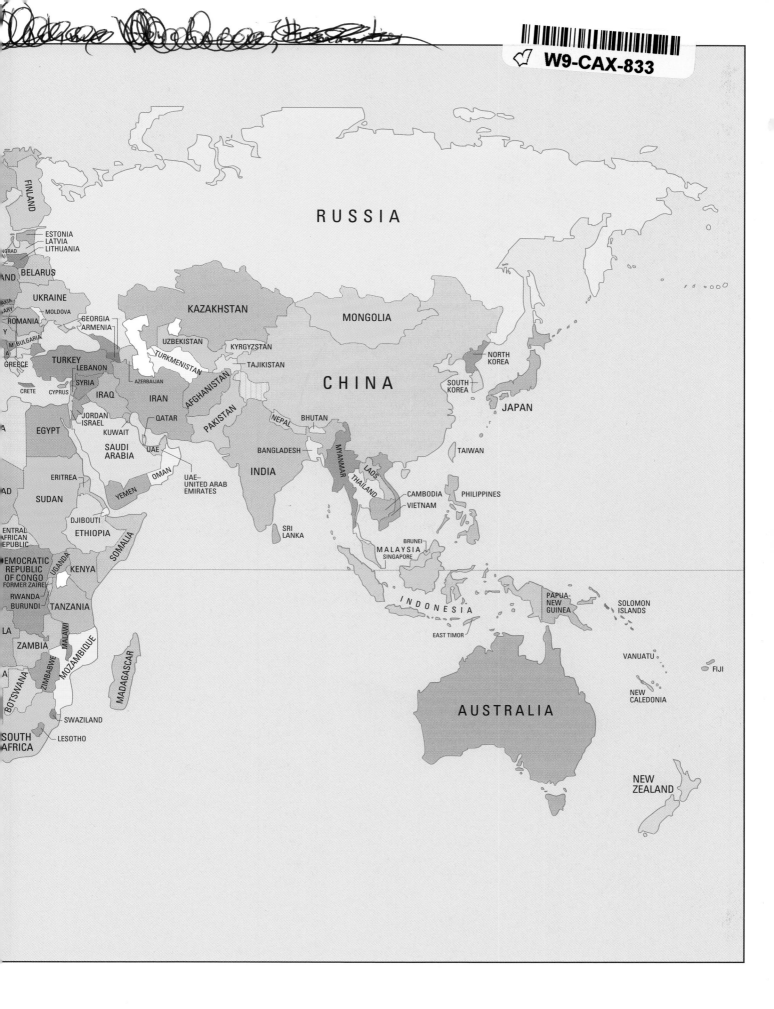

human geography

FIFTH EDITION

human geography

FIFTH EDITION

william norton

OXFORD
UNIVERSITY PRESS
1904 ✦ 2004
100 YEARS OF
CANADIAN PUBLISHING

OXFORD
UNIVERSITY PRESS

70 Wynford Drive, Don Mills, Ontario M3C 1J9
www.oup.com/ca

Oxford University Press is a department of the University of Oxford.
It furthers the University's objective of excellence in research, scholarship,
and education by publishing worldwide in

Oxford New York

Auckland Bangkok Buenos Aires Cape Town
Chennai Dar es Salaam Delhi Hong Kong Istanbul Karachi
Kolkata Kuala Lumpur Madrid Melbourne Mexico City
Mumbai Nairobi São Paulo Taipei Tokyo Toronto

Oxford is a trade mark of Oxford University Press
in the UK and in certain other countries

Published in Canada by Oxford University Press

National Library of Canada Cataloguing in Publication

Norton, William, 1944–
Human geography/William Norton.—5th ed.

Includes bibliographical references and index.
ISBN 0-19-541908-1

1. Human geography. I. Title.

GF41.N67 2004   304.2   C2003-907330-0

Cover and text design: Joan Dempsey

Cover photo: Hugh Sitton/Getty Images

Text composition: Valentino Sanna, Ignition Design and Communications

This book is printed on permanent (acid-free) paper ∞.

Printed in Canada

CONTENTS

Figures, Tables, and Boxes vii

PREFACE xi
Fifth Edition: Special Features xi
To the Student xii
To the Instructor xii
Acknowledgements xiii

INTRODUCTION 1
Three Recurring Themes 1
Why Human Geography? 3
The Goal of Human Geography 3
The Human Geographer at Work 4
About This Book 9
About Being a Human Geographer 12
Summary 12
Further Explorations 13
On the Web 13

1 WHAT IS HUMAN GEOGRAPHY? 15
Preclassical Geography 16
Classical Geography 17
The Fifth to Fifteenth Centuries: Geography in
    Europe, China, and the Islamic World 19
The Age of European Exploration and Discovery 22
Geography Rethought 25
Institutionalization: 1874–1903 30
Prelude to the Present: 1903–1970 33
Contemporary Geography 35
Postscript 38
Summary 40
Links to Other Chapters 41
Further Explorations 41
On the Web 42

2 STUDYING HUMAN GEOGRAPHY 45
Philosophical Options 46
Human Geographic Concepts 55
Techniques of Analysis 65
A Concluding Comment 75
Summary 76
Links to Other Chapters 77
Further Explorations 78
On the Web 79

3 THE EARTH: A HUMAN ENVIRONMENT 81
Planet Earth 82
Physical Processes 83
Making the Physical Landscape 86
Global Environments 90

Life on Earth 94
Human Origins 95
Understanding Human Evolution: Three Myths 99
Conclusion 100
Summary 101
Links to Other Chapters 102
Further Explorations 102
On the Web 103

4 THE EARTH: A FRAGILE HOME 105
A Global Perspective 106
Humans as Simplifiers of Ecosystems 107
Environmental Ethics 110
Human Impacts 113
Earth's Vital Signs 130
Sustainability and Sustainable Development 132
Summary 135
Links to Other Chapters 136
Further Explorations 137
On the Web 138

5 THE HUMAN POPULATION: HISTORY
   AND CONCEPTS 141
Fertility and Mortality 142
History of Population Growth 163
Explaining Population Growth 166
The Need for Better Explanations 170
Summary 172
Links to Other Chapters 174
Further Explorations 174
On the Web 175

6 THE HUMAN POPULATION:
   AN UNEQUAL WORLD 177
Distribution and Density 178
Migration 181
Refugees 192
The Less Developed World 196
Summary 212
Links to Other Chapters 213
Further Explorations 214
On the Web 215

7 CULTURES: THE EVOLUTION AND
   REGIONALIZATION OF LANDSCAPE 217
A World Divided by Culture? 218
Humans in Groups 218
The Evolution of Culture 221
Cultural Regions 224
The Making of Cultural Landscapes 229

Language   232
Religion   240
Cultural Globalization   247
Summary   250
Links to Other Chapters   251
Further Explorations   252
On the Web   253

8   CULTURES: SYMBOLIC AND SOCIAL
     LANDSCAPES   255
Rethinking Culture   256
Types of Society   257
Social Theory   259
Landscape as Place   264
Vernacular Regions   266
Ethnicity   268
Landscapes and Power Relations   270
Folk Culture and Popular Culture   277
Group Engineering in a Globalizing World?   281
Summary   282
Links to Other Chapters   283
Further Explorations   284
On the Web   285

9   THE POLITICAL WORLD   287
State Creation   288
Geopolitics (and *Geopolitik*)   296
The Stability of States   298
The Role of the State   312
Elections: Geography Matters   315
The Geography of Peace and War   317
Our Geopolitical Future?   320
Summary   326
Links to Other Chapters   327
Further Explorations   328
On the Web   329

10   INTERACTION AND ECONOMIC
      GLOBALIZATION   331
Distance Friction and Frictionless Distance   332
Concepts of Distance and Space   333
Diffusion   339
Transportation   343
Trade   348
Towards One World   351
Economic Globalization   354
A Global Village?   360
Summary   361
Links to Other Chapters   362
Further Explorations   362
On the Web   363

11   AGRICULTURE   365
Inventing and Reinventing Economic Geography   366
The Agricultural Location Problem   371
Distance, Land Value, and Land Use   379
World Agriculture: Origins and Evolution   382
World Agriculture: Types and Regions   391
A Marxist Perspective on Agriculture   396
Political Ecology and Agriculture   401
Consuming Food   403
Summary   405
Links to Other Chapters   407
Further Explorations   407
On the Web   409

12   SETTLEMENT   411
Rural Settlement   412
Origins and Growth of Cities   419
Urban Locations I: Historical Explanations   424
Urban Locations II: Central Place Theory   426
Cities in the More Developed World   431
Cities in the Less Developed World   446
Summary   452
Links to Other Chapters   454
Further Explorations   455
On the Web   457

13   INDUSTRY AND REGIONAL DEVELOPMENT   459
The Industrial Location Problem   460
The Industrial Revolution   465
World Industrial Patterns   469
Recreation and Tourism   486
The Geography of Uneven Development   491
Summary   495
Links to Other Chapters   497
Further Explorations   498
On the Web   500

CONCLUSION: THE HUMAN GEOGRAPHY
OF THE FUTURE…AND THE FUTURE OF
HUMAN GEOGRAPHY   501
Changing Human Landscapes   502
Changing Human Geography   506
Being Human Geographers: Where We Began   508
Summary   509
Further Explorations   509
On the Web   510

References   *511*
Glossary   *520*
Credits   *529*
Index   *531*

FIGURES, TABLES, AND BOXES

FIGURES

Intro.1 World political divisions, 2000   4
Intro.2 World political divisions, 1938   5
1.1 The world according to Eratosthenes   17
1.2 The world according to Ptolemy   18
1.3 An example of a T-O map   19
1.4 An example of a Portolano chart   19
2.1 The scientific method   48
2.2 The site of Winnipeg   57
2.3 The situation of Winnipeg within North America   57
2.4 Clustered, random, and uniform point patterns   60
2.5 A typical distance decay curve   60
2.6 Impact of spatial scale on descriptions of point patterns   61
2.7 Urban centres in Manitoba   61
2.8 A typical S-shaped growth curve   62
2.9 Images of North America in 1763   63
2.10 Mapping at a scale of 1:250,000   66
2.11 Mapping at a scale of 1:50,000   67
2.12 Schematic representation of a dot map   67
2.13 Schematic representation of a choropleth map   67
2.14 Schematic representation of an isopleth map   67
2.15 Mercator projection   68
2.16 Mollweide projection   68
2.17 Robinson projection   68
2.18 Two topographic maps of the Love Canal Area, Niagara Falls, New York   69
3.1 Revolution of the Earth around the Sun   83
3.2 Moving continents   84
3.3 Major ocean currents   85
3.4 Global cordilleran belts   87
3.5 Global distribution of soil types   88
3.6 Global distribution of natural vegetation   89
3.7 Global distribution of climate   89
3.8 Generalized global environments   90
4.1 Chemical cycling and energy flows   107
4.2 Some consequences of human-induced vegetation change   116
4.3 Location of tropical rain forests   118
4.4 Protecting Australian agriculture from rabbits and dingoes   120
4.5 The global water cycle   123
4.6 Impact of sea-level change on Bangladesh   128
4.7 Global distribution of some major environmental problems   130
5.1 World distribution of crude birth rates, 2002   148
5.2 Generalized J-shaped curve of death rates and age   149
5.3 World distribution of crude death rates, 2002   150
5.4 World distribution of life expectancy, 2002   151
5.5 World distribution of rates of natural increase, 2002   155
5.6 Age and sex structure in China, 1990   158
5.7 Age structure of populations   159
5.8 Age–sex structure in Brazil: 1950, 1970, 1987   160
5.9 Age–sex structure in Canada: 1861, 1921, 1981, 2036   162
5.10 World population growth   165
5.11 The demographic transition model   168
6.1 World population distribution and density   178
6.2 Mental maps   185
6.3 The push–pull concept and relevant obstacles   186
6.4 Major world migrations, 1500–1900   187
6.5 Refugee numbers, 1960–2002   192
6.6 Refugee outflows by origin, 1997–2001   194
6.7 Refugee population by country of asylum, 2001   195
6.8 The Horn of Africa   197
6.9 The Third World   198
6.10 North and South   198
6.11 Groups of economies   200
6.12 The world system   202
6.13 The construction of entitlements: selected factors and mechanisms at different scales   206
7.1 Civilizations in the ancient world   224
7.2 Cultural regions of the world   225
7.3 Europe defined   226
7.4 Regions of North America   227
7.5 Cultural regions of the United States   228
7.6 Regions of Canada   228
7.7 Core, domain, and sphere   230
7.8 World distribution of language families   233
7.9 Initial diffusion of Indo-European languages   235
7.10 Diffusion of Indo-European languages into England   235
7.11 Four official languages in Switzerland   236
7.12 Dutch, French, and German in Belgium   236
7.13 French and English in North America   237
7.14 Hearth areas and diffusion of four major religions   240
7.15 World distribution of major religions   241
8.1 Components of place as a historically contingent process   261
8.2 North American vernacular regions   267
8.3 Apartheid on the national scale in 1975   273
9.1 The British Empire in the late nineteenth century   290

9.2   Principal elements in the process of exploration   292
9.3   Territorial expansion of the United States   293
9.4   Jones field theory   294
9.5   Mackinder heartland theory   297
9.6   African ethnic regions   300
9.7   African political areas in the sixteenth, eighteenth, and nineteenth centuries   301
9.8   Areas with nationalist/separatist movements in Europe   302
9.9   The former Yugoslavia   303
9.10 The former USSR   306
9.11 Some areas of conflict in South Asia   307
9.12 European ethnic regions   311
9.13 World distribution of state types   314
9.14 The original 'gerrymander'   316
9.15 Congressional districts in 1960s Mississippi   316
9.16 World civilizations   321
10.1 A non-straight-line, shortest-distance route   333
10.2 A one-way system   333
10.3 Time distance in Edmonton   334
10.4 Toronto in physical space and time space   334
10.5 Logarithmic transformation of distances from Asby, Sweden   336
10.6 The London underground system   337
10.7 Distribution of innovativeness   341
10.8 Effects of pre-emption on the adoption curve   343
10.9 Diffusion of transport innovations in Britain, 1650–1930   344
10.10 Evolution of a transport network   345
10.11 The road network of Martinique   347
10.12 Conventional international trade and intra-firm trade within a transnational corporation   354
10.13 The contemporary geo-economy   356
11.1 Relationship between mean annual rainfall and wheat yield, 1909   371
11.2 Crop and livestock associations, US–Canada border   374
11.3 Supply and demand curves   375
11.4 Rent-paying abilities of selected land uses   375
11.5 Economic rent lines for three crops and related zones of land use   376
11.6 Agricultural land use in the isolated state   377
11.7 Relaxing a von Thünen assumption: a navigable waterway   378
11.8 Relaxing two von Thünen assumptions: multiple markets and a transport network   378
11.9 Scatter graphs, best-fit lines, and r values   380
11.10 Agricultural land use in Uruguay: (a) as predicted by von Thünen theory; (b) actual   381
11.11 World agricultural regions   392
11.12 Percentage of labour force in agriculture by country   397
11.13 The food supply system   400
11.14 Global dietary patterns   403
12.1 Growth of urban population relative to growth of world population   413

12.2 Township survey in the Canadian prairies   414
12.3 Examples of nucleated rural settlement patterns   415
12.4 Approved urban areas in the regional municipality of Niagara, 1989   419
12.5 Percentage urban population by country, 1996   422
12.6 British new towns   425
12.7 A triangular lattice   429
12.8 Theoretical trading areas   429
12.9 A central place system: the marketing principle   430
12.10 A central place system: the transportation principle   430
12.11 A central place system: the administration principle   431
12.12 Urban land values   433
12.13 Three models of the internal structure of urban areas   433
12.14 Three images of Los Angeles   441
12.15 A model of postmodern urban structure   444
12.16 Cities with 10 million or more population   444
12.17 World cities and spheres of influence   446
12.18 Cities in the less developed world with more than 2 million population   448
12.19 Cities in China   449
13.1 A locational triangle   462
13.2 A simple isotim map   463
13.3 An isodapane map   463
13.4 Transport cost and distance   464
13.5 Stepped transport costs   464
13.6 Major oil trade movements (million tonnes)   471
13.7 Major world industrial regions   473
13.8 Export-processing zones in Asia   476
13.9 Special economic zones in China   478
13.10 Relationship between economic growth and distribution of employment   482
13.11 Percentage of labour force in industry by country   483
13.12 Percentage of labour force in services by country   484
13.13 The tourism system of place construction   491

TABLES

Intro.1   Selected demographic data   7
3.1   Basic chronology of life on earth   95
3.2   Basic chronology of human evolution   97
4.1   Shallow and deep ecology compared   115
4.2   Global deforestation: estimated areas cleared (000s km^2)   116
5.1   Contraceptive use, by regions, 2002   145
5.2   Population data, Canada, 2002   149
5.3   Countries with highest levels of HIV/AIDS prevalence, 2001   153
5.4   Countries with highest rates of natural increase (3.0 and above), 2002   154
5.5   Countries with lowest rates of natural increase (less than 0.0), 2002   155
5.6   Projected population growth, 2002–2050   156

5.7   Global aging, 1950–2050   161
5.8   Major epidemics, 1500–1700   164
5.9   Adding the billions: actual and projected   165
5.10 Anticipated population declines greater than
      five million, 2002–2050   166
6.1   World population distribution by major area
      (percentages): actual and projected   178
6.2   The ten most populous countries, current and
      projected   179
6.3   Population densities of the ten most populous
      countries, 2002   179
6.4   Some push and pull factors   182
6.5   Some typical moorings   186
6.6   Refugees and persons of concern to UNHCR,
      worldwide, 1980–2001   193
6.7   Persons of concern to UNHCR, 2001   194
6.8   Refugee numbers by country of origin and country
      of asylum, 2001: the ten largest groups   195
6.9   Countries of asylum with more than 100,000
      refugees, 2001   196
6.10 Extremes of human development, 2000   200
6.11 Population densities, selected countries, 2002   204
6.12 Natural events and human disasters, regional data,
      1947–1967 and 1969–1989   209
7.1   A unilinear evolutionary model of culture   223
7.2   Basic chronology of early civilizations   224
7.3   Language families   234
7.4   Numbers of speakers, major languages   234
7.5   Major world religions: numbers of adherents
      (thousands), 2001   241
8.1   Gender-related development index   272
9.1   Ethnic groups in the former Yugoslavia   303
9.2   Ethnic groups in the former USSR   306
9.3   Disputes and conflicts involving the UN, 1945–1990
      319
10.1 From protectionism to free trade: a chronology   355
10.2 Sales data for transnationals compared to GDP for
      selected countries, 1997   357
10.3 Net foreign direct investment flows: selected
      countries and groupings, 1989 and 1997   358
11.1 Contrasting farming types in Illinois   373
11.2 Average distances from London, England, to regions
      of import derivation (miles)   382
11.3 Multinationals and crops in the less developed world
      399
11.4 Determinants of peasant–herder conflicts in northern
      Ivory Coast   402
12.1 Some definitions of urban centres   412
12.2 Percentage change in rural population by region,
      Canada, 1981–1986   418
12.3 Factors influential in the decision to move to
      exurban locations around Woodstock, Ont.   419
12.4 Seven stages of premodern urban development
      426
12.5 Three global cities   445
12.6 Five tiers of global cities   445

13.1 Outlook for world total primary energy supply,
      2010 and 2020 (%)   470
13.2 Regional shares of energy production by type,
      1973 and 2000 (%)   470
13.3 Location of R&D facilities and plants of nine
      Japanese electronics firms, 1975 and 1991   481
13.4 Mineral production in the less developed world   483
13.5 Summary of world employment   485
13.6 Labour markets: from Fordism to post-Fordism   486
13.7 Characteristic tendencies: conventional mass
      tourism vs alternative tourism   490
Concl.1 Disappearing peoples   506

BOXES

Intro.1 Germany: a new country   5
Intro.2 Less-developed Canada?   7
Intro.3 This book and some subdisciplines of human
         geography   11
1.1   Did China 'discover' the world in 1421?   20
1.2   Evaluating our place in the world   22
1.3   Exploration or invasion?   23
1.4   The southern continent   25
1.5   Varenius   26
1.6   Humboldt   28
1.7   Ritter   29
1.8   Ratzel   30
1.9   Vidal   32
1.10 Taylor   33
1.11 Sauer   34
1.12 Hartshorne   35
2.1   Environmental determinism   47
2.2   Positivistic human geography   49
2.3   The Schaefer–Hartshorne debate   50
2.4   Humanistic human geography   51
2.5   Marxist thought and practice   53
2.6   Marxist human geography   55
2.7   Map projections   68
2.8   The power of maps   69
2.9   Using a GIS   71
2.10 Some qualitative resources   74
3.1   Earthquakes and volcanic eruptions   85
3.2   The geomorphology of Canada   86
3.3   Gaia   96
3.4   Species and races   99
3.5   A history of racism   100
4.1   Lessons from Easter Island   108
4.2   The tragedy of the commons or collective
      responsibility?   114
4.3   The threatening forest   117
4.4   Defeating desertification   119
4.5   Unwanted guests   121
4.6   A tale of three water bodies   124
4.7   Chloro-fluorocarbons and the atmosphere   126
4.8   Rising sea levels   128
4.9   The skeptical environmentalist   131
4.10 Clayoquot sound: the case for sustainability   133

5.1   Fertility in tropical Africa   144
5.2   Primitive abortion   146
5.3   Declining fertility in the less developed world   147
5.4   Declining fertility in the more developed world   149
5.5   Problems in central Asia and Russia   151
5.6   Fertility in Romania, 1966–1989   157
5.7   Population in China   158
5.8   The prospect of underpopulation   166
5.9   Fertility, population growth, and the changing status of women   171
6.1   The 1991 Nigerian census   180
6.2   The Ravenstein laws   183
6.3   'A new system of slavery': Indian indentured labour in Mauritius   188
6.4   Scandinavian migration to North America   189
6.5   Migration and ethnic diversity in Canada   190
6.6   Restrictive immigration policies   191
6.7   Refugees in the Horn of Africa   197
6.8   The less developed world: Ethiopia   199
6.9   The less developed world: Sri Lanka   201
6.10   The less developed world: Haiti   203
6.11   The Grameen Bank, Bangladesh   208
6.12   Flooding in Bangladesh   210
6.13   The best or the worst of times?   211
7.1   The concepts of culture and society   219
7.2   The superorganic concept of culture   220
7.3   Defining 'civilization'   222
7.4   Europe as a cultural region   226
7.5   The Mormon landscape   231
7.6   Linguistic territorialization in Belgium and Canada   237
7.7   The Celtic languages   238
7.8   The Mithila cultural region   245
7.9   Religion and 'irrational' choices   247
7.10   Religious landscapes: Hutterites and Doukhobors in the Canadian west   249
8.1   The concept of alienation   258
8.2   An application of structuration theory   261
8.3   A positivist response to postmodernism   262
8.4   A Marxist response to postmodernism   263
8.5   The iconography of landscape   265
8.6   Psychogeography: the sense of Oklahomaness   268
8.7   The origins of apartheid   273
8.8   The geography of fear   276
8.9   Consigned to the shadows   278
8.10   Geophagy   279
9.1   Remnants of empire   291
9.2   Laws of the spatial growth of states   294
9.3   The Jewish state   295
9.4   A different Canada?   296
9.5   The plight of the Kurds   299
9.6   Tribes, ethnicity, political states, and conflict in Africa   301
9.7   Conflicts in the former Yugoslavia   303
9.8   The collapse of the USSR   306
9.9   Regional identities and political aspirations   308
9.10   The Maghreb union   312
9.11   Some traditional rivalries   318
9.12   Naturally aggressive?   322
10.1   Ekistics: a science of human settlements   332
10.2   The tyranny of distance   338
10.3   Cholera diffusion   340
10.4   Port system evolution   346
10.5   Railways and economic growth   347
10.6   Explaining commodity flows: Ullman   349
10.7   Looking back at the twentieth century   353
10.8   The World Trade Organization   356
11.1   Human geography and economics   366
11.2   The economic operator concept   368
11.3   Marxist political economy   369
11.4   Government and the agricultural landscape   374
11.5   Calculating economic rent   376
11.6   Correlation and regression analysis   380
11.7   Agricultural core areas, c. 500 BCE   384
11.8   Frontier agriculture: subsistence or commerce?   385
11.9   The pastoral frontier   386
11.10 Organic farming   388
11.11 The green revolution in India   389
11.12 Pastoral nomadism in the Sahara and Mongolia   393
11.13 Canadian farmers: fewer and older   396
11.14 Agricultural change in Bhutan   398
11.15 Peasant–herder conflict in Ivory Coast   402
12.1   Rural settlement in the Canadian Prairies   414
12.2   Exurbanization in southwestern Ontario   419
12.3   Theory construction by Christaller   427
12.4   Nearest-neighbour analysis   428
12.5   Testing central place theory   431
12.6   Modern and postmodern urban theory   432
12.7   A Marxist interpretation of the urban experience in a capitalist world   438
12.8   Slums and the cycle of poverty   439
12.9   Suburbia and postsuburbia   440
12.10 Urbanization in the less developed world: Brazil   450
13.1   Factors related to industrial location   461
13.2   Testing Weberian theory   464
13.3   Locational interdependence: the Hotelling model   465
13.4   The period of the industrial revolution   466
13.5   Geographical change and industrial growth: the Tyneside coalfield   468
13.6   Oil-producing countries   471
13.7   An energy 'crisis': New Zealand in 1992   472
13.8   Explaining local industrial change   473
13.9   Industry in Canada   474
13.10 Industry in Japan   475
13.11 Deindustrialization and reindustrialization   480
13.12 Tourism in Sri Lanka   489
13.13 The Rostow model of economic growth   492

The discipline of human geography is evolving today at an almost alarming pace. Its links with physical geography are attracting renewed attention, especially in the context of relations between humans and land. Technological advances, particularly in the areas of remote sensing and geographic information systems, continue at an ever-increasing rate. New concepts and approaches are emerging through dialogue with other disciplines. Above all, perhaps, there is a growing awareness of how much the traditional human geographic concepts of space, place, and distance really do matter to our understandings of the world and our actions in it, especially as these understandings and actions relate to an increasingly interconnected and globalizing world. Together, these developments make human geography not only one of the most intellectually challenging of academic disciplines, but also one of the most relevant, both environmentally and socially.

This book attempts to capture both the spirit and the practical value of our contemporary human geography. Like the discipline itself, it encompasses an extraordinarily broad range of subject matter—including history, biology, philosophy, demography, sociology, politics, cultural studies, and economics—and no single approach or methodology dominates. Loosely organized around three recurring themes—relations between humans and land, regional studies, and spatial analysis—this text emphasizes how human geography has developed in response to society's needs, and how the discipline continues to change accordingly. It also stresses the links between human geography and other disciplines, not only in order to clarify the various philosophies behind different types of human geographic work, but also to encourage students to apply their human geographic knowledge and understanding in other academic contexts.

Fifth Edition: Special Features

Of course, it is not only the academic discipline of human geography that changes: even more obviously, the subject matter of human geography changes. Environmental issues, population circumstances, the political world, and economic landscapes are subject to ongoing and sometimes unexpected transformations. Especially significant today are the many and varied consequences of the processes associated with globalization. Accordingly, much of the detailed factual content in this fifth edition is new. Among the specific changes are the following:

- Much of the Introduction has been rewritten to reflect ongoing changes both in the academic discipline of geography and in the specific geographies discussed here.
- Chapters 1, 2, and 3 all contain new material. Especially notable is the account in Chapter 2 of globalization as an overarching concept
- Substantial additions and amendments to Chapter 4 take into account changing understandings of human impacts on the earth. In particular, attention is drawn to diverse, sometimes contradictory, views concerning the nature of those impacts at the global scale.
- The Chapter 5 account of human population has been significantly revised, not only to update data, but also to reflect new information about and understandings of such important topics as aids and aging populations.
- Similarly, the data in Chapter 6 have been updated and the discussions of differing perspectives extended.
- Chapters 7 and 8 both include new material, especially relating to cultural globalization.
- Chapter 9 includes new and revised content reflecting several recent geopolitical events, most notably the 2001 terrorist attacks on New York and Washington and their aftermath. There is also an overview of conflicts in South Asia, along with a substantial new account of possible geopolitical futures, including political globalization.
- The sequence of the final four substantive chapters has changed, the former Chapter 13 moving up to become Chapter 10. This shift places one of the chapter's key themes—the declining importance of physical distance in an era of globalization—prior to the chapters on agricultural, settlement, and industrial geographies (now renumbered as 11, 12, and 13).
- The new Chapter 10, on the closely related topics of interaction and economic globalization, includes a comprehensive account of the details and implications of economic globalization.

- Chapter 11, on agricultural geography, includes substantial new material, much of it linked to the discussion of economic globalization in the new Chapter 10. There is also a new conceptual discussion at the beginning of the chapter that provides insights helpful to this and the following two chapters. Other additions focus on nitrogen fertilizers, the green revolution, and organic farming. There is also an extended account of the Marxist perspective and the insights it suggests, a new regional overview of change on the Canadian prairies, and a new section on food consumption.
- The Chapter 12 account of settlement geography includes new material on the links between capitalism and urban growth and change; urban models; and urban land uses such as retailing and housing. Many of these new discussions build on the economic globalization material in Chapter 10.

Flowing directly from the account of settlement geography, the Chapter 13 discussion of industry includes new material on the ongoing processes of industrial restructuring and the transition from a Fordist to a post-Fordist economy and society. In addition, the discussion of energy production and consumption has been substantially developed.

Finally, the entire text has been revised for clarity and readability, and a number of errors have been corrected.

To the Student

A package of resources specifically for students—including links to human geography websites, chapter summaries and objectives, and short-answer exercises—is available on a companion website. For details, please visit www.oup.com/ca/he.

This book has five general goals:

- to introduce the discipline of human geography;
- to highlight the value of human geography in understanding our complex and ever-changing world;
- to provide a solid foundation for further courses in human geography;
- to help you develop the habit of thinking logically; and
- to encourage you to read not just widely but critically: in other words, to question whatever you read—including this book.

As a student, of course, you have your own goals. Whatever they may be, in order to achieve them, you need to think seriously about your learning strategy. Although we all develop our own distinctive ways of studying, the following general advice should be helpful to anyone.

First, organize your time. This is not as simple as it sounds. Most of us find it only too easy to become distracted from important tasks. The key to sound time management is to schedule all your activities—academic and social—on a daily basis. Allocate the appropriate amounts of time for each of the subjects you are studying—and for going to the movies, or playing sports, or visiting friends—and stick to your plan. Study human geography when you are scheduled to study it.

Second, develop efficient reading and studying habits. How? Many students find it helpful to read and study a text chapter in five relatively discrete stages.

1. *Survey*: use headings, subheadings, the end-of-chapter summary, and the glossary to get a broad overview of the material.
2. *Question*: Skim through the chapter again, this time looking for 'facts' and arguments that raise questions in your mind. Instead of passively accepting what the author says, be an active reader.
3. *Read*: Only now, once you have a sound general feel for the chapter, should you start your careful reading. Again, read actively—taking notes, underlining important words, highlighting key passages.
4. *Rephrase*: State the major ideas out loud, in your own words.
5. *Review*: Go over the chapter one more time to confirm your grasp of its content.

Believe it or not, this five-step approach will save you time and energy in the long run (particularly at the end of term) by ensuring that your understanding of the material grows and develops in an integrated way. Enjoy learning.

To the Instructor

A generous package of ancillary materials is available to adopters of this text:

- Instructor's manual, available on-line at www.oup.com/ca/he: Includes chapter outlines for use as lecture resources, suggestions for small-group assignments, recommended audio and visual resources for classroom use, and suggested topics for essay assignments.
- PowerPoint® slides, also on-line at www.oup.com/ca/he: Hundreds of slides (including graphics and tables from the text) that can be edited to suit the individual instructor's needs.
- Test-item file, on CD-ROM: Approximately 1500 questions in a variety of formats (multiple-choice, short answer, true/false, and essay), with a wide range of options for sorting, editing, importing, and distribution.

For more information, please contact your local Oxford University Press sales representative or visit the website www.oup.com/ca.

A good atlas is another invaluable accompaniment to the text: the *Canadian Oxford World Atlas* (fifth edition) is highly recommended. Of course, current demographic and political data, as well as general information on global change, are always useful; instructors should have little difficulty using their own examples to complement the material in this volume.

At the end of the Introduction, a section entitled 'About this Book' provides a broad overview of the contents and outlines the reasoning behind the sequence in which topics are presented; for a concise index of subdisciplinary topics, see Box Intro.4. I hope most instructors will find that the sequence chosen represents an appropriate course organization.

Acknowledgements

Textbook authors owe many debts of thanks to the authors, past and present, who have contributed to their discipline. There is little that is original in this volume. I hope that it is a fair reflection of human geography as it exists today, and that, having read it, students will have an appropriate background for more advanced study.

Many of the substantial revisions in this fifth edition have been prompted by the insightful comments of fellow human geographers. Hilary Janzen identified some of the new data incorporated in this edition, and I am grateful for her support. Yet again, I thank Barry Kaye, friend and colleague, who never ceases to provide me with helpful material and ideas.

Douglas Fast prepared the new and revised graphics in this fifth edition and I appreciate his professionalism and good judgement. I am most grateful to the staff at Oxford University Press—Sally Livingston, Laura Macleod, Euan White, and Phyllis Wilson—who supported this textbook and myself with just the right combination of professionalism and friendship. Finally, and as always, thanks to Pauline.

As a young man, my fondest dream was to become a geographer. However, while working in the customs office I thought deeply about the matter and concluded that it was far too difficult a subject. With some reluctance, I then turned to physics as a substitute.

—Albert Einstein

You know that this book is an introduction to human geography—but what does that term mean? Note down in a few key words your idea of what human geography is. It should be interesting to compare your initial perception with your understanding after you have read these introductory remarks.

Consider the word 'geography'. Its roots are Greek: *geo* means 'the world', *graphei* means 'to write'. Literally, then, what geographers do is write about the world. The sheer breadth of this task has demanded that it be divided into two relatively distinct disciplines. *Physical geography* is concerned with the physical world, and *human geography* is concerned with the human world.

So human geographers write about the human world. But how do they approach their work? What methods do they employ? Not surprisingly, human geographers have diverse approaches and methods from which to choose, and the later chapters of this book reflect that diversity. Nevertheless, three themes are central to any study of the human world: relations between humans and land, regionalization, and spatial analysis.

Three Recurring Themes

Keeping these three themes in mind from the start will help you to follow and understand the complex connections among the many aspects of human geography. The intellectual origins and evolution of these themes are outlined in Chapter 1, and a philosophical justification for each theme is introduced in Chapter 2.

Humans and Land

The human world is not in any sense preordained. It is not the result of any single cause such as climate, physiography, religion, or culture. Rather, it is the ever-changing product of the activities of human beings, as individuals and as group members, working within human and institutional frameworks to modify pre-existing physical conditions. Thus human geographers often focus on the evolution of the human world with reference to people, their cultures, and physical environments. We usually describe the human world as a **landscape**. There are two closely related aspects to 'landscape' in this sense:

1. First, landscape is what is there as a result of human modifications of physical geography. This aspect of landscape includes crops, buildings, lines of communication, and other visible, material features.
2. Second, landscape has significant symbolic content—in other words, meaning. It is hard to view a church, a statue, or a tall

office building without appreciating that it is something more than a visible, material human addition to the physical geography of a place: such features are expressions in landscape of the cultures that produced them.

As human geographers, we are interested in landscape both for what it is and for what it means to live in it: we interpret landscape as the outcome of particular relationships between humans and land. For some geographers, both physical and human, the study of humans and land together *is* geography.

Regional Studies

To facilitate their task of writing about the world, human geographers often divide large areas into smaller areas that exhibit a degree of unity—that share one or more features in common. These smaller areas are **regions**. To regionalize is to classify on the basis of one or more variables. The fact that we are able to regionalize tells us that human landscapes make sense; they are not random assemblages of features. Groups of people occupying particular areas over a period of time create regions: human landscapes that reflect their occupancy and that differ from other landscapes. Much contemporary regional study focuses on this social organization of space and the impact of region creation on social and economic life.

Building on a long tradition, contemporary human geography considers regions at a wide range of scales—from the local (a suburban neighbourhood, for example) all the way through to the global (the world as a single region). In acknowledging the relevance of different spatial scales of analysis, human geography reveals the importance of place in all aspects of our lives.

Spatial Analysis

Understanding the human world requires that we explain **location**: why things are where they are. Typically, a geographer using a spatial analysis approach tackles this question through theory construction, models, and hypothesis testing, using quantitative methods. For example, we may find that towns in an area are spaced at relatively equal distances apart, or that industrial plants are located close to one another. The goal of spatial analysis is to explain such locational

regularities. Thus we have already identified two types of spatial organization: first, the one that is reflected in the creation of regions; second, the one that is reflected in distinct patterns of locations.

Central to this spatial analysis is the realization that all things are related—and for geographers, to say that things are related is usually to say that they interact. Perhaps the best examples of interaction in the contemporary world have to do with what is often described as **globalization**—a complex set of processes with economic, political, and cultural dimensions.

A Common Thread

Our three recurring themes reflect three separate but overlapping traditions (to be outlined in the next chapter). One common thread among them, however, is the fact that the human world is always changing. Thus human geography, regardless of the specific focus, typically incorporates a time dimension. Change can be a response to either internal or external factors. Some changes are rapid, some gradual. Landscapes change in content and meaning; regions gain or lose their distinctive features; and locations adjust to changing circumstances.

One compelling example of changing geographies can be found in the tragic events of September 11, 2001, when a series of concerted terrorist attacks destroyed the twin towers of the World Trade Center in New York City and part of the Pentagon in Washington, DC, killing more than 3000 people. The consequences of those attacks, some immediate, some longer-term, have served to reshape both local and global geographies. The skyline of New York City will never be the same again. Nor will the pattern of global politics, since the United States responded to the attacks by invading first Afghanistan, where the ruling Taliban had supported terrorist groups, and then Iraq. Many commentators suggested that global politics would henceforth be characterized by a 'clash of civilizations', and there was widespread acknowledgement that fighting terrorism would require in-depth understanding of local geographies, both physical and human.

In identifying our three principal themes, we explicitly acknowledge the legitimacy of multiple approaches to our subject matter: the human world. It is also important to note how the application of these approaches changes over time. Such change is equally legitimate—and necessary—in any dynamic discipline.

Why Human Geography?

Although many geographers have striven to make their field a truly integrated one, combining physical and human components, human geography and physical geography are relatively distinct disciplines. Contemporary human geographers do not deny the relevance of physical geography to their work, but they do not insist that their studies always include a physical component. Because human geography studies human beings, it has close ties with social sciences such as history, economics, anthropology, sociology, psychology, and political studies. One of the particular strengths of human geography is that it considers the human world in multivariate terms, incorporating physical and human factors as necessary. Where appropriate, then, this book introduces aspects of physical geography. For example, to understand world population distributions we obviously need to know about global climates. However, we also need to know about human perceptions of those climates. Thus human geography is related to but separate from physical geography in much the same way as it is related to but separate from the other disciplines that study human beings.

The central subject matter of human geography is *human behaviour* as it affects the earth's surface. Expressed in this way, the subject matter of human geography is very similar to that of the various other social sciences, all of which focus on human behaviour in some specific context.

Defining Human Geography

Providing a meaningful one-sentence definition of any academic discipline is a real challenge. In the case of human geography, however, the American geographer Charles Gritzner has suggested a useful definition in the form of a question: 'What is where, why there, and why care?' (Gritzner, 2002). Every exercise in human geography, regardless of which theme it highlights, begins with the spatial question: 'where?' Then, once the basic environmental, regional, and spatial facts are known, the geographer focuses on

understanding, or explaining, why things are where they are. Finally, the third part of the question, 'why care?', draws attention to the pragmatic nature of human geography: geographic facts matter because they affect human life. This is true of any geographic fact, whether it is something as seemingly mundane as the distance between my home and the nearest convenience store or something as far-reaching or serious as a drought or a civil war that causes peasant farmers in Ethiopia to lose all their crops.

The Goal of Human Geography

Writing about the human world to increase our understanding of it has been the goal of human geography at least since the time of the ancient Greeks. Two twentieth-century statements of this continuing goal help to explain why it remains so important:

> The function of geography is to train future citizens to imagine accurately the conditions of the great world stage and so help them to think sanely about political and social problems in the world around (Fairgrieve 1926:18).

> Geography's *raison d'être* should be to develop appreciation of the great variety of cultures that comprise the contemporary world, and to show how in each society these have evolved—and are evolving—as specific responses to environment, to place and to people (Johnston 1985:334).

Although written some sixty years apart, these two statements express essentially the same view. Human geography is a practical and socially relevant discipline that has a great deal to teach us about the world we live in and how we live in the world. The goal of this particular book is to provide a basis for comprehending the human world as it is today and as it has evolved. If it succeeds, it will have made a contribution to the more general goal of advancing a just global society. Students of human geography are in an enviable position to accomplish this more general goal. A holistic discipline, it is not restricted to any narrowly defined subject or single theme that might limit our ability to appreciate what we might call the interrelat-

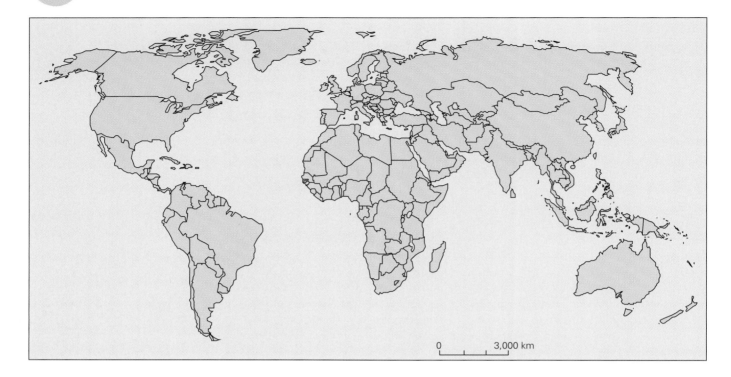

Figure Intro.1 (above)
World political divisions, 2003.

Figure Intro.2 (opposite)
World political divisions, 1938.

edness of things. Traditional links with physical geography clearly play a key role, but human geographers do not hesitate to draw on material and ideas from many other disciplines. This book explicitly acknowledges the controlled diversity of human geography.

The Human Geographer at Work

What do human geographers actually do? What questions do they ask and what problems do they strive to solve? These questions raise others. What is the nature of the world we live in? How do we live in the world? The following four vignettes offer some preliminary insights into the concerns and activities of human geographers, insights that will be further expanded in later chapters. The first two vignettes focus on the ways in which our world is divided; the second two look at some of the ongoing transformations of global life.

A World Divided 1: Political Divisions

Figure Intro.1 is a simple map showing how the land surface of the earth is divided into countries—one example of the diversity of human life. What do these divisions tell us? Have they always existed? Are the divisions changing today—possibly even between the time I write and the time you read these

words? Is the total number of countries increasing or decreasing? What do these political divisions reflect—population distribution, cultures, development? Is it possible to aggregate countries? Are there meaningful groupings? Do basic aggregations such as north versus south, or east versus west, have any value? These are the sorts of questions that human geographers ask.

Figure Intro.2 is similar to Figure Intro.1, except that it depicts the political divisions that existed in 1938. To say that the two maps are different seems elementary, but it is a crucial point. The human world does indeed change—substantially. The major changes evident when we compare the two figures reflect the decolonization of Africa and Asia, the political consequences of the Second World War, and the post-1989 political transformation of Europe and the former USSR (Box Intro.1). The most general consequence is a significant increase in the number of countries, from about seventy in 1938 to more than 190 in 2003. There is little reason to believe that a political world map in 50 years' time will closely resemble that of today. The contemporary political world is characterized by two tendencies in particular: first, a tendency for regions to begin a process of integration, as in the case of the European Union, and, second, a tendency for portions

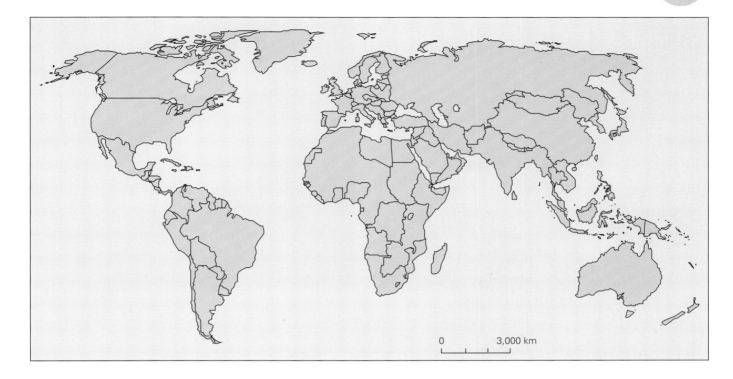

0   3,000 km

GERMANY: A NEW COUNTRY

The new Germany was born on 3 October 1990 as a result of the unification of West Germany and East Germany. That this unification was achieved peacefully is remarkable, given the political tensions that had plagued Europe since 1945. A united Germany was one product of the sweeping changes that occurred throughout eastern Europe beginning in 1989.

Germany first appeared as a country on the map of Europe in 1871, following the Franco-Prussian War. The specific territory changed in accord with a series of military activities (most notably, of course, the two world wars), and in 1945 Germany was occupied by the four Allied powers and divided into four zones: American, British, French, and Soviet. In 1949 the three western zones linked to create West Germany. The Soviets chose not to cooperate and instead created East Germany. Gradually, during the 1950s, links between the two diminished and an 'Iron Curtain' was put in place separating not only East and West Germany but eastern (communist) and western (democratic) Europe. The most notorious indicator of this division was the Berlin Wall, initially set up in 1961, which divided the then international city of Berlin. Berlin, although located in East Germany, was a microcosm of Germany, with American, British, French, and Soviet sectors. In 1989, however, the wall (literal and figurative) dividing Germany began to crumble.

The changes in eastern Europe began in Poland and quickly spread, removing communist governments in favour of democratic political systems. For Germany, the remarkable outcome was a decision, agreed to in February 1990 at a meeting in Ottawa, to unify Germany. Technically, East Germany disappeared some eight months later, absorbed into West Germany, and the united country is now known simply as Germany.

Germany is again a major European power with a population of 82 million (as of 2002) and a new capital in Berlin, but numerous problems remain to be faced. Economic problems in the former East Germany include the need to make industry more competitive and to modernize agriculture. In the former West Germany, economic recession and inexpensive imports have given rise to an economic and social crisis in the major industrial region of the Ruhr. Major environmental problems, including high levels of pollution and related industrial hazards, are apparent in the east. It is suspected, for example, that large areas of forest have died because of sulphur dioxide pollution, and that groundwater is substantially polluted. There are serious social problems, especially in the urban areas, resulting from the differences between the former states and the immigration of labour from elsewhere. The collapse of communism in the east caused mass unemployment, which has prompted a tightening of the rules for admitting refugees. Right-wing nationalists exploit social discontent and direct racist violence against foreign workers, especially Turks and Jews. In 1992, 80 Jewish cemeteries were desecrated, roughly the same number as in the five-year period from 1926 to 1931. Despite all these problems, however, the new Germany is clearly playing a key role in the Europe of the twenty-first century.

Urban street scene, Canada
(above)
(Dick Hemingway, 1999).

Rural family in El Salvador (right)
(International Fund for Agricultural
Development photo/ Louis Dematteis).

of some countries to seek a separate political identity, as in the case of Quebec within Canada. (These ongoing trends are explicitly addressed in Chapter 9.)

Human geographers are interested in these issues for at least three general reasons. First, the political partitioning of the world is an example of one of our recurring themes: regionalization, or the organization of space. Second, political units usually play a major role in determining such matters as the numbers of births and deaths and the availability of paid employment. Third, many political divisions reflect differences in ethnicity and/or culture (for instance, in language or religion). Human geographers are able to identify relationships between culture and political divisions, often with the help of maps. These and related issues are addressed specifically in Chapter 9.

A World Divided 2: Economic Divisions

The world is divided not only into political units but also into larger areas to which labels such as 'more developed' and 'less developed' are often applied. Our human world is full of diversity: languages, religions, ethnic identities, standards of living, all vary from place to place. There are differences between individuals and also more general differences between groups of individuals. Similarly, landscapes vary. Each particular location is unique, of course, different from every other—but it is often more useful to note the more general differences that exist

between regions. The photographs on this page suggest a basic distinction in our contemporary world. In the first we see middle-class people in a prosperous area of a modern western urban centre; we can probably assume that most of them are employed, own or rent comfortable homes, and have surplus ('disposable') income for purposes such as recreation. In the second image we see a family in a rural area of what is often called the less developed world: small farmers with little more than what they need for subsistence. Pictures like these suggest that the human world we are studying is in reality at least two worlds.

It is important to realize that major differences exist not only between major world regions but also between other spatial entities, such as regions inside countries (Box Intro.2). Even within a single city, the quality of life can differ enormously between the outer suburbs and the inner-city skid row. For more on this theme, see especially Chapters 6, 8, and 12.

A World Transformed 1: The Impact of Modernization

Table Intro.1 provides basic sample information on population totals, births and deaths, and life expectancy for various years from 1700 to 2002. Again you will notice that things have changed. Population totals have increased, the birth rate has declined, the death rate has declined, and life expectancy has increased. Why have these changes taken place? What do they mean for general

Table Intro.1 **SELECTED DEMOGRAPHIC DATA**

(a) World Population Totals (millions)		(b) Birth Rates and Death Rates (per 1,000), Finland			(c) Life Expectancy at Birth (number of years), France	
Year	Total	Year	Birth Rate	Death Rate	Year	Life Expectancy
1700	680	1755	45.3	28.6	1750	27
1800	954	1805	38.4	24.7	1825	41
1900	1,600	1909	31.0	17.7	1905	49
2002	6,214	2002	11.0	10.0	2002	79

Notes:
1. Finland and France are selected because both countries have better than average records.
2. Birth rate refers to the total number of live births in one year for every 1,000 people living.
3. Death rate refers to the total number of deaths in one year for every 1,000 people living.

human well-being? A process that we might label 'modernization' has been taking place. Birth rates decline primarily because of changing cultural conditions; death rates decline and life expectancy increases primarily because of technological advances related to sanitation and health. Population totals increase because of differences between birth and death rates. Clearly, however, as we already seen, this process of modernization has not had the same effects in all areas of the world.

Intro.2

LESS DEVELOPED CANADA?

Although Canada is a part of the developed world, as measured by various cultural and economic criteria, the benefits are not distributed equally among all Canadians. Throughout Canada, Aboriginal people live different lives from most other Canadians. Aboriginal communities are characterized by more youthful populations, shorter life expectancies, and higher birth rates that result in a high rate of natural increase (about double the national average and comparable to rates in many less developed countries). More generally, Aboriginal people in Canada are disadvantaged; they are more likely than other Canadians to be unemployed, to rely on social assistance, to live in poor-quality housing, to have limited access to medical services, and to have little formal education.

One of the more notorious settings was Davis Inlet, a community of about 680 people that lacked even running water and had deep-seated social problems. In late 2002 and early 2003, the community moved—at a cost to the federal government of some $152 million—15 kilometres across the Labrador Sea to the new village of Natuashish, in the hope that a new place would allow for a fresh start. The situation on reserves in the north is also deplorable. A survey of housing on northern Manitoba reserves estimated that 1,800 (29 per cent) of the 7,200 units were in need of major repair and that an additional 1,400 units were required. More than 50 per cent of reserve residents in Saskatchewan are dependent on welfare, and on some isolated reserves the rate is close to 90 per cent. Further, many communities lack piped water and sewage systems—a situation that has clear implications for health. For residents of these reserves, the rate of tuberculosis is eight times the provincial average. Frustration, hopelessness, depression, and apathy are common, with high rates of violent death, family breakdown, and substance abuse. That such conditions should persist in relatively close proximity to some of the most affluent communities in the world is a scandal; yet most Canadians are still able to ignore it, because Aboriginal people are typically marginalized, both spatially and socially.

Explaining this state of affairs is far from easy. Writing with general reference to the Canadian north, Bone (1992:197) noted that 'the notion of "ethnically blocked mobility" arises; that is, Native Canadians may value educational and occupational achievements less than other Canadians because of cultural differences such as the sharing ethic and because of perceived or experienced discrimination.' Other important reasons have to do with the history of relations between indigenous people and European Canadians and the nature of capitalism as a social and economic system.

United Nations headquarters, New York. Founded in 1951 with an initial membership of 51 countries, the UN now has 191 members. Its goals are to maintain international peace and security; develop friendly relations among nations; cooperate on solving international economic, social, cultural, and humanitarian problems and promoting respect for human rights and fundamental freedoms; and help to harmonize members' efforts those ends. The United Nations family of organizations is made up of the UN Secretariat, programs and funds such as the UN Development Programme (UNDP) and UN Children's Fund (UNICEF), and various specialized agencies. Individual programs, funds, and agencies have their own governing bodies and budgets, and set their own standards and guidelines. Together they provide technical help and other practical assistance in virtually all areas of economic and social endeavour (UN/DPI).

Modernization involves much more than these demographic changes; cultural and economic aspects of landscape and life have also been transformed over the past three centuries, and human geographers employ their various perspectives—those suggested in our three recurring themes—to describe and explain these changes. Questions of population growth and change are discussed in Chapter 5.

One aspect of modernization that is especially evident today in advanced economies, and that is causing changes both in landscape and in way of life, is the rise of postindustrial society. Among the changes associated with this process are an increase in the importance of information technologies, an economic transformation from a goods-producing to a service economy, the rise of professional and technical workers as a new middle class, and the related decline of the working class. It is possible that these ongoing changes will have consequences for our social and economic identities that will be just as dramatic as those of the industrial revolution. In North American and European cities, for example, there is evidence that traditional industries are declining, the profes-

sional workforce is growing, and the process of decentralization is accelerating. Such postindustrial cities include large numbers of people who are disadvantaged, unemployed, or 'working poor' (many of the latter are part-time workers).

There is a new geography of employment, a new geography of housing, and a new geography of neighbourhoods. Here again, the human geographer uses her particular approach (any one of our three recurring themes) to identify these ongoing trends, the reasons for them, and their character, and to analyze their impacts on landscapes and ways of life. These and related topics are discussed in detail in Chapters 12 and 13.

A World Transformed 2: Globalization

Our fourth vignette addresses what many observers consider to be the most important change—or, more accurately, set of changes—evident in the early twenty-first century. Put simply, our long history of creating distinctive lives and landscapes in essentially separate parts of the world is ending. More and more, we are seeing groups of people and the places they occupy changing in

Kensington Market, Toronto. In the Kensington neighbourhood, shops open out onto the streets, forming an old-fashioned market. Local markets offer a pleasant alternative to the more common urban experience of indoor malls and major grocery chains. Kensington's appeal is enhanced by its ethnic diversity. In earlier years many Jewish and Italian immigrants settled here, but eventually they moved on to more affluent neighbourhoods, and today the dominant cultures are Portuguese and Caribbean. Close by the market are two other distinctive areas: the city's original Chinatown and the College Street strip of upscale cafés and clothing stores (Dick Hemingway).

accord with larger global (as opposed to regional or local) forces. These global forces are something more than the movement between places that began in the fifteenth century as Europeans moved overseas, and are affecting all aspects of people and place. There is no single reason for these changes, but human geographers are always quick to point out that a key factor is humans' increasing ability to overcome what we call the friction of distance. Today we can move our goods and ourselves from one place to another much more rapidly than ever before, while ideas, words, and capital can travel anywhere in the world almost instantaneously. We will consider some detailed implications of this point in Chapter 10.

Culturally, as we will see in Chapters 4, 5, 6, 7, and 8, a particular set of attitudes, beliefs, and behaviours is spreading that reflects a more ecocentric environmental ethic, increasing insistence on gender equality, and a better understanding of the linguistic, religious, and ethnic bases for differences between groups of people. More generally, we are seeing the extension of some American characteristics to other parts of the world—what has been described as 'coca colonization'. Politically, as discussed in Chapter 9, many countries are working together for strategic reasons—the United Nations is the prime example of such a grouping. Economically, as discussed in Chapters 10, 11, 12, and 13, the connections between different parts of the world are constantly increasing, in terms both of product movement and of capital investment; corporate headquarters in major cities now control economic activities throughout the world.

But this description is incomplete, and therefore misleading. It is tempting to jump on the globalization bandwagon and begin interpreting all human geographic change in those terms. But even a brief reflection on our personal lives or a cursory glance at newspaper headlines will make it clear that the local still matters to us. Think of your own life, of the places you frequent and the people who matter to you. For most of us, it is our home and local community, our family, friends, and neighbours, that give shape and meaning to our lives. Distance has not become irrelevant, and the local has not been submerged under a tidal wave of globalization. Today there are many groups of people who want to see their specific culture recognized and their particular region attain some autonomous status; in the Canadian context alone, we can point to several separatist movements, the political aspirations of First Nations, and the tribal urge sometimes evident in the behaviours of some cultural groups. Certainly the forces associated with globalization are powerful, and their impacts are undoubted, but they are not the only forces at work today.

About This Book

The diversity of human geography means that we have much subject matter and many methods to introduce. Chapters 1 and 2 address two basic issues.

First, in Chapter 1 we investigate the nature of human geography. Our discipline has a long and distinguished history, involving many different groups of people and many noteworthy individuals working in different places at different times. The close ties between human and physical geography are evident. By the time you reach the end of this chapter, the origins of our three recurring themes should be clear.

Second, in Chapter 2, we look at the many ways in which human geography is studied today. This subject naturally includes many technical matters. Before turning to them, however, we will look at the philosophies behind each of our three recurring themes. Many of the ideas introduced in Chapter 2 may seem too abstract on your first reading, but as you read further you will find that many of them are central to the issues discussed in later chapters.

Chapter 3 provides basic factual information concerning the earth as a physical environment and as the home of humans. This material is an essential prologue to subsequent chapters that discuss how we have chosen to live in this home, but it is especially relevant to Chapter 4, which draws on physical geography to offer an overview of humans' global impact on and adjustment to the physical environment. Surveying humans' occupation of the earth through time, we consider the use and abuse of resources, and related environmental issues—issues of crucial importance to contemporary life. The theme of the relationship between humans and land is especially evident in this chapter.

The next two chapters are closely related: both deal with population issues, specifically questions of growth through time, both globally and in major world regions. Chapter 5 describes and suggests explanations for the exponential growth in global population that has occurred especially since about 1650, and identifies some of the consequences of that growth. Understanding population growth is essential if we are to understand many of the other issues discussed in this book. The use of population data as indicators of regional inequalities is discussed in Chapter 6, which provides a fuller account of the idea, introduced in Chapter 4, that the countries of the world can be divided into two general groups, less developed and more developed. All three of our recurring themes come into play in these two chapters.

Much human geography is best studied on a group scale, employing the concepts of culture and society as they relate to human landscapes. Chapter 7 focuses on the evolu-

tion and regionalization of landscapes, and includes discussions of such cultural factors as language and religion. Chapter 8 introduces some current theoretical concepts in order to clarify the relevance to landscape of social factors such as gender, ethnicity, and class. Like Chapter 6, Chapter 8 encourages us to think about various examples of inequality in the human experience from place to place. Two of our three recurring themes, humans and land and regional studies, are central to these discussions.

Political groupings of people in space are the subject of Chapter 9, which looks at the reasons behind the current partitioning of the world into territories. Because division into political units is the most fundamental of the ways in which humans divide the world, this chapter includes discussions of the significance of these divisions for our way of life. Perhaps the most compelling aspect of this chapter is its discussion of the considerable evidence indicating that many of the tensions in our contemporary world—tensions that are reported regularly in the news media—result when politics and culture disagree on the way space should be organized. Here again, as in the two preceding chapters, two themes—humans and land and regional studies—are crucial.

One feature of the contemporary world that will become obvious in Chapters 4 through 9 is the increasing necessity for human geographers to discuss topics at a global (as opposed to merely regional or national) scale. To point out this trend is one way of saying that our world is becoming more and more integrated and interconnected. Located at a pivotal point in this textbook, Chapter 10, 'Interaction and Economic Globalization', builds on much of the content of Chapters 5 through 9 and anticipates much of the content of Chapters 11 through 13. Discussions of the means by which humans overcome distances through transportation and trade lead logically to consideration of globalization processes. At the same time the opportunity is taken to reinforce some of the fundamental human geographic concepts introduced in Chapter 2.

The next three chapters share a common thrust in that each focuses on a particular human activity and includes a good deal of spatial analysis—our third recurring theme—

while also reflecting our other two themes. Chapter 11, on agriculture, asks why specific agricultural activities are located where they are; describes how the various major agricultural regions around the world have evolved; discusses contemporary agricultural landscapes in both the less and the more developed worlds, using political economy and political ecology perspectives; and considers issues related to food consumption.

Chapter 12, on settlement, discusses the reasons behind settlement patterns, both rural and urban; the differences between rural and urban experiences; and the origins, growth, and contemporary circumstances of urban centres. Chapter 13, on industrial activity, looks at the reasons behind industrial

location decisions; the period known as the industrial revolution; contemporary industrial circumstances in a global context; the complex processes of industrial restructuring (possibly related to a new postindustrial phase); and the general tendency for countries to exhibit uneven industrial development. This last discussion is closely linked to the earlier discussions of inequalities in Chapters 6 and 8. Finally, the Conclusion discusses the future of human geography and the human geography of the future—if only tentatively!

For a summary of the structure of this book, indicating the chapters of most relevance for the various subdisciplines of human geography, see Box Intro.3.

THIS BOOK AND SOME SUBDISCIPLINES OF HUMAN GEOGRAPHY

Intro.3

This book has 13 substantive chapters. Although each of them can stand alone, many issues are discussed in more than one chapter—a reflection of the fact that the book could have been structured in several different ways. Particularly important links between chapters are pointed out at the end of each chapter under the heading 'Links to Other Chapters'. It is hoped that these links will encourage you to appreciate the interrelatedness of issues that, for organizational reasons, are discussed in different parts of the book. Although the present arrangement was judged to be the most appropriate, there are other ways of presenting an introduction to human geography. For example, we might have structured the book around a single central theme, such as less developed/more developed, rural/urban, or global/local distinctions, or perhaps focused on one such theme within the existing structure. The basic advantage of the present structure is the flexibility that it offers both students and instructors.

The sequence of chapters invites grouping into five sections:

Section	Chapters
History and methods of human geography	1 and 2
Physical geography and human activities	3 and 4
Population geography	5 and 6
Cultural, social, and political geography	7, 8, and 9
Economic geography	10, 11, 12, and 13

Following an introductory course in human geography, it is usual for departments of geography to offer a series of courses

on systematic subdisciplines, regional topics, and techniques and methods. Because this book aims to introduce the entire spectrum of human geography, it should provide a background for all such advanced courses. It provides the basic understanding required for courses on regional topics, such as the geography of Canada, as well as the background required for courses on techniques and methods (in Chapter 2).

There is no universal agreed-upon list of systematic subdisciplines, of course, but some typical examples are identified below, along with the chapters in which they are discussed:

Systematic subdiscipline	Chapters
Geography of natural resources	3 and 4
Geography of energy	4 and 13
Population geography	5 and 6
Cultural geography	7 and 8
Social geography	8, 12, and 13
Political geography	9
Economic geography	10, 11, 12, and 13
Transportation geography	10
Geography of trade	10
Agricultural geography	11
Settlement geography	12
Rural geography	11 and 12
Urban geography	12 and 13
Industrial geography	13
Geography of tourism	13
Regional development and planning	13

As you are beginning to realize, human geographers seek to understand both the world we live in and how we live in the world. To achieve these aims, this book incorporates not only a wide variety of facts about humans and their world, but a number of different approaches to those facts. Our understanding of human geography is enhanced by acknowledging the links between human geography and physical geography, as well as other social science disciplines. Where appropriate, these links are highlighted.

About Being a Human Geographer

Human geography is worth studying not only for the importance of the subject matter itself but also for the career training that it offers. Many graduating students find that a background in human geography, especially in conjunction with physical geography, leads to diverse employment opportunities in such areas as education, business, and govern-ment. Geography students are employed as cartographers, geographic information specialists, demographers, land-use and environmental consultants, social service advisers, and industrial and transportation planners, to mention only a few possibilities.

There are three distinctive features of an education in geography that make graduates particularly valuable to employers:

1. subject matter, physical and human, that provides the geographer with a specific perspective and knowledge base;
2. technical experience with data collection and analysis; and
3. recognition that any attempt to understand our contemporary world must take into account the crucial roles played by space and place.

Human geography, then, is both a scholarly pursuit that deepens our understanding of the world and a practical area of study that leads to varied career opportunities.

SUMMARY

Geography
Literally, writing about the world.

Three recurring themes
1. Humans and land: the human world, a landscape, is continuously changing because of human actions within institutional and physical frameworks.
2. Regional studies: describing the earth and dividing the whole into parts—regions—in which one or more variables exhibit some uniformity.
3. Spatial analysis: explaining the locations of geographic phenomena using abstract arguments and quantitative procedures.

Because the human world is constantly changing, human geography typically includes a time dimension.

Human geography/physical geography
Related disciplines, both traditionally and for logical reasons. Best introduced separately at the introductory level because physical geography is a physical science, while human geography is a human science.

Our subject matter
Human behaviour as it affects the earth's surface.

Goal of the book
To provide a basis for comprehending the human world as it is today and as it has evolved.

This book is about
The world we live in and how we live in the world.

About being a human geographer
As a human geographer, you will have an understanding of our world and a sound training for a variety of careers.

FURTHER EXPLORATIONS

AGNEW, J., D.N. LIVINGSTONE, and A. ROGERS, eds. 1996. *Human Geography: An Essential Anthology*. Oxford: Blackwell.

An interesting collection of readings for both new and advanced students of human geography.

DEMKO, G.J., J. AGEL, and E. BOE. 1992. *Why in the World: Adventures in Geography*. Toronto: Anchor Books.

Aimed at the popular market, this entertaining volume asks and answers a multitude of questions to demonstrate the practical relevance of geography.

DOUGLAS, I., R. HUGGETT, and M. ROBINSON, eds. 1996. *Companion Encyclopedia of Geography: The Environment of Humankind*. London: Routledge.

A comprehensive volume centred on the idea that geography reflects the interdependence of humans and environment.

FIELDING, G.J. 1974. *Geography as Social Science*. New York: Harper and Row.

A human geography textbook in the spatial analysis tradition (the third of our three recurring themes).

HANSON, S.E., ed. 1997. *Ten Geographic Ideas That Changed the World*. New Brunswick, NJ: Rutgers University Press.

A cleverly conceived book written in a style accessible to introductory-level students and containing statements by leading practitioners of the discipline; a good read.

JACKSON, W.A.D. 1985. *The Shaping of Our World: A Human and Cultural Geography*. New York: Wiley.

A human geography textbook with an especially strong emphasis on relations between humans and land (the first of our three recurring themes).

JOHNSTON, R.J. 1988. 'There's a Place for Us'. *New Zealand Geographer* 44:8–13.

An article, written by one of the more perceptive commentators on contemporary human geography, that provides an enlightening perspective on what human geographers do and why they do it.

———, ed. 1993. *The Challenge for Geography: A Changing World, A Changing Discipline*. Oxford: Blackwell.

This book demonstrates that the world is changing rapidly, argues that the discipline of human geography needs to respond accordingly, and asks whether human geographers should influence the direction of change.

ROGERS, A., H. VILES, and A. GOUDIE, eds. 1992. *The Student's Companion to Geography*. Oxford: Blackwell.

A clear and comprehensive student guide to both physical and human geography comprising over fifty entries on such diverse topics as the history of geography, the art of interviewing, world libraries, and careers for geographers.

ON THE WEB

http://atlas.gc.ca/
An excellent resource for learning about the geography of Canada; useful throughout this book.

http://www.canadainfolink.ca/geog.htm
Numerous useful resources for teachers and students of Canadian geography.

http://www.geography.about.com/science/geography/?once=true&
A site designed to attract a wide audience; addresses basic popular geographic questions.

http://www.geography.com.sg/
An online resource for younger geography students that aims to introduce interesting concepts and issues in an educational style.

http://geography.pinetree.org/
A virtual library for geography; includes a list of websites on geography and geographical issues.

http://www.schoolnet.ca/
A publication of the Canadian government that offers a wide range of materials intended for teachers; includes some excellent human geographic material.

What Is Human Geography?

This chapter provides an overview of the origins and evolution of human geography, describing advances in geographic knowledge through time and noting how this knowledge was gradually organized to form a new academic discipline. The approach is chronological: from the major contributions of the Greek, Chinese, and Islamic civilizations, to the period of European overseas exploration, which began in the fifteenth century and led to the development of new techniques of mapping and improved descriptions of the world, to the ambitious writings of the great nineteenth-century geographers such as Humboldt and Ritter.

Beginning with the attempts by seventeenth- and eighteenth-century writers such as Varenius and Kant to organize geographic knowledge, the formal discipline of geography emerged in European and North American universities during the late nineteenth and early twentieth centuries. The chapter concludes with a look at the changing character of contemporary geography, the links between physical and human geography, and the current status of human geography as a social science.

A detail from Nicolas de Fer's map 'America', 1698 (National Archives of Canada, NMC–026825-1).

To understand contemporary human geography, it is essential to understand how it has changed through time. Indeed, there would be cause for great concern if such change did not occur. *Geography, like most other academic disciplines, functions to serve society. As society and societal requirements change, so does geography.* Geographers have always had a consistent purpose: to describe and explain the world. It is only the manner in which they approach this task that changes.

For example, for many years geographers were principally involved in discovering, describing, and explaining an increasingly better-known world. As unknown areas became known, geographers worked feverishly either to fit new facts into established knowledge or to propose radically new knowledge bases. Their most important tool was the map. More recently, since the nineteenth century, geographers have reoriented their activities. Once basic global descriptions were in place, the emphasis shifted to developing better explanations and clearer understandings of geographic facts.

Full of drama and intriguing individuals, the history of geography is a fascinating subject in its own right. At the same time, it provides the background that is essential if we are to fully understand contemporary geography.

Preclassical Geography

It seems likely that the earliest geographic descriptions took the form of maps—a simple but effective means of communicating spatial information. No doubt maps have been created, temporarily at least, by all human groups. Roughly sketched with a stick in sand or soil, or carefully scratched into rock or wood, maps could be used to show the location of water, game, or a hostile group. To the extent that geography is about maps, humans have always been geographers. Lack of hard evidence need not prevent us from recognizing the centrality of geography to our human existence. Our ancestors could not function without maps any more than we can today.

The world's first civilization emerged in Mesopotamia (now southern Iraq) some time after 4000 BCE. Surviving Mesopotamian maps, drawn on clay tablets, typically show local areas, reflecting limited knowledge of areas beyond the immediate environment. Geographical knowledge was probably similarly limited for all the civilizations

The GA.SUR or Nuzi map. This early map was engraved on a clay tablet c. 2200 BCE and discovered in 1930–1 by archaeologists digging at an ancient site in what is now north-eastern Iraq. The original tablet has since been lost. Measuring just 7.6 x 6.5 cm, the map was oriented with the east at the top and showed an area bounded by two ranges of hills and bisected by a stream. It seems to have been intended to indicate the location of a parcel of land to the west of the stream, perhaps for purposes of defining ownership or assessing taxes (Semitic Museum, Harvard University).

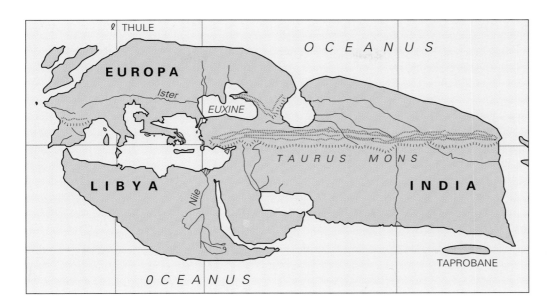

Figure 1.1 The world according to Eratosthenes.

before that of classical Greece: namely, the civilizations in the Nile Valley, the Indus Valley, China, Crete, the Greek mainland (Minoan and Mycenaean), and southeast Mexico (Olmec).

Classical Geography

With the emergence of classical Greece, shortly after 1000 BCE, geographic understanding—and hence maps—for the first time began to cover more than small local areas. The reason was that the Greeks were the first civilization to become geographically mobile and to establish colonies. Greek scholars initiated two major geographic traditions.

The first tradition is *literary*, involving written descriptions of the known world. Many Greek scholars contributed to this tradition, although relatively few of them were geographers. Herodotus (484–*c.*425 BCE), for example, is best known as the first great historian, but he was also an accomplished geographer. His descriptions of lands and peoples—based on observations made during extensive travels to such areas as Egypt, Ukraine, and Italy—make it clear that he saw geography as a necessary background to history and vice versa. Possible relationships between latitude, climate, and population density were noted by Aristotle (384–322 BCE), who also speculated about the ideal locations for cities and the conflicts between rich and poor groups.

Eratosthenes (*c.*273–*c.*192 BCE)—often considered the father of geography because he coined the word—contributed to the literary tradition by writing a book describing the known world, though unfortunately no copy of this work has survived. He also mapped the known world at the time: the Mediterranean region and adjacent areas in Europe, Africa, and Asia (Figure 1.1). We know about the contributions of Eratosthenes and others because the literary tradition they began was summarized by Strabo (64 BCE–20 CE) in a multi-volume work that has survived, entitled *Geographia*. An encyclopedic description of the entire world known to the Greeks, it consists of two introductory books, eight books on Europe, six on Asia, and one on Africa. Much of what we know today about Greek geographers before Strabo comes from his detailed summaries of their work.

The second tradition is *mathematical*. Thales (*c.*625–*c.*547 BCE), one of its originators, took a scientific view of the world and successfully predicted an eclipse of the sun in 585 BCE. By the fifth century BCE the Greeks knew that the earth was a sphere, and in the second century BCE Eratosthenes calculated the circumference of the earth. Shortly after that, Hipparchus devised a grid system of imaginary lines on the surface of the earth based on the poles and the equator—these, of course, were the lines of **longitude** and **latitude**, and they made

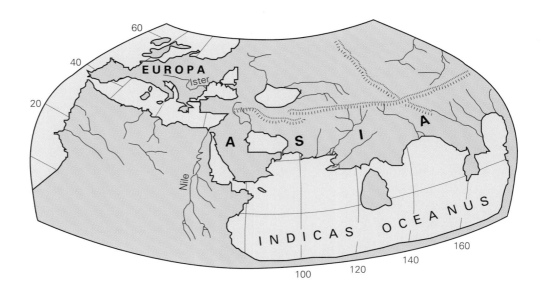

mapping considerably more accurate. Determining the exact position of any location, however, remained difficult. Latitude was relatively easy to calculate by observing the angle of the sun's shadow using an early version of a sundial, but longitude continued to require estimation because there was no way to measure time precisely, especially at sea. In addition, Hipparchus was the first person to tackle a problem that remains with us today: how to map the curved surface of the earth on a flat surface.

Much of this mathematical tradition was summarized by the Alexandrian Ptolemy (90–68 CE) in his eight-volume *Guide to Geography*. Using the grid system devised by Hipparchus, Ptolemy produced the first index of places, or gazetteer, of the world. Although his gazetteer is sometimes fanciful (for instance, he believed that a great continent must exist in the southern ocean; see Box 1.4, p. 25), and includes numerous errors because latitude was not carefully measured and longitude was necessarily estimated, Ptolemy also produced a world map (Figure 1.2) that includes a grid system and that in its details is generally a clear improvement over Eratosthenes' map (Figure 1.1).

Together, the written works of Strabo and Ptolemy and the world map of Ptolemy provide a useful indication of the achievements of Greek civilization in geography, for they make it clear that geography emerged and evolved in response to larger societal requirements. The civilization of ancient Rome added little to geographic knowledge, despite the expansion of the Roman empire. Apart from some geographic guidebooks published in response to exploratory and expansionist needs, there was little development of either the literary or the mathematical tradition begun by the Greeks. Although Roman civilization continued until the fifth century CE, classical geography effectively ended with Ptolemy, and it seems that Roman leaders continued to rely on Ptolemy's work.

Thus the classical geography of the Greeks is the true beginning of our contemporary discipline. Geography in the classical world was concerned with the locations and interconnections of places on the surface of the earth. In addition to the word 'geography', the Greeks introduced 'cosmography', 'chorography' and 'topography'. **Cosmography** refers to the universe, heavens and earth. **Chorography** refers to places smaller than the earth, such as countries, while **topography** refers to local areas within countries. The Greeks' introduction of these terms indicates that they recognized the importance of differences in scale. As we have seen, the Greeks also produced relatively accurate maps using a sophisticated grid system. What is perhaps most significant is that the mapping procedures devised by Ptolemy have persisted to the present. Both the basic style and the language of the maps that we use today originated with Ptolemy.

The Fifth to Fifteenth Centuries: Geography in Europe, China, and the Islamic World

The period from the fifth to the fifteenth century was one of only sporadic and limited geographic work in Europe. Elsewhere, however, especially in China and the Islamic world, geography flowered. Again we find a close relationship between expanding societies and a thirst for geographic knowledge.

The European Decline

The word 'geography' did not enter the English language until the sixteenth century. Medieval Europeans knew little beyond their immediate environment. Over the centuries, much of the knowledge gained by the Greeks was lost, and the only place where any kind of geographic work continued was in the monasteries. The general assumption, inside and outside the monasteries, was that God had designed the earth for humans (this doctrine is called **teleology**). In effect, during the Middle Ages geography as such no longer existed.

The clearest evidence of decline can be seen in maps of the period. The ancient Greek maps had been drawn by scholars with expertise in astronomy, geometry, and mathematics. The medieval European map-makers, by contrast, were more interested in symbolism than scientific facts. Stylizing geographic reality in order to arrive at a predetermined structure, they produced maps that are less detailed and accurate than those produced 1,500 years earlier by the Greeks. The best

examples are the 'T-O' maps produced between the twelfth and the fifteenth century. Consisting of a T drawn within an O (Figure 1.3), they show the world as a circle divided by a T-shaped body of water. East is at the top of the map; above the T is Asia; below left is Europe, and below right is Africa. The cross of the T is the Danube–Nile axis, the perpendicular part is the Mediterranean, and the map is centred on Jerusalem. In these maps depicting scriptural dogma—what Christians were expected to believe—symbols triumphed over facts. Another type of medieval map divided the world into climatic zones, largely hypothetical, on either side of the equator. Others still were lavishly decorated: the Ebstorf *Mappemonde* (*c.*1284) had as background a picture of the crucifixion, while the Hereford map (*c.*1300) was really an encyclopedia. Perhaps the only medieval maps that served a practical purpose were the ones known as Portolano charts (Figure 1.4). Dating from

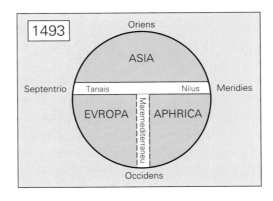

Figure 1.3 An example of a T-0 map.

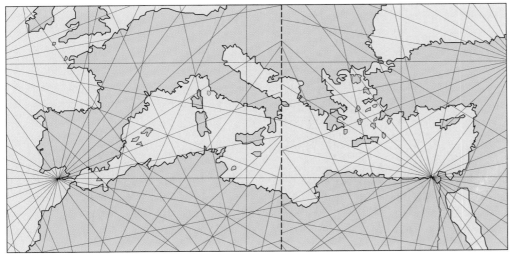

Figure 1.4 An example of a Portolano chart.

about 1300, these maps depicted a series of radiating lines. The lines did not serve to locate positions on the map, nor did the maps use any projection. Nevertheless, they often succeeded in locating coastlines accurately.

Medieval Europeans made very few contributions to geographic knowledge. Although Norsemen sailed to Greenland and North America, and Christian Europeans embarked on a series of crusades and military invasions to the Holy Land, the results, as far as geographic knowledge was concerned, were minimal. The most significant exploratory journey was that of Marco Polo (1254–1323), a Venetian who visited China and wrote a description of the places he visited. He was unable to add to Greek knowledge, however, because he was largely unaware of it. The distinction is not always an easy one to grasp, but Marco Polo was an explorer, not a geographer.

It was not in Europe but in China and the Islamic world that the major geographic advances took place during the period after the Greeks and before the fifteenth century.

Geography in China

In China a great civilization—clearly the ancestor of contemporary China—developed before 2000 BCE. The longest-lasting civilization in the world, China inevitably made important contributions to geographic knowledge. Writings describing the known world of the Chinese date back to at least the

fifth century BCE. Later, the Chinese also explored and described areas beyond their borders; in 128 BCE, for example, Chang Chi'en discovered the Mediterranean region, described his travels, and initiated a trade route. Other Chinese geographers reached India, central Asia, Rome, and Paris. Indeed, Chinese travellers reached Europe before Marco Polo reached China (Box 1.1).

There is one important respect in which early Chinese geography differed from the European equivalent. It is a difference of geographic perspective—a different way of looking at the world. Traditionally, Chinese culture has viewed the individual as *a part of* nature, whereas Greek and subsequent European culture have typically viewed the individual as *apart from* nature. This distinction reflects the differing attitudes underlying Confucianism (which dates from about the sixth century BCE) and Christianity. Given their view of humans and land as one, it was natural for Chinese descriptive geographers to integrate human and physical description.

Maps were central to geography in China, as elsewhere, and there is evidence that a grid system was in use during the Han dynasty (third century BCE to third century CE). It appears that the first Chinese map-makers were civil servants who drew and revised maps in the service of the state. Their maps were symbolic statements, asserting the state's ownership of some territory.

1.1

DID CHINA 'DISCOVER THE WORLD' IN 1421?

China was a great naval power in the early fifteenth century— so great that Chinese navigators may have reached North America 70 years before Columbus, circumnavigated the globe 100 years before Magellan, and reached Australia 350 years before Cook. We know that Emperor Zhu Di sponsored several voyages during his reign (1403-24) for the purposes of exploration, mapping, and collecting tribute from other peoples. We also know that in 1421 the great admiral Zheng He set sail with many ships across the Indian Ocean to the east coast of Africa. However, a recent book— *1421: The Year China Discovered the*

World (Menzies 2002)—suggests that the expedition continued around the southern capes of Africa and South America and across the Pacific before arriving back in China in 1423. With the emperor's death the following year, the country entered a long period of isolation, and presumably this was why the story of the voyage remained unknown in Europe. Menzies' theory— based in part on the existence of European maps from as early as 1428 depicting regions that Europeans themselves had not yet seen—is fascinating, although cartographers are not likely to rewrite their histories without some additional evidence.

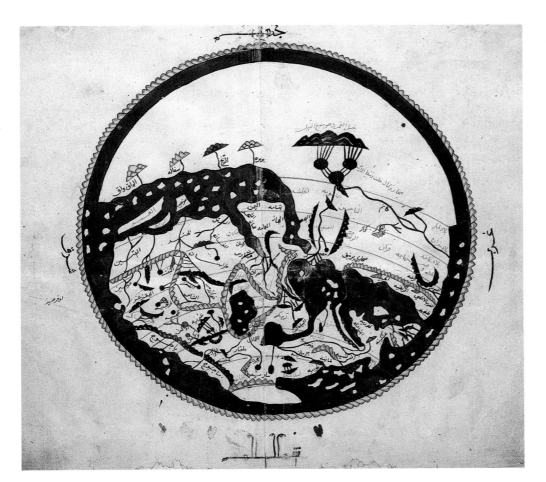

Al-Idrisi's circular world map: a copy (dated 1456) of the twelfth-century original. As early as the ninth century Muslim traders had travelled east to China by land and by sea; south along the Indian Ocean coast of Africa; north into Russia; and west as far as the Atlantic.

Al-Idrisi combined the knowledge acquired through these journeys with that available from the works of Greek and Persian geographers to produce the geographic description of which this map was a part. Depicting Europe and Africa as well as Asia, it is oriented with the south at the top (Bodleian Library, Oxford Pococke Manuscript 375, fols. 3c-4r).

Geography in the Islamic World

The second contribution to geography outside Europe came from the Islamic world. As the religion of Islam—founded in the seventh century CE by the prophet Muhammad (d. 632)—spread, it served as a unifying force, bringing together previously disparate tribes. Consequently, at the time when Europe was immersed in the Dark Ages, civilization flowered in Arabia. As Islamic conquests spread beyond the Arab region, the geographic knowledge base expanded to include north Africa, the Iberian peninsula, and India.

By the ninth century Islamic geographers were recalculating the circumference of the earth. From then until the fifteenth century, they and their successors produced a wealth of geographic writings and maps based on earlier Greek work as well as Islamic travels. Among the most notable contributions were those of al-Idrisi (1099–1180), whose book on world geography corrected many of Ptolemy's errors. Perhaps the best-known traveller was ibn-Batata (1304–c.1368), who

journeyed extensively in Europe, Asia, and Africa. A third major addition to geography came from ibn-Khaldun (1332–1406), a historian who wrote extensively about the relations between humans and the environment. Maps produced by Islamic geographers, including al-Idrisi, centred on Arabia.

An important eleventh-century Arabic atlas, previously unknown to modern geographers, was discovered in a private collection in 2002, and is now housed in a library in Oxford, England. Widely regarded as a missing link in the history of cartography, this two-volume, 96-page manuscript includes 17 maps, two of them depicting the world as it was known at the beginning of the second millennium. The fact that some of the maps show travel routes suggests that, unlike earlier Greek and later European maps, they were intended not to represent actual landscapes but to serve a practical purpose as memory aids for travellers.

Chinese and Islamic geographies prior to the fifteenth century were roughly comparable to Greek geography. Two traditions, mathematical and literary, developed, and map-making (see Box 1.2) was central to most geographic work. In all cases, the geographers' work reflected the knowledge and needs of particular societies. As Box 1.2 suggests, maps in particular are perhaps best understood as one component of the gradual development of human knowledge about the world and ourselves.

The Age of European Exploration and Discovery

By 1400, then, geographic knowledge had grown considerably, but the 'known world' on which it was based was still limited to a small portion of the earth. In the fifteenth century European scholars began to recognize this fact, and the impact on geography was dramatic and significant. All three components of the geography discussed so far—mathematical, literary, and cartographic—underwent rapid change. Between the fifteenth and seventeenth centuries, Europeans embarked on a period of unprecedented exploratory activity that happened to coincide with a decline in Chinese and Islamic explorations.

Many factors, among them the desire to spread Christianity and to establish trade routes, contributed to the surge in European exploration, but there were two interesting additional factors. Printing technology, which was first applied to maps by the Chinese in 1155, was first used for map production in Europe in 1472. In 1410, editions of the work of Ptolemy were printed; editions including maps (redrawn using new projections) appeared in 1477. Printing allowed information, including that contained on maps, to diffuse rapidly. The second additional factor, which reflected the period's thirst for knowledge, was the establishment of what might be called centres of geographic analysis. The first, initiated by Prince Henry of Portugal in 1418, focused on key geographic questions such as the size of the earth and the suitability of tropical environments for habitation. In addition, techniques of navigation were taught there, and Prince Henry set in motion a series of explorations along the western coast of Africa. Interestingly, the maps produced following these voyages did not always reflect new discoveries. It seems that a 1459 map by Fra Mauro deliberately concealed new information in order to maintain secrecy.

1.2

EVALUATING OUR PLACE IN THE WORLD

Maps are useful indicators not only of what we know about the world, but also of how we see ourselves in the world. It might be said that while we create maps, maps in turn recreate us. T-O maps, for instance, served to reinforce the Christian world-view, while the ability to map an area suggests power—perhaps even ownership rights—over it. Maps can be, and often are, used to convey messages and are thus powerful symbolic representations. Nevertheless, most early maps were explicitly practical, whether they were used by mariners for navigation or by state officials as aids to tax collection.

The long history of map-making might be seen as one of slow intellectual growth and gradual spatial expansion. We begin with knowledge of our own bodies before moving on to our immediate surroundings, gradually progressing to a partially known world, and finally to detailed knowledge of the world. Throughout this process, the unknown is usually seen

as dangerous, and often it is only with great reluctance that we allow newly discovered fact to triumph over familiar fiction. Herodotus did not accept Phoenician accounts of a circumnavigation of Africa about 600 BCE, and when Pytheas sailed out of the Mediterranean to England and Iceland in about 320 BCE he was branded a liar by Strabo because he was describing places presumed not to exist. In both of these cases, evidence contradicting existing maps was rejected because it did not fit the story that was accepted and understood.

Even today, maps are both enabling and constraining. They enable us to see what our own vision does not permit, but the fact that they necessarily simplify complex realities constrains our understanding. For more on these themes, see Chapter 2, especially Boxes 2.7 (on map projections) and 2.8 (on maps that are intentionally misleading). *The maps that we create in turn recreate us.*

Exploration

Exploration is not, of course, geography. Yet it was exploration that furnished new facts and provided the basis for new maps and new descriptive geographies. The relationship between geography and exploration is symbiotic, with maps and books encouraging exploration and exploration in turn generating new maps and books. The major explorations were led by Bartolomeu Dias around southern Africa (1486–7); by Christopher Columbus to North America (1492–1504); by Vasco da Gama to India (1497–9); and by Ferdinand Magellan, who reached Asia by sailing west (1519–22). The last in this line of great explorers was James Cook, who made three voyages into the Pacific (1769–80). It was Cook who corrected one of Ptolemy's greatest errors, revealing that the supposed southern continent did not exist. By 1780 the basic outlines of the world map were established, adding considerably to Europeans' knowledge of the world—and thus to the geographer's task (Box 1.3).

Exploratory activities became easier as several basic hurdles were overcome. Most important was the absence, recognized a thousand years earlier by the Greeks, of an accurate method of determining location. As we have seen, establishing latitude was no problem, but an instrument for establishing longitude at sea was not available until 1761 and was not used on a major voyage until Cook's second voyage, in 1772–5.

Mapping

It was during the age of exploration that science in general changed from being a practice controlled by the church to one concerned with the acquisition of knowledge. In response to the demands of sea travellers, maps in this period returned to the model developed by the Greeks—a model in which facts triumphed over imagination. Unlike medieval maps, with their mixture of fantasy and dogma, the typical map from the fifteenth century onwards was functional. Maps showing the grid of latitude and longitude replaced T-O maps centred on Jerusalem. One of the pioneers in these endeavours was the Polish scientist Nicolaus Copernicus (1473–1543), who produced maps of Poland that were used to settle a variety of boundary disputes.

The demand for geographic knowledge, and hence for maps, coincided with the rise of printing to encourage new developments. Gerardus Mercator (1512–94) was undoubtedly the most influential of the new mapmakers. He tackled the crucial problem of projection: how to represent a sphere on a flat surface. His answer was the famous 1569 Mercator projection, still used extensively today. This projection, which showed the earth as a flat rectangle with a grid of latitude and longitude lines, was enormously useful to sea travellers. By the early seventeenth century, Mercator's map had replaced all earlier charts used at sea, including the Portolanos.

1.3

EXPLORATION OR INVASION?

The overseas expansion of Europe that began in the fifteenth century resulted in much more than new knowledge and new tasks of description and cartography. Although that expansion has characteristically been interpreted as simply a matter of exploration and discovery, another interpretation is possible, less innocent and less Eurocentric. As a consequence of expansion, Europeans were able to diffuse their power and authority throughout much of the world—a process that was a principal cause of the inequalities that currently prevail.

Indeed, alternative terms for 'exploration' and 'discovery' are 'invasion' and 'conquest'.

The various European activities involved in expansion gave birth and, eventually, credibility to the idea that the European world was distinctive from other worlds; Europe was seen as superior, powerful, enlightened, and civilized, while the rest of the world was seen as inferior, powerless, ignorant, and barbaric. It was against this intellectual backdrop that the discipline of human geography and the various other social sciences began to emerge.

Mercator's map of the North Pole, 1595 (National Archives of Canada, NMC-016097).

Another map-maker, Abraham Ortelius (1527–98), produced the first modern atlas in 1570; such was the demand for geographic knowledge that it ran into 41 editions by 1612. From Mercator onwards, map-making has steadily improved and consistently reflected new geographic knowledge.

Geographic Description

The awakening of Europe and the burst of exploratory activity had an impact not only on map-making but also on geographic description. European scholars had to make sense of the 'new world'. Once again, society turned to geography to provide answers to a multitude of questions concerning the shape of the earth, the location of places, physical processes, and human lifestyles. Just as earlier Greek, Chinese, and Islamic geographers had described the worlds known to them, so the European geographers of the sixteenth century onwards needed to describe the world they knew—one that was constantly expanding and changing. Now geographers faced an enormous task: writing about all aspects of the entire world.

The resulting flurry of geographic writing was of mixed quality. Some works of fiction were regarded as factual. Imaginative works by an author of uncertain identity, Sir John Mandeville, were reprinted three times in 1530 alone. Other authors perpetuated Ptolemy's

error concerning the 'great southern continent'; as late as 1767 a Scottish geographer, Sir Alexander Dalrymple (1737–1808), wrote that almost all of the unknown areas between the equator and 50°S were land (Box 1.4).

Other geographers, however, largely reflected available knowledge and excluded content that was essentially surmise. Their works, along with developments in map-making, provide a clear indication of the progress of geography and geographic understanding. Among them was Peter Apian (1495–1552), a map-maker and writer who in 1524 published a book that divided the earth into five zones (one torrid, two temperate, and two frigid), provided notes on each continent, and listed major towns. Sebastian Münster (1488–1552), a contemporary of Apian, produced *Cosmography* in 1544, the first major work following the initial burst of European expansion activities that included descriptions of the major regions of the earth. The book is a careful compilation of existing knowledge, but is relatively weak for areas other than Europe. Nevertheless, it remained the standard geographic work for at least 100 years and was printed in 46 editions and six languages.

The period from the early fifteenth century to the mid-seventeenth was one of enormous growth in geographic fact and fiction alike. Geography itself continued in the tradition begun by the Greeks. The process of discovering and making sense of ourselves and our world meant questioning old ideas as well as exploring. For geography, the result was a dramatic increase in factual content. Map-makers and writers strove to keep pace with these questionings and explorations and to encourage more. By the mid-seventeenth century, geography's main concerns were still mapping and world description.

Geography Rethought

Varenius

Contemporary geography continues to be concerned with map-making and description, but it also addresses many other questions. The first indication of these new questions appeared in 1650 with the publication of *Geographia Generalis* by Bernhardus Varenius (1622–50) (Box 1.5), which remained the standard geographic text for at least a century; the last English-language edition appeared in 1765. What made this work so distinctive and important was that in it Varenius provided an explicit definition of geography:

> Geography is that part of mixed Mathematics, which explains the State of the Earth, and of its Parts, depending on Quantity, viz. its Figure, Place, Magnitude, and Motion, with the Celestial Appearances, etc. By some it is taken in too limited a Sense, for a bare description of the several Countries; and by others too extensively, who along with such a Description would have their Political Constitution.

Clearly, Varenius was interested in establishing an appropriate place for geography in the system of knowledge as a whole. He went on to clarify the definition above:

1.4

THE SOUTHERN CONTINENT

The Greeks were the first to suggest that a large land mass should exist in the Southern hemisphere, to balance the large known areas of land in the North. Along with so many other provocative ideas, this notion was lost during the medieval period; the medieval belief in a flat world did not require any such symmetry. The first knowledge of the basic distribution of land and sea in the Southern hemisphere was gained with the beginnings of European overseas exploration in the fifteenth century. Early voyages were often prompted by incorrect information and sometimes they returned with fiction as well as fact. For example, in the early sixteenth century Gonneville claimed to have discovered a tropical paradise in the south, inhabited by an easygoing, contented people. All records of this voyage were lost, but the claim clearly influenced French exploration for the next 200 years. Subsequent French and other explorers sought to rediscover this idyllic 'Gonneville land'.

When Magellan, on his voyage of 1518, discovered the Straits of Magellan, separating the island of Tierra del Fuego from the mainland of South America, it was thought that the island was the edge of the huge southern continent. Cartographers such as Orontius, in 1531, responded with maps that included the continent. By 1578, however, Drake had shown that Tierra del Fuego was not a large continent. Exploration then turned east with the discovery of the islands of South Georgia by de la Roche in 1675. The French continued to search for Gonneville's lost land. When, in 1772, Kerguelen reported that he had discovered a place promising all the resources the mother country desired—wood, minerals, diamonds, and rubies—he called his discovery South France, and many thought he had finally located the long-sought Gonneville land. Find Kerguelen Island in your atlas and you will see just how wrong Kerguelen was: the island named after him is in an isolated, far from tropical location in the south Indian Ocean. Clearly Kerguelen discovered what he wanted to discover, just as Gonneville had almost 200 years earlier.

The distribution of land and sea in the Southern hemisphere became clearer in the late eighteenth century, even as Dalrymple was unequivocally asserting that almost all of the area between the equator and 50° S was land. In 1773 Cook spent much of a long voyage searching for the southern continent, sailing farther south than any previous explorer; but instead of land, all he discovered was pack ice. Although, on his 1775 voyage, Cook discovered numerous small islands, all were inhospitable environments. The coastline of Antarctica was finally mapped during the nineteenth century.

This intriguing example is not necessarily atypical. Once a false idea is established, considerable effort may be needed before the true facts are known. False ideas can and do influence human behaviour.

We call that *Universal Geography* which considers the whole Earth in general, and explains its Properties without regard to particular Countries: But *Special* or *Particular Geography*, describes the Constitution and Situation of each single country by itself which is twofold, viz. *Chorographical*, which describes Countries of a considerable Extent; or *Topographical*, which gives a View of some place or small Tract of the Earth.

It is worth noting that the universal/special distinction noted by Varenius was an important intellectual issue for all branches of knowledge at the time. Universal geography, more typically called *general geography*, was seen as crucial. It was not adequate merely to provide detailed descriptions of particular places; it was also necessary to develop an understanding of laws that applied to all places. In geography as in any other subject, this is a basic distinction. Varenius saw the two as mutually supportive. Interestingly, as we shall see, twentieth-century geographers sometimes failed to appreciate the need for both types of study, arguing for one at the expense of the other.

Varenius confirmed geography as the study of the earth's surface, both physical and human. Detailed description—special geography—is essential, as is the development of generalizations—general geography. There was little that was really new here; the Greeks, for example, had contributed to both of these areas. But the explicit recognition of two different approaches—a recognition prompted by the remarkable increases in knowledge and the prevailing intellectual trends at the time—did mark a new phase in geography's evolution.

Broadening Vistas: 1650–1800

The one and a half centuries, 1650 to 1800, following the publication of Varenius's major work witnessed major advances in a wide range of geographic issues. Exploratory activity continued apace, as did map-making. In addition, geographic questions were asked that, while often prompted by exploration, had little to do with the acquisition of new facts and the mapping of newly discovered places. Questions were asked about such fundamental issues as the role of the physical environment as a cause of the growth of civilization, the unity of the human race, and the relationship between population density and productivity.

These were broad questions that generated a variety of responses. Most of the individuals

1.5

VARENIUS

Bernhardus Varenius was born in northern Germany in 1622. Following an education in philosophy, mathematics, and physics in Hamburg, he pursued further studies in Königsberg and Leiden before accepting employment as a private tutor in Amsterdam. In this stimulating environment, Varenius first applied his studies in a geographic context by responding to the unquenchable thirst for knowledge about overseas areas. In a book published in 1649 he provided Amsterdam merchants with information about Japan and southeast Asia, both areas with which the merchants were doing business. This was practical geography responding to the needs of society.

However, it appears that Varenius was aware that descriptive works of that kind made only a limited contribution to intellectual life. Accordingly, he began work immediately on his *Geographia Generalis*. Varenius died just six months later, but his last work was a landmark contribution to geography, at least in part because of its explicit acknowledgement that regional descriptions needed to be placed in a larger conceptual context if they were to be regarded as contributions to knowledge. *Geographia Generalis* was translated into several languages and was revised by the great mathematician and physicist Isaac Newton in 1672 and 1681 for students at Cambridge University.

Although his early death left much planned geographic writing undone, Varenius had rethought the nature of geography, beginning the process that would culminate in the late nineteenth century with the formal recognition of geography as a university discipline.

who tackled them were not geographers as such—in fact, knowledge had not yet been divided into the disciplines that we know today. Nevertheless, the views of J. Bodin (*c*.1528–96) and C. Montesquieu (1689–1755) on the role of the physical environment, of G.-L. Buffon (1707–88) on the unity of the human race, and of T.R. Malthus (1760–1834) on population and food supply were all eminently geographic. Each of these issues will be considered later in this textbook. For now, the point to note is simply that all of them were raised before 1800.

Other geographic writing reflected the current trends. Thus several important geographies were published after 1650 that stressed encyclopedic description. Most such studies concentrated on describing political units. Both A.F. Büsching (1724–93), in a six-volume description of Europe published in 1792, and J.C. Gatterer (1727–99), who saw geography as earth description and description of humans using the earth, focused almost entirely on regional description.

Immanuel Kant

Immanuel Kant lived in Königsberg, Germany, from his birth in 1724 until his death in 1804. For most of his adult life he taught at the University of Königsberg, becoming a professor of logic and metaphysics in 1770. Through his writings, Kant has had a substantial influence over subsequent philosophy, but he was also interested in the natural sciences. His early publications were in astronomy and geophysics, and he lectured in physical geography (broadly interpreted to include much human geography) for 40 years beginning in 1756. He has been described as 'the outstanding example in western thought of a professional philosopher concerned with geography' (May 1970:3).

To introduce his lectures on physical geography, Kant emphasized that the subject involved the description or classification of facts in their spatial context. As with Varenius, this focus was not particularly novel. However, it seems that Kant has been interpreted by some later geographers, especially Alfred Hettner, as an explicit advocate of geography as a regional study. This interpretation is based on Kant's argument that geography is description according to space and that history is description according to time. In *Physische Geografie*, published in 1802,

Kant asserted that geography and history together comprise all knowledge.

Although, again, there was nothing especially new about these ideas, it is possible that Kant contributed heavily to the notion of geography as a descriptive regional science. In an important sense, his view of geography as special, not general, represents a departure from Varenius. At the same time, there is little doubt that Kant was very much in accord with prevailing late eighteenth-century thought.

Universal Geography: 1800–1874

The first three-quarters of the nineteenth century saw the decline of the geography associated with exploratory activity and centred on mapping and description. This period was one of radical change in the way knowledge was organized, in which the academic disciplines that we know today formally emerged at universities. The first geography departments were established in 1874, at several Prussian universities. Simply put, geography could not continue to be primarily mapping and description—the late nineteenth century required a much more academic discipline—and it was necessary that mid-nineteenth-century scholars lay the groundwork for the formal discipline. The years 1800–74 witnessed the culmination of all that we have discussed so far and the beginnings of much that has happened since.

The first major work of this period was published by a Danish geographer, Conrad Malte-Brun (1775–1826), between 1810 and 1829. Following in the established tradition, this all-embracing study includes both general and special geography as defined by Varenius. Mathematical, physical, and political principles were discussed along with physical phenomena, including animals and plants, and human matters, including race, language, beliefs, and law. Describing all areas of the known world, Malte-Brun succeeded in producing a complete geography.

Humboldt and Ritter

Geography in the first half of the nineteenth century was dominated by two German scholars, Alexander von Humboldt (1769–1859) and Carl Ritter (1779–1859) (Boxes 1.6 and 1.7). Humboldt's greatest published work was *Cosmos*, a five-volume work published between

1845 and 1862; the title is the Greek term for an orderly universe (as opposed to chaos). Like earlier geographers, Humboldt wanted to describe the universe, but that was not all. As he wrote, 'my true purpose is to investigate the interaction of all the forces of nature.' For Humboldt, humans were part of nature—a major shift of emphasis in the European world. In addition to conventional regional descriptions, Humboldt strove to offer a complete account of the way all things are related. General concepts were carefully blended with precise observation.

Ritter too was concerned with relationships and argued for coherence in describing the way things are located on the earth's surface. Like Humboldt, he expressed interest in moving from description alone to description and laws. This, of course, was exactly where Varenius had directed our attention. These interests are paramount in *Die Erdkunde*, an only partially complete world geography comprising 19 volumes published between 1817 and 1859. This outline of the contributions of Humboldt and Ritter makes it clear that all three of our recurring themes are evident in their work. The study of humans and land is central to their conception of geography, there is a focus on regions, and the concerns of spatial analysis are anticipated in their interest in the formulation of concepts as general statements that aid in the understanding of specific facts.

Together, Humboldt and Ritter represent a fitting conclusion and an appropriate beginning. Although much of the world was still unknown to Europeans—Africa was largely a mystery, as was Asia—these two men produced two great works of geography. Both recognized the need for complete geographic descriptions and yet neither could truly fulfil that aim. By the year of their deaths, 1859, there was simply too much geography. A complete, encyclopedic description of the world was now beyond the reach of a single scholar. In this sense they represent a conclusion. Yet in another sense they represent a beginning. Humboldt and Ritter were the first geographers to pay full attention to concept formulation, that is, the derivation of

1.6

HUMBOLDT

Alexander von Humboldt was born in Berlin in 1769 into the land-owning aristocracy. Before attending university, he had the opportunity to mix with a diverse group of liberal intellectuals, and this experience, coupled with an enquiring mind, meant that he was interested in a wide range of academic and related topics. Following brief spells at two other universities, he commenced studies at the university in Göttingen and there met George Foster, who had recently returned from a world voyage with Cook. Apparently it was this encounter that started Humboldt along the geographic path. His studies continued at Freiburg, and when he inherited a substantial income, he decided to use it to support travel and related publication.

Humboldt (Metropolitan Toronto Reference Library).

From 1796 until his death in 1859, Humboldt travelled extensively and published prolifically. His major travels were to Central and South America from 1799 to 1804. His American travels, which culminated in a visit to Washington and a series of meetings with Thomas Jefferson, resulted in a thirty-volume work that was published between 1805 and 1834 and in many languages.

The writing of *Cosmos*—one of the most important and successful pieces of scientific writing ever accomplished—occupied Humboldt from the late 1820s until his death. In it he strove to investigate the interaction of all natural and human forces, drawing on all the sciences to explain everything from the Milky Way galaxy to microscopic organisms. In many areas of thought he was well ahead of his time; he insisted, for example, that all races had a common origin and that no one race was inherently superior to any other.

Although he made no single great discovery, Humboldt was one of the most admired men of the nineteenth century and might justifiably be regarded as one of the last universal men. Today, as when he lived, Humboldt is acknowledged as a great geographer. His conception of geography as an integrating science and his insistence on seeing humans as a part of nature, not separate from it, remain key ideas for many people.

general statements from the detailed factual information available. Their interest in the relations between things and their inclusion of humans as part of nature established the basis for much twentieth-century geography. Thus they represent both an end to the long road begun by the Greeks and a beginning to a geography combining regional description and concept formulation.

Map-Making, Exploration, and Geographical Societies

Despite considerable new thinking about geography, there was also a steady continuing interest in some of the traditional geographic concerns. Map-making was considered so important that governments began to assume responsibility for the task (you will recall that in China civil servants were responsible for map-making, probably as early as the third century CE) and in England the Ordnance Survey was founded in 1791. Large-scale topographic maps, showing small areas in considerable detail, became possible with the development of exact survey techniques in eighteenth-century France. Atlases also became much more sophisticated. Exploration continued and was

greatly assisted by the founding of geographical societies in Paris (1821), Berlin (1828), London (1830), St Petersburg (1845), and New York (1851). The ever-growing interest in overseas areas encouraged prosperous individuals as well as governments to support these organizations.

The home of the Royal Geographical Society in London. The Royal is the largest geographical society in Europe, and one of the largest in the world. In the nineteenth century, such societies helped geographers to organize their field of study, and promoted recognition of their work. Most also encouraged exploration and implicitly —if not explicitly —supported colonial activity (Nick Ivans/Royal Geographical Society).

1.7

RITTER

Carl Ritter was born in 1779, ten years after Humboldt, and died in 1859, the same year as Humboldt. Although he never became as well known in the larger scientific world, Ritter proved to have a greater influence on German geography.

Ritter had the good fortune to be selected at age five for involvement in a new system of education that encouraged understanding rather than conventional rote learning. One

Ritter (Metropolitan Toronto Reference Library).

of his teachers was a geographer who emphasized that humans were part of nature, not some separate entity—a viewpoint that would be central to Ritter's own work. He was able to fund his studies at universities including Halle and Göttingen by working as a private tutor. Becoming professor of history at Frankfurt in 1819, in 1820 he obtained the post he would hold for the rest

of his life: the first chair of geography in Germany at the University of Berlin.

As early as 1811, Ritter published a two-volume textbook on Europe; the first volume of his most important work, *Die Erdkunde,* was published in 1817. At the time of his death *Die Erdkunde* had reached nineteen volumes, most of which focused on Asia; he had not even begun to discuss Europe in this series.

For Ritter, geography was clearly an empirical and descriptive science reflecting the unity of humans and nature. Ritter had first met Humboldt in 1807 and the two established an important relationship. There was a clear agreement regarding the importance of the unity of all geographic facts. Unlike Humboldt, Ritter did not undertake any of the exploratory activity that led to the discovery of new facts, but he was an inspiring lecturer who influenced many subsequent geographers. In both his lectures and his writings, Ritter employed a conventional regional focus, one that continued to dominate in late nineteenth-century German geography.

Institutionalization: 1874–1903

The year 1874 marks the formal beginning of geography as an institutionalized academic discipline. In 1903, the first North American university department of geography was established. This 30-year period was one of dramatic developments in geography.

First, once geography became legitimized in institutions of higher learning, it was necessary that university geographers clearly define their discipline and distinguish it from the various other disciplines emerging at the same time. Not surprisingly, not all geographers immediately agreed on the discipline's content and methods. Second, the mid- to late nineteenth century was a period of major advances in various areas of knowledge that had direct effects on geography, when a mechanistic view of science, emphasizing cause and effect, prevailed. Third, the publication of Charles Darwin's (1809–82) *On the Origin of Species* in 1859 had repercussions throughout the scientific world. Fourth, Europe now had a 400-year history of expansion and movement into regions clearly different from the home area. Finally, the rapid growth of physical science encouraged the development of topics that were traditionally geographic in a number of other disciplines, notably geology.

At the time when geography was institutionalized, then, the intellectual environment was complex and changing. Not only did geographers have to consider the appropriate weight to give the many aspects of established geographic traditions—mapping, description, general and special geography, regions, human–land relations—but now they also had to take into account ongoing developments in the physical sciences, the social sciences, and mechanistic science, as well as the hugely controversial issues raised by Darwinian thought and the moral questions associated with expansion overseas. In brief, from 1874 onward, geographers had a great deal to accommodate. How did they respond?

Germany

In 1874 the Prussian government established departments of geography in all Prussian universities. Why? Possibly because the time was ripe; possibly because the value of geog-

1.8

RATZEL

Friedrich Ratzel was born in Karlsruhe in 1844. Following university studies in Heidelberg and Jena, he fought in the Franco-Prussian War (1870–1). After another brief spell at the university in Munich, he travelled in Europe and then in North and Central America in 1874–5. The American travels in particular impressed upon him the importance of humans as makers of landscapes and inspired him to pursue a career in the new field of geography.

Ratzel (Metropolitan Toronto Reference Library).

He taught geography in Munich from 1875 until 1886 and then succeeded Richthofen as professor of geography at Leipzig, where he remained until his death in 1904. The first volume of Ratzel's *Anthropogeographie*, published in 1882, was specifically designed to be a study of the effects of physical landscapes on history. The second volume, published in 1891, focused attention on relations between humans and land, with a primary emphasis on the role of humans. These two volumes together represent a major contribution to human geography. The first volume proved to be especially influential in the United States. Ratzel also made major contributions in political geography, regarding states as spatial organisms.

In many respects, several of Ratzel's ideas were used somewhat indiscriminately by later geographers. Volume 1 of *Anthropogeographie* was seen as a powerful argument for viewing human landscape as secondary to—even caused by—physical landscapes, and the political geographic ideas that Ratzel expressed were used by Nazi geographers in the 1930s to justify spatial aggression. Nevertheless, Ratzel remains an important figure, largely because he was the first influential geographer to explain clearly the concept of human landscapes as ever-changing additions to original physical landscapes—an idea integral to the thinking of later German, French, and American geographers.

raphy was especially evident following the Franco-Prussian war of 1870–1; possibly because of a belief that geographic knowledge would facilitate political expansion (see Box 1.1). Whatever the reasons, it became the responsibility of these new departments to clarify what geography entailed. Several leading scholars attempted this task.

For Ferdinand von Richthofen (1833–1905), geography was—as it had been earlier—the science of the earth's surface. Research centred on field studies and observation. Richthofen maintained the distinction between special and general, first noted by Varenius, and further argued that the two could be combined to form a chorological (regional) approach. The subject matter of this regional geography (or **chorology**) included human activities, but only in relation to the physical environment.

Friedrich Ratzel (1844–1904) focused on human geography, or what he termed 'anthropogeography' (Box 1.8). His major work was published in two volumes (1882 and 1891). Volume 1 had as an overriding theme the influence of physical geography on humans, while Volume 2 focused on humans using the earth. Interestingly, it was Volume 1 that had the greatest immediate impact, possibly because its inherently simple cause-and-effect logic was in accord with mechanistic views and helped to assert the importance of physical geography. It is appropriate to regard Ratzel as a founder of human geography, since he was probably the first to focus on human-made landscape.

Alfred Hettner (1859–1941) was the most influential follower of Richthofen. In his methodological and descriptive work, geography was unequivocally regarded as the chorological science of the earth's surface, a view that had first been clearly enunciated by Kant. A persuasive advocate of regional geography, Hettner was highly influential in the United States. However, his work excluded studies of human–land relations, as well as studies involving time.

Thus German geography following 1874 consisted of three quite different interpretations. First, there was geography as chorology (Richthofen and Hettner); second, there was geography as the influence of physical geography on humans (Ratzel, Volume 1); and, third, there was geography as the study of the human landscape (Ratzel, Volume 2). These differences would continue well into the twentieth century.

France

Geography in France followed a route independent of German developments, although it reflected many of the same ideas expounded by Ratzel in Volume 2. One influential scholar, Pierre Le Play (1806–82), was not a geographer but a sociologist; he focused on human–land relations and the impact of technology on social groupings. Elisée Réclus (1830–1905), along with Le Play, paved the way for later French geography. This geographer and anarchist, who was barred from France and imprisoned at various times, published a descriptive systematic geography of the world and a 19-volume universal geography. His work was very much in the Ritter tradition. Interestingly, another well-known anarchist, Peter Kropotkin, a former Russian prince, was also a leading geographer in the late nineteenth century.

The most important French geographer, Paul Vidal de la Blache (1845–1918) (Box 1.9), followed both Ratzel and Le Play and succeeded in introducing a well-articulated geographic method. His overriding concerns were the relations between humans and land, the evolution of human landscapes, and the description of distinctive local regions. For Vidal, geography should consider both physical geographic impacts on humans and human modification of physical geography. The parallels with Humboldt, Ritter, and Ratzel (Volume 2) are clear. Vidal and his many followers continue to exert enormous influence on human geography.

Britain

The first British geography department was established in 1900 at Oxford, largely as a result of efforts by the Royal Geographical Society. The dominant influence at first was Halford J. Mackinder (1861–1947). Interestingly, he determined to become the first European to climb Mount Kenya, in Africa, to ensure that his geographic views would be favourably received; to be respected as a geographer, he felt, one had to be a successful explorer! For Mackinder, geography and history were closely related: a global geographic perspective was essential, and, following

Richthofen, physical geography was a prerequisite for human studies. In Britain, then, the traditionally close association between geography and mapping, description, and exploration continued: the emphasis was on physical rather than human geography.

United States

The establishment of the first North American department of geography, at the University of Chicago in 1903, came at a time when American geography was influenced largely by German scholars. Physical geography (physiography) was dominant, possibly because of the powerful personality of William Morris Davis (1850–1934), a geologist who promulgated the views (imported from Germany) that physical geography influenced human landscapes and that geography was essentially a regional science. Ellen Churchill Semple (1863–1932) followed the early Ratzel to become a leading proponent of the 'physical influences' school of thought, while many American geographers quickly adopted a regional focus. At the time there

was no evidence of a distinctive school analyzing the relations between humans and physical geography. Neither the general legacies of Humboldt and Ritter nor the views of Vidal were in evidence, although George Perkins Marsh (1801–82), an American geographer and congressman, had earlier focused on the same issues.

Geography in 1903

Once again we have reached an important turning point. By 1903 geography was a full-fledged academic discipline in many European countries and in the US. Elsewhere geography was institutionalized somewhat later, and largely as a consequence of academic contacts with these pioneering countries. In Canada, for example, a partial department of geography was created at the University of British Columbia in 1923, 12 years before the first full department, at Toronto, was established in 1935. Today Canada has more than 40 geography departments.

In 1903 the general subject matter was clear—there had been no real change since

1.9

VIDAL

Paul Vidal de la Blache was born in 1845, studied history and geography in Paris, and then taught successively at Nancy, Paris, and the Sorbonne until his death in 1918. Vidal laid down ideas that, once accepted and amplified by others, established a dominant French geographic tradition, *la tradition Vidalienne*. This tradition was not, of course, established in an intellectual vacuum, as Vidal had been introduced to the great works of Humboldt

Vidal (Metropolitan Toronto Reference Library).

and Ritter during his schooling and also travelled often to Germany, meeting with such geographers as Richthofen and Ratzel.

The first clear methodological statement by Vidal was in his Sorbonne inaugural address in 1899; other articles appeared in the journal he founded in 1891, *Annales de Géographie*. Vidal rejected the idea that the physical landscape determined the human. Rather, he believed that geography should focus on the

reciprocal relationship between humans and land. Central to his blueprint for geography were the ideas of *genre de vie* ('way of life', or culture), *milieu* (local area), and the regional concept of *pays* (natural region). Vidal's conception of geography involves chorological analysis, but with a distinctive humans-and-land focus that was not clearly formulated by either Richthofen or Hettner. There are similarities between Vidal and the Ratzel of *Anthropogeographie*, Volume 2, and also between Vidal and the German geographer Otto Schlüter (1872–1952).

In France, Vidal's followers included such geographers as Jean Brunhes (1869–1930) and Albert Demangeon (1872–1940). The absence of any alternative French conception of geography did not mean, however, that Vidal's view was unchallenged. There was an ongoing debate between Vidal and the great French sociologist Emile Durkheim (1858–1917). Simply put, Durkheim was critical of Vidal and geography in general for emphasizing physical landscapes at the expense of an adequate emphasis on society. Vidal died in 1918, having founded a distinct school of geographic thought.

Greek times—and a number of different approaches were advocated. Many people studied physical geography in its own right; some focused on physical geography as the cause of human landscapes; others concentrated on regional description; while still others centred their attention on humans and nature combined. These general threads continued through much of the last century. Geography also continued its association with mapping and, to a lesser extent, exploration.

Prelude to the Present: 1903–1970

The period from 1903 to 1970 was characterized by several different approaches to geographic subject matter. As in the past, academic geography continued to emphasize three principal areas of study—physical geography as cause, humans and land, and regional studies—to which was added a fourth, relatively novel, approach: spatial analysis. The physical-geography-as-cause approach proved relatively short-lived; the other three, of course, are our recurring themes.

Physical Geography as Cause

Much of the attraction of the emphasis on physical geography as cause (most clearly stated in Ratzel's Volume 1) lay in its relative simplicity. Arguing that human landscapes and cultures resulted primarily from physical geography reduced the need to think about economics, politics, societies, and so forth. Among the well-known geographers who favoured this approach were Semple, Ellsworth Huntington (1876–1947), and Griffith Taylor (1880–1963) (Box 1.10). Nevertheless, the principal significance of the physical-geography-as-cause perspective lies in the fact that it was simply taken for granted as a self-evident truth. Much twentieth-century geography that did not explicitly take this perspective—also known as **environmental determinism**—did so implicitly. The greatest problem with this approach is the tendency to use it in isolation. There are very important and often intimate relations between physical and human geography, but these are never so straightforward as to allow us to assume that the former is the cause and the latter the effect. Fortunately for geography, environmental determinism, although popular in the early twentieth century and occasionally taken to extremes, never became the focus of a formal school of geography, and is now discredited.

Humans and Land

The view of geography as the study of all things physical and human, specifically as the study of relationships between physical and

1.10

TAYLOR

Griffith Taylor lived from 1880 to 1963. Born in Britain, he was responsible for introducing geography as a university discipline to Australia and Canada. A member of the Scott expedition to the Antarctic, 1910–12, Taylor had been trained as a geologist, and it was this not altogether unusual background that led him into geography. He first taught geology at the university in Melbourne and was appointed the first professor of geography at the university in Sydney in 1920.

Taylor was a dedicated and effective teacher with strong opinions on the influence of physical geography on human activities. His dim view of the settlement potential of the arid interior of Australia—diametrically opposed to the propaganda of the Australian government—brought him into considerable disrepute.

At least partly as a result of this controversy, in 1928 he moved to Chicago, where he taught until 1935. In that year he moved to Toronto, where he was responsible for the founding of the first full Canadian department of geography and remained until his retirement in 1951.

Always a stimulating writer, Taylor produced some twenty books and 200 articles. In Canada, as in Australia, he was involved in discussions about settlement of environmentally difficult areas, in this case the Canadian north. Interestingly, when Taylor returned to Australia at the age of seventy, he was treated as a national hero, his earlier views having been vindicated. At the time of his death, Taylor was being considered in Australia for a knighthood.

human facts, has a long tradition. Humboldt, Ritter, and Ratzel (Volume 2) all took this view, and their specific legacy is evident in much subsequent geography. However, such a vast subject required a more solid foundation than the physical-geography-as-cause school could provide. In particular, it needed a better definition of its subject matter and a formal methodology. These were provided by three scholars.

Vidal (Box 1.9) developed a distinctive French school, *géographie Vidalienne*, beginning about 1899. Otto Schlüter (1872–1952) founded a German school of *Landschaft-skunde*—'landscape science', or 'landscape geography'—in 1906, which provided a clear definition of subject matter. The third scholar was Carl Sauer (1889–1975), an American who, in 1925, effectively introduced the various European ideas to North America and elaborated on them (Box 1.11). The **landscape school** that grew out of his work focuses on the transformation of the physical geographic landscape by human cultural groups over time.

Together, the ideas of Vidal, Schlüter, and Sauer provide us with a clearly defined approach to our subject matter. Landscape geography explicitly rejects any suggestion of physical-geography-as-cause by rejecting environmental determinism in favour of **possibilism**: the view that human activity is determined not by physical environments, but rather by choices that humans make.

Regional Studies

Regional geography, or chorology, proved to be the most popular focus during the first half of the nineteenth century. Pioneering work by German geographers such as Richthofen and Hettner (in his later arguments) was carried over into English-language geography. In Britain a view emerged that the ultimate task of geography was to delimit regions. In America this view was attractive for physical geographers in particular, and Davis, for example, produced a map of regions as early as 1899. In 1905 the British geographer A.J. Herbertson (1865–1915) proposed an outline of the world's natural regions. These developments culminated with the 1939 publication of *The Nature of Geography* by the American Richard Hartshorne (1899–1992) (Box 1.12). This substantial contribution to geographic scholarship argued forcefully for geography as the study of regions—what Hartshorne often called **areal differentiation**.

1.11

SAUER

For students of English-language human geography, Carl Sauer is a major figure. Born in 1889 in Missouri, Sauer obtained a doctorate from the University of Chicago and taught at the University of Michigan from 1915 to 1923, but he is most closely associated with the department of geography at the University of California at Berkeley, where he taught from 1923 to 1957. He died in 1975. Proponents of Sauer's conception of geography are often labelled the 'Berkeley school'.

Sauer articulated that conception in his inaugural address at Berkeley. Entitled 'The Morphology of Landscape', the address presented a highly original version of earlier European ideas. Sauer saw geography as a chorological discipline, but one largely concerned with the manner in which humans modify physical landscapes. Humboldt, Hettner, Schlüter, and the core French school of Vidal are all evident influences. Sauer explicitly opposed the popular view that physical geography was a cause of human landscapes. He employed the term 'landscape' as the object of geographic study in preference to 'region', which he felt was too closely associated with overly detailed descriptions. Although Sauer's view of geography was largely derivative of others, it proved to be highly influential, and has contributed significantly to the development of historical and cultural emphases.

Sauer succeeded in establishing a landscape geography that continued from the 1920s to the present with relatively little change until the 1980s. This was no mean achievement, and it reflects both the quality of his methodological and research writings and the legitimacy of his basic arguments. In addition to launching the study of landscape geography, Sauer produced highly original studies of early America and the origins of agriculture.

Geography in the North American world by 1953 was thus characterized by two related but different emphases, on (1) analysis of the relationship between humans and land and (2) regional studies. Both were the products of long geographic traditions, and both have continued to change. In 1953, however, a new focus was forcefully introduced.

Spatial Analysis

A paper by F.W. Schaefer, published in 1953 (see Box. 2.3, p. 50), is usually seen as the beginning of an approach called spatial analysis. This approach, as we saw in the Introduction, focuses on explaining the location of geographic facts. Schaefer argued that geographers should move away from the description characteristic of regional studies to a more explanatory framework based on scientific methods such as the construction of theory and the use of quantitative methods. Here again we can recognize the special/general distinction identified by Varenius. Schaefer objected to what he saw as the overly special focus of regional geography. Regardless of the specific merits of Schaefer's arguments, it is clear that his views struck home to many. What followed might be called a revolution, in which the emphasis shifted to quantitative studies. The decade of the 1960s was characterized by phenomenal

growth in detailed analyses based on the proposal and testing of hypotheses by means of quantitative procedures.

By 1970, regional geography was receding in popularity and spatial analysis had found a niche alongside somewhat modified versions of both the landscape and regional approaches.

Contemporary Geography

Since 1970, geography has seen a number of important revisions and additions to traditional interests and approaches. Thus geography today, while clearly the product of the earlier developments discussed above, also incorporates some relatively new components. There is the welcome addition of a wide range of new (for the human geographer) philosophies that are prompting major shifts in the traditional approaches (human–land relations and regional studies). There is also the ever-increasing impact of technological advances in data collection and analysis.

To conclude this chapter on the evolution of human geography, and to whet your appetite for the next chapter, detailing what human geographers are really doing today, this final section will provide a brief summary of seven major trends in contemporary human geography:

1.12

HARTSHORNE

Richard Hartshorne was born in 1899, received a doctorate from the University of Chicago, and taught at the university level in Minnesota from 1924 to 1940 and in Wisconsin from 1940 to 1970. In 1939 he published a book entitled *The Nature of Geography*, and in 1959 a revision entitled *Perspective on the Nature of Geography*. The first of these, considered a landmark in the history of writing about geography, was based on extensive research in Europe. It is impossible to do justice to the significance of this methodological statement in a few words, but it is important to note that Hartshorne based his work closely on earlier German writings, such as those by Hettner, that argued for geography as a chorological science. Geography's unique role was the study of 'the world, seeking

to describe, and to interpret, the differences among its different parts, as seen at any one time, commonly the present time' (Hartshorne 1939:460).

Although both Sauer and Hartshorne based many of their ideas on earlier German geographic thought, they clearly arrived at different conclusions, and each became identified with a particular approach to geography: Sauer with the landscape school and Hartshorne with regional studies. Sauer was critical of Hartshorne's conception of geography as the study of physical and human facts in a regional context, especially because it ignored change over time, but Hartshorne's view was very much in accord with the prevailing American opinion and continued to dominate geography until well into the 1950s.

1. an increasing separation of the physical and human components of geography;
2. a revitalized landscape approach;
3. a revitalized regional geography;
4. an ongoing interest in spatial analysis;
5. recognition of the need for a global perspective;
6. an increasing concern with applied matters; and
7. an increasing emphasis on technical content.

Physical and Human Geography

Our discussion so far has not made any formal attempt to distinguish between physical geography and human geography. As we have seen, geography from the Greeks onwards has had both physical and human components. Prior to Humboldt and Ritter, however, there was a clear tendency to treat them separately. Descriptions of the earth typically dealt first with physical and second with human matters. Humboldt and Ritter, by contrast, focused closely on integrating the two. Subsequently, some geographers saw physical geography as paramount (the physical-geography-as-cause, or environmental determinist, school), some saw the human landscape as the relevant subject matter (the landscape school), and some saw the two as separate (the regional school). It was not unusual for geographers to assert a unity between physical and human geography, but there was not much evidence of such unity. Very few geographers deny the relevance of physical to human geography, but equally few see the one as necessary to the other. Today, therefore, we tend to teach and research the two separately.

The separation of physical from human geography and the recognition of human geography as a separate discipline is thus relatively recent. Of course, some regional accounts and some specific issues require consideration of both dimensions. Chapter 4 of this book is a clear instance of the value of a geography that is both physical and human. And a basic understanding of global physical geography is obviously relevant to much work in human geography. Overall, though, the two are no longer seen as so closely related that we need to discuss them as one. Contemporary human geography is a social science, but one with special and valuable ties to physical sciences, especially physical geography.

Contemporary Landscape Geography

At least since the writings of Vidal, Schlüter, and Sauer, landscape geography has maintained a consistent place in human geography alongside the regional approach. Also known as social geography (in Britain) and cultural geography (in North America), landscape geography typically involves studies of human ways of life, cultural regions and related landscapes, and relationships between human and physical landscapes. But the landscape approach has recently been enhanced by the inclusion of new conceptual concerns. Since about 1970, ideas associated with humanism and Marxism, along with a generally increased awareness of advances in other social science disciplines, have enriched landscape studies.

Today landscape geography is concerned with both visible features (such as fields, fences, and buildings) and symbolic features (such as meaning and values). Thus many contemporary landscape geographers focus on the human experience of being in landscape; here landscapes are regarded as relevant not simply because of what they are but also because of what we think they are. Further, there is explicit acknowledgement that landscapes, like regions, both reflect and affect cultural, social, political, and economic processes.

Over time, studies in human–land relations have become increasingly subtle and profound, and they are now closely tied to conceptual developments in disciplines such as psychology, anthropology, and sociology.

Contemporary Regional Geography

The rise of spatial analysis came largely at the expense of the regional geography ('areal differentiation') articulated by Hartshorne. Since 1970, however, regional geography, like landscape geography, has resurfaced in a somewhat different guise and has once again become a central perspective. Most human geographers accept that the earlier regional geography is no longer appropriate; there is a clear need to move beyond regional classification and description. Yet the regional approach is obviously a valuable and distinctive geographic device. One geographer described regional geography as 'the highest form of the geographer's art' (Hart 1982:1).

Today regional geography emphasizes the understanding and description of a particular region and what it means for different peo-

Lunenburg, Nova Scotia, from the harbour. Geographers often suggest that landscapes can speak to us, communicating information about the people who live there, what they do, and the values they hold. Established by the British in 1753 as their second outpost in Nova Scotia (the former French colony of Acadia), Lunenburg was first settled by Protestants specifically recruited from Germany, Switzerland, and France. Farming, fishing, ship-building, and ocean-based commerce provided the foundations of a vibrant economy, and today tourism is a major source of income. This photograph reflects elements of both the town's economic history and its character. For example, there is no evidence of the ongoing physical change—whether construction or demolition—so typical of most urban areas. Rather, the presence of many carefully maintained old buildings suggests permanence and stability (Gerald Hallowell photo).

ple to live there. This emphasis reflects at least three general concerns, each of which is tied to a particular set of conceptual constructs. First, there is a concern with regions as settings or locales for human activity. Second, there is a concern with uneven economic and social development between regions, including a focus on the changing division of labour. Third, there is a concern with the ways in which regions reflect the characteristics of the occupying society and in turn affect that society. The underlying concepts for each of these three regional geographic concerns are humanism and Marxism; both of these are introduced in Chapter 2. Looking at regional geography in this light, we can see it as increasingly satisfying the requirement that human geography serve society by addressing a range of economic and social problems.

The new directions in regional geography and landscape geography are sufficiently similar that these two approaches now share several common interests.

Contemporary Spatial Analysis

The spatial analytic approach that first became a prime interest of human geographers in the mid-1950s was very influential until about 1970. Since that time it has continued to be an integral part of human geography, but it is now generally regarded as only one among several widely accepted approaches. A central concern is that the theoretical constructs it uses to explain locations are somewhat limited, and hence that it tends to emphasize generalizations at the expense of specifics. The topics studied by spatial analysts are typically economic (for example, the locations of industrial plants), as opposed to political, cultural, or social. This tendency reflects the influence of various economic location theories.

There is a clear distinction between the spatial analytic approach and the contemporary regional and landscape approaches. Indeed, some geographers believe that spatial analysis is overly concerned with spatial issues, perhaps to the point of seeing space as a cause of human landscapes, and as a result ignores the full range of human variables as causes of landscapes. Contemporary regional and landscape geographers explicitly argue that space is important only when it is analyzed in terms of the use that humans make of it. This is an important philosophical question that will be discussed more fully in the next chapter.

Global Issues

Geography has always been a global discipline. Much of the work of the early Greek, Chinese, and Islamic geographers, for example, was concerned with describing the limits of the known world and identifying links between places. European geography, as it developed from about 1450 onwards, focused on understanding where places were, mapping routes, and describing what was previously unknown. Much contemporary human geography continues in these traditions.

Nevertheless, there is something else behind our current concern with global issues. The fact is that people and places in the modern world are increasingly connected and interconnected. Today we are interested in the global movements of people, products, ideas, and capital. We are interested in how the many and diverse peoples and places scattered throughout the world are increasingly associated one with another. Human impacts on the earth, population growth, the spread of diseases such as AIDS and SARS, international migration and refugee problems, food shortages, vanishing languages, the spread of democracy, agricultural change, urban growth, and industrial restructuring are just a few of the major issues in the world today that require thinking on the global scale. More than ever before, human geographers are finding it necessary to look at the big picture—the world—in order to understand the detail at the local level.

Applied Geography

Because geography is an academic discipline that serves society, it has always had an applied component in which geographical skills are used to solve problems. Geography as exploration is a prime example of such applied work. In this century, geographers have played a major role in land-use studies. For example, the studies of the British Land Utilization Survey directed by L.D. Stamp in the 1930s proved enormously valuable during the Second World War, and American geographers made important contributions in both land-use and military matters.

Contemporary human geography responds to the wide range of social and environmental issues that confront us today with regular contributions on such topics as peace, energy supply and use, food availability, population control, and social inequalities. All these issues are covered in this book, for human geographers recognize that geography has responsibilities to society and the world. Much recent geography of this type reflects the influence of other disciplines (for example, in the use of Marxist analysis). There are also close links between applied geography and both spatial analysis and technical advances.

Technical Advances

Human geography today is greatly aided by a variety of technical advances in areas such as navigation-assisted exploration. Among the technical advances that have facilitated data acquisition are aerial photography and both infrared and satellite imagery. Similarly, advances in computer technologies and geographic information systems have facilitated mapping and data analysis. These important geographic techniques will be discussed in detail in the next chapter.

Postscript

As befits geographers, we have travelled a long way in this chapter. From the earliest map-makers to the present, we have seen the evolution of geography as an academic discipline. We know that human geography has a rich heritage and exciting contemporary developments. But we have not yet travelled far enough. The next chapter presents fuller discussions of the various contemporary developments that have received only brief mention so far.

Human geography today is a responsible social science with the basic aim of advancing knowledge and serving society. In this respect it is no different from Greek, Chinese, Islamic, or later European geographies. Our subject matter is clear and our methods various. Our key goal—to provide a basis for comprehending the human world as it is today and as it has evolved—is of crucial importance to contemporary world society. As a brochure produced by the Association of American Geographers expresses it, 'Without geography you're nowhere.'

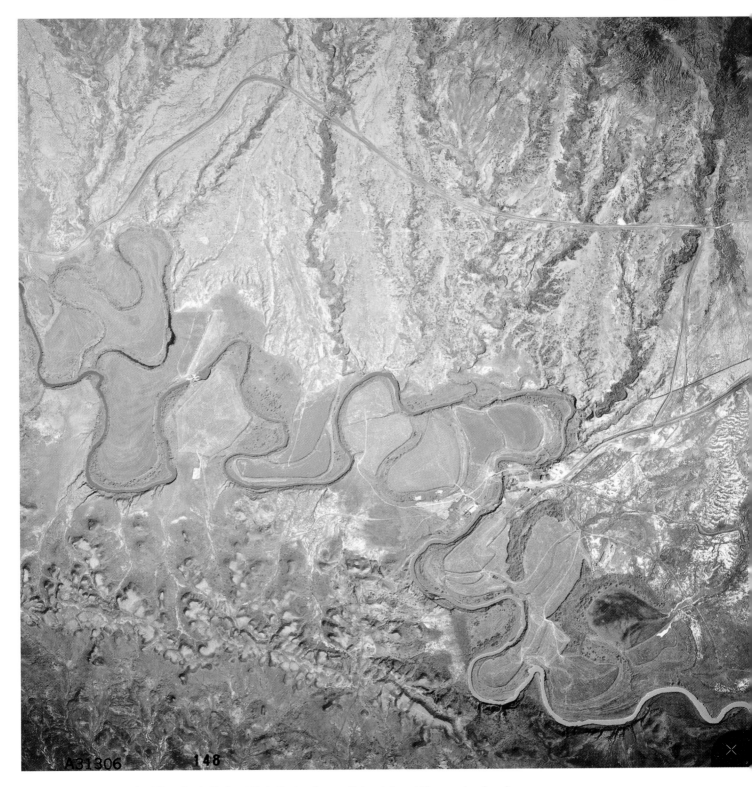

Infrared aerial photograph of Grasslands National Park, Saskatchewan. Colour infrared film was developed during the Second World War to detect painted targets camouflaged to look like vegetation. In this image healthy grass and tree leaves appear red (A31306-148, © 1982 Her Majesty the Queen in right of Canada, reproduced from the collection of the National Air Photo Library with permission of Natural Resources Canada).

Chapter One Summary

A raison d'être for geography
To serve society.

Preclassical geography
The earliest geographic descriptions were maps. The earliest surviving maps, from ancient Mesopotamia, showed only local areas.

Classical geography
Because the Greeks were geographically mobile, they began a descriptive literary tradition that included the mapping of the entire world as they knew it. Major figures in this tradition include Herodotus, Eratosthenes, and Strabo. The Greeks also initiated a mathematical tradition; Eratosthenes measured the circumference of the earth and Hipparchus devised a grid system for maps. Ptolemy summarized the mathematical tradition. Roman civilization added little to geographic knowledge.

The fifth to fifteenth centuries
From the fifth to the fifteenth century, only sporadic and limited geographical work was accomplished in Europe. Christian dogma determined the contents of T-O maps. Only Portolano charts represented a development in mapping. Thus geographic study largely disappeared. In China from c.2000 BCE onwards, and in the Islamic world from the seventh century CE on, geography flourished; significant progress was made in exploration, mapping, description, and mathematical work.

A European resurgence
Beginning in the fifteenth century, Europe expanded overseas; cartography was transformed with new map projections, especially that of Mercator; geographic description reappeared with major works by Apian and Münster. By the mid-seventeenth century, geographers had added significantly to our knowledge of ourselves and our world; geography itself continued to be concerned with mapping and describing the earth.

Rethinking geography
Varenius was the first scholar to formally define geography and distinguish between general and special geography (in 1650). Kant contributed further to the understanding of geography by identifying it as the science that describes or classifies things in terms of area (c.1765). Between roughly 1650 and 1800, geography broadened to include discussion of such issues as physical-geography-as-cause, the growth of civilizations, the unity of the human race, and population density and productivity relationships. Descriptive geography and mapping continued.

Universal geography
The mid-nineteenth century was the high point for geography as mapping and description. By this time considerable geographic knowledge had been accumulated, and specialist sciences were emerging. Major nineteenth-century geographies include those by Malte-Brun, Humboldt, and Ritter. The latter two made novel contributions by discussing human–land relations in addition to providing comprehensive descriptions; both considered geography to require both general and special studies.

Institutionalization
Geography was firmly established as a university discipline in Prussia in 1874. Other countries rapidly followed suit. Programmatic statements about geography were plentiful: Richthofen and Hettner formulated geography as chorology; Ratzel introduced anthropogeography and Vidal followed his lead; Mackinder focused on a world view; and Davis pioneered physiography. The first North American department of geography was established in Chicago in 1903.

Prelude to the present
Four principal emphases were evident between 1903 and 1970. Physical-geography-as-cause flowered initially. Sophisticated human and land views developed under Vidal, Schlüter, and Sauer and have persisted to the present. Regional geography dominated North American work until the 1950s. Spatial analysis appeared in the 1950s and flourished in the 1960s, but is now regarded as just one of several valid approaches.

Human geography today

Contemporary human geography has close ties to the long and illustrious history of geography and is affected by several relatively new influences. Today human geography is easily distinguished from physical geography, includes revitalized regional and landscape approaches, takes an ongoing interest in spatial analysis, is increasingly concerned with applied issues in the global context, and has a continually expanding technical component.

Conclusion

Knowledge of geography is essential to the education of all humans. Its fundamental goal is still to serve the society of which it is a part.

LINKS TO OTHER CHAPTERS

- Because this chapter traces the roots and development of human geography over time, it has links with all later chapters.

- Philosophies, concepts, and techniques: Chapter 2.

- Humboldt, Ritter, Vidal, and Sauer:
  Chapter 4 (human impacts on land)
  Chapter 7 (approaches to landscape study).

- The history of geography:
  Chapter 8 (changing types of society)
  Chapter 9 (European political expansion).

FURTHER EXPLORATIONS

DICKINSON, R.E. 1969. *The Makers of Modern Geography*. Boston: Routledge and Kegan Paul.

    A traditional history of geography that emphasizes German and French scholars and contains much detailed information.

EDSON, E. 1997. *Mapping Time and Space: How Medieval Mapmakers Viewed Their World*. London: The British Library.

    For students especially interested in the history of mapmaking, this is an excellent and often entertaining survey of the years 500 to 1500; stresses the meaning and purpose of maps following the logic that the hand of the map-maker is responding to a particular spatial and temporal context.

HARRIS, N. 2002. *Mapping the World: Maps and Their History*. San Diego: Thunder Bay Press.

    A well illustrated overview of the history and purpose of mapping; tells the story of our quest to understand the earth and our place in it.

HARTSHORNE, R. 1939. *The Nature of Geography: A Critical Survey of Current Thought in the Light of the Past*. Lancaster: Association of American Geographers.

    The leading methodological statement arguing for regional geography; worth a look, but difficult to read in depth.

HARVEY, D.W. 1969. *Explanation in Geography*. London: Arnold.

    A comprehensive and pioneering statement about geography as a science that, like the Hartshorne book, is not an easy read for the beginning geographer, but rewards even a cursory inspection.

HOLT-JENSEN, A. 1988. *Geography: Its History and Concepts,* 2nd ed. Totowa: Barnes and Noble.

    A stimulating evaluation of geography as a discipline that is especially readable for the new geographer.

HUNT, A. 2000. '2000 Years of Map Making'. *Geography* 85:3-14.

A brief, informative, and very readable overview of the history of maps and map-making during the past 2000 years.

JOHNSTON, R.J. 1997. *Geography and Geographers: Anglo-American Human Geography Since 1945,* 5th ed. London: Arnold.

An overview of English-language geography since 1945; thoughtful debate of the role of geography in society.

LIVINGSTONE, D.N. 1992. *The Geographical Tradition*. Oxford: Blackwell.

A fascinating account of the last 500 years of geography; comprises a series of discussions of intellectual history, the first of which is entitled 'Should the History of Geography be X-Rated?'

————, AND C.W.J. WITHERS, eds. 1999. *Geography and Enlightenment*. Chicago: University of Chicago Press.

A series of critical essays that together stress the need to understand the nature of geography in terms of the particular meanings that the subject has had at different times and in different places.

MARTIN, G.J., and P.E. JAMES. 1993. *All Possible Worlds: A History of Geographical Ideas*, 3rd ed. New York: Wiley.

The best comprehensive history of geography, with detailed discussions of geographic developments and early geographers.

MINSHULL, R. 1967. *Regional Geography: Theory and Practice*. London: Hutchinson.

A clear account of the regional approach, focusing on current issues in method and practice.

MORRILL, R.L. 1984. 'The Responsibility of Geography'. *Annals, Association of American Geography* 74:1–8.

A thoughtful assessment of the obligations of geographers, including those to society and humanity.

SAUER, C.O. 1925. 'The Morphology of Landscape'. *University of California Publications in Geography* 2:19–53.

The landmark article introducing the landscape concept to North American geography.

SEMPLE, E. 1911. *Influences of Geographic Environment*. New York: Henry Holt.

A classic example of the physical-geography-as-cause interpretation of human geography; very readable today.

TAAFE, E.J. 1974. 'The Spatial View in Context'. *Annals, Association of American Geographers* 64:1–16.

A brief and effective summary of major geographic approaches prior to the 1970s.

UNWIN, T. 1992. *The Place of Geography*. Harlow, Essex: Longman.

A readable and critical account that places particular emphasis on the consequences of the current division between human and physical geography.

WOOLDRIDGE, G.W., and W.G. EAST. 1951. *The Spirit and Purpose of Geography*. London: Hutchinson.

An overview co-written by a physical geographer and a human geographer arguing for geography as a link between physical and human sciences.

ON THE WEB

http://www.aag.org/

Association of American Geographers. The most important American site for professional geographers; a valuable resource that offers information on many topics, including careers for geographers.

http://cag-acg.ca/en

Canadian Association of Geographers. Useful information on geography in Canada; intended primarily for professional geographers and those whose employment is closely related to their geographic education.

http://www.cangeo.ca/

The website for *Canadian Geographic* magazine; an informative source that deals with a wide range of geographical issues for Canada.

http://geowww.uibk.ac.at/geolinks/simple.html
A listing of geography departments around the world.

http://www.nationalgeographic.com/main.html
As is to be expected from the National Geographic Society, an excellent site, often full of newsworthy items.

http://www.ncge.org/
National Council for Geographic Education. A site for professional geography educators that includes a link to an excellent tutorial on the National Geography Standards developed for high-school instruction in the US.

Studying Human Geography

The broad subject matter of human geography has remained relatively constant over several thousand years and in various cultural settings. Approaches to that subject matter have changed, however, in response to the specific needs of time and place. Today human geography comprises a wide—some might say bewildering—array of questions, general concepts, and techniques of analysis.

In order to make sense of such a complex subject, this chapter is divided into three parts: philosophical, conceptual, and analytical. First, we consider the philosophical component: why geographers ask the questions they do. The answer may seem self-evident, but it is not. Different geographers ask different questions, and we need to know why. Second, we consider the conceptual element: what general concepts human geographers use to assist their research activities. Once again, we find several core concepts that are central to almost any piece of geographic research, and several other concepts that tend to be associated with particular types of questions. Third, we look at the many analytical techniques that are available to help human geographers answer the questions they ask. In fact, the three parts of this chapter are all intimately related.

The space station Mir over New Zealand (World Perspectives/Getty Images).

Charles Darwin and the title page from his landmark work, *On the Origin of Species*, 1859 (Thomas Fisher Rare Book Library).

Why is philosophy important to the beginning geographer? Our discussion of the discipline's evolution made it clear that the specific content of human geography is essentially the product of three distinct eras: a preinstitutional phase, before the first university geography departments were established in 1874; the period when the discipline was institutionalized, and geographers needed to articulate a specific content for it; and more recent developments. We can fully appreciate our content—the way we analyze and conceptualize problems—only if we first understand our philosophies. It is our philosophical perspectives that explain our specific content, concepts, and techniques of analysis.

In human geography, as in other disciplines, facts, concepts, and techniques are logically interrelated. But it is not enough simply to accept that these are related: we need to know *why*. This means we must understand the philosophical viewpoints that serve as the 'glue'. Much (though not all) geographic work is guided by a particular philosophical viewpoint. Before we embark on such work, we need to learn about our philosophical options—to understand why we ask the questions we do, why we conceptualize as we do, and why we use particular techniques. Philosophy is at the heart not only of this chapter but of much of this book. We begin with an overview of our philosophical options.

Philosophical Options

First, it may be helpful to realize that in fact we have already started discussing such matters, if only indirectly, in the Introduction and Chapter 1. To clarify this point, let us return to the concept of 'physical geography as cause', or environmental determinism.

This term, as we have seen, implies that the physical environment is the principal determinant of all human matters, including human landscapes. It refers to what Ratzel (Volume 1) called anthropogeography and was subsequently used by many others, notably Semple, Huntington, and Taylor. But we have already seen that there are other views. Possibilism implies that the physical environment does not determine human matters, but simply offers various possibilities. This is the cultural landscape view variously introduced by Vidal, Schlüter, and

Sauer. Immediately, then, we find that what appeared to be merely two alternative views about human geography are in fact the tip of a massive philosophical iceberg: determinism versus free will.

Determinism is a philosophical concept postulating that all events, including human actions, are predetermined. Free will, on the other hand, postulates that humans are able to act in accordance with their will. Environmental determinism is a specific version of the larger concept of determinism, while possibilism is a specific variant of the larger concept of free will—a variant that does not necessarily require acceptance of free will itself. It is not difficult to understand why environmental determinism proved attractive to geography:

1. According to the prevailing late nineteenth-century view, it was the job of science to explain phenomena in terms of cause and effect. Thus a geography centred on environmental determinism was a geography with scientific credibility. Specifically, its emphasis on cause and effect linked it, philosophically, to Darwin's landmark work, *On the Origin of Species,* published in 1859.
2. There was a long history of environmental determinist thought beginning with the Greeks and including such philosophers as Bodin and Montesquieu. We might argue that Ratzel, in 1882, was merely introducing to geography a long-established and appropriate concept.
3. When geography was in its institutional infancy, any approach that explicitly asserted the importance of geography to human affairs would enhance the new discipline's status.
4. The apparent logic of the environmental determinist argument was attractive in itself. As Taylor once wrote, 'as young people we were thrilled with the idea that there was a pattern anywhere, so we were enthusiasts for determinism' (cited in Spate 1952:425).

Instead of engaging in any substantial determinism-versus-free will debate, those geographers influenced by Ratzel's Volume 1 simply accepted the logic of determinism. It was not until the 1950s that essentially philosophical cases for environmental determinism and possibilism were made (Box 2.1).

Possibilism emerged at much the same time as environmental determinism, for three reasons:

1. A substantial literature already existed that centred on humans and land and allowed humans to make decisions not explicitly caused by the physical environment. (Humboldt and Ritter were its two most notable authors.)

2. Geographers could point to many instances in which different human landscapes were evident in essentially similar physical landscapes.

3. Possibilism corresponded to the historically popular view to the effect that every event is the result of individual human decision-making.

This brief discussion of environmentalism and possibilism reminds us that almost every geographic question we ask has philosophical

---

2.1

ENVIRONMENTAL DETERMINISM

'In the early part of this century, a major stumbling block to meaningful objective research in American geography was the assumption that the physical environment largely determined the cultural landscape' (Dohrs and Sommers 1967:121).

There is no denying the long scholarly pedigree of this view in Western thought. Both Plato and Aristotle regarded Greece as having an ideal climate for government. Montesquieu contended that areas of high winds and frequent storms were particularly conducive to human progress. As the French philosopher Cousin put it:

> Yes, gentlemen, give me the map of a country, its configuration, its climate, its waters, its winds and all its physical geography; give me its natural productions, its flora, its zoology, and I pledge myself to tell you, a priori, what the man of this country will be, and what part this country will play in history, not by accident but of necessity, not at one epoch but at all epochs (quoted in Febvre 1925:10).

Environmental determinism was not only in accord with the prevailing views of science in the late nineteenth century, but was seen by some as offering a distinct and necessary basis for the new discipline of geography.

Nevertheless, in retrospect it is still surprising that so many geographers accepted environmental determinism in a relatively uncritical manner. Davis, for instance, considered any statement to be geographic if it linked physical factors to some human response, while Semple (1911:1) wrote, 'Man is a product of the earth's surface.' Most significantly, until the 1950s environmental determinism was implicitly accepted in much geography, including much of the regional geography that was the dominant approach at the time.

Taken out of context, many of the apparently extreme statements of this view, such as those quoted from Cousin and Semple above, are easily misinterpreted. However, careful reading suggests that such dramatic statements are not substantiated in the larger body of their work—see, for example, G.R. Lewthwaite (1966:9). Indeed, Huntington, a human geographer usually regarded as an archetypal determinist, wrote, 'Physical environment never compels man to do anything: the compulsion lies in his own nature. But the environment does say that some courses of conduct are permissible and others impossible' (Huntington 1927:vi).

Debates about the relative merits of determinism and alternative interpretations began by the late 1940s and became common in the 1950s. Environmental determinists posited that it was essential to talk in terms of cause and effect, and therefore it was necessary to be a determinist. The basic counter to this assertion was that environmental determinist statements were incapable of being tested; they were merely working assumptions, useful only for suggesting research directions. According to this logic, the fundamental error of much environmental determinism is that it simply assumes a certain cause and effect to be true, when in actuality there is no basis for any such assumption.

Many other social scientists regarded geography with some dismay because of what they perceived as an overemphasis on the role of physical factors in discussions of human behaviour. Durkheim even objected to the inclusion of physical factors in the French school of geography initiated by Vidal, a school that itself opposed environmental determinism.

Regardless of where one stands on the relative importance of physical and human factors, it is always an oversimplification to treat physical geography as a cause of human behaviour.

underpinnings. The only aspects of geography that are essentially independent of philosophy are exploration and mapping. Our concern now is to identify the major philosophies that play a part, implicitly or explicitly, in contemporary geography: empiricism, positivism, humanism, and Marxism.

Empiricism

Much human geography, even today, may appear to be aphilosophical. Certainly most human geographers before the 1950s ignored philosophical issues; they simply conducted research that was considered appropriate in the light of the historical development of the discipline. Such work, particularly in regional and cultural studies, made no claim to have a philosophical base; human geographers were simply doing what human geographers did. Nevertheless, it can be argued that there was an implicit philosophy in such regional and cultural work. This is the philosophy of **empiricism**, which contends that we know through experience, and that we experience only those things that actually exist. Empiricism typically sees knowledge acquisition as an ongoing process of verifying and, as necessary, correcting factual statements. In practical terms, an empiricist approach allows the facts to speak for themselves. By definition, empiricism rejects any philosophy that purports to be an all-embracing system.

Empiricism in turn is rejected by most other philosophies. It is, however, a fundamental assumption of positivism.

Positivism

For some human geographers **positivism** is a very attractive philosophy because it is rigorous and formal. In principle a clear and straightforward philosophy for human geography, positivism makes the following arguments:

1. Human geography needs to be objective; the personal beliefs of the geographer should not influence research activity. Do you agree with this principle? If so, do you believe that it is possible to research human geography without being affected by your personal beliefs? According to humanism and Marxism (outlined in the two following sections) objectivity is not only undesirable but in fact not possible.

2. Human geography can be studied in much the same way as any other science. For the positivist, there is really no such thing as a separate geographic method; all sciences rely on the same method. Specifically, positivism first found favour in the physical sciences, and its applications in human geography reflect the belief that humans and physical objects can be treated in a similar fashion. Once again, humanism and Marxism reject this assumption, believing that it dehumanizes human geography.

3. The specific method that positivism sees as appropriate for all sciences, physical and human, is known as the **scientific method** (Figure 2.1): reflecting the empiricist character of the philosophy, research begins with facts; a **theory** is derived from those facts, together with any available laws or appropriate assumptions; a **hypothesis** (or a set of hypotheses) is derived from the theory; and that hypothesis becomes a **law** when verified by the real world of facts. Thus the scientific method consists of the study of facts, the construction of theory, the derivation of hypotheses, and the related recognition of laws (Box 2.2). For the positivist, any science rests on the twin pillars of facts and theory, and a disciplinary focus on one at the expense of the other is wrong.

Figure 2.1 The scientific method. Science stands on the twin legs of facts and theory; theoretical work depends on and stimulates factual work.

Source: Adapted from J.G. Kemeny, *A Philosopher Looks at Science* (Englewood Cliffs: Prentice Hall, 1959):81.

Positivist philosophy has roots in the work of the French social philosopher Auguste Comte (1798–1857) and was introduced into human geography relatively late (1953). The principal impetus behind Comte's initial argument for positivism was to distinguish between science on the one hand and metaphysics and religion on the other. In human geography, positivism was closely associated with quantitative methods and theory development during the 1960s. Positivism is an integral part of what we described earlier as the spatial analysis approach. Its introduction to geography was controversial because it directly challenged the regional approach dominant at the time. Box 2.3 offers a brief sketch of that controversy.

Humanism

Compared to positivism, **humanism** is a much more loosely structured set of ideas. It comes in several significantly different versions. Contemporary humanistic geographers argue for intellectual origins in earlier geographic work, such as that of Vidal and Sauer; in fact, however, this earlier work was essentially aphilosophical, whereas the contemporary work is typically rooted in some philosophical ground.

Humanistic geography developed from about 1970 onwards, initially in strong opposition to positivism. It focuses on humans as individual decision-makers, on the way humans perceive their world, and emphasizes subjectivity in general. Thus there are substantial differences between the two philosophies. Positivists argue that the world exists as an objective reality, independent of the mind that perceives it, whereas humanists believe it is impossible to separate the two—and hence that research is inevitably influenced by the particular qual-

---

POSITIVISTIC HUMAN GEOGRAPHY

2.2

An example of a positivistic approach to research in human geography will demonstrate the value of working within this philosophical framework. This research, analyzing the retailing and wholesaling activities in urban centres, proceeds through five conventional stages (Yeates 1968:99–107)

1. *Statement of problem*: The purpose of the study is to determine the relationships between the numbers of retailing and wholesaling establishments and the size of urban centres in a part of southern Ontario. This problem is suggested by earlier work, both factual and theoretical, suggesting that urban land uses and activities may be explained in terms of variables such as size.

2. *Hypotheses:* Three specific hypotheses are identified on the basis of available human geographic theory (the theory in question is known as central place theory and is discussed in Chapter 12). These are: (1) the number of retail establishments and the population size of an urban centre are directly related; (2) the number of retail and service functions in an urban centre increases with the size of that centre, but at a decreasing rate; (3) the number of wholesale establishments and the population size of an urban centre are directly related.

3. *Data collection*: Population, retail, and wholesale data are collected for all urban centres of 4,000 or more population

(1966). In addition to fieldwork, the data sources used include city directories, telephone directories, and censuses. Note that the use of a minimum population of 4,000 is an example of an operational definition (a way of describing how an individual object can be identified as belonging to a general set).

4. *Data analysis:* This is the stage in the research when information is subjected to appropriate analysis to formally test the stated hypotheses (the verification stage). In this study, the data are graphed (for example, the number of establishments against the population of urban centres), and a statistical procedure known as correlation analysis is used to calculate the extent of the relationship between the two sets of data.

5. *Stating conclusions:* In standard positivistic research, the formal testing of a hypothesis results in a statement concerning the validity of the hypothesis. In this study, all three of the stated hypotheses are confirmed by the statistical analysis; such analysis increases our confidence in the theory that generated the hypotheses and also provides insights into the specific landscape investigated.

As this example shows, it is no exaggeration to describe positivism as a rigorous and formal philosophy.

ities of the individual researcher. Positivists seek to generalize, whereas humanists focus on the individual and specific. Positivists base their research on theories, test hypotheses, and develop laws; humanists have no such package of standard, reproducible methods. Rather, humanistic research is essentially personal, the product of the individual geographer, involving intuition and interpretation.

Not surprisingly, positivistic and humanistic geographers usually study different types of research problems. Humanists tend to focus on humans and land, landscapes, and regions, whereas positivists tend to focus on spatial analysis. These differences will become apparent in later chapters.

Discussions of humanism easily become complicated because there are several human-istic philosophies, among them pragmatism and phenomenology:

1. The philosophical approach known as **pragmatism** developed in the United States at the end of the nineteenth century and was especially influential before 1945. Pragmatism centres on the idea that human actions are structured by our subjective interpretations of the world we live in. Moreover, those interpretations have a marked practical—pragmatic—content. The central argument is that every human action taken in the world is based on human perceptions and observation of the apparent conditions for, and practical consequences of, previous actions—in short, on experience. Thus pragmatism is closely associated with empiricism. There is little evidence to

2.3

THE SCHAEFER–HARTSHORNE DEBATE

Because overviews of the evolution of an academic discipline, particularly with respect to philosophical changes, are supposed to be dispassionate, they can rarely do justice either to the individuals involved or to the events in which they participate. On this occasion, however, we will depart from tradition.

By the mid-1950s, geography seemed ripe for change. The combination of regional geography and environmental determinism was coming to be seen as unproductive, while the landscape school led by Sauer appealed only to a minority, albeit a substantial one. The transition to spatial analysis was not, however, easy. One important element in that transition was the Schaefer–Hartshorne debate.

An account of this debate can be found in Martin (1989). In 1952, Fred Schaefer submitted a manuscript entitled 'Exceptionalism in Geography: A Methodological Examination' to a major geographical periodical, the *Annals of the Association of American Geographers*. In the winter of 1952–3, Schaefer suffered a heart attack. On 6 June 1953, at the age of 48, he died of a second attack, and his paper (Schaefer 1953) was published shortly thereafter. In it he refuted the claim that geography was necessarily a regional discipline, and advocated the adoption of a positivist approach and related methods. Specifically, Schaefer objected to the argument that he saw as implicit in Richard Hartshorne's work, to the effect that geography studied unique places or regions and was therefore unlike other disciplines.

Hartshorne first saw the article in October 1953. He corresponded with the *Annals* editor several times in that month, complaining that Schaefer's article was 'a palpable fraud, consisting of falsehoods, distortions and obvious omissions' (quoted in Martin 1989:73), and later published a series of substantial rebuttals (Hartshorne 1955, 1959).

One geographer (Bunge 1968) wrote a biographical sketch of Schaefer's life that was later condensed (Bunge 1979). As a socialist in Germany during Hitler's rise to power, Schaefer was 'immediately subjected to terrorism. . . . He was under constant police surveillance' (1979:128). He became a political refugee in England and subsequently moved to Iowa City, where he joined the university's geography department. Bunge notes the harassment that Schaefer faced in the early 1950s, when a US Senate committee headed by Senator Joseph McCarthy was attempting to find and punish suspected communists in the US. The article, the greatest achievement of his life, faced strong opposition, which further drained his energies. 'Hartshorne, in a commanding position in geography, turned his fury against Schaefer. With sheer brilliance of argument as his only weapon, Schaefer fought to be heard. Fire engine sirens vividly reminded him of Hitler's terrorism. . . . McCarthyism repulsed and sickened Schaefer.' At the time of his death, Schaefer's 'article—his "reason for being"—had not yet been published' (Bunge 1968:20–1).

date of explicit uses of this philosophy in human geography, although it is basic to some of our qualitative research techniques.

Unlike some other humanistic philosophies, pragmatism has a marked focus on the group or society (as opposed to the individual). One influential offshoot of pragmatism sees individual behaviour as comprehensible only by reference to the behaviour of the appropriate larger groups to which the individual belongs. This offshoot is known as interactionism.

2. The philosophy of **phenomenology** has many variations, but all of them are based on the idea that knowledge is subjective. To understand human behaviour, therefore, it is necessary to reconstruct the worlds of individuals. A central component of phenomenology is the idea that researchers need to demonstrate *verstehen*, or sympathetic understanding of the

issue being researched. The distinction with respect to positivism is clear. Phenomenology seeks an empathetic understanding of the lived worlds of individual human subjects, whereas positivism seeks objective causal explanations of human behaviour, without reference to individual human differences. One especially thoughtful geographer has written a series of books and articles that are phenomenological in focus (Box 2.4).

Several other humanistic philosophies, such as **existentialism** and **idealism**, have been advocated by geographers, but they have not yet proved to be highly influential.

Our brief discussion of humanism has identified two general issues. First, the basic distinction between positivism and humanism is one of objectivism versus subjectivism. Positivism contends that the study of human

HUMANISTIC HUMAN GEOGRAPHY

2.4

What questions does humanistic geography ask? What types of research issues does it address? What kinds of answers does it look for? Although the 'humanistic' name covers many different approaches, one way to think about these questions is to consider the work of a leading practitioner, Yi-Fu Tuan.

Tuan was born in China in 1930 and educated at London and Oxford universities in England and at Berkeley in California. An initial interest in physical geography soon expanded into a broad focus on questions of humans and nature. Today, Tuan is a leading humanistic geographer and one of the first to make explicit reference to philosophy, although he does not claim to subscribe to any single approach: 'My point of departure is a simple one, namely, that the quality of human experience in an environment (physical and human) is given by people's capacity—mediated through culture—to feel, think and act...' (Tuan 1984:ix).

With that aim in mind, Tuan is critical of much geography for ignoring individuals and the relationships that they have with one another. For Tuan, there is a need to be comprehensive, like a novelist: 'comprehensiveness—this attempt at complex, realistic description—is itself of high intellectual value; places and people do exist, and we need to see them as they are even if the effort to do so requires

the sacrifice of logical rigour and coherence' (Tuan 1983:72).

Three examples of Tuan's work, all well received within and beyond our discipline, illustrate the broad range of his interests: a discussion of landscapes of fear, proposing that fear is a component part of some environments and that, in turn, it helps create environments (Tuan 1979); a discussion of gardens as an example of our attempts to dominate environments (Tuan 1984); and a discussion of the links between territory and self (Tuan 1982). In the latter, Tuan contends that the presence of segmented spaces—for instance, the many discrete spaces in any urban area—leads to segmented societies. Humans retreat into segmented social worlds in response to the difficulties of coping in a complex society. There is a clear distinction between a small settlement in the less developed world, which has few human-constructed barriers and few spaces assigned for exclusive activities, and a modern city with numerous physical and social barriers. The human experiences in these two environments are very different, with significant implications for individual and group behaviour.

The differences between this humanistic geography and the positivistic variety are striking. The two approaches tackle different problems; they raise different questions; and they reach different answers.

phenomena can be objective; humanism says that it cannot. This argument is an acknowledged, long-established, and unresolved issue in the social sciences. Three questions are relevant here:

1. Is there an interaction between the researcher and the researched that invalidates the information collected?
2. Does the researcher have a personal background that effectively influences his or her choice of problem, methods, and interpretation of results?
3. If we view humans objectively, does this mean we see them as objects? If so, is this approach dehumanizing?

Answering 'no' to these three questions means that you have positivistic tendencies; 'yes' means that you have humanistic tendencies. If your answer is 'don't know', that is perfectly understandable at this stage in your studies.

The second issue that is becoming apparent is that of social scale. Do we as human geographers study individuals or groups of people? Traditionally we have focused on groups, but you will have noted that the classic humanistic philosophies place some emphasis on individuals. For many human geographers, this emphasis on individuals is inappropriate, and much humanistic work in geography has been done on a group scale. Typically, groups are defined by culture, religion, language, ethnicity, or class.

Marxism

The ideas of Karl Marx (1818–83) need to be taken very seriously. Unfortunately, Marx and his collaborator, Friedrich Engels, are both difficult writers to summarize. Nowhere are his ideas presented in a clear, unequivocal fashion, and any interpretation of **Marxism** is bound to give rise to disagreement. In broad terms, then, Marx was a political, social, economic, and philosophical theorist who worked to construct a body of social theory that would explain how society actually worked; he was also a revolutionary who aimed to facilitate a change in the economic and political structure of society, from **capitalism** to communism.

Before reading about Marx's ideas, it is important to recognize the enormous influence they have had on recent world history. Before the collapse of the former USSR in 1991 and related changes in eastern Europe, approximately one-third of the world's population lived in societies that traced their origins to Marx's ideas (Box 2.5).

What is it about those ideas that has led so many countries to adopt them in practice—and at the same time has attracted so much academic elaboration by so many different schools of thought (see Box 2.5)? The simple answer is that Marx has something relevant to say, both pragmatically and intellectually.

The Marxist perspective is often described as **historical materialism**. This term refers to Marx's concern with the material basis of society and his effort to understand society and social change by referring to historical changes in social relations. Marx summarized the character of society as follows. First, there are **forces of production**: the raw materials, implements, and workers that actually produce goods. Second, there are **relations of production**: the economic structures of a society, that is, the ways in which the production process is organized. The most important relations are those of ownership and control. In a capitalist society, for example, those relations are such that workers are able to sell their labour on the open market. Thus, although it is the forces of production that produce goods, it is the relations of production that determine the way in which the production process is organized.

Together, these forces and relations make up what is called the **mode of production**. This concept is the key to understanding the composition of society. Examples of modes of production include slavery, feudalism, capitalism, and socialism. It is helpful to think of the mode of production as a stage in a process.

Marx used these ideas to sketch a history of economic change that traces the transitions from one mode of production to another—for example, from feudalism to capitalism to socialism. Marx was especially critical of the capitalism that flourished in his lifetime because of his belief that one class (owners) was able to, and did, exploit the other (workers) in order to maximize its own profits. Exploitation itself was not new; slavery and serfdom were both obvious forms of exploitation. What Marx recognized was that under capitalism, the same kind of exploitation continued, except that now it was disguised.

It is for this fundamental reason, the exploitation of workers by owners, that Marx called for a socialist revolution in the form of a class struggle: a struggle to overturn one mode of production and replace it with another mode of production under the control of a different class—workers themselves.

Two more terms are important to our human geographic understanding of Marxism. **Infrastructure (base)** is simply another word for the relations of production; **superstructure** is the level of appearances, that is, the legal and political system, the social consciousness, and, for our purposes, the larger human geographic world. These terms are important to us because they are connected: the infrastructure is seen as the cause of the superstructure. Thus a primary concern of Marxist human geography is to understand the crucial role played by the infrastructure—in this case the economic structure of capitalist society—in determining the superstructure of the human geographic world. The human geographer who works within Marxism thus analyzes superstructures, at any location at any time, as they

2.5

MARXIST THOUGHT AND PRACTICE

To understand the broad outlines of Marxist thought, it may help to look at how that thought has been applied from the mid-nineteenth century to the present, and how it has been interpreted and modified by others. Three overlapping stages can be identified in the history of Marxism:

1. During the first period, from the publication of the *Communist Manifesto* in 1848 to roughly 1917, followers of Marx focused on four classic concerns: (1) explaining the means by which capitalists exploited workers; (2) advocating a change from private ownership to state control of property and industry; (3) denouncing religion as the 'opium of the people'; and (4) preparing for the inevitable revolution that, when it came, would eliminate the class system.

2. In the second period Marx's ideas were put into practice in some countries, beginning with Russia. Following the Russian Revolution of 1917, in which communist (or Bolshevik) forces overthrew the czar, communism expanded into eastern Europe, parts of the less developed world, and, by 1949, China. Until the collapse of communist systems in the former USSR and eastern Europe in the early 1990s, about one-third of the population of the world lived in a political and social system inspired by Marxism. (In the early twenty-first century, the principal communist country is China, with about one-fifth of the world population—1.3 billion of a total of 6.2 billion in 2002.) It is important to point out that in none of these cases was the system that Marx described fully implemented. In the former USSR (Union of Soviet Socialist Republics), and those countries influenced by it, it was

Karl Marx.

the government—controlled by a communist political party—that controlled economic activity. Contrary to Marx's own expectations, in the USSR workers had minimal influence on economic policies. In fact, the system in place has sometimes been described as state capitalism rather than communism. The Chinese variant of communism was closer to Marx's ideas until the 1976 death of Mao Zedong; since then the government, though still communist, has made substantial moves towards a capitalist economy (some of the important agricultural and industrial changes since 1976 are addressed in Chapters 11 and 13 of this book).

3. The third stage, which might be called western academic Marxism, began in Europe in the 1920s but became influential in North America only in the 1960s. This academic Marxism developed in opposition to the Soviet version and includes numerous variants. Two major schools of thought in this Western tradition, which you may encounter if you read further in this area, are the Frankfurt school of critical theory, which flourished in the 1960s, and the structural Marxism developed by Althusser.

By 1970, when 'Marxist' ideas were first introduced to human geography, the term itself could refer to a multitude of related but often quite different ideas.

Manchester cotton mills, c. 1940. Similar landscapes developed in many parts of northern England during the industrial revolution. Marx's collaborator, Friedrich Engels, wrote of Manchester a century before this photograph was taken: 'If any one wishes to see in how little space a human being can move, how little air—and such air!—he can breathe, how little of civilisation he may share and yet live, it is only necessary to travel hither' (*The Condition of the Working-Class in England in 1844* [London, 1892, p. 53]). In 'dark satanic mills' like these, workers manufactured cotton goods, spinning the yarn and weaving it on huge looms. Some factories were attached to orphanages that offered children board and lodging in return for their labour. Small children—employed to clear out cotton fluff from the looms—were sometimes crushed and killed by the fast-moving machines. In the nineteenth century, mill owners who used child labour were often praised for their philanthropy in keeping the children off the streets and providing them with food and shelter. Today some mills remain as museum pieces, cleaned up and Disneyfied for tourist consumption (Hulton Archive/Getty Images).

derive from the relevant infrastructures (essentially sets of economic processes).

Marxism differs profoundly from both positivism and humanism in that it sees human behaviour as being constrained by economic processes. Marx held human societies alone responsible for the conditions of life within them. In other words, he believed that social institutions were created by humans—and that when those institutions no longer served a society's needs, they could and should be changed.

Thus, as Box 2.5 suggests, Marxism has a significant revolutionary dimension. Working within Marxism typically implies that one is striving both to understand the human world and to change it. Box 2.6 summarizes one example of the Marxist concern with the global spread of capitalist culture. The long-term aim of such work is to facilitate a social and economic transformation from capitalism to socialism. In the shorter term, Marxism is an attractive philosophy for many human geographers who feel strongly that their work should focus explicitly on social and environmental ills and contribute to solutions to those ills. Other philosophies, such as positivism, are seen as merely describing the superstructure—which they implicit-

ly accept as appropriate. It is worth noting that Marxist geography is not the only kind of geography that strives to contribute solutions to environmental and social ills; however, it is fair to say that that focus is most explicit in Marxist work.

The ideas of Marx live on in many different interpretations and elaborations. His writings and those of his followers are essential reading for any social scientist, especially those who are critical of our current societies.

Philosophy and Human Geography

It was aphilosophical human geography, or empiricism, to which positivism objected in the mid-1950s. In turn, by about 1970 positivism was facing criticism from humanism and Marxism—philosophies that also object to certain aspects of one another. Not surprisingly, human geographers continue to welcome further additions to an already broad range of options. Among the current developments in social thought that you will encounter later in this book are feminism and postmodernism.

First, however, there is one more important distinction to be drawn: between idiographic and nomothetic methods. An

idiographic approach is concerned with individual phenomena; traditional empiricist regional geography—Hartshorne's 'areal differentiation'—was idiographic in focus, as is much humanistic geography. A **nomothetic** approach, on the other hand, formulates generalizations or laws. Research in a positivist tradition is nomothetic, as is much Marxist human geography.

The philosophical diversity of contemporary human geography reflects the diversity of our subject matter, namely human behaviour in a spatial context. It also means that we draw on a wide variety of concepts and methods. Some aspects of the human geography presented in this book are clearly aphilosophical or empiricist, some are positivist, some are humanist, some are Marxist, and some reflect the influence of other aspects of current social thought. That some may be difficult to pin down clearly is a reflection of the fact that—as we will see in Chapter 8—the philosophies outlined in this chapter, although they are the principal ones, are by no means the only ones relevant to human geography.

Human Geographic Concepts

Like any other academic discipline, human geography involves two basic endeavours. First, there is the need to establish facts. To know where places are and to be aware of their fundamental characteristics is certainly a vital starting point. This is what we might call *geographic literacy*. Second, there is the need to understand and explain the facts—to know why the facts are the way they are. This is what we might call *geographic knowledge*. Understanding and explaining require that we ask intelligent questions; this in turn requires the conscious adoption of an appropriate philosophical stance.

But there is much more involved in the acquisition of geographic knowledge. Philosophy guides us, but it does not necessarily provide us with all the tools we need. Our task now is to identify those tools, first in the form of concepts and then, in the next section, in the form of techniques. Some of these concepts and techniques are relevant regardless of philosophical bent, while others are philosophy-specific. Where

2.6

MARXIST HUMAN GEOGRAPHY

Much contemporary human geography employs—implicitly or explicitly—a Marxist philosophical framework. The following discussion of the global spread of a capitalist culture may help to clarify what doing Marxist research really means (Peet 1989).

What is culture? How is it formed? Why has capitalist culture been able to spread across the globe and destroy previously existing cultures? A Marxist response to these questions begins with the idea of the consciousness of a culture—that is, the general way that humans think about things. Put simply, different human experiences in different regions of the world result in different regional consciousnesses and cultures. Before the introduction of capitalism as a new economic and cultural system (a new mode of production), the world comprised a variety of regional cultures, each with its own distinct kind of consciousness and its own specific means and modes of production. One way to explain these regional variations is to recognize that human culture was closely related to physical geography (this is not an environmental determinist assertion: simply a recognition that humans with limited technology are especially closely related to their environment).

To understand cultural change as a result of the rise of capitalism, a Marxist acknowledges that human history is a story of domination by forces beyond human control. Initially, as we have seen, the dominating force was nature—physical geography—but over time capitalism came to replace nature as the dominating force.

Like nature before it, capitalism has the ability to cause human wealth or poverty, success or failure. It is a powerful and all-pervading economic and cultural system that can overwhelm and replace earlier regionally based economies and cultures. Thus capitalist culture is now a world culture, while regional non-capitalist cultures in the less developed world either have been or are being dramatically transformed. In explicitly Marxist terminology, this process can be seen as reflecting change in the mode of production.

Typically, a human geographer with a Marxist perspective regards these developments as undesirable because the regional cultures that are disappearing had adapted to deal with a wide range of environmental problems, while the new culture, with its new ways of life and thought, is unsuited to handling those issues.

appropriate, concepts are linked to the parent philosophy. Concepts and techniques that are not tied to particular philosophies typically focus on factual matters, such as determining where things are on the surface of the earth; those tied to a philosophy typically focus on understanding or explaining why things are where they are.

Space

We have already had occasion to use the word **space** several times in this book. In fact, it is not uncommon to describe human geography as a spatial discipline, one that deals primarily with space. To understand the meaning of this term, it is helpful to distinguish between absolute and relative space. *Absolute space* is objective: it exists in the areal relations among phenomena on the earth's surface. This conception of space is at the heart of map-making, chorology, and spatial analysis, and it is central to the ideas of Kant, who saw the geographer as ordering phenomena in space. *Relative space*, by contrast, is perceptual: it is socially produced, and therefore, unlike absolute space, it is subject to continuous change.

Some human geographers, notably those with a primarily humanistic perspective, are critical of spatial analysis for what they see as its **spatial separatism**, or 'spatial fetishism': they believe that a human geography based on spatial analysis focuses on space alone as an explanation of human behaviour. Such critics argue that space itself is devoid of content, and becomes important only in the context of human life.

Location

Regardless of philosophical persuasion, every human geographer begins with one central question: where? **Location** is the basic concept; it refers to a particular position within space, usually (but not necessarily) a position on the earth's surface. As we have just noted with regard to space, there are two ways to describe location. *Absolute location* identifies position by reference to an arbitrary mathematical grid system such as latitude and longitude; for example, Winnipeg, Canada, is located at 49° 53'N latitude, 97°09'W longitude. Such mathematically precise statements are often essential. Sometimes, though, they may not be as meaningful as a statement of *relative*

location: the location of one place relative to the location of one or more other places. Absolute locations are unchanging, whereas some relative locations do change. The location of Winnipeg, relative to lines of communication such as roads and air routes, changes over time.

There are other means by which the location of a geographic fact can be identified. A location can be described simply by reference to its place name or toponym: for instance (with increasing degrees of exactness), Canada, Manitoba, Winnipeg, Portage Avenue.

Geographers also distinguish between site and situation. **Site** refers to the local characteristics of a location, whereas **situation** refers to a location relative to other locations. Figures 2.2 and 2.3, respectively, illustrate the site and the situation of Winnipeg.

Place

The term **place** has developed a special meaning in human geography. It refers not only to a location, but also and more specifically to the values that we associate with that location. We all recognize that some locations are in some way distinctive to us or to others. A place, then, is a location that has a particular identity. Our home, our place of worship, and the shops we frequent are all examples. Clearly, for one person or another, many locations on the surface of the earth qualify as places. The term **sense of place** was popularized in the 1970s by humanistic geographers with philosophical roots in phenomenology. It refers to the attachments that we have to locations with personal significance, such as our home, and also to especially memorable or distinctive locations. It is possible to be aware that a particular location evokes a sense of place without necessarily visiting that location; Mecca, Jerusalem, Niagara Falls, and Disney World are all examples.

A closely related concept in sociology, now also utilized by humanistic and other geographers, is that of **sacred space**. This term refers to landscapes that are particularly esteemed by an individual or a group, usually for a religious but possibly also for some political or other comparable reason. Mormons, a Protestant religious group based in Utah, regard a number of spaces as sacred. All their temples are sacred, because they are open only to Mormons and are considered to be locations where communication between

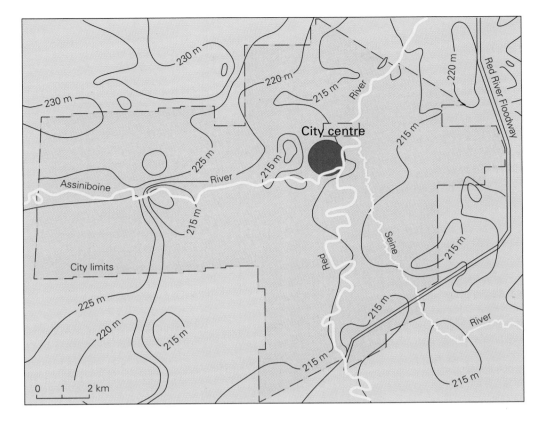

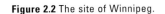

Figure 2.2 The site of Winnipeg.

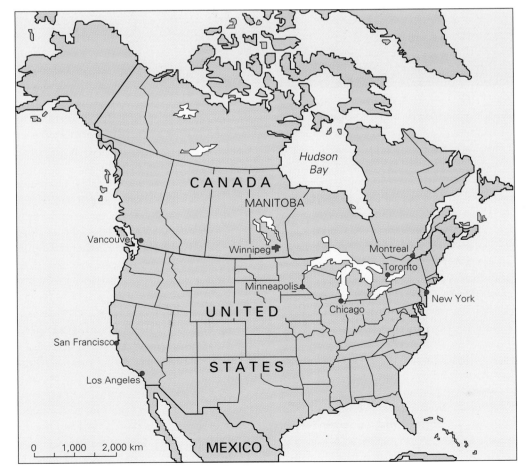

Figure 2.3 The situation of Winnipeg within North America.

'Honeymoon capital of the world': Many North Americans would easily recognize Niagara Falls from this label alone. One of the most readily identifiable places in the world, Niagara Falls is a distinctive physical landscape with a distinctive cultural image. The reason for the association with honeymoons is debated, but there is no doubt that the falls have traditionally represented a 'trip of a lifetime' destination (Andrew Leyerle).

earth and heaven occurs; another sacred Mormon space is a location near Palmyra in New York state, where they believe their founder received God's message (Jackson and Henrie 1983). By contrast, mundane space is occupied by humans but has no special quality. Humanistic geographers use the related concept of **placelessness** to identify landscapes that are relatively homogeneous and standardized; examples include many tourist landscapes, urban commercial strips, and suburbs (Relph 1976). There is an interesting parallel here with the Marxist perspectives outlined in Box 2.6: places may be thought of as especially characteristic of precapitalist cultures, whereas placelessness is more evident in the modern industrial world.

Tuan, the humanistic geographer whose work is discussed in Box 2.4, introduced the concept of **topophilia**, literally 'love of place', to human geography. This term refers to the positive feelings that link humans to particular environments. By contrast, the less frequently used term **topophobia** refers to dislike of a landscape that may prompt feelings of anxiety, fear, or suffering.

These concepts add valuable refinement not only to the more general concept of location, but any human geographic work that aspires to understand people and their relationships with land. Indeed, we can think of places as emotional anchors for much human activity.

Region

Region is one of the most useful and yet at the same time one of the most confusing of geographic concepts:

> So much geography is written on a regional basis that the idea of the region and the regional method is as familiar and as accepted as is Mercator's map in an atlas. Yet as with so many other familiar

ideas which we use every day and take for granted, the concept of the region floats away when one tries to grasp it, and disappears when one looks directly at it and tries to focus (Minshull 1967:13).

Chapter 1 identified regional geography as a traditional focus. In the simplest sense, geographers have long recognized that geographic accounts cannot encompass the earth as a whole; hence the logical tendency to subdivide. In similar fashion, historians have traditionally subdivided the span of human history into periods. By the early twentieth century, however, regional geography had come to a formal definition of region as 'a device for selecting and studying areal groupings of the complex phenomena found on the earth. Any segment or portion of the earth surface is a region if it is homogeneous in terms of such an areal grouping' (Whittlesey 1954:30). Such a definition can accommodate many types of region, both physical and human.

Dividing a large area into regions, or **regionalization**, is a process of classification in which each specific location is assigned to a region. Human geographers recognize various types of regions, most notably the **formal** (or uniform) **region**, an area with one or more traits in common, and the **functional** (or nodal) **region**, an area with locations related either to each other or to a specific location. Thus an area of German-speaking people qualifies as a formal region, while the sales distribution area of a city newspaper qualifies as a functional region. Delimiting formal regions is notoriously troublesome. Consider, for example, how you might delimit a wheat-growing area in the Canadian prairies. Would you delimit it according to the percentage of wheat farmers per township? the percentage of acreage under wheat? the percentage of farm income derived from wheat? These are only some of the possibilities. Clearly, the choice of measure involves considerable subjectivity. Another problem inherent in region delimitation is the implication that regions are geographically meaningful: in fact, a prairie wheat-growing region may or may not be a significant portion of geographic space.

In the 1960s, with the rise of spatial analysis, the traditional concept of formal regions declined and the concept of functional regions increased in popularity. Since roughly 1970 humanistic geographers have argued for a revitalized regional geography involving **vernacular regions**: regions perceived to exist by people either within or outside them. The 'Bible belt' of the American south and midwest is one example. This region is readily identifiable to most Americans as an area with a particularly strong commitment to various conservative Protestant religions. For many of those living in the region, the name is a source of pride; Oklahoma City declares itself to be the 'buckle on the Bible belt'.

The concept of region is fundamental. We have identified four separate applications:

1. Regionalization is a valuable simplifying device, comparable to periodization for historians; the exercise of classifying in itself may be a valuable aid to the understanding of landscapes.
2. Delimitation of formal regions was central to the chorological approach that dominated during much of the first half of the twentieth century.
3. Delimitation of functional regions was an important aspect of spatial analysis.
4. Many contemporary geographers see vernacular regions as crucial to our understanding of human landscapes.

Distance

Since our first three concepts—location, place, and region—all refer to portions of the earth's surface, logically we are also interested in the **distance** between locations, between places, and between or within regions. Not surprisingly, because distance is quantifiably measurable, spatial analysts tend to be particularly interested in it. Distances may be measured in many ways, but most measuring systems typically involve some standard unit such as the kilometre. Human geographers also measure distances in terms of time and cost, or—rather more difficult to measure—in terms of cognitive or social variables.

The **distribution** of geographic facts—such as churches or towns—can often be explained by reference to the distance between them and other geographic facts. Geographers often talk about distributions, patterns, or forms in reference to the mapped appearance of spatial facts (see Figure 2.4).

Figure 2.4 Clustered, random, and uniform point patterns. A clustered pattern is typically produced when the geographic facts benefit from close proximity —for example, specialty retail outlets such as jewellery stores. A uniform, or dispersed, pattern is produced when geographic facts benefit from separation because they compete for surrounding space— for example, urban centres in a region. A random pattern may be the result of a random process; may reflect the influence of more than one process; or may represent a temporary stage during changing circumstances.

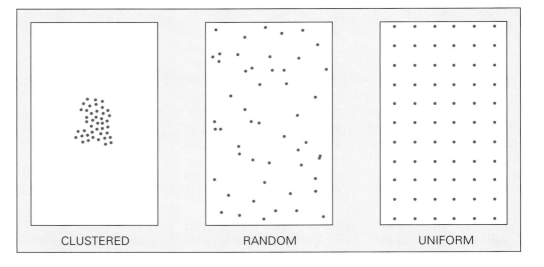

CLUSTERED   RANDOM   UNIFORM

It is often the case that such maps reveal some sort of order. Specialty retail stores (jewellers and shoe stores, for example) tend to locate near one another. Similarly, financial institutions prefer to be close to one another. It is useful to characterize such distributions as resulting from a clustering process. Urban centres, on the other hand, typically develop at some distance from one another, since one of the principal reasons for a particular location is to serve surrounding rural populations. Urban centres also locate at least partly in response to competition. Other geographic facts locate apart because they involve the provision of services without involving competition; hospitals, community centres, and recreation areas are examples.

Much of what has been said about distance so far is evident in what W. Tobler (1970) referred to as the first law of geography: *everything is related to everything else, but near things are more related than distant things.* This is the notion of **distance decay**, graphically depicted in Figure 2.5 and sometimes referred to as the effect or friction of distance; typically both time and cost are involved in overcoming distance. This concept lies at the heart of much spatial analysis (for example, see Box 2.2) and is subject to criticism from humanistic geographers who consider it to be an example of spatial separatism, in that distance is seen as having causal powers.

The distance concept is related to several other ideas. **Accessibility** refers to the relative ease with which a given location can be reached from other locations and therefore indicates the relative opportunities for contact and interaction; it is a key concept in the agricultural, settlement, and industrial location theories that we will encounter later in this text. **Interaction** refers to the act of movement, trading, or any other form of communication between locations. **Agglomeration** describes situations in which locations (usually of activities related to production or consumption) are in close proximity to one another. Conversely, **deglomeration** refers to situations in which those locations are characterized by separation from one another.

Distance is conventionally measured as absolute distance using some standard unit. Distance measured in economic, temporal, cognitive, or social terms is relative. Such distances can be, and frequently are, quite different from corresponding absolute distances. Minimum-cost distances between two locations often involve greater absolute

Figure 2.5 A typical distance decay curve. The specific slope is a function of the particular variable being plotted against distance. The slope usually becomes less steep over time, reflecting decreasing distance friction as a result of constantly improving technology.

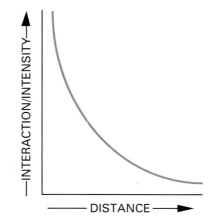

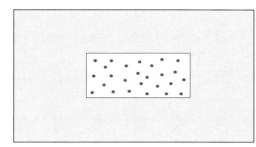

Figure 2.6 The impact of spatial scale on descriptions of point patterns. Whether we describe this point pattern as clustered or dispersed depends on the area within which it is contained. Using the inner boundary, the pattern is dispersed; using the outer boundary, it is clustered.

distances than do straight-line distances because of the lower cost of water compared to land transport, or because of physical or political barriers. Distance may also be measured in terms of the time required: times are usually shorter in more developed than in less developed areas; they also tend to decrease with increasing levels of technology. Cognitive and social distances vary according to individual and group circumstances; such differences often reflect differences in knowledge of the areas between the two locations.

The concept of distance, together with the various closely related concepts discussed above, often lies at the heart of geographic research. Indeed, much of world economic history can be interpreted as involving the gradual overcoming of distance as obstacle. One leading Australian historian talked about Australian development in terms of the 'tyranny of distance' (Blainey 1968).

Scale

One of the first decisions made in any piece of geographic research relates to the selection of appropriate scales—spatial, temporal, and social. It is possible to study a large area or a small area, a long period of time or a particular moment in time, a great many people or a single individual. The choice of **scale** is usually determined by the questions posed. But it is important to recognize that different scales can generate different answers.

Geographers use the concept of *spatial scale* in two distinct ways. First, there is a technical meaning associated with the use of maps; in this case, scale is the ratio of distance on the

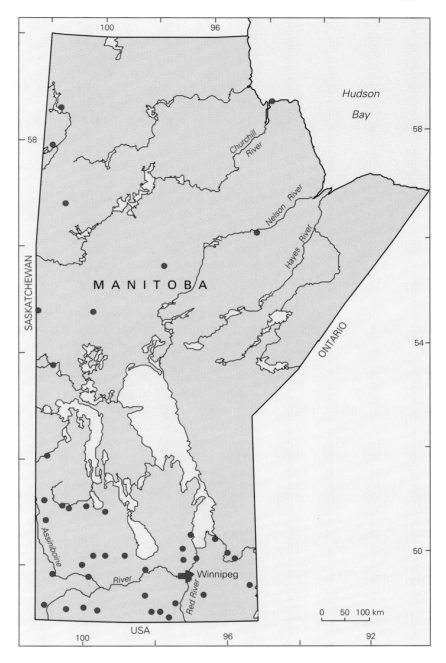

map to distance on the ground. It is in this sense that we might describe the world maps in this book as being at a 'small scale', the maps of regions as being at 'intermediate scales' and maps of local areas as being at a 'large scale'. Note that maps of large areas are at a small scale and maps of small areas are at a large scale.

Second, if we ask whether the locations in a given set are clustered together (agglomerated) or dispersed (deglomerated), the answer will vary with the area selected (Figures 2.6 and 2.7). Spatial scale also needs to be carefully identified whenever we make state-

Figure 2.7 Urban centres in Manitoba. This real-world pattern encompasses two relative distinct regions: a sparsely settled centre and north and a densely settled south.

ments about density. In Canada, for example, a single statistic masks enormous variations; although the average population density for Canada as a whole is 3.25 persons per square kilometre, for an urban centre the density may be 10,000 per square kilometre.

The choice of *temporal scale* is also important. If questions are asked concerning the evolution of landscape, then a temporal perspective is essential; but if questions are asked concerning the manner in which a given area functions, then a temporal perspective may be quite irrelevant. Historical and cultural geography typically emphasize time, while chorological work and spatial analysis tend to emphasize the present. Contemporary human geography uses whatever temporal scale is most appropriate.

The relevance of *social scale* has emerged as a major concern with the rise of humanistic perspectives in human geography. Although the focus has traditionally been on groups, typically delimited by reference to culture, some humanistic geographers today favour an individual scale of analysis. Selecting the scale that is most appropriate to the particular question posed, and that will therefore lead most directly to the correct answer, is not simple. Scales must be selected with care, and properly justified, if we are to be confident of our results.

Diffusion

Landscapes, regions, and locations—our three recurring themes from Chapter 1—are all subject to change. **Diffusion**—the spread of a phenomenon over space and growth through time—is one way change occurs. The migration of people, the movement of ideas (for example, the spread of a religion), and the expansion of land use (for instance, wheat cultivation) are all examples of diffusion. Diffusion-centred research has long been central to cultural geography because of the need to understand landscape evolution. Probably the greatest impetus for such research was the work of a pioneering Swedish geographer, T. Hagerstrand, who developed a series of diffusion-related concepts in 1953; his work was beginning to become known in the English-language world by the early 1960s, but was not translated *in toto* until 1967. Hagerstrand and the work that he inspired, which was largely

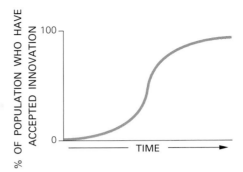

Figure 2.8 A typical S-shaped growth curve. Adoption of the new idea begins slowly, then proceeds rapidly, only to slow down as the adoption rate reaches 100 per cent.

positivistic in character, introduced three important ideas.

First is the *neighbourhood effect*. This term describes situations where diffusion is distance-biased: that is, where a phenomenon spreads first to individuals or groups nearest its place of origin. Other situations, however, involve a *hierarchical effect*. In this case, the phenomenon first diffuses to large centres, then to centres of decreasing size. One geographer noted that the 1832 cholera epidemic in North America diffused in a neighbourhood fashion, whereas the 1867 epidemic diffused hierarchically. Why the difference? Because North America in 1832 had a limited number of urban centres and a limited communication network, whereas North America in 1867 had many urban centres and a well-developed communication network. A third contribution of Hagerstrand's focused on the well-established fact that most diffusion situations proceed slowly at first, then very rapidly, ending with a final slow stage to produce an *S-shaped curve* (Figure 2.8).

The diffusion concept is perhaps best described as a process that prompts changes in landscapes, regions, and locations.

Perception

In 1850, Humboldt noted that, 'in order to comprehend nature in all its vast sublimity, it would be necessary to present it under a twofold aspect, first objectively, as an actual phenomenon, and next subjectively as it is reflected in the feelings of mankind' (quoted in Saarinen 1974:255–6). Despite this pioneering statement, geographers paid relatively little attention to subjective matters, especially the perceived environment, until the late 1960s. We now recognize that all humans relate not to some real physical or social environment but rather to their **perception** of that environment—a perception that varies with knowledge and is closely related to cultural and social considerations. Humanistic geographers in particular discuss the mental images of places and other people, and seek to describe and understand the mental maps that we carry in our heads. These images and maps are important for at least five reasons:

1. Mental **images** of other places and other people are always changing. We all live in a shrinking world. We are becoming increasingly aware that when we pollute a specific location, we are in fact polluting the world.

2. Research into **mental maps** demonstrates that humans have varying perceptions of environments. These perceptions help explain population movements. In the less developed world, the most obvious migrations are from rural to urban areas; in the US, there has been a continuing movement to the south, especially since 1950.

3. The mental maps of particular individuals are of great importance—for example, those of people who make decisions about where to locate industrial plants. The location of a plant that will employ perhaps 1,000 workers will often reflect the biases and values of a single person or small group. Today the traditional industrial areas of the developed world are rarely favoured by such decision-makers. Instead, many prefer to locate close to recreational amenities and/or large cities. In the US the south and west are generally favoured over the north and east; and in England the London region is favoured over the midlands and north. In Canada there is an additional factor, as many locational decisions reflect the biases and values of Americans, not Canadians. The mental maps of decision-makers are especially important because so many industries today are footloose—not tied to specific raw materials or power sources.

4. Serious problems can arise when people in positions of power have distorted mental maps. We may find it humorous when an American president mistakes one country for another, but the consequences could be disastrous. Research in the 1980s showed that the mental maps of the typical American military officer completely confused the sizes of areas and relative populations, and tended only to see one threatening country (the former USSR).

5. Mental maps do change. In Brazil, the movement of the capital from Rio de Janeiro to Brasilia completely changed the image of the interior from one of a worthless wasteland to one of a region with huge potential. In Canada, the image of the north changed with the beginning of oil exploration, and the image of Quebec changes in response to changes in the political fortunes of the separatist movement.

6. Mental maps of relatively unknown areas are especially subject to error. Figure 2.9 is an interesting composite of the image of North America held by Europeans in the mid-eighteenth century.

Clearly, we acquire our mental maps from a variety of sources, but one is of particular relevance to us: human geography teaches us about the world, about where

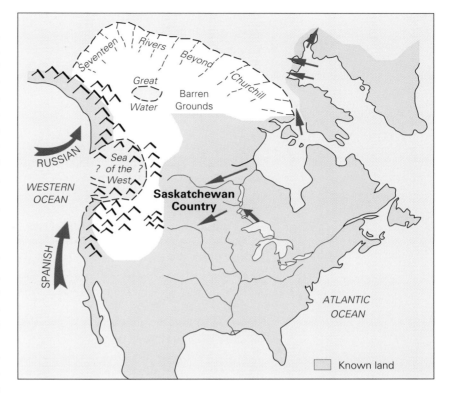

Figure 2.9 Images of North America in 1763. Limited knowledge resulted in many inappropriate decisions. As reality became better understood, spatial behaviour became increasingly appropriate.
Source: R.I. Ruggles, 'The West of Canada: Imagination and Reality', *Canadian Geographer* 15 (1971):237.

things are located, why they are there, and what they really are. By the time you have completed this course, you may notice a significant improvement in your own mental maps and images!

Development

In their studies of the landscapes created by humans, human geographers recognize that any one area changes through time, and that different areas have different landscapes. Such conditions are often interpreted in terms of **development**. The general meaning of this term involves measures of economic growth, social welfare, and modernization. It is possible to define, for any given area at any given time, the level or state of development. Following this logic, certain areas are qualified as more developed and others as less developed. (This distinction will be pursued in some detail in Chapter 6.)

Such basic identifications can be of real value, but we need to use caution in applying them. It is important that human geographers highlight spatial disparities in economic well-being—but it is also important that we interpret variations with reference to cultural and social considerations. If, for example, we use income levels as a measure of development, then we must not neglect to recognize that income level may reflect both economic success and cultural values.

Contemporary human geographers analyze development while remaining fully aware of the risks of oversimplification. A Marxist might view underdevelopment as a consequence of the rapid diffusion of the capitalist economic and social system, arguing that areas brought into the expanding capitalist system become dependent. Another Marxist perspective might suggest that a capitalist system tends to create depressed areas within any given country, thus prompting uneven development. Such variations are evident throughout what we generally label the more developed world. In Canada, for example, an economically advantaged core area, the St Lawrence lowlands of Quebec and Ontario, is surrounded by a number of relatively disadvantaged peripheries, such as the maritime area and the north.

Discourse

The root meaning of the word 'discourse' is 'speech'. In the social sciences, however, the term also refers to a way of communicating, in speech or in writing, that serves to identify the person communicating as a member of a particular group. (Note that language, dialect, or accent can serve the same function.) For example, the technical vocabulary introduced above—'space', location', and so on—is part of the discourse of human geography, and serves to identify those who use that vocabulary as members of the group of human geographers.

'Discourse' also has a more profound meaning, however, derived from the work of the French social theorist Michel Foucault. Foucauldian theory was introduced into the literature of human geography in the 1980s, as one aspect of contemporary social theory. According to Foucault, the history of ideas is a history of changing discourses in which (a) there is a fundamental connection between power and knowledge, and (b) truth is not absolute but relative, dependent on the power relations within the societies that construct it. This more complex meaning of 'discourse' is briefly pursued in Chapter 8, in connection with feminism and postmodernism. The latter are bodies of social theory that challenge established discourses because those discourses are seen as products of people in positions of academic power who are able to define the truth in their terms, to the exclusion of the concerns of other, usually marginalized, groups. See Box 1.3 (p. 23) for an example of the difference that one's choice of words can make.

Globalization

Our final concept integrates—some might even say replaces—several of the concepts introduced above: namely space, location, place, region, and distance. So far, we have had occasion to refer to globalization (see glossary) in three contexts: in the Introduction, as the subject of a preliminary vignette; in Chapter 1, as one of the directions taken by contemporary human geography; and earlier in this chapter, as a focus of concern among those who approach the study of cultures from a Marxist perspective. Those earlier discussions suggested that our complex and varied human worlds are coming, however unevenly, to look more and more like a single world. Accordingly, we now identify globalization as an overriding metaconcept, providing human geographers with a body of ideas that may facilitate analysis of environmental, cultural, political, and economic topics.

The term 'globalization' came into widespread use only in the 1980s. In general, it refers to the idea that the world is becoming increasingly homogenized—economically, politically, and culturally. Globalization is both a result and a cause of an ever-increasing connectedness of places and peoples as economic, political, and cultural institutions, flows, and networks all combine to bring previously separated peoples and places together. Among the most obvious components of globalization are advances in communications technologies and the increasing dominance of **transnationals** (corporations whose activities are not confined to a single state). Among the most obvious consequences is the fact that, because we are generally able to overcome distances much more easily today than we were in the past, distance no longer plays the critical role it once did in promoting the development of separate human geographic worlds.

Globalization increases both the quantities of goods, information, and people that move across national boundaries and the speed with which they do so. Most readers of this book are accustomed to having significant world happenings—sports competitions, concerts, political events—beamed directly to their living rooms, to receiving information and goods from faraway places, and to seeing local decisions affected by larger global circumstances. For many who regularly travel to other countries, the sense of place-to-place difference is vanishing as a few dominant brands become ever more ubiquitous—in the area of information (CNN, the BBC) no less than that of consumer goods (McDonald's, Coca Cola, Benetton, Nike). Many of us have also had direct experience of financial homogenization: today we can purchase goods or local currency almost anywhere in the world using only one major credit card. Virtually all of these developments have come about since 1980.

Some observers go so far as to predict the emergence of a global village, a global culture. Although such a development may still be far in the future, we do appear to be entering a new era involving the creation of global landscapes, in which places separated by great distances are nevertheless very similar in character. Think of shopping malls, major hotels, fast food restaurants, and airports. In these instances, similar settings may be patronized by people with similar lifestyles and similar aspirations. At the centre of all this increasing interaction and homogenization is the capitalist mode of production.

We will see evidence of the trend towards globalization throughout this book—in our discussions of human impacts on the environment (Chapter 4), population growth (Chapter 5), migration and food issues (Chapter 6), language and religion (Chapter 7), ethnic identities and popular culture (Chapter 8), and political transitions (Chapter 9), as well as agricultural, urban, and industrial change (Chapters 11, 12, and 13). Most tellingly, we will see evidence of globalization in our discussions of distance, transportation, trade, and economic integration (Chapter 10).

Concepts: A Summary

This discussion of human geographic concepts has introduced the terms that are central to understanding how human behaviour affects the earth's surface—space, location, place, region, distance, scale, diffusion, perception, development, discourse, and globalization—along with various concepts that can be conveniently subsumed under those headings. In most cases, as we have seen, a concept can be understood and used in multiple ways, depending on philosophical approach; think especially about region, distance, scale, and development. All the terms introduced here will reappear in subsequent accounts of the substantive work of human geographers; all the concepts will be put into practice.

Techniques of Analysis

Now that we have a basic understanding of why human geographers ask the research questions they ask, what their philosophical

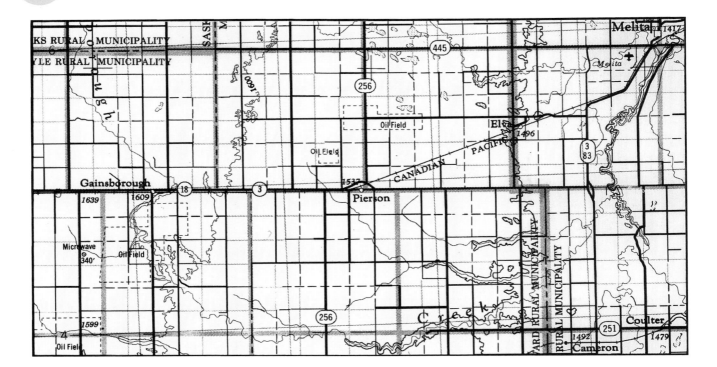

Figure 2.10 Mapping at a scale of 1:250,000.

options are, and the concepts they use to structure their research activity, we are ready to look at the techniques they use to collect, display, and analyze data.

The first three techniques discussed are cartography, computer-assisted cartography, and geographic information systems. Each of these closely related techniques is inherently geographic, having been largely developed within geography, and each involves inputting, storing, analyzing, and outputting spatial data. A group of techniques that relate primarily to the collection of data is discussed next, namely remote sensing. Finally, we consider both qualitative and quantitative methods of collecting and analyzing data.

Although many of these techniques are philosophically neutral, this section is not entirely philosophy-free. Data collection by fieldwork clearly includes some procedures that are positivist and others that are humanist. We will return to the question of underlying philosophies—and what they mean for the 'human' in 'human geography'—in the conclusion of this chapter.

Cartography

Chapter 1 highlighted the centrality of maps to the geographic enterprise. The science of map-making is known as **cartography**.

Until the 1960s, cartography was essentially limited to map production, following data collection by surveyors and preceding analysis by geographers. There was much emphasis on manual skills. The primary purpose of such maps was to communicate information. Quite simply, maps are an efficient means of portraying and communicating spatial data. Today, however, cartography is less dependent on manual skills and is closely integrated with analysis.

As communication tools, maps describe the location of geographic facts. As analytical tools, they can be used to clarify questions and suggest research directions. In the production of maps, cartographers need to decide on questions that can significantly affect map appearance and quality: scale, type, and projection.

As we have already noted, map scale relates to the area covered and the detail presented. Large-scale maps portray small areas in considerable detail, while small-scale maps portray large areas with little detail. A map of a city block is at a large scale and a map of the world at a small scale. Scale is always indicated on a map, whether as a fraction, a ratio, a written statement, or a graphic scale. Canada has a National Topographic System, with maps at a scale of

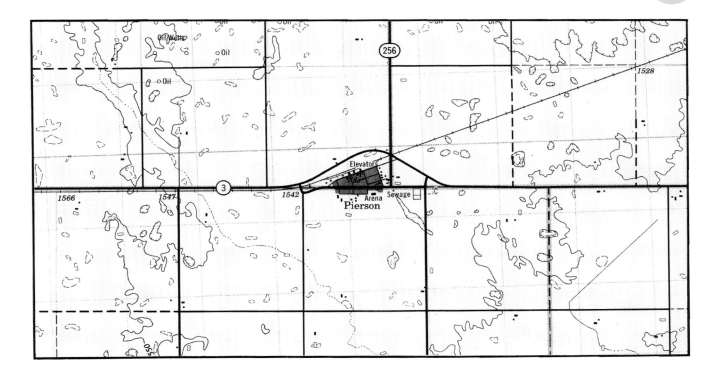

Figure 2.11 Mapping at a scale of 1:50,000.

1:250,000 covering approximately 15,539 km² (6,000 square miles) and maps at a scale of 1:50,000 covering 1,036 km² (400 square miles) in greater detail. On a 1:50,000 map, farm buildings are located and contours (lines of equal elevation) are shown at 7.5-m (25-ft) intervals. Figures 2.10 and 2.11 provide examples of each type.

The type of map constructed depends on the information being presented. Dot maps are useful for data such as towns, wheat farming, cemeteries, incidence of diseases, and so forth. Typically, each dot represents one occurrence of the phenomenon being mapped (Figure 2.12). A **choropleth** map displays data using tonal shadings that are proportional to the density of the phenome-

na in each of the defined areal units (Figure 2.13). Choropleth maps sacrifice detail for improved appearance. An **isopleth map** consists of a series of lines, isopleths or iso-lines, that link points having the same value (Figure 2.14). Examples include contour maps, isochrome maps showing lines of equal time, and isotim maps showing lines of equal transport cost.

Small-scale maps (of large areas such as the earth) raise a fundamental question: how best to represent a nearly spherical earth on a flat surface (a process of conversion known as a **projection**). No satisfactory answer to this question has yet been found, although hundreds of projections have been developed (Box 2.7).

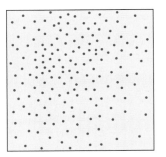

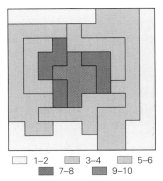

1–2 3–4 5–6
7–8 9–10

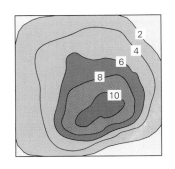

Figure 2.12 (left) Schematic representation of a dot map.

Figure 2.13 (middle) Schematic representation of a choropleth map.

Figure 2.14 (right) Schematic representation of an isopleth map.

2.7

MAP PROJECTIONS

'Not only is it easy to lie with maps, it's essential' (Monmonier 1991:1). Monmonier is referring to the fact that even the best-intentioned cartographers are forced to distort reality because they must always use a map projection: a systematic two-dimensional transformation of the three dimensions of a sphere. This issue is most important when the map is of a large area (small scale).

A projection attempts to retain one or more of the following characteristics of a sphere: shape, size, direction, and distance. The correct projection to use is the one that best serves the objectives of the map. The main decision to be made concerns the choice between a *conformal* projection, which depicts shape correctly, or an *equal area* projection, which maintains the relative areas of all parts of the earth.

Figures 2.15, 2.16, and 2.17 show three important projections. The Mercator projection, introduced in 1569, greatly exaggerates the size of high-latitude land masses (Figure 2.15). Note that Alaska and Brazil appear to be of similar size; yet in reality Brazil is some six times larger. The Mercator is a conformal projection. The great advantage of the Mercator projection is that it can be easily used for navigational purposes; one has only to draw a straight line between two locations on the projection and that line gives the necessary compass direction. The Mollweide projection, introduced in 1805, depicts all regions in correct relative size, but compresses high-latitude regions (Figure 2.16). Mollweide is an equal-area projection. The third projection was introduced in 1963 by the geographer Robinson and was adopted in 1988 by the National Geographic Society (Figure 2.17). The Robinson projection has distortions both at the equator and at the poles and is accurate only at latitudes 38°N and 38°S. Because this projection offers a good compromise between geographic accuracy and aesthetic, it is the one used in this text. *Even so, whatever projection a cartographer uses, getting one thing right will always mean getting something else wrong.*

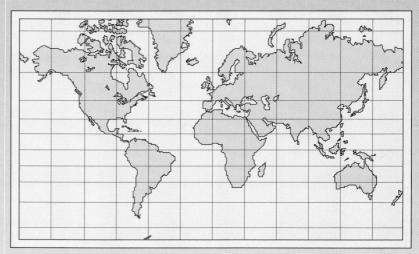

Figure 2.15 Mercator projection.

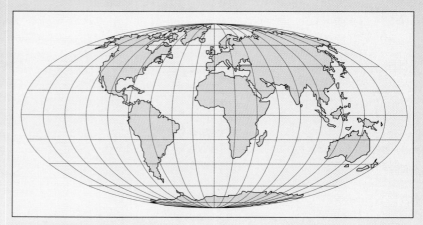

Figure 2.16 Mollweide projection.

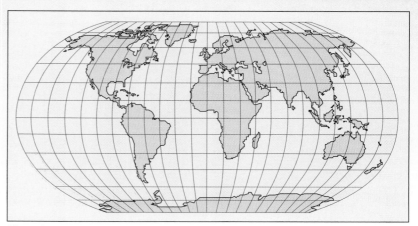

Figure 2.17 Robinson projection.

Map users need to understand the cartographic design process, including the significance of the chosen scale, the types of symbols, and the projection, in order to interpret a map correctly (Box 2.8).

Computer-Assisted Cartography

Computer-assisted cartography, sometimes called digital mapping, is covered separately from traditional cartography because it represents much more than just another evolution in production techniques, although technical advances have significantly reduced the need for manual skills.

Computers allow us to amend maps by incorporating new and revised data; they enable us to produce various versions of the mapped data to create the best version;

THE POWER OF MAPS

'The authoritative appearance of modern maps belies their inherent biases. To use maps intelligently, the viewer must understand their subjective limitations' (Wood 1993:85). While Box 2.7 outlines the unavoidable errors in maps that result from the need to employ a projection, this discussion centres on how maps reflect the assumptions of their creators, and how they may be deliberately employed to convey specific misleading messages.

Recall the T-O maps introduced in Chapter 1; these medieval European maps clearly reflect Christian values, with Jerusalem placed at the centre of the world. Other maps from that place and time show *terra incognita* to the south and make extensive use of decorative pictures. Nineteenth-century European maps reflect the political and economic concerns of their creators: colonial possessions, for example, are prominently displayed. These maps, building on Ptolemaic traditions, established the conventions that most of us take for granted today: north at the top, 0° longitude running through Greenwich, England, and the map centred on either North America or western Europe. These conventions are totally arbitrary—and yet their influence is enormous, because in effect they tell us that certain countries are world centres and others are outliers.

Other maps may convey a biased message either deliberately or because the map-maker has some ulterior motive. As Monmonier (1991) suggests in his book *How to Lie with Maps*, there are many ways to deceive. Political propaganda maps represent an extreme example. Less extreme but still significant are maps that simply ignore important features: for example, maps of the Love Canal area in Niagara Falls, New York (Figure 2.18), that simply fail to include extremely important information. The message here is that 'a single map is but one of an indefinitely large number of maps that might be produced for the same situation or from the same data' (Monmonier 1991:2).

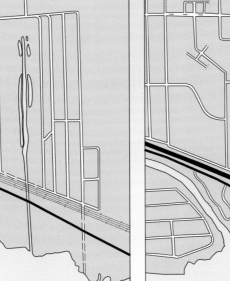

1946

1980

Figure 2.18 Two topographic maps of the Love Canal area, Niagara Falls, New York. The 1946 map shows the canal (the long vertical feature in the centre of the map), but does not show the use made of it, beginning in 1942, as a dump for chemical waste. The 1980 map does not show the filed-in canal, nor does it indicate that dumping continued until 1954. In 1978 the New York health commissioner declared a state of emergency and relocated 239 families.
Source: M. Monmonier, *How to Lie with Maps* (Chicago: University of Chicago Press, 1993):121-2. 1946: US Army Map Service, 1946, Tonawanda West, NY, 7.5-minute quadrangle map. 1980: US Geological Survey, 1980, Tonawanda West, NY, 7.5-minute quadrangle map.

through the use of mapping packages, they diminish the need for artistic skills and allow for desktop map creation. Computer-assisted cartography has introduced maps and map analysis into a wide range of new arenas: for example, businesses can now use cartography to realign sales and service territories. Computer-generated maps facilitate decision-making, and are becoming important in both academic and applied geography.

As a means of upgrading cartographic method and practice, as opposed to speeding up the production process, computer-assisted cartography is one component of geographic information systems.

Geographic Information Systems

A **geographic information system** (GIS) is a computer-based tool that combines several functions: storage, display, analysis, and mapping of spatially referenced data. A GIS includes processing hardware, specialized peripheral hardware, and software. The typical *processing hardware* is a personal computer or workstation, although mainframe computers may be used for especially large applications. *Peripheral hardware* may be used for data input—for example, digitizers and scanners—while printers or plotters produce copies of the output. *Software* production is now a major industry, and numerous products are available for the GIS user; examples include IDRISI, a university-produced package designed primarily for pedagogic purposes, and ARC/INFO, a package developed by the private sector that is widely used by governments, industries, and universities.

The origins of contemporary GIS can be traced to the first developments in computer-assisted cartography and to the Canada GIS of the early 1960s. These developments centred on computer methods of map overlay and area measurement—tasks previously accomplished by hand. Since the early 1980s, however, there has been an explosion of GIS activity related to the increasing need for GIS and the increasing availability of personal computers.

The roots of GIS are clearly in cartography, and maps are both its principal input and its principal output. But computers are able to handle only characters and numbers—not spatial objects such as points, lines, and areas. Hence GISs are distinguished according to the methods they use to translate spatial data

into computer form. There are two principal methods of translation. The **vector** approach describes spatial data as a series of discrete objects: points are described according to distance along two axes; lines are described by the shortest distance between two points; and areas are described by sets of lines. The **raster** method represents the area mapped as a series of small rectangular cells known as pixels (picture elements); points, lines, and areas are approximated by sets of pixels, and the computer maintains a record of which pixels are on or off. Box 2.9 summarizes the sorts of activities involved in using a GIS.

What is the value of a GIS? What applications are possible? The short answer is that GISs have numerous and varied applications in any context that may be concerned with spatial data. We will encounter several of these in later chapters. For the moment, we need only identify two general applications: measuring and analyzing spatially distributed resources and managing spatially distributed facilities. Here are a few specific areas of application: biologists analyze the effect on wildlife of changing land-use patterns; geologists search for mineral deposits; market analysts determine trade areas; and defence analysts select sites for military installations.

The common factor throughout is that the data involved are spatial. After handling spatial data for more than 2,000 years, geographers now have important new capacities. In brief, a GIS achieves a whole new range of mapping and analytical capabilities—additional ways of handling spatial data.

Remote Sensing

No map can be produced without data. GISs and analytical methods in general also require data. Where do these data come from? Early map-makers obtained their data from explorers and travellers. The contemporary geographer collects different data in different ways.

One particularly important group of collection methods focuses on gathering information about objects from a distance. The term **remote sensing** describes the process of obtaining data using both photographic and non-photographic sensor systems. In fact, we all possess remote sensors in the form of eyes, and it has been one of humanity's ongoing aims to improve their ability to acquire information. Improving our eyes themselves (using telescopes), improving our field of vision by gaining altitude (climbing hills or using balloons and aircraft), and improving our recording of what is seen (using cameras) are all advances in remote sensing. Today, most applications of remote sensing rely on electromagnetic radiation to transfer data from the object of interest to the sensor. Using satellites as far as 36,000 km (22,370 miles) from the earth, we are now able to collect information both about the earth as a whole and about what is invisible to our eyes. Electromagnetic radiation occurs naturally at a variety of wavelengths, and there are specific sensing technologies for the principal spectral regions (visible, near-infrared, infrared, and microwave).

The conventional camera was the principal sensor used until the introduction of earth orbital satellites in the 1960s. Aerial photography is still used for numerous routine applications, especially in the visible and near-infrared spectral regions. The near-infrared spectral region has proved particularly useful for acquiring environmental data. Today, the emphasis is on satellite imagery, especially since the launching of Landsat (initially called ERTS—Earth Resources Technology Satellite) in 1972 as the first unmanned satellite. Satellite scanners numerically record radiation and transmit numbers to a receiving station; these numbers are used to computer-generate pixel-based images. Satellite images are not photographs.

2.9

USING A GIS

A GIS can be used to perform as many as eight different functions (Raper and Green 1989):

1. *Data capture:* Vector data are recorded by manual tracing or electronic digitizing of the spatial data. Raster data are recorded by a scanner that uses a light sensor to record the spatial data as 'on' pixels.
2. *Editing:* Once captured, the data are edited to correct errors: for example, to relocate misplaced points in vector data or to remove stray pixels in raster data.
3. *Structuring:* This involves storing the data in a form suitable for rapid spatial retrieval; a number of technical procedures are available to achieve this goal.
4. *Restructuring:* Changing the level of map detail or converting between vector and raster data types.
5. *Manipulation:* One of the greatest advantages of using a GIS is the ease with which spatial data can be manipulated and revisions made.
6. *Search or retrieval:* Another major benefit of working with a GIS is the ease of identifying and retrieving information if necessary.
7. *Analysis:* The ability to analyze, not merely represent, spatial data is perhaps the most important activity that a GIS is able to perform, and is the clearest distinction between GIS and computer-assisted cartography. The quality of GIS software is often judged primarily in terms of this activity.
8. *Integration:* One output possible from a GIS is the integration of two or more maps of an area to form a single map; in the simplest terms, a map overlay procedure is used to create a new composite map.

Several of these activities are technical and are both prompting and responding to further technical advances. Not surprisingly, given the difficulty of understanding the internal workings of a GIS, the mapped output must always be interpreted and used with care.

There are some important differences between traditional maps and a GIS. In fact, some are so significant that they are changing the nature of geographic information and the role that this information plays in society. Whereas a traditional map is a single product, a GIS can be 'customized' by individual users, who may change data, update details, and superimpose multiple scales. Further, a GIS challenges one of the most basic of geographic conventions, namely, the idea that regions are separated by lines. Of course, we know that this is not the case, but the very act of drawing maps with neat boundaries effectively reifies both boundaries and the regions they circumscribe—it misleads us into thinking of them as fixed realities. A GIS is much more capable than a conventional map of depicting continuous spatial change, and thus requires users to think about geographic worlds without precise boundaries. This is an example of how a new technology can generate new ideas.

Interestingly, there is an ongoing debate concerning what some human geographers see as the philosophical limitations of GIS, which has close associations with positivism but no connection at all to any of the post-positivist philosophies such as humanism and Marxism. Today, as critical analyses of the social implications of GIS technology begin to appear, there are encouraging signs that GIS may yet play a bigger role in the broader practice of human geography.

When geographers collect data in the field, they typically need to specify the geographic locations at which they take measurements. These locations are then mapped to facilitate identification in a corresponding remote sensing image. Today, precise field locations can be established using a global positioning system (GPS). Originally developed by the US military, a GPS involves a group of satellites circling the earth in precisely known orbits and transmitting signals that are recorded by ground-based, usually hand-held, receivers like the one shown. The US GPS consists of 24 satellites. There is also a Russian GPS, and a European system is in development. (Courtesy of Thales Navigation).

What are the principal advantages of satellite remote sensing? First, repeated coverage of an area facilitates analysis of land-use change. Thus the most recent Landsat satellite covers most of the earth's surface every 16 days; these data are homogeneous and comprehensive. Second, because the data collected are in digital format, rapid data transmission and image manipulation are possible. Third, for many parts of the globe, these are the only useful data available. Finally, remote sensing allows the collection of entirely new sets of data; it was satellite data that first alerted us to the changing patterns of atmospheric ozone in high-latitude areas.

For the human geographer, remote sensing is especially valuable in aiding understanding of human use of the earth; much of the discussion in Chapter 4 relies on remotely sensed data. Remote sensing is less useful if we are concerned with underlying economic, cultural, or political processes.

Most satellites have been launched by either the US or the European Space Agency. Canada launched a satellite, Radarsat, in 1996. The Earth Observing System platforms are scheduled to be placed in orbit from 1997 to 2010. A recent substantial achievement—indeed a breakthrough in the science of remote sensing—was the remarkably detailed mapping of the earth's surface in 2000 by a manned NASA space shuttle. This mission involved a partnership among the military, intelligence-gathering, and environmental communities, and resulted in a topographic map of the earth's land mass between 60°N and 56°S that is about thirty times as precise as the best maps available before the mission.

There is another means, besides remote sensing, by which some data can be collected. In the early 1990s geographers began to make use of another new digital geographic technology, the global positioning system (GPS). A GPS is an instrument (either hand-held or installed in a personal computer) that uses signals emitted by satellites to calculate location and elevation. Along with remotely sensed data, data from a GPS can be integrated into a GIS.

Qualitative Methods

Human geographers collect and analyze data using a broad range of **qualitative methods** —a term that is widely used in other social sciences and refers to social research with a focus on attitudes and behaviour. Qualitative methods are a part of **ethnography**, a general approach that requires researcher involvement in the subject studied.

Much **fieldwork**—a traditional term for the methods that geographers use to obtain primary data (for example, observation and questionnaires)—is qualitative in character. Observation of landscapes was central to much early regional and cultural geography, but decreased in importance around 1960, as geographers began to focus more on secondary data. Recently, however, new types of fieldwork have appeared in response to humanistic concerns, and human geographers now use a range of qualitative methods for collecting and analyzing data (Box 2.10). Early fieldwork was not philosophically motivated, although it was implicitly empiricist because it assumed that reality was present in appearance. By contrast, contemporary qualitative fieldwork is by nature humanistic—a response to the humanistic requirement that human geography strive to understand the nature of the social world.

For the humanist, qualitative methods that involve a researcher's observation of and involvement in everyday life are central to understanding humans and human landscapes. Thus **participant observation**, a standard method in anthropology and sociology, is now a popular geographic approach. The principal advantage of this method is its explicit recognition that people and their lives do matter.

To conduct research using qualitative methods requires considerable skill. A subjective procedure such as participant observation does not provide any means for the researcher to objectively control the relationship between observer and observed. You will recall that this was one of the key issues in our discussion of the differences between humanism and positivism. It may be the case that the researcher, who is often of a higher social status, is ethnocentric (**ethnocentrism** is the presumption that one's own culture is normal and natural and that other cultures are inferior); or the researcher may identify with the group being observed and become their advocate. Other disadvantages of qualitative proce-

Aerial photograph of the Ottawa region (A31736-180 ©1996 Her Majesty the Queen in right of Canada, reproduced from the collection of the National Air Photo Library with permission of Natural Resources Canada).

SOME QUALITATIVE RESOURCES

The term 'qualitative resources' is used to refer to both methods and data sources. Holland et al. (1991) have identified nine of the many such resources available. Here is their list, accompanied by some comments that point to the breadth of humanistic research today:

1. *Talking and listening:* The typical quantitative interview uses a carefully structured questionnaire to ensure that multiple respondents all answer a specific set of questions. By contrast, a qualitative interview seeks to gain insight into the human worlds of individuals. It is particularly important to talk with and listen to respondents when the human world under investigation is different from the interviewer's own. In planning such research, you should always try to take into account the respondents' likely perspectives and be ready to change your questions, depending on what they say.

2. *A feminist perspective:* Feminist research focuses on the social significance of gender and especially the unequal distribution of power between women and men. A feminist perspective will influence the questions posed as well as the techniques used to carry out research.

3. *Literature:* Why do humans change landscapes as we do? Some geographers have turned to literature to answer this question. Creative writing often reflects cultural attitudes.

4. *Cartoons:* Although, like works of literature, cartoons are usually the work of an individual, they often reflect a broad social consensus on political, environmental, or economic issues.

5. *Photographs:* Historians frequently turn to photographs for information about the past. Photos can be invaluable not only as records of historical landscapes but as evidence of cultural attitudes. Choice of subject matter (including what the photographer omits or ignores) and manner of presentation can be extremely revealing.

6. *Art:* Many works of art have been interpreted as reflecting the distinctive character of a particular time and place.

7. *Popular music:* Music is often closely associated with place—just think of blues and the American south. Today much popular music is a direct reflection of inner-city urban experiences, highlighting social and environmental ills.

8. *Buildings:* Buildings are among the most important features of human landscapes, especially in urban areas. Different styles and functions reflect larger geographies, past and present.

9. *Monuments, memorials, and cemeteries:* Careful analysis of these landscape features can tell us a lot about the cultural identities of their creators.

These are only a few examples of the many qualitative possibilities open to the skilled researcher.

"Damn! Somebody's pinched our spot!"

Cartoon by Chic (Reproduced with permission of Punch Ltd).

dures include the risk that the researcher will begin with a biased or otherwise inappropriate idea about the data to be collected, or that the subjects of the study may not be sufficiently representative to provide an accurate picture. You will have an opportunity to assess the quality of some of this work, especially in Chapters 8 and 12.

Quantitative Methods

Some fieldwork is explicitly quantitative in character, notably the use of a **questionnaire** to survey people. Like the traditional qualitative fieldwork procedure of observation, a questionnaire is part of an empiricist research activity. Unlike qualitative fieldwork, however, a questionnaire asks all individuals the same questions in the same way. The value of questionnaire results depends on the response rate achieved and the way potential respondents are selected, namely the **sampling** method. Proper sampling methods, based on statistical sampling theory, allow the results of a sample to be treated as representative of the population, within certain error limits. The most common technique used for selecting respondents is random sampling.

During the 1960s, **quantitative methods** in general developed extensively in association with the spatial analysis school and the general acceptance of a positivist philosophy. The principal methods used were statistical, and the purposes were to describe data (by calculating a mean and a standard deviation, for example) and to test hypotheses generated by theory (using correlation tests, e.g., as in Box 2.2, point 4). We will encounter applications of these and other tests in Chapter 11.

The spatial analysis school recognized early on that models could play a much greater role in analyzing data. A **model** is an idealized, simplified representation of the real world that highlights its key properties and eliminates incidental information. Many of the earliest spatial models employed by human geographers were based on generalizations about the relationships between the distribution of geographic facts and distance. (Some of these classic spatial models will be discussed in Chapters 10, 11, 12, and 13.)

Geographers use quantitative techniques for a wide variety of purposes, but especially for analyzing relationships between spatial patterns and for classifying data. Describing relationships is fundamental in producing explanations and is usually broached by proposing a functional relationship such that one variable is dependent on one or more other variables. The relationship specified is, ideally, derived from appropriate theory in accordance with the scientific method outlined earlier. Classifying imposes order on data, and a number of techniques facilitate that activity.

A Concluding Comment

We conclude this chapter with a provocative, if somewhat crude, distinction. Positivism and the quantitative procedures associated with it have a tendency to exclude the individual human element from research, preferring to focus on aggregate data; critics consider such work to be dehumanized human geography. We might call it a geography *of* people. Humanism and the qualitative procedures associated with it, on the other hand, emphasize the integration of researcher and researched, thus generating a geography *for*, and possibly even *with*, people.

We do not, however, need to debate the relative merits of differing research strategies; we simply need to acknowledge that contemporary human geography is an exciting and diverse discipline incorporating a variety of useful philosophies, concepts, and techniques that complement each other and combine to offer an enviable range of procedures.

CHAPTER TWO SUMMARY

The importance of philosophy

The subject matter of human geography, the questions asked, and the aids used in obtaining answers are best understood by reference to various underlying philosophies. Most human geography is explicitly or implicitly guided by one particular philosophical viewpoint.

Philosophical options

Contemporary human geography draws on four principal philosophies. Empiricism gives priority to facts over generalizations; traditional regional geography, although typically presumed to be aphilosophical, was empiricist in character. Positivism argues for use of the scientific method, as in the physical sciences, and purports to be objective. The approach known as spatial analysis follows a positivistic logic. Humanism is a set of subjective philosophies, including pragmatism and phenomenology, with a focus on humans. Early cultural geography had some humanistic content, but humanist philosophies did not begin to play a major role in human geography until about 1970. Marxism is a loosely structured philosophy that aims to understand and change the human world; work inspired by Marxist thought tends to centre on social and environmental problems.

Human geographic concepts

In order to increase geographic knowledge, a great many concepts are used. We constantly refer to locations (where a geographic fact is present), place (the quality of a given location), region (a grouping of similar locations or places), and distance (the interval, however it is measured, between locations or places). Geographic research is conducted at specific spatial, temporal, and social scales. Explaining and understanding landscape change is often enhanced by considering both diffusion—the spread and growth of geographic facts—and perception—the way in which landscapes are viewed by individuals and groups. Analyses of landscapes frequently distinguish different degrees of development. The technical vocabulary of human geography—for example, conceptual terms such as 'place' and 'location'—is an example of discourse. Most notably, human geographers identify globalization as an overriding metaconcept, providing a body of ideas that may facilitate analysis of environmental, cultural, political, and economic topics.

Techniques of analysis

Geographic data are collected and handled by a variety of techniques. Maps are used to store, display, and analyze data, and can be produced manually or by computer. Geographic information systems are computer-based tools that analyze and map data, and have a wide range of applications. An important method of collecting data involves remote sensing techniques, such as aerial photography and satellite imagery. Fieldwork, both observation and interviewing, is an essentially qualitative procedure for the collection and analysis of data and is favoured by humanists. The use of models and quantification is associated with positivistic spatial analysis and involves abstraction and empirical testing of ideas.

LINKS TO OTHER CHAPTERS

- Philosophies, concepts, and techniques:
    Chapter 1 (history).

- Environmental determinism and possibilism:
    Chapter 4 (natural resources and
        environmental ethics)
    Chapter 7 (cultural landscapes)
    Chapter 11 (agricultural regions).

- Positivism:
    Chapters 11, 12, and 13 (location theories
        of von Thünen, Christaller, and Weber,
        respectively).

- Humanism:
    Chapters 7 and, especially, 8 (landscapes)
    Chapter 12 (contemporary urban experience).

- Marxism:
    Chapter 5 (Marxist theory of population
        growth)
    Chapter 6 (world systems approach)
    Chapter 8 (unequal landscapes)
    Chapter 12 (contemporary urban experience).

- Concepts of space, location, and distance:
    Chapters 10, 11, 12, and 13 (location
        theories).

- Concept of place:
    Chapters 7 and 8 (understandings of
        landscape).

- Concept of region:
    Chapter 3 (global environment)
    Chapters 5 and 6 (demographic maps)
    Chapter 7 (cultural regions)
    Chapter 11 (agricultural regions).

- Spatial scale:
    Tables and maps generally, especially in
        Chapters 5 and 6 (population).

- Concept of diffusion:
    Chapter 6 (migration)
    Chapter 10 (diffusion processes).

- Concept of perception:
    Chapter 6 (migration)
    Chapters 7 and 8 (group identities)
    Chapter 12 (urban areas).

- Concept of development:
    Chapter 6 (more and less developed worlds)
    Chapter 13 (uneven spatial development).

- Concept of discourse:
    Chapter 8 (feminism and postmodernism).

- Globalization:
    Chapter 4 (ecosystems and global impacts)
    Chapter 5 (population growth, fertility
        decline)
    Chapter 6 (refugees, disease, more and less
        developed worlds)
    Chapter 7 (cultural globalization)
    Chapter 8 (popular culture)
    Chapter 9 (political globalization)
    Chapter 10 (economic globalization)
    Chapter 11 (agriculture and the world
        economy)
    Chapter 12 (global cities)
    Chapter 13 (industrial restructuring)

- Map projections:
    World maps (all chapters).

- GIS and remote sensing:
    Data collection and analysis (all chapters).

- Qualitative methods:
    Chapters 8 and 12.

- Quantitative methods:
    Chapters 10, 11, 12, and 13 (spatial analysis).

FURTHER EXPLORATIONS

BIRD, J.H. 1989. *The Changing Worlds of Geography: A Guide to Concepts and Methods*. New York: Oxford University Press.

A comprehensive account of concepts and methods, many of which are discussed in great detail.

CLOKE, P., C. PHILO, and D. SADLER. 1991. *Approaching Human Geography: An Introduction to Contemporary Theoretical Debates*. New York: Guilford.

A useful text with good accounts of humanism and Marxism; also useful for the conceptual content of Chapter 8.

EYLES, J., ed. 1988. *Research in Human Geography: Introductions and Investigations*. Oxford: Blackwell.

An entertaining series of statements by leading geographers concerning research inspirations and practices; includes examples of humanistic and applied work.

GOODCHILD, M.F. 1988. 'Geographic Information Systems'. *Progress in Human Geography* 12:560–6.

A brief history and assessment of geographic information systems that helps to place this major growth area in an appropriate perspective; avoids technical terminology.

GOULD, P., and R. WHITE. 1986. *Mental Maps*, 2nd ed. Boston: Allen and Unwin.

A highly readable and well-researched discussion of the topic; the inspiration for this original research is one of the stories told in Eyles (1988).

HAGERSTRAND, T. 1967. *Innovation Diffusion as a Spatial Process*. Translated by A. Pred. Chicago: University of Chicago Press.

A highly influential analysis with an excellent review of diffusion studies by the translator; this is a classic cultural geographic problem approached for the first time in an essentially positivist manner.

HONDERICH, T. 1995. *The Oxford Companion to Philosophy*. Toronto: Oxford University Press.

An excellent guide to specific philosophical ideas and terms that will be useful for students wishing to appreciate the larger intellectual con-

text for some of the philosophies introduced in this chapter.

JOHNSTON, R.J. 1986. *Philosophy and Human Geography: An Introduction to Contemporary Approaches*, 2nd ed. London: Arnold.

A clear, readable statement of our various philosophies; employs a tripartite division into positivism, humanism, and Marxism.

———, D. GREGORY, and D.M. SMITH, eds. 2000. *The Dictionary of Human Geography*, 4th ed. Oxford: Blackwell.

A remarkably comprehensive, detailed, and clear dictionary with major entries often treated in an almost encyclopedic fashion; highly recommended.

LIMB, M., and C. DWYER. 2001. *Qualitative Methodologies for Geographers: Issues and Debates*. New York: Oxford University Press.

An innovative text that covers many methodologies as they are used in practical research contexts; covers research design, interviewing, group discussions, participant observation and ethnography, interpretative strategies, and writing.

NORTHEY, M., and KNIGHT, D.B. 2001. *Making Sense: A Student's Guide to Research and Writing*, 2nd ed. Don Mills: Oxford University Press.

A valuable guide for students that provides insights and guidelines concerning all aspects of the research and report-writing process from a student perspective.

PEET, R. 1998. *Modern Geographical Thought*. Oxford: Blackwell.

An overview of human geographic thought from about 1970 to the present; also useful for the conceptual content of Chapter 8.

PICKLES, J. 1985. *Phenomenology, Science and Geography*. New York: Cambridge University Press.

A detailed and thoughtful appraisal of phenomenology as applied in geographic research.

QUANI, M. 1982. *Geography and Marxism*. Oxford: Blackwell.

A useful introduction that focuses primarily on European rather than North American work; somewhat dated now, but still a valuable source of information and ideas.

RELPH, E.C. 1976. *Place and Placelessness*. London: Pion.

A good example of writing from a phenomenological perspective.

ROBINSON, A.H., and B.B. PETCHENIK. 1976. *The Nature of Maps: Essays towards Understanding Maps and Mapping*. Chicago: University of Chicago Press.

A highly readable book full of useful ideas. The first author devised the projection used for world maps in this text.

SCHMITT, R. 1987. *Introduction to Marx and Engels: A Critical Reconstruction*. Boulder: Westview Press.

Although not written especially for human geographers, this is an exceptionally clear and comprehensive introduction to a challenging school of thought.

STODDARD, R.H. 1982. *Field Techniques and Research Methods in Geography*. Dubuque: Kendall Hunt.

A comprehensive book covering such topics as data collection and analysis; provides clear guidelines concerning how to conduct positivist research.

WATSON, J.W. 1955. 'Geography: A Discipline in Distance'. *Scottish Geographical Magazine* 71:1–13.

A landmark statement anticipating the spatial analysis of the 1960s and highlighting the 'distance factor'.

WATSON, M.K. 1978. 'The Scale Problem in Human Geography'. *Geografiska Annaler* 6OB:36–47.

This is a valuable read for all geographers.

ON THE WEB

http://www.cig-acsg.ca/page.asp
Home page of the Canadian Institute of Geomatics.

http://www.ccrs.nrcan.gc.ca/
Canada Centre for Remote Sensing. A wide range of information on remote sensing technologies and applications; includes sample images.

http://www.esri.com
The home page of esri, the creator of two of the most widely used gis software packages, ArcView and ArcInfo; includes useful general information on gis technologies.

http://www.geoplace.com/
Highlights new and interesting information about GIS; good global coverage.

http://sciwebserver.science.mcmaster.ca/gislab/index.html
A McMaster University website focusing on GIS, with links to other GIS websites.

http://www.utm.edu/research/iep/
An Internet encyclopedia of philosophy that includes details on all of the philosophical terms introduced in this chapter.

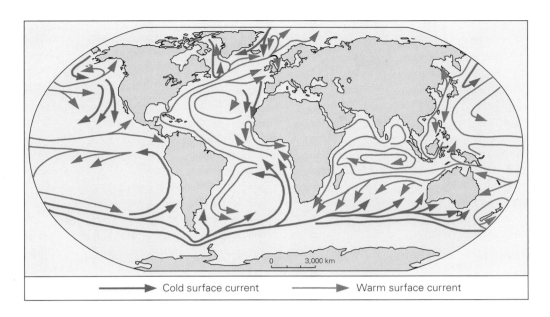

Figure 3.3 The major ocean currents: The prevailing winds and the rotation of the earth combine to cause gyres (large whirlpools). These gyres move warm water away from the equator and cold water towards it. There are five major gyres and thirty major currents.

→ Cold surface current   → Warm surface current

3.1

EARTHQUAKES AND VOLCANIC ERUPTIONS

If we were able to peel away the thin shell of air, water, and solid ground that is the outer layer of the earth, we would find that our planet is a furnace. Many of our major cities are a mere 35 km (22 miles) above this furnace, and in the deep ocean trenches, the solid crust above the furnace is as little as 5 km (3 miles) thick. Earthquakes and volcanic eruptions, not surprisingly, are regular occurrences. There are usually between 200 and 300 earthquakes each year in Canada alone, while several major volcanoes have recently made headlines (Mount Pinatubo in the Philippines and Mount Unzen in Japan, both in 1991). Perhaps the most dramatic such event in recent times was the 1963 birth of an island, Surtsey, near Iceland in the north Atlantic. Surtsey was the product of a massive eruption of molten rock through a split in the crust. Over the course of a few months, 300 million m³ (almost 400 million cubic yards) of lava, twice as much ash, and large amounts of carbon dioxide and steam literally poured out of the sea.

Needless to say, contemporary scientists are continually attempting to improve our ability to predict earthquakes and volcanic eruptions. We know that most earthquakes occur at the margins of tectonic plates, when the plates move against each other; we also know some of the signs that can indicate a quake is likely in a particular area. To date, however, more specific predictions—predictions that might save lives—remain elusive.

Predicting volcanic eruptions is equally difficult. Although we know where volcanoes are located, we do not know when an eruption will occur. One of the best-known eruptions was that of Vesuvius in 79 CE, when the town of Pompeii was covered by

3 m (10 ft) of hot cinders, pumice stone, and burning rock. Since then Vesuvius has erupted many times, most recently in 1906 and 1944. The 1980 eruption of Mount St Helens in Washington state was successfully predicted (Arkell 1991).

A resident of a poor neighbourhood in Mexico City is shown in April 2003 in her home—a room of a house that was damaged in the earthquake of 1985. Nearly twenty years after the disaster, poor people were still living in more than 400 such buildings. Attention was drawn to the problem in 2002, when a building that had been home to 60 people collapsed (AP photo/Jaime Puebla).

Making the Physical Landscape

The physical landscape in which human activities take place (and which is continually modified by those activities) is a product of a series of processes. Crustal movement and the effects already noted are followed by weathering and gradational processes that combine to produce specific *landforms*. Weathering is the reduction in size of surface rock; gradation is the movement of surface material under the influence of gravity and typically involves water, ice, or wind. In addition to creating landforms, these processes combine with the activities of living organisms to create soil. Plant and animal life in turn are closely related in a circular fashion to the landscapes created.

Weathering occurs when rock is mechanically broken and/or chemically altered. Temperature change is an important cause of mechanical weathering; water is the principal cause of chemical weathering. Once rocks are weakened in this way, they become especially susceptible to gradational processes. Water, ice, and wind can erode, move, and deposit material. Many of the earth's distinctive physical landscapes are clear evidence of these processes at work.

Water is probably the most important cause of gradation. A continuous transfer of water from sea to air to land to sea is taking place. River erosion creates V-shaped valleys and waterfalls, while river deposition creates flood plains and deltas. Sea water also affects land surfaces, and coastlines undergo constant change; cliffs are the result of erosion, beaches the result of deposition. Other parts of the earth's surface show the effects of ice and snow; during the most recent glacial period, ice covered an area approximately three times greater than that covered today, including

3.2

THE GEOMORPHOLOGY OF CANADA

Canada is the product of three primary geologic developments. First, the Canadian or Laurentian Shield was formed about 2,000 mya as a result of the collision and welding together of seven ancient microcontinents. Much of the shield consists of Archaean rocks 2,500 million years old (mostly granite or granite gneiss). Second, there are three regions where mountains developed around the shield margins as a result of sediment accumulation in long, narrow basins: the Cordilleran mountain system in western Canada, which has a long and complex history related to the presence of an active continental margin (including volcanic and earthquake activity); the Appalachian–Acadian system in eastern Canada, which is older than the Cordilleran and has been substantially lowered by erosion; and the Innuitian system, across the High Arctic from Alaska to northern Greenland. The third geologic development was the deposition of chemical and organic sediments in shallow seas (often called epeiric seas) in the intervening areas, about 500 mya.

This structural, or tectonic, history helps us to understand the various physiographic regions of Canada: the shield, the three mountain areas, and the several lowlands and plains that resulted from deposition in epeiric seas (such as the prairies and the St Lawrence lowland). But the landscapes of each of these regions also developed through time as weathering and gradational processes, closely related to climate,

affected the physical surface. The most important of these processes, for much of Canada, were the glacial periods; almost all of Canada was under ice at some time during the last 1.5 my (only 1 per cent is under ice today), and Canada was the site of the largest ice sheet in the Northern hemisphere during the most recent glacial period, which ended about 10,000 ya. As a result, many of Canada's land-forms are primarily glacial in origin (Trenhaile 1990).

Landscape of the Canadian Shield showing spruce/fir forests and small lakes (Philip Dearden).

northern Europe and Canada. Areas former-ly covered by ice show distinctive evidence both of erosion (e.g., U-shaped valleys) and of deposition (e.g., the low hills called drum-lins). Wind erosion results either from wind-blown materials colliding with stable materi-als (as in sandblasting) or, more often, the wind's removal of loose particles from a sur-face. Sand dunes are among the best-known products of wind deposition of material.

Each of these processes is affecting parts of the earth's surface at all times (Box 3.2). Physical landscapes, then, are subject to con-tinuous modification—almost fine-tuning. In principle, the eventual outcome, without crustal movement or climatic change, would be an almost featureless physical landscape.

Our physical environment is, of course, much more than the distribution of land-forms. Soils, vegetation, and climate are addi-tional components of a physical environ-ment. Figures 3.4, 3.5, 3.6, and 3.7 provide general outlines of global landforms, soils, vegetation, and climate respectively.

Figure 3.4 maps the major *cordilleran belts*—the backbones of the continents. The slope and ruggedness of local landforms in particular are major considerations in human decisions about the location of eco-nomic activities, such as agriculture, indus-try, and transport.

Soil types are distributed in a systematic fashion (Figure 3.5). There are close general links to both climate and natural vegetation; these are two of the most important determi-nants of soil type. Soils are important for both their agricultural potential and their capacity to support buildings. Much of the European migration to the New World was closely tied to the availability of soils suitable for com-mercial agriculture; in the late nineteenth century, many settlers found suitable grassland soils in Canada, Argentina, and Australia.

Figure 3.6 maps the global distribution of *natural vegetation*. The four principal types are grassland, forest, desert scrub, and tundra. Much of the history of agriculture is intimate-ly tied to their distribution. In circumstances of limited agricultural technology, for example, forested areas can be a significant barrier.

Climate is the characteristic weather of an area over an extended period of time and can be usefully summarized by reference to mean monthly and annual statistics of temperature and precipitation. These means reflect lati-tude. In Figure 3.7, it is clear that climatic distributions are generally linked to latitude. Three low-latitude, two subtropical, three

Figure 3.4 Global cordilleran belts. The backbones of the continents, these cordilleran belts also extend through ocean areas.
Source: Adapted from R.C. Scott, *Physical Geography* (St Paul: West, 1989):295. © 1989 West Publishing Co. Used by permission of Wadsworth Publishing Co.

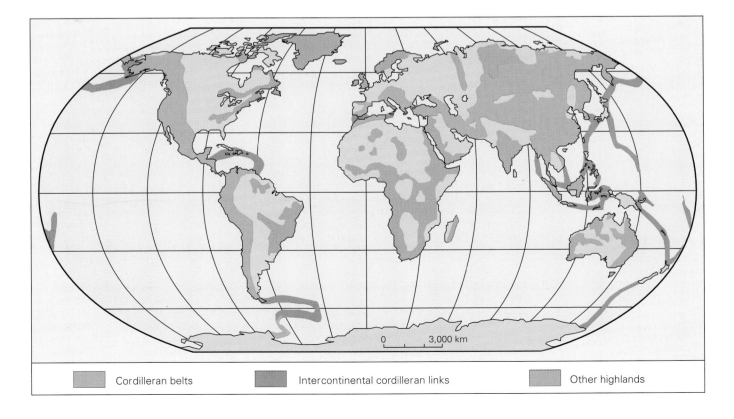

Cordilleran belts	Intercontinental cordilleran links	Other highlands

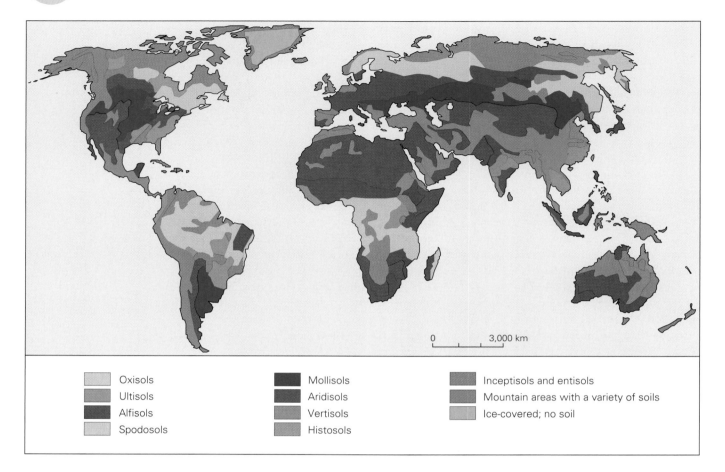

	Oxisols		Mollisols		Inceptisols and entisols
	Ultisols		Aridisols		Mountain areas with a variety of soils
	Alfisols		Vertisols		Ice-covered; no soil
	Spodosols		Histosols		

Figure 3.5 (above) Global distribution of soil types. Oxisols are typical of tropical rain forests, and are deep soils but not inherently fertile: if the natural vegetation is removed, the soil is soon impoverished. Ultisols, in the humid tropics, are low in fertility. Alfisols, dispersed throughout low- and middle-latitude forest/grassland transition areas, are of moderate to high fertility. Spodosols are associated with the northern forests of North America and Eurasia; low fertility is typical. Mollisols, associated with middle-latitude grasslands, are often very fertile. Aridisols, located in areas of low precipitation, lack water. Vertisols, limited to areas of alternate wet and dry seasons, may be very fertile. Histosols are most common in northern areas of North America and Eurasia; they are composed of partially decayed plant material, usually waterlogged, and typically fertile only for water-tolerant plants. Inceptisols and entisols are immature soils found in environments that are poorly suited to soil development.
Source: Adapted from R.C. Scott, *Physical Geography* (St Paul: West, 1989):276–7. © 1989 West Publishing Co. Used by permission of Wadsworth Publishing Co.

Figure 3.6 (upper right) Global distribution of natural vegetation. Differences in forest types are related to climate. Forests become less dense as they move from the equator, and have fewer species. Deciduous trees dominate the middle latitudes, evergreens the higher latitudes. Savanna, steppe, and prairie are composed chiefly of grasses. Desert vegetation is closely linked with an arid climate. Tundra vegetation is the most cold-resistant.
Source: Adapted from R.C. Scott, *Physical Geography* (St Paul: West, 1989):232–3. © 1989 West Publishing Co. Used by permission of Wadsworth Publishing Co.

Figure 3.7 (lower right) Global distribution of climate. These types of climate, distinguished on the basis of moisture and temperature, are derived from the Köppen classification. Tropical wet, tropical wet and dry, and low-latitude dry climates may be grouped as low-latitude climates; dry summer subtropic, humid subtropical, mid-latitude dry, marine, humid, and continental climates as middle-latitude climates; and subarctic, tundra, and polar climates as high-latitude climates.
Source: Adapted from R.C. Scott, *Physical Geography* (St Paul: West, 1989):158–9. ©1989 West Publishing Co. Used by permission of Wadsworth Publishing Co.

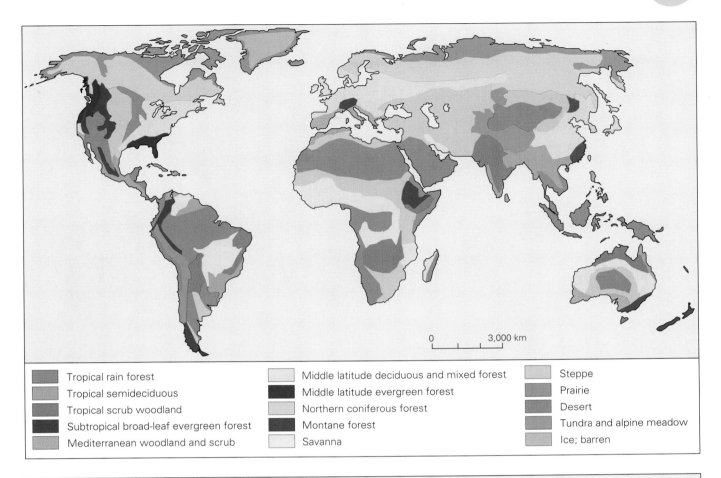

Tropical rain forest
Tropical semideciduous
Tropical scrub woodland
Subtropical broad-leaf evergreen forest
Mediterranean woodland and scrub

Middle latitude deciduous and mixed forest
Middle latitude evergreen forest
Northern coniferous forest
Montane forest
Savanna

Steppe
Prairie
Desert
Tundra and alpine meadow
Ice; barren

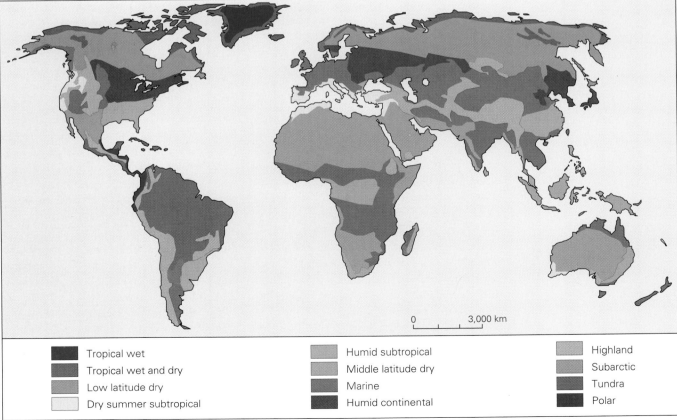

Tropical wet
Tropical wet and dry
Low latitude dry
Dry summer subtropical

Humid subtropical
Middle latitude dry
Marine
Humid continental

Highland
Subarctic
Tundra
Polar

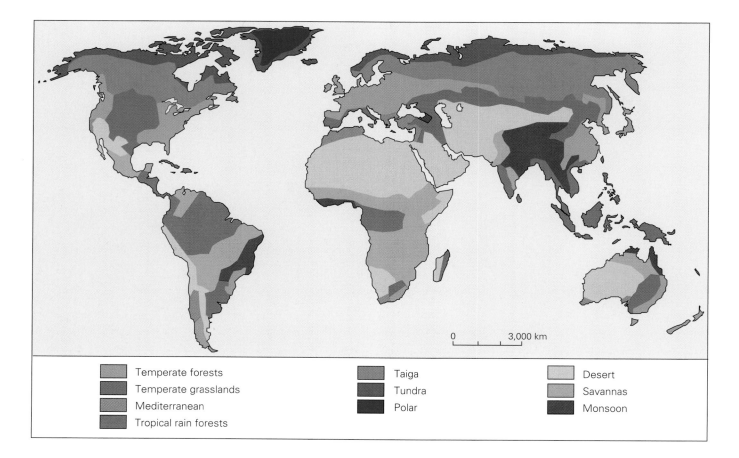

Temperate forests		Taiga		Desert
Temperate grasslands		Tundra		Savannas
Mediterranean		Polar		Monsoon
Tropical rain forests				

Figure 3.8 Generalized global environments.

mid-latitude, and three high-latitude climates are mapped, along with one highland climate. Low-latitude climates are characteristically hot; the distinguishing feature of each of the three is precipitation. The two subtropical climates are less hot, and again the distinguishing feature is precipitation. The three low-latitude climates lack any significant seasonality. The subtropical and mid-latitude climates vary according to the degree of continentality (distance from ocean) and, of course, latitude. All five of these climates have seasonal variations. Proximity to oceans tends to increase precipitation and reduce temperature variations. The three high-latitude climates are cold most of the year and have low precipitation. Finally, highland climates display many local variations, since temperature and precipitation are affected by altitude in addition to latitude.

Climate is clearly a key consideration in many human decisions. But *weather* is also important. Weather can have a major impact on agricultural yields. Atypical or violent weather can cause floods or droughts, and hurricanes—vast tropical storms—can cause serious damage to populated land areas.

Global Environments

The preceding section has highlighted a wide variety of physical processes and environments. Their relationship to human activities is often intimate, but these physical factors do not in any sense determine our human activities. Although we need to be aware of physical factors at all levels, we must not assume any causal relationships. As the remainder of this book will demonstrate, we humans make the decisions—albeit with many physical factors in mind! This section presents a brief summary of the major global environments (Figure 3.8), integrating much of the material above in a more applied human context.

The notion of global environmental regions is useful here. Indeed, those regions are closely correlated with climatic regions in particular. Humans have favoured certain environments and have made changes to all areas they have settled in.

Tropical Rain Forests

These areas are located close to the equator and experience high rainfall—typically exceeding 2,000 mm (78 in.) per annum—and high temperatures—24°C to 30°C (75°F to 86°F). They lack seasonal variation. Broadleaf trees dominate, and the vegetation is dense and highly varied. Nevertheless, once the vegetation is removed, the soil is characteristically shallow and easily eroded. Tropical rain forests have not become major agricultural regions. Among the economically valuable products of rain forests are mahogany and rubber, as well as various starchy food plants, such as manioc, yams, and bananas.

Monsoon Areas

Monsoon areas are characterized by marked seasonality, especially an extended dry season and heavy rainfall. Rainfall may be as high as 2,000 cm (780 in.) in some areas. The typical natural vegetation is deciduous forest, but all monsoon areas have been changed dramatically as a result of human activity. Agriculture has been practised for a long time, with rice as the basic crop in many such areas. Population densities are usually high.

(left) Tropical rain forest, Costa Rica (© Mark Hobson).

(below) Monsoon area, Bali (United Nations photo 153860/P.S. Sudhakaran).

(above) Savanna with baobab, **Malawi** (Victor Last, Geographical Visual Aids).

(right) Desert, Death Valley, **California** (Victor Last, Geographical Visual Aids).

Savannas

Savannas are located to the north and south of the tropical rain forests. They have a distinctive climate: a very wet season and a very dry season. The natural vegetation is a combination of trees and grassland. Africa has the most extensive savanna areas, which are rich in animal species. To date, most savannas are relatively undeveloped economically, but they offer great promise for increases in animal and crop yields. Important grain crops include sorghum in Africa and wheat in Asia.

Deserts

A desert area is one that has low and infrequent rainfall, perhaps as low as 100 mm (4 in.) per annum and perhaps no rain for ten years. Areas are typically defined as arid if they receive less than 255 mm (10 in.) of rain per annum and as semiarid if they receive less than 380 mm (15 in.). There are hot, cold,

sandy, and rocky deserts. In all deserts, water is a crucial consideration in human activities. Irrigation and nomadic pastoralism are two human responses to desert environments.

Mediterranean Areas

There are only a few Mediterranean environments. They are characterized by long, hot, dry summers and warm, moist winters. The typical location is on the western side of a continent in a temperate latitude. Evergreen forests are the usual natural vegetation, although human activity means that scrub land prevails today. Important crops, especially in the Mediterranean proper, are wheat, olives, and grapes.

Temperate Forests

Much of Europe and eastern North America has a temperate climate with a distinct seasonal cycle, fertile soils, and a natural vegetation composed primarily of mixed deciduous trees. The seasonal cycle gives this area its distinctive character. The many animals and plants have adapted to the variations in heat, moisture, food, and light. Annual loss and growth of leaves is an especially striking adaptation that affects all other forest life-forms. Particularly since the beginnings of agriculture, these areas have been attractive to and significantly altered by humans, and they are now dominated by agriculture, industry, and urbanization. These changes began in Europe about 7,000 ya, in Asia about 6,000 ya, and in North America about 1,000 ya. Indeed, the removal of forest cover has been so dramatic that today there is a general recognition of the need to preserve and even expand the remaining forests.

Mediterranean area, near St-Tropez, France (Victor Last, Geographical Visual Aids).

Temperate Grasslands

North American prairies, Russian steppes, South African veld, and Argentine pampas are four of the large temperate grasslands. Located in continental interiors, they have insufficient rainfall for forest cover and tend to show great annual temperature variations. Such grasslands were once occupied by particular large migrating herbivores such as the bison (North America), antelopes, horses, and asses (Eurasia), and pampas deer (South America). After remaining almost devoid of humans until the late seventeenth century, these grasslands became the granaries of the world in the twentieth century. Their soils are typically suited to grains; wheat became dominant in most areas.

Boreal Forests

Boreal forests, often called taiga, are located only in the Northern hemisphere—in North America and Eurasia. Their winters are long and cold, their summers brief and warm, and precipitation is limited. The natural vegetation is evergreen forest. Agriculture is very limited, and the major economic importance of taiga consists in forest products and minerals.

Tundra

Located north of the boreal forest, this area experiences long, cold winters; temperatures average –5° C (23° F) and there are few frost-free days. Perennially frozen ground—permafrost—is characteristic and presents a severe challenge to human activities. Only a

Temperate forest, Cathedral Grove, Vancouver Island
(Philip Dearden).

(above left) Temperate grassland, Head Smashed In Buffalo Jump, Alberta
(Victor Last, Geographical Visual Aids).

(below left) Boreal forest, Terra Nova National Park, Newfoundland
(Victor Last, Geographical Visual Aids).

(above) Tundra, Herschel Island, Yukon (Philip Dearden).

(above right) Polar region, Ellesmere Island (Philip Dearden).

few plants and animals have adapted to this environment because the land has been exposed for only about 8,000 years, since the last retreat of ice-caps. There has been little time for even soil to form.

Polar Areas

Polar conditions prevail in both hemispheres and are best represented by the Greenland and Antarctic ice sheets. Antarctica is a continental land mass under an ice sheet that reaches thicknesses of 4,000 m (more than 13,000 ft). These areas lack vegetation and soil and have proven hostile to human settlement and activity. Indigenous inhabitants of these areas lived in close relationship to the physical environment, subsisting on hunting and fishing products. In recent years, non-indigenous inhabitants have greatly increased in number, particularly because of the discovery of oil.

These brief descriptions of ten global environments make no attempt to describe all areas. Nor do all areas within a particular environment have uniform characteristics. Regions merge into one another; the tropical rain forest merges into savanna and then into desert, for example. Rarely are there clear, sharp divisions. As a result, many areas are difficult to classify. What this outline has emphasized is that the earth comprises many environments, with which humans have to cope. Some have been greatly affected by human activities; others, hardly at all. All environments need to be considered in any attempt to comprehend our human use of the earth.

Life on Earth

Air-breathing life cannot exist without oxygen, which was not a part of the original atmosphere. However, life in the form of primeval bacteria and algae evolved without oxygen; these life-forms consumed carbon dioxide and nitrogen, which were in the original atmosphere, and emitted oxygen as a waste. In addition to adding oxygen to the atmosphere, this process also formed the ozone layer, which filters out harmful ultraviolet radiation from the sun. The first life-forms evolved in the seas. As indicated in Table 3.1, there is evidence of life-forms as early as 3,500 mya, of an ozone layer 2,500 mya, and of a breathable oxygenated atmosphere 1,700 mya. Such early life-forms were not affected by the absence of an ozone layer because they lived below the surface of the water. Oxygen-breathing life, initially single- and later multicelled, appeared following the creation of a suitable atmosphere. Soft-bodied animals, comparable to jellyfish, evolved 650 mya, and shelled animals about 70 million years later.

About 400 mya, fishes diversified considerably and gradually some moved into areas of shallow water that was subject to drying up. Amphibious **species** appeared shortly thereafter, moving onto land already occupied by plants and insects, especially close to water. Much of the earth at this time comprised swamp environments, which were well suited to plant growth and provided food for amphibians. One group of amphibians evolved into reptiles: animals with waterproof skins capable of remaining out of the water indefinitely. These reptiles were able to move away from water and soon colonized large areas, dominating much of the earth for about 135 million years. During this period mammals and birds appeared; and when the reptiles vanished, 65 mya, the mammals and birds remained. The cause of the extinction of reptiles remains uncertain (Lewin 1982).

With the absence of reptilian competition, mammals and birds flourished, expanding and diversifying, filling the ecological gaps left by the dinosaurs. Among the new evolutionary lines that began at this time were the primates; today, humans are one of about 200 species of living primates. Apes, a new group of primates, evolved about 25 mya. Some 10 mya, mammals reached their greatest diversity. Most available evidence suggests that the human evolutionary line separated from that of the apes about 6 mya (Feder and Park 1997:120).

Three factors need to be highlighted in this account of the emergence and spread of life on earth. First, the earth was a changing environment in this period. Climatic change, especially a series of glacial periods, was ongoing. Second, the land masses were moving. The most recent movements began with the initial breakup of Pangaea (see Figure 3.2) some 225 mya. The spread of life has been greatly affected by both of these factors, as land routes around the world have changed. Third, the relationship between the physical earth and the emergence and evolution of life is not fully understood (Box 3.3).

Human Origins

Although our knowledge continues to increase, there is still much that we do not know about human origins and early history (Table 3.2). Current evidence suggests that the large apes evolved as one primate line in Africa about 25 mya and then split into several relatively distinct evolutionary lines. One of these in turn split into two further lines, chimpanzees and humans. We can suggest with reasonable confidence, using a combination of fossil, geological, and genetic evidence, that this split occurred—that is, human ancestors diverged from the ape line—at least 6 mya, and the 2001 discovery of a skull in Chad, known as *Sahelanthropus*,

raises the possibility that pre-human ancestors may date as far back as 7 mya. The most compelling evidence of a common origin is that humans and chimpanzees differ in only about 1 per cent of their genes; this means that our two species could not have been evolving separately for much more than about 6 million years (Feder and Park 1997:122).

Our Earliest Ancestors: *Australopithecus*

What was it that distinguished the earliest human ancestors from other primates? Unlike all earlier primates, they were bipedal—they walked upright. This was a

| Table 3.1 | **BASIC CHRONOLOGY OF LIFE ON EARTH** |

Time	Event
Physical Origins	
10–20 billion years ago?	Creation of universe
5,000 mya	Solar system forms
4,500 mya	Earth forms
Origins of Life	
3,500 mya	Oldest life-forms
3,200 mya	Oxygen-creating bacteria
2,500 mya	Ozone shield forms
2,200 mya	Oxygen in atmosphere
1,700 mya	Atmosphere breathable
Origins and Evolution of Life-forms	
1,400 mya	Oxygen-using animals
800 mya	Multicellular life
650 mya	Soft-bodied animals
580 mya	Shelled animals
500 mya	First fishes
430 mya	First land plants
360 mya	First insects
350 mya	First amphibians
280 mya	First reptiles
225 mya	Breakup of Pangaea
220 mya	First mammals
200 mya	Reptiles dominant for next 135 my
190 mya	First birds
65 mya	Dinosaurs extinct; birds and mammals diversify; first primates
25 mya	Grasslands become widespread; first apes
10 mya	Maximum diversity of mammals
6 mya	Emergence of earliest human ancestors

Note:
mya = million years ago

Olduvai Gorge, Tanzania. Located in the Ngorongoro crater, in the rift valley of East Africa, Olduvai has been the site of so many important fossil and tool discoveries that it has become known as the 'cradle of mankind'. The first fossilized remains were found here in 1911 by a German scientist searching for butterflies, but the gorge became famous through its association with the paleontologists Mary and Louis Leakey. The earliest remains at the Olduvai site have been carbon-dated to approximately 1.75 million years. Among the hominid species found there are *Australopithecus boisei* (originally named *Zinjanthropus*; the first major discovery, made by Mary Leakey in 1959), *Homo habilis*, and *Homo erectus* (Copyright 2001 by Philip Martin. All rights reserved).

3.3

GAIA

'Gaia' is the Greek name for the goddess of the earth. Today the term is used to denote a self-regulating system, with all components of the system, or ecosphere, in a stable balance. This remarkable concept was first introduced in a book by James Lovelock (1979). In brief, Gaia is seen as a self-regulating entity that keeps the environment relatively constant and comfortable for life; earth and all life on it have evolved together as one. Specifically, the Gaia hypothesis supposes 'that the atmosphere, the ocean, the climate, and the crust of the Earth are regulated at a state comfortable for life because of the behaviour of living organisms' (Lovelock 1988:19). According to Gaia, the earth itself is 'alive'. This revolutionary idea contradicts the standard view in which life and the earth evolved separately and life had to adapt to conditions on earth.

The Gaia concept obliges us to assume a global perspective. It is the planet as a whole that matters, not any particular species—including humans. If we humans continue to change the earth, then we may precipitate our replacement. If, on the other hand, we act as part of a living organism, then perhaps we will remain on earth for a long time.

The Gaia concept has helped to explain some crucial scientific issues. For example, the surface temperature of the earth has remained relatively constant since life first appeared, even though the heat from the sun has increased by about 25 per cent. How has this occurred? The level of CO_2 in the atmosphere has decreased over this period, reducing the natural greenhouse effect and allowing surface temperatures to remain relatively constant. Life, specifically photosynthesis, is the cause of the long-term CO_2 decline. Without life, CO_2 would probably increase once again. (These comments should help to place our present greenhouse crisis, discussed in the next chapter, in perspective: by adding CO_2 and other greenhouse gases to the atmosphere, we are countering a 3.5-billion-year trend towards CO_2 reduction.) A second example that supports the Gaia concept concerns the salinity of sea water. The fact this has barely changed over the long term is puzzling, given that salt is continually being added to the sea via rivers. How is salt being removed from the seas? The Gaia response is that various life-forms in the seas promote the segregation of sea water in lagoons, allowing the sun to evaporate the water and remove the salt.

Both of these examples illustrate the meaning of Gaia as a self-regulating entity, keeping the environment fit for life. The existence of Gaia is not a proven fact; indeed, many scholars consider the hypothesis too general to be tested (Kirchner 1989). Nevertheless, it is both scientifically stimulating (in 20 years, Gaia has gained considerable scientific legitimacy) and valuable for the global perspective it represents.

critical adaptation that permitted early humans to move over great distances while carrying objects at the same time; fossil evidence has been found throughout eastern and southern Africa. The first known **hominid** group is called *Australopithecus*. Most authorities currently agree that the common ancestor of all hominids was *Australopithecus afarensis*; fossil evidence of this species, discovered in east Africa, has been dated to more than 3.5 mya. *A. afarensis* had a small brain of about 440 ml (modern human brains average 1,450 ml).

Towards Modern Humans: *Homo habilis*

The next critical evolutionary event in the human lineage occurred about 3 mya when the hominid line split into two types, one of which became extinct and one of which evolved into modern humans. The first representative of the line leading to modern humans appears to be the species *Homo habilis*. Both *Australopithecus* and *Homo habilis* were probably—though not certainly—restricted to Africa, notably the east and south African savanna areas, where they lived by hunting small animals and gathering plants. But *H. habilis* had a larger brain (about 680 ml), which made possible the development of an important new cultural behaviour: the fabrication of stone tools. Tool-making was a major technological advance. Brain size distinguished *Homo habilis* from other previous hominids and from other hominid lines living at the same time; some evidence even suggests that *habilis* possessed the capacity for speech.

Why did *Australopithecus* evolve, perhaps as long ago as 6 million years? Why did *Australopithecus afarensis*, our common ancestor, split into two types about 3 mya? Why did one of these become extinct about 1 mya? Although we are uncertain of the answers, there is evidence to suggest that global climatic change was a causal factor. Declines in global temperatures correspond with each of these evolutionary events. Such declines prompted drying trends in Africa: drying reduced the extent of tropical forest, increasing that of open woodland and dry savanna. It is in this way that evolutionary changes can be seen as adaptations to changing physical environments.

Out of Africa: *Homo erectus*

Beginning about 1.8 mya, *Homo erectus* appeared. This new species was distinguished from *Homo habilis* primarily by a brain size of about 1,000 ml—roughly the lower limit of modern human brain size. *Homo erectus* first appeared in Africa and subsequently spread over much of Africa and into the warm temperate areas of Europe and Asia. Currently this hominid species is thought to have been the first to move out of Africa. Some recent evidence suggests that *Homo erectus* may have reached Asia not long after emerging in Africa. The reasons for this movement are not known, but they probably had to do with climatic change, population increases, and the associated search for food. Again, the basis of subsistence was hunting and gathering.

Table 3.2   BASIC CHRONOLOGY OF HUMAN EVOLUTION

6	mya	Probable earliest date for evolution of first hominids, the bipedal primate *Australopithecus*, in Africa
3.5	mya	First fossil evidence of *Australopithecus afarensis* in east Africa
3	mya	Probable appearance of *Homo habilis* in east Africa; evidence of first tools
1.8	mya	First appearance of *Homo erectus* in east Africa
1.5	mya	Evidence of *Homo erectus* in Europe; evidence of chipped stone instruments in Africa; beginning of Pleistocene
700,000	ya	First evidence of chipped stone instruments in Europe
400,000	ya	*Homo erectus* uses fire; develops strategies for hunting large game; first appearance of archaic *Homo sapiens*
100,000	ya	Appearance of *Homo sapiens sapiens* (in Africa only, according to favoured replacement hypothesis)
80,000	ya	Beginnings of most recent glacial period
40,000	ya	Arrival of humans in Australia
25,000–15,000	ya	Arrival of humans in America
12,000	ya	First permanent settlements
10,000	ya	End of most recent glacial period; beginning of Holocene period

Note:
mya = million years ago

After the appearance of *Homo erectus*, human biological evolution stabilized for some time. But cultural developments, including the introduction of clothing and shelters, was considerable. Different cultural adaptations were devised for different environments. By about 400,000 ya, hunting strategies had progressed to include the use of fire and possibly the planned hunting of larger game. An innovation at this time was the introduction of relatively permanent pair bonding—monogamy. Human language use also increased substantially, encouraging the gradual development of human culture. All of these adaptations took place against a background of changing physical environments, as the geological period beginning about 1.5 mya, the **Pleistocene**, was one of considerable fluctuation in global temperatures. Several extremely cold periods, when large areas of the earth were covered in ice, were punctuated by periods with average temperatures comparable to those of today—an alternation of glacial and interglacial phases. These fluctuations continued until as recently as 10,000 ya and may indeed be continuing; it is possible that we are in an interglacial period at present.

Modern Humans:
Homo sapiens sapiens

Some 400,000 ya, a new hominid evolved from *Homo erectus*. This species was similar to modern humans, but because there are physical differences between them and modern humans, they are often called *archaic Homo sapiens*. Their mean brain size of 1,220 ml was about 85 per cent that of modern humans. The best-known subset of these archaic humans were the Neanderthals, who first appeared about 130,000 ya and lasted until about 30,000 ya. In the early twentieth century, following the first fossil discoveries, Neanderthals were regarded as subhuman in intellect, but by the 1950s it was not uncommon for them to be regarded as virtually modern in appearance and behaviour. The current interpretation is a compromise—neither brutish apes nor modern humans.

We have now reached a critical point: the emergence of anatomically modern humans, officially called *Homo sapiens sapiens*. Where and how did this come about? There are two rather different explanations. The *multiregional* hypothesis proposes that modern humans evolved independently from archaics, at more or less the same time in a number of different geographic areas.

However, all the currently available palaeontological, genetic, and archaeological evidence favours a second hypothesis. This *replacement* hypothesis proposes a single evolution in a limited geographic area and a subsequent spread to replace archaics elsewhere. Specifically, modern humans, *Homo sapiens sapiens*, evolved in a single place, in Africa, from a group of archaics before 100,000 ya. This hypothesis received additional confirmation with the 2003 discovery in Ethiopia of anatomically modern human skulls dated 160,000 ya. These humans spread throughout the world, replacing existing archaic groups because of some adaptive advantage. From that time onwards, human evolution has been cultural, not biological. Thus acceptance of the favoured replacement hypothesis, centred on a single exodus from Africa, has enormous implications for our understanding of ourselves. Simply put, we are all related. Modern humans settled much of Africa, Asia, and Europe, and reached Australia 40,000 ya and America perhaps 25,000 ya, but certainly more than 15,000 ya. These movements of humans were, of course, related to the continuing climatic changes associated with the glacial and interglacial periods. Modern humans began to evolve and spread across the earth during a period of harsh climate. Evidence suggests that favoured locations were at the boundaries of geographic regions, where the first permanent settlements appeared about 12,000 ya. The most recent glacial period ended 10,000 ya, marking the end of the Pleistocene and the beginning of the **Holocene** period.

The fact that early humans effectively settled most areas of the earth is particularly significant. It suggests that humans had some adaptive advantage over other primates, something that allowed them to adjust and survive in harsh new environments where other species could not; as we will see in later chapters, this advantage was probably *cultural*. Migration to previously unsettled areas enlarged the overall resource base and allowed slow but steady increases in human populations. We can only estimate the number of people in these early stages. *Australopithecus* was restricted to Africa, and population estimates range from 70,000 to 1 million. *Homo erectus* covered much more of the globe and possibly attained a population of 1.7 million. A figure of 4 million seems reasonable for the human population about 12,000 ya.

Understanding Human Evolution: Three Myths

Today, scientific research is finally dispelling three long-standing myths about human evolution. The *first myth* concerns the idea of human evolution, or evolution generally, as some sort of ladder of progress. As the above account suggests, evolution is not a history of steady progress to a finished product; it is not correctly seen as some sort of ladder. Rather, it is helpful to borrow a metaphor popularized by the distinguished evolutionary biologist Steven Jay Gould. For Gould, evolution is best seen as a bush with endlessly branching twigs, each twig representing a different species. In the terms of this metaphor, the human species is 'a fragile little twig of recent origin' (Gould 1987:19).

A *second myth* concerns the supposed existence of several different human **races** (Gould 1985; Montague 1964). As modern humans, *Homo sapiens sapiens*, moved from Africa to different physical environments and splintered into distinct spatial groupings, adaptations developed not only in culture but in certain body features or **phenotypes**. Thus relatively distinct phenotypes emerged: different skin colours, head shapes, blood groups, and so forth. Certain characteristics became dominant in certain areas, either because they were environmentally appropriate or as a result of chance genetic developments. The resulting variations in human physical characteristics are quite minor and do not in any way represent different types of humans (Box 3.4).

3.4

SPECIES AND RACES

There is only one species of humans in the world, namely *Homo sapiens sapiens*. That all humans are members of the same species is biologically confirmed by the fact that any human male is able to breed with any human female. All species, including humans, display variation among their members. Indeed, many plant and animal species exhibit sufficient variation that they can be conveniently divided into *subspecies* or races: geographically defined aggregates of local population. Different races emerge within a species if groups are isolated from one another, in different environments, for the necessary length of time. As a consequence of selective breeding (the result of isolation), together with various adaptations to their respective environments, each group will eventually exhibit some genetic differences from every other group.

To what extent is it possible to delimit distinct races—geographically defined aggregates of local populations—in the human species? To answer this question, we need to identify three stages:

1. We know that *Homo sapiens sapiens* evolved at least 100,000 years ago, probably only in Africa.
2. Members of this species then moved across the surface of the earth, replacing archaic *Homo sapiens* and gradually separating into various groups; as a consequence of these movements and groupings, some selective breeding and adaptation to different environments occurred, giving rise to some genetic differences.

3. As a result of population increases and further migrations, groups that had been temporarily isolated began to mix with other groups.

It was during the second stage that the groups of humans often known as Negroid, Mongoloid, Caucasoid, and Australoid appeared. But are these groups races? *The answer is no, because none of these human groups was ever isolated long enough or completely enough to allow separate independent genetic changes to occur.* Use of the word 'race' is therefore mistaken, because genetically speaking, humans cannot be divided into clear-cut stable subspecies.

Because race is a biological concept referring to a genetically distinct subgroup of a species, there are today no such things as races within our human species. There are genetic differences between members of the human species as a consequence of the earlier period of isolation of groups in different environments. Biologically, however, these differences are not sufficient to justify the identification of separate races.

We are a biologically variable species, and our variability results from the same processes that produce variability in all living things: namely, natural selection in different physical environments. These biological facts are of basic importance. They immediately inform us, for example, that it is incorrect to equate a cultural group with a racial group. Thus there is a Jewish religion, but not a Jewish race. Similarly, there are Aryan languages, but not an Aryan race. There are no subspecies—races—within our human species.

A *third myth* is the idea that races not only exist but can be classified according to their level of intelligence. We are fortunate today in that we have clear scientific evidence of the unity of the human race. Such knowledge was not available until relatively recently. The characteristic response to physical variations in humans, particularly different skin colours, has been to regard these as evidence of different types of humans with different levels of intelligence. Box 3.5 summarizes the history of the erroneous idea that the human population can be divided into distinct and unequal groups, usually called races. The reality is that all humans are members of the same species; the only differences within the species are differences of secondary characteristics, such as skin colour (Kennedy 1976).

Conclusion

The physical environments that are now occupied by humans are diverse; they are also changing (change tends to be localized in the short term, global in the long term). This chapter has identified some major global characteristics of the earth, some key physical processes that affect physical landscapes, and some characteristic landscapes. It has also commented on the types of life-forms that appeared, and outlined the current understanding of human evolution.

The physical geographic content in this chapter is much more than mere background or stage setting for the material to follow. To say that human activities are closely related to physical environments is not to validate environmental determinism. Rather, it is to acknowledge a central and pervasive relationship. Humans make decisions, as individuals and as groups, but they make them with a multitude of physical factors in mind. Success in adaptation—that is, coping—means adapting to both the physical environment and relevant human variables.

In a book dealing with geography and social theory, the following statement appears:

> The journey along Mulholland Drive, atop the Hollywood Hills, provides one

3.5

A HISTORY OF RACISM

Box 3.4 explained the unity of the human species: there are no distinct subspecies, races, within that species. Notwithstanding these biological facts, the concept of distinct racial groupings has long been popular in both lay and scientific circles. Ideas about the existence of races and the relative abilities of the supposed races have been central to many cultures.

Racism is the belief that human progress is inevitably linked to the existence of distinct races. Notions of racial variations in ability appear to have been common in most cultures. They flowered especially in Europe beginning about 1450 with the overseas movement of Europeans, which brought them into contact with different human groups, and later new theories were developed according to which the different groups observed came from different origins. This notion of multiple origins is incorrect, as we saw in Box 3.4.

By the eighteenth century, racist explanations for the increasingly evident variety of global cultures were standard. Philosophers such as Voltaire and David Hume perceived clear differences in ability between what were commonly seen as racial groups. Undoubtedly one of the key racist thinkers was Joseph Arthur de Gobineau, whose *Essay on the Inequality of Human Races* was published in 1853–5. Gobineau ranked races as follows: Whites, Asians, Negroes. Within the Whites, the Germanic peoples were seen as the most able. A second influential racist was Houston Stewart Chamberlain, but racist logic was also apparent among many of the greatest scientists. Darwin wrote of a future when the gap between human and ape would increase because such intermediaries as the chimpanzee and Hottentot would be exterminated (see Gould 1981:36). As we have seen in Box 3.4, the racial categories employed in all such discussions have no biological meaning.

One explanation for the popularity of racist thought is the fact that most cultures classify other cultures relative to themselves. The consequences of racist thinking are varied and considerable. Belief in the inferiority of specific groups has led to mass exterminations, slavery, restrictive immigration policies, and, most generally, unjust treatment.

of the world's great urban vistas. To the south lies the Los Angeles basin, a glittering carpet. To the north, the San Fernando Valley (still part of the City of Los Angeles) unfolds in an equivalent mass of freeways, office towers, and residential subdivisions. There is probably no other place in North America where such an overpowering expression of the human impact on landscape can be witnessed. And yet, the physical landscape cannot be denied. Even in this region of almost 12 million people, the landscape still contains and molds the city (Dear and Wolch 1989:3).

The 12 million people in the Los Angeles region today represent about three times the number on the entire earth 12,000 ya. The story of human biological and early cultural evolution before this date is a complex and sometimes uncertain one. The account provided in this chapter emphasizes the known facts of the story; where the facts are uncertain, this is acknowledged, and current consensus thought is reflected.

Chapter Three Summary

The earth

A habitable environment, the earth is a sphere rotating on its axis; one rotation defines a day. The earth orbits the sun and one revolution is a year. Revolution, combined with an axial tilt, produces seasons and variations in length of daylight hours. Diurnal, seasonal, and annual cycles are basic to much human activity. Locations on the earth can be identified by reference to an imposed geographic grid: the lines of latitude and longitude.

The earth's crust

The present distribution of land and water is the consequence of a long and ongoing process of the lithosphere's movement known as continental drift. These movements are also responsible for major mountain chains, such as the Himalayas and Appalachians, and for earthquakes.

Physical landscape makers

The details of a physical landscape result from ongoing weathering and gradational processes. Once rock is weathered, either mechanically broken or chemically altered, water, ice, or wind can erode, move, and deposit the weathered material. Water, ice, and wind each produce distinctive landforms. In addition, our physical environments are characterized by particular soils, vegetation, and climates. Each of these three is distributed in a relatively systematic fashion, related to latitude and to each other.

Global environments

The general correspondence between soils, vegetation, and climate permits classification of the earth's surface into typical environments. Ten such environments are noted and the typical physical content identified. Human activities are often closely related to these physical environments, but they are not caused by them. Although global environments can be distinguished, there are rarely clear divisions between different environments; any one environmental type can include a wide range of physical conditions.

Life on earth

Life on earth evolved 3,500 mya and has undergone a series of additions, deletions, and gradual changes since that time. The first life-forms emerged under water, but their presence led to the development of an oxygenated atmosphere that could support air-breathing life and of an ozone layer that acts as a protective shield from the sun's ultraviolet rays. During the long history of life on earth, numerous physical changes have taken place, including the movement of continents and climatic change.

Human origins

The earliest humans, the first bipedal primates, were *Australopithecus* (6 mya), who emerged in east Africa. Subsequent species had increasingly larger brains. *Homo habilis* (3 mya) also evolved in east Africa; these early humans used tools. *Homo erectus* (beginning

1.8 mya) moved out of Africa to Europe and Asia. Archaic *Homo sapiens* emerged 400,000 ya. Modern humans, *Homo sapiens sapiens*, appeared in Africa no later than 100,000 ya. These modern humans reached Australia 40,000 ya and North America possibly 25,000 ya; they settled in environmentally suitable locations.

The unity of the human race

Humans are members of one species—*Homo sapiens sapiens*. Early spatial separations of groups of humans facilitated the development of physical variations. These are of minimal relevance to our understanding of either people or places. The concept of race among humans has no basis in fact.

LINKS TO OTHER CHAPTERS

- Physical processes, physical landscapes, global environments:
    Chapter 4 (human impacts).

- Physical regions:
    Chapter 2 (concept of regions).

- Global environments:
    Chapter 6 (population distribution and density)
    Chapter 9 (reasons for European overseas expansion)
    Chapter 11 (distribution of agricultural regions).

- Human origins:
    Chapter 5 (population geography)
    Chapter 7 (evolution of culture).

- The idea of race, racism:
    Chapter 6 (Box 6.6, on restrictive immigration)
    Chapter 8 (landscapes and power relations; ethnicity).

FURTHER EXPLORATIONS

FEDER, K.L., and M.A. PARK. 1997. *Human Antiquity: An Introduction to Physical Anthropology and Archaeology*, 3rd ed. Toronto: Mayfield.
    An introductory physical anthropology textbook that contains a detailed discussion of the origin of humans; a balanced presentation.

GORE, R. 1997. 'The First Steps'. *National Geographic* 191, no. 2:72–99.
    Articles on human evolution are published regularly in this popular and accessible magazine; always well illustrated and readable; a good source of information.

MACDONALD, G. 2003. *Biogeography: Space, Time, and Life*. New York: Wiley.
    One chapter of this book offers an excellent overview of the relationships between biogeography and human evolution, including an account of the geographic expansion of modern humans.

MCKNIGHT, T.L., AND D. HESS. 2000. *Physical Geography: A Landscape Appreciation*, 6th ed. Toronto: Prentice Hall.
    A detailed and well-written textbook that is especially strong on questions of human impacts.

PASSINGHAM, R.E. 1982. *The Human Primate*. Oxford: Freeman.

> An excellent book that discusses humans as an animal species.

ROBERTS, N., ed. 1993. *The Changing Global Environment*. Oxford: Blackwell.

> Includes details of global environmental change, with emphasis on climate, oceans, and water; also very useful for Chapter 4.

———. 1997. *The Holocene,* 2nd ed. Oxford: Blackwell.

> A clear account of the past 12,000 years in the history of the earth, including details of climate change; also very useful for Chapter 4.

SCOTT, R.C. 1989. *Physical Geography*. New York: West.

> A physical geography textbook with an especially clear account of our planetary setting.

TRENHAILE, A.S. 1998. *Geomorphology: A Canadian Perspective*. Toronto: Oxford University Press.

> A textbook presenting a systematic explanation of the landforms of Canada.

ON THE WEB

http://www.nrcan.gc.ca/gsc

The home page of the Geological Survey of Canada, offering a wide variety of information on physical geography and links to relevant provincial and territorial websites.

http://www.usgs.gov

This site is published by the United States Geological Survey and includes clear overviews of physical geographic landscapes; offers a good account of the application of GIS technologies.

The Earth: A Fragile Home

Most human impacts on the earth are the result of efforts to improve our well-being. Yet, ironically, sometimes these efforts have quite the opposite effect. To understand their impacts, we need to keep in mind five facts:

1. Everything in nature is related; hence to change one thing is to change many things.
2. In addition to basic physiological needs, humans have many culturally based wants.
3. Humans' ability to affect the earth increases with increasing levels of technology, increasing use of various sources of energy, and increasing numbers of people.
4. Different cultures have different attitudes towards the environment; also, attitudes within a single culture change over time.
5. There is evidence that a shift is under way in contemporary western culture, from an anthropocentric (human-centred) world view to an ecocentric (environment-centred) one.

Following a general discussion of these five points, specific human impacts—on vegetation, animals, land and soil, water, and climate—are outlined both globally and with reference to regional and local examples. Finally, a discussion of the current state of the earth raises a provocative question that will resurface in various forms in later chapters: Is a global disaster looming, or will human ingenuity and technology be able to solve all our problems? We conclude with a look at the progress currently being made towards a sustainable world.

Water pollution caused by the forest industry, Shawinigan, Quebec (Victor Last, Geographical Visual Aids).

For most of the time that humans and their ancestors have been using the earth, their impacts on the planet have been slight, for two reasons: technologies were limited and populations were small. The overview of human evolution in Chapter 3 noted a few technological advances, such as the use of stone tools by *Homo habilis* some time after 3 million years ago and the use of fire as a hunting strategy by *Homo erectus* before 400,000 years ago. But these technologies were used by a limited number of people (the world's total human population was probably as small as 4 million as recently as 12,000 years ago), and the few environmental changes they brought about were temporary and restricted to a local scale. Today, with a substantial array of agricultural and industrial technologies and a population of 6.2 billion (2002), the situation is vastly different. Our interactions with and impacts on the environment are now numerous, relatively permanent, and often on a global scale.

A Global Perspective

Our topic is the impact of human groups upon land. What is the value of a global perspective on this question? The short answer is that *everything is related to everything else; one cannot change one aspect of nature without directly or indirectly affecting other aspects*. In principle, then, human activity in any one area has the potential to affect all other areas; and in practice the evidence that it does so is now overwhelming. One way to approach these issues is to consider the question of how a civilization survives. Smil (1987:1) answered, 'It survives by harnessing enough energy and providing enough food without imperilling the provision of irreplaceable environmental services. Everything else is secondary.'

In exploring how harnessing energy and producing food have changed our environment, and whether the changed environment threatens human survival, many human geographers find that the most valuable concepts are those of systems and ecology.

Systems, Ecology, and Ecosystems

In principle, the concept of the **system** is simple and widely applicable. It is surprising, therefore, that it has not been more widely used in studies of the relationship between humans and land. Systems can be usefully defined as sets of interrelated parts. The attraction of the systems concept is its ability to describe a wide range of phenomena and offer a simplified description of what is usually a complex reality. Descriptions of systems typically focus on distinctions between the parts of the system and the relationships between the parts. Open systems interact with elements outside the system, necessitating the study of inputs and outputs of energy and matter. Closed systems—lacking inputs and outputs of matter—are less common. Finally, relationships between the parts of a system, or between a system and some external elements, are often described as feedbacks. Positive feedback reinforces some change that is occurring; negative feedback counters some change.

The roots of the term '**ecology**' are Greek: 'eco' comes from *oikos*, 'house' or 'place to live', and the common suffix '-logy', meaning 'study of', comes from *logos*, 'word' or 'reason'. Thus ecology is the study of organisms in their homes. The concept of an ecosystem, integrating systems and ecology, was formally developed by an English botanist, A.G. Tansley, in 1935.

Ecosystems can be identified on a wide range of scales. The global ecosystem (often called the ecosphere or biosphere) is the home of all life on earth. It is a thin (about 14-km/8.5-mile) shell of air, water, and soil. Ecosystems can also be identified within the larger ecosphere: any self-sustaining collection of living organisms and their environment is an ecosystem. Thus the ecosystem concept refers to distinct groupings of things and the relationships between these things.

On the global scale, the three basic parts of the ecosystem are the atmosphere, the hydrosphere, and the lithosphere: air, water, and land. This global ecosystem has one key input: the sun. It is the sun that is the source of the energy that warms the earth, providing energy for photosynthesis and powering the water cycle to provide fresh water. In order to be self-sustaining, our global ecosystem depends on the cycling of matter and the flow of energy. Figure 4.1 outlines the cycling of critical chemicals and the one-way flow of energy. Matter must be cycled; the law of conservation of matter states that matter can be neither created nor destroyed, only changed in form. Energy flows through the

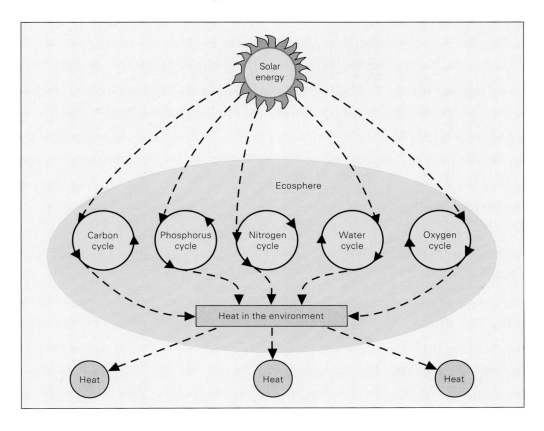

Figure 4.1 Chemical cycling and energy flows.

system because of the second law of thermodynamics, that energy quality cannot be recycled. We call the circular pathways of chemicals biogeochemical cycles; 'bio' refers to life, 'geo' to earth. Our global ecosystem, indeed every ecosystem, is dynamic. As we are about to find, one of the most important—and least understood—causes of ecosystem change is the human population.

Humans as Simplifiers of Ecosystems

Any human change to an ecosystem is typically a simplification, and a simplified ecosystem is usually vulnerable (Box 4.1). Ecologists agree that, if the global ecosystem is to survive, systems modified by humans must be balanced by systems that have not been modified by humans. For example, imagine a cultivated lowland ecosystem. Such a system depends on an upland forest system nearby that releases water and minerals to it. If humans remove the forest system, the cultivated area will suffer. This small example illustrates the interrelatedness of local ecosystems and underlines the point that reality is complex—that in fact (as the basic principle of ecology states) all things are related. Today,

one of the most urgent issues facing us is the need to change the way we live inside our ecosystems—to change our longstanding habits of domination for new ones of cooperation. Nothing illustrates this domination better than our use of energy and our high valuation of technology.

Energy and Technology

Humans have basic physiological needs for food and drink; we also have a host of culturally based wants that have no upper limit. Needs and wants are satisfied largely as a result of humans' using their energy to harness other forms of energy. **Energy** is the capacity to do work. The more successfully we can utilize other energy, the more easily we can fulfil our needs and wants and the more profoundly we can affect the environment. We become aware of other energy sources and acquire the ability to use those sources through the development of new technology.

All forms of **technology** represent ways of converting energy into useful forms. An important early technological advance was the human use of fire to convert inedible plants into an edible form for human use. In domesticating plants, humans gained control over a natural energy converter: the plants that convert solar energy into organic material via photosynthesis. Similarly, in domesticating animals, humans took control over another natural converter: the animals that change one form of chemical energy (inedible plants, usually) into another form usable by humans (such as animal protein). In this sense, the domestication of plants and animals, sometimes known as the first **agricultural revolution**, was an early example of how humans have used technology in order to tap new energy sources.

Between the agricultural revolution (approximately 10,000 years ago) and the eighteenth century, many other technological changes permitted increasing human control of energy sources: new plants and animals were domesticated and new tools and techniques were invented. In addition, three new energy converters were developed to utilize the energy in water and wind: the water mill, the windmill, and the sailing craft. But it was only with the **industrial revolution** of the eighteenth century, most particularly the steam engine, that we began using inanimate converters on a large scale to tap into new energy sources: coal in the second half of the eighteenth century, oil and electricity in the second half of the nineteenth century, and nuclear power in the middle of the twentieth century.

This brief account raises two key points. First, energy sources such as coal and oil are not literally 'new': they are 'fossil fuels', formed from ancient organic matter, and they are not renewable. Second, we are using these sources in ways that are often harmful to our ecosystem. Environmental degradation resulting from **pollution** is now com-

4.1

LESSONS FROM EASTER ISLAND

One of the most remote inhabited places on earth, Easter Island, in the Pacific Ocean, is 3,747 km (2,328 miles) off the coast of South America and 2,250 km (1,400 miles) southeast of Pitcairn Island, the nearest inhabitable land. When Europeans first visited Easter Island on Easter Sunday, 1722, they encountered a population of about 3,000 living in a state of warfare, with minimal food available from limited resources and a treeless landscape (Ponting 1991:1–7). They also encountered clear evidence of a once-flourishing earlier culture: between 800 and 1,000 huge stone statues, or *moai*. Because visiting Europeans believed that these statues could not have been carved, transported, and erected by the impoverished local population, it was long assumed that Easter Island must have been visited by a more culturally and technologically advanced group.

There is a simpler but more disturbing answer to this 'mystery' of Easter Island. Radiocarbon dating of evidence suggests that Easter Island was settled by the seventh century, and possibly as early as the fifth century; the first settlers were likely Polynesians, not South Americans. They arrived at an island with few species of plants and animals and limited opportunities for fishing, but with considerable areas of woodland. Their diet consisted of sweet potatoes (an easy crop to cultivate) and chicken. The considerable amount of free time available allowed the Easter Islanders to engage in elaborate rituals and construct the *moai*. Agricultural activities

required the removal of some trees, but most of the deforestation was carried out for the purpose of moving statues. It is likely that statue construction involved competition between different groups. By about 1550, the population peaked at roughly 7,000. Deforestation, which would, over time, have led to soil erosion, reduced crop yields, and a shortage of building materials for both homes and boats, was probably complete by 1600. It was the Easter Islanders impoverished by the consequences of deforestation that the Europeans encountered.

It appears that what began the collapse of the Easter Island culture and economy was total deforestation. What is the significance of this lesson? Bahn and Flenley offer this suggestion:

> We consider that Easter Island was a microcosm which provides a model for the whole planet. Like the Earth, Easter Island was an isolated system. The people there believed that they were the only survivors on Earth, all other land having sunk beneath the sea. They carried out for us the experiment of permitting unrestricted population growth, profligate use of resources, destruction of the environment and boundless confidence in their religion to take care of the future. The result was an ecological disaster leading to a population crash (Bahn and Flenley 1992:212–13).

monplace in industrial societies. According to Smil (1989:10), 'Environmental pollution, previously a matter of regional impact, started to affect more extensive areas around major cities and conurbations and downwind from concentrations of power plants as well as the waters of large lakes, long stretches of streams and coastlines, and many estuaries and bays.' Much pollution can be attributed to twentieth-century developments such as the thermal generation of electricity and the mass use of automobiles, plastics, nitrogenous fertilizers, and pesticides.

Energy use is spatially variable. Canadians lead the world in the per capita use of energy, followed by Americans. Although Europeans and Japanese use less energy than we do in North America, countries in the **more developed world** generally use more energy per capita than do countries in the **less developed world**. Energy sources are also spatially variable. Globally, oil accounts for 40 per cent of energy used, coal for 23 per cent, natural gas for 22 per cent, electricity for 5 per cent, nuclear power for 4 per cent, and other sources for 6 per cent. In many parts of the less developed world, the primary energy sources are not oil, coal, and gas but the traditional **biomass** sources of wood, crop waste, and animal dung. And energy use is increasing. It has been estimated that in one year humans use an amount of fossil fuel that took one million years to produce. Clearly we cannot continue indefinitely in this fashion, because we are draining the supply of non-renewable energy sources. We must realize that human survival depends on our retaining access to appropriate energy sources—and we must act on that realization.

Natural Resources and Human Values

Humans are continually evaluating physical environments. As human culture (especially technology) changes, so do those evaluations. Thus something becomes a 'resource' only if humans perceive it as useful in some way: technological, political, economic, or social. Different groups do not necessarily agree on what is and is not a resource. Some people may see an area of wetland in prairie Canada as a valuable scientific or recreational landscape; others will see it as potential farm or building land. Different interest groups use different evaluation criteria, and hence may hold radically different views.

Traditionally, geographers have divided resources into two types. **Stock resources** include all minerals and land and are essentially fixed, as they take a very long time, by human standards, to form. **Renewable resources** are those that are continually forming, such as air and water. This simple distinction is generally useful, but it does blur some key issues. There are, of course, a whole series of resources that lie somewhere between the two extremes of stock and renewable and whose continuing availability depends on how we manage those resources. Obvious examples include game populations for hunting economies and fish populations for much of the contemporary world. In both cases, there may be a perceived need for conservation, but also some compelling reasons to continue depleting the resources. In Canada, the east coast fishery has been especially susceptible to controversies over such issues.

Renewable Energy Sources

Perhaps the biggest technological challenge of the early twenty-first century is to decrease our reliance on non-renewable fossil fuels (coal, oil, natural gas) as sources of energy and find ways of generating more energy from renewable sources (Chapter 13 includes an overview of current global production and consumption). There are many reasons to favour increasing use of renewable energy. Environmental reasons include the impact on the atmosphere of burning fossil fuels and the global warming that is believed to be a consequence (discussed later in this chapter). Political reasons include the pressure exerted on governments by voters who are increasingly aware of environmental issues; the need for national economies to be self-sufficient in energy; and the need to diversify in order to reduce dependence on particular suppliers of energy sources. Economic reasons centre on the fact that coal, oil, and natural gas will become more expensive as supplies become more scarce. Many of these arguments first came to the fore as a result of the energy crisis of the early 1970s, although interest waned when energy costs fell again in the 1980s.

Ships emerge from the five-stage lock at the Three Gorges dam on the Yangtze River in June 2003. More than two kilometres wide and 185 metres high, on completion (scheduled for 2009) the dam will create a reservoir extending nearly 650 km into the country's interior. Ocean-going vessels will have access to agricultural and manufactured products from a vast region, while the dam's hydropower turbines will generate as much electricity as 18 nuclear power plants. The dam is the largest construction project in China since the Great Wall (AP photo/Greg Baker).

Today, renewable energy sources are tapped in only a few areas, on a relatively local scale. The main potential sources are the sun, the wind, biomass (e.g., wood), geothermal energy (the natural heat found inside the earth), and water (rivers, waves, and tides). Nuclear power is also considered renewable, since uranium reserves are expected to last about 1,000 years if an efficient process is used.

To date, the principal technology for generating renewable energy has involved the damming of rivers. Many large dams have been constructed in parts of the less developed world, often with the financial support of the World Bank. Unfortunately, although such projects do make effective use of a renewable energy source, they can cause serious damage to the environment. The massive Three Gorges project in China was refused World Bank funding, partly because of the anticipated human and environmental impacts. Nevertheless, in June 2003 the raising of the world's third largest river, the Yangtze, began. By the time the project is completed, in 2009, some 1.8 million people will have been displaced and many ancient fortresses, temples, and tombs submerged under water; it is also feared that the eventual reservoir may become a giant cesspool filled with sediment washed down from the deforested mountains that surround it. Another Chinese megaproject, the principal purpose of which is to divert water from the south to the north, is similarly subject to criticism.

Nuclear power does not pollute, but it has not yet established itself as a source of safe and inexpensive energy. Several serious nuclear accidents have occurred, the most devastating being the 1986 Chernobyl explosion in northern Ukraine (then part of the USSR), as a result of which as many as 7,000 people died and up to 3.5 million have suffered from diseases related to the release of radioactive material. There are more than 400 nuclear reactors in the world, most of which use the basic nuclear fission process, although future reactors are likely to use the more efficient 'fast breed' process.

Solar power has been widely discussed, and, although it remains very expensive, there is great potential for use of solar panels in private homes as well as businesses. Several countries in the more developed world are now making effective use of wind power; Denmark is a leader in wind farm technology, and the numbers of such farms in Canada and the United States are increasing. Several countries have experimented with wave and tidal power, but these have not established themselves as viable sources. Biomass sources, such as wood, remain important in the less developed world. Geothermal energy is a minor contributor at the present time.

How the energy picture will change in the twenty-first century is difficult to predict, at least partly because the amounts of energy available from renewable sources can be affected by weather conditions. In the United States, for example, consumption of hydroelectric power from renewable sources fell in 2001 because of widespread drought conditions.

Environmental Ethics

Western Environmental Concern before 1900

Use or abuse? It is not always easy to distinguish the two; they are, after all, relative terms. Yet there is considerable evidence to suggest that we are currently causing damage—perhaps irreparable damage—to our environment. Concerns about the consequences of human activities were raised by the ancient Greeks; Plato himself noted the detrimental effects of agricultural activities on soil. Despite such early observations, this general question received relatively little attention in the western world until the eighteenth century. Before that time, geographers were most interested in the earth as a home for humans made by God (teleology) and the land as a cause of human activity (environmental determinism). Significantly, the Europeans' general failure to appreciate the potential dangers of certain human activities appears to have been unique: many other cultures have recognized the necessity of protecting natural 'resources'.

The eighteenth-century origins of western environmental concern reflected the overseas movement of Europeans, particularly their colonization of tropical areas that, in Europe, had long been regarded as pristine utopias. It soon became obvious that European activity in those areas was environmentally destructive. Scientists, typically employed by commercial groups such as the East India Company, responded by cataloguing the newly recognized flora and fauna. In Mauritius the French introduced a number of conservationist measures, particularly concerning the location and amount of forest that could be removed, and these measures were imitated by the British in several Caribbean islands.

The general question of human impact on the land first received scholarly attention from Buffon (1707–88) in discussions of the contrasts between settled and unsettled areas and of the human domestication of plants and animals. Buffon believed that humans inhabit the earth in order to transform it. Malthus (1760–1834) established the terms of the current debate by focusing attention on the relationship between available resources and numbers of people. On a more specific level, Humboldt, during his travels in South America, explicitly identified lowered water levels in lakes as human impacts, and explained that they were caused by deforestation for the purpose of agricultural activities.

Probably the earliest systematic work on human impacts was done by G.P. Marsh (1801–82), an American geographer and congressman. His book *Man and Nature, or Physical Geography as Modified by Human Action* (1864, with revised editions in 1874 and 1885) was intended

to indicate the character and, approximately, the extent of the changes pro-

duced by human action in the physical conditions of the globe we inhabit; to point out the dangers of imprudence and the necessity of caution in all operations which, on a large scale, interfere with the spontaneous arrangements of the organic and of the inorganic worlds; to suggest the possibility and the importance of the restoration of disturbed harmonies and the material improvement of wasted and exhausted regions; and, incidentally, to illustrate the doctrine that man is, in both kind and degree, a power of a higher order than any of the other forms of animated life, which, like him, are nourished at the table of bounteous nature (Marsh [1864] 1965:3).

This powerful statement seems more reminiscent of the 1960s than the 1860s. The late nineteenth-century western world, heavily involved in colonial expansion, became concerned about environmental change only when that change had negative impacts on human economic interests. 'If a single lesson

In September 2003 ENMAX Corporation and Vision Quest Windelectric announced the official opening of the $100-million McBride Lake Wind Farm in southern Alberta. With 114 turbines, this facility will generate approximately 235,000 megawatts of electricity a year—enough energy to power more than 32,500 homes. Similar projects are underway in many parts of the world, but local responses are not always favourable. In the Lake District of northwest England, for example, opponents object that turbines are an eyesore in a landscape widely considered to be idyllic (Vision Quest Windelectric Inc.).

can be drawn from the early history of conservation, it is that states will act to prevent environmental degradation only when their economic interests are shown to be directly threatened. Philosophical ideas, science, indigenous knowledge and people and species are, unfortunately, not enough to precipitate such decisions' (Grove 1992:47).

The Current Debate: Origins

A concern with the environment is one of the great developments of the second half of the twentieth century, and many students know a good deal about it, from various sources. Environmental concerns are for the first time at the forefront of public consciousness. In 1989 *Time* magazine declared the earth to be the 'Planet of the Year'; many throughout the world celebrate Earth Day each 22 April, and many people make conscious efforts to conserve and recycle.

The first signs of this real shift in our appreciation of human impacts on ecosystems came in the 1960s with the publication of Rachel Carson's *Silent Spring* (1962), which revealed the dangers associated with indiscriminate use of pesticides. A few years earlier, in 1956, a seminal academic work, totalling 1,194 pages and entitled *Man's Role in Changing the Face of the Earth* by Thomas et al., had come to many of the same conclusions. The 1960s also saw increasing pressure from advocates of wilderness preservation, and new scientific evidence about worsening air pollution. Two popular explanations for the environmental 'crisis' pointed to, first, the Judaeo-Christian belief that humans had been placed on earth to subjugate nature and, second, the failings of capitalism. The first of these explanations is too simplistic, and ignores the complexity of Christian attitudes. The second explanation is one aspect of an ideological approach that stresses the links between different parts of the world: 'Clearly there are problems, many of them—and all of them intertwined in the operations of a capitalist world economy, which is hellbent on annihilating space and place. Those problems are severe now at a global scale, and life-destroying in some places' (Johnston and Taylor 1986:9). Today, our awareness of environmental impacts outside the capitalist world economy, in eastern Europe and the former Soviet Union prior to the major political changes that began in 1989, should lead us to question such assertions, although the basic idea of interrelatedness is sound.

The Current Debate: Political Overtones

Environmental issues entered the political arena in the early 1970s with the creation in the United States of the Environmental Protection Agency, while the first major international meeting, the United Nations Conference on Human Environment, was held in Stockholm in 1972. By 1980 the western world was becoming increasingly aware of environmental and related food supply problems in the less developed world, with food shortages in India and droughts in the Sahel region of Africa. A series of disasters and discoveries during the 1980s ensured that the environment was always in the news. These included the 1984 leak of methyl isocyanate from a pesticide plant in Bhopal, India, that killed perhaps as many as 10,000 and disabled up to 20,000; the 1986 nuclear disaster in Chernobyl, Ukraine, that killed up to 7,000 and will cause many more to die of radiation poisoning or related cancer; the 1985 discovery of a seasonal ozone hole over Antarctica; recognition of both the rapidity and the consequences of tropical rain forest removal, especially in Brazil; and, more generally, an increasing concern for numerous local environmental problems.

By the late 1980s the environment was on the national agenda of many countries, as well as the international political agenda. At the *national* level, green political parties first appeared in West Germany in 1979 and were present in most countries in the more developed world by 1990. Further, many countries have some form of green plan; an encyclopedic survey of the Canadian environment is available as a part of Canada's Green Plan (Supply and Services Canada 1991).

International agreement is the best way to solve those environmental problems that transcend national boundaries, some of which may have literally global impacts. There have been many calls for the creation of international institutions and policies, most notably by the Brundtland Commission (World Commission on Environment and Development 1987). Not surprisingly, although many governments agree on the need for international policies, most are unwilling to sacrifice their sovereignty; before they can reach agreements, countries need to work together to resolve their conflicting goals and priorities. Major international developments include the 1987 Montreal protocol aimed at the reduction and eventual elimination of chloro-fluorocarbons (CFCs), which are one cause of global warming; the 1992 United Nations Conference on Environment and Development (the Rio Earth Summit); the 1997 Kyoto protocol designed to reduce emissions of greenhouse gases; and the 2002 United Nations World Sustainable Development Summit held in Johannesburg. Both of the UN-sponsored meetings were attended by thousands of delegates in addition to various world leaders. As of 2003, the Kyoto protocol has been only partially implemented, and the United States still has not agreed to it.

The Current Debate: Three Contentious Issues

Before we address the impacts that humans are currently having on the environment, three issues should be identified. The first concerns relationships between the environment and the economy. Market forces are unlikely to solve environmental problems; market-based decisions are rarely in the environment's interest, even at the national level. A detailed analysis of economic decision-makers in Canada showed that only 6 per cent gave significant consideration to the environment. Such findings suggest that the integration of economic and environmental concerns will be a major challenge in the future (Gale 1992).

Second, environmental problems are increasingly affecting relationships between countries—not only because of the international implications of many human impacts, but also because environmentalists in, for example, North America are increasingly anxious to impose their standards on other countries. Payment is one solution: the 1987 Montreal protocol included a fund to assist those countries most likely to suffer economically as a result of the protocol. International disapproval is another possible solution: Britain eventually agreed to cease dumping sewage sludge in the North Sea because of the political costs of the dumping. Finally, trade policies are among the few weapons available to one national government to per-

suade another national government to amend its environmental behaviour.

The third issue, closely related to the second, concerns the behaviour of individuals as group members (Box 4.2). The ecophilosopher Arne Naess argues that humans need to develop a new world view recognizing how we are all connected; that we need to work with and not against nature; and that a central goal of human activity is the preservation of ecosystems. 'Deep ecology' is a term sometimes used to describe this viewpoint (see Katz et al. 2000). Table 4.1 provides a comparison of deep and shallow views of ecology. Both represent improvements over some traditional attitudes towards land—for instance, the notions that humans are the source of all value, that land exists for human use, and that energy and other resources are unlimited.

To the extent that deep ecology calls for significant changes to contemporary lifestyles, it finds parallels in the concepts of sustainability and sustainable development. Before we examine these important concepts in detail, however, we need to take a closer look at various specific human impacts on the earth, so that we can understand why our current social and economic systems must change if the environment is to be safeguarded for future generations.

Human Impacts

Human history is a history of impacts on land; we must have an impact on land to survive. Only recently have we realized that some of our impacts actually threaten our continued survival:

1. Small, often insignificant, changes to the environment can have major impacts if they are repeated often enough. Arable and pastoral activities can lead, over time, to major environmental problems.
2. Technological changes related to demands for energy continually change the environment.
3. The lifestyles promoted by technological changes also work to change the environment.
4. Increasing human populations are a threat to the environment.
5. Increasing connections between different regions of the globe mean that human activities that used to have merely local or

regional consequences are now more and more likely to be global in their impact. (The most obvious example is global warming; see p. 126 below.)

Our ability to collect and analyze information about current human impacts is greatly enhanced by technologies such as remote sensing and GIS (introduced in Chapter 2). It is possible to use different wavebands of the electromagnetic spectrum to gather information on particular land covers, such as areas under crop, pasture land, forest, and wetlands. It is also possible to measure such properties of ecosystems as soil surface moisture and temperature, to map soils, and to 'sense' (as in remote sensing) bio-geochemical cycles. Much of the detailed information in this discussion of human impacts is derived from remote sensing by satellite and analyzed in a GIS.

Impacts on Ecosystems in General

Hunter gatherers affect ecosystems only in specific areas and on a short-term basis. Their environmental impacts are limited, both spatially and temporally, by the small numbers of people involved, low levels of technology, and

---

4.2   THE TRAGEDY OF THE COMMONS OR COOPERATIVE RESPONSIBILITY?

Imagine that you and a group of friends are dining at a fine restaurant with an unspoken agreement to divide the check evenly. What do you order? Do you choose the modest chicken entrée or the pricey lamb chops? The house wine or the Cabernet Sauvignon 1983? If you are extravagant you could enjoy a superlative dinner at a bargain price. But if everyone in the party reasons as you do, the group will end up with a hefty bill to pay. And why should others settle for pasta primavera when someone is having grilled pheasant at their expense? (Glance and Huberman 1994:76)

This lighthearted scenario accurately depicts the clash between individual and collective attitudes that, in the context of human use of the environment, was so forcefully put forward by Hardin (1968) as follows:

• a group of graziers use an area of common land;
• they continually add to their herds so long as the marginal return from the additional animal is positive, even though the common resource is being depleted and the average return per animal is falling;
• indeed, individual graziers are obliged to add to their herds because the average return per animal is falling;
• clearly, efficient use of the common resource requires restricted herd sizes;
• but individuals will not reduce herd sizes on the common land unless all other group members similarly reduce their herd numbers;
• hence the metaphor 'the tragedy of the commons'.

Individual rational behaviour does not result in a collectively prudent outcome when individuals have access to common resources.

Both examples, lighthearted and serious, prompt the same question: How do we ensure that individuals act for the common good rather than for personal gain? With reference to the environment, three solutions have been proposed (Johnston 1992):

1. Resources may be privatized, with the private owners implementing strategies for environmental preservation not available to group owners.
2. The group owners may be able to devise an agreement about the use of the common resource that they are able to implement themselves. According to some recent work in social theory, the success of local and regional recycling programs depends on such group cooperation (Glance and Huberman 1994:80).
3. The common resource may be subject to some external control. However difficult it may be to implement, it is the third proposed solution that appears most necessary for ensuring the reduction of deleterious human impacts on the environment.

In fact, efforts to implement these three possible solutions, all of which involve cooperative responsibility, are increasingly evident. They may be a part of a general transformation of environmental ideology in the more developed world away from an anthropocentric (human-centred) to an ecocentric (environment-centred) world view. General evidence to support this suggestion includes the rise of political green parties, the current proliferation of various ecophilosophies, and the fact that environmental issues, particularly those involving environmental abuse, are becoming increasingly important factors in decisions about national security and foreign policy for some states.

some deliberate conservation strategies. It seems probable that most ecosystems recover from hunter–gatherer activities. Natural energy flows are not greatly altered except in cases of regular use of fire or animal overkill.

Agriculturalists, both cultivators and pastoralists, affect far more extensive areas over longer time periods. Cultivators change normal energy flows, often permanently, and often consciously direct new flows. Pastoralists have less effect on energy flows, as domesticated animals may merely replace previously wild populations.

Industrialists have affected virtually the entire surface of the earth. Energy flows have substantially increased with the use of fossil fuels and nuclear-based power. Today, however, oil supplies are often restricted not only by the limited amounts available, but by political circumstances. Ecosystem stability or instability is now very much a product of human decision-making rather than natural processes. We have now 'advanced' sufficiently to cause global instabilities, primarily through our depletion of the ozone layer and the increasing concentration of atmospheric carbon dioxide.

Impacts on Vegetation

When we consider human impacts on vegetation, animals, land and soil, water, and climate, it is appropriate to discuss vegetation first, since modification of plant cover results in changing soils, climates, geomorphic processes, and water: 'Indeed, the nature of whole landscapes has been transformed by man-induced vegetation change' (Goudie 1981:25). Figure 4.2 summarizes some of the consequences of vegetation change. Table 4.2 summarizes deforestation through time by major world regions, showing that most clearing has occurred since 1850, and that Europe, Asia, and the former USSR were the first regions to experience any considerable impact. It is only relatively recently that the tropical and subtropical areas of Central and South America have been subjected to significant deforestation.

Fire

For at least 1.5 million years, humans used fire deliberately to modify the environment. Initially, vegetation removal probably resulted in increased animal numbers and greater mobili-

ty for human hunters. Fire also offered security and a social setting at night, and encouraged movement to colder areas. For later agriculturalists, fire was a key method of clearing land for

Table 4.1	SHALLOW AND DEEP ECOLOGY COMPARED
Shallow Ecology **(Spaceship earth)**	**Deep Ecology** **(Sustainable earth)**
Views humans as separate from nature	Views humans as part of nature
Emphasizes the right of humans to live (anthropocentrism)	Emphasizes the idea that every life-form has in principle a right to live; recognizes that we have to kill to eat, but that we have no right to destroy other living things without sufficient reason based on ecological understanding
Concerned with human feelings (anthropocentrism)	Concerned with the feelings of all living things; deep ecologists feel sad when another human or a cat or dog feels sad and grieve when trees and landscapes are destroyed
Concerned with the wise management of resources for human use (anthropocentrism)	Concerned about resources for all living species
Concerned with stabilizing the population, especially in less developed countries	Concerned not only with stabilizing the human population worldwide, but also with reducing the size of the human population to a sustainable minimum without revolution or dictatorship
Either accepts by default or positively endorses the ideology of continued economic growth	Replaces this ideology with that of ecological sustainability and preservation of biological and cultural diversity
Bases decisions on cost-benefit analysis	Bases decisions on ethical intuitions about how the natural world really works
Bases decisions on short-term planning and goals	Bases decisions on long-range planning and goals and on ecological intuition when all facts are not available
Tries to work within existing political, social, economic, and ethical systems	Questions these systems and looks for better sytems based on the way the natural world works

Source: G.T. Miller, Jr, *Living in the Environment*, 4th ed. (Belmont, CA: Wadsworth, 1982):456.

Figure 4.2 Some consequences of human-induced vegetation change.
Source: Adapted from A. Goudie, *The Human Impact: Man's Role in Environmental Change* (Oxford: Blackwell, 1981):25.

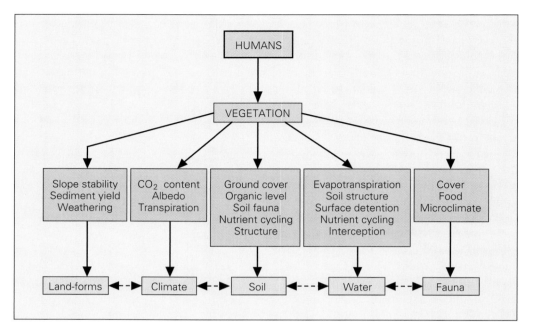

agriculture and improving grazing areas; fire continues to serve these and similar functions today. Indeed, deforestation by fire or other means has been prompted largely by the need to clear land for agricultural activities, both pastoral and arable. In Europe, large-scale deforestation continued from about the tenth century onwards, and in temperate areas of European overseas expansion it was carried out largely in the nineteenth century. Box 4.3 describes the European attitude towards forests in one part of the 'new world'. Together, deliberate burning and the natural fires that are much more common—it has been estimated that lightning strikes some 100,000 times each day (Tuan 1971:12)—have drastically modified vegetation cover.

Fire has played a major role in creating some of the vegetation systems discussed in the previous chapter—savannas, mid-latitude grasslands, and Mediterranean scrub lands are prime examples—and has probably affected all vegetation systems except tropical rain forests. Areas significantly affected by fire typically possess considerable species variety.

Plant domestication

Domestication is a process whereby a plant is modified in order to fulfil a specific human desire; once domesticated, the plant is permanently different from the original. This process is ongoing and is an important part of agricultural research today. Associated with plant domestication has been the labelling of plants that are not domesticated as weeds—and their removal. Once again, such human activity contributes to ecosystem simplification. Early domesticates included wheat, barley, oats (southwest Asia), sorghum, millet (west Africa), rice (southeast Asia), yams (tropical areas), potato (Andes), and manioc and sweet potato (lowland South America). Other domesticates include such pulses as peas and beans and such trees/shrubs as peach and grape. Domestication involves the

Table 4.2	**GLOBAL DEFORESTATION: ESTIMATED AREAS CLEARED (000s KM^2)**				
Region	**Pre-1650**	**1650–1749**	**1750–1849**	**1850–1978**	**Total**
North America	6	80	380	641	1,107
Central America	1530	40	200	285	2055
Latin America	15	100	170	637	922
Oceania	4	5	6	362	377
Former USSR	56	155	260	575	1,046
Europe	190	60	166	81	497
Asia	807	196	601	1,220	2,824
Africa	161	52	29	469	711
Total	**1,254**	**678**	**1,652**	**4,185**	**7,769**

Source: Adapted from M. Williams, 'Forests', in *The Earth as Transformed by Human Action: Global and Regional Changes in the Biosphere over the Past 300 Years*, edited by B.L. Turner et al. (Cambridge: Cambridge University Press, 1990):180.

Stretching from Prince George in central British Columbia south to central Idaho, the world's largest remaining temperate rain forest includes at least fifteen tree species—western red cedar, western hemlock, mountain hemlock, ponderosa pine, Douglas fir, western larch, lodgepole pine, western white pine, subalpine fir, western yew, trembling aspen, paper birch, and three species of spruce—and is a priceless habitat for a wide range of life forms.

In the background of this photo is a hillside that has been 'clear-cut'. Such indiscriminate logging destroys both the trees themselves and the biologically diverse ecosystem that they support. It can also have drastic effects on streams like this one when logging debris accumulates after flooding (Philip Dearden).

introduction and perpetuation of human-induced selection at the expense of natural selection. The choice of domesticates reflects a combination of physical and cultural (preference) variables. Pastoral activities resulting from animal domestication can increase species diversity, but are characteristically detrimental to vegetation cover, causing overall soil deterioration and erosion.

Remote sensing is a versatile and effective tool for monitoring forestry operations. In Canada, Landsat and other imagery allow forest managers to collect data on forest inventory, depletion, and regeneration in areas as small as 2 ha (5 acres) and with boundary accuracy within 25 m (27 ft). Such data are invaluable in the accurate mapping of, for example, clear-cut areas.

4.3

THE THREATENING FOREST

Europe was once heavily wooded, but by the eighteenth century it was a largely agricultural region with few remnants of the once dominant forest. For eighteenth-century and later Europeans, cleared land represented progress and the triumph of technology.

When Europeans moved to temperate areas overseas, they confronted a very different environment. Eastern North America in particular was densely wooded. Aboriginal populations typically cleared only small areas and then moved elsewhere, allowing the forest to regrow. Among the nineteenth-century British in Ontario, the prevailing attitude towards the forest was antagonistic: 'settlers stripped the trees from their land as quickly as possible, shrinking only from burning them as they stood. They attacked the forest with a savagery greater than that justified by the need to clear the land for cultivation, for the forest smothered, threatened and oppressed them' (Kelly 1974:67).

Forests were seen as indicative of a lack of progress. They were removed not simply because cleared land was needed for agriculture, but also because, for the majority of settlers, the forest was oppressive and threatening—a symbol of nature's domination over humans.

Deforestation in Ontario proceeded apace as humans established their dominance over the land. By the 1860s, settlers were becoming aware of the disadvantages of deforestation, such as lack of shelter belts, fuel, and building materials. But by then the damage was done.

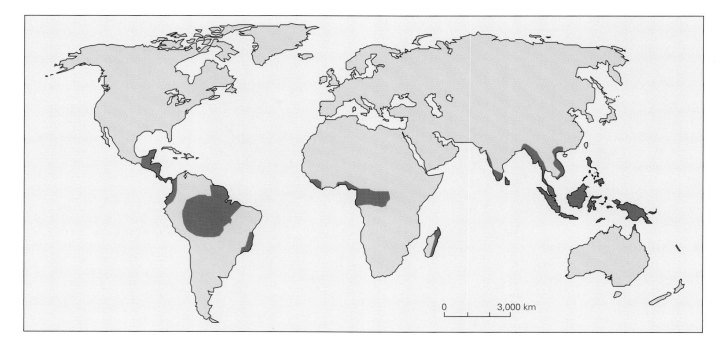

Figure 4.3 Location of tropical rain forests.

Tropical rain forest removal

Human removal of vegetation—particularly of tropical rain forest—and desertification continue to be of major concern. Without human activity, forests would cover most of the earth's land surface. Large-scale deforestation accompanied the rise of the Chinese, Mediterranean, and western European civilizations, as well as the nineteenth-century expansion of settlement in North America and Russia. Today, deforestation is concentrated in the tropical areas of the world. Viewed in this historical perspective, the current removal of rain forest may not seem excessive, but there are important differences between temperate and tropical deforestation. Tropical forests typically grow on much poorer soils that are unable to sustain the permanent agriculture now practised in temperate areas. Also, tropical rain forests now play a major role in the health of our global ecosystem—a fact that has been significantly acknowledged only in recent years. Figure 4.3 shows the distribution of tropical rain forests. These rain forests cover about 8.6 million km² (3.3 million square miles).

The current rate of depletion and the loss to date are both open to dispute. The rate of depletion may be as high as 100,000 km² (38,500 square miles) per year; a rate that, if sustained, will result in complete elimination of rain forest in the first half of this century. Fortunately, remote sensing using Landsat,

together with other satellite data, permits some objective assessment of rates of clearance. These data show that annual clearance rates in Brazil and elsewhere were several times greater than the early 1980s estimates made by the United Nations Food and Agricultural Organization (Repetto 1990). On the other hand, the World Bank's 1989 estimate of a 12 per cent loss of Brazilian rain forest was shown to be about twice the actual loss. Clearly, there are still uncertainties about the details of rain forest removal. Why are the rain forests being removed, and what are the ecological impacts?

Rain forests are located primarily in less developed areas of the world, such as Bolivia, Brazil, Colombia, Venezuela, Gabon, Zaire, Indonesia, and Malaysia, but the more developed areas of the world are a leading cause of deforestation because of their enormous appetite for tropical timber and the inexpensive beef produced in areas cleared of tropical timber. In addition, poor people in the less developed world use cleared land for some subsistence farming. Often such farming is possible only for a few years because of cultivation techniques that rapidly deplete the soil of key nutrients. Cattle ranching also quickly becomes less profitable, because rain forest soil supports grazing for only a few short years. Nevertheless, as already noted, from the perspective of the countries experiencing rain forest clearance, such activity can be seen as a means of reducing population pressure else-where and as generally equivalent to the resource and settlement frontier of, say, North America during the past 200 years and of Europe during the past 1,000 years.

Rain forest removal has two principal eco-logical consequences. First, it is a major cause of species extinction because the rain forests are home to at least 50 per cent of all species (the total has been estimated at 30 million). It is not easy to appreciate what such a statis-tic means. One way might be to recognize that, as E.O. Wilson puts it:

> The human species came into being at the time of greatest biological diversity in the history of the earth. Today as human populations expand and alter the natural environment, they are reducing biological diversity to its lowest level since the end of the Mesozoic era, 65 million years ago. The ultimate conse-quences of this biological collision are beyond calculation and certain to be harmful. That, in essence, is the biodiver-sity crisis (Wilson 1989:108).

From a strictly utilitarian viewpoint, many tropical forest species are (or may prove to be) important to humans as foods, medicines, sources of fibres, and petroleum substitutes.

The second principal ecological conse-quence of rain forest removal involves global warming. Carbon is stored in trees, and when burning occurs, the carbon is transferred to the atmosphere as carbon dioxide. In addi-tion, soil is a source of carbon dioxide, methane, and nitrous oxide, all of which are released into the atmosphere as a result of forest removal and farming. Each of these gases contributes to what we now call the greenhouse effect—a topic that we will con-sider shortly.

Desertification

Desertification is land deterioration caused by climatic change and/or by human activi-ties in semiarid and arid areas: 'It is the process of change in these ecosystems that can be measured by reduced productivity of desirable plants, alterations in the biomass and the diversity of the micro and macro fauna and flora, accelerated soil deteriora-tion, and increased hazards for human occu-pancy' (Dregne 1977:324). The significance of desertification lies not simply in the clear-ing of vegetation but also in the conse-quences of clearing, which include soil ero-sion by wind and water and possible alter-ations of the water cycle.

Deserts are natural phenomena, but deser-tification is the expansion of desert areas. The human causes are complex, but typically involve removal of vegetation as a result of overgrazing, fuel-gathering, intensive cultiva-tion, and waterlogging and salinization of

4.4

DEFEATING DESERTIFICATION

The Kenyan Green Belt Movement had its origins in a 1974 Nairobi tree-planting scheme that focused on the value of working on a community level and, especially, with women (Agnew 1990). The objectives of the movement are many and varied, as befits any effort to solve so difficult a problem as desertification.

The central objective is to reclaim land lost to desert and to guarantee future fertility. Specific objectives include conserv-ing water, increasing agricultural yields, limiting soil erosion, and increasing wood supplies. The ideas of working on the local level and involving women are crucial. Reclaiming land is important to the local people, and their direct involvement makes the exercise much more meaningful to them, while women (the principal wood-gatherers in Kenya and in most of the Sahel) are increasingly aware of future needs. Planting crops around trees improves yields—an important and direct consequence in a subsistence economy.

Tree-planting to combat desertification is far from a panacea, but it is proving to be a positive development. Other parts of the world are adopting the strategies of the Kenyan movement, particularly its community-based focus. The par-allels with the Grameen Bank (Box 6.11) are intriguing.

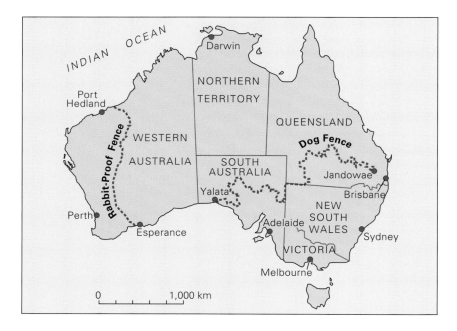

Figure 4.4 Fences built to protect Australian agriculture from rabbits and dingoes (wild dogs). Stretching 1,833 km (1,139 miles), the rabbit fence was built in the 1890s in an attempt to confine the introduced species to the central desert, but it was too late; rabbits had already entered western Australia. The dingo fence in eastern Australia stretches 5,321 km (3,307 miles) and was built to protect sheep; dingoes are thought to have been introduced into Australia by Asian seafarers more than 3,500 years ago.

irrigated lands. The reasons behind these activities are usually population pressure and/or poor land management. Fortunately, the technology to combat desertification is available; unfortunately, the will to use it is rare (Box 4.4). A major international effort to combat desertification, the 1977 United Nations Plan of Action, was a failure. There were two reasons for this failure: (1) technical solutions were applied to areas where the key causes were economic, social, and political, and these underlying causes were not addressed; and (2) local populations were not involved in the search for solutions. A potentially important development is discussion concerning an international convention to combat desertification as first proposed at the 1992 Earth Summit in Rio de Janeiro.

Estimates of the spatial extent of desertification in 1992, by the United Nations Environment Program, arrived at a figure of 3.5 million ha (8,648,000 acres) affected. Since remote-sensing imagery and local area surveys have not confirmed this figure, there is considerable confusion concerning the actual spatial extent. The most publicized area experiencing desertification is the Sahel zone of west Africa, an area that was first brought to world attention following the 1968–73 drought. Here, desertification is caused by population pressure, inappropriate human activity, human conflict, and periods of drought. Elsewhere, desertification similarly has multiple causes and no simple solution.

Many of the human causes of desertification may be related to the pressures placed on local people by the introduction of capitalist imperatives into traditional farming systems.

As is the case with the tropical rain forests, areas subject to desertification are home to poor people who are not necessarily able to adopt appropriate remedies and lack the necessary political influence. A proper solution requires that the dry-land ecosystems be treated as a whole—land management is needed. Further, population pressures need to be reduced, land needs to be equitably distributed, and greater security of land tenure is required.

Impacts on Animals

Animal domestication serves many purposes, providing foods such as meat and milk (cows, pigs, sheep, goats), as well as draft animals (horses, donkeys, camels), and pets (dogs, cats). Once domesticated, animals have often been moved from place to place, both deliberately and accidentally. Some deliberate introductions, such as that of the European rabbit to Australia, have had drastic ecological consequences. Whalers and sealers were probably the first to introduce rabbits into the Australian region, but the key arrival was in 1859, when a few pairs were introduced into southeast Australia to provide so-called sport for sheep-station owners. Following this introduction, rabbits spread rapidly across the non-tropical parts of the continent, prompting a series of 'unrelenting, devastating' (Powell 1976:117) rabbit plagues. Rabbits consume vegetation needed by sheep and remain a problem today despite the introduction of the disease myxomatosis and rabbit-proof fences (Figure 4.4). The European rabbit in Australia has probably caused more damage than any other introduced animal anywhere in the world—but it is only one of many unwanted guests (Box 4.5).

As human numbers and levels of technology have increased, so also have animal extinctions. It is possible that humans have caused species extinction since 200,000 BCE, although the evidence is uncertain. What is certain is that hunting populations have caused extinctions. After Europeans arrived in New Zealand, all moa species became extinct, and there is much evidence to suggest that animal extinctions in North

Desertification outside Benguela city, Angola, 1991. Agriculture in the area was disrupted by war, and the soil, left untended, became desert. The traces of former crop furrows are still visible (CIDA photo/Bruce Paton).

4.5

UNWANTED GUESTS

Beginning in 1788, the European colonization of Australia ended a long period of isolation and introduced numerous animals and plants. Many of these species, whether introduced deliberately or accidentally, proliferated in their new environment, where the natural controls on their numbers (such as predators and diseases) were absent. Inevitably, the presence of introduced species reduces the populations of native animals that evolved for millions of years in isolation.

Rats and mice that arrived in the holds of ships multiplied rapidly after being accidentally introduced. Sheep and cattle were brought over to satisfy British economic needs, while animals brought over to satisfy the 'sporting' habits of British settlers included deer, fox, and, of course, rabbits. The interior deserts prompted the introduction of pack animals such as burros, asses, and camels, and soon there were more camels in Australia than in all of Arabia. Many varieties of introduced livestock have reverted to a feral (wild after previously being domesticated) existence, including pigs, horses, camels, and water-buffalo.

Feral water-buffalo damage forest ecosystems by destroying trees and eroding soil. More generally, because all feral species require a supply of that most precious resource in the interior, water, they reduce the water available for native species. Many of the native animals of Australia have disappeared because of the presence of the introduced species. Of the smaller marsupials, 17 have become extinct and another 29 are considered endangered.

A recent introduction that is causing considerable damage is the cane toad, first brought to Queensland in the 1930s to combat the cane grub, which was damaging the sugar cane crop. Unfortunately, cane toads do not eat only cane grubs—they eat almost anything. They are also capable of much more rapid reproduction than native toad species, with females producing up to 40,000 eggs a year; they compete effectively with the native species; and they have no natural predators. Cane toads have now moved far beyond the sugar cane area, and there are no signs of an end to their movement.

The Carolina parakeet was one of the many species that humans have hunted to extinction. The last flock, numbering 30 birds, was sighted in Florida in 1920. From John James Audubon's *Birds of America* (Metropolitan Toronto Reference Library).

world are dying off as humans remove or destabilize their environments. This loss of biodiversity is irreversible by any reasonable human standards of time, and future generations are certain to live in a world that is biologically impoverished.

Most commentators explain biodiversity loss by reference to population growth and increasing consumption of energy and resources. With more and more people around the world aspiring to the lifestyle of the affluent minority, the human ecological footprint is outgrowing the resources needed to support it. Wilson extends this argument in a more controversial direction to suggest that the impulse towards environmental destruction is innate, hard-wired into us—in short, that clearing forests and killing animals is instinctual for humans. Furthermore, he says, we are unable to see things in the long term. Whether or not Wilson is correct, the evidence of biodiversity loss is compelling; however, as much of the content of this chapter makes clear, the significance of this loss is still hotly debated.

Impacts on Land and Soil

Because we humans live on the land and use soil extensively, we are geomorphic agents that change land and affect that thin and vulnerable resource, soil.

Most human activities actually create landforms. Excavation of resources such as common rocks (limestone, chalk, sand, gravel), clays (stoneware clay, china clay), minerals (dolomite, quartz, asbestos, alum), precious metals (gold, silver), and fossil fuels (oil, coal, peat) can have major impacts: changing ecosystems, lowering land surfaces, flooding, building waste heaps, creating toxic wastes, and leaving scenic scars. These are not the only human activities that make us geomorphic agents: when we modify river channels, for example, sand dunes are affected; other effects include coastal erosion and coastal deposition. There are many causes and consequences of human impact on land.

Degradation and loss of arable land are occurring throughout the world as a result of population increases, industrialization, and improper agricultural practices. Smil (1993:67) reported that the average annual loss of farmland in China between 1957 and 1980 was a mammoth 1 million ha (2.5 million acres).

America coincided with human arrival. The 1859 publication of Darwin's *On the Origin of Species* helped to place the extinction of plant and animal species in context and was followed by protectionist legislation in several of the British colonial areas; an early example was the 1860 law in Tasmania to protect indigenous birds.

Today our role in animal extinction is clearer. When natural animal habitats are removed, as in the case of tropical rain forest, extinction follows. This is one component of what we may regard as a major threat to biodiversity.

Biodiversity loss

As suggested earlier, E.O. Wilson, the best-known Darwinian thinker today, believes that the earth is entering a new evolutionary era involving the greatest mass extinction since the end of the Mesozoic era, 65 million years ago. Humans are the cause of the current extinction phase, in which species around the

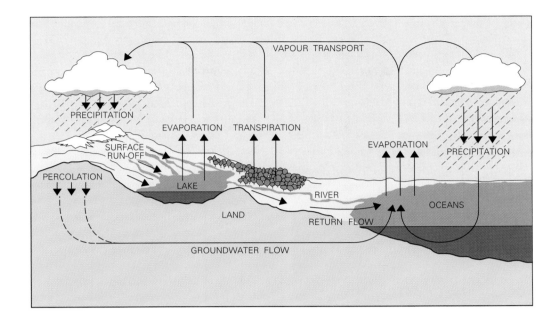

Figure 4.5 The global water cycle.

By its very nature, soil is especially susceptible to abuse, and humans use and abuse soils extensively. Agricultural activities have the greatest impact on soil, not only through erosion but through chemical changes. Humans increase the salinity (salt content) of soil largely through irrigation—and yet increased salinity has a negative effect on plant growth. Humans increase the laterite content of soil (laterite is an iron- or aluminum-rich duricrust naturally present in tropical soils) by removing vegetation—and yet laterite is essentially hostile to agriculture.

Soil erosion is associated with deforestation and agriculture. Forests protect soil from run-off and roots bind soil. Probably the best-known example of human-induced soil erosion is the 'Dust Bowl' of the 1930s in the North American mid-latitude grasslands. Among the causes of this phenomenon were a series of low-rainfall years, overgrazing, and inappropriate cultivation procedures associated with wheat farming. The 'black blizzards' that were the result led many people to leave the prairies.

Impacts on Water

Water is an essential ingredient of all life. We know this and yet we choose to ignore it. Rather than carefully safeguarding water quantity and quality, we cause shortages and continually contaminate it. The two issues of scarcity and contamination dominate our consideration of the human impact on water.

The global water cycle

How much water is available? Figure 4.5 outlines the global water cycle, identifying three principal paths—precipitation, evaporation, and vapour transport. Total annual global precipitation is estimated at 496,000 km³ (119,000 cubic miles), most of which (385,000 km³/92,000 cubic miles) falls over oceans and cannot be easily used. Water returns to the atmosphere via evaporation from the oceans (425,000 km³/102,000 cubic miles) and from inland waters and land along with transpiration from plants (71,000 km³/17,000 cubic miles combined). In addition, some of the precipitation that falls on land is transported to oceans via surface run-off or groundwater flow (41,000 km³/9,800 cubic miles), and some water evaporated from the oceans is transported by atmospheric currents and subsequently falls as precipitation over land (again, some 41,000 km³/9,800 cubic miles). In principle, 41,000 km³ (9,800 cubic miles) of water are available each year globally, but this figure is reduced by 27,000 km³ (6,500 cubic miles) lost as flood run-off to the oceans and by another 5,000 km³ (1,200 cubic miles) flowing into the oceans in unpopulated areas. Perhaps 9,000 km³ (2,200 cubic miles) are readily available for human use.

This total of 9,000 km³ (2,200 cubic miles) is possibly enough water for 20 billion people. However, some states have a plentiful amount and others have an inadequate amount. Where water is abundant, it is treated as though it were virtually free; where it is scarce, it is a precious resource. Thus the average citizen of the US annually consumes 70 times more water than does the average citizen of Ghana.

Using and polluting water

Agriculture is the world's main user of water, consuming 73 per cent of global supplies, often highly inefficiently. In second place is industry, which consumes perhaps 10 per cent of global supplies. The bulk of what is left goes to supply basic human needs. As each of these uses increases in any given area, whether through agricultural and industrial expansion or population increases, water quantity may become a problem. Shortages are common in many areas, some continuous and some periodic. Bahrain, for example, has virtually no fresh water and relies on the desalinization of sea water. Groundwater depletion is common in the US, China, and India, and water levels have fallen in both Lake Baikal and the Aral Sea.

The second major issue, after water quantity, is water quality. As water passes through the cycle described in Figure 4.5, it is polluted in two ways. Organic waste (from humans, animals, and plants) is biodegradable, but can still cause major problems in the form of oxygen depletion in rivers and lakes and in the form of water contamination, causing such diseases as typhoid and cholera. Much industrial waste is not easily degraded (paper, glass, and concrete are exceptions), and such wastes are a major cause of deteriorating water quality. These pollutants enter water via pipes from industrial plants, diffuse sources (run-off water containing pesticides and fertilizers), and the atmosphere (**acid rain**).

4.6

A TALE OF THREE WATER BODIES

Until the 1960s, the *Aral Sea*, on the border between the central Asian states of Kazakhstan and Uzbekistan (both former parts of the Soviet Union), was the fourth largest inland water body in the world. By the early 1990s, however, remote-sensing imagery showed that the area covered by the sea had been reduced by more than 40 per cent. The explanation was that, since the 1950s, much of the water that formerly flowed into the sea had been diverted to irrigate cotton. Although data for the late 1990s show some signs of recovery—water levels have risen as a result of dam construction and reductions in water use—the Aral Sea remains vulnerable to a range of human activities. Tragically, because cotton requires large quantities of defoliants, pesticides, and fertilizers, the Aral Sea has also become dangerously polluted, and in the local area there are high rates of infectious disease, cancer, miscarriage, and fetal abnormalities. These problems are not likely to be solved any time soon.

The *Rhine River*, one of the most important international waterways in the world, is 1,320 km (820 miles) in length, and has a catchment area of 185,000 sq km (71,428 sq miles) populated by about 50 million people in six European countries. The river receives pollutants from several heavily industrialized areas in addition to domestic sewage and shipping discharges. By the late 1980s, as a result of these human activities, the quality of Rhine water was so compromised that treatment was necessary before the water could be used for drinking or crop irrigation in the Netherlands, and local ecosystems were changed, destroying much animal and plant life. However, considerable success in combating these problems has been achieved through the activities of the International Commission for the Protection of the Rhine against Pollution, established in 1950. Today the Rhine is in relatively good health; one recent indicator was the return of salmon in 2000.

When Europeans first reached *Lake Erie*, the water was clear and supported a large fish population. By the 1960s, however, Erie was one of the most seriously polluted large water bodies in the world. With some 13 million people and much heavy industry in its watershed, over time the lake was polluted by domestic sewage, household detergents, agricultural wastes, and industrial wastes. Clear water became green, fish perished, and algae growth was enhanced by the addition of phosphates and nitrates. These problems were acknowledged and the Canadian and American governments, urged to take action by newly formed environmental groups, initiated a comprehensive program to solve these problems in the 1970s. Known as the Great Lakes Water Quality Agreement, this program has been remarkably successful, and today Lake Erie has been rehabilitated.

These three examples demonstrate that, given effective management strategies, seriously polluted water bodies can recover; however, such strategies are more likely to be employed in the more developed than in the less developed world.

Both inland waters and oceans are suffering the consequences of pollution (Box 4.6).

Notorious examples include the Aral Sea and the toxic chemical waste in waters at Love Canal in the US (refer to Figure 2.18). Both Lake Erie and the Rhine River, especially in the Netherlands, are suffering from more general urban industrial pollution; such surface water pollution is reversible, as evidenced by the results of the Rhine Action Plan. Acid rain is a more difficult issue because the pollution itself may occur far away from its source. Acid rain is one general consequence of urban and industrial activities that release large quantities of sulphur and nitrogen oxides into the atmosphere. The effects of acid rain are not entirely clear, but there is no doubt about its negative impacts on some aquatic ecosystems. One of the most severe problems involved in reducing water pollution is the difficulty of securing the necessary international cooperation. The Rhine Action Plan involves only four countries; efforts to combat acid rain, especially in Europe, require cooperation among many more countries.

Nowhere is the need for international agreement more evident than in combating ocean pollution. Many states exploit oceans, but no state is prepared to assume responsibility for the effects of human activities. More than 50 per cent of the world's population live close to the sea, and most of the world's ocean fish are taken from coastal waters. Water quality in the oceans, especially coastal zones, is seriously endangered, and many ocean ecosystems have been damaged. Remote sensing enables objective assessments of water pollution; images of the Mediterranean Sea show a marked contrast between the northern shore, heavily polluted by water from major European rivers and coastal towns, and the southern shore. Remote sensing permits effective monitoring of oil spills, as in the *Exxon Valdez* incident off the Alaskan coast in 1989. Once again, we do not know enough about the consequences of our activities, but we do know that restoring the quality of ocean water is likely to be much more difficult than is the case with surface inland water.

Impacts on Climate

Nowhere is the concept of interrelations better demonstrated than in a consideration of human impacts on climate. Human impacts on other

Spanish air force personnel formed a human chain to clean oil from the rocks on the northwest coast of Spain in December 2002. The oil was spilled when the tanker *Prestige* sank with a cargo of fuel oil estimated at 77,000 tons (AP photo/Carmelo Alen).

components of the ecosystem also affect climate, and many environmentalists feel that our most damaging impacts of all are on global climate. Unfortunately, as in our discussions of other human impacts, we are confronted with two general areas of uncertainty: the role played by humans (as opposed to physical factors) and the extent of any human-induced change.

Physical causes of climatic change

Not all climatic changes are caused by humans; in fact, such changes can occur for various reasons, and on various time scales. It appears that some changes result from external variables, such as variation in the input of solar radiation. Yet it also seems that climate oscillates regardless of such factors; it has recently been determined, for example, that long-term changes are related to the changing distribution of land and sea, with the current distribution conducive to the emergence of glacial periods in North America and Eurasia. The most recent glacial period ended some 12,000 years ago. Other climatic changes occur at time scales of a few hundred to a few thousand years, and these changes are not well understood. During our current interglacial period (see Chapter 3), there was a cool phase 10,000–11,000 years ago, a distinctly warm phase 6,000 years ago, and another warm period about 1,000 years ago. Historical evidence shows that, before the fourteenth century, England was warm enough for wine cultivation, and that Europe in general was unusually cold in the sixteenth and seventeenth centuries.

It is important to consider human impacts in the above perspective. We know that we are capable of changing climate, but we are not sure of precisely how we may affect future climates, because we do not know how our activities may interact with other, non-human variables. We know that we are moving towards another glacial period, but we do not know what shorter-term changes may naturally occur prior to the onset of a new ice age.

The natural greenhouse effect

Temperatures on the earth's surface result from a balance between incoming solar radiation and loss of energy from earth to space. If the earth had no atmosphere, then the average surface temperature would be about $-19°C$ ($-3°F$), but the presence of an atmosphere results in an actual average surface temperature of $15°C$ ($60°F$). The atmosphere causes an increase in surface temperatures because it prevents about one-half of the outgoing radiation from reaching space; some is absorbed and some bounces back to earth. This natural 'greenhouse' effect is not related to human activity but is the result of the presence in the atmosphere of water vapour, carbon dioxide (CO_2), ozone (O_3), and other gases. These greenhouse gases are only a fraction of the atmosphere—nitrogen and oxygen make up 99.9 per cent of it (excluding the widely varying amounts of water vapour)—but their impact is never-theless considerable. Today we are increasing the greenhouse effect by adding CO_2 and some other gases that perform a similar function, such as sulphur dioxide (SO_2), nitrous oxide (N_2O), methane (CH_4), and a variety of chloro-fluorocarbons (CFCs). How are we doing this?

Human-induced global warming

In the most general sense, the human additions to the natural greenhouse effect are the product of our increasing numbers and advancing technology. More specifically, they are the result of burning fossil fuels, increased fertilizer use, increased animal husbandry, and deforestation. Until recently, much carbon was stored in the earth in the form of coal, oil, and natural gas. Burning these resources releases CO_2, water vapour, SO_2, and other gases that are then added to the atmosphere. Estimates suggest that the concentration of CO_2 in the atmosphere has increased from 260 ppm (parts per million) 200 years ago to 350 ppm today and might be 400 to 550 ppm in 2030. Burning wood also adds CO_2 to the atmosphere. Further, soil contains large quantities of organic carbon in the form of humus, and agricultural activity speeds up the process by which this carbon adds CO_2 to the atmosphere. Agricultural activity also adds CH_4 and N_2O to the atmosphere. CH_4 is increasing particularly as a result of paddy rice cultivation.

4.7

CHLORO-FLUOROCARBONS AND THE ATMOSPHERE

Chloro-fluorocarbons (CFCs) exemplify the devastating environmental impact of many efforts to satisfy human demands through the use of technology. CFCs were not even synthesized until the late 1920s; at that time they represented a remarkable advance. They are ideal as coolants because they vaporize at low temperatures and also serve well as insulators. Most important, they are easy and therefore inexpensive to produce: hence their widespread use since the Second World War as coolants in refrigerators, as propellant gases in spray cans, and as ingredients in a wide range of plastic foams.

Recognition of the impact that CFCs have on the atmosphere, especially on ozone, prompted 24 countries to gather in Mon-treal in 1987 and agree to limit CFC production by reducing production by 35 per cent by 1999. Most environmental experts argued that this target was inadequate. Subsequently, a 1990 London agreement set the goal of eliminating CFC production by the year 2000. Neither of these goals was achieved. Because CFCs have a long atmospheric lifetime, peak levels will not be reached until roughly a decade after production ceases.

What we really need, of course, is a substitute for CFCs. So far, however, any possible substitutes are more expensive to produce, and world economies, whether capitalist or socialist, are not keen to sacrifice short-term economic benefits for long-term planetary stability.

Each of the greenhouse gases noted so far (CO_2, N_2O, SO_2, CH_4) is naturally present in the atmosphere; human activities simply increase the natural concentrations of them. However, we are also adding new greenhouse gases, the most important being two of the CFCs—$CFCL_3$ and CF_2CL_2. These gases are added primarily by our use of aerosol sprays, refrigerants, and foams (Box 4.7).

It is widely accepted that this human-induced global warming will raise the average temperature of the earth during the present century. The 2001 report of the UN's Intergovernmental Panel on Climatic Change predicted a minimum increase of 2°C and a maximum of 5.8°C, although some scientists think the increase will be less dramatic because the panel overestimated the rate of economic growth in the less developed world. The principal response to the fact of global warming and its probable consequences has been an effort to implement policies that will reduce the emission of greenhouse gases. Most notably, the Kyoto protocol establishes goals that, if met, will slow the rate of global warming.

Some facts and some uncertainties

The best way to detect global temperature change is to use satellites to measure ocean temperatures (because ocean temperatures are not subject to the same diurnal and seasonal fluctuations that air temperatures are). These data indicate that global temperatures have increased by about 0.5°C (1.0°F) since the late nineteenth century.

However, it is difficult to be certain that the global temperature rise can be directly attributed to human activities. 'In this century, instrumental data show no warming trend globally in the period 1940–80; indeed between 1950 and 1976 the trend if anything was towards cooler conditions, but the period since then shows a strong warming of the order of 0.3°C since 1970' (Chambers 1998:267). Some scientists have proposed that the recent temperature rise has more to do with the length of the solar cycle than with CO_2 emissions.

It has even been suggested that recent climatic changes are not exceptional. The majority view is that humans are the cause of the current warming trend. But the minority view—that current warming can be

A harp seal pup stranded in the Gulf of St Lawrence, 21 March 2000, after unusually warm weather—perhaps caused by global warming—led to premature break-up of the ice that adult females need to give birth and nurse their young. Pups prematurely separated from their mothers rarely survive (CCN International Fund For Animal Welfare).

accommodated within the longer-term history of natural physical change—needs to be acknowledged. For our purposes, these complex debates highlight one simple point: namely, that scientists are uncertain of the extent to which humans are responsible for recent climatic changes.

Further, it is important to appreciate that many of the popular accounts of global warming, and some of the scientific ones too, make little if any reference to the uncertainties involved in predicting either the magnitude or the consequences of the human-induced greenhouse effect. For example, there is now strong evidence suggesting that the higher precipitation and related melting of snow and ice in northern areas that is likely to accompany global warming will, in turn, lead to increased river discharge into the Arctic and North Atlantic oceans. The resulting decrease in ocean salinity may disturb the balance of water flow (known as the thermohaline circulation) that involves a warm northward flow of surface water and a returning southward flow of deepwater. A possible consequence of reduced thermohaline circulation might be to flip the North Atlantic back into a glacial mode and to substantially lower temperatures in northwest Europe (Anderson 2000). This scenario is simply one example of the many uncertainties involved in attempting to anticipate the consequences of global temperature changes.

One of the few predictions that scientists feel relatively confident in making is that, as warming occurs, the earth's polar regions will be the most seriously affected. By the late 1990s, it was clear that such warming was already underway, and not all the evidence to this effect came from scientific sources. In northern Canada, Inuit elders and hunters reported that glaciers were receding and coastlines eroding, that the fall freeze was arriving later, and winters becoming less severe. The combination of scientific data and traditional ecological knowledge makes for a compelling argument. Although at present the evidence of warming in lower latitudes is less clear, it is generally agreed that the next several decades will see a poleward retreat of cold areas, a corresponding expansion of forests and agricultural areas, changes in the distribution of arid areas, and, of course, rising sea levels as the ice-caps melt (see Box 4.8 and Figure 4.6).

4.8

RISING SEA LEVELS

Global warming is expected to affect sea levels. A rise of as much as 1.5 m (almost 5 ft) has been predicted for 2050. If this occurs, the consequences for many currently populated areas may be catastrophic. Up to 15 per cent of Egypt's arable land would be at risk, and many coastal cities such as New York and London would be below sea level. The consequences for two of the most densely populated areas in the world—coastal Bangladesh and the Netherlands—will be disastrous unless we are able to adapt.

The Netherlands has successfully adapted to a situation where much of the current area is already almost 4 m (13 ft) below sea level. In principle, the Netherlands model—construction of levees and dikes—can be followed. Venice is already pursuing a similar strategy, constructing a flexible seawall to protect the city against Adriatic storms. But what are the costs of such projects? One estimate places the cost at approximately US $300 billion to protect only major areas, not including coastal margins.

In theory, there is an alternative to adaptation: moving away from the threatened areas. But for many people in the less developed world, this is hardly an option, and for many in the more developed world, it is culturally and economically unthinkable at the present time.

On the other hand, our prediction of a sea-level rise is at best uncertain. In 1985, one major United States scientific committee predicted a rise of 1 m (3 ft) by 2100, and subsequently, in 1989, amended the prediction to a rise of .3 m (1 ft). Even the higher of these two is less than the prediction referred to earlier. Why such uncertainty? The principal unknown factor is the effect of warming on the Antarctic ice sheet. Instead of increasing ice melt, it is possible that warmer weather could increase snowfall, which would in turn help build up the ice sheet.

The problem is serious. We need to adapt soon to an unknown situation. Dikes and population movements may or may not be needed. If they are needed, we do not know the extent of the need. In this way global uncertainty becomes local political uncertainty.

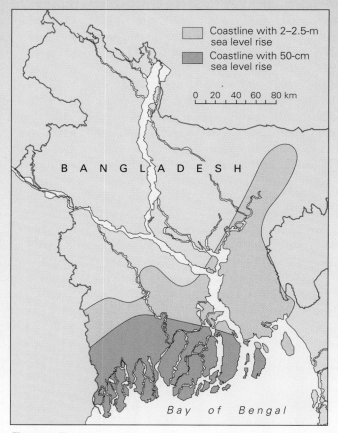

Figure 4.6 The impact of sea-level change on Bangladesh.
Source: G.A. McKay and H. Hengeveld, 'The Changing Atmosphere', in *Planet Under Stress*, edited by C. Mungall and D.J. McLaren (Toronto: Oxford University Press, 1990):66.

Damaging the ozone layer

A second atmospheric concern is the ozone layer. Ozone (O_3) is a form of oxygen that occurs naturally in the cool upper atmosphere. It serves as a protective sunscreen for the earth by absorbing the ultraviolet solar radiation that can cause skin cancer, cataracts, and weakening of the human immune system; ultraviolet radiation is also damaging to vegetation. Ozone depletion was first recognized in 1985 by scientists with the British Antarctic Survey. An ozone hole over Antarctica appears to be about one-half the size of Canada. The principal culprits are the CFCs already noted and some greenhouse gases. As these rise into the atmosphere, chemical reactions occur and ozone is destroyed.

Urban climates

Any discussion of human impacts on climate must also consider urban climates. Not surprisingly, any urban development affects local climate. Built-up areas store heat during the day and release it at night; they also generate artificial heat, creating an 'urban heat island'. Built-up areas also affect cloud formation and precipitation, but these impacts are more difficult to determine and measure. Perhaps the most obvious impact of all involves the creation of smog (smoke-fog). Chemical smog—largely the product of automobile emissions—is common in areas such as Los Angeles.

Globalization

Many human impacts on environment are exacerbated by current trends towards globalization, especially the activities of transnationals (see Chapter 2). It seems clear that transnationals are among the principal actors in many environmental issues. For example, fast food corporations have been blamed for the destruction of tropical rain forests because they use meat from animals raised in areas that have been cleared specifically to create pasture. Similarly, oil companies that explore for and produce oil in such environmentally sensitive areas as the Amazon basin, the Canadian Arctic, and the mangrove swamps of the Niger delta are particularly condemned by environmentalists.

According to a 2003 Amnesty International report, a planned oil and gas pipeline

"Hey, let's not underestimate science! I bet they're working on a replacement to the ozone layer right now."

Cartoon by J.B. Handelsman, from *Punch*, 14 Sept. 1990 (Reproduced with permission of Punch Ltd).

from the Caspian Sea port of Baku in Azerbaijan to the Mediterranean Turkish port of Ceyhan will infringe on the human rights of thousands of people and cause massive environmental damage. The route of the pipeline—to be constructed by a BP-led consortium—is designed to avoid passing through Russian or Iranian territory. As a result, according to Amnesty, 30,000 villagers will be forced to give up their land rights. In addition, Amnesty claims that the health and safety precautions in place are inadequate, and that any local protesters face state oppression. BP, for its part, argues that the project follows the highest international standards, that appropriate compensation has been paid, and that there will not be any environmental damage.

Transnational companies have sometimes been charged with actively perpetuating local social injustices. Talisman Oil, the largest independent Canadian oil and gas producer, has operations in Canada, the North Sea, and Indonesia. It is currently conducting explorations in Algeria and Trinidad, and between 1998 and 2003 was also active in Sudan. Its Sudanese operations were severely criticized by social justice and church groups in Canada because of claims that Talisman actively supported the government of Sudan in an ongoing civil war. That government, headed by the National Islamic Front, has

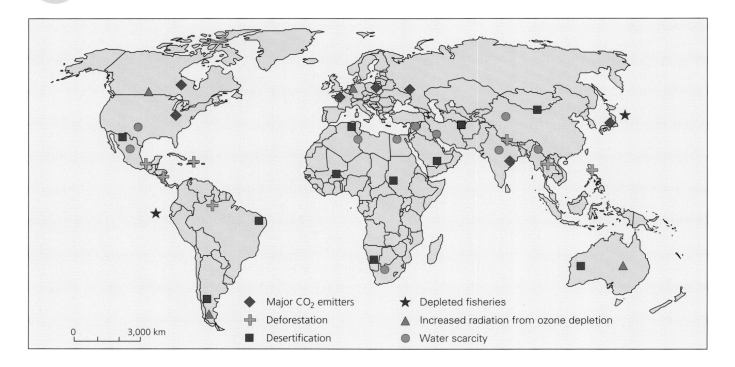

Figure 4.7 Global distribution of some major environmental problems. This map shows the general regional occurrence of six current environmental stresses. Symbols are not intended to identify specific locations, or country, but rather to indicate general regions affected by particular kinds of stress.

Source: Adapted from *Current History* 95, no. 604 (November 1996).

consistently been identified as violating basic human rights. The oil fields in question were located in southern Sudan—traditionally a rebel stronghold—and there was evidence that the government forcibly removed people in order to make the area safe for Talisman. However, the company denied complicity in any such social injustices and in 1999, in response to a request from the Canadian government, formally adopted the International Code of Ethics for Canadian Business, developed in 1997 by a group of Canadian companies with transnational operations. It was at least partly because of the difficult political and social environment in Sudan that Talisman sold all its interests there in 2003.

Earth's Vital Signs

As the preceding account demonstrates, there is no doubt that our current impacts on ecosystems, from the global to the local, are greater than ever before, and are increasing as a general result of the growth of population and technology (Figure 4.7). Further, although there is reason to believe that the situation is not as desperate as some would suggest, international cooperation will certainly be required to address many of the problems that human activities have caused (see Box 4.2); as already noted, the first such

international agreement was the 1987 Montreal protocol limiting the production of CFCs. Debate continues, however, concerning the present condition of the environment and the probable future scenario. As Smil (1993:35) put it, 'Confident diagnoses of the state of our environment remain elusive.'

Apocalypse Now, Deferred, or Never?

In fact, there are many different opinions about the impact of human activities on the environment. At one extreme are **catastrophists** who view the current situation and future prospects in totally negative terms—a view recently articulated by R.D. Kaplan (1994, 1996). At the other extreme are **cornucopians** who believe that the gravity of current problems has been greatly exaggerated and that human ingenuity and technology will overcome the moderate problems that do exist (Simon and Kahn 1984). This is a broad and complex debate, which we will revisit in Chapters 5, 6, and 9.

A balanced view of such an emotional topic is not easy. Certainly the environment is sustaining considerable damage. Yet, as V. Smil (1993:5) convincingly argues, since the 1960s the environment has been the subject of many ill-informed commentaries and predictions: 'Once the interest in environmental degradation began, the western media, so diligent in search of catastrophic happenings, and scientists whose gratification is so often achieved by feeding on fashionable topics, kept the attention alive with an influx of new bad news.' As our discussion of human impacts has demonstrated, it is difficult to separate the many very real problems from the numerous exaggerated claims (Box 4.9).

Responding to Uncertainty

In view of the many uncertainties that surround environmental questions, Smil writes,

> The task is not to find a middle ground: the dispute has become too ideological, and the extreme positions are too unforgiving to offer a meaningful compromise. The practical challenge is twofold. First, to identify and to separate the fundamental long-term risks to the integrity of the biosphere from less important, readily manageable concerns. The second task is to separate effective solutions to such problems from unrealistic paeans to the power of human inventiveness (Smil 1993:36).

The first challenge is continually being addressed by geographers and other environmental scientists. Given our current understanding of global environmental problems, there are three possible general responses to the second challenge:

1. We can attempt to develop new technologies to counter our deleterious impacts; for example, some scientists have suggested that dust might be deliberately spread in the upper atmosphere to reflect

4.9

THE SKEPTICAL ENVIRONMENTALIST

An impressive body of scientific research indicates that the environment is in bad shape. For example, a 2002 study by the World Wildlife Fund warned that the Earth is being plundered at a rate that outstrips the capacity to support life, and that unless drastic action is taken, humans will have to colonize other planets by the year 2050. Nevertheless, at several points in this chapter we have had occasion to note uncertainties concerning both the facts of environmental change and the consequences of such change. Interestingly, comparable uncertainties will be evident in the two chapters that follow concerning the related matters of population growth and food supply, and also in Chapter 10, concerning the consequences of economic globalization.

Perhaps nothing better demonstrates these uncertainties than the sharply divergent responses sparked by the 2001 book *The Skeptical Environmentalist: Measuring the Real State of the World*, by Bjorn Lomborg. Some view Lomborg's massive compendium of facts as confirmation that the prophets of environmental doom and gloom are either exaggerating or just plain wrong, while others view his work as preposterous; one critic expressed his opinion on the matter by throwing a pie at the author. In short, conservative and business-oriented commentators hailed Lomborg as the voice of sanity, while environmentalists and most scientists rejected his views in no uncertain terms.

Bjorn Lomborg is a Danish statistician who wrote *The Skeptical Environmentalist* following a failed attempt to disprove the late Julian Simon's claim that the state of the environment was not getting worse, but rather was improving all the time. Unable to find any flaws in either Simon's data or his logic, Lomborg conducted his own analyses and then wrote his book endorsing Simon's claims. According to Lomborg, we are not running out of energy or natural resources; the world's species are not disappearing at an alarming rate; air and water supplies are becoming less polluted; and warnings of a possible 5°C temperature rise this century are unrealistic. Lomborg even goes so far as to suggest that climate change may offer positive benefits, including improvements to human health and increases in crop yields. Lomborg's critics, including the World Wildlife Fund, maintain that he is wrong on all these counts; that he mischaracterizes the environmental movement and commits precisely those sins for which he attacks environmentalists, namely exaggeration, gross generalization, presentation of false choices, selective use of data, and factual errors. For his part, Lomborg suggests that bad news attracts more attention than good news, and that scientists are more likely to receive funding for research into identifiable problem areas.

Whatever the rights and wrongs of Lomborg's claims, debating them will help us to reach the best possible understanding both of what is happening now and of what may happen in the future.

sunlight, replacing the depleting ozone layer. Overall, though, it seems inappropriate to rely too much on such technological solutions. Despite massive evidence that we are changing the earth's climate on a variety of scales—local to global—we still know too little to change it deliberately in ways we desire. The results of attempts at rain-making, hurricane modification, and fog dispersal have been mixed.

2. We can acknowledge that environmental impacts are inevitable and emphasize our own need to adapt to such changes. In the case of possible climatic change, adaptation might involve new water-supply systems and coastal defences.

3. The third response, probably the most popular and most logical, involves the conservation of resources and the prevention of harmful impacts. **Conservation** refers generally to any form of environmental protection. Prevention involves limiting the increase of greenhouse gases, reducing the use of certain materials, and reusing and **recycling**. Recycling often makes sense economically as well as environmentally.

Prevention is central to many current moves to protect environments. It can be argued that economic systems, including capitalism, do not properly reward the efficient use of resources, and that the physical environment has been seen as irrelevant in the final economic accounting. Only recently have we begun to understand that price and value are not equal. But how do we assign a monetary value to a wilderness landscape, for example? One answer, albeit an imprecise one, is to assert that certain ecosystems or landscapes are sufficiently distinct as to merit protection or preservation. The largest protected ecosystem today is probably that of Antarctica. Other protected landscapes include wilderness regions such as Canada's national and provincial parks. (The first areas to be preserved as national parks were Yosemite [1864] and Yellowstone [1872] in the US.)

To improve the health of the earthly patient of which we are all a part, solutions are needed at all spatial and social scales. Above all, perhaps, what we require is educa-tion about and understanding of the need for reducing harmful human impacts.

Sustainability and Sustainable Development

To conclude this chapter, we return to the key concepts of sustainability and sustainable development that were briefly introduced on p. 113. Clearly, we are transforming the earth in ways that we do not intend. And, equally clearly, we need to manage the earth along appropriate pathways. Management requires us to understand what kind of earth we want, to reach consensus, and to find an appropriate balance between the values of economic development and conservation. Although these aims will be difficult to achieve, if only because people in different areas live in very different circumstances and have very different value systems, it is imperative that a balance be established: the relationship between humanity and the land is such that environment and economics must both be central concerns.

The term 'sustainability' was introduced in the late 1970s to refer to the idea that our current way of life, based on ever-increasing consumption of resources, could not continue indefinitely and that we would have to find a more sustainable way of life. It was difficult to argue against the desirability of sustainability, but there was considerable disagreement as to what changes were needed to achieve the desired state. The debate on this theme advanced significantly with the introduction of the concept of **sustainable development**, loosely defined as economic development without harm to the environment. This term was introduced by the influential Brundtland Report, published by the World Commission on Environment and Development in 1987 under the title *Our Common Future*.

The term was originally defined in *Our Common Future* as follows: 'Sustainable development is development that meets the needs of the present without compromising the ability of future generations to meet their own needs.' A more elaborate definition was put forward in 1989: 'Sustainability is the nascent doctrine that economic growth and development must take place, and be main-

CLAYOQUOT SOUND: THE CASE FOR SUSTAINABILITY

Clayoquot (pronounced Klak-wat) Sound, on the west coast of Vancouver Island in British Columbia, is one of the largest (about 3,500 km²/1,351 square miles) areas of coastal temperate rain forest in the world. It is a distinctive ecosystem, with the highest biomass (weight of organic matter to land area) of any forest type. It also contains at least 4,500 known plant and animal species, and possibly several thousand other insects and micro-organisms. Several of the species, such as the sea otter, are on the Canadian endangered species list. Aboriginal groups, including the Nuu-chah-nulth (meaning 'all along the mountains'), have lived in the area for more than 8,000 years without causing environmental degradation, but in recent decades the area has become attractive to lumber companies, and clear-cut logging is posing real threats to the ecosystem.

There are, then, several conflicting interests at stake. Conservationists, including groups such as the Sierra Club, argue for preservation of the ecosystem; Aboriginal people ask that their values and lifestyles, which are intimately related to the forest, be recognized and respected; and lumber companies seek to make profits in a province that has 20 per cent of its economy in the forestry sector. In 1984 the first open confrontation occurred when local Aboriginal and other residents set up a blockade to prevent logging. The provincial government established a Wilderness Committee in 1986, then three mediating forums in 1986, 1989, and 1992. Meanwhile, in 1988, a second confrontation began in another location.

Discussions at the various forums failed to arrive at an agreement satisfactory to all interest groups, and in 1993 the provincial government announced the Clayoquot Sound Land Use Decision, permitting logging in two-thirds of the area and protecting one-third. A massive civil disobedience movement followed, with about 900 people arrested by the end of 1993. In 1994 the forum that was set up in 1992 (the Commission on Resources and the Environment) proposed arrangements for the area that would result in the loss of about 900 logging jobs. This prompted a mass demonstration by loggers.

In 1995 the provincial government endorsed a report of the Clayoquot Sound Scientific Panel, another advisory forum that it had established following the 1993 protests; the report made a series of recommendations that included significant reductions in the rate of cutting but did not call for permanent protection. The recommendations represented a substantial change from previous reports, as they explicitly acknowledged the importance of maintaining the rain forest ecosystem. This change of attitude appears to be continuing. In 1996, at their annual meeting, members of the World Conservation Union, including the BC government, endorsed a proposal supporting

In the summer of 1993 more than 800 people were arrested for blockading a logging road into Clayoquot Sound. This non-violent civil disobedience campaign was designed to attract media coverage and pressure the provincial and federal governments into putting a stop to clearcut logging in the area (Philip Dearden).

the designation of Clayoquot Sound as a United Nations Biosphere Reserve. Further, in 1999 the provincial government reached agreements with MacMillan Bloedel concerning compensation for the company's loss of cutting rights following the creation of new parks; and, in a move welcomed by environmental groups, the company announced new forest management policies involving the phasing out of clear-cutting, increased conservation of old growth, and independent validation of their forest practices. In November 1999 MacMillan Bloedel was acquired by Weyerhaeuser to create the largest North American forest products company. Currently, the annual total logged ranges between zero and 100,000 cubic metres—a massive reduction from the late 1980s' high of almost 1 million cubic metres per year. Nevertheless, environmentalists have expressed concern over some of the current activities of International Forest Products Limited, a logging and manufacturing company.

This example highlights many important aspects of current human impacts on land. Above all, it suggests that governments find it hard to appreciate the arguments for a sustainable economy—not surprising, given that governments need political support from both industry and trade unions and also benefit financially from economic activities (in this case from the sale of logging licences). But it also indicates that public education campaigns and the intense public pressure they generate, combined with international market campaigns against large companies, can have an impact on government attitudes and policies.

tained over time, within the limits set by ecology in the broadest sense—by the interrelations of human beings and their works, the biosphere and the physical and chemical laws that govern it' (Ruckelshaus 1989:167). This second definition is reminiscent of the deep ecology perspective summarized in Table 4.1. The concept of sustainability can also be explained by reference to the systems concept we looked at earlier. The earth can be regarded as a closed system in that, although energy enters and leaves the system, matter only circulates within it. This type of system can reach a state of dynamic equilibrium—one that involves optimal energy flow and matter cycling in such a way that the system does not collapse. In a sense, sustainable development would represent a similar state of dynamic equilibrium.

Once the term was introduced, the idea of sustainable development became a key focus at the United Nations Conference on Environment and Development held in Rio de Janeiro in 1992. The essence of sustainable development is the attempt to blend into one process what are often seen as opposites, namely sustainability and development, and for some critics the phrase is an oxymoron—a contradiction in terms. Certainly it is true that in order to limit damage to environments, it might be necessary to limit growth. This contradiction is most easily resolved, in principle if not in practice, by suggesting that rich countries should strive for sustainability by limiting their own growth, while poor countries, which have not reached an adequate level of economic development and should have a chance to do so, should strive for sustainable development.

How do we move towards a sustainable world, a world where environmental changes are in accord with sound ecological principles? Any such move clearly requires a significant—and deliberate—shift towards a new attitude. Four principles are essential to that new attitude:

1. We need to recognize that humans are a part of nature. To destroy nature is to destroy ourselves.
2. We need to account for environmental costs in all our economic activities.
3. We need to understand that all humans deserve to achieve acceptable living standards. A world with poor people cannot be a peaceful world.
4. We need to be aware that even apparently small local impacts can have global consequences. This is one of the basic themes of ecology, and it becomes ever more obvious as globalization processes unfold.

There are, of course, vast differences between acknowledging the need for a new attitude, seeing it adopted globally, and, finally, putting it into practice.

Many strides have been made in the right direction. Clean air acts, environmental impact assessments, and departments of the environment are now standard in many countries. Although it would be an exaggeration to assert that solutions are in sight, or even that they are being sought in all cases (Box 4.10), environmental issues are increasingly recognized as relevant at all levels, from individuals to governments. But this is not equally the case for all countries. Thus it is the clear responsibility of the more developed world to demonstrate environmental concern by example:

> in creating the consciousness of advanced sustainability, we shall have to redefine our concepts of political and economic feasibility. These concepts are, after all, simply human constructs; they were different in the past, and they will surely change in the future. But the earth is real, and we are obliged by the fact of our utter dependence on it to listen more closely than we have to its messages (Ruckelshaus 1989:174).

CHAPTER FOUR SUMMARY

Use or abuse?

Today there is considerable evidence that we are damaging our home—the earth—and increasing recognition of how fragile that home is.

Scale

The value of a global perspective in any discussion of human use of the earth is clear: everything is related to everything else.

The increasing human impact

Two variables—sheer numbers of people and increasing use of technology and energy—explain why human activities today have such great impact on the earth.

Ecosystems

Combining systems logic and ecological principles produces the concept of the ecosystem: any self-sustaining collection of living organisms and their environment is an ecosystem. A whole series of ecosystems together make up the global ecosystem—the ecosphere or biosphere—which is the home of all life on earth. The ecosphere includes air, water, land, and all life.

Ecosystem simplification

When humans change an ecosystem, the result is usually a simplification. We cause changes largely because of our numbers, our technology, and our energy use. Today we number in excess of 6.2 billion, and our technological ability—that is, our ability to convert energy into forms useful to us—is constantly increasing. Examples of technological change include the deliberate use of fire; animal and plant domestication; and the use of fossil fuels associated with the industrial revolution.

As human culture (especially technology) changes, so does the value attributed to resources. At the same time, different interest groups evaluate resources according to different criteria and hence have differing views. Stock resources are finite in quantity, where-as renewable resources are relatively unlimited in quantity.

Our simplification of ecosystems is now sufficiently evident that a new environmental ethic is needed—one involving cooperation with nature, not domination over it.

Human impacts

The presence of humans on the earth changes the planet. Some of the changes brought about by humans are inevitable because they are necessary to human survival. Other changes, however, are the result of inappropriate actions motivated by human greed and misunderstanding. Over time, humans' ability to affect ecosystems at all levels has increased. Although it may be inappropriate to focus on individual parts of an ecosystem—which is by definition an inseparable whole—it is convenient to consider the impact of human activities on specific parts of ecosystems such as vegetation, animals, land and soil, water, and climate. In each case, we have caused many changes.

Vegetation change, for example, is closely related to the human use of fire, domestication of grazing animals, and deforestation. Two important issues today are tropical rain forest depletion and desertification; in both cases, human activity is at least partially explained by the problems experienced by poor people in less developed countries. Impacts on animal life include those associated with domestication and species extinction. Extensive changes to land surfaces result from a host of human activities such as resource extraction and the creation of urban industrial complexes. Soil is especially susceptible to abuses involving salinization, laterization, and erosion. Water, an essential ingredient for life, is typically used in ways that result in shortages and contamination. Even the oceans are being polluted—a problem that urgently requires international cooperation.

The ecosystem concept of interrelatedness is especially useful with respect to climate.

We are having a significant effect on the atmosphere by increasing the quantity of greenhouse gases and damaging the ozone layer; to fully appreciate the human impact in this area, however, we also need to understand more about climatic change that is not caused by humans.

The health of the planet

Although debates continue concerning the consequences of human activity for the health of the planet, most agree that human-induced ills are many and serious. Probably the best solution is prevention—an argument advanced by many environmentalist groups. One way to encourage prevention would be to offer rewards for responsible environmental behaviour.

Sustainable development

Sustainable development is development that serves our present needs, but does not compromise the ability of later generations to meet their needs.

LINKS TO OTHER CHAPTERS

- Globalization:
    Chapter 2 (concepts)
    Chapter 5 (population growth, fertility decline)
    Chapter 6 (refugees, disease, more and less developed worlds)
    Chapter 7 (cultural globalization)
    Chapter 8 (popular culture)
    Chapter 9 (political globalization)
    Chapter 10 (economic globalization)
    Chapter 11 (agriculture and the world economy)
    Chapter 12 (global cities)
    Chapter 13 (industrial restructuring)

- The interrelatedness of all things:
    All chapters.

- Relations between humans and land:
    Chapter 1 (Humboldt, Ritter, Vidal, and Sauer)
    Chapter 2 (environmental determinism; possibilism).

- Energy use and technology:
    Chapter 13 (especially the industrial revolution and current circumstances).

- More developed and less developed regions:
    Chapter 6.

- 'Agricultural revolution':
    Chapter 7 (beginnings of civilization)
    Chapter 11 (agriculture).

- Natural resources and human values:
    Chapter 2 (possibilism).

- Human impacts:
    Chapters 10, 11, 12, 13 (economic geography)
    Chapter 11 (agriculture)
    Chapter 13 (industrial activity).

- The catastrophist–cornucopian debate:
    Chapters 5, 6, 9.

FURTHER EXPLORATIONS

BROWN, L.R., et al., eds. 2003. *State of the World, 2003*. New York: W.W. Norton.
> An annual publication by the Worldwatch Institute on the progress being made—or not being made—towards a sustainable society.

CIPOLLA, C.M. 1974. *The Economic History of World Population*, 6th ed. Harmondsworth: Penguin.
> A survey that focuses on the importance of energy sources in understanding economic change.

CUTTER, S.L. 1994. 'Environmental Issues: Green Rage, Social Change and the New Environmentalism'. *Progress in Human Geography* 18:217–26.
> Insightful discussion focusing on the usefulness of a variety of new environmental philosophies.

FRIDAY, L.E., and R.A. LASKEY, eds. 1988. *The Fragile Environment*. Cambridge: Cambridge University Press.
> A series of essays by specialists in various disciplines addressing many of our basic themes, but always with an ecological focus.

GOODIN, R.E. 1992. *Green Political Theory*. Oxford: Polity.
> A novel and stimulating book arguing that environmental policy needs to be guided by an appreciation of the value that humans sense of themselves in relation to nature.

GOUDIE, A. 1993. *The Human Impact on the Natural Environment*, 4th ed. Oxford: Blackwell
An excellent general textbook by a well-known geographer.

———, ed. 1997. *The Human Impact Reader: Readings and Case Studies*. Oxford: Blackwell.
> A reader intended to be used with the above textbook; includes sections on geomorphological and surface impacts, soil impacts, water impacts, climatic and atmospheric impacts, and biological impacts, with each section comprising between five and ten case studies.

HUGHES, J.D. 2001. *An Environmental History of the World: Humankind's Changing Role in the Community of Life*. New York: Routledge.
> A history of the world organized around the concept of ecology that treats humans as part of the environment and assumes that population is out-stripping resources; includes a series of interesting and informative case studies.

JOHNSON, C. 1991. *The Green Dictionary: Key Words, Ideas and Relationships for the Future*. London: McDonald Optima.
> A useful source for definitions, and sometimes discussions, of a wide variety of terms relevant to human impacts on the environment.

JOHNSON, D.L., ed. 1977. 'The Human Face of Desertification'. *Economic Geography* 53:317–432.
> A theme issue that contains nineteen articles, many of which offer specific examples of desertification in a wide range of spatial, cultural, and technological contexts.

KEMP, D.D. 1998. *The Environment Dictionary*. New York: Routledge.
> A remarkably comprehensive dictionary, written by a geographer.

MACAULEY, D., ed. 1996. *Minding Nature: The Philosophers of Ecology*. New York: Guilford.
> One of the better sources to aid understanding of the often complex ideas that underlie our contemporary attitudes to human and land issues.

MANNION, A.M. 1991. *Global Environmental Change*. New York: Wiley.
> A substantial book that reviews environmental change over the last 3 million years.

MATTHEWS, J.A. et al., eds. 2003. *The Encyclopaedic Dictionary of Environmental Change*. London: Arnold.
> A comprehensive survey of both natural and human-induced environmental change; an invaluable source of facts and definitions.

MUNGALL, C., and D.J. MCLAREN, eds. 1990. *Planet under Stress*. Toronto: Oxford University Press.
> Produced for the Royal Society of Canada; an excellent up-to-date account of numerous issues; includes a full-colour pictorial introduction.

PAEHLKE, R., ed. 1995. *Conservationism and Environmentalism: An Encyclopedia*. New York: Garland Publishing.
> A comprehensive and informative encyclopedia that defines and discusses a wide range of relevant terms and issues.

PARK, C. 2001. *The Environment*, 2nd ed. New York: Routledge.

> A comprehensive text on the environment and human interactions with it; covers all the topics addressed in this chapter, and includes numerous detailed regional examples.

PHILLIPS, D.E., A. WILD, and D.S. JENKINSON. 1990. 'The Soil's Contribution to Global Warming'. *Geographical Magazine* 62, 4:36–8.

> A brief popular piece that succinctly summarizes the important links between soils and the greenhouse effect.

PORTER, G., and J.W. BROWN. 1991. *Global Environmental Politics*. Boulder: Westview.

> A good discussion of the links between environmental issues and other aspects of public policy; includes evaluations of how well national governments and international organizations are responding to environmental challenges.

SIMMONS, I.G. 2000. 'The Environment at 2000: "A Tremendous Number of Layers"'. *Geography* 85:97–105.

> A summary article outlining human links with nature at the beginning of a new century.

———. 1996. *Changing the Face of the Earth*, 2nd ed. Oxford: Blackwell.

> A detailed text that gives coherence to a vast array of factual information with an ecological viewpoint by focusing on energy flows through ecosystems.

SMIL, V. 1990. 'Planetary Warming: Realities and Responses'. *Population and Development Review* 16:1–29.

> A clear and objective account of global warming that concludes that warming may lead to desirable changes in economic and population policies.

———. 1993. *Global Ecology: Environmental Change and Social Flexibility*. New York: Routledge.

> A detailed, informative, thoughtful, and thought-provoking book that develops many of the ideas introduced in this chapter.

———. 2002. *The Earth's Biosphere: Evolution, Dynamics, and Change*. Cambridge: MIT Press.

> A readable, thoughtful overview of the biosphere, focusing on facts rather than political issues; emphasizes the gaps in our understanding that make it impossible to diagnose accurately the current state of the world.

ON THE WEB

http://www.arkive.org/

This website, launched in May 2003, is intended to serve as a 'digital ark' for the planet's flora and fauna, bringing together pictures, sounds, and information. Speaking at its launch, the distinguished British naturalist Sir David Attenborough noted that humanity is on the verge of causing a catastrophic extinction rivalling any in geological history. The site includes the only known film of the now-extinct Tasmanian tiger and the last known film of the Amazonian golden toad.

http://www.corpwatch.org/

A site established to be a self-proclaimed 'watchdog on the web' that provides critical evaluations of the environmental impacts that may result from the activities of some leading corporations.

http://earthwatch.unep.net/

Earthwatch is a broad UN initiative aimed at coordinating, harmonizing, and encourage environmental observation activities among all UN agencies; very comprehensive.

http://www.ec.gc.ca/
Environment Canada offers a comprehensive over-view of the Canadian environment and of environ-mental issues.

http://www.grida.no/
Probably the most comprehensive, informative, and reliable website on global environmental issues; published by the United Nations Environmental Programme (UNEP).

http://www.worldwatch.org/
The Worldwatch Institute offers a regularly updated survey of human impacts on environment.

http://www.sdri.ubc.ca/
The website of the Sustainable Development Research Initiative located at the University of British Columbia; looks at many issues concerning the environment and development.

CHAPTER **5**

The Human Population: History and Concepts

Together, Chapters 5 and 6 present an overview of the subdiscipline of population geography. This chapter focuses on the fundamentals of human population growth through time, at both global and regional scales. The two factors on which all changes in population size depend are fertility and mortality. This chapter provides a brief introduction to the various measures used to track fertility and mortality rates; some of the factors that influence those rates; and the temporal and spatial variations that occur in both cases. We also look at recent evidence indicating that fertility is declining not only in the more developed parts of the world but in the less developed parts as well—even though populations there are still growing much more rapidly than those in the more developed world.

This fundamental difference between the two worlds is a direct reflection of the many political, social, and other human inequalities that will be discussed in detail in Chapter 6. We conclude this chapter with a look at six theories that have been proposed to explain the history of population growth and to predict possible scenarios for the future.

Dhaka, Bangladesh, August 1994 (AP photo/FILE/John Moore).

At the time of writing in 2003, the world population is estimated at 6.2 billion. By the time you read these words, that figure will be much higher; the United Nations estimates 7.9 billion by the year 2025, 9.1 billion by 2050, and a stable population of about 10 billion by 2200. More significant even than this numerical increase, however, is the fact that the greatest increases are occurring in those parts of the world that are currently least capable of supporting increased numbers. Hence an ever-present theme in this chapter and the one following is the uncertainty that such increases imply for human well-being. A number of human geographers and other scholars have suggested links between population numbers and such problems as famine, disease, and, as discussed in the previous chapter, environmental deterioration. Others, however, are more optimistic about our ability to cope with increasing numbers, arguing that past increases in population have been accompanied by improvements in human well-being through technological change.

Much of this chapter draws on measures and procedures developed in **demography**: the science that studies the size and make-up of populations (according to such variables as age and sex), the processes that influence the composition of populations (notably fertility and mortality), and the links between populations and the larger human environments of which they are a part.

Fertility and Mortality

At the global level, all changes in population size can be understood by reference to two factors: fertility and mortality. Thus:

$$P_1 = P_0 + B - D$$

where

P_1 = population at time 1
P_0 = population at time 0 (before time 1)
B = number of births between times 0 and 1
D = number of deaths between times 0 and 1

Fertility and mortality rates vary significantly according to time and location, and both rates are affected by many different variables.

At the sub-global level a third factor becomes relevant: migration into and out of that particular subdivision of the world. Thus:

$$P_1 = P_0 + B - D + I - E$$

where

I = number of immigrants to area between times 0 and 1
E = number of emigrants from area between times 0 and 1

The effects of migration are discussed in Chapter 6.

Fertility Measures

The simplest and most common measure of **fertility** is the *crude birth rate* (CBR): the total number of live births in a given period (usually one year) for every 1,000 people already living. Thus:

$$CBR = \frac{\text{number of live births in one year}}{\text{mid-year total population}} \times 1,000$$

The CBR for the world in 2002 was 21. Historically, measures of CBR have typically ranged from a minimum of 15 (the theoretical minimum is, of course, 0) to a maximum of 55, which is a rough estimate of the biological maximum. Clearly, the CBR is a very useful statistic, but it may be somewhat misleading because births are related to the total population and not to that subset of the population that is able to conceive (it is called 'crude' for this reason). To more accurately reflect underlying fertility patterns by reference to the biological concept of female **fecundity**—the ability of a woman to conceive—two other, more sophisticated measures are used: general fertility rate and total fertility rate. Both relate births as directly as possible to that segment of the population responsible for them, excluding those unable to conceive, namely males, and those unlikely to conceive, namely very young girls and women over the age of 50.

First, the *general fertility rate* (GFR) refers to the actual number of live births per 1,000 women in the fecund age range: those years in which a woman has the ability to conceive, typically defined as ages 15 to 49 (sometimes 15 to 44). It is calculated as follows:

$$CBR = \frac{\text{number of live births in a one-year period}}{\text{mid-year number of females}} \times 1{,}000$$
$$\text{aged } 15\text{–}49 \text{ years}$$

Second, and more useful, the *total fertility rate* (TFR) is the average number of children a woman will have, assuming she has children at the prevailing age-specific rates, as she passes through the fecund years. This is an age-specific measure of fertility that is useful because child-bearing during the fecund years varies considerably with age. It is calculated as follows:

$$TFR = 5 \sum_{A=1}^{7} \frac{\text{number of births to women in age group A in a given period}}{\text{mid-year number of females in age group A}}$$

where A refers to the seven five-year age groups of 15–19, 20–4, 25–9, 30–4, 35–9, 40–4, and 45–9. The 5 preceding the summation sign is necessary because each age group covers five years. The world TFR for 2002 was 2.8, meaning that the average woman has 2.8 children during her fecund years.

Although there are various other measures of fertility, with different measures used in different circumstances and with different aims in mind, we will concentrate on two measures, the CBR and the TFR. The CBR is a factual measure reflecting what has actually happened in a given time period—*x* number of children per 1,000 members of the population were born. The TFR, on the other hand, reflects an assumption that as a woman reaches a particular age, she will have the same number of children as do the women of that age today; hence it is a predictive measure.

The impact of a particular CBR or TFR on population totals is related to mortality. Generally, a TFR of between 2.1 and 2.5 is considered replacement-level; that is, it maintains a stable population. The lower figure, 2.1, applies to areas with relatively low levels of mortality and the higher figure, 2.5, to areas with relatively high levels of mortality.

Factors Affecting Fertility

Fertility, the reproductive behaviour of a population, is affected by biological, economic, and cultural factors. It is not difficult to identify many of these factors; it is difficult, however, to assess their specific effects.

Biological factors

Age is the key biological factor in fecundity. Fecundity begins at about age 15 for females, reaches a maximum in the late 20s, and terminates in the late 40s. The pattern for males is less clear, with fecundity commencing at about age 15, peaking at about age 20, and then declining, but without a clear termination age. Some females and males are sterile—incapable of reproduction; data are not readily available, but it is probably not unrealistic to say that in 10 per cent of all married couples in the more developed world there is one sterile partner.

Reproductive behaviour is also affected by nutritional well-being; populations in ill health are very likely to have impaired fertility. Thus periods of famine reduce population growth by lessening fertility (as well as by increasing mortality). Box 5.1 reviews the situation in contemporary tropical Africa, a region with generally high birth rates but some particular areas of low fertility.

A third biological factor affecting fecundity is also related to diet. Women with low levels of body fat tend to be less fecund than others. Members of nomadic pastoral societies are particularly likely to have low levels of body fat because they eat a low-starch diet, and they are characterized by low fertility.

Economic factors

Another group of variables affecting fertility are principally economic. Indeed, until recently, fertility changes in developed countries were essentially one part of the process of economic development. With increasing industrialization, fertility declines. This economic argument suggests that traditional societies, in which the family is a total production and consumption unit, are strongly *pronatalist*: that is, they favour large families. By contrast, modern societies emphasize small families and individual independence; in economic terms, whereas children were once valued for their contributions to the household, today they represent an expense. The economic argument is that the decision to have children is essentially a cost-benefit decision. In the traditional (often extended-family) setting, children are valuable both as productive agents and as sources of security for their parents in old age—hence large families. Neither of these factors is important in modern societies.

This economic argument, which implies that any reductions in fertility are essentially caused by economic changes, is central to the demographic transition theory—one of the six explanations of population growth through time to be discussed later in this chapter.

Cultural factors

A host of complex and interrelated cultural factors also affect fertility; in fact it is now often suggested that the reasons behind current reductions in fertility are primarily cultural rather than economic; this suggestion is central to the fertility transition theory,

another of the six models of population growth through time to be discussed later.

Most cultural groups recognize *marriage* as the most appropriate setting for reproduction. There are several measures of **nuptiality**, the simplest of which is the nuptiality rate:

$$\text{nuptiality rate} = \frac{\text{number of marriages in one year}}{\text{mid-year total population}} \times 1,000$$

The age at which females marry is important because it may reduce the number of effective fecund years. A female marrying at age 25, for example, has 'lost' ten fecund years. Until recently, perhaps the most obvious instance of such loss was Ireland. In the 1940s, the average age at marriage for Irish

5.1

FERTILITY IN TROPICAL AFRICA

Fertility is typically high throughout tropical Africa—in several countries the total fertility rate (TFR) is between 6 and 7—but the national statistics mask some remarkable variations within countries. For some areas, TFRS are as low as 2 to 5. Two geographers working at Syracuse University identified a belt of low rates from southwestern Sudan, through the Central African Republic, and into the Democratic Republic of Congo (formerly Zaire). Other low-fertility areas include parts of Cameroon and Gabon, the Lake Victoria area, isolated locations in the savanna zone of west Africa, parts of Namibia and Botswana inhabited by San peoples, parts of Ethiopia inhabited by nomadic pastoralists, and parts of the east African coast (Doenges and Newman 1989). In each of these cases, there appears to be a close correlation between a low TFR and a particular ethnic group.

There are various explanations for the fact that fertility rates in some areas are impaired (abnormally low for the region). Four will be noted here. First, there are cultural variations in the length of time a baby is breast-fed; breast-feeding limits ovulation and extends the period of infertility following birth from about two months to eighteen months. The second cause of localized areas of low fertility is the impact of diseases such as gonorrhea and syphilis, which can cause sterility and also tend to reduce the likelihood of sexual intercourse. Poor nutrition is the third cause: fertility declines following famine and is consistently low in areas experiencing chronic undernutrition. The fourth cause relates to marriage. Although almost all women marry, and the age at first marriage is usually in the mid-teens, there are exceptions. For example,

Urban street scene, Cameroon (CIDA photo/Roger Lemoyne).

among the Rendille of northern Kenya, cultural practices result in one-third of the women not marrying until their mid-thirties; and among nomadic pastoralists, periods of prolonged spousal separation are not uncommon.

Doenges and Newman (1989:111) concluded their study of tropical African fertility as follows: 'If in the future impaired fertility ceases to be a significant concern for African populations, deliberate birth control will have a better chance of becoming an accepted social norm.' In this respect, impaired fertility acts very much like high infant mortality: both are important limitations on social well-being that require resolution before the state of regulated fertility can be reached.

women was 28 and for Irish men 33; in addition, high proportions of the population lost all their fecund years as a result of not marrying—as many as 32 per cent of women and 34 per cent of men. Late marriage and non-marriage are usually explained by reference to social organization and economic aspirations. Interestingly, since about 1971 the Irish people have tended to marry earlier and more universally. Other cultures actively encourage early marriage. Some Latin American countries, such as Panama, have legal marriage ages as low as 12 for females and 14 for males. Overall, delayed marriage and celibacy are uncommon throughout Asia, Africa, and Latin America. Some of the recent success in reducing fertility in China can be explained by the government requirement that marriage be delayed until age 25 for females and 28 for males.

Another cultural factor that affects fertility is *contraceptive use*. Attempts to reduce fertility within marriage have a long history. The civilizations of Egypt, Greece, and Rome all used contraceptive techniques. Today contraceptives are used around the world, but especially in developed countries. The less developed countries typically have lower rates of contraceptive use, although most evidence shows that the rates are increasing. Data for 2002 contraceptive use are shown in Table 5.1. Two statistics are included for each area: the percentage of married women practising contraception and the percentage of married women using modern contraceptive methods such as the pill, IUD, and sterilization. It is noteworthy that for many countries, and for many of the larger regions of the world, data are not available. The practice of contraception is closely related to government attitudes and religion. Today most of the world's people live in countries that actively encourage limits to fertility; yet as recently as 1960 only India and Pakistan had active programs to reduce fertility.

Another important cultural factor affecting fertility is *abortion*: the deliberate termination of an unwanted pregnancy. Abortions outnumber deaths today, and for every two births in the world, at least one pregnancy is deliberately terminated. Abortion is an even more complex moral question than contraception (Box 5.2). Although it is a long-standing practice, it is still subject to widespread condem-

nation, usually on religious and moral grounds. The reality, however, is that as many as 60 million abortions are performed each year, only about half of which are legal.

In many countries, governments actively try either to promote abortion or to discourage it, for their own pragmatic reasons. China has a liberal abortion law that reflects the need for population control; Sweden has a liberal abortion law on the grounds that abortion is a human right. In other countries, however, especially those where the dominant faith is either Catholicism or Islam, access to abortion is either denied or limited to instances in which the woman's life is threatened. In some other countries, such as Canada, abortion is one of the most hotly debated biological, ethical, social, and political issues.

Attitudes towards contraception and abortion often reflect religious beliefs, and it is frequently assumed that religion is a major influence on fertility in general. For example, accounts of fertility in Pakistan and India often note how much higher the fertility rate is in Pakistan (4.8 in 2002) than in India (3.2 in 2002) and attribute the disparity to religious differences between Hindus and Muslims. But this simple cause-and-effect relationship is clearly flawed, as a more detailed regional analysis shows. For example, the Indian state that has been most successful in reducing fertility—Kerala—is second only to Kashmir in the proportion of its residents who are Muslim. Indeed, throughout India,

Table 5.1 **CONTRACEPTIVE USE, BY REGION, 2002**

	% Married Women Using Contraception	
	All methods	**Modern methods**[1]
World	61	55
More developed world	68	58
Less developed world	60	54
Less developed world (excluding China)	48	41

Source: Population Reference Bureau, *2002 World Population Data Sheet* (Washington, DC: Population Reference Bureau, 2002).

Note 1: 'Modern' methods include the pill, IUD, condom, and sterilization.

A family-planning fieldworker talks about contraceptives with village women in Bangladesh (CIDA photo/Roland Perker).

5.2

PRIMITIVE ABORTION

It appears that abortion has long been practised to limit population numbers or, more specifically, family numbers. A passage from one of the first (1925) realist Canadian novels shocked not only the central character of the novel but also the Canadian reader of the time:

So, from many things that were said and from some that I saw I inferred that mother expected shortly to have another child and that that greatly worried her; but even more did it worry my father. He began to speak still more curtly to my mother; and he treated her as if she were at fault and had committed a crime. He prayed even more than before, both more frequently and longer. Gradually my mother began to get into a panic about her condition.

So, one day, taking me along, she went over to see that woman in the potato-patch once more.

A number of things were said back and forth which I remember with great distinctness but which have nothing to do with my story.

At last Mrs Campbell laughed out loud. Of course, she said, it's plain to be seen by now. It's a curse. But I can tell you I wouldn't be caught that way. Not I! I'm wise.

But what can you do? my mother exclaimed. He comes and begs and says that's what God made them male and female for. And if you want to hold your man . . .

I was only ten years old. But I tell you, I knew exactly what they were talking about. And right then I vowed I should never marry. I was furious at the woman and afraid of her.

You're innocent all right, she said at last contemptuously. I don't mean it that way, child. But when I'm just about as far gone as you are now, then I go and lift heavy things; or I take the plow and walk behind it for a day. In less than a week's time the child comes; and it's dead. In a day or two I go to work again. Just try it. It won't hurt you. Lots of women around here do the same.

So, when we came home, my mother took some heavy logs, dragged them to the saw-buck, and sawed them. I begged her not to do it; but even I could see that she was desperate.

Next day she was very sick. I was sent to the house of the German preacher in the village. And when I was allowed to come back, my mother was at work again on the land. She looked the picture of death; but she was cheerful. My father prayed more than ever.

Source: F.P. Grove, *Settlers of the Marsh* (Toronto: McClelland and Stewart [1925] 1966):108–9.

fertility rates among Muslims are usually closer to those of neighbouring Hindu groups than they are to those of Muslims in Pakistan. Similarly, the fertility rate of 3.3 in Muslim Bangladesh is very close to the Indian rate. So what does explain these differences in fertility rates? As Box 5.3 suggests, and as the account of the fertility transition later in this chapter makes clear, regional differences in fertility reflect differences not in religion but rather in the empowerment of women and hence in gender inequality.

Variations in Fertility

Spatial variations in fertility today correspond closely to spatial variations in levels of economic development, a fact that offers some support for the importance of the economic factors outlined earlier. It appears that, especially since the onset of the industrial revolution in the mid-eighteenth century, modernization and economic development have prompted lower levels of fertility. Thus in 2002, at the broad regional level, the more developed world had a CBR of 11 and a TFR of 1.6, while the less developed world had a CBR of 24 and a TFR of 3.1. More specifically, in the more developed world, Spain was among the group of countries with low fertility, with a CBR of 9 and a TFR of 1.2, while in the less developed world, Niger had a CBR of 55 and a TFR of 8.0. Figure 5.1 maps CBR by country. In general, variations of this type reflect economic conditions.

5.3

DECLINING FERTILITY IN THE LESS DEVELOPED WORLD

The less developed world is currently experiencing a decline in fertility that began in the 1970s and appears to have little to do with economic factors and everything to do with cultural factors (Robey, Rutstein, and Morris 1993). Birth rates are falling rapidly without any prior improvement in economic or living conditions.

The magnitude of this decline is evident in Thailand, which had a 1975 TFR of 4.6 and a 2002 TFR of 1.8; similar reductions have occurred in many other countries in Asia and Latin America, including Indonesia, Turkey, Colombia, and Morocco. The fact that several sub-Saharan African countries—Kenya, Botswana, Zimbabwe, and Nigeria—are also experiencing reduced fertility is especially significant, as this area has always appeared to be resistant to any such trend.

The explanation for these declines appears to be related to the availability and acceptance of new contraceptive technologies, the success of family-planning programs, and the educational power of the mass media. Traditional methods of contraception, such as periodic abstinence and withdrawal, are being replaced by modern methods. Female sterilization is the favoured method of contraception in Asia and Latin America, but is less popular in Africa and the Middle East.

Although this decline in fertility in the less developed world apparently began in the 1970s, the reasons behind it were not apparent until recently because data on such matters were not readily available. The information now available—derived largely from 44 surveys of more than 300,000 women carried out during the 1990s—are substantial and well documented. These fertility declines are in addition to the case of China, where government policies, often compulsory, have resulted in a below-replacement fertility level of 1.8 only 25 years after the initial recognition of the problem.

One detailed analysis of Botswana (VanderPost 1992) showed that, geographically, the fertility decline there is proceeding in an uneven manner; as is so often the case, national statistics disguise local details. Indeed, although in some areas of Botswana fertility is decreasing, in other areas it is actually increasing. The difference is essentially an urban/rural one: in 1988, the TFR in urban areas was 4.1 and in rural areas 5.4.

The importance of fertility declines in the less developed world is difficult to exaggerate, both practically and in terms of our understanding of fertility change. Because family planning is now practised throughout much of the less developed world (typically it appears without any substantial prior improvement in economic circumstances), many earlier accounts of anticipated population growth are being revised. The conventional wisdom in the 1970s was well expressed by Demeny (1974:105): countries in the less developed world 'will continue their rapid growth for the rest of the century. Control will eventually come through development or catastrophe.' It now appears that rapid growth is slowing down, but not for either of these two reasons: instead, cultural changes mean that people are becoming more willing to employ modern contraceptive methods in order to have smaller families.

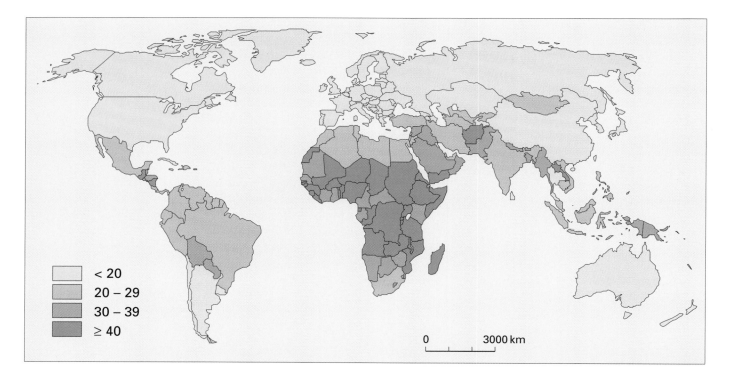

Figure 5.1 World distribution of crude birth rates, 2002.
Source: Population Reference Bureau, 2002 *World Population Data Sheet* (Washington, DC: Population Reference Bureau, 2002).

Legend:
< 20
20 – 29
30 – 39
≥ 40

0   3000 km

Current evidence, however, indicates that although fertility rates in the less developed world are still high, they are decreasing, and that, as already suggested, the reasons are more cultural than economic. As recently as 1990 the less developed world had a CBR of 31 and a TFR of 4—considerably higher than the 2002 rates of 24 and 3.1. There is also evidence of declining fertility in several countries in the more developed world that may be prompted by cultural factors (Box 5.4).

Fertility also varies significantly within any given country. There is, for example, usually a clear distinction between urban areas with relatively low fertility and rural areas with relatively high fertility. This distinction can be found in all countries, regardless of development levels. Similarly, fertility within a country is higher for those with low incomes and for those with limited education.

Mortality Measures

Like fertility, mortality may be measured in a variety of ways. The simplest, which is equivalent to the CBR, is the *crude death rate* (CDR): the total number of deaths in a given period (usually one year) for every 1,000 people living. Thus:

$$CDR = \frac{\text{number of deaths in one year}}{\text{mid-year total population}} \times 1,000$$

The CDR for the world in 2002 was 10. Measures of CDR have typically ranged from a minimum of 5 to a maximum of 50.

The CDR does not take into account the fact that the probability of dying is closely related to age (it is called 'crude' for this reason). Usually, death rates are highest for the very young and the very old, producing a characteristic J-shaped curve (Figure 5.2). Other mortality measures take into account the age structure of the population. The most useful of these for our purposes is the *infant mortality rate* (IMR): the number of deaths of infants under 1 year old per 1,000 live births in a given year. Thus:

$$IMR = \frac{\text{number of infant deaths under one year old}}{\text{number of births in that year}} \times 1,000$$

The world IMR for 2002 was 54, although figures for individual countries ranged from as low as 3.0 for Iceland to as high as 153 for Sierra Leone. The IMR is sensitive to cultural and economic conditions, declining with improved medical and health services and better nutrition; reductions in the IMR usually precede overall mortality decline.

Although not a mortality measure, another useful statistic that reflects mortality is *life expectancy* (LE), the average number of years to be lived from birth. World life expectancy in 2002 was 67, although for Japan and some European countries the LE was about 80, while for several African countries it was in the low 40s.

Factors Affecting Mortality

Humans are mortal. Whereas it is possible in principle for the human population to attain a CBR of 0 for an extended period, the same cannot be said for the CDR. Unlike fertility, which is affected by many biological, cultural, and economic variables, mortality is relatively easy to explain— despite the fact that the World Health Organization has recognized some 850 specific causes of death! As the CDR and LE figures cited above suggest, mortality generally reflects socio–economic status: high LE statistics are associated with

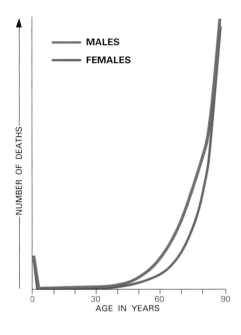

MALES
FEMALES

NUMBER OF DEATHS

AGE IN YEARS
0   30   60   90

Figure 5.2 Generalized J-shaped curve of death rates and age.

5.4

DECLINING FERTILITY IN THE MORE DEVELOPED WORLD

The lowest fertility rates in the world today are found in several European countries; the TFR is as low as 1.1 in Ukraine and the Czech Republic, and 1.2 in Latvia, Romania, and Slovakia. The extraordinarily low figures for much of Europe reflect both a desire to postpone starting families and a desire for smaller families. For many Europeans, the ideal family now includes only one child; a 1989 survey in France showed that 19 per cent favoured a one-child family, compared to only 3 per cent in 1979. Such preferences may be prompted by uncertain economic circumstances and the growth of employment opportunities for women (Hall 1993; King 1993).

The figure for Germany, 1.3, reflects an unprecedented trend in the former East Germany, which appears to have come as close to a temporary suspension of child-bearing as any large population in the human experience. Eastern Germans have virtually stopped having children. The explanation is not an increase in abortions, which have also fallen abruptly. One possible explanation is the trauma associated with the transition from communism to capitalism and, specifically, concern about employment opportunities for future generations in a region with high levels of unemployment. It may be appropriate, then, to interpret the low fertility rate in Germany—actually a form of demographic disorder—as one temporary outcome of the transition from the 'old' to the 'new' political order. This suggestion appears to have some merit, as fertility also declined between 1989 and 1993 by 20 per cent in

Poland, 25 per cent in Bulgaria, 30 per cent in Romania and Estonia, and 35 per cent in Russia (*The Economist* 1993:54).

Whether the low fertility evident in many European countries will continue at or close to the current low values remains to be seen. It may be that if economies boom and employment opportunities increase, then fertility will rise. As of 2002 there was little indication that such a rise was occurring, and the current rates typically imply net losses in population, since the replacement level in such countries is a TFR of about 2.1.

Although Canada does not belong to the group of lowest-fertility countries, both CBR and TFR statistics have been declining or stable in recent years. Many European countries have slightly lower CB and TF rates than Canada, but in the United States they are slightly higher (CBR is 15 and TFR is 2.1). Table 5.2 summarizes the current demographic situation in Canada.

Table 5.2 POPULATION DATA, CANADA, 2002

Total population	31.3 million
CBR	11
CDR	7
RNI	0.3%
IMR	5
TFR	1.5

Source: Population Reference Bureau, *2002 World Population Data Sheet* (Washington, DC: Population Reference Bureau, 2002).

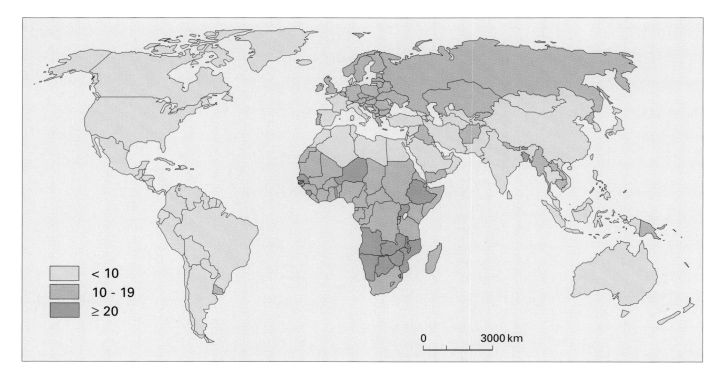

Figure 5.3 World distribution of crude death rates, 2002.
Source: Population Reference Bureau, *2002 World Population Data Sheet* (Washington, DC: Population Reference Bureau, 2002).

high-quality living and working conditions, good nutrition, good sanitation, and widely available medical services—and vice versa.

Variations in Mortality

The world pattern for CDR shows much less variation than does that for CBR (Figure 5.3). This is a reflection of the general availability of at least minimal health care facilities throughout the world. Figure 5.4 maps LE by country. In this case, major variations are evident: as already noted, LE figures are more sensitive to availability of food and health-care facilities than are death rates. The map of LE is a fairly good approximation of the health status of populations. Low LE statistics are found in tropical countries in Africa and in south and southeast Asia. High LE statistics are particularly common in Europe and North America.

Mortality measures also vary markedly within countries. In countries such as Canada, the United States, and Australia, certain groups have higher CDRs and IMRs and lower LEs than the population as a whole. These differences reflect what we might call the social inequality of death. Two examples suffice. In Australia in 1980, the IMR for Aboriginal people was 32.7, compared to 10.2 for non-Aboriginals. In the US, the national IMR in 1993 was 8.6, but Washington, DC, which has a very large

Black population, reported a figure of 21.1, while for both South Dakota (with a large Indian population) and Alabama (largely Black) the figure was 13.3. Black babies in the US are almost twice as likely to die in their first year as white babies—a state of affairs that has remained essentially unchanged for the last fifty years.

Box 5.5 provides two examples of LE and IMR data that clearly reflect environmental and health problems at least partly attributable to inefficient political systems.

Twenty-First-Century Nightmare?

A most disturbing trend is the increasing mortality caused by AIDS (Acquired Immune Deficiency Syndrome). AIDS itself was first identified in 1981, and HIV (the Human Immunodeficiency Virus) was recognized as its cause in 1984. The immune system of an individual infected by HIV weakens over time, leaving the body less and less able to combat infection. The majority of HIV-infected people develop AIDS within a few years. Today, in the more developed areas of the world, drugs make AIDS relatively manageable, but the worst-affected regions are in less developed areas—notably sub-Saharan Africa—where drugs are largely unavailable, and it is not unusual for a person developing AIDS to die just six months after infection.

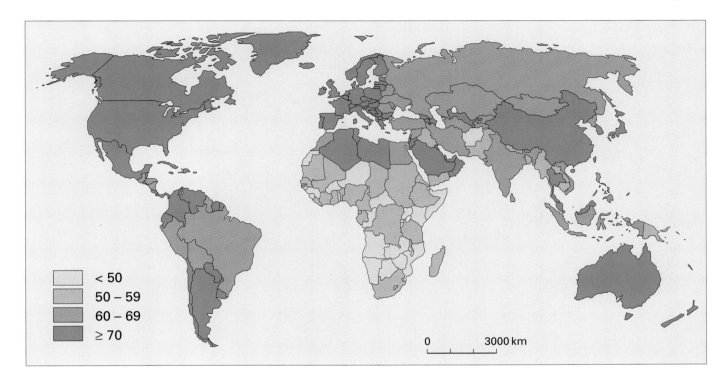

Figure 5.4 World distribution of life expectancy, 2002.

Source: Population Reference Bureau, *2002 World Population Data Sheet* (Washington, DC: Population Reference Bureau, 2002).

AIDS is the greatest threat to human health in the world. As of 2002, it was estimated that some 35 million people were infected with HIV, and approximately 20 million have died of AIDS since the early 1980s. Given these numbers, and the fact that AIDS has now spread to all parts of the world, the disease is appropriately described as a **pandemic**. Although it is especially prevalent in sub-Saharan Africa and among the poorest members of individual populations, AIDS honours no social or geographical boundaries. It is therefore very difficult to predict its spatial spread. Because AIDS is most often transmitted

5.5

PROBLEMS IN CENTRAL ASIA AND RUSSIA

Measures of infant mortality and life expectancy are especially useful indicators of quality of life. In both cases, statistics reflect living conditions with respect to the quality of the diet available, the health care system, public sanitation, and disease control. Two examples that indicate serious problems are outlined here, both in the former USSR.

In one area on the shores of the Aral Sea in central Asia, the IMR is estimated to be as high as 111 (see Box 4.6). This high rate reflects a number of basic health problems—a contaminated water supply, an abominable sewage system, and unsanitary hospitals. These problems are exacerbated in the summer as a consequence of increased temperatures and related disease problems. These basic health problems, in turn, reflect a host of social problems resulting from years of neglect by political leaders.

A remarkable example of an unexpected decline in LE is that of Russia. In 1987, the LE was 67 for Russian males and 74 for Russian females, but since then, there have been significant decreases. The 2002 data show that LE has fallen to 59 for males and to 72 for females. These figures are below those of all major industrialized countries; comparable Canadian data for 2002 are 74 for males and 80 for females. The explanation appears to involve a combination of issues. The transition from communism to capitalism may have created a new class of wealthy entrepreneurs, but it has also resulted in economic problems for other Russians. For many, diet is worsening because of high prices for meats, fruits, and vegetables; the industrial policies of the previous communist regime, which led to pollution of the air and drinking water, contributed to the high IMR noted earlier and are undoubtedly affecting the health of many Russians; there is an epidemic of alcohol abuse; and the high rate of abortions may have left many females unable to bear children.

Much of the effort to combat HIV/AIDS focuses on education. Here are two posters from Africa. The one at the right, from Namibia, was produced by the local Catholic church and highlights the links between AIDS and other illnesses (Charlotte Thege/DAS FOTOARCHIV). The poster above, from Lesotho, stresses the need for people to behave responsibly and inform themselves (Friedrich Stark/DAS FOTOARCHIV).

sexually, the most vulnerable population is the 15-to-49 age group—the very group that ought to be most highly productive, and that is most likely to have dependents, both young and old.

Sub-Saharan Africa

The area of the world most devastated by AIDS is sub-Saharan Africa. In 1991 a UNICEF-sponsored study forecast an AIDS crisis in several African countries, predicting that 5.5 million children on the continent would lose their mothers to the disease in the 1990s. By the end of the 1990s it was clear that this projection was a tragic underestimate; in fact, approximately 10 million African children had lost their mothers to AIDS since 1991. A 2002 UN prediction paints an even grimmer picture, with 40 million AIDS orphans in sub-Saharan Africa by 2010.

Since the early 1980s, some national death rates have doubled; rates of infant mortality have increased sharply; and life expectancy has been reduced by as much as 23 years. Table 5.3 provides data on the ten countries with the most adults living with HIV/ AIDS. These rates are devastatingly high, and a 2002 UN prediction anticipated that a further 68

million Africans would die of AIDS by about 2020. Comparisons with more developed parts of the world are especially telling. In Malawi—one of the poorest of African countries, with a population of less than 11 million—the number of people infected with HIV is about the same as in the United States, a rich country with a population of 287 million. In sub-Saharan Africa, HIV/AIDS is most common in two age groups: infants and adults aged between 20 and 40 years. Women are at greater risk of contracting AIDS than men, by a ratio of about 1.5:1 (Daniel, 2000:47), and often infect their infants, either before birth or after, through breastfeeding. No amount of data can do justice to the human suffering caused by AIDS. Nevertheless, a particularly poignant estimate is that a 15-year-old boy in Botswana today has an 80 per cent chance of dying of AIDS. Writing about one clinic in Malawi, Toolis (2000:29) stated:

If you have no money, then there is no choice; you must queue in the morning heat of the Boma's outpatients clinic and hope to be admitted. Demand outstrips supply. There are 130 beds, and every morning another 250 would-be patients appear at the gates. All but the sickest are turned away, as well as the medically hopeless, who are sent back home to die. There is not enough staff, not enough drugs, not enough needles, not enough anything.

It seems clear that poverty is the biggest single factor contributing to the spread of AIDS. Poverty means that infections remain untreated, greatly increasing the risk of further transmission. Poverty also keeps children out of school, increasing the likelihood that the sale of sex will be a primary means of supporting themselves. But of course AIDS only serves to exacerbate poverty; a 2002 estimate suggests that by 2010 the economic output of Botswana will decline by 32 per cent because of it. Yet in many countries the disease is still seen as so shameful that it cannot even be named; in Mozambique, for example, death certificates for AIDS patients often bear the words 'cause unknown'. As a result of these attitudes, the frank discussion needed to raise public awareness is still lacking in many cases.

Russia, China, and India

Most accounts of AIDS today focus on sub-Saharan Africa, but the disease may soon have major impacts elsewhere. According to Eberstadt (2002:24), 'it is quite possible that the center of the global HIV/AIDS crisis, in terms of absolute numbers, will shift from Africa to Eurasia over the coming generation.' A 2002 report by the National Intelligence Council in the United States suggested that Russia, India, and China (along with Nigeria and Ethiopia) were especially likely to see substantial increases in the numbers of persons infected.

In all three cases the data are unreliable. Nevertheless, it appears that the infection rate in Russia may be three times that in the United States, in part because of the social and economic changes that have been underway since the fall of communism. Together, increasing poverty, increasing social freedom, and (especially) increasing illegal drug use are making Russians particularly vulnerable to infection.

In the case of China the data are even less reliable, but a 2002 UN report suggested that the number of people with HIV was between 800,000 and 1.5 million, and some other estimates are much higher. In a country with a total population of more than a billion, these numbers represent a rate of infection no higher than that in the United States, but they are still disturbing, as is the speed with which the disease appears to be spreading. The Chinese epidemic seems to be transmitted primarily through heterosexual sex, although one secondary cause is the sale of blood by impoverished peasants to illegal brokers (*The Economist* 2002:75).

Finally, India may have as many as 8 million people infected with HIV. Most of the current epidemic there is concentrated in major cities and is associated with heterosexual behaviour. Because there is little open discussion of AIDS, the risks of infection are not widely recognized.

In all three countries it is likely that the disease will soon spread into the general population, largely because governments are unwilling to take the necessary public health measures. Eberstadt (2002:34-38) painted a dire picture of the spread and growth of AIDS, with an 'intermediate epidemic' in all three countries resulting in 259 million new HIV cases and 155 million deaths by 2025. Such an epidemic would represent a humanitarian crisis of staggering proportions, with significant impacts both on population growth and on national economies.

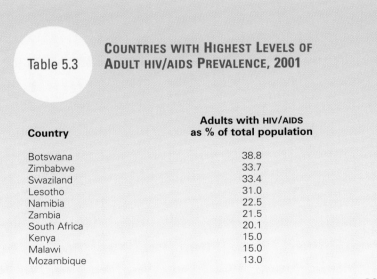

Table 5.3	COUNTRIES WITH HIGHEST LEVELS OF ADULT HIV/AIDS PREVALENCE, 2001	
Country		**Adults with HIV/AIDS as % of total population**
Botswana		38.8
Zimbabwe		33.7
Swaziland		33.4
Lesotho		31.0
Namibia		22.5
Zambia		21.5
South Africa		20.1
Kenya		15.0
Malawi		15.0
Mozambique		13.0

Source: Population Reference Bureau, *2002 World Population Data Sheet* (Washington, DC: Population Reference Bureau, 2002).

On the eve of world AIDS Day (1 December) 2002, this student at a government high school in Gauhati, India, pinned a red ribbon—the symbol of AIDS awareness—on a classmate. Along with education, peer support is an important part of the fight against AIDS
(AP photo/Anupam Nath.)

The Future of AIDS

So many factors can influence the course of the battle against AIDS that it is impossible to predict the future with any certainty. However, some positive signs are emerging.

Overall, the fact that impacts are being kept at a relatively low level in more developed areas suggests that negative consequences can be contained. Some of the differences between the United States and Canada may be instructive. The infection rate in Canada is half that in the US—a significant difference that is attributed to Canada's having more open sex education programs, easier access to condoms, better and more widely available needle exchange programs, and free treatment provided by its universal health care system. Elsewhere, countries such as Thailand, the Philippines, and Brazil have been fighting the disease with much success since the 1990s through prevention programs designed to educate people about such matters as condom use.

Perhaps most notably, there are positive signs in parts of sub-Saharan Africa. Attitudes towards AIDS are changing and governments are actively supporting safer sex practices. The greatest success story appears to be Uganda, where the adult infection rate has dropped from 30 per cent in 1992 to about 5 per cent in 2002. Perhaps surprisingly, this reduction is attributed not to increased use of condoms, but to a government campaign stressing abstinence and fidelity. In many other countries, including South Africa, people are increasingly well educated concerning the causes of AIDS, and there are clear signs that attitudes towards sexual behaviour are changing. In Botswana, some mining companies that employ large numbers of adult males are providing hospital facilities for those already infected and working to prevent AIDS not only through education but by making free condoms readily available.

Another important factor in the worldwide fight against AIDS is the response of the more developed world, and here too the signs are increasingly positive. A 2001 UN summit agreed to set up a global fund to fight AIDS, along with tuberculosis and malaria. Despite initial funding problems, there is now reason to believe that the errors of the 1990s, when the more developed world quite literally looked away, are now being corrected. In 2003 the United States committed up to $15 billion over five years to combat AIDS in Africa and the Caribbean. It is far too early to be confident of success, but changing attitudes in both less and more developed countries may well help to limit the spread of AIDS in the coming years.

Natural Increase

The *rate of natural increase* (RNI) is determined by subtracting the CDR from the CBR; thus it measures the rate (usually annual) of population growth. In 2002 the world CDR was 9 and the CBR was 21, producing an RNI of 13 per 1,000; this figure is typically expressed as a percentage of total population—hence 1.3 per cent. This percentage has been relatively constant in recent years. Because RNI data take into account only mortality and fertility, not migration, they generally do not reflect the true population growth of any area smaller than the earth. Not surprisingly, the highest RNI figures are for those countries in

Table 5.4	COUNTRIES WITH HIGHEST RATES OF NATURAL INCREASE (3.0 AND ABOVE), 2002

Country	Natural increase (%)	Population (millions)
Niger	3.5	11.6
Chad	3.3	9.0
Yemen	3.3	18.6
DR of Congo	3.1	55.2
Liberia	3.1	3.3
Burkina Faso	3.0	12.6
Congo	3.0	3.2
Eritrea	3.0	4.5
Madagascar	3.0	16.9
Mali	3.0	11.3
Uganda	3.0	24.7

Source: Population Reference Bureau, *2002 World Population Data Sheet* (Washington, DC: Population Reference Bureau, 2002).

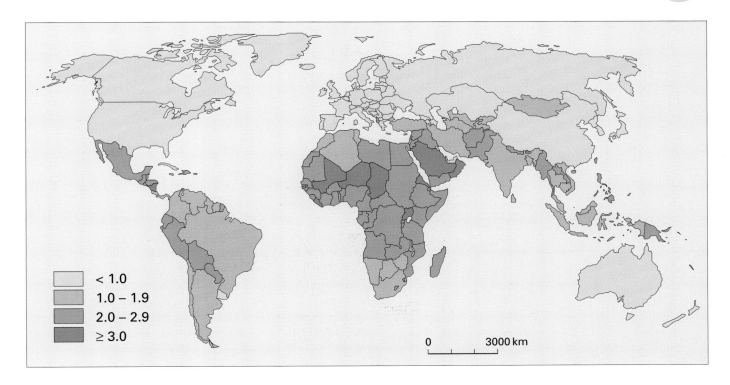

the less developed world. Tables 5.4 and 5.5 show RNI data for two groups of countries, while Figure 5.5 maps RNI data for individual countries.

Together, these three sets of information provide a useful commentary on the present world population. Low RNI statistics are essentially restricted to Europe, especially some former communist countries and several bordering on the Mediterranean. High RNI statistics are concentrated in the western Asian Islamic world and tropical Africa (countries that have temporarily inflated RNI statistics because of refugee in-movement are excluded from Table 5.4).

Although, as suggested in Boxes 5.3 and 5.4, fertility is declining in parts of both the less and the more developed worlds, the number of females of reproductive age continues to rise. Thus the total world population continues to grow rapidly.

In 2002 the world's population increased by about 79 million people. The size of the annual increase is declining each year—or, to put it another way, *world population is still increasing, but at a decreasing rate.* The reduction in annual increases began about 1990, when the annual increase reached 87 million—and the reason behind it is that the world TFR is declining. This decrease in the size of the annual increase is very significant.

Regional Variations

Of course, to use world numbers in this way is to ignore important regional differences. Even a cursory glance at Figure 5.5 shows that much of the current growth is concentrated in Africa, where TFR rates are still relatively high. It is also important to recognize that large populations grow faster than small ones even when they have lower fertility rates—simply because the base population is so much larger.

Figure 5.5 World distribution of rates of natural increase, 2002. Source: Population Reference Bureau, *2002 World Population Data Sheet* (Washington, DC: Population Reference Bureau, 2002).

Table 5.5	**COUNTRIES WITH LOWEST RATES OF NATURAL INCREASE (LESS THAN 0.0), 2002**	
Country	**Natural increase (%)**	**Population (millions)**
Ukraine	−0.8	48.2
Russia	−0.7	143.5
Latvia	−0.6	2.3
Belarus	−0.5	9.9
Bulgaria	−0.5	7.8
Estonia	−0.4	1.4
Hungary	−0.4	10.1
Lithuania	−0.3	3.5
Croatia	−0.2	4.3
Czech Republic	−0.2	10.3
Romania	−0.2	22.4
Germany	−0.1	82.4
Moldova	−0.1	4.3

Source: Population Reference Bureau, *2002 World Population Data Sheet* (Washington, DC: Population Reference Bureau, 2002).

At present, China and India are home to almost 38 per cent of the world's population; hence even a small increase in fertility in one of these countries will mean a significant increase in the total world population. (Such a development is more likely in India, which appears to have ended a period of fertility decline and is again growing rapidly.)

Table 5.6, showing projected data for selected world regions/countries, highlights the fact that by far the greatest portion of the growth predicted between now and 2050 will occur in African countries and India. Indeed, of the anticipated world growth of 2889 million, Africa will contribute 1005 million and India 570 million, together accounting for 55 per cent of the world's total growth. Most of the rest of the growth will be in other high-population Asian countries. More generally, almost 99 per cent of the growth projected by 2050 will take place in the less developed world. These regional projections also have implications for the larger arena of political power and future developments in international affairs. For example, Africa has less than 14 per cent of world population today, but will have more than 20 per cent in 2050. Conversely, Europe has almost 12 per cent today, but will have only 7 per cent in 2050.

Doubling time

Related to the rate of natural increase is the useful concept of doubling time: the number of years needed to double the size of a population, assuming a constant RNI. For example, at the current 1.3 per cent growth rate, the world population will double in approximately 54 years. Relatively minor variations in RNI can significantly affect the doubling time. An RNI of 0 (where CDR and CBR are equal) results in a stable population with an infinite doubling time; on the other hand, an RNI of 3.5 per cent produces a doubling time of a mere 20 years.

Government Policies

So far we have paid only incidental attention to governments' impact on population. Most governments attempt to influence various aspects of population, such as total numbers and spatial distribution. Governments can also attempt, directly or indirectly, to control deaths, births, and migrations.

All death-control policies have the same objective: to reduce mortality. Such policies, usually adopted for both economic and humanitarian reasons, include measures to provide medical care and safe working conditions. Despite the near universality of such policies, many governments actively raise mortality levels at specific times—for instance, in times of war. Further, many governments do not ensure that all members of their population have equal access to the same quality of health care. This is true throughout the world, regardless of levels of economic development.

Unlike death-control policies, those related to birth control have varying objectives. Many governments choose not to establish any formal policies, either because of indifference to the issue or because public opinion is divided.

Other governments are actively *pronatalist*. Such governments are typically found in countries dominated by a Catholic or Islamic theology (e.g., France, Iran), in countries where the politically dominant ethnic group is in danger of being numerically overtaken by an ethnic minority (e.g., Israel), and in countries where a larger population is perceived as necessary for economic or strategic reasons (see Box 5.6).

Two countries in Asia, both of which belong in the Asian group of newly industrialized countries (see Chapter 13), began to actively encourage fertility increases in the 1980s (Dwyer 1987). Both Singapore and

Table 5.6	PROJECTED POPULATION GROWTH, 2002–2050		
Region or Country	Population, 2002 (millions)	Projected population, 2050 (millions)	Amount of projected increase (millions)
World	6215	9104	2889
Africa	840	1845	1005
Europe	728	651	−77
US and Canada	319	450	131
China	1281	1394	113
India	1050	1628	578

Source: Population Reference Bureau, *2002 World Population Data Sheet* (Washington, DC: Population Reference Bureau, 2002).

Malaysia had succeeded in lowering their TFRs by following an antinatalist line; but in 1984 both governments decided that new policies were required to reverse the trend and increase fertility again. In Singapore this reversal was related to the economic difficulties perceived to be the result of reducing the TFR to 1.6 by 1984; there was a perceived need to provide a larger market for domestic industrial production and a larger workforce. The Malaysian case was similar in that the motivation was economic, although the TFR remained above replacement level in 1984 at 3.9. The extent to which such pronatalist policies are successful is debatable; the 2002 TFR for Singapore was only 1.4 and for Malaysia 3.2 (both decreases since 1984). More recently, Japan has introduced financial incentives with the aim of encouraging a baby boom. The concern in this case is explicitly economic, as it is feared that an aging population supported by fewer working people would keep the country in a state of permanent recession. In addition, day-care facilities are being expanded and employers are making their companies more family-friendly.

In addition to such direct attempts to increase fertility, many countries indirectly support increased fertility through 'baby bonuses', establishment of day care centres, and generally widespread high-quality housing; Canada is an example. It seems likely that more and more countries will adopt pronatalist policies if their fertility declines to the point where not only the economic but the political and cultural consequences come to be seen as unacceptable—for instance, if it is feared that a smaller population will mean a loss of national identity.

Today, however, the most common policies relating to births are still *antinatalist*. Since about 1960, many of the less developed countries have initiated policies designed to reduce fertility. The argument behind such policies is that overpopulation is a real danger and that **carrying capacity** (the maximum population that can be supported by a given set of resources) has been (or will soon be) exceeded. Regardless of whether this threat is real, there is a common assumption that certain areas are too densely populated and that the best solution is reduced fertility.

5.6

FERTILITY IN ROMANIA, 1966–1989

In the October 1986 issue of the German magazine *Der Spiegel*, the president of Romania was quoted as saying that 'The fetus is the socialist property of the whole society. Giving birth is a patriotic duty, determining the fate of our country. Those who refuse to have children are deserters, escaping the laws of national continuity.'

Alarmed by a birth rate of 16 in 1966, the communist government in Romania ended all legal access to abortion and set a goal of increasing the national population to between 24 and 25 million by 1980 (a 30 per cent increase). Contraceptives were banned, and all Romanian employed women under 45 had to take a monthly gynecological examination. Unmarried people over 15 and married couples without children or a valid medical reason were assessed an additional 30 per cent income tax. The total fertility rate increased from 1.9 in 1966 to 3.6 in 1968, but since that time the rate has steadily dropped, and in 2000 it stood at 1.3. There is reason to believe that many women had illegal abortions despite the establishment of a special unit within the State Security Police whose job it was to combat abortion. Interestingly, all the strategies devised to increase the fertility rate were negative: no positive inducements, such as maternity leave or family income supplements, were offered.

The government's principal reasons for attempting to increase fertility were probably to strengthen national security (more people equals more strength) and improve productivity (by alleviating labour shortages).

In general, it seems reasonable to say that Romania before 1989 had the most stringent fertility policy in the world. Among other previously communist countries, the former USSR encouraged higher fertility, but at the same time probably had the highest national legal abortion rate in the world (more than one abortion for every birth).

Directly or indirectly, most states have an influence on fertility. What types of policies are evident in Canada? Does the Canadian government encourage or discourage births? What role do churches play?

As we noted earlier, birth rates are indeed falling in many of the less developed countries as a result of changes in attitudes, government programs, and such general social advances as improvements in literacy and rural development.

India was the first country to intervene actively to reduce fertility. Beginning in 1952, the Indian government introduced a series of programs designed to encourage contraceptive use and sterilization. The first program offered financial incentives; others have relied

5.7

POPULATION IN CHINA

Data from the 1990 census and more recent estimates provide a relatively sound factual description of the Chinese demographic situation (Jowett 1993). As of 2002, China continued to enjoy a declining CBR, one that is low by Asian standards (13), a CDR that is among the lowest in the world (6), an RNI of 0.7, and an LE that is close to the average for the more developed world (71).

A useful way to depict the recent history of Chinese population growth is with a population pyramid (Figure 5.6). The indentations in the pyramid show that there have been several changes in fertility and mortality over the past forty years. The 1960s were a period of generally high fertility and low mortality, reflected by the bulge for the ten-year cohort aged 18–27 in 1990; these were people born in the high-fertility period, 1963–72. In 1964 more than 40 per cent of the population were under age 15. The high fertility of 1972 was followed by dramatic decreases, beginning in 1973, that resulted in a narrowing of the pyramid, and the 1990 percentage under 15 was reduced to 28.

The broadening at the base of the pyramid reflects the fact that fertility is highest for women in their late twenties; if the number of births to married women of this age remains constant, the CBR will remain high. As Jowett (1993:405) noted, 'China is currently faced with the "echo" effect from demographic developments that occurred 20–25 years ago. Thus the large-small-large-small cohorts of population aged eight, six, three and zero are a condensed form of the fluctuations from the previous generation aged 32, 29, 27 and 23.'

The 1990 census also provides details of spatial variations in the demographics of the Chinese population that are masked by the national data. For example, the western and southwestern interior includes ethnic minorities exempt from the one-child policy, and the resultant high fertility means that their populations are relatively youthful. From 1964 to 1990, the majority Han group (92 per cent of the population) increased by 60 per cent, while the ethnic minorities increased by 128 per cent. These contrasting figures are explained by the differing levels of fertility and by the fact that some Han have chosen to be reclassified as members of one of the minority groups—probably in order to gain exemption from the one-child policy.

The census also indicates a strong negative correlation between fertility on the one hand and literacy, income, industrialization, and urbanization on the other. For example, fertility varies substantially between urban areas, where the TFR is 1.7,

comparable to many European countries, and rural areas, where the TFR is 2.7. Overall, the information provided by the 1990 census reflects various government policies and priorities that typically favour urban rather than rural areas, the coastal rather than the interior areas, men rather than women, and the majority Han rather than the various ethnic minorities.

The most recent data for China highlight two developments. First, it seems that if current fertility trends continue, China's population will begin to decline about the year 2042. This astounding prospect prompted authorities to informally relax the one-child policy, and it is now common, especially in rural China, for couples to have two children. The policy was formally abandoned in 2001, when urban couples were permitted a second child. The second significant development is a marked disparity in the sex ratio of children born in recent years. The natural ratio of males to females among newborns is 105 to 107, but recent Chinese data show a ratio of 117 males to 100 females. These numbers suggest that increasing numbers of female fetuses have been aborted: apparently parents prefer sons because a daughter is responsible for her birth family only until she marries, whereas a son remains responsible after his marriage. The only other country with a similar pattern of female abortion is India, where many parents still have to offer dowries in order to find husbands for their daughters.

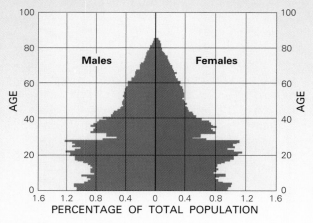

Figure 5.6 Age and sex structure in China, 1990.
Source: J. Jowett, 'China's Population: 1,133,709,738 and Still Counting', *Geography* 78 (1993):405.

on coercion, others still on education. Fertility rates have declined, but with a 2002 RNI of 1.7 per cent, it is clear that the various programs have not achieved the desired result.

China, on the other hand, has made great strides towards reducing fertility. The 2002 RNI was 0.7 per cent and the doubling time 100 years. Much of this success can be attributed to the programs introduced, mostly in the late 1970s, by China's communist government. Families were restricted to having one child, and marriage was prohibited until the age of 28 for men and 25 for women. Contraception, abortion, and sterilization were free. There were financial incentives for families with only one child and penalties for those with more than one. As noted in Box 5.7, this policy was terminated in 2002.

It is interesting, if a little perplexing, to note that China's strict antinatalist policy did not reflect the conventional Marxist position. According to classical Marxism, human problems such as apparent limits to resources and apparent overpopulation are actually caused by inappropriate modes of production and social organization; therefore the solution to population issues is not to reduce fertility, but to correct the economic and social conditions at the root of those problems. This does not mean that Marxism is pronatalist—simply that it sees social change as the first priority; fertility is only secondary. This view does not preclude the possibility that an antinatalist policy may be a step in the right direction.

Governments may also attempt to influence migration, especially between countries. Most countries formally restrict immigration, and until about 1989 some countries, especially communist countries in eastern Europe, limited emigration.

Age Structure

Fertility and mortality vary significantly with age. Inevitably, then, the growth of a population is affected by the age composition of the population. Age composition is dynamic. As we have seen in Box 5.7, it is usual to represent age and sex compositions with a **population pyramid**. Three general patterns can be suggested (Figure 5.7). First, if a population is rapidly expanding, a high proportion of the total population will be in the younger age groups. Because fertility is high, each successive age group is larger than the one before it. Second,

Virtually all the children in this photo of a school playground in Shinian, China, in June 2001, are boys. A few months later, census figures were reported to show that the number of girls born in China the previous year was nearly 900,000 short of the figure that would be expected from natural sex ratios among newborns (AP photo/Martin Fackler).

if a population is relatively stable, then each age group, barring the older groups that are losing numbers, is similarly sized. Third, if a population is declining, then the younger groups will be smaller than the older groups.

The three generalizations in Figure 5.7 are useful, but specific age and sex composition pyramids are more revealing. Usually, population pyramids distinguish males and females and divide them into five-year categories. Each bar in the pyramid indicates the percentage of the total population that a particular group (such as 30- to 34-year-old females) makes up. Figure 5.8 presents the population of Brazil for the period 1950 to 1987 in a series of three pyramids.

Global Aging

The year 2000 was a watershed: the first year in which people under 14 were outnumbered by people over 60. The process by which older individuals come to make up a proportionally larger share of the total popu-

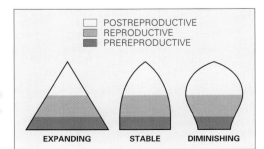

EXPANDING   STABLE   DIMINISHING

POSTREPRODUCTIVE
REPRODUCTIVE
PREREPRODUCTIVE

Figure 5.7 Age structure of populations. Expanding populations have a high percentage in the pre-reproductive age group. Stable populations have relatively equal pre-reproductive and reproductive age groups. Diminishing populations have a low percentage in the pre-reproductive age group.

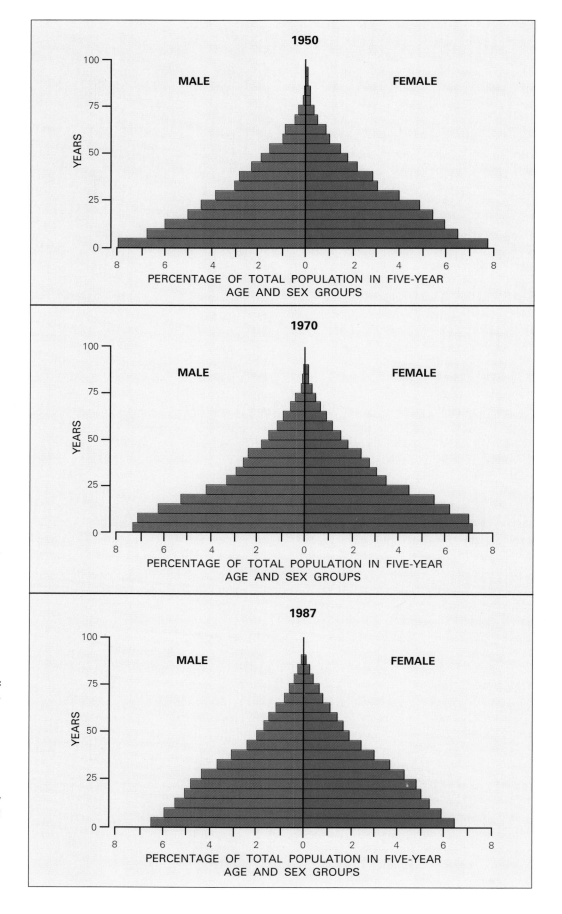

Figure 5.8 Age–sex structure in Brazil: 1950, 1970, 1987. Age–sex pyramids show the changes in Brazil's population over a 37-year period. Wide bases and gently sloping sides indicate high rates of both births and deaths. In the first pyramid much of the population was under 20, ensuring that births will continue to be high in the future. Later pyramids show slight reductions in numbers of births, and decreases in deaths: the first stage in the demographic transition. Throughout this period, the population increases rapidly because birth rates are high and death rates are dropping. The 1987 pyramid indicates that Brazil's population growth has not yet reached its maximum acceleration—a fact that does not bode well for a nation already experiencing difficulties supporting its current population.
Source: United Nations Demographic Yearbooks, various years.

lation is known as population aging, and it is occurring in most countries around the world, albeit at different rates.

The data are compelling. In 1900, less than 1 per cent of the world's population was over age 65; by 2002 this figure had increased to about 7 per cent, and the estimate for 2050 is 20 per cent. Furthermore, global aging is now occurring at an increasing rate: the older population is growing faster than the total population in most parts of the world, and the difference between the two growth rates is increasing. Table 5.7 provides data on the global aging trend, while Figure 5.9 shows the expected impact of aging on the Canadian population.

Causes of population aging

Population aging is the result of ongoing changes in two areas: fertility and mortality (United Nations, 2002).

1. Fertility is declining, both in more and in less developed countries. This is the primary reason for population aging: the fewer the young people in a population, the larger the proportion of elderly people. Globally, the TFR has declined by about one half in the past 50 years. As fewer babies are born, the base of the population pyramid narrows, producing a relative increase in the older population.
2. At the same time, mortality is declining—and hence life expectancy is increasing, (particularly for women)—everywhere as well. Globally, life expectancy has increased by about 20 per cent in the past 50 years, with females expected to live 69 years and males 65 years. Japan had the highest life expectancy in 2002, with the female rate at 85 years and the male rate at 78 years. Life expectancy is increasing because of ongoing improvements in health care and living conditions.

As a result of these changes, the median age of the world's population is expected to increase from 27 years in 2002 to about 37 years in 2050. Among specific countries, Italy and Japan have the highest median age today (40.2 years), but it is estimated that by 2050 there will be ten countries with a median age above 50, with Spain heading the list at 50.4 years.

Although population aging is indeed a global phenomenon, there are significant regional differences, most obviously related to differences between the more and less developed worlds in their patterns of fertility decline and increasing life expectancy. In the case of fertility, the decline started later in the less developed world but is proceeding more rapidly. In the case of life expectancy, the increases are lower throughout the less developed world, and in some places they are plummeting because of AIDS.

Consequences of population aging

One commentator has suggested that the greying of the population in the less developed world may do more to reshape our collective future than even the proliferation of chemical and other weapons, terrorism, global warming, and ethnic tribalism (Peterson 1999:42). The reasoning behind this prediction has to do with changes in the ratio between what might be described as the dependent elderly population and those of working age. Estimates suggest that this ratio will double in much of the more developed world, and triple in much of the less developed world.

Perhaps most notably, an aging population will place increasing stress on retirement, pension, and related social benefits, necessitating radical changes in social security programs. Further, global aging will lead to quite different patterns of disease and disability; for

Table 5.7   GLOBAL AGING, 1950–2050

Region	Median age (years)			% aged 60 or older		
	1950	**1999**	**2050**	**1950**	**1999**	**2050**
World	23.5	26.4	37.8	8.1	9.9	22.1
More developed world	28.6	37.2	45.6	11.7	19.3	32.5
Less developed world	21.3	24.2	36.7	6.4	7.6	20.6

Source: Population Division of the Department of Economic and Social Affairs of the United Nations Secretariat, *The World at Six Billion*, 1999.

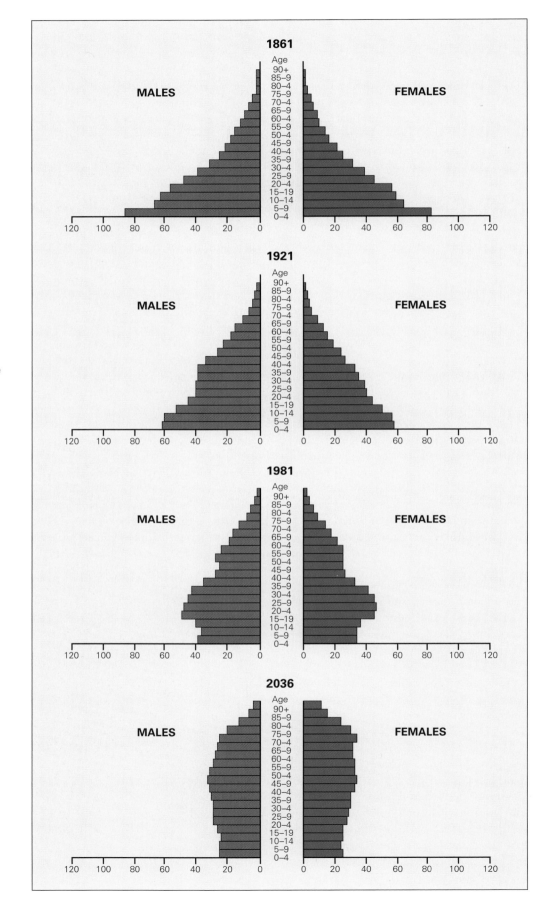

Figure 5.9 Age–sex structure in Canada: 1861, 1921, 1981, 2036. This series of four population pyramids provides considerable information about the population composition of Canada through time. The numbers along the base of each pyramid indicate the number of people in each age and sex group per 1,000 of the total population. In 1861 immigration and high fertility combined to produce a pyramid that shows cohorts steadily decreasing in size with age. The 1921 pyramid shows the effects of reduced fertility in a narrowing of the base, but the effects of immigration are still apparent. In 1981 the Canadian population is characterized by extremely low rates of both birth and death; the age–sex structure exhibits a 'beehive' shape. The pyramid bulge between the ages of 15 and 35 represents the postwar baby boom, which accelerated population growth. The birth rate has long since returned to its normal low; without the baby boom, the pyramid would have had almost vertical sides. A continuing low birth rate and an aging population are reflected in the projection for 2036; this age–sex structure may well represent the future for most countries in the more developed world.

Note: As Figures 5.6, 5.8, and 5.9 demonstrate, population pyramids may be presented in various styles.

Source: Statistics Canada, *Report on the Demographic Situation in Canada*, 1992, Catalogue 91-209E Annual (Ottawa: Statistics Canada, 1992).

example, degenerative diseases associated with aging, such as cancer, heart problems, and arthritis will become increasingly common. The need for adjustments to national health care programs is apparent.

At the same time, national economies will face enormous strain as the numbers of workers available to support the growing population of non-working elderly gradually declines. It is possible that a decrease in the number of dependent young will free up some resources for the dependent elderly; however, the amount of public spending required to support an elderly person is two to three times that required to support a young person. Countries with a strong tradition of children supporting elderly parents, such as China, will face particular stress as increasing life expectancy makes this more and more difficult.

Inevitably, the problems of aging will be exacerbated in those parts of the world that already lack financial and other resources. The fact that the more developed world became rich before aging, whereas the less developed world is aging first, before it has a chance to get rich, is a very significant difference between the two. The implications of population aging for poorer countries are considered sufficiently serious that a 2002 UN conference vowed to defend the rights of the old through improvements in education, employment, pension guarantees, housing, health care, and human rights.

History of Population Growth

Like all animal populations, early human populations increased in some periods and declined in others. The principal constraints on growth were climate and the (related) availability of food. Unlike other animals, humans gradually increased their freedom from such constraints as a consequence of the development of culture. Cultural adaptation has enabled humans to increase in numbers and—so far—to avert extinction. For most of our time on earth, however, our numbers have increased very slowly.

Among the cultural changes that permitted human numbers to increase before about 12,000 years ago were the evolution of speech, which facilitated cooperation in the search for food; the introduction of monogamy, which increased the chances that children would

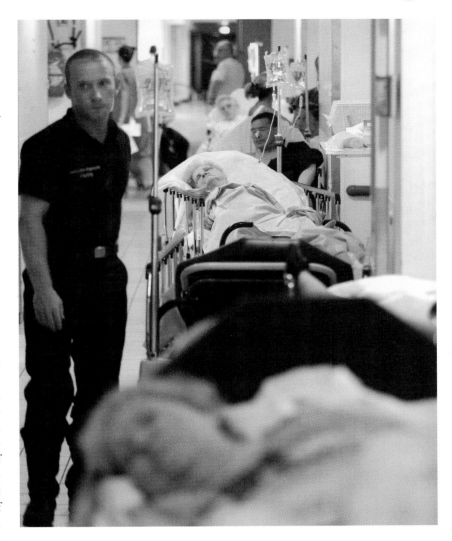

survive; and the use of fire and clothing, which together made it possible to move into cooler areas. Nevertheless, the cumulative effect of these advances was not great; 12,000 years ago, the human population totalled perhaps 4 million. The relatively rapid growth since that time has not been regular. Rather, there have been several relatively brief growth periods and longer intervals of slow growth. Each growth period can be explained by a major cultural advance.

The first such advance, about 12,000 years ago, was the so-called agricultural revolution: the gradual process, over thousands of years, by which humans domesticated animals and plants. As agriculture spread, the first human economic activities, hunting and gathering, became marginal. This 'revolution' took place in various centres, beginning in the Tigris and Euphrates valleys of present-day Iraq. Some 9,000 years ago the first region of high

In August 2003, elderly people sickened by a brutal heat wave lay in the corridors of a Paris hospital. More than 11,000 deaths were attributed to the heat. Among the factors that may have contributed to the toll were hospital understaffing and a lack of family support for frail elders during the traditional vacation period (AP photo/Franck Prevel).

population density appeared, stretching from Greece to Iran and including Egypt. By 4000 BCE agriculture and related population centres had developed on the Mediterranean coast and in several European locations, as well as Mexico, Peru, China, and India. Around the beginning of the Common Era new centres arose throughout Europe and Japan, and the total world population reached an estimated 250 million—a dramatic increase in the 10,000 years since the beginnings of agriculture.

Before the cultural innovation of agriculture, population numbers had changed little, increasing or decreasing depending on cultural advances and physical constraints. Birth and death rates were high, about 35 to 55 per 1,000, and life expectancy was short, about 35 years. Following the introduction of agriculture, birth rates remained high but death rates fluctuated; as the epidemics listed in Table 5.8 suggest, humans still had relatively little control over the environment. This state of affairs continued until the seventeenth century. The 250 million at the beginning of the Common Era had reached 500 million by 1650, but in the absence of any major cultural advance, the pace of growth remained slow.

From about 1650 onwards, however, the world population increased rapidly in response to the second major cultural advance, the industrial revolution. Originating in England, the industrial revolution initiated a rapid growth and diffusion of technologies, and industry replaced agriculture as the dominant productive sector. The agricultural revolution had involved more effective use of solar energy in plant growth. This second revolution involved the large-scale exploitation of new sources of energy—coal, oil, and electricity. The result was a rapid reduction in death rates, with a delayed but equally significant drop in birth rates. World population increased to 680 million in 1700, 954 million in 1800, and 1.6 billion in 1900.

The Current Situation

The growth initially spurred by industrialization has largely ceased in the more developed world, but it continues in the less developed world. As a result, the total world population continues to increase substantially. Table 5.9 places the recent and expected growth in context by noting the number of years taken to add each additional one billion people, and Figure 5.10 displays these data graphically. What is clear from these data is that the world population is no longer growing at an increasing rate—indeed, the largest annual increments to the world population occurred in the late 1980s, when about 86 million people were added each year. By 2002 this annual increment had fallen to less than 79 million. The principal reason for this reduction is the current fertility decline that is occurring in the less developed world; details of, and reasons for, this decline are discussed later in this section.

To summarize the current situation: the world population growth rate—the rate of natural increase—has fallen from a late 1960s high of 2.04 per cent per year to the current 1.3 per cent per year. Because of this decline, the world population will grow less rapidly in the current century than it did in the twentieth century, although it will continue to increase substantially. Specifically, there will be continued rapid growth until about 2050, followed by much slower growth after that date and, perhaps, a stable population by 2200.

Table 5.8	**MAJOR EPIDEMICS, 1500–1700**

Date	Event
1517	Smallpox in Mexico, death of one-third of population, followed by other diseases in Mexico and other American countries
1522–34	Series of epidemics in western Europe resulted in significant decrease of population
1563	Plague in Europe
1580–98	Epidemics in France
1582–3	Epidemics and famines in southern Sudan
1588	Epidemics and famines throughout China
1618–48	German population declined from 15 million to 10 million because of epidemics and war
1628–31	Disastrous plague epidemics in Germany, France, and northern Italy
1641–4	Epidemics, famines, and conflicts reduced Chinese population by 13 million
1648–57	Plague in Spain; 1 million died
1650–85	Plague and conflict in Britain; 1 million died
1651–3	Plague and conflict in France; 1 million died
1690–4	Food shortages and epidemics throughout Europe

Of course, future increases in population will not be uniformly distributed. Because of spatial variations in fertility, especially between the more and less developed worlds, the percentages of the world population in the various major regions are expected to change dramatically by the mid-twenty-first century, with the most notable changes being an increase in Africa and a decrease in Europe. These changes in distribution were noted earlier in this chapter and are detailed at the beginning of the next.

This brief account explains the basis of population growth. Numbers have increased in response to cultural, specifically technological, advances. The dramatic increases since about 1650 reflect the fact that the death rate began to fall before the birth rate did. In what is now the more developed world, populations thus grew rapidly after about 1650; elsewhere numbers increased rapidly in the last century, especially since the 1940s.

Population Projections

Predicting population growth is hazardous. In the early 1920s, Pearl predicted a stable population of 2.6 billion persons by 2100. A second example is the 1945 prediction by an eminent American demographer, Frank Rotestein, of 3 billion by 2000. These are just two of many examples. But despite some unimpressive precedents, there is good reason to suggest that we are now in a position to make better forecasts. The principal reason is that both fertility and mortality rates now lie within narrower ranges than was previously the case. In the more developed world, fertility and mortality rates are relatively low. In the less developed world, however, both rates remain relatively high.

Current United Nations 'medium' projections suggest a world population of 9.1 billion for 2050. This projection assumes that the mortality transition will be complete at that time, such that the CDR is approximately equal throughout the world. A look further ahead might see the fertility transition that is currently taking place in the less developed world complete by the year 2100, with a relatively stable world population of about 10 billion by 2200.

What these population projections do not take into account is the possibility that there may be **limits to growth** (this phrase

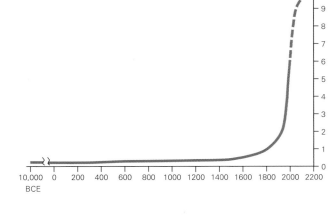

Figure 5.10 World population growth.

became popular after it was used as the title of a report published by the Club of Rome [Meadows et al. 1972]; we mentioned this issue briefly in Chapter 4 under the heading 'Earth's Vital Signs'). Many environmentalists and ecologists argue cogently that there are definite limits to the growth both of populations and of economies. The earth is finite, and many resources are not renewable. The classic argument in this genre is, of course, that of Malthus (see p. 167), but the 1972 Club of Rome report broadened the debate to include natural resources and environmental impacts. Judging by trends in the early 1970s, the authors of the report argued that world population was likely to exceed world carrying capacity, leading to a collapse of both population and economy. A well-publicized

Table 5.9	ADDING THE BILLIONS: ACTUAL AND PROJECTED	
Billions	**Year reached**	**Years taken**
1	1804	
2	1927	123
3	1960	33
4	1974	14
5	1987	13
6	1999	12
7	2013	14
8	2028	15
9	2054	26
10	2183	129

Source: Population Division of the Department of Economic and Social Affairs of the United Nations Secretariat, *The World at Six Billion*, 1999.

earlier work, *The Population Bomb* (Ehrlich 1968), similarly anticipated widespread famine, raging pandemics, and possibly nuclear war by about 2000 as a result of worldwide competition for scarce resources.

As we saw in Chapter 4, this thesis is typically articulated by catastrophists, but there is another, more positive view, typically articulated by cornucopians. According to this second view, technological advances will make new resources available as old resources are depleted: 'there are limits to growth only if science and technology cease to advance, but there is no reason why such advances should cease. So long as technological development continues, the earth is not really finite, for technology creates resources' (Ridker and Cecelski 1979:3–4).

Indeed, some believe that people themselves are the ultimate resource. Japan, for example, has few resources apart from people, and that resource is slowly decreasing (see Box 5.8). According to some commentators, a slow-down in population growth will lead to a slow-down in economic growth—and therefore is to be avoided. The debate between the catastrophists and the cornucopians is far from over. Perhaps the best we can do at present is to recognize the wide range of views on the subject and to weigh the evidence on both sides with extreme care.

Explaining Population Growth

Making sound predictions about future population numbers is not easy; nor is it easy to explain past population growth. In this section six different theories and models are described and evaluated.

The S-Shaped Curve Model

All species, including humans, have a great capacity to reproduce, which is regulated by constraints such as space, food supplies, disease, and social strife. Our first explanation proposes a simple and natural statement of population growth known as the S-shaped curve (see Figure 2.8, p. 62).

5.8

THE PROSPECT OF UNDERPOPULATION

It is possible that the population explosion will prove to be a short-term phenomenon, a period of perhaps 250 years in which life expectancy increased dramatically as a result of improvements in health and living conditions—all very positive developments. As our discussion of population projections notes, it seems likely that the explosion is now over in the sense that the world population, although continuing to increase, is doing so at a decreasing rate. Some commentators are even beginning to raise the spectre of underpopulation rather than overpopulation. But we should not be misled by suggestions of this kind. The shrinking populations that those commentators worry about are all in the more developed world. Meanwhile, serious problems continue in many parts of the less developed world, largely as a result of imbalances between population and resources. Table 5.10 identifies the six countries that are expected to experience population losses of more than 5,000,000 between 2002 and 2050. In total, 32 countries are expected to lose population in that time, most of them in Europe.

It is important to recognize that the numbers in Table 5.10 are susceptible to even slight changes in fertility rates; therefore the predictions for 2050 are especially uncertain. Consider, for example, what will happen to the population of Japan if the current fertility rate trends continue. According to the Japanese government, the country's total population—127 million in 2002—will fall to only 500 by the end of the current millennium, and to a single person by the year 3500. This is an exercise in fantasy, of course, but the numbers do provide a hint of what the future may hold in the more developed world if fertility rates stabilize below the replacement level of about 2.1.

Table 5.10 **ANTICIPATED POPULATION DECLINES GREATER THAN FIVE MILLION, 2002–2050**

Country	Population decline 2002–2050 (millions)
Russia	41.8
Japan	26.8
Germany	14.7
Ukraine	9.8
Italy	5.9
Romania	5.3

Source: Population Reference Bureau, *2002 World Population Data Sheet* (Washington, DC: Population Reference Bureau, 2002).

This curve is produced under carefully controlled experimental conditions. The growth process begins slowly, then increases rapidly (exponentially), and finally levels out at some ceiling. It seems probable that such growth curves never actually occur in nature. It may not be unusual for growth to be first slow, then rapid, but it does appear to be unusual for any population (plant or animal) to remain steady at some ceiling level. A more characteristic final stage would involve a series of oscillations. Various scientists nevertheless predicted that world human population growth would correspond to the S-shaped curve. We have already referred to Pearl's prediction in the 1920s of a stable population of 2.6 billion by 2100; clearly, using the curve as a predictive tool is risky. Simply put, the curve model does not take into account the variety of cultural and economic factors that affect human populations. As we have already ascertained in our discussion of fertility, stable populations result not from some natural law but from human decisions to control birth rates knowingly and willingly. On the other hand, it is still possible, as suggested in Figure 5.10, that an S-shaped curve will be evident by 2200.

Malthusian Theory

Our second model, unlike the S-shaped curve, is rooted in real-world facts. Malthusian theory was first presented in 1798 and remains relevant today. Thomas Robert Malthus (1766–1834) was born in England at a time of great technological and other changes associated with industrialization and rapid population growth. A humane man, he was greatly concerned about the welfare of the growing numbers of poor people. He opposed the prevailing school of economic thought, mercantilism, which was explicitly pronatalist; for mercantilism, more births meant more wealth, since a large labour force was needed both in England to increase national productivity and overseas to increase English strength in the colonies.

Malthus expressed his views in a book titled *An Essay on the Principle of Population*, first published in 1798 with a final, seventh, edition in 1872. Basic Malthusian logic is straightforward. It can be usefully presented in terms of two axioms and a hypothesis:

Axiom 1: Food is necessary for human existence; further, food production increases at an arithmetic rate, i.e., 1, 2, 3, 4, 5 . . .

Axiom 2: Passion between the sexes is necessary and will continue; further, population increases at a geometric rate, i.e., 1, 2, 4, 8, 16 . . .

Given these two axioms, which Malthus regarded as reasonable assumptions, the following hypothesis is deduced:

Hypothesis: Population growth will always create stress on the means of subsistence.

According to Malthus, this conclusion—the inevitable result of the different growth rates of food supplies and population—applied to all living things, plant or animal. As rational beings (that is, with culture), humans in principle had the capacity to anticipate the consequences and therefore avoid them by deliberately reducing fertility through adoption of what Malthus called *preventive checks*: means such as moral restraint and delayed marriage. However, he believed that in practice humans were incapable of voluntarily adopting such checks: they would do so only under the pressure of extreme circumstances such as war, pestilence, and famine—what he called *positive checks*. Hence the human future would be one of famine, vice, and misery.

Malthus's central concern was imbalance between population and food. His arguments are intriguing, if not especially prophetic. In considering his theory, we must first note that there is no particular justification for the specific rates of increase proposed in the two axioms. Indeed, Malthus himself acknowledged in later editions of his work that the rates of population and food increase could not be definitely specified. Malthus was not able to anticipate that food supplies could be increased not only by increasing the supply of land (which he correctly saw as finite) but also by improving fertilizers, crop strains, and so forth. Nor could he have predicted that contraception would become normal and accepted. He was aware of contraception, of course, but saw it as an immoral preventive check (homosexuality was another), unlike the moral preventive checks of delayed marriage and self-restraint, of which he approved.

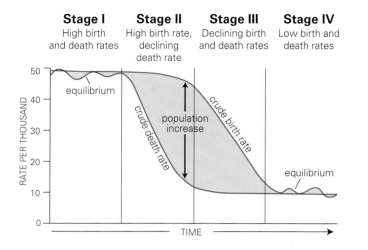

Stage I	Stage II	Stage III	Stage IV
High birth and death rates	High birth rate, declining death rate	Declining birth and death rates	Low birth and death rates

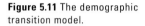

Figure 5.11 The demographic transition model.

Events have proven Malthus incorrect in his predictions, and his theory lost favour in the mid-nineteenth century as birth-rate reductions and emigration eased population pressures in Europe. Today what might be called neo-Malthusian theory is often purported to be relevant in the less developed world. China is the prime example, but because its political philosophy actively encourages state intervention in matters of everyday life, it has been able to implement a rigorous population policy.

Marxist Theory

One of the earliest and most powerful critics of Malthusian theory was Marx, who objected to the rigorous axioms and hypothesis used by Malthus and believed that population growth must be considered in relation to the prevailing mode of production in a given society. For Marx, Malthus represented a bourgeois viewpoint in which the primary aim was to maintain existing social inequalities. Malthus saw population growth as the primary cause of poverty, whereas Marx saw the capitalist system as the primary cause. The difference is fundamental: for Marx the problem was not a population problem at all, but a resource distribution problem caused by capitalism.

A related distinction can be made between the concepts of overpopulation and surplus population. These terms are not synonyms. Malthus was concerned with overpopulation: he believed that a society could be said to be overpopulated when food and other necessities of human life were in such short supply that life-threatening circumstances arose. For Marx, by contrast, the key con-

cept was surplus population: those surplus —unemployed—workers who represent a 'reserve' labour force. According to Marx, capitalism depends on the existence of those 'surplus' workers in order to keep wages low and profits high; surplus population is therefore an inevitable consequence of capitalism. Both concepts are valuable to us as geographers.

Boserup Theory

A fourth theory is available in the writings of Ester Boserup on historical changes in agriculture in subsistence societies (Boserup 1965). According to this argument, subsistence farmers select farming systems that permit them to maximize their leisure time, and they will change these systems only if population increases and it becomes necessary to increase the food supply accordingly.

Boserup argued that in subsistence societies, population growth requires that farming be intensified so that the supply of food is increased to feed the additional population. Thus population growth has positive effects, improving human welfare. However, though population increase prompts an increase in gross food output, food output per capita decreases. Further, a series of negative changes ensues because most of the agricultural areas with high and increasing populations are already areas of poverty and often limited agricultural technology.

The contrast with Malthus is clear. Malthus saw population as dependent on food supply; Boserup reversed this relationship by proposing population as the independent variable. Evidence suggests that this theory is applicable in a subsistence context, but not in the more developed world, where technology plays a much larger role and agricultural populations are declining.

The Demographic Transition

So far in our discussion of theories about population growth, we have considered one natural law (the S-shaped curve) and three influential writers (Malthus, Marx, and Boserup). Each has something to offer; each is at least partially flawed. Clearly, population growth is not easily explained. Our fifth explanation is in a somewhat different category.

The demographic transition model is a descriptive generalization; it simply describes changing levels of fertility and mortality, and hence of natural increase, over time in the contemporary more developed world. Because it is based on known facts rather than specific axioms or general assumptions, it has a major advantage: although simplified, it is clearly more realistic than most of the other explanations we have looked at. Inevitably, it also has a major disadvantage: because it is descriptive, it does not offer a bold, provocative perspective or hypothesis. Figure 5.11 presents the demographic transition in conventional graphic form.

1. The first stage is characterized by a high CBR and a high CDR—the two rates are approximately equal. The CDR fluctuates in response to war and disease. This stage involves a low-income agricultural economy. Population growth is limited.
2. In the second stage, there is a dramatic reduction in CDR as a result of the onset of industrialization and related medical and health advances. This reduction is not accompanied by a parallel reduction in CBR; hence the RNI is high and population growth is rapid.
3. The principal feature of the third stage is a declining CBR, the result of voluntary decisions to reduce family size—decisions facilitated by advances in contraceptive techniques. The evidence suggests that voluntary birth control is related to increased standards of living. The RNI falls during this third stage as the CBR approaches the already low CDR.
4. In the fourth (and final?) stage, the CBR and CDR are once again, as in the first stage, approximately equal—but now, instead of being wastefully high, both rates are low, as is the RNI.

Growth rates are similar in the first and fourth stages, but the ways in which the rates are generated are very different. The transition also involves two central stages during which birth and death rates are unequal and a high rate of increase prevails. All three population pyramids in Figure 5.8 (p. 160) exhibit the characteristics of a country at the onset of the transition. The four population pyramids in Figure 5.9 (p. 162) show a country that has passed through the transition, with the added complications

A family planning poster in Barbados (Victor Last, Geographical Visual Aids).

of the baby boom and an aging population.

The demographic transition model accurately describes in a simplified form the experience of the more developed world. Most of that world reached the fourth stage during the first half of the twentieth century—but what about the less developed world? Throughout much of Asia, Africa, and Latin America, the situation resembles the third stage of the model. Can we assume that it is only a matter of time before economic development guarantees passage through the third stage and entry into the stable fourth stage? To put the question another way: Is the demographic transition a valuable predictive tool? The answer appears to be yes, but for rather different reasons.

As we have already seen, there is compelling evidence to suggest that the fertility decline evident in the third and fourth stages of the demographic transition is indeed occurring in the less developed world (see Box 5.3, p. 147).

The Fertility Transition

Since about 1970, fertility has declined throughout much of the less developed world and is continuing to decline. To help understand this change, it is appropriate to refer not to the experience of the more developed world, which is summarized in the demographic transition model, but to an explanation based on the current circumstances of the areas being affected. Although the current fertility declines match the pattern described by the demographic transition model, they are occurring more rapidly than they did in the more developed world, and under very different social and economic circumstances. In fact, the currently available data for the less developed world appear to conform more closely to what is called the fertility transition model (Robey, Rutstein, and Morris 1993).

During the period since 1970, the most powerful influence on fertility in the less developed world has been the extent to which modern contraceptive methods are employed. If family planning increases, fertility drops. In the less developed world (not including China), about 48 per cent of married females now practise family planning, and about 85 per cent of these employ modern as opposed to traditional methods of contraception. Although better-educated females are most likely to limit family size, education is not a prerequisite for using contraception, as is evidenced by a decreasing correlation between education and use of contraceptives. Information about family planning is widespread because of the influence of the mass media.

According to the fertility transition model, large families have fallen out of favour because of the obvious problems—such as pressure on agricultural land and poor quality of urban life—associated with rapid and substantial population increases. The evidence also suggests that females—who are generally experiencing a rise in social status—favour later marriage, smaller families, and more time between births.

The principal reason fertility is declining so rapidly in the less developed world is probably that today effective contraception is available to meet the demand as soon as the necessary cultural impetus is in place; this was not the case when the more developed world was first culturally predisposed to smaller families, a time when abstinence, withdrawal, and abortion were the only techniques available. The fertility decline that is occurring today may be best described as a 'reproductive revolution' because it is rapid and substantial.

Economic development does help to create a climate conducive to reductions in fertility, but the key point of the fertility transition argument is that these reductions are caused by a new cultural attitude, the willingness to

employ modern contraceptive methods, and by the ready availability of these methods. Although it has been said that development is the best contraceptive, in fact it is not: 'contraceptives are the best contraceptive' (Robey, Rutstein, and Morris 1993:65) (Box 5.9).

The Need for Better Explanations

How do the six theories and models above aid our understanding of population growth? In general, human geographers are far from satisfied with their explanatory and predictive power. Moreover, as we saw in Chapter 2, the varied philosophies of individual human geographers give rise to a wide range of interpretations.

Evaluating the Available Explanations

An *empiricist* philosophy focuses on facts, not theory. Remember that much traditional regional and cultural geography was—at least implicitly—empiricist. Much the same can be said of population geography; its focus is on facts, to the general exclusion of interpretation, understanding, or explanation. Both the demographic and fertility transitions are empiricist descriptions based on available data, although both also suggest explanations for the trends shown by the data. It is now reasonably clear that the demographic transition model, which centres on changing economic circumstances, provides a good account of the post-1650 experience of what is now the more developed world and is a useful predictive tool for the less developed world, albeit for different specific reasons. The fertility transition model suggests that the changes currently occurring in the less developed world can be explained by changing cultural preferences.

A *positivist* philosophy focuses on theory construction, related hypothesis and law formulation, and statistical analysis. The S-shaped curve model, Malthusian theory, and the Boserup thesis all have positivistic overtones in that they explicitly relate cause and effect. In the S-shaped curve, population size is related to time and the availability of space and food. In Malthusian theory, population growth is similarly explained, while the Boserup thesis sees population as cause and food supplies as effect. All three contributions are useful, but they are far from universally valid. The S-shaped curve is of limited use; Malthusian theory has clear limitations and the Boserup argument applies only to subsistence societies. A problem with the use of a positivist philosophy is that it requires data to test hypotheses statistically—data that are rarely available at the full range of spatial and temporal scales in question.

A *humanistic* perspective has rarely been employed in analyses of population growth. Yet there are some obvious applications, given the humanistic emphasis on individual experience. What, for example, was the reaction of Romanian women to the strict pronatalist policies of their country prior to 1989? Or how did Chinese peasant couples respond to the one-child policy? The fertility transition model seems to be one case where a humanist approach might well be useful.

A *Marxist* philosophy applied to population issues sees population growth as an outcome of the society's particular mode of production. Fertility behaviour, for example, is seen as the product of society rather than of free decision-making. It seems that the value of Marxist theory is essentially restricted to the period of early industrial capitalism.

Alternative Explanations?

It is worth noting that in these discussions we have rarely chosen to subdivide human populations on the basis of age, ethnicity, class, religion, gender, or other relevant variables. The decision to treat humans as a cohesive whole reflects our favoured scale of analysis, which is large: either the world as a whole or large areas within it. (Material focusing on population subgroups will be introduced in later chapters.) Nevertheless, we must acknowledge that the human experience is not unitary: in fact, it is closely tied to a number of variables. One important variable that merits discussion at this time is gender.

Feminist geography has developed as an area of interest in human geography since the 1970s. Like Marxist geography, it is concerned with criticizing and changing established procedures. Specifically, feminist geography addresses the question of gender inequality in our human experience. The aim is not only to recognize that there are different gender geographies but, much more fundamentally, to develop gender-specific theory.

How might feminist geography contribute to our discussions of population? Answers are many and varied. We will briefly consider

---

5.9 FERTILITY, POPULATION GROWTH, AND THE STATUS OF WOMEN

The fertility transition undoubtedly marks a significant change in patterns of population growth. Perhaps the most important aspect of this change is the changing status of women in the less developed world. Two major conferences held in the mid-1990s addressed this subject.

The critical importance of fertility to population growth was a central theme at the third United Nations-sponsored International Conference on Population and Development, held in Cairo in 1994; the previous conferences had been held in Romania in 1974 and Mexico City in 1984. The Cairo meeting was the first to recognize the centrality of women to debates about ways to limit future population growth. In several discussions of a very sensitive nature, the Vatican and various Islamic countries debated controversial issues such as contraception and abortion with representatives of European countries and feminist groups. The program that was eventually approved in Cairo represented a compromise.

These discussions continued at the United Nations-sponsored Fourth World Conference on Women held in Beijing in 1995. When, at this meeting, the UN Secretary General stated that 'No progress is possible without the full and equal participation of women and men,' he implicitly underlined the importance of improving the social and economic status of women, especially in the less developed world.

The program of action approved at the meeting included several references to population matters, notably contraception, abortion, different forms of family, and sexual rights. From the perspective of feminist groups and liberal countries generally, the program of action approved moved well beyond the one agreed to at Cairo a year earlier. Together, these two meetings have helped to focus attention on the importance of improving the status of women in the less developed world. Such improvements will undoubtedly encourage further reductions in fertility and help to slow the rate of population growth.

gender differences in the more developed world as an example. The principal change currently occurring involves women's role: from wife and mother to wife, mother, and worker outside the home. In the more developed world, the typical family structure remains the conjugal variety.

The family has responsibility for reproduction, and the changing role of women in society is clearly linked to fertility. In all the more developed countries, fertility rates are low and women's participation in the labour force is increasing. Thus women and men alike now have links with a world outside the family—a fact that should, in principle, foster greater commonality of experience between men and women. Whether this is actually happening is debatable.

A brief review of philosophies and related attempts at explaining changes in fertility and mortality cannot do justice to them. Still, it does appear that current explanations, though useful, are far from adequate. What would a more complete explanation need to include?

Regardless of philosophical affiliation, new explanations need to explain reality at a variety of spatial, social, and temporal scales. As an example, let us consider the reality of the demographic transition. How are we to explain changing death and birth rates? An appropriate theory needs to incorporate economic circumstances (such as industrialization) and cultural issues (such as desirable family size). But we also need to consider the behaviour of individual males and females and the motivations for their behaviour. We also need to recognize that the transition was not the same in every region. In France, for example, the decline in fertility was underway by the 1830s, but in England it did not become evident until the 1890s. Thus in reality there are significant spatial variations in the timing and details of what we presented above as a uniform transition. Detailed analysis of such variations might facilitate the construction of a more adequate theory by identifying specific causal variables. A full theory to explain the transition is still wanting.

CHAPTER FIVE SUMMARY

How many?
A population of 6.2 billion in 2002.

Fertility
There are various measures of fertility; these include the crude birth rate, the general fertility rate, and the total fertility rate. In 2002 the world birth rate was 21 (21 live births per 1,000 members of the population), while the total fertility rate was 2.8 (the average woman has 2.8 children). A total fertility rate of about 2.1 to 2.5 is sufficient to maintain a stable population. Fertility is affected by many variables: age and related fecundity, nutrition, level of industrialization, age at marriage, celibacy, governmental policies, contraceptive use, abortion, and empowerment of women. Spatial variations in fertility are closely related to level of development; the total fertility rate in the more developed world is 1.6; in the less

developed world it is 3.1. At present there is evidence of declining fertility in much of the less developed world (related to widespread acceptance of family planning) and in much of the more developed world (possibly related to uncertain economic prospects).

Mortality
Three mortality measures are the crude death rate, the infant mortality rate, and life expectancy. In 2002 the death rate was 10 and life expectancy was 67 years. Major causes of death are old age, disease, famine, and war. Around the world, death rates shows much less variation than birth rates; central Africa is the last region of high mortality. Data on infant mortality and life expectancy are good indicators of health. Many countries exhibit significant internal variations in mortality patterns.

The AIDS pandemic

Increased mortality and reduced life expectancy as a result of AIDS are a major humanitarian and economic concern in much of sub-Saharan Africa, and there are fears that incidence of the disease will increase significantly in some other countries, including China, India, and Russia. AIDS is usually seen as the greatest threat to human health in the world. As of 2002, it was estimated that approximately 35 million people around the world were infected with HIV, and that some 20 million people had died of AIDS since the early 1980s.

Natural increase

In 2002 the rate of natural increase was 1.3 per cent. This rate will result in a doubling of world population in about 54 years. Natural increase is affected by the age composition of a population.

Population aging

The age structure of the world's population is changing significantly as the proportions of elderly people increase relative to other age groups. This change is the result of declines in fertility and increases in life expectancy. The year 2000 was a watershed: the first year in which people under 14 were outnumbered by people over 60. The social and economic implications of this shift are considerable.

Government intervention

Most governments intervene directly and/or indirectly to influence growth. Death control policies are normal and designed to reduce mortality. Birth control policies can be *laissez-faire*, pronatalist, or antinatalist. A pronatalist approach may be related to a dominant religion or economic and strategic motives. Antinatalist policies, of varying degrees of success, are common in many of the less developed countries. The two most populous countries, China and India, have employed different approaches; China has been highly successful, India less so.

Predicting growth

Past forecasts have often erred seriously. As of 2002, it is estimated that population will continue to increase rapidly for about another fifty years and then increase more slowly. The current projection is for an increase to about 9.1 billion by 2050 and a relatively stable population of about 10 billion by 2200. There is much uncertainty concerning the relative merits of two arguments. The limits-to-growth, or catastrophist, thesis sees definite limits to population and economic growth because the earth is finite. The cornucopian thesis sees technology as continually making new resources available and hence enabling the earth to accommodate increasing numbers of people.

World population growth

The world's population increases in response to cultural change. An agricultural revolution began about 10,000 BCE and prompted increases from about 4 million to 250 million by the beginning of the Common Era. A second major change, the industrial revolution, made it possible for the world's population to increase from 500 million in 1650 to 1.6 billion in 1900.

Explaining growth

The S-shaped curve is a useful biological analogy, but is clearly not applicable to the current situation. Malthus, Marx, and Boserup each contributed usefully to our understanding of population growth—Malthus saw population as limited by food supplies; Marx saw population as a response to a particular social and economic structure; Boserup saw population increases as prompting food supply increases—but none is of general applicability. Malthus is the most discussed theorist, and a contemporary version of his theory is known as neo-Malthusianism. The demographic transition model is a summary of birth and death rates over the long period of human history in what is now the developed world; for the less developed world an accurate description is provided by the fertility transition model. There is a clear need for more and better theory—whether it be positivist, humanist, or Marxist.

LINKS TO OTHER CHAPTERS

- Population growth—especially the exponential growth that has taken place over the last 250 years:
    Chapter 6 (population distribution and density; migration).

- Demographic measures (CBR, LE, etc.):
    Chapter 6 (inequalities).

- Human origins:
    Chapter 3.

- Human impacts on earth:
    Chapter 4.

- Inequalities between the more and less developed worlds:
    Chapter 6.

- Changes in social and cultural landscapes:
    Chapter 7
    Chapter 8.

- The political world today:
    Chapter 9.

- Agricultural change:
    Chapter 11.

- Settlement:
    Chapter 12.

- Industry:
    Chapter 13.

- Transportation and trade:
    Chapter 9.

- The catastrophist–cornucopian debate:
    Chapter 4 (environmental futures)
    Chapter 6 (refugees, the less developed world)
    Chapter 8 (landscapes, power relations).

- Explaining population growth:
    Chapter 2 (philosophical approaches)
    Chapter 8 (feminism).

- Globalization: Chapter 2 (concepts)
    Chapter 4 (ecosystems and global impacts)
    Chapter 6 (refugees, disease, more and less developed worlds)
    Chapter 7 (cultural globalization)
    Chapter 8 (popular culture)
    Chapter 9 (political globalization)
    Chapter 10 (economic globalization)
    Chapter 11 (agriculture and the world economy)
    Chapter 12 (global cities)
    Chapter 13 (industrial restructuring)

FURTHER EXPLORATIONS

BARRETT, H. 2000. 'Six Billion and Counting.' *Geography* 85:107–120.
    An excellent overview of the world population situation, including both fertility and mortality trends, at the beginning of a new century; numerous tables and figures.

CIPOLLA, C.M. 1974. *The Economic History of World Population*, 6th ed. Harmondsworth: Penguin.
    A stimulating and provocative overview of world population growth emphasizing our use of energy sources; dated but still interesting.

DAUGHERTY, H.G., and K.C.W. KAMMEYER. 1995. *An Introduction to Population.* New York: Guilford.
    A comprehensive overview of population studies; covers the various topics addressed in this chapter in much greater detail.

DAY, L.H. 1992. *The Future of Low Birth-Rate Populations.* London: Routledge.
    An assessment of the demographic situation, policy alternatives, and likely future changes in an aging and stable or declining population.

KALIPENI, E. 1995. 'The Fertility Transition in Africa'. *Geographical Review* 85:286–300.
    Clearly written analysis showing that fertility is declining in many African countries as a result of improved education, the changing status of women, and increased contraceptive use.

LUTZ, W., S. SCHERBON, and A. VOLKOV, eds. 1994. *Demographic Trends and Patterns in the Soviet Union before 1991.* London: Routledge.
    A compendium of demographic research on the former USSR; detailed reviews of fertility,

mortality, age, marriage, and the family; contrasts the developed west and north with the less developed central Asian region.

NEWMAN, J.L., and G.E. MATZKE. 1984. *Population: Patterns, Dynamics and Prospects*. Englewood Cliffs: Prentice-Hall.

This population geography textbook can be used by students with minimal previous study of the topic.

OMRAN, A.R. 1992. *Family Planning in the Legacy of Islam*. London: Routledge.

An examination of the Islamic view of marriage, family formation, and child-rearing.

PACIONE, M., ed. 1986. *Population Geography: Progress and Prospect*. London: Croom Helm.

An advanced book with good chapters on, among other topics, theory, fertility, and mortality.

PETERS, G.L., and R.P. LARKIN. 1983. *Population Geography: Problems, Concepts and Prospects*, 2nd ed. Dubuque: Kendall Hunt.

An excellent textbook in population geography covering all the topics in this chapter in detail;

includes clear accounts of the various measures of fertility and mortality.

POPULATION REFERENCE BUREAU. 2000. *2000 World Population Data Sheet*. Washington, DC: Population Reference Bureau.

An invaluable factual document, published annually; a comprehensive source of basic demographic data.

SCIENTIFIC AMERICAN. 1974. *The Human Population*. San Francisco: Freeman.

A collection of articles all originally published in *Scientific American*. The data are somewhat dated, but most articles remain of value, especially those dealing with the genetics of human populations.

WOODS, R. 1979. *Population Analysis in Geography*. London: Longman.

A very useful and straightforward text on population geography.

————. 1981. *Theoretical Population Geography*. London: Longman.

A stimulating and critical overview of conceptual issues.

ON THE WEB

http://www.census.gov/
Home page of the United States Census Bureau.

http://www.census.gov/cgi-bin/ipc/popclockw
This website, published by the US Census Bureau, provides the estimated total world population on a daily basis.

http://www.prb.org/
The most fundamental and reliable of many websites dealing with population matters. The Population Reference Bureau publishes an annual Population Data

Sheet that provides basic demographic data for most countries in the world; these data are posted on the website approximately one year after they are first available in printed form. This site also includes commentaries on specific issues.

http://www.statcan.ca/start.html
Statistics Canada maintains this comprehensive site, which includes recent census data.

http://www.unfpa.org/
Home page for the United Nations Population Fund; covers a wide range of topics and regularly updated.

The Human Population: An Unequal World

In the preceding chapter we saw that the current rate of population growth is a significant—and possibly temporary—deviation from the rates that prevailed during most of human history; that the number on which this rate is based is the highest in human history (6.2 billion in 2002); and that there are significant spatial variations in growth rates. In this chapter we examine several general questions about population patterns:

1. Where do we live—and why?
2. Why have so many people, past and present, moved from one location to another?
3. Why are so many people today effectively without a home country?
4. Why is the world demographically divided between a stable population in the more developed world and a rapidly increasing population in the less developed world?
5. What are the consequences of this division?

Not only are all these questions related to one another, but all of them relate to one crucial question: Why is the world divided into more and less developed areas? This is a complex question, and to answer it we must consider a wide range of factors, from the quantity and quality of food supplies in different parts of the world to the spatial distribution of debt, from the regional selectivity of disasters to the measures employed to distinguish levels of development—not to mention the political and economic relationships between the world's countries.

An Afghan woman begging on the street in Kabul, 2002 (AP photo/Amir Shah).

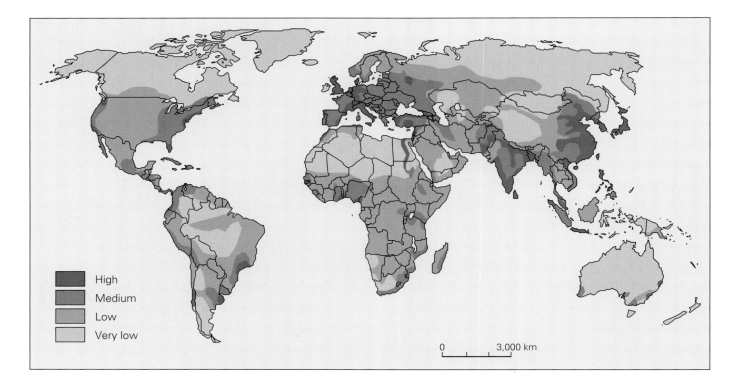

Figure 6.1 World population distribution and density.

Distribution and Density

Determining population distributions and densities—where people are located and in what numbers—is one of the human geographer's central concerns. Establishing these basic facts is not as easy as you might think. There are three related problems. First, the required data are not available for all countries—let alone all regions within countries—and even the data that are available may not always be reliable. Second, the data that are available have been collected for various purposes, and hence may reflect different divisions from the ones we are interested in. Third, since distribution and density statistics are closely related to spatial scale (recall Figure 2.6, p. 61), they can often suggest an inaccurate picture of reality.

Measuring Density

Distribution refers to the spatial arrangement of geographic facts; **density** refers to the frequency of occurrence of geographic facts within a specified area. Conventionally, population density is arithmetic density: the total number of people in a unit area. Maps of population often combine two characteristics: distribution and density. Figure 6.1, showing population density and distribution for the world, is a good example.

Many data sources use a single statistic to identify the population density of a country, but that practice can be seriously misleading. Canada is a prime example: because its population is not evenly distributed—in some areas the density is very high, in others it is very low—a single density statistic for the entire country would be meaningless. We mentioned this problem in Chapter 2, in reference to spatial scale. For Canada the 2002 population density was 3.14 people per square kilometre—but this single statistic is merely an average produced by combining

Table 6.1	WORLD POPULATION DISTRIBUTION BY MAJOR AREA (%): ACTUAL AND PROJECTED	
Region	**2002**	**2050**
Africa	13.5	20.3
Asia	60.6	58.2
Europe	11.7	7.2
Latin America and Caribbean	8.5	9.0
North America	5.1	4.9
Oceania	0.5	0.5

Source: Population Reference Bureau, *2002 World Population Data Sheet* (Washington, DC: Population Reference Bureau, 2002).

large areas that are virtually unpopulated and relatively small areas that are densely populated. In most cases, the local scale is the only one where density measures are appropriate; the smaller the area, the less likely it is to include significant spatial variations.

More refined density measures relate population to some other measure, such as cultivated or cultivable land; for example, **physiological density** is the relation between population and that portion of the land area deemed suitable for agriculture.

There is one exception to the general rule about density measures and scale. Simple density measures employed at the national or multinational scale can be useful for showing the broad outlines of world population patterns. In Table 6.1, for instance, which presents 2002 population data by continent, the use of a simple density measure highlights the disparity between Asia (61.0 per cent of the world's population) and Europe (only 12.0 per cent). Similarly, as we noted in Chapter 5, the projections for 2050 in the same table point to some quite dramatic changes at the large spatial scale: a significant increase in the percentage of world population located in Africa; a slight increase in the Latin American/Caribbean region; and a significant decrease in Europe, with moderate decreases in North America and Asia.

Table 6.2 identifies the ten countries with the largest total populations in 2002. China ranks first, followed by India; these two countries have much larger populations than any other country. Four other Asian countries—Indonesia, Bangladesh, Japan, and Pakistan—are also in the top ten. (For insight into some of the factors that, in the past, have made it difficult to arrive at accurate population figures, especially in the less developed world, see Box 6.1, on the Nigerian census of 1991.)

Table 6.2 also includes projections for 2050. Although China and India continue to dominate, India has overtaken China as the world's most populous country. African countries are also more prominent by 2050, with Nigeria moving up from tenth to sixth place and the Democratic Republic of the Congo and Ethiopia occupying ninth and tenth place respectively. Two countries in the top ten for 2002 have slipped off the list by 2050: Russia and Japan are the two countries expected to lose the most popula-

tion between 2002 and 2050, because of low fertility. In 2002 only the ten countries listed in Table 6.2 and Mexico had populations of more than 100 million, but by 2050 there are expected to be five more (Philippines, Vietnam, Russia, Egypt, and Japan). Table 6.3 shows the number of people per square kilometre in each of the top ten countries; perhaps surprisingly, the

| Table 6.2 | THE TEN MOST POPULOUS COUNTRIES: CURRENT AND PROJECTED |

2002 Country	Population (millions)	2050 Country	Population (millions)
China	1281	India	1628
India	1050	China	1394
United States	287	United States	413
Indonesia	217	Pakistan	332
Brazil	174	Indonesia	316
Russia	144	Nigeria	304
Pakistan	144	Brazil	247
Bangladesh	134	Bangladesh	205
Nigeria	130	DR of Congo	182
Japan	127	Ethiopia	173

Source: Population Reference Bureau, *2002 World Population Data Sheet* (Washington, DC: Population Reference Bureau, 2002).

| Table 6.3 | POPULATION DENSITIES OF THE TEN MOST POPULOUS COUNTRIES, 2002 |

Country	Population density per square kilometre
Bangladesh	927.6
Japan	337.1
India	319.3
Pakistan	180.3
Nigeria	140.5
China	134.0
Indonesia	113.9
US	29.7
Brazil	20.5
Russia	8.5

Source: Population Reference Bureau, *2002 World Population Data Sheet* (Washington, DC: Population Reference Bureau, 2002).

countries with the largest populations are not those with the highest densities. Of the ten most populous countries, Bangladesh is by far the most densely populated, followed by Japan and India.

Mapping World Population

Although the world map of population distribution and density (Figure 6.1) is valuable, it is a static picture of a dynamic situation, providing no indication of how the distribution developed or how it might change in the future. It shows three areas of population concentration: eastern Asia, the Indian subcontinent, and Europe. Both of the Asian areas are long-established population centres, locations of early civilizations and early participation in the agricultural revolution. Today, both include areas of high rural population density, especially in the coastal, lowland, and river valley locations and large urban centres. In Europe densities are high, but lower than in the high-density areas of Asia, and are related primarily to urbanization. In all three areas, population densities are clearly related to land productivity.

There are other scattered areas of high density: northeastern North America; around large cities in Latin America; the Nile Valley; and parts of west Africa. Perhaps the most compelling impression conveyed by Figure 6.1, however, is that a large proportion of the earth's surface is only sparsely populated.

Explaining the Map of World Population

Physical variables

There is a simple—but potentially misleading—correlation between Figure 6.1 and basic physical geography. Certainly humans have favoured some physical environments and not others. Even a cursory comparison of Figure 3.8 (our map of global environments) and Figure 6.1 suggests that three particular environments—monsoon, Mediterranean, and temperate forest areas—are associated with high population densities, while three others—desert, tundra, and polar areas—are associated with very low densities. Once again, though, it is important to remember that physical geography does not cause human geography; the correlation between the two simply reflects the fact that humans have recognized the relative attractiveness of certain

6.1

THE 1991 NIGERIAN CENSUS

The principal sources of information about the demographic characteristics of a country are its censuses. The enumeration of populations has a long history, but the first modern censuses were conducted in some European countries in the eighteenth century. Relatively reliable data are available for most European countries from at least the early nineteenth century. The united Province of Canada conducted its first comprehensive **census** in 1851. Almost all countries have conducted censuses during the second half of the twentieth century. In order to be useful, censuses need to be comprehensive (that is, to include all members of the population) and taken at regular intervals (in many countries, at least once every ten years).

In 1991 Nigeria conducted its first official census since 1963 (Porter 1992). The long delay between counts reflected both the cost involved (at least Can. $150,000) and, more important, two related political issues. First, because the spatial distribution of Nigerian government spending on services and amenities was linked to population numbers, some areas inflated their returns in the 1963 census, which as a result contained many errors; ten years later, the 1973 census was actually declared invalid

because the results, both total and regional, were clearly false. Second, since Nigeria gained independence from Britain in 1960, different ethnic groups have used population numbers as arguments in the ongoing debate over which group will exercise political control; the principal conflict is between a primarily Muslim north and a primarily Christian south.

During the early 1990s, the Population Reference Bureau, based in the United States, estimated the total population of Nigeria at 119 million, a figure based on the recorded 1963 total and assumptions about subsequent natural increase rates. The 1991 census therefore provided a major surprise when it reported a total population of only 88.5 million. Based on this census information, the 1992 estimate for Nigeria was 90 million and the 2002 estimate, as detailed in Table 6.2, is a total population of 130 million. Although the possibility of significant errors in the 1991 census cannot be discounted, there are sound reasons to be reasonably confident of the quality of the data.

This example offers an important lesson: that all data have to be carefully assessed for their reliability, and that any uses made of data need to acknowledge possible limitations.

areas, especially with respect to productivity.

At the global scale, four physical variables are relevant: temperature, availability of water, relief (the physical contours of the land), and soil quality. High-density areas typically have temperatures that permit an agricultural growing season of at least five to six months a year. Water is essential, whether it comes as precipitation or in the form of irrigation water. Optimum temperature levels and precipitation amounts are not easy to specify, since they vary with technology, but it is clear that extremely high temperature and precipitation together, as in tropical rain forests, are not associated with dense population.

Cultural variables

A second set of factors related to global population densities has to do with cultural organization. In many of the high-density areas shown in Figure 6.1, a form of state organization was established relatively early; China, India, southern Europe, Egypt, Mexico, and Peru were all centres of early civilizations. A strong state organization facilitates the concentration of population; conversely, the collapse of such organization (usually with the onset of war) works against continued concentration. However, not all areas of high density today experienced early state development: in western Europe and northeastern North America, high densities developed much later, in conjunction with the industrial revolution and the urbanization associated with it.

Together, the distribution and density of world population today represent the dynamic outcome of a long historical process. Is it possible that the pattern is stabilizing today? We know that certain areas are continuing to experience high rates of natural increase and are therefore likely to experience density increases. We also know that population continues to move from one location to another; indeed, some areas, such as northeastern North America, have grown largely as a consequence of migration.

Migration

Humans have always moved from one location to another—early pre-agricultural humans moved out of Africa to populate all major areas of the world except Antarctica. Such movements expanded the resource base available, facilitated overall population increases, and stimulated cultural change by requiring ongoing adaptations to new environmental circumstances. Migration has continued to be an important feature of the human occupation of the earth.

For our purposes, migration may be defined as a particular kind of mobility that involves a spatial movement of residence. Thus we do not regard the journey to work, or the trip to the store, or the seasonal movements of some pastoralists and agriculturalists as migration. Focusing on those movements that involve a change of residence, we are particularly concerned with the distance moved, the time spent in the new location, the political boundaries crossed, the geographic character of the two areas involved, the causes of the migration, the numbers involved, and the cultural and economic characteristics of those moving.

Why People Migrate

To ask why people migrate is not to suggest that there is a single reason behind the multitude of migrations that have taken place in human history. We may, however, propose a useful generalization.

Push–pull logic

People move from one location to another because they consider the new location to be more favourable, in some crucial respect, than the old location. This is the key idea of inequality from place to place. How much more favourable the new location needs to be is a matter of individual judgement.

In mid-nineteenth-century Ireland, for example, there was virtually unanimous agreement that some overseas location such as the United States, Canada, or Australia was preferable to Ireland. In this case, the advantages of a new location appeared to be so overwhelming that the decision to stay behind may have been just as difficult as the decision to leave, if not more so. In other cases, though, only a small percentage of the population will decide to migrate, suggesting that the perceived difference between old and new locations is not so substantial. If we rank each area (however defined) of the world on an attractiveness scale from, say, 1 to 10—1 being the least attractive and 10 being the most attractive—who will want to move where? The answer is that those living in

Eastern European immigrants to North America on board ship, c.1910–14 (National Archives of Canada c–68842).

low-ranked areas will wish to move to high-ranked areas; thus there will be large-scale movements from areas ranked low to areas ranked high. This scenario is a simple way of saying that migration decisions can be conceptualized as involving a push and a pull. Being located in an unattractive area is a push; being aware of an attractive alternative area is a pull. Table 6.4 summarizes some common push and pull factors.

Typical push–pull factors can be sorted into three categories: economic, political, and environmental. A simple *economic* explanation is that migration is a consequence of differences in wages, with people moving from low- to high-wage areas. Relatively low wages are a push; relatively high wages are a pull. Another economic explanation involves the relative availability of agricultural land. From perhaps the sixteenth century to the twentieth century, land was in shorter supply in Europe than in temperate areas overseas. A third economic explanation, and perhaps the most fundamental, applies in situations where life itself is threatened because of inadequate food supplies; in this case, any alternative area will be more attractive.

Another respect in which areas can be seen as unequal has to do with *politics*. The post-1945 desire of many eastern Europeans to migrate reflected assessment of the relative merits of communist and democratic political orders. In some extreme cases, people have felt obliged to seek refuge in a country other than their own. In recent years, the political environments in Afghanistan, Ethiopia, the former Yugoslavia, and Rwanda have all prompted massive refugee movements. (This topic is so important to human geographers that it will be addressed separately in the following section.)

Relative attractiveness can also be measured *environmentally*. Migration may be

Table 6.4	SOME PUSH AND PULL FACTORS

Push Factors	Pull Factors
Localized recession because of declining regional income	Superior career prospects, and increased income
Cultural or political oppression or discrimination	Improved personal growth opportunities
Limited personal, family, career prospects	Preferable environment (climate, housing, medical care, schools)
Disaster, such as floods, earthquakes, wars	Other family members or friends

Source: After D.J. Bogue, *Principles of Demography* (New York: Wiley, 1969):753–4.

induced by flooding and desertification, for example, and the historical experience indicates that most migrations have been towards the temperate climatic areas.

As obvious as push–pull reasoning may seem, it does have a serious limitation: it assumes that people invariably behave in a theoretically logical way. Specifically, it is unable to explain why some people decide to stay in an unfavourable area when a more favourable alternative is available. In other words, the concept of push–pull factors assumes that the decision to migrate is beyond the control of the individual.

Laws of migration

The logic behind the push–pull concept is areal inequality. Although this logic is valuable as a general explanation, applicable in many different circumstances, clearly it is not especially profound to assert that people move from A to B because B is in some way preferable to A.

A classic attempt to formulate specific 'laws' of migration was made in the late nineteenth century by E.G. Ravenstein (1876, 1885, 1889). Based on analyses of population movements in Britain, Ravenstein's laws are not laws in a formal, positivistic sense, but rather generalizations with varying degrees of applicability. The eleven generalizations that Ravenstein developed (Box 6.2) are still among the most valuable concepts we have for understanding migration. However, like push–pull logic generally, they too fail to take into account individual differences.

6.2

THE RAVENSTEIN LAWS

More than a century ago, E.G. Ravenstein wrote three articles on migration that have been highly influential in much subsequent research on the subject. On the basis of information contained in the British censuses of 1871 and 1881, Ravenstein identified eleven 'laws'—or, more correctly, generalizations. In the following list, each generalization is followed by a brief comment.

1. The majority of migrants travel only a short distance. The idea of distance friction, noted in Chapter 2, is one of the most fundamental geographic concepts; for more on the 'tyranny of distance', see Chapter 10, especially Box 10.2.
2. Migration proceeds step by step. Thus a migrant from Europe to Canada might go first to a port city such as Montreal and then to rural Quebec.
3. Migrants moving long distances generally head for one of the great centres of commerce or industry. This reflects the fact that large centres are usually better known than small ones to people from far away.
4. Each current of migration produces a compensating countercurrent. Any such countercurrent is usually relatively small.
5. The natives of towns are less migratory than those of rural areas. This law reflects the frequency of rural-to-urban migration.
6. Females are more migratory than males within their country of birth, but males more frequently venture beyond. Females often move within a country in order to marry. International migrants are usually young males.
7. Most migrants are adults. Families rarely migrate out of their country of birth.
8. Large towns grow more by migration than by natural increase. Remember that Ravenstein was writing at a time of dramatic industrial and urban growth.
9. Migration increases in volume as industries and commerce develop and transport improves. Such developments make urban centres more attractive and reduce distance friction.
10. The usual direction of migration is from agricultural areas to centres of industry and commerce. This is still the most common direction in the early twenty-first century, as evidenced by the ongoing rural depopulation in the Canadian prairies.
11. The major causes of migration are economic. This law too reflects Ravenstein's time: the present text includes several examples of migrations undertaken for social or political reasons.

Although these eleven statements have been modified by subsequent research, they have not been disproven. Clearly, they are somewhat time-specific, but all of them have some validity in different places and at different times (Grigg 1977). Probably Ravenstein's principal limitations are his neglect of various forms of forced migration and failure to account for the current exodus from large cities (see Chapter 12).

What is your impression of these laws? Do they make sense to you, both intuitively and with reference to the ideas introduced in this chapter? Or are they too general?

The mobility transition

A third effort to explain why people migrate was made by Zelinsky (1971). This theory is labelled the mobility transition. The term is derived from the concept of demographic transition (see Chapter 5). Zelinsky proposed five phases of temporal changes in migration:

1. the premodern traditional society;
2. the early transitional society;
3. the late transitional society;
4. the advanced society; and
5. a future superadvanced society.

Each of these five societies has particular migration characteristics. In Phase 1 there is minimal residential migration and only limited human mobility. Zelinsky sees this phase as temporally parallel to the first stage of the demographic transition (high birth rates, high and fluctuating death rates). In Phase 2, numerically significant migration begins in the form of rural to urban and overseas movements. This phase is temporally parallel to the second stage of the demographic transition (continuing high birth rates, rapidly falling death rates) and includes the mass movements associated with industrialization and European overseas expansion. In Phase 3, rural-to-urban mobility declines somewhat but remains numerically significant, while overseas migration is drastically reduced. This phase is temporally parallel to the third stage of the demographic transition (declining birth rates, low death rates). In Phase 4 residential mobility continues apace; rural-to-urban movement lessens but continues; urban-to-urban movement is significant; and international migration increases, with both unskilled and skilled workers moving from the less developed to the more developed world. This phase is temporally parallel to the fourth stage of the demographic transition (low birth rates, low death rates). Finally, in Phase 5—Zelinsky's attempt to predict future trends—most migration is between urban centres. There is no equivalent stage in the demographic transition model.

Zelinsky's mobility transition theory is a very useful summary of temporal changes in mobility. Essentially, it proposes two causal factors: first, the demographic context and, second, technology. Although it does not explain any specific migration, it offers a useful classi-fication scheme that helps us to generalize about migration and migration trends. More-over, the links it suggests with the demographic transition model are intriguing.

But Zelinsky's theory is not without its crit-ics. The mobility transition can be regarded as an example of **developmentalism**—as a 'geography of ladders' in which 'the world is viewed as a series of hearth areas out of which modernisation diffuses, so that the Third World's future can be explicitly read from the First World's past and present in idealised maps and graphs of developmental social change' (Taylor 1989:310). An alternative way of look-ing at the mobility history of a particular country would be to see it as reflecting global processes as well as those taking place within the country. The same issue arose, somewhat less formally, in our discussion of the demo-graphic transition; we will pursue it in a more general context later in this chapter.

A behavioural explanation

Another way of approaching the question of why people migrate is from a more humanis-tic perspective. None of the three approaches outlined so far has focused primary attention on the people themselves. Push–pull logic, the eleven laws of Ravenstein, and the mobility transition model all have positivistic over-tones: they all seem to imply that each indi-vidual human responds in an identical fashion to various external factors. Of course, any human geographer knows that this is not so, but many of us have been willing to sacrifice some reality for the sake of simplicity. Our fourth and final attempt to explain why peo-ple migrate shifts attention to the people themselves and away from the forces pre-sumed to be affecting their decisions.

For convenience, we will label this approach the behavioural view of migration, since it centres on the behaviour of individ-uals rather than on aggregate, usually large-group, behaviour. Interestingly, behavioural approaches usually employ a version of push–pull logic, but typically do so at the level of the individual.

Place utility is a measure of the extent to which an individual is satisfied with particu-lar locations. Typically, the place utility that people attribute to their current place of res-idence is much better grounded in fact than the place utility they attribute to other loca-

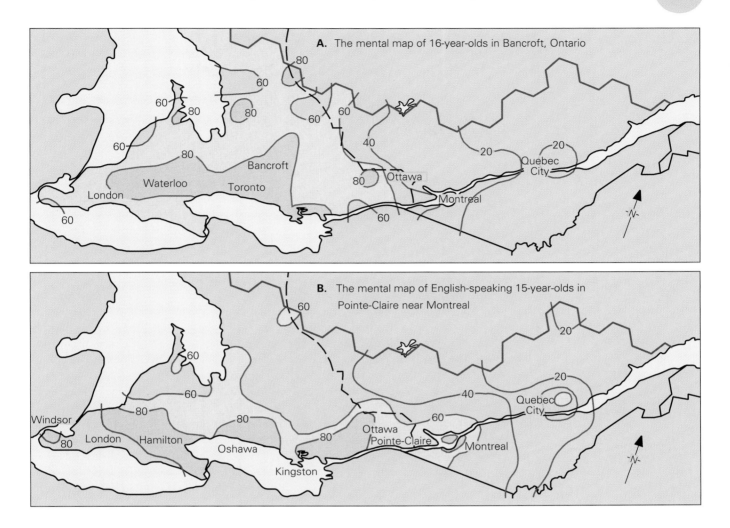

Figure 6.2 Mental maps; numbered 'isolines' represent spatial preferences on a scale from 1 to 100. These particular examples demonstrate that, in some cases, cultural considerations are more important than distance. Map A represents the mental map of 16-year-olds in English-speaking Bancroft, Ontario. Note the preference for larger urban areas and, especially, the sharp drop in ratings across the border with Quebec. Map B represents the mental map for English-speaking 15-year-olds in Pointe-Claire, near Montreal, in the predominantly French-speaking province of Quebec. Instead of favouring their local area, these young anglophones favour various areas in English-speaking Ontario. Source: Adapted from P. Gould and R. White, *Mental Maps*, 2nd ed. (Boston: Allen and Unwin, 1986):77-9.

tions. Place utility is thus an individually focused version of push–pull logic. This concept was first introduced by Wolpert (1965), who argued that it was necessary to research an individual's **spatial preferences**.

All too clearly, such preferences are based on perceptions, not objective facts. All people have what we call 'mental maps' (see Chapter 2): mental images of various places, images that contribute to migration decisions. To determine what these maps are, geographers typically use questionnaires asking individuals for their perceptions of different areas. The pioneering work in this area was done by Gould and White (1986), who gathered the data on group perceptions (spatial preferences) mapped in Figure 6.2. In principle, we can use maps such as these to predict migration behaviour.

The selectivity of migration

Migration is a selective process. Objectively speaking, as our discussion of place utility emphasized, areas in themselves are neither

unattractive or attractive. Relative attractiveness is a matter of subjective, individual perception: what is attractive or unattractive to one person may not be so to another, and migration decisions are, in the final analysis, made by individuals. We are not all equally affected by the general factors prompting migration. Who moves and who stays?

Among the factors that appear to influence individual decisions about migration are age (most migrants are older adolescents or young adults); marital status (today, as in the past, most of the people migrating from the less developed world are single adults); gender (males are typically more migratory, but there are many exceptions to this generalization); occupation (higher-skilled workers are most likely to move); and education (migrants have higher levels of education than non-migrants). In the most general sense, there is a useful relationship between **life cycle** and the likelihood of individual migration.

In addition, there is often a substantial difference between what people would *like*

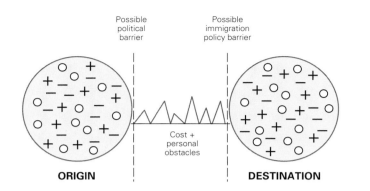

Figure 6.3 The push–pull concept and relevant obstacles.

to do and what they are *able* to do. People who want to migrate may be unable to leave their home for political reasons, and people who want to move to a specific new area may not be able to do so because of the immigration policies of that area. Similarly, potential migrants need to consider the economic and personal costs; some may be unable to pay for the move; some may not be able to move because of health, age, or family circumstances.

These ideas can be usefully conceptualized as shown in Figure 6.3. Origins and destinations alike have good (+), bad (–), and neutral (0) attributes. Sometimes movement

away from the place of origin is prevented by political barriers; sometimes movement to the destination is prevented by immigration policies. Between these two external barriers to migration are various costs and personal obstacles (Lee 1966).

Moorings

The studies in migration that we have looked at here are useful, but they are not conceptually sophisticated by the standards of much contemporary human geography; indeed, 'migration research is in danger of being left behind by recent developments in social theory' (Halfacree and Boyle 1993:337). Perhaps the greatest limitation, particularly in push–pull logic, the laws of migration, and the mobility transition theory, is the tendency to base ideas on inflexible assumptions about human behaviour. Even though the choice to migrate is typically a major life decision, determined to a large extent by individual personality and the culture of the group to which the individual belongs, much migration research focuses exclusively on the material aspects of the decision.

One way of advancing studies in this area is to acknowledge that migration reflects a personal decision made within a larger political and economic framework—for example, by drawing on theories of human motivation as developed in social psychology and putting greater emphasis on the cultural influences on migration. An example is the idea of 'moorings'—issues through which individuals give meaning to their lives—suggested by Moon (1995). Table 6.5 identifies some typical 'moorings'.

The moorings approach centres on the idea that individuals' perception of their current location, and hence the likelihood of their either remaining there or migrating to another location, depends on the value they place on their various moorings. It seems probable that future efforts to explain why people migrate will increasingly focus on conceptual formulations that incorporate both the personal and the cultural aspects of the migration decision.

Types of Migration

One of the most useful attempts to classify migration was made by the sociologist Petersen (1958), who identified four classes of migration:

Table 6.5　**SOME TYPICAL MOORINGS**

Life-Course Issues
　　household/family structure
　　career opportunities
　　household income
　　educational opportunities
　　care-giving responsibilities

Cultural Issues
　　household wealth
　　employment structure
　　social networks
　　cultural affiliations
　　ethnicity
　　class structure
　　socio-economic ideologies

Spatial Issues
　　climate features
　　access to social contacts
　　access to cultural icons
　　proximity to places of recreation interest

Source: B. Moon, 'Paradigms in Migration Research: Exploring "Moorings" as a Schema', *Progress in Human Geography* 19 (1995):515.

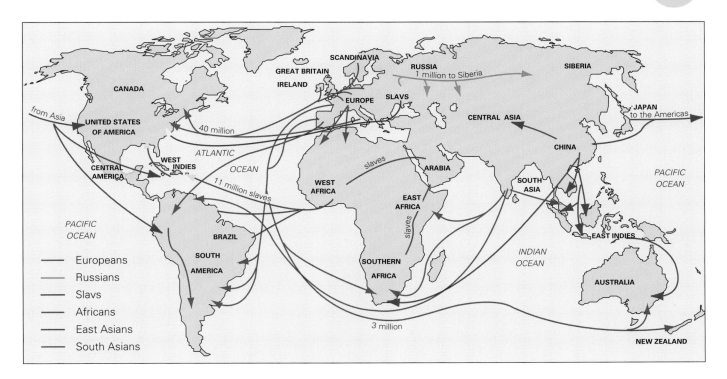

Figure 6.4 Major world migrations, 1500–1900.

1. primitive migration, associated with preindustrial peoples and caused by some ecological necessity;
2. forced migration, in which people have little or no alternative but to move, usually as a result of political circumstances;
3. free migration, in which people decide to move or stay on the basis of place utility; and
4. illegal migration, which can take two forms: illegal exit occurs when a country prohibits out-movement; illegal entry occurs when people enter a country without official approval.

Figure 6.4 provides an overview of the principal forced and free migrations that have taken place at the global scale since about 1500.

Primitive migration

Primitive migration is really a specific instance of adaptation to environment, in which people respond to an unfavourable environment by leaving it for a more favourable one. Preindustrial societies tend to make such adaptation decisions on a group rather than an individual basis. Thus hunting and gathering groups might migrate regularly as the resources of an area are depleted or as game animals move on; some agricultural groups might move as soil loses fertility. Another instance of primitive migration occurs when populations increase in size so that additional land is needed. In preindustrial societies, primitive migration was a normal part of the human search for appropriate environments— the search that was responsible for the human occupation of most the earth. As we noted in Box 4.1, by the fifth century CE, even Easter Island, one of the most isolated locations in the world, was settled.

Forced migration

Forced migration has a long history. **Slavery** was an indispensable institution in early civilizations such as those of Greece and Rome. It appears that slaves forced to migrate from areas occupied by the Romans made up the largest portion of Rome's population at its peak. The Europeans who colonized the Caribbean and the warm coasts of North and South America—areas not conducive to large-scale European migration—also relied on slave labour. Perhaps as many as 11 million slaves were moved out of Africa between 1451 and 1870.

A second example of forced migration can be seen in the late nineteenth century, when workers from China, Java, and India were shipped to the new European-controlled plantations of Malaysia, Sumatra, Burma, Sri Lanka, and Fiji. These workers were supposed to be engaged voluntarily, on the basis of contracts, but in fact force was often used (Box 6.3). A third and quite differently motivated example was the post-1938 movement of Jewish populations in areas controlled by Nazi Germany. In each of these instances, movement was literally forced on people.

A variant of forced migration is the situation in which the migrant has some voice, however small, in the decision-making process. Examples of such *impelled* migration would include the many cases in which people have chosen to flee oppressive political regimes, war zones, and areas of famine. The dividing line between forced and impelled is not clear, nor is the line between impelled and free. Most contemporary refugee movements would qualify as impelled.

Free migration

In free migration the person has the choice either to stay or to move. Historically, free migrations moved from densely settled countries to less densely settled ones, as in the period between 1800 and 1914, when some 70 million people migrated from Europe to temperate areas such as the United States, Canada, Australia, New Zealand, South Africa, and Argentina (see Box 6.4). As we saw in our discussion of the mobility transition, this migration was closely related to the demographic and technological changes that began in Europe after about 1650.

Thus emigration relieved Europe of at least some of the population pressures that came with the second stage of the demographic transition. In 1800, people of European origin totalled 210 million; by 1900, the total was 560 million—a 166 per cent increase—and one in every three people in the world was of European origin. Most European countries participated in the nineteenth-century wave of migration. Irish, English, Scottish, Germans, Italians, Scandinavians, Austro-Hungarians, Poles, and Russians all moved overseas in large numbers; at the same time, many other Russians moved east to the Caucasus and Siberia.

6.3   'A NEW SYSTEM OF SLAVERY': INDIAN INDENTURED LABOUR IN MAURITIUS

From a broad global and historical perspective, unfree labour has been normal: free labour, in which the labourer has the right to choose an employer, did not become usual in Europe until the late eighteenth century. Indeed, in parts of Europe, slavery persisted into the nineteenth century; serfdom (an essential social relationship in feudalism that involved the legal subjection of peasants to a lord) was not abolished in Poland until 1800 and in Russia until 1861. Forms of unfree labour were also central to the evolution of the European colonial world. To produce tropical products and precious metals overseas, European powers relied first on slavery. Following the abolition of slavery by the British in 1834, the French in 1848, and the Dutch in 1863, these powers turned to indentured (that is, contracted) labour. The migration of indentured labourers became a key component in the global system by which Europeans combined cheap labour and abundant land in order to make big profits in their colonial areas.

The majority of indentured labourers—perhaps as many as 1.5 million—came from India between the years 1830 and 1916 (Tinker 1974). The British were able to establish and maintain this system as a result of their penetration into the Indian economy and society, a penetration that brought commercialization of agriculture, payment of rents in cash rather than kind, a decline in traditional crafts, and discriminatory taxation. Togeth-er, these changes had a severe impact on the lower agricultural classes and served to make the prospect of emigration attractive. In the course of the nineteenth century more than 500,000 indentured labourers left India for the island of Mauritius (a British colony in the Indian Ocean), where most worked on sugar cane plantations.

In 1837 British legislation laid down the rules to be followed in organizing the emigration of indentured labourers. In principle, the labourers contracted to work for a certain period (normally five years) in exchange for their passage. In reality, however, forced banishment, kidnapping, and deception were common parts of the emigration process. Once overseas, indentured workers were not slaves, but their freedom was severely constrained. Movement outside the plantations was restricted, and the opportunities available after contracts expired were limited by measures such as taxes and vagrancy laws. Overwork, low wages, poor food, illness, low-quality housing, and inadequate medical and educational facilities were usual. Typically, any attempt at resistance on the part of workers was met by stringent labour legislation. Although one of the conditions of indenture was to offer all workers transport back to India after ten years, this responsibility was neglected in Mauritius after 1851. Very few indentured labourers ever returned home, and their descendants are the majority population in Mauritius today.

This historically brief period of movement had massive and wide-ranging effects in areas of both origin and destination, redistributing a large number of people and bringing into contact many previously separate groups. Indeed, migration lies at the root of some of the most difficult political issues in the world today, especially those involving the territorial claims of various minority groups: in Quebec, for example, non-francophone immigrants were blamed for the failure of the 'yes' side in the 1995 referendum on separation. Box 6.5 considers some of these issues in the context of the Canadian immigration experience.

Free migration continues today, but now all of the more developed countries are popular destinations, including the European states that so many earlier migrants left behind. This situation confirms our observation that individuals' migration decisions largely reflect their assessments of relative place utility.

Many decisions to migrate, however, cannot be carried out, because most developed countries have implemented restrictive immigration policies. In the early twentieth century, the immigration policies of countries such as the United States, Canada, Australia, New Zealand, and South Africa were explicitly racist (see Box 6.6). Today, the reasons for restricting immigration are more likely to be economic.

In fact, much free migration today takes place within rather than between countries. In the United States, the south and west are attractive destinations for a combination of reasons having to do with climate and job opportunities; the situation is similar in England, where the southeast is the most attractive area. In many countries, people continue to move from rural to urban areas and from central to suburban zones within urban areas. Because these movements tend to be selective, reflecting the migrants' stage in the life cycle, gender, and ethnic background, they often lead to significant changes in the structure and composition of local populations. Most migration continues to be primarily economic in motivation, being related to employment, income potential, and the housing market.

Illegal migration

The term 'illegal migration' covers a wide variety of situations. Although it is obviously not possible to determine the exact number of people involved, it is certainly significant. The most important category of illegal immigrants consists of those who consciously violate immigration laws, but other immigrants become 'illegal' through no fault of their own, as a result of policy changes or the complexities of maintaining legal residency. (One example of a policy change that creat-

6.4

SCANDINAVIAN MIGRATION TO NORTH AMERICA

Significant migration from Denmark, Norway, Sweden, and Finland to North America began in 1825. For several decades, considerable movement had taken place within Scandinavia, especially from rural to new urban areas, and also from Scandinavia to elsewhere in Europe, especially the Netherlands. But the overseas migrations were far greater in number than the earlier movements within Europe. A small number of migrants overseas went to Australia and South Africa, but the favoured destinations were, first, the United States and, second, Canada. Overall, Scandinavian migrants ranked second to the Irish in total number.

The movements from Scandinavia were prompted by both push and pull factors (Ostergren 1988). In many cases migration reflected dissatisfaction with authority. For example, some religious dissidents moved because they rejected the state church, and Scandinavians, especially Danes, were among the very first Mormon settlers in Utah. Increasing populations and related problems of rural landlessness in Scandinavia were more general push factors, affecting large numbers of people. The principal pull factors were the perceived opportunities to work and to obtain land.

Within North America, Scandinavian settlers favoured the land-rich areas of the American midwest and the Canadian prairies. Arriving in various waves, migrants from Norway settled mostly in Illinois, Wisconsin, Iowa, Minnesota, South Dakota, and North Dakota; those from Finland in Minnesota and Michigan; those from Sweden in Minnesota, South Dakota, and North Dakota; and those from Iceland in Manitoba. Most Scandinavian migrants came in family groups, often as part of a larger kinship network, and the new communities encouraged the maintenance of established traditions.

ed illegal immigrants involved several million west Africans who had legally immigrated to Nigeria during the 1970s and early 1980s; their status was changed after a new government changed the immigration regulations.) Other migrants, however, do deliberately violate immigration laws. The simple explanation sees such illegal movement as the result of desperate push factors (overpopulation, political turmoil, economic crises) combined with irresistible pull factors (high wages, plentiful job opportunities). This explanation is accurate in many cases, but it is not adequate. Indeed, much illegal movement takes place between more developed countries; for example, a 1993 analysis concluded that Italians made up the largest group of illegal immigrants in New York City. Most illegal immigrants are young, clustered in urban areas, and involved in industries such as construction and hospitality. The 1990 illegal alien population in the US was conservatively estimated at 3.4 million.

One of the best-known cases of illegal immigration is that of Mexicans moving into the US; every night, the border sees several

6.5

MIGRATION AND ETHNIC DIVERSITY IN CANADA

The population of Canada has changed dramatically since the beginning of the twentieth century, both in total numbers and in ethnic composition. In 1901 the country had 5.3 million people, most of British origin; in 2002 Canadians numbered 31.3 million, and made up one of the most ethnically diverse populations in the world. Four relatively distinct waves of migration can be identified during the twentieth century. The years 1901 to 1921 brought dramatic growth; many of these immigrants came from the new source areas of eastern and southeastern Europe and headed for the prairies. For some years during this period, the annual population growth rate was almost 3 per cent, but economic recession abruptly ended this period in 1921, and lower annual population growth rates of about 1.4 per cent were usual until the end of the Second World War.

The next major increase in immigration, between 1951 and 1961, joined with high fertility to produce an average annual growth rate of 2.7 per cent. In this period, during which the Canadian government was anxious to fill labour shortages in both the agricultural and industrial sectors, it nevertheless continued to show a strong preference for British and other European settlers. Potential immigrants from Africa, the Caribbean, and Asia were subject to quotas (see Box 6.6 for the larger context of such policies).

Since 1961, both immigration and fertility in Canada have declined, resulting in annual growth rates of about 1.3 per cent. In 1962, Canada's immigration rules dropped all references to race and nationality, and in 1967 a points system was introduced that reflected economic needs and gave particular weight to education, employment qualifications, English- or French-language competence, and family reunification. Thus policies in this most recent phase have been relatively liberal, permitting significant diversification in the ethnic mix of immigrants. Immigrants from Europe and the United States accounted for about 95 per cent of the total in 1960, but make up only about 20 per cent today. Application of the points system has resulted in an

Fifty-three new Canadians taking the citizenship oath in a ceremony at the Museum of Civilization in Gatineau, Quebec, on 1 July 2002 (CP photo/Jonathan Hayward).

increase in the number of qualified professionals, many from Asia and the Caribbean. More generally, there have been huge increases in immigrant numbers from less developed countries. A new immigration act in 1978 had three principal objectives: to facilitate family reunification, to encourage regional economic growth, and to fulfil moral obligations to refugees and persecuted people. Although subsequent acts have modified details, especially concerning refugee claimants and illegal immigration, their overall thrust has remained the same.

Canada's liberal immigration policies over the last four decades have benefited the country in many ways. Yet increasing ethnic diversity has made the establishment of a coherent cultural identity an elusive goal. Moreover, some Canadians' reactions against that diversity make it clear that we still have a long way to go if we are to free ourselves of racist attitudes.

thousand attempts at illegal immigration. But there are numerous other examples. Approximately 10,000 illegal Chinese immigrants enter the US each year; there is also a significant illegal Chinese movement into Canada. Australia is experiencing overstays of legal temporary admissions; the total number of overstays recorded by 1992 was 81,500—a significant number when compared with the legal immigration in that year of 35,100. The countries of the European Union may have as many as 3 million illegal immigrants from Africa, eastern Europe, and Asia. Major movements of illegal immigrants occur within Asia, where the favoured destinations are Japan and Singapore. Some of this movement occurs between more developed countries (for example, from South Korea to Japan). Before 1997, when it was returned to China, Hong Kong received many illegal immigrants from the mainland, while Singapore, despite very tight policies, has received many illegal immigrants from Malaysia.

These examples are only the best-known. Although the motives for illegal immigration in many cases appear obvious, efforts at

6.6

RESTRICTIVE IMMIGRATION POLICIES

Discriminatory immigration policies were typically introduced in the nineteenth and early twentieth century in those areas experiencing large-scale immigration. Although in some cases certain European groups were also targeted, in general such policies were specifically intended to limit immigration from Asia.

As we saw in Box 6.3, the termination of a slave trade prompted the introduction of an indentured labour system, initially involving Indian labourers on the island of Mauritius but soon involving others as well in the Caribbean, East Africa, South Africa, and some Pacific Islands, notably Fiji. The success of this system, as measured by the plantation owners, resulted in its expansion to include Chinese indentured labourers, most of whom went to the Caribbean and South America.

In the mid-nineteenth century, Chinese also moved to North America and Australia as free migrants, prompted in both cases by the discovery of gold. Many worked in transcontinental railroad construction, first in the United States in the mid-1860s and then in Canada in the 1880s. But their presence was quickly opposed by local populations motivated by some combination of racist (see Box 3.5) and economic fears.

In countries that were part of the British empire, the agreement of the British government was required before restrictive immigration policies could be implemented. Had this not been the case, it is probable that in some colonies restrictive policies would have been in place much earlier. In Australia, restrictions were imposed following the immigration of Chinese labourers and gold field workers. The first legislative action, in Victoria in 1855, was followed by similar legislation in other states. But the most explicit restriction was introduced in the South African province of Natal in 1897. This was a requirement that all new immigrants be proficient in a European language. Although, in principle, such a restriction is 'colour blind', in practice it is not. New Zealand introduced a restrictive immigration policy in 1881 but replaced it with a literacy requirement in 1899. In Australia, the in-movement of all coloured peoples was similarly limited:

the requirement that all immigrants be literate in a European language marked the beginning of the 'White Australia Policy'.

In Canada, similar tactics were used in British Columbia, where Chinese immigrants had at first been welcomed as cheap labour; discrimination began in 1885 with the introduction of a head tax, and in 1923 Chinese immigration was virtually prohibited by an act that was not repealed until 1947. Immigration from India began about 1900 but was largely restricted in 1908 by the requirement that immigrants arrive by a 'continuous voyage'—this at a time when there were no direct voyages between India and Canada. Japanese immigration, however, was not severely curtailed during the period of the Anglo-Japanese Alliance (1902–22).

In the United States the Chinese Exclusion Act was in place from 1882 until 1943 and restrictions were imposed on most other Asian groups after 1917. These policies were greatly expanded in 1924 to include some European groups, especially from eastern Europe, because of widespread concerns about political radicalism as well as the various nationalisms and economic dislocations of the postwar period. The 1924 act included an intentionally discriminatory system and set quotas on the numbers of immigrants from specified groups. One term used to identify the outlook that prompted such policies is **nativism**: the protection of the interests of the native-born population against those of 'foreign' minorities. It was not until after the Second World War that these restrictions were gradually relaxed.

By the time of the First World War, then, many of the countries of European overseas settlement had in place restrictive immigration policies directed specifically at Asians, policies that continued for several more decades. Of the motivations behind such policies, racist attitudes are the most obvious; British immigrants were typically regarded as the most desirable. In addition, however, there was often fear of the economic competition that Asian immigrants—as an abundant source of cheap labour—would represent.

In August 1999 members of the Canadian Forces, Canadian Coast Guard, and RCMP rescued 190 Chinese migrants from a rusting unmarked ship off the coast of Vancouver Island. Suspected of intending to enter Canada illegally, they were taken first to Gold River, BC, where this photo shows them wrapped in blankets, waiting on the dock, and eventually to Esquimalt for processing (CP photo/Chuck Stoody).

Figure 6.5 Refugee numbers, 1960–2002.
Source: Updated from United Nations High Commissioner for Refugees, *The State of the World's Refugees: The Challenge of Protection* (London: Penguin, 1993).

explaining other cases are complicated by the absence of precise data. What we can say with some certainty is that international migration, legal and illegal, reflects globalization processes, specifically a growing interdependence among the world's countries and a web of international relations that is becoming ever more complicated.

Refugees

In general, population migrations can be seen as efforts by humans to settle the surface of the earth in a rational fashion. Once politically unconstrained, contemporary migrations between countries are now subject to a host of restrictions. While political considerations often play a role in international migration, intranational movements of populations—migrations within a given country—are not usually determined by political

matters. Thus in most cases we can interpret intranational migration as a good reflection of relative place utilities.

Refugee Movements: A Growing Problem

The first major movements of refugees after the Second World War took place in response to changing political circumstances. Immediately after 1945, about 15 million Germans relocated. The partition of India in 1947, which created the Muslim state of Pakistan, caused the movement of about 16 million people—8 million Muslims fled India for the new state, while 8 million Hindus and Sikhs fled Pakistan for India. A rather different migration took place prior to the construction of the Berlin Wall in 1961, when some 3.5 million moved from what was then communist East Germany to democratic West Germany. Even after the wall was in place, about 300,000 succeeded in fleeing west before the country was reunited in 1989.

From about 1960 to the mid-1970s, the annual total number of refugees in the world was relatively low: between 2 and 3 million (Figure 6.5). However, in 1973 the end of the Vietnam War gave rise to one of the first large transcontinental movements of refugees; some 2 million people fled Vietnam, necessitating a major international relief effort. Because of its involvement in the war, the US received about half of these refugees. The flow of refugees from Vietnam continued until the early 1990s. More generally, the 1980s saw a huge increase in numbers of refugees for various political, economic, and environmental reasons. Since then the total number of refugees has fallen from a high of 17.8 million in 1992 to 12 million in 2001. Table 6.6 shows both the number of refugees and the total number of persons of concern to UNHCR for the period 1980 to 2001.

The Problem Today: Numbers and Causes

It is generally agreed that refugees are people forced to migrate, usually for political reasons. There is, however, no agreement on how to determine the legitimacy of refugee claims; there are also considerable logistical difficulties in counting refugees, and governments often have a vested interest in providing incorrect information on numbers.

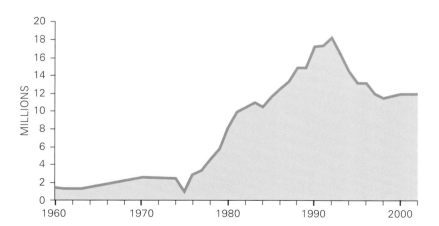

Accordingly, estimates of total numbers of refugees vary.

The most reliable source of information on refugee numbers today is the United Nations High Commission for Refugees (UNHCR). A few definitions are important at the outset of this discussion. According to the 1951 Convention Relating to the Status of Refugees, a refugee is a person who 'owing to a well-founded fear of being persecuted for reasons of race, religion, nationality, membership in a particular social group, or political opinion, is outside the country of his nationality, and is unable to, or unwilling to avail himself of the protection of that country.' Although most countries in the world offer protection to their citizens, some do not—either because they choose not to or because they are unable to do so—and it is these countries that prompt refugee movement. Today, most refugee movements are prompted by civil wars and forms of ethnic conflict. A country that receives refugees is known as a country of asylum.

In addition to refugees, the UNHCR includes three other classes of persons under the heading 'persons of concern to UNHCR':

1. Asylum seekers are people who have left their home country and applied for refugee status in some other country, usually in the more developed world.
2. Returnees are refugees who are in the process of returning home (the desired goal for most refugees); in recent years the UNHCR has been involved in major repatriations in many places, including Afghanistan, Iraq, Myanmar, Cambodia, and several African countries.
3. Internally displaced persons (IDPs) are people who flee their homes but remain within their home country; unlike refugees, they do not cross an international boundary.

Table 6.7 presents some very disturbing data at the scale of global regions. For 2001, the total number of persons of concern to UNHCR globally was about 19.8 million: one in every 300 of the world's people. The principal problem areas for both refugees and IDPs were Africa, Asia, and Europe. Most asylum seekers were located in countries in Europe and North America, and most of the

returnees in Africa. But even a figure of 19.8 million seriously underestimates the problem. According to UNHCR estimates, there are probably as many as 50 million people worldwide who have been forced to leave their homes—the difference between that figure and the official figure of 19.8 million reflects many distressing situations for which

Displaced people fleeing from Totota, Liberia (about 100 km from Monrovia), September 2003. Tens of thousands of civilians fled the refugee camps in response to sounds of gunfire and mortars and rumors of attacks nearby (AP photo/Pewee Flomoku).

Table 6.6	REFUGEES AND PERSONS OF CONCERN TO UNHCR, WORLDWIDE, 1980–2001	
Year	Refugees	Persons of concern
1980	8,439,000	
1981	9,696,000	
1982	10,300,000	
1983	10,602,000	
1984	10,710,000	
1985	11,844,000	
1986	12,614,000	
1987	13,103,000	
1988	14,319,000	
1989	14,706,000	
1990	17,370,000	
1991	16,829,000	
1992	17,802,000	
1993	16,242,000	23,033,000
1994	15,637,000	27,419,000
1995	14,855,000	26,103,000
1996	13,312,000	22,729,000
1997	11,966,000	22,376,000
1998	11,430,000	21,460,000
1999	11,626,000	22,257,000
2000	12,062,000	21,814,000
2001	12,051,000	19,783,000

Source: United Nations High Commissioner for Refugees.

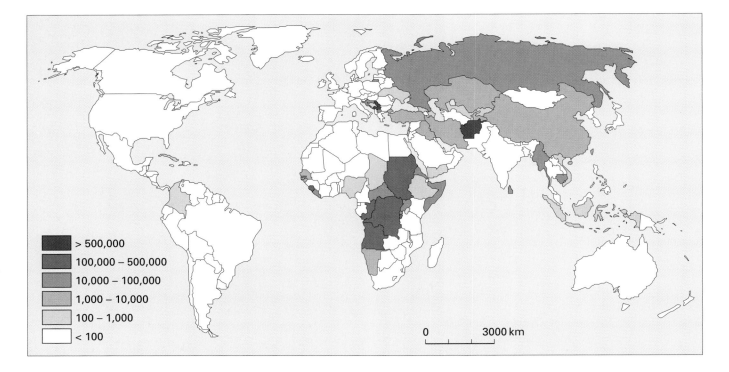

Figure 6.6 Refugee outflows by origin, 1997–2001.

reasonably detailed data are simply not available, as well as a variety of situations that are more difficult to categorize. There are, for example, people in refugee-like circumstances in Bangladesh, where about 258,000 Pakistanis have been stranded since the partition of 1947. Similarly, there are about 800,000 Iranians in Turkey who are not recognized as refugees.

Table 6.8 provides more detailed data on the ten largest refugee groups by country of origin. Afghanistan is the home country of the most refugees—more than 3.8 million in 2001. Figure 6.6, mapping refugee outflows by origin for the period 1997–2001, highlights the key problem areas (east and central Africa and west Asia). Not included in the table or the map are an estimated 3.9 million Palestinian refugees because, as of 2003, they remain without a territory of their own. Numerically less notable but nevertheless locally significant refugee or related problems also exist in places such as Sri Lanka and several Latin American countries.

It is often suggested that the vast majority of refugees and other persons of concern to UNHCR are women and children, but this is not the case. Globally, of the total of 19.8 million persons of concern, 48.1 per cent are female and 11.6 per cent are under the age of five. These percentages do not differ radically from standard gender and age distributions. It seems that when a population is displaced *en masse*, its demographic structure remains relatively balanced.

Most refugees move to an adjacent country, and this can result in some very complicated regional patterns when the same country is both a country of asylum for some refugees and the home from which

Table 6.7	PERSONS OF CONCERN TO UNHCR, 2001				
Region	**Refugees**	**Asylum seekers**	**Returned refugees**	**IDPs and others of concern**	**Total**
Africa	3,305,070	107,159	266,804	494,494	4,173,527
Asia	5,770,345	33,111	49,246	2,967,964	8,820,666
Europe	2,227,900	335,675	146,457	2,145,645	4,855,377
Latin America and Caribbean	37,377	7.878	194	720,000	765,449
North America	645,077	441,681			1,086,758
Oceania	65,361	15,587		313	81,251
Total	12,061,120	940,791	462,701	6,328,426	19,783,028

Source: United Nations High Commissioner for Refugees.

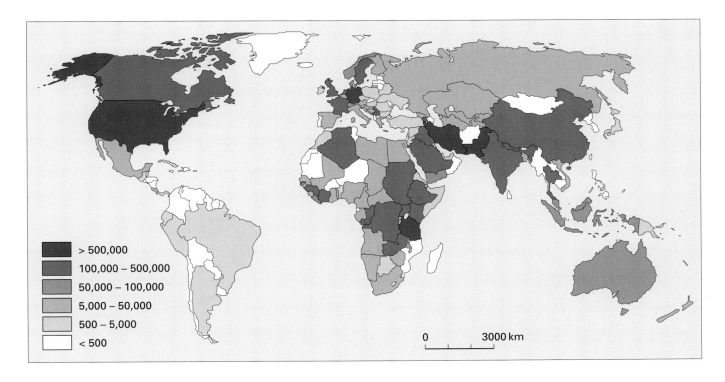

others have fled. Such situations often arise in the context of ethnic conflict and civil war, and those affected are usually in serious need of aid and support. As of 2002, examples included the east African countries of Sudan, Somalia, Ethiopia, and Djibouti; the central African countries of Burundi, Rwanda, Uganda and DR of Congo; and the west Asian countries of Afghanistan, Iran, and Iraq. Some other refugees are able to move to countries in the more developed world. Table 6.9 lists countries of asylum with more than 100,000 refugees, and Figure 6.7 shows refugee population by country of asylum. Together, this table and figure confirm that, although most refugees are in countries adjoining their homelands, many others have obtained legal status in the more developed world, whether in the United States, Canada, Germany, or the United Kingdom. Most countries in the more developed world receive locally significant numbers of refugees.

Refugee and related problems are greatest, however, in the less developed world, where few asylum countries have the infrastructure to cope with the additional pressures that refugees bring. In the more developed world, the collapse of communism in several European countries in the late 1980s, the breakup of the former USSR in 1991, and the conflict in the former Yugoslavia after 1991, all gave rise to mass refugee movements.

Solutions?

Different refugees have different reasons for moving, and there are no simple ways to resolve their many different cases. The UNHCR has traditionally proposed three

Figure 6.7 Refugee population by country of asylum, 2001.

Table 6.8	REFUGEE NUMBERS BY COUNTRY OF ORIGIN AND COUNTRY OF ASYLUM, 2001: THE TEN LARGEST GROUPS

Country of origin	Main countries of asylum	Number
Afghanistan	Pakistan, Iran, Netherlands	3,809,600
Burundi	Tanzania	554,000
Iraq	Iran	530,100
Sudan	Uganda, DR of Congo, Ethiopia, United States	489,500
Angola	Zambia, DR of Congo, Namibia	470,600
Somalia	Kenya, Ethiopia, Yemen, United States	439,900
Bosnia and Herzegovina	Yugoslavia, United States, Sweden, Denmark	426,100
DR of Congo	Tanzania, Congo, Zambia, Rwanda, Burundi	392,100
Vietnam	China, United States	353,200
Eritrea	Sudan	333,100

Source: United Nations High Commissioner for Refugees.

solutions: voluntary repatriation, local settlement, and resettlement. More recently, the focus has turned to attacking the underlying causes of refugee problems. But all attempts at solutions are problematic.

The UNHCR tends to favour *voluntary repatriation*, but this is not possible for most refugees, since in most cases the circumstances that caused them to leave have not changed.

Local settlement is difficult in areas that are poor and lack resources. Whatever the reason behind a refugee movement—political, environmental, or otherwise—nearby areas often face similar problems, and hence are rarely able to offer solutions. People in areas that are already poor and subject to environmental problems, such as drought in parts of eastern and southern Africa, will be hard-pressed to provide refugees with the food, water, and shelter they need (Box 6.7).

Resettlement in some other country is an option for only a few. No country is legally obliged to accept refugees for resettlement,

and only about twenty countries do so on a regular basis (see Table 6.9). Attacking the root causes of refugee problems is a mammoth challenge. Not only are those causes often a complex mixture of political, economic, and environmental issues for which no simple solutions are available, but refugee problems are greatest in parts of the less developed world that already face huge challenges.

The Less Developed World

Many of the topics introduced in earlier chapters are directly relevant to the emergence of what we are calling the less developed world. As we saw in Chapter 1, the growth and institutionalization of geography in the nineteenth century was largely spurred by the perceived importance of the discipline as a source of information about the parts of the world 'discovered' by Europeans during the period of overseas exploration. The concept of race introduced (and debunked) in Box 3.5 played an important part in justifying Europeans' colonization and exploitation of those parts of the world inhabited by 'inferior races'. Chapter 4 pointed out that a key distinction between more and less developed countries concerns energy: both the quantities used and the sources exploited. Finally, Chapter 5 painted a vivid picture of global differences in such basic demographic measures as birth rates, death rates, rates of natural increase, life expectancy, and infant mortality. These and related ideas are the subject of the present section. They will also reappear in later discussions of the political world, agriculture, settlement, industry, and, most generally, globalization processes.

What Is the Less Developed World?

The term 'Third World' was first used in the early 1950s, in the context of suggestions that former colonial territories might follow a different economic route from either the capitalist 'First' or the socialist 'Second' World. By 1960, 'Third World' was being used to designate a group of African, Asian, and Latin American countries (see Figure 6.9, p. 198) that in 1969 would be described by Prime Minister Lee Kuan Yew of Singapore as 'poor, strife-ridden, [and] chaotic'. Like all groupings based on broad generalizations, the 'Third World' contained many more variations than it did similarities. Nevertheless, there is some

Table 6.9	COUNTRIES OF ASYLUM WITH MORE THAN 100,00 REFUGEES, 2001

Country of asylum	Refugees
Pakistan	2,198,797
Iran	1,868,000
Germany	903,000
Tanzania	668,107
United States	515,853
Yugoslavia	400,304
Congo	362,012
Sudan	349,209
China	295,325
Zambia	284,173
Armenia	264,337
Saudi Arabia	245,268
Kenya	239,221
Uganda	199,736
Guinea	178,444
India	169,549
Algeria	169,422
Ethiopia	152,554
Netherlands	152,338
United Kingdom	148,550
Sweden	146,551
France	131,601
Nepal	130,945
Canada	129,224
Iraq	128,142
Ivory Coast	126,239
Congo	119,147
Thailand	110,711

Source: United Nations High Commissioner for Refugees.

value in the classification, and other terms have been used in much the same way.

For example, the Brandt Report, *North–South: A Programme for Survival* (Brandt 1980) explicitly distinguished between 'north' and 'south', 'rich' and 'poor' (Figure 6.10). Other common terms are 'developed' and 'underdeveloped', or 'developed' and 'developing'. Boxes 6.8, 6.9, and 6.10 provide capsule commentaries on three case-studies, one each from Africa, Asia, and Latin America.

The principal advantage of the terms 'more developed' and 'less developed' is that they are frequently used in reports by organizations such as the United Nations, the World Bank, and the Population Reference Bureau. The more developed world comprises all of Europe and North America, plus Australia, New Zealand, and Japan. All other countries in the world are classed as less developed. Thus the less developed world is essentially the Third World as delimited by Dickenson et al. (1996) or the south as delimited by the Brandt Report (1980).

In general, countries in the less developed world have relatively high levels of mortality and fertility and relatively low levels of literacy and industrialization; in addition, they are often beset by political problems stemming from ethnic or other rivalries. The basic demographic data are often unreliable; not only do the poorest countries have limited capital to conduct censuses, but low literacy levels may affect the quality of the data collected, and—as Box 6.1 pointed out—data may even be falsified for political reasons.

6.7

REFUGEES IN THE HORN OF AFRICA

The Horn of Africa (Figure 6.8) comprises three countries: Somalia, Ethiopia, and Djibouti. Today, along with neighbouring Sudan, the Horn is a land of refugees. As of 2001 the estimated numbers were as follows:

- 324,500 Eritrean refugees in Sudan;
- 62,100 Somali refugees in Ethiopia, 144,300 in Kenya, 67,500 in Yemen, and 21,700 in Djibouti, with smaller numbers in Eritrea, Egypt, and Libya; and
- 35,600 Ethiopian refugees in Sudan, with smaller numbers in Kenya and Djibouti, and also some Djibouti refugees in Ethiopia.

The reasons behind this situation are a tragic mix of human and physical geographic factors. In 1962 Ethiopia annexed Eritrea, beginning a bitter war that has continued ever since. Ethiopia sees it as a secessionist war, while Eritrea sees it as a fight for self-determination; the two positions seem irreconcilable. When, for several years in the 1980s, Eritrea and the neighbouring Ethiopian province of Tigray experienced severe droughts, the Ethiopian government refused to allow international aid efforts access to these areas.

Politics is also a factor in the Somalian conflict. Even though it is one of Africa's poorest countries, Somalia places economic and social development second to the political ideal of integrating all Somali people into one nation. Since independence in 1960, this has involved conflict with Ethiopia (in the Ogaden) and Kenya especially. A full-scale war between Somalia and Ethiopia in 1977–8 coincided with drought. As of 2003, there is a serious risk of another severe famine, which would exacerbate an already tragic scenario.

Neither of these political conflicts is merely regional. Both are closely related to European colonial policies and, subsequently, the involvement of the United States and the USSR (more recently Russia).

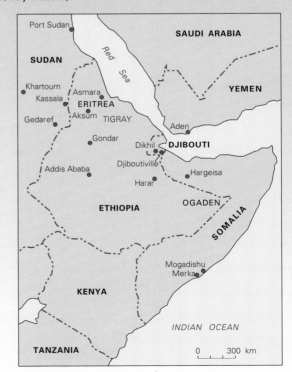

Figure 6.8 The Horn of Africa.

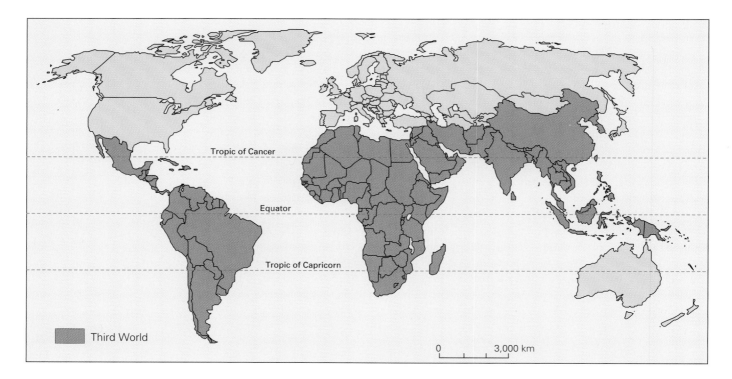

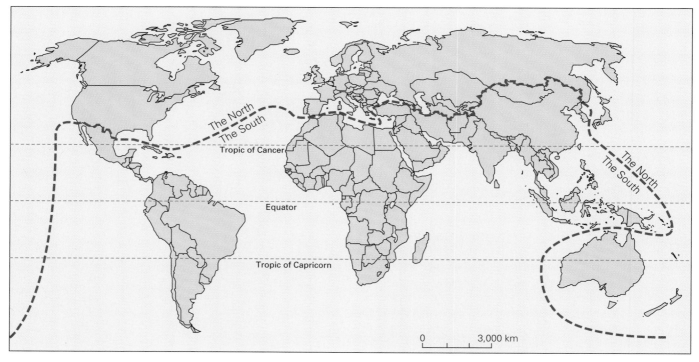

Figure 6.9 (above) The Third World.
Source: J. Dickenson et al.,
A Geography of the Third World
(New York: Routledge, 1996):2.

Figure 6.10 (below) North and South.

Development: Problems of Defining and Measuring

Traditionally, economic and social development have been measured by reference to **gross domestic product** (GDP) per capita or **gross national product** (GNP)—now usually called **gross national income (GNI)**—per capita on the grounds that such macroeconomic indicators not only provide reliable data for comparing the economic performance of various countries but also serve as reliable surrogate measures of social development in the areas of health, education, and overall quality of life. Others, however, believe that such measures are inappropriate because they do not take into account either the spatial distribution of economic benefits or the real-life conditions that less developed countries face: for example, population displacement, inadequate food supplies, and vulnerability to environmental extremes. It can be argued that for the less developed countries, GDP or GNI may indicate how the minority wealthy population are progressing, but tell us nothing about the poor majority. The different opinions on measures of development reflects a lack of agreement on what 'development' itself means.

One problem is that definitions of development are often ethnocentric: thus the standard view in the more developed world equates development with economic growth and modernization.

Measuring development

The *World Development Report* is an annual publication of the World Bank that measures development on the basis of selected economic criteria, grouping countries into three categories—low-income, middle-income, and high-income—according to GNI (formerly GNP) per capita. Figure 6.11 illustrates this classification of national economies. For 2000, the global GNI was US$31,315 billion, with the 66 low-income

countries accounting for only $997 billion, the 90 middle-income countries for $5,319 billion, and the 52 high-income countries for $24,994 billion. The problem with this way of ranking countries is that it reflects a **developmentalist** bias, suggesting that as countries become more technologically advanced they can—and should—increase their GNI. Moreover, as the World Bank itself has admitted, this measure 'does not, by itself, constitute or measure welfare or success in development. It does not distinguish between the aims and ultimate uses of a given product, nor does it say whether it merely offsets some natural or other obstacle, or harms or contributes to welfare' (World Bank 1993:306–7).

Measuring human development

An annual *Human Development Report* that first appeared in 1990 is intended to complement GNI measures of development. There are three distinctive characteristics to this publication. First, the concept of development underlying it focuses on the satisfaction of basic needs, gender inequality, and environmental issues. Second, it uses a wide variety of data to construct a Human Development Index (HDI) based on three

6.8

THE LESS DEVELOPED WORLD: ETHIOPIA

As we saw in Box 6.7, Ethiopia is one of several countries in the Horn of Africa that are currently suffering as a result of a tragic combination of human and environmental factors.

Ethiopia was first settled by Hamitic peoples of north African origin. Following an in-movement of Semitic peoples from southern Arabia in the first millennium BCE, a Semitic empire was founded at Axsum that became Christian in the fourth century CE. The rise of Islam displaced that empire southwards and established Islam as the dominant religion of the larger area; the empire was overthrown in the twelfth century. Ethiopia escaped European colonial rule, but its borders were determined by Europeans occupying the surrounding areas, and the Eritrea region was colonized by Italy from the late nineteenth century until 1945. Attempts by the central government in Addis Ababa to gain control over both the Eritreans and the Somali group in the southeast generated much conflict. In 1974–5, a revolution displaced the long-serving ruler, Haile Selassie, and established a socialist state.

Ethiopia is slightly larger than Ontario—1.2 million km^2 (463,400 square miles). Much of the country is tropical highlands, typically densely populated because such areas have good soils and are free of many diseases. In 2002 the population was estimated at 67.7 million; the CBR was 40, the CDR 15, and the RNI 2.5. There is no clear national population policy, nor any family-planning program. Population growth is highly uneven, with the central, western, and southern areas growing at the expense of the northern and eastern areas. Low urbanization rates are characteristic; urban growth has actually declined since 1975 because of socialist land reform policies and the low quality of life in urban areas.

The majority of Ethiopians have little or no formal education, especially among the (predominant) rural population. The key social and economic unit is the family, in which women are subordinate, first to their fathers, then to their husbands, and then, if widowed, to adult sons. Health care varies substantially between urban and rural areas.

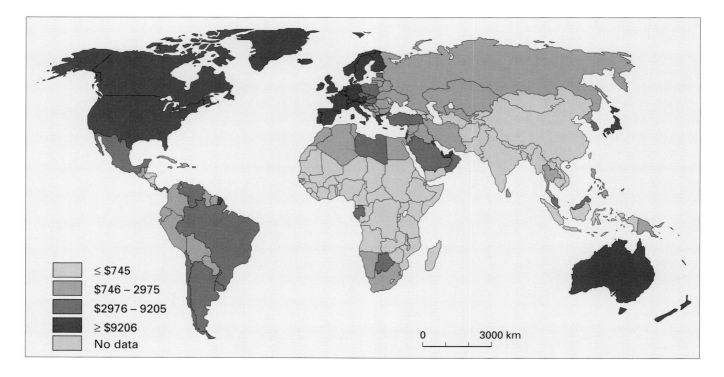

≤ $745
$746 – 2975
$2976 – 9205
≥ $9206
No data

0   3000 km

Figure 6.11 Groups of economies. Low-income countries have a GNI per capita of less than US$745, lower middle-income countries have from US$746 to US$2975, upper middle-income countries from US$2976 to US$9,205, and upper-income countries US$9,206 and above.

goals of development: LE, education, and income. Third, it is explicitly concerned with how development affects the majority poor populations of the less developed world, and recognizes that there is a need to enlarge the range of individual choice. The HDI does not measure absolute levels of human development but ranks countries in relation to one another.

Table 6.10 presents data on 20 countries: the ten with the highest HDI values and the ten with the lowest HDI values in 2000 (out of the 173 countries for which data are available). Again, as with GNI per capita data, African countries are the least developed. The HDI has a maximum value of 1.000 and a minimum value of 0.000; thus Norway, with a score of .942, has a shortfall in human development of roughly 6 per cent, whereas Sierra Leone, with a score of .275, has a shortfall of 72 per cent. It is also important to note that, although there is clearly a general relationship between economic prosperity and human development, there is no direct link, since some countries are more successful than others in translating economic success into better lives for people. For example, although Spain and Singapore have similar HDI levels, Spain's GDI per capita is only about half that of Singapore; similarly, Lesotho and Morocco have comparable HDI levels even though Lesotho's GDI is only about half that of Morocco.

The HDI data in Table 6.10 paint a distressing picture of global inequalities. There is a positive side to the picture, however, in that most countries have managed over time to reduce their shortfall from the maximum value of 1.0. Of the 79 countries for which comparative data are available for the period from 1975 onwards, only Zambia experienced a decline in its HDI (the reason was the impact of AIDS).

| Table 6.10 | **EXTREMES OF HUMAN DEVELOPMENT, 2002** | | | | |

Top Ten			Bottom Ten		
Country	**Rank**	**HDI value**	**Country**	**Rank**	**HDI value**
Norway	1	0.942	Sierra Leone	173	0.275
Sweden	2	0.941	Niger	172	0.277
Canada	3	0.940	Burundi	171	0.313
Belgium	4	0.939	Mozambique	170	0.322
Australia	5	0.939	Burkina Faso	169	0.325
United States	6	0.939	Ethiopia	168	0.327
Iceland	7	0.936	Guinea Bissau	167	0.349
Netherlands	8	0.935	Chad	166	0.365
Japan	9	0.933	Central African Republic	165	0.375
Finland	10	0.930	Mali	164	0.386

Source: United Nations Development Programme, *Human Development Report, 2002: Deepening Democracy in a Fragmented World.* (New York: Oxford University Press, 149–52).

Relations with the More Developed World: World Systems Theory

Perhaps the single most important factor in explaining the plight of the countries in the less developed world is their relationship with more developed countries. Most of the less developed countries have a colonial history; even those (such as China, Thailand, Liberia, Saudi Arabia, Iran, and Afghanistan) that have not been colonies of European countries have been affected by Europe's world dominance between about 1400 and 1945. Why is this relationship so important?

First, on the world scale, **colonialism** has led to **dependence**. In the past, many former colonies became economically dependent on the more developed countries; more recently, aid intended to promote development has served to encourage increased dependence. Second, the indigenous cultures and social structures of former colonies have been largely relegated to secondary status, their place taken by European structures; thus, in the broadest sense, the less developed countries lack power, including the power to control and direct their own affairs.

An exciting contribution to current human geography that addresses the issue of dependence is the notion of world systems proposed by Wallerstein (1979). Describing the dynamic capitalist world economy from 1500 onward, it examines the roles that specific states play in the larger set of state interrelationships. It can be briefly summarized as follows. Capitalism emerged gradually from feudalism in the sixteenth century, consolidated up to 1750, and expanded to cover the world in the form of industrial capitalism by 1900; in 1917, however, the capitalist system entered a long period of crisis that may eventually bring the world closer to a socialist system. Although the changes that have taken place, especially in Europe, since 1989 appear to make the prospect of further movement towards socialism less likely, this in no way detracts from Wallerstein's general argument.

The contemporary result of this historical process is a world divided into three principal zones: core, semiperiphery, and periphery (Figure 6.12). The *core* states benefit from the current situation, as they receive the surpluses produced elsewhere. The principal core states are Britain, France, the Netherlands, the US, Germany, and Japan—the countries where world business and financial matters are centred. The *semiperiphery* consists of states that are partially dependent on the core: for example, Argentina, Brazil, and South Africa.

THE LESS DEVELOPED WORLD: SRI LANKA

6.9

Sri Lanka (Ceylon until 1972) is an island located in the Indian Ocean, south of India. Mostly low-lying, it has a tropical climate and a limited resource base. Minerals are in short supply, as are sources of power. The dominant economic activity is agriculture; about 36 per cent of the country is under cultivation.

In 2002 Sri Lanka had a population of 18.9 million, with a CBR of 18, a CDR of 6, and an RNI of 1.2. Although these figures are good by the standards of the less developed world, in 2002 Sri Lanka had a per capita gross national income of US $820—well below that of countries such as Singapore and Malaysia, with which its per capita GNI was on par in the 1960s.

Three factors account for Sri Lanka's continuing problems. First, although it has now effectively passed through the demographic transition (along with only a few other low-income countries such as China), in the past its population did increase rapidly.

Second, Sri Lanka has a long history of colonization: by the Portuguese from 1505 to 1655, the Dutch until 1796, and the British until independence was achieved in 1948. Problems of colonial dependency developed largely in the nineteenth century, when the ruling British established a system of plantation agriculture benefiting themselves rather than the local population. By 1945, tea plantations covered about 17 per cent of the cultivated area.

Third, Sri Lanka continues to suffer severe ethnic conflict between the Sinhalese (74 per cent of the population), who moved down from north India and conquered the island in the sixth century BCE, and the Tamils (18 per cent of the population), who arrived from south India in the eleventh century. Today the Tamils still have neither citizenship status nor voting rights, and in addition to seeking recognition as an indigenous people (in order to protect themselves from persecution by the Sinhalese majority) are engaged in ongoing guerrilla activity. This conflict is discussed in a broader South Asian context in Chapter 9.

Contemporary Sri Lanka is a far cry from the tropical Indian Ocean island called 'Serendip' by the first Arab visitors and 'Paradise' by many later Europeans.

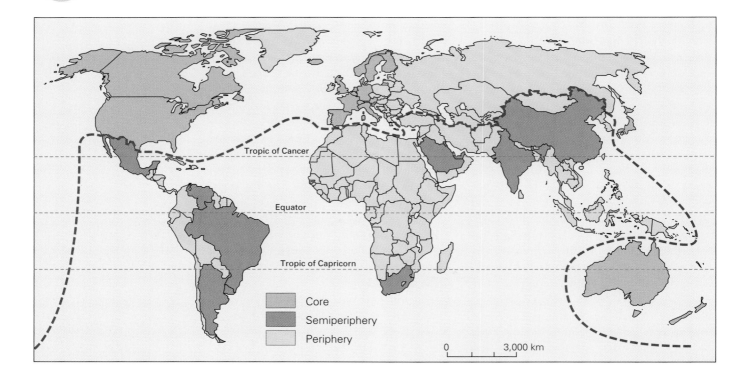

Figure 6.12 The world system: core, semiperiphery, and periphery. Source: Adapted from P. Knox and J. Agnew, *The Geography of the World Economy*, 2nd ed. (London: Arnold, 1994):2.

The *periphery* consists of those states that are dependent on the core and are effectively colonies; all the countries that we regard as less developed belong in this group.

Although this world system is dynamic, it is extremely difficult for a state to move out of peripheral status because the other states have vested interests in maintaining its dependency. Later, in Chapter 11, we will explore a more sophisticated conceptual basis for world systems theory, derived from Marxism.

Population and Food

Undernutrition and malnutrition

Until the nineteenth century, hunger and malnutrition were not uncommon in Europe. Today, however, large-scale food problems are for the most part limited to the less developed world. There is a world food problem, despite evidence that if the world's food production were equally divided among the world's population, nobody would go hungry.

A diet may be deficient in quantity, quality, or both. Requirements vary according to age, sex, weight, average daily activity, and climate, but an insufficient quantity of food (or calories) results in **undernutrition**. The best-known cases of acute undernutrition are the famines that attract the attention of the world's media.

An adequate diet includes protein to facilitate growth and replace body tissue, as well as various vitamins. A diet that is deficient in quality results in **malnutrition**, usually a chronic condition.

Among the health problems that can be caused by undernutrition or malnutrition are kwashiorkor (too few calories), poor sight (inadequate vitamin A), poor bone formation (inadequate vitamin D), and beriberi (inadequate vitamin B_1). The most extreme consequence of undernutrition and malnutrition is death.

The extent of the problem

How much food is lacking? How many people suffer from undernutrition or malnutrition? These are not easy questions to answer. We all know that various parts of the world, especially in Africa, seem to be especially vulnerable to hunger and famine. What may not be so well known is that world food supplies are increasing faster than world populations. In the period 1961 to 1980, food production in the less developed world increased 3.1 per cent per year, while population increased only 2.4 per cent. Clearly, improvements at the world level do not translate into improvements at the regional level.

The World Bank estimates that over 1 billion people receive insufficient nourishment to support normal activity and work. The highest levels of undernutrition are in Somalia, Mozambique, Sierra Leone, Bangladesh, and Bolivia.

Food aid

Food aid provided by the more developed world has not proven to be a solution to famine. First, such aid tends to be directed to urban areas, even though the greatest need is usually in rural areas; indeed, much donated food goes to governments, which then sell it for profit. Second, food aid tends to depress food prices in the receiving country, thus reducing the incentive for it to grow crops and increasing its dependence. Third, in many cases food aid is not effectively distributed, whether because of inadequate transport or because undemocratic governments in the receiving countries control the food supply and feed their armies before anyone else.

6.10

THE LESS DEVELOPED WORLD: HAITI

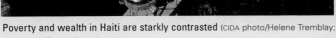

Poverty and wealth in Haiti are starkly contrasted (CIDA photo/Helene Tremblay; CIDA photo/Benoit Aquin).

The 'nightmare republic' of author Graham Greene, Haiti is a third example of a politically troubled country in the less developed world (Barberis 1994). The basic demographic data for 2002 show a very high CBR (33), a high CDR (15), and a high RNI (1.7). The use of modern contraceptive techniques is low— although there are good reasons to believe that demand for them is high. The total population is 7.1 million, with a high population density of 234 per square kilometre.

Haiti has had a long history of political turmoil since gaining independence from France in 1804 following a twelve-year rebellion. After 30 years under the rule of the Duvalier family, the first free election was held in 1990, but the victor was overthrown in 1992 and the country was again ruled by a military despot until democracy was re-established in 1994.

Economic disparities are extreme, not only between the poor (95 per cent of the population) and the rich (5 per cent) but also between the capital city of Port-au-Prince and the rural areas.

A 1990 estimate suggested that the wealthiest 1 per cent of the population held 44 per cent of the wealth. The rural population practise subsistence agriculture on soils that are generally poor, leading to erosion and declining soil fertility. Access to clean drinking water is difficult, hunger is widespread, and rates of infant mortality, tuberculosis, and HIV are all high.

Population and food: The future

As we saw in Chapters 4 and 5, there is a wide range of opinions on the number of humans that the earth can support. The same is true of the world food problem. Neo-Malthusians anticipate catastrophe, while cornucopians argue that food supplies can easily be increased sufficiently to feed a growing population.

The problem is complex. It is true that the number of undernourished people in the world is increasing. But it is also true that enough food is being produced today to feed all 6.12 billion people alive today. There seem to be good reasons to believe that current food shortages are not caused by inadequate supplies, and that supplies will be adequate in the future (Bongaarts 1994; Smil 1987). Rather, like several other current problems (for instance, the problem of refugees), the food problem appears to be associated with specific areas and a multitude of complex interrelated causes—physical, political, economic, and cultural.

Explaining the World Food Problem

Until recently, most explanations of the world food problem have focused on three factors:

1. *Overpopulation:* The root cause of the food problem is often said to be the sheer number of people in the world. Yet there is evidence to suggest that in fact there are not too many people in the world. 'Overpopulation' is a relative term, and areas that are densely populated are not necessarily overpopulated. Table 6.11 confirms that countries with dense populations, such as the Netherlands and South Korea, are not necessarily overpopulated, while less densely settled countries, such as Ethiopia and Mexico, may very well be relatively overpopulated. A compelling example is China, which experienced regular famines when it had a population of 0.5 billion but is now, with a population of 1.3 billion, essentially free of famine.

2. *Inadequate distribution of available supplies:* Most countries have the transportation infrastructure to guarantee the intranational movement of food, but other factors prevent satisfactory distribution.

3. *Physical or human circumstances:* A 1984 drought in central and eastern Kenya is estimated to have caused food shortages for 80 per cent of the population. Other causes in specific cases include flooding (Bangladesh) and war (Ethiopia). Circumstances such as these obviously aggravate existing problems, but they cannot be considered root causes.

Political and economic explanations

More recent theories have tended to focus on the political and economic aspects of the food problem. For example, the *world systems* perspective introduced earlier in this chapter suggests that global political arrangements are making it increasingly difficult for many in the less developed world to grow food for themselves. Not only is the percentage of the population involved in agriculture in the less developed world declining (from more than 80 per cent in 1950 to less than 60 per cent in 2002), but the vast majority of those remaining in agriculture are incapable of competing with the handful of commercial agriculturalists who are able to benefit from technological advances. Indeed, the vast majority have lost control over their own production because of larger global causes. For example, farmers in Kenya are actively encouraged to grow export crops such as tea and coffee instead of staple crops such as maize; peasant farmers in many countries are losing their freedom of choice because credit is increasingly controlled by large corporations; and many governments make it their policy to provide cheap food for urban populations at the expense of peasant farmers.

Table 6.11	POPULATION DENSITIES, SELECTED COUNTRIES, 2002

Country	Population density per square kilometre
South Korea	487.3
Netherlands	395.0
Ethiopia	61.4
Mexico	52.1

Source: Calculated from data in Population Reference Bureau, *2002 World Population Data Sheet* (Washington, DC: Population Reference Bureau, 2000).

The essential argument here—and it is an argument, not a fact—is that the capitalist mode of production is affecting peasant production in the less developed world in such a way as to limit the production of staple foods, thus causing a food problem.

In the same way, global economic considerations make it increasingly difficult for people in the less developed world to purchase food. Throughout the world, food is a commodity; hence, throughout the world, production is related to profits. In other words, food is produced only for those who can afford to buy. In 1972, a year when famine was widespread in the Sahel region of Africa, farmers in the US were actually paid to take land out of production in order to increase world grain prices.

Both of these arguments suggest that the cause of the world food problem is the peripheral areas' dependence on the core area. Yet any attempt to correct that problem (apart from clearly humanitarian action) is likely doomed to failure, because the more developed world is not about to initiate changes that would lessen its power and profits. If this argument is followed to its logical conclusion, the world food crisis can only get worse, regardless of technological change, because the cause of the problem lies in global political and economic patterns.

The fact is that feast and famine live side by side in our global village. There are rich and poor countries, and in the poor countries especially, there are rich and poor people. Increasingly, agriculture is a business, and it is in the nature of business that not all participants compete equally well. Food problems are not typically caused by overpopulation, and population density is a poor indicator of pressure on resources. We will not solve food problems merely by decreasing human fertility.

What is needed is a concerted and cooperative international effort to improve the quality of peasant farming and to reorient it towards the production of staple foods. But there is little indication that this will be achieved in the foreseeable future. Meanwhile, undernutrition and malnutrition are not disappearing, and famines, typically prompted by events such as droughts and wars, will continue to occur. The less developed world is not about to develop, whatever criterion of development is employed.

Severely malnourished children at a camp for refugees, returnees, and displaced persons in the Ogaden region of Ethiopia in 1991. Many factors contributed to the Ethiopian famine of the early 1990s, including drought in some regions; rains elsewhere that came at the wrong time, destroying the crops; returnees from the war in Somalia (many had gone to Somalia a few years before to escape the war in Ethiopia); and the escalation of the civil war in Ethiopia, during which the Mengistu government siphoned off government resources that could have helped famine victims (CIDA photo/Roger LeMoyne).

Level	Factors	Mechanisms
International	historical	• processes of integration: accumulation of power and resources
	economic structure	• institutions: World Bank. International Monetary Fund. General Agreements on Tariffs and Trade. • transnational corporations • gendered macro-economic policy
	political/ideological structures	• development philosophies • Structural Adjustment Policies • interpretations of hunger poverty
National	historical	• processes of integration: character and capacity of the state
	economic structure	• export orientation • infrastructure (transport financial health education)
	political/ideological structures	• public policy • development strategies • gender relations
Regional	historical	• processes of integration: regional and ethnic disparities • environmental legacies
	economic structure	• limited infrastructural development • vulnerable income base
	political/ideological structures	• public policy • discrimination
Household	historical	• social relations
	political/ideological structures	• public policy • gender relations

Figure 6.13 The construction of entitlements: selected factors and mechanisms at different scales.
Source: L. Young, 'World Hunger: A Framework for Analysis', *Geography* 81 (1996):100.

What makes the world systems perspective particularly valuable is that it emphasizes the need to focus on structural causes—not just the relatively obvious immediate causes, such as drought and crop failure, that typically receive the attention of the media. But the basic world systems logic can be usefully applied at various spatial scales below that of the world as a whole. Indeed, the role of power, in the sense of the politics that govern access to food, is relevant at various scales.

A case in point is the explanation of world food problems suggested by Young: 'Patterns of food distribution may be examined with reference to people's entitlements reflected in their ability to *command* food' (Young 1996:99). *Entitlements* are the factors and mechanisms that explain people's ability to acquire food in terms of their power. In addition to entitlements at the international scale (addressed by world systems theory), this argument identifies entitlements at three smaller scales: national, regional, and household (see also Figure 6.13):

1. At the international scale, as world systems logic suggests, among the historical legacies of colonialism is an emphasis on export production at the expense of local production and hence a vulnerability to

global market changes. In other words, small changes in the price of a commodity can significantly affect entitlements. Especially in the 1980s, high levels of debt led the International Monetary Fund and the World Bank to require that countries in the less developed world adopt structural adjustment policies—typically involving even greater emphasis on export production—before additional loans would be issued or existing loans restructured. Contemporary globalizing trends are aggravating this situation as the demand for food continues to increase in parts of the world that are already poor and already lack power. At the national scale, governments may not be committed to ensuring that economic growth is accompanied by the elimination of food shortages. Because of the distribution of power, national governments may support urban activities and commercial agriculture at the expense of the rural peasant sector.

2. At the regional scale, governments may neglect areas inhabited by relatively powerless minority ethnic groups. Further, problems of various kinds are often regionally distinct: some areas may be more subject to conflict, or environmental problems, than other areas.

3. At the household scale, the most vulnerable family groupings are those that are poorest, that include many dependents, that are isolated, and that are powerless. Even within households, there are differences in ability to command food; females and the elderly are often the most vulnerable.

This multi-scale approach to understanding the world food problem stresses the inequalities that exist in people's ability to acquire food. Food shortages are placed in context using broad geographic and historical frameworks with emphasis on entitlements and the related ability of people to command food. We will introduce a conceptual basis for this political economy approach in Chapter 11. Those global areas, countries, regions within countries, households, and even individuals within households whose entitlements are most limited are the ones least likely to command adequate amounts of food. The basic equation is this: *lack of power equals lack of food*.

In a similar vein, the Indian economist Amartya Sen argues that widespread hunger has nothing to do with food production and everything to do with poverty—which in turn is closely related to political governance. Sen points out that famine does not occur in democratic countries, because even in the poorest democracy famine would threaten the survival of the ruling government. Even a cursory review of current global famine areas confirms that famine seems to have two principal causes: bad weather and bad government. As of 2003, famine threatened four countries—Angola, Zambia, Zimbabwe, and Mozambique—and in all of them both of those conditions could be seen to apply. Not everyone agrees with Sen's argument, however. In 2002, for example, at the Johannesburg Earth Summit, the main subjects of discussion in talks aimed at addressing world poverty and hunger were colonialism and the obligations of the more developed to the less developed world. Sen does not deny the role played by history, and agrees that the more developed world should do much more to help the less developed world, but he emphasizes that the central problems of governance can be addressed internally.

If this argument is correct, then the prospects for alleviating or avoiding famine in much of Africa are clearly improving. In recent years, many African countries—including Ghana, Niger, Mali, Gambia, Malawi, and Kenya—have shown enthusiasm for democratic institutions and increasing personal freedom. A 1999-2001 survey of more than 21,000 people in 12 African countries showed that more than 70 per cent of respondents favoured democracy over other forms of government. As of 2003, the repressive regime of Robert Mugabe in Zimbabwe is more the exception than it is the norm.

Further support for the connection between famine and politics can be found in the historical record. Between 1875 and 1914, some 30 to 50 million people died in a series of famines in India, China, Brazil, and Ethiopia. Describing these famines, Davis (2001) recognizes the role played by meteorological conditions, but also points to the complex politics of colonialism and capitalism as essential causes.

Feeding the world

How many people is the earth capable of feeding? This question has no meaningful answer. The fact is that, in a world that believed in consumption equity and nutritionally adequate diets for all the world's people, there would be no problem in feeding many more than ten billion people—although even the current six-plus billion people could not be fed if the North American diet became the norm worldwide.

How best can we ensure adequate food for the nine billion people who are expected to occupy the earth by the middle of this century? On the basis of a scientifically detailed analysis of the complete food cycle, Smil (2000:315) arrived at a conclusion he described as 'encouragingly Malthusian'. In other words, the future may not be as bright as we might hope, but neither is it totally bleak. Thus, with reference to the competing catastrophist and cornucopian positions noted in Chapters 4 and 5, we might suggest that neither extreme is a realistic portrayal of our future.

The World Debt Problem

The long-term indebtedness of the less developed world—money owed to international lending agencies and commercial banks in the more developed world—has increased from US$59.2 billion in 1970, to 445.3 in 1980, to 1167.9 in 1990, and to more than 2000 by 2001. The 1970s was a decade of high lending to less developed countries by the commercial banking sector, development agencies, and governments in the more developed world. These loans, intended for the establishment and support of economic and social programs, had long payback terms, because it was generally expected that, in the long run, the less developed economies would boom and eventually provide good returns on the loan investments.

Unfortunately for all concerned, the 1980s brought a recession, in the course of which interest rates rose, world trade declined, and debt steadily increased. As a result, many countries are now so poor and owe so much that they need to borrow more money just to keep up the interest payments on their existing debt—a situation that first arose in 1982 in Mexico (Sowden 1993).

The significance of debt to a country depends on the country's economy. The US is the biggest net debtor in the world, but the effect of the debt on the economy is minimal because the US is a high-income country with a large volume of exports. Other countries such as South Korea, which borrowed heavily to finance industrialization, have been able to repay their loans because of their industrial and export success. In many less developed countries, however, the cost of servicing the foreign

6.11

THE GRAMEEN BANK, BANGLADESH

A remarkable transformation is occurring in parts of Bangladesh, a transformation brought about by a credit program that focuses exclusively on improving the status of landless and destitute people—especially women.

Bangladesh is one of the poorest countries in the world. It has a large population of 133.6 million, an exceptional population density of 928 per square kilometre, a CBR of 30, a CDR of 8, and an RNI of 2.2. The per capita gross national income is one of the lowest in the world at US $350.

The Grameen Bank program began in 1976 as a research project at the University of Chittagong investigating the possibility of providing credit for the most disadvantaged of the Bangladeshi people. In 1983 the project became an independent bank, which by 1988 had 571 branches covering 17 per cent of the villages in Bangladesh. Essentially, the Grameen program seeks to enhance the social and economic status of those most in need, especially women, by issuing them loans. The bank requires that loan applicants first form groups of five prospective borrowers and meet regularly with bank officials. Two of the five then receive loans and the others become eligible once the first loans are repaid. The focus is on the collective responsibility of the group. As the recovery rate—a remarkable 98 per cent—shows, the Grameen Bank's innovative approach works.

This successful technique is now being replicated elsewhere in the world (see Mahmud 1989; Todd 1996).

debt accounts for up to 30 per cent of all income from exports. For those countries, foreign debt is such a crushing burden that some, such as Colombia, have actually exported food to help repay it, even though their own populations are malnourished; not surprisingly, there is a concentration of severely indebted low-income countries in Africa.

Thus loans intended to help impoverished countries can have quite the opposite effect. Box 6.11 outlines one alternative approach, an innovative program in which loans are provided directly to poor individuals themselves, rather than to their country as a whole. It is also possible to rethink the whole question of world debt and argue that the debt owed by the less developed world is only one side of the coin. For example, a 1999 report published by Christian Aid, a leading British charity, asserted that the more developed world owes a huge debt to the less developed world because of the disproportionate environmental damage that the more developed world causes. Similarly, discussions at the 2002 World Conference Against Racism, Racial Discrimination, Xenophobia, and Related Forms of Intolerance, sponsored by the United Nations and held in Durban, South Africa, included arguments that rich countries should pay poor countries compensation for the abuses (such as slavery) inflicted on them in the past.

The Selectivity of Disasters

Human geographers have long used the term 'natural disasters' to refer to physical phenomena such as floods, earthquakes, and volcanic eruptions and their human consequences. In recent years, however, we have recognized that this label is inappropriate. Certainly such events are natural in the sense that they are part of larger physical processes—but not all such events become disasters. To understand why a natural event becomes a human disaster, we need to understand the larger cultural, political, and economic framework. In short, some parts of the world and some people are more vulnerable than others to natural events. Just as, on average, the rich live longer than the poor because they are better able to afford good nutrition and medical care, so some parts of the world are better able to protect themselves against natural events.

People take shelter on the roof of a house on the outskirts of Dhaka, Bangladesh, in July 1998. Floods caused by torrential monsoon rains affected more than three million people and claimed at least 72 lives (AP photo/Pavel Rahman).

The United Nations designated the 1990s as a decade to focus on reducing the human disasters that so often accompany natural events, especially in the less developed world. Between the 1960s and the 1980s, the number of major disasters increased fivefold, fatalities increased considerably, and regional disparities between the more and less developed worlds were widening. Table 6.12 provides data for three regions on the average loss of life associated with disasters; there is clear evidence that the less developed world is especially vulnerable and that its vulnerability is increasing over time.

As with food supplies and national debts, again the less developed world suffers the most. Adverse cultural, political, and eco-

Table 6.12	**NATURAL EVENTS AND HUMAN DISASTERS, REGIONAL DATA, 1947–1967 AND 1969–1989**

Region	Average loss of life per disaster
1974–67	
North America	38
Western Europe	230
Less developed world	984
1969–89	
North America	19
Western Europe	99
Less developed world	2,066

Source: M. Degg, 'Natural Disasters: Recent Trends and Future Prospects', *Geography* 77 (1992):201.

nomic conditions combine to place increasing numbers of people at serious risk in the event of any environmental extreme. 'The problems of staggering population/urban growth and crippling overseas debt that face many Third World countries have often become manifested in poor construction standards, poor planning and infrastructure, inadequate medical facilities and poor education—all of which exert a direct influence on vulnerability to natural hazards' (Deg 1992:201). Overall, there is a close relationship between vulnerability and poverty. Box 6.12 provides details on a poor and densely populated area of the world that is regularly subject to devastating floods, while Box 6.13 returns to the debate, introduced in Chapter 4, between catastrophists and cornucopians, this time in the context of development.

Striving for Equality

The World Bank has identified eight 'millennium development goals':

- *Eradicate extreme poverty and hunger.* Data for recent years show that GNI per capita in the less developed world is increasing and that the number of people living on less than US$1 per day is declining. These improvements are most evident in Asia, especially China. Most regions are reducing the proportion of underweight children, and malnutrition rates in the less developed world have fallen from 46.5 per cent in 1970 to 27 per cent in 2000.

- *Achieve universal primary education.* Education is essential both to the reduction of poverty and inequality and to the building of successful economies and democratic institutions. Data for 2000 suggested that there were 120 million primary-age children not in school, mostly in south Asia and sub-Saharan Africa.

- *Promote gender equality and empower women.* Here again, education is the key—yet in the less developed world, girls are even less likely than boys to attend school and more likely to drop out. Education allows people to improve their economic circumstances by helping them to find better jobs. As we saw in Chapter 5, empowering women is essential to reduce fertility rates.

6.12

FLOODING IN BANGLADESH

Bangladesh is a low-lying country (mostly only 5–6 m/16–20 ft above sea level) that lies at the confluence of three large rivers, the Ganges, the Brahmaputra, and the Meghna. Floods are normal in this region; indeed, they are an essential part of everyday economic life, as they spread fertile soils over large areas. During the monsoon season, however, floods often have catastrophic consequences, especially if they coincide with tidal waves caused by cyclones in the Bay of Bengal (see Figure 4.6, p. 128). Three factors contribute to flooding:

1. Deforestation in the inner catchment areas that results in more run-off;
2. Dike and dam construction in the upstream areas that reduce the storage capacity of the basin; and
3. Coincidental high rainfall in the catchment areas of all three rivers.

Regardless of specific causes, floods are difficult to control because of their magnitude and because the rivers often change channels.

The most extreme consequence of flooding is death; in a 1988 flood, over 2,000 died. Impelled migration is another serious consequence; also in 1988, over 45 million people were uprooted and forced to flee. Disasters occur on a regular basis. In a normal year, over 18 per cent of Bangladesh is flooded, and even these normal floods result in shifting of river courses and erosion of banks that, in turn, result in population displacement. In an already poor country, such displacement aggravates problems such as landlessness and food availability. Most displaced persons move as short a distance as possible for cultural and family reasons. During these impelled migrations, the poorest suffer the most, and women-headed households are especially vulnerable. Some migrants opt to move to towns, hoping to become more economically prosperous. Typically, however, such rural-to-urban migrants become further disadvantaged, clustering together in squatter settlements that are generally regarded as unwelcome additions to the established urban setting.

- *Reduce child mortality*. Deaths of infants and children have fallen dramatically in recent years, largely because of improved health care. Still, the majority of children who die before the age of five do so because of disease and/or malnutrition. These deaths are preventable.
- *Improve maternal health*. Complications related to pregnancy and childbirth are a leading cause of death among women of reproductive age in the less developed world, where a rate of 1 maternal death per 100 births is not unusual. The corresponding figure in the more developed world is about 1 in 100,000. Again, these deaths are preventable.
- *Combat disease*. Many areas suffer from malaria and tuberculosis in addition to AIDS.

6.13

THE BEST OR THE WORST OF TIMES?

Much of this chapter's discussion of the less developed world paints a grim picture. Yet there are also promising signs. What will our global future be? The answer is far from clear—and, as our discussion of world systems analysis demonstrates, it is at least partly dependent on ideological perspective. As we have seen, there is an ongoing debate between pessimists and optimists or, as they are often known, catastrophists and cornucopians. Similarly antithetical perspectives on the future of the world have existed for centuries. In the mid-1700s, just a few years before Malthus predicted a future of famine, vice, and misery, a Berlin vicar and statistician named Johann Peter Süssmilch claimed that the earth could feed 10 billion people.

The differences of opinion in the contemporary debate often seem almost equally extreme. Why is this so? The fundamental answer is that pessimists and optimists base their forecasts on very different kinds of assumptions. Pessimists such as Ehrlich and the Club of Rome (p. 165) usually assume an existing set of conditions, 'other things being equal'. Optimists, by contrast, argue that 'other things' are not equal; they maintain that circumstances change, and in particular that human ingenuity is always increasing, finding new ways to cope with problems. As we saw in Box 4.9's account of 'The Skeptical Environmentalist', the best-known such authors in recent decades are the economist Julian Simon and the statistician Bjorn Lomborg.

One writer who has focused attention on specific problem regions in several African and Asian countries is Kaplan (1994, 1996). Reporting in a style that one reviewer described as 'travel writing from hell' (Ignatieff 1996:7), he details a nightmare scenario of population growth, ecological disaster, disease, refugee movements, ethnic conflict, civil war, weakening of government authority, empowerment of private armies, rural depopulation, collapsing social infrastructure, increasing criminal activity, and urban decay, and identifies West Africa as the region in the most severe trouble. Kaplan writes convincingly about the interrelatedness of problems such as population growth, environmental degradation, and ethnic conflict—all characteristic in many parts of the less developed world. The evidence that he finds, particularly concerning refugees and food supply, leads him to predict an immediate future of instability and chaos.

Others disagree with both the details of Kaplan's findings and his conclusions. Pointing to steady improvements in health, education, life expectancy, and nutrition, both globally and specifically in the less developed world, Gee (1994:D1) asserts that 'By almost every measure, life on earth is getting better.' It is certainly true that, by some measures, life is improving. Throughout much of the less developed world, fertility is declining, life expectancy is increasing, and adult literacy is increasing. One especially interesting development concerns Botswana, a country that has seen rapid growth in income per person over the past 35 years. Attempts to explain this surprising fact are varied, but it appears that Botswana has succeeded in aligning the interests of the elite with those of the masses, and has used aid money to improve the legal and court system and to fight disease. Elsewhere in Africa, Uganda is educating nearly all of its children of elementary school age.

Critics like Gee have compared Kaplan to such earlier doomsayers as Malthus and the Club of Rome, whose most dire predictions have not come to pass. But there are differences, and they may be significant. Unlike most earlier predictors of nightmare futures, Kaplan was looking at the relatively short term—perhaps two decades—and basing his judgments on specific circumstances in the mid-1990s. As of 2003, the overall picture looks brighter than a pessimist might have predicted. For example, the proportion of the world's people living in extreme poverty fell from 29 per cent in 1990 to 23 per cent in 1999. On the other hand, specific regional circumstances may well confirm the worst fears. Perhaps the 2002 UN Human Development Report (United Nations Development Programme 2002:19) is a fair summary of the current situation: 'The level of inequality worldwide is grotesque. But trends over recent decades are ambiguous. The range of economic performance across countries and regions means that inequality has increased between some regions and decreased between others.'

- *Ensure environmental sustainability*. Wherever resources are depleted, it is the poorest who suffer first and most severely. Sustainable practices are especially urgent with regard to fresh water, forests, and soils.
- *Develop a 'global partnership for development'*. An open and nondiscriminatory trading and financial system must take into account the special needs of the less developed world. It is crucially important to address the debt problem faced by many countries.

For each of these general goals, the World Bank has established specific indicators of success and estimated the likelihood that they will be achieved. Overall, success depends on the growth of national economies, which in turn depends on the full and equal participation of all people. As the final goal suggests, it is now widely accepted that achieving these goals is a global responsibility. International aid is most effective when it goes to countries with good economic policies and sound governance.

CHAPTER SIX SUMMARY

Distribution and density
'Distribution' refers to the spatial arrangement of a phenomenon; 'density' refers to the frequency of occurrence of a phenomenon within a specified area. Population maps frequently combine the two characteristics. Asia has almost 61 per cent of the world's population; China is the most populous country, followed by India. A third area of population concentration is Europe. The Asiatic regions include some areas with very high rural population densities. Distribution and density of population are related to land productivity and cultural organization.

Causes of human mobility
Human populations have always been mobile. Migrations are often explained in terms of the relative attractiveness of different locations. This idea can be expressed by reference to simple push and pull logic, 'laws' such as Ravenstein's, or the concept of place utility. Another explanation for migration, related to the demographic transition, is called the mobility transition.

Types of migration
'Primitive migration' is the normal process by which humans originally moved over the surface of the earth, gradually adapting to new environments. 'Forced migration' occurs when people have little or no choice but to move; slaves and refugees are examples of forced migrants. Refugees are prompted to move for human and/or environmental reasons; major

problem areas today include Ethiopia and neighbouring countries (Box 6.7), Israel and neighbouring countries, Iraq, and various Asiatic countries. 'Free migration' is the result of a decision made following evaluation of available locations. 'Illegal migration' is increasingly common in a world where many countries have restrictive immigration policies intended to ensure that only members of desired groups are admitted.

The less developed world
'Less developed' countries may be identified according to various criteria: some, such as GDP or GNI, are solely economic; others reflect a number of measures of human development. All these countries have similar relationships with the more developed world. Many have been colonies and most are in a dependent situation today; virtually all have massive debt loads; and most are vulnerable to environmental extremes such as droughts, floods, and earthquakes.

World systems theory
Wallerstein's world systems approach posits a world divided into three zones: core, semi-periphery, and periphery. The world is organized in such a way that core countries benefit at the expense of peripheral countries. (All the countries in the less developed world are peripheral.) It has been argued that the world food problem can only get worse, regardless of technological change, because the causes of the problem lie in global political and economic patterns.

The world food problem

Undernutrition is caused by insufficient food; malnutrition is caused by low-quality food lacking the necessary protein and vitamins. One explanation of the global food problem is over-population (too many people for the carrying capacity of the land); another is inadequate distribution of food. Some people have argued that the world food problem is a consequence of the imposition of a capitalist mode of production, which has made it impossible for increasing numbers of people in the less developed world either to grow or to purchase food. There is also much evidence to suggest that food shortages and related problems can often be traced to a lack of democratic institutions.

Our global future

Debate continues between the pessimists who foresee deepening misery and the optimists who foresee ongoing improvements in human circumstances.

LINKS TO OTHER CHAPTERS

- Globalization:
    Chapter 2 (concepts)
    Chapter 4 (ecosystems and global impacts)
    Chapter 5 (population growth, fertility decline)
    Chapter 7 (cultural globalization)
    Chapter 8 (popular culture)
    Chapter 9 (political globalization)
    Chapter 10 (economic globalization)
    Chapter 11 (agriculture and the world economy)
    Chapter 12 (global cities)
    Chapter 13 (industrial restructuring)

- Population distribution and density:
    Chapter 2 (environmental determinism and possibilism);
    Chapter 3 (physical geography, especially Fig. 3.8: global environment regions)
    Chapter 5 (population growth).

- Explanations of migration:
    Chapter 2 (positivism and humanism).

- Forced migration, immigration to Canada, restrictive immigration policies:
    Chapter 3 (racism, Box 3.5)
    Chapter 8 (types of society; landscapes and power relations; ethnicity).

- Free migration within countries:
    Chapter 12 (rural settlements in transition; rural–urban fringe).

- Illegal migration:
    Chapter 9 (globalization).

- Refugees:
    Chapter 9 (political geography, especially Box 9.6).

- The less developed world (including Box 6.13):
    Chapter 4 (energy and technology; earth's vital signs
    Chapter 5 (fertility and mortality)
    Chapter 9 (colonialism)
    Chapter 11 (agriculture)
    Chapter 12 (settlement)
    Chapter 13 industry).

FURTHER EXPLORATIONS

ATKINS, P., AND I. BOWLER. 2001. *Food in Society: Economy, Culture, Geography*. New York: Oxford University Press.

>Covers a wide range of topics, including food supply and famine.

BLACK, R., and V. ROBINSON, eds. 1993. *Geography and Refugees: Patterns and Processes of Change*. New York: Belhaven Press.

>A collection of detailed case-studies including examples from both the more and the less developed worlds.

BRADLEY, P.N. 1986. 'Food Production and Distribution—and Hunger'. In *A World in Crisis: Geographical Perspectives*, edited by R.J. Johnston and P.J. Taylor, 89–106. New York: Blackwell.

>A readable and provocative article explaining the world food problem in terms of global economic patterns.

BROHMAN, J. 1996. *Popular Development: Rethinking the Theory and Practice of Development*. Oxford: Blackwell.

>Critical evaluation of approaches to development, relations to policy, and results of applications in the less developed world.

CHAMPION, A., and A. FIELDING, eds. 1992. *Migration Processes and Patterns, Volume 1*. New York: Belhaven Press.

>A series of detailed analyses of contemporary migration trends; strong emphasis on links to cultural and economic change; focus on British examples.

COHEN, R. 1995. *The Cambridge Survey of World Migration*. Cambridge: Cambridge University Press.

>A comprehensive survey made up of 95 contributions covering a multitude of migration topics, from the sixteenth century to the present; broad geographical coverage, including discussions of refugees; an invaluable source of information.

DEMENY, P. 1989. 'World Population Trends'. *Current History* 88:17ff.

>A factually based discussion of trends and projections.

DESAI, V., AND R.B. POTTER. Eds. 2002. *The Companion to Development Studies*. New York: Oxford University Press.

>Many short pieces on a wide range of topics; a very useful source of basic factual information.

DICKENSON, J., et al. 1996. *A Geography of the Third World*, 2nd ed. New York: Routledge.

>An excellent overview that employs a developmentalist perspective to explore population and economic issues; relevant also for Chapters 10, 11, and 12; succinct summaries of many key topics and numerous insightful case-studies.

DYSON, T. 1996. *Population and Food: Global Trends and Prospects*. New York: Routledge.

>Good overview of this fundamental topic.

GRIGG, D.B. 1977. 'Ravenstein and the "Laws of Migration"'. *Journal of Historical Geography* 3:41–54.

>An excellent summary and discussion of the topic, which helps our understanding of the contemporary relevance of Ravenstein specifically as well as the value of laws generally.

———. 1993. *The World Food Problem*, 2nd ed. Oxford: Blackwell.

>A comprehensive survey emphasizing history and regional issues.

KAPLAN, R.D. 1996. *The Ends of the Earth: A Journey at the Dawn of the 21st Century*. New York: Random House.

>As noted in Box 6.13, Kaplan has been called a master of 'travel writing from hell'; having travelled a number of less developed countries in Africa and Asia, he argues that our global future is fraught with difficulties and is being shaped at 'the ends of the earth'.

LEWIS, G.J. 1982. *Human Migration*. New York: St Martin's Press.

>A useful book covering the many facets of the topic.

LOESCHER, G. 1993. *Beyond Charity: International Cooperation and the Global Refugee Crisis*. Toronto: Oxford University Press.

>A powerful book that argues for a reform and strengthening of organizations such as UNHCR and for a concerted strategy by the industrial nations to tackle refugee problems effectively.

NEWLAND, K. 1994. 'Refugees: The Rising Flood'. *World Watch* 7, no. 3:10–20.

A brief, readable account that covers basic data; information on political, environmental, and economic causes of specific problems; discussions of ethnic tensions, human rights violations, preventive measures, and possible responses.

POOLEY, C.G., and I.D. WHYTE. 1991. *Migrants, Emigrants and Immigrants: A Social History of Migration*. New York: Routledge.
   Includes a series of case-studies with emphasis on Britain and areas of British settlement overseas.

STILLWELL, J., P. REES, and P. BODEN, eds. 1992. *Migration Process and Patterns, Volume 2: Population Redistribution in the United Kingdom*. New York: Belhaven Press.

A series of empirical analyses of the migration component of population change between and within the various regions of the United Kingdom.

YOUNG, L. 1996. 'World Hunger: A Framework for Analysis'. *Geography* 81:97–110.
   Clear and concise review of debates about world hunger with an emphasis on the political context.

ZELINSKY, W. 1971. 'The Hypothesis of the Mobility Transition'. *Geographical Review* 61:219–49.
   Links mobility to the demographic transition and technological change.

ON THE WEB

http://www.amnesty.org/
Amnesty International is an international organization with a vision of a world in which every person enjoys all of the human rights enshrined in the Universal Declaration of Human Rights and other international human rights standards. With more than a million members, Amnesty is independent of any government, political persuasion, or religious creed

http://www.ercomer.org/
A useful source of data and general information on migration topics especially as these relate to ethnic identities.

http://www.undp.org/
The home page of the United Nations Development Programme; includes past issues of the Human Development Report.

http://www.unhcr.ch/
The United Nations High Commission on Refugees maintains this regularly updated site; provides a wealth of information and commentary.

http://www.unicef.org/uwwide/
Created by the United Nations General Assembly in

1946 to assist children in Europe after the Second World War, UNICEF today helps children around the world get the care and stimulation they need in the early years of life; it encourages families to educate girls as well as boys; strives to reduce childhood death and illness; and works to protect children in the midst of war and natural disaster. UNICEF supports young people in making informed decisions about their own lives, and strives to build a world in which all children live in dignity and security.

http://www.wfp.org/
The home page for the United Nations World Food Programme; a substantial resource on matters of hunger and famine in a global context.

http://www.who.int/
This home page of the World Health Organization includes useful material on world health issues.

http://www.worldbank.org/
Another valuable web resource with comprehensive information on the policies and activities of the World Bank.

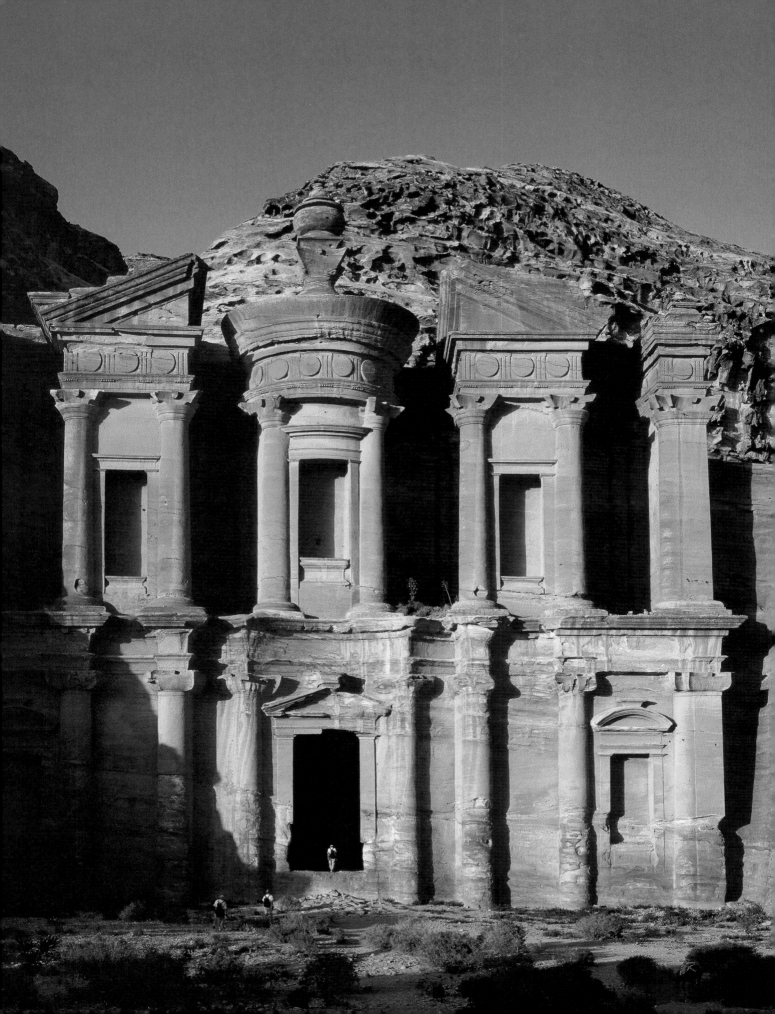

CHAPTER 7

Cultures: The Evolution and Regionalization of Landscape

Both this chapter and the one following focus on culture, and both are closely related to the political geography that is the subject of Chapter 9. It may be helpful, therefore, to view the three chapters as an integrated unit. More specifically, Chapters 7 and 8 focus on three closely related subdisciplines of human geography—historical, cultural, and social geography—that share a concern with the behaviour of humans, both as individuals and as group members, and with the landscapes that humans create. The differences in focus between this chapter and Chapter 8—here, the geographic expression of culture in landscape; there, the spatial constitution of culture, its symbolic expression and social significance—are reflected in different approaches: whereas the emphasis here is strongly empirical, in the next chapter it will be much more theoretical and conceptual.

We begin with some provocative ideas about cultural identity in the contemporary world (a subject we will revisit in Chapters 8 and 9). Following a brief introduction to the terms 'culture' and 'society' as they are used by human geographers, we look at cultural evolution, the origins of 'civilization', cultural geographic regions—a good example of our second recurring theme, regional studies—and the making of cultural landscapes. We continue with discussions of the two cultural variables that are perhaps most important with respect to both human identity and the human creation of landscape: language and religion. The concluding material reviews the evidence for cultural globalization.

'The Monastery', Petra, Jordan. The capital of the Nabataean kingdom from 400 BCE until 106 CE, when it was annexed by Rome, Petra sat at the crossroads of several major trade routes (Andrew Leyerle).

A World Divided by Culture?

Humans have 'brought into being mountains of hate, rivers of inflexible tradition, oceans of ignorance' (James 1964:2).

In 1964 the eminent American geographer Preston James published an introductory geography textbook with an inspired title: *One World Divided*. We have already recognized some physical divisions (climate, relief, soil, vegetation, etc.) and some human divisions (demographic differences). We have also identified, described, and attempted to explain the fundamental division of the world into two regions: more developed and less developed. Our concern now is the human divisions that derive from differences in culture. In fact, the barriers created by human variables are often far more difficult to cross than any physical barrier. The divisions we noted in our accounts of the human population are only the tip of the iceberg.

The world we live in today is one of tremendous cultural diversity—but it is also one of increasing interaction between cultures. As a result, in many areas traditional groups are struggling to protect their established ways of life against 'foreign' influences; in others, the often uncontrolled passions that surround language, religion, and ethnicity are erupting in cultural and sometimes political conflicts. Ironically, it is our greatest human achievement—our culture—that has been responsible not only for erecting barriers but for sparking conflict between peoples.

Introducing Culture
Humans differ from all other forms of life in that they have developed not only biologically but culturally. Other forms of life, limited to biological adaptation, have become so highly specialized that most are restricted to particular physical environments; indeed, in many cases environmental change has resulted in species extinction. If, so far, humans have avoided such a fate, it is primarily because of our culture: our ability first to analyze and then to change the physical environments that we encounter. Unlike other animals, humans can form ideas out of experiences and then act on the basis of these ideas; we are capable not only of changing physical environments, but of changing them

in directions suggested by experience. This is one of the meanings of the term 'culture'.

Why is it, then, that this remarkable human ability has tended to have the unfortunate consequences noted by James? A large part of the reason is that each cultural change, each new idea prompted by experience, brings with it new knowledge and responsibilities that, at least initially, do not fit easily into the cultural framework of existing attitudes and behaviours. The value that humans generally continue to place on tradition and the past means that our cultural frameworks are rarely conducive to the efficient use of changes in culture. Once in place, differences in attitude and behaviour tend to be self-perpetuating and can be resolved only with the development of new, overarching values. Developing and establishing such new values is a challenging task, as it involves human engineering, literally changing ourselves in accord with these new values.

We can summarize this important introductory argument as follows:

1. Our world is divided, especially because of spatial variations in culture.
2. By culture, we mean the human ability to develop ideas from experiences and subsequently act on the basis of those ideas.
3. Once cultural attitudes and behaviours are in place, there is an inevitable tendency for them to become the frame of reference—however inappropriate—within which all new ideas (developments in culture) are placed and evaluated.
4. One task humans may choose to tackle today is to create new sets of values—in effect, to re-engineer ourselves.

We will return to the idea raised by James—that culture acts as a barrier between peoples at the world scale—in the concluding section of Chapter 9, where we discuss possible geopolitical futures.

Humans in Groups
The terms 'culture' and 'society' are two of the most awkward employed in human geography, if not in all social science. Both terms are primarily associated with disciplines other than human geography—culture with anthropology, society with sociology—both of which study human behaviour

at the group scale (the individual scale is the realm of psychology). Anthropology and sociology evolved as separate disciplines in the nineteenth century, the former focusing on non-western and rural issues, the latter on western and urban issues. As recently as 1958, however, eminent representatives of the two disciplines found it necessary to clarify what they meant by the terms 'culture' and 'society' (Box 7.1).

Human Geography and Culture

Following in the tradition pioneered by earlier geographers, especially Humboldt, Vidal treated humans as part of nature, not separate from it. As we saw in Chapter 1, he focused on the relations between humans and land, with emphasis on culture ('*genre de vie*'). A proponent of the possibilist view, he explicitly opposed the idea of determinism, whether environmental or social.

A view comparable to Vidal's did not develop in North America until the 1920s, when Sauer put forward a series of ideas that would become central to the 'landscape school' (see Box 1.11). These ideas, derived from possibilism, centred on the creation of cultural landscapes from earlier physical landscapes. Sauer emphasized that the object of

geographic study was the landscape itself, not the culture that created it.

There have been suggestions that Sauer was influenced by a school of thought in American anthropology espoused by Alfred Kroeber and known as the **superorganic** (Box 7.2). The evidence on this count is contradictory. There is little doubt, however, regarding the impact of the landscape school on North American human geography—it has survived largely intact from the 1920s to the present.

By now it will be apparent that **culture** is a complex concept in geography. Earlier, we suggested that 'culture' referred to our unique human ability to deliberately change physical environments in directions suggested by experience—a process that over time led to the development of distinct ways of life. But this is an extremely broad, general statement. To gain a more specific understanding, it is useful to note that culture can be divided into two categories, non-material and material (Huxley 1966; Zelinsky 1973:72–4). *Non-material* culture has two components: (1) key attitudinal elements or values, such as language and religion, known as **mentifacts**, and (2) the norms involved in group formation, such as rules about family

---

7.1

THE CONCEPTS OF CULTURE AND SOCIETY

The emerging disciplines of anthropology and sociology focused on 'culture' and 'society' respectively, but their definitions of these concepts do not differ in any significant way. The classic early definition of culture proposed in 1871 by Tylor—'that complex whole which includes knowledge, belief, art, morals, law, customs and any other capabilities and habits acquired by man as a member of society' (see Friedl and Pfeiffer 1977:288)—and developed by Boas around the turn of the twentieth century, seems very close to the idea of society presented by the great early sociologists: Comte, Spencer, Durkheim, and Weber. Thus the core concepts of late nineteenth- and early twentieth-century anthropology and sociology were not clearly distinguished. Rather, the distinction between the disciplines was operational: anthropologists applied those concepts largely to non-western groups, while sociologists applied them to western ones.

In 1958, distinguished representatives of both disciplines collaborated to clarify the terms, proposing that 'culture' be reserved for a narrower sense than it had been previously, to refer to the 'transmitted and created content and patterns of values, ideas and other symbolic-meaningful systems [that are] factors in the shaping of human behavior and the artifacts produced through behavior'. 'Society' or 'social system', on the other hand, would refer to the 'specifically relational system of interaction among individuals and collectivities' (Kroeber and Parsons 1958:583).

Although the distinction proposed by Kroeber and Parsons has been highly influential, today anthropologists and sociologists alike continue to use the two terms in various ways. Traditionally, human geographers have had their closest links with anthropology, and hence with the culture concept, but now they are turning to sociology and the society concept for new inspirations.

structure, known as **sociofacts**. *Material culture* comprises all the elements related to people's livelihood, known as **artifacts**, and thus includes the human landscape.

Traditionally, then, human geographers have followed Vidal and Sauer (but especially the latter) in taking as their prime object of study the landscapes created by humans through their cultures. In our study of human landscapes, both mentifacts and sociofacts facilitate our understanding. Anthropologists study human cultures, whereas human geographers have traditionally paid attention to the landscape expressions of these cultures.

Human Geography and Society

Simply put, society is to sociology what culture is to anthropology: the central concept of the discipline. Like 'culture', 'society' may be defined in various ways, depending on theoretical perspective. According to the traditional view, the term **society** refers to a cluster of institutionalized ways of doing things. Whereas 'culture' refers to the way of life of the members of a society, 'society' refers to the system of interrelationships that connects those individuals as members of a culture (Giddens 1991:35). Thus 'society' refers to recurrent attitudes and behaviours among a particular group. Clearly, to the extent that 'society' includes non-material culture—languages, religions, and social structures are all institutionalized ways of doing things—the term overlaps with 'culture'.

Let us turn to the concept of society as it is used in human geography. North American geographers who adhere to Sauer's landscape school have employed the term 'culture' rather than 'society' in their analyses of the human landscape; British and European geographers have done the reverse, using 'society' rather than 'culture'. In fact, there is no clear and absolute distinction between these two terms, either in social science generally or in human geography specifically. Not surprisingly, current research indicates that human geographers now recognize the need to integrate North American cultural and European social geography. Relatively nominal differences in emphasis cannot disguise a basic unity. Human geographers are able to use the two related concepts of culture and society to analyze landscape.

7.2

THE 'SUPERORGANIC' CONCEPT OF CULTURE

Sauer regarded culture as 'the impress of the works of man upon the area' (Sauer 1925:30). His highly influential view was derived from earlier French and German perspectives, but it also incorporated a rather different dimension. Unlike Vidal, who rejected Durkheim's emphasis on social determinism, Sauer appears to have accepted the popular North American view of culture as cause: cultural determinism.

One of North America's leading anthropologists, Alfred Kroeber, proposed that culture was 'superorganic'—at a higher level than the individual—and that it constrained individual behaviour. Kroeber taught anthropology at Berkeley, and Sauer, who taught geography there, sent most of his students to study with him. The suggestion that Sauer accepted Kroeber's view of culture is supported by his various references to culture as the 'agent' and the 'shaping force' (Sauer 1925, 46). But the evidence is far from unequivocal on this point, for Sauer does not downplay the role of physical geography, and his own prolific writings are definitely not examples of the application of the superorganic concept.

Whatever Sauer himself may have thought, there are indications that many of his students subscribed to Kroeber's view, and they were a highly influential group in mid-twentieth-century human geography. It is therefore reasonable to suggest that human geography's focus on landscape, especially the visible landscape, is a direct reflection of the discipline's general acceptance of culture as a causal variable.

The issues raised in this box are not straightforward. For some scholars Kroeber is the archetypal cultural determinist; for others he is not a cultural determinist at all; for some Sauer followed Kroeber, while for others Kroeber played only a minor role in Sauer's ideas; for some Sauer's legacy is an inflexible cultural determinism, while for others that legacy is rich and varied. At this point, all that we can usefully conclude is that Sauer's landscape school paid much attention to landscape as a geographical expression of the impact of culture, and little attention to culture itself.

The Importance of Human Scale

As human geographers, we are particularly aware of the importance of selecting appropriate spatial scales of analysis when conducting research. Our concern here is with a second choice of scale, human scale.

As students of human populations, we may in principle choose to work at any scale, from the individual to the world population, through the intermediate scales of nuclear families, extended families, friendship circles, voluntary associations, involuntary associations, institutions, and nations. Selecting the appropriate scale is a function of the particular type of study being conducted. In order to improve understanding of our divided world, human geographers have typically focused on cultures. Thus we study groups delimited on the basis of common operating rules (or what we might call common mentifacts or common sociofacts). For human geographers, an appropriate social scale typically allows delimitation of a group that is meaningful given the aims of the work. This point is actually more important than it may appear.

Philosophical emphasis and human scale

Choosing the most appropriate human scale is a thorny issue in social science. Those with a humanistic focus recognize the need to study both individuals and groups; in human geography, then, we need to study individual humans, their actions, and intentions. Those with a Marxist perspective typically focus on groups because they believe that individuals cannot be understood without reference to the appropriate larger cultural context. Most traditional cultural geography has favoured a group scale because that is the scale best suited to the typical geographic interest in the world or regions of the world. Finally, most contemporary social theory favours the group scale, largely on the grounds that individual actions are determined by ideas and beliefs rooted in groups defined on the basis of interaction and communication.

The Evolution of Culture

Early Human Groups

Our knowledge regarding the earliest cultural attainments of humans and the earliest organization of human groups is limited.

Evidence of both language and tool-making dates from about 2.5 million years ago. Fire was first used perhaps 1.5 million years ago, prompting the identification of specific locations as home bases. Language appears to have expanded substantially about 400,000 years ago. Together, these developments facilitated the formation of social groups. Early human groups probably numbered between 10 and 30 (occasionally as many as 100) people linked by kinship. Groups of this size prevailed until the agricultural revolution.

No doubt many factors encouraged language, tool-making, and group formation, among them the requirements of basic subsistence, mating, child care, cooperative hunting, information sharing, and minimizing conflict. In early human groups—typically labelled 'bands'—there was probably little labour specialization (since economic activities were limited to the basic search for food) or variation in status. Age and gender were likely the two principal means by which individuals were differentiated, and any authority was probably exercised by heads of families, not separate political authorities. Similarly, religion did not have a separate existence; any beliefs or activities that we today might see as religious were simply a part of everyday life, centred on the search for food within a particular physical environment.

As a way of life, hunting and gathering was relatively undemanding, without either long-term food shortages or major outbreaks of disease. Nevertheless, it has been abandoned throughout the world with only a few exceptions, such as the Kazakhs of central Asia or the Inuit of northern Canada. The cultural development that led to its abandonment was the agricultural revolution.

The Beginnings of Civilization

According to Smil, 'the maintenance of a suitable environment . . . remains a critical precondition of human existence, [but] much more is needed for the development of civilizations':

The essence of these additional requirements is the appropriation of energies, first merely by better management of human labour, later by extensively harnessing various renewable transformations of solar radiation . . . , finally by extracting fossil fuels and generating electricity (Smil 1987:1).

It was in Chapter 4, during our discussion of human impacts on environment, that we first noted the importance of humans' use of energy sources. In this section we will examine various theories about where, how, and why some human groups began the 'appropriation of energies' essential to the development of 'civilization'.

Prior to the agricultural revolution, as we have seen, the religious, political, and economic dimensions of life were so rudimentary that any variations in culture between bands were intimately related to the physical environment. It was only with the agricultural revolution that what we might now term **civilization** appeared: a particular type of culture that includes a relatively sophisticated economy, political system, and social structure (Box 7.3). What specific changes gave rise to the beginnings of civilization?

There are many possible responses to this question. Racist explanations such as that proposed by Gobineau (see Box 3.5) were common in the nineteenth century, but are now discredited. Explanations based on environmental determinism, such as Huntington's (1924), are no longer tenable because they are not supported by facts. The nineteenth-century anthropologist Morgan proposed that all cultures evolve in the same basic way, through successive levels of material achievement (Table 7.1); but this explanation is also inadequate, because it has nothing to say about what brought about the necessary changes in material achievement. Three other suggested explanations, however, all of which presuppose an increasing population, merit more detailed comment.

The hydraulic hypothesis

Two theories centre on water. According to Childe (1951), the introduction of irrigation significantly increased agricultural productivity. More specifically, according to the German historian Wittfogel (1957), an expanding population would have required more agricultural land—but in a semiarid area, the only way to achieve such expansion would have been through irrigation. In turn,

---

7.3

DEFINING 'CIVILIZATION'

In many respects 'civilization' is an awkward term, since it can be interpreted as implying that members are 'civilized' and non-members are not. Indeed, there is good reason to believe that this was precisely the way the early 'civilizations' of China, India, Greece, Rome, Persia, and the Middle East saw themselves in relation to others. Until the early twentieth century, when the idea of multiple civilizations emerged, the concept was almost by definition exclusionary. There is no single accepted definition of civilization. However, the word itself derives from the Latin *civis*, citizen, and we generally associate 'civilization' with an urban way of life.

Some anthropologists, such as Kroeber, have defined 'civilization' in terms of internal complexity (for example, the nature of social relations and economic activities): for them, civilizations are more internally complex than other forms of human organization. Others, such as Morgan, have considered complexity less important than type of organization. A third approach is to define civilization by reference to a particular set of characteristics. Childe (1951) identified ten characteristics, subsequently organized by Charles Redman as follows:

Primary
1. city settlement
2. labour specialization
3. concentration of surpluses
4. class structure
5. state organization

Secondary
6. monumental public works
7. long-distance trade
8. standardized monumental artwork
9. writing
10. arithmetic, geometry, astronomy

The primary characteristics are aspects of human organization, while the secondary characteristics are features of material culture.

The first civilizations in the world evolved in areas where agriculture also evolved. For example, in Mesopotamia—the area between the Tigris and Euphrates rivers, which today is part of southern Iraq—irrigated agriculture, water management, and land reclamation were probably being practised as early as 5,000 years ago. The fact that this 'cradle of civilization', as it is often known, introduced regulations on water use indicates the complexity of its social organization, and the city of Babylon became a great administrative, cultural, and economic centre.

irrigation would have required the construction of canals and aqueducts—which in turn would have required that the people work cooperatively, under the direction of some central authority. In short, a new form of cultural organization based on central control arose in response to the need for irrigation.

Coercion

Another theory supposes that the necessary social changes, particularly the imposition of central control, preceded agricultural development, as groups and their territories increased in size. This second suggested explanation proposes that the development of agriculture and civilization may have resulted from increases in population pressure in an area too small to support more people. In this case the population growth would have been accommodated not through intensification of agriculture (as in the hydraulic hypothesis) but through territorial expansion by means of force—the same kind of expansion that, over time, creates empires. According to Carneiro (1970:734), 'only a coercive theory can account for the rise of the state.'

Marxism

A Marxist view sees civilization as the product of class differences, which in turn developed as a logical consequence of the wealth accumulated by those individuals possessing domesticated animals. Once such class differences develop, class conflict follows, as it becomes necessary for those with wealth to defend themselves from those without wealth. At the same time, exchange begins between those with and those without animals, and as population increases, a class of specialists become involved in the exchange process. In this way people not directly involved in production begin to control economic life, adding a second dimension to the emerging class conflict. Such class conflict in turn generates central control as a means for the more dominant class to maintain and even enhance its dominance. Civilization follows from these beginnings in the form of a culture consisting of rulers and ruled.

Early Civilizations

Whatever the specific reasons behind them, early civilizations represented major cultural advances and brought many changes. Although

Irrigated terraces at the ancient Incan site of Pisac, Peru
(Philip Baird).

each one had its own distinctive character, they shared several characteristics. Agriculture and urbanization are the two most obvious developments that have impacts on land, but civilization also involves changes in non-material

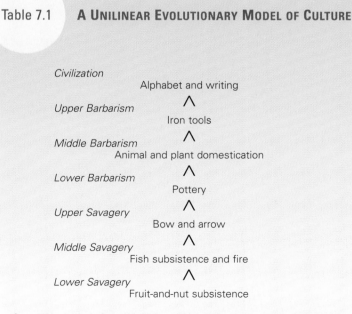

Table 7.1   **A UNILINEAR EVOLUTIONARY MODEL OF CULTURE**

Civilization
                    Alphabet and writing
                           ∧
Upper Barbarism
                        Iron tools
                           ∧
Middle Barbarism
                Animal and plant domestication
                           ∧
Lower Barbarism
                         Pottery
                           ∧
Upper Savagery
                      Bow and arrow
                           ∧
Middle Savagery
                  Fish subsistence and fire
                           ∧
Lower Savagery
                  Fruit-and-nut subsistence

Source: *Human Antiquity* by K.L. Feder and M.A. Park with permission of Mayfield Publishing Co. Copyright © 1993, 1989 Mayfield Publishing Co.

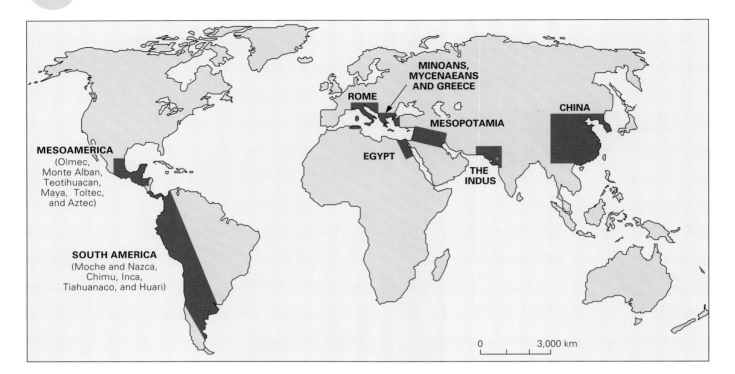

Figure 7.1 Civilizations in the ancient world.

culture: the rise of powerful elites and the disappearance of the egalitarian family-based group, the growth of bureaucratic systems, and the institution of private land ownership. In many cases religion was the basis of early power, but all civilizations became increasingly secular. The great early civilizations are mapped in Figure 7.1; for a general time scale, see Table 7.2. As the table makes clear, almost all the great early civilizations—sets of cultural advances—have been temporary; only the civilization of China has lasted to the present day. All others have collapsed, whether because of natural disasters (such as earthquakes or floods) or because of conflict (internal or external). Earlier cultural groupings might have been able to adapt and survive crises such as these, but the size, density, and organizational complexity of civilizations made them vulnerable.

Cultural Regions

The areas occupied by early civilizations are examples of cultural regions. In this section we will examine how the concept of **cultural regions** is used in contemporary human geography.

Traditionally, human geographers have been concerned not with humans themselves or human cultures, but with human impacts on landscape. Yet these impacts can vary considerably, depending on the specific characteristics of the human group in question—in other words, depending on culture. Because different cultures have emerged in different areas and because the earth is a diverse physical environment, there is a wide variety of human landscapes. In addition to describing and explaining these landscapes, human geographers have sorted them into cultural regions.

Delimiting cultural regions requires decisions on at least four basic points:

Table 7.2	BASIC CHRONOLOGY OF EARLY CIVILIZATIONS
Egypt	3000–332 BCE
Minoan	3000–1450 BCE
Indus	2500–1500 BCE
Mesopotamia	2350–700 BCE
China	2000 BCE–present
Mycenaean	1580–1120 CE
Olmec	1500–400 BCE
Greece	1100–150 BCE
Rome	750 BCE–375 CE
Monte Alban	200 BCE–800 CE
Moche and Nazca	200 BCE–700 CE
Teotihuacan	100–700 CE
Maya	300–1440 CE
Tiahuanaca	600–1000 CE
Toltec	900–1150 CE
Chimo and Inca	1100–1535 CE
Aztec	1200–1521 CE
Benin	1250–1700 CE

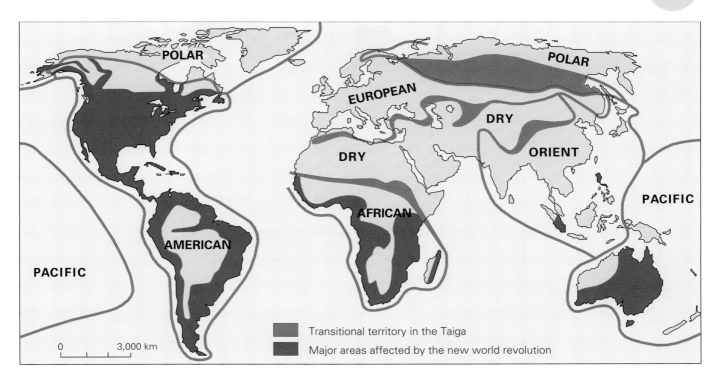

Transitional territory in the Taiga

Major areas affected by the new world revolution

Figure 7.2 Cultural regions of the world.

Source: Adapted from R.J. Russell and F.B. Kniffen, *Culture Worlds* (New York: Macmillan, 1951).

1. criteria for inclusion;
2. date or time period (since cultural regions change over time);
3. spatial scale; and
4. boundary lines.

These four issues are all interrelated, and they are not easy to resolve.

World Regions

On the world scale, many notable attempts have been made to delimit a meaningful set of cultural regions. The historian A.J. Toynbee (1935–61), for example, identified a total of 26 civilizations as responses to environment, past and present: 16 'abortive' and 10 surviving. Three of the latter—the Polynesian, Nomad, and Inuit civilizations—Toynbee described as 'arrested' because their overspecialized response to a difficult environment left them unable either to expand into different regions or to cope with environmental change. He identified the remaining seven living civilizations as Western Christendom, Orthodox Christendom, the Russian offshoot of Orthodox Christendom, Islamic culture, Hindu culture, Chinese culture, and the Japanese offshoot of Chinese culture; at the time when he was writing, Toynbee considered most of Africa to be 'primitive'.

Although its limitations are clear, Toynbee's schema is instructive. His principal criterion

is obviously religion, since he named most of his civilizations for their religions, but in a broader sense the most important feature of each is the manner in which it has responded to the environment—hence Toynbee's recognition of 'arrested' and 'abortive' civilizations.

The first geographers to attempt to delimit world regions were Russell and Kniffen (1951), who began by recognizing various cultural groups and then related them to areas so as to delimit seven cultural regions, each a product of a long evolution of human–land relations (Figure 7.2; the term 'new world revolution' in the legend refers to the diffusion of European ideas and technologies with European expansion). With the addition of one 'transitional' area, these regions effectively cover the world. Russell and Kniffen's regions are well justified, and their classification is a valuable contribution to our general understanding of the world. Nevertheless, it is important to emphasize that world regionalizations of this type do not represent the best use of the culture concept in geography. The central problem is scale: the larger the area to be divided, the more likely it is that the regions identified will be either too numerous or too superficial to be helpful. Like regionalizations based on physical variables (see Chapter 3), cultural regionalizations are more appropriately seen as useful classifications than as insightful applications of geographic methods. The difficulties involved in delimiting Europe as a cultural region are outlined in Box 7.4.

Cultural regions and the distinctive landscapes associated with them are more effectively considered on a more detailed scale. If, for example, we take North America as one example of a world region, we find that subdividing greatly enhances our appreciation of the relationship between culture and landscape.

North American Regions

North America is a relatively easy region to subdivide because the development of its contemporary cultural regions is recent enough that we are able to trace their origins. Figure

7.4 delimits 16 regions in North America on the basis of several criteria, such as physical and economic differences, not simply culture. Each of these regions is likely to be generally recognizable, and the authors of the map attempt to convey the 'feeling' of each region in their discussions. This regionalization is interesting because it focuses on broad regional themes rather than specific criteria, and recognizes that the border between the United States and Canada is not a fundamental geographic division.

A regionalization of the United States

Figure 7.5 is a regionalization of the United States explicitly based on the interesting con-

7.4

EUROPE AS A CULTURAL REGION

Although clearly an area of considerable diversity, Europe is generally seen as a single world region—'a *culture* that occupies a *culture area*' (Jordan 1988:6). Before the sixteenth century it was widely believed that Europe was a continent physically separated from Asia, and even after this error had been corrected, the image of a separate continent was so powerful that a new divide was needed: thus the idea became generally accepted that the Caucasus, the Urals, and the Black Sea represented a meaningful boundary. The image of a separate Europe, an area that was culturally distinctive, was allowed to remain intact.

Using the political map of the mid-1980s, Europe was defined on the basis of twelve traits measured for each country (Figure 7.3):

- majority of population speak an Indo-European language;
- majority of population have a Christian heritage;
- majority of population are Caucasian;
- more than 95 per cent of population are literate;
- an IMR of less than 15;
- an RNI of 0.5 per cent or less;
- per capita income of US $7,500 or more;
- 70 per cent or more of population are urban;
- 15 per cent or less are employed in agriculture and forestry;
- 100 km (62 miles) of railway plus highway for 100 km² (39 square miles);
- 200 or more kg (441 lbs) of fertilizer are applied annually per ha of cropland; and
- free elections are permitted.

Figure 7.3 implies a regionalization; to proceed further would require a more explicit focus on variables such as language or religion. One particularly innovative delimitation of Europe, by Jordan (1988:402), combines three approaches: a north/south distinction, an east/west distinction, and a core/periphery distinction.

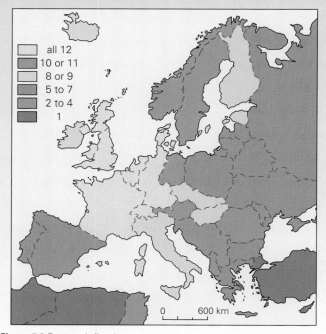

Figure 7.3 Europe defined.
Source: T.G. Jordan, *The European Culture Area: A Systematic Geography*, 2nd ed. (New York: Harper & Row, 1988):14.

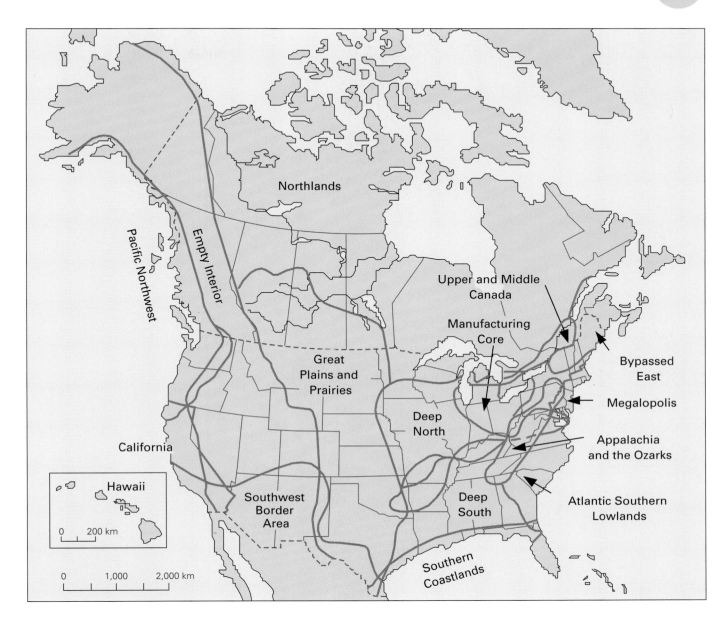

California

Hawaii

0   200 km

0   1,000   2,000 km

Pacific Northwest

Empty Interior

Northlands

Great
Plains and
Prairies

Southwest
Border
Area

Upper and Middle
Canada

Manufacturing
Core

Deep
North

Deep
South

Southern
Coastlands

Bypassed
East

Megalopolis

Appalachia
and the Ozarks

Atlantic Southern
Lowlands

cept of **first effective settlement**: 'Whenever an empty territory undergoes settlement, or an earlier population is dislodged by invaders, the specific characteristics of the first group able to effect a viable, self-perpetuating society are of crucial significance for the later social and cultural geography of the area, no matter how tiny the initial band of settlers may have been' (Zelinsky 1973:13).

Accordingly, Zelinsky demarcates five regions—west, middle west, south, midland, and New England. Only New England has a single major source of culture (England); the other four regions are further divided into various subregions, all of which have multiple sources of culture. Thus the midland is

divided into two subregions, each of which has multiple sources; the south is divided into three subregions, each of which has multiple sources; the middle west is divided into three subregions, each of which has multiple sources; and the west is divided into nine subregions, each of which has multiple sources. In addition, a number of sub-subregions are noted, along with three regions of uncertain status (Texas, peninsular Florida, and Oklahoma). As a cultural regionalization, this example is especially useful because of the single clear variable employed. Yet the simplicity of the approach does not disguise the complexity of the cultural landscapes revealed by the many subregions.

Figure 7.4 Regions of North America. Overlapping boundaries are included to emphasize the uncertain status of areas that might be seen as belonging to more than one region. This is another way of acknowledging that many of the boundaries between regions are zones of transition rather than firm divisions.

Source: Adapted from S.S. Birdsall and J.W. Florin, *Regional Landscapes of the United States and Canada*, 2nd ed. (New York: John Wiley & Sons, 1981):18. Reprinted by permission of John Wiley & Sons, Inc.

Figure 7.5 Cultural regions of the United States.
Source: Adapted from W. Zelinsky, *Cultural Geography of the United States* (Englewood Cliffs: Prentice-Hall, 1973):118.

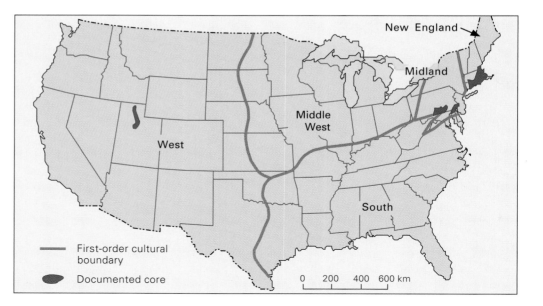

Figure 7.6 Regions of Canada.
Source: Adapted from J.L. Robinson, *Concepts and Themes in the Regional Geography of Canada*, rev. ed. (Vancouver: Talon Books, 1989):17.

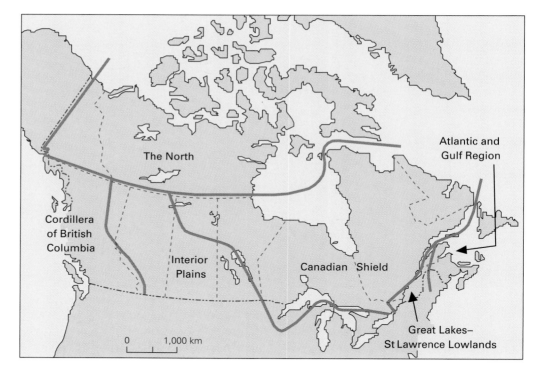

A regionalization of Canada

In our third North American example, Canada is divided into six regions (Figure 7.6). This regionalization employs several variables, but a basic cultural division is evident. The north is delimited according to political boundaries and can be easily subdivided into a northwest that is relatively forested and has a mixed Native and European population and an Arctic north that is treeless, is populated mostly by Inuit people, and has limited resource potential by southern Canadian standards. The British Columbia region is characterized by major differences in physical and human geography and typically has a resource-based economy; the same is true of the Atlantic and Gulf region, which in addition has the problems of low incomes and high unemployment rates. The interior plains are an agricultural region (cattle, wheat, and mixed farming), and its dispersed farmsteads and often regularly spaced towns reflect government surveying and planning before settlement. The core region of Canada is the

Great Lakes–St Lawrence lowlands, with most of the people, the largest cities, major industry, and intensive agriculture; this is the cultural and economic heartland of Canada. The largest region is the Canadian Shield, a sparsely inhabited area in which most of the settlements are resource towns.

The Making of Cultural Landscapes

So far, our discussion of cultural regions has been descriptive rather than explanatory. Now we will consider in more detail how regions of distinctive cultural landscape arise, beginning with a look at cultural variations over space and the ability of culture to affect landscape.

Although all cultures share certain basic similarities—all need to obtain food and shelter and to reproduce—they differ in the methods used to achieve those goals. As humans settled the earth, culture initially evolved in close association with the physical environment. Over time, however, cultural adaptations brought increasing freedom from environmental constraints. Gradually, as humans' ties to the physical environment loosened, their ties to their culture increased. One of the most important changes in human history was the shift (associated with the industrial revolution) to a capitalist mode of production, which largely destroyed our ties to the physical environment but created a new set of ties—to culture itself. Some human geographers see this change as part of an ongoing process that is effectively homogenizing world culture by minimizing regional variations; this view is derived most clearly from Marx (see Box 2.6, p. 55).

Cultural Adaptation

As humans settled the earth, different cultures evolved in different locations. Despite their many underlying similarities, each one developed its own variations on the basic elements of culture: language, religion, political system, kinship ties, and economic organization. As we seek to understand how different cultures create different landscapes, we must also ask how sound relationships develop between humans and their physical environment. This question is central to human geography: witness environmental determinism, possibilism, and related viewpoints. One answer centres on the concept of **cultural adaptation**.

Humans are continually adapting to the environment, genetically, physiologically, and culturally. Cultural adaptation may take place at an individual or a group level—human geographers are equally interested in both—and consists of changes in technology, organization, or ideology of a group of people in response to problems, whether physical or human. Such changes can help us to deal with problems in any number of ways: by permitting the development of new solutions or improving the effectiveness of old ones, by increasing general awareness of a particular problem or enhancing overall adaptability. In other words, cultural adaptation is one of the essential processes by which sound relationships evolve between humans and land.

We noted earlier the idea that culture represents the human ability to develop ideas from experiences and subsequently to act on the basis of those ideas. Cultural adaptation thus includes changes in attitudes as well as behaviour. One current example of attitudinal change is our increasing awareness of the importance of environmentally sensitive land-use practices; an example of behavioural change would be the implementation of such practices. Needless to say, behavioural changes do not always accompany attitudinal changes—or even follow them.

Human geographers have proposed a number of theories to explain more precisely what cultural adaptation is and how it works. Among them are the following.

Culturally habituated predisposition

In their analysis of the evolution of agricultural regions, Spencer and Horvath (1963:81) suggested that such regions could be seen as the 'landscape expression of . . . the totality of the beliefs of the farmers over a region regarding the most suitable use of land'. According to this view, cultures are particular beliefs, psychological mindsets, that result in a culturally habituated predisposition towards a specific activity and hence a specific cultural landscape; for example, the corn landscape of the early European American midwest reflected the commercial aims of incoming settlers.

Cultural preadaptation

According to another view, a cultural group moving into a new area may be preadapted for that new area: in other words, conditions

Figure 7.7 Core, domain, and sphere.
Source: Adapted from D.W. Meinig, 'The Mormon Culture Region: Strategies and Patterns in the Geography of the American West, 1847–1964', *Annals, Association of American Geographers* 55 (1965):214.

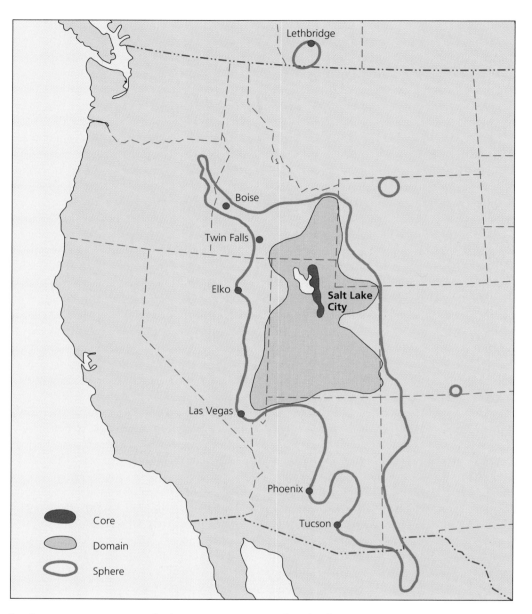

in the source area are such that any necessary adjustments have already been made prior to the move or are relatively easy to make thereafter. Following this logic, Jordan and Kaups (1989) proposed that the American backwoods culture had significant northern European roots, specifically in Sweden and Finland.

Core, domain, and sphere

Perhaps the most useful explanation is the core, domain, and sphere model (Meinig 1965). According to this view, a cultural region or landscape can be divided into three areas—core (the hearth area of the culture), domain (the area where the culture is dominant), and sphere (the outer fringe)—and

cultural identity decreases with increasing distance from the core. Figure 7.7 shows what most observers would agree is one of the easiest regions to delimit: the Mormon region of the United States (Box 7.5). But Meinig has successfully extended these ideas, in modified form, to other regions such as Texas and the American southwest. Few attempts have been made to apply these ideas outside North America.

Each of the three views outlined above is, however indirectly, very much in the Sauer landscape school tradition. There is a consistent concern with the landscapes created by cultural groups and particularly with the material manifestations of culture: the visible landscape.

THE MORMON LANDSCAPE

Members of the Church of Jesus Christ of Latter-day Saints, commonly known as Mormons, first arrived near Salt Lake in 1847 and subsequently settled in an extensive area that included part of southern Alberta. Most decisions concerning settlement expansion and landscape activities were made by the church leadership and reflected the fundamental character of the Mormon religion: decisions were made in response to the requests of church leaders and as one part of the larger experience of being a Mormon. Thus settlement followed an orderly sequence, in accord with larger church concerns and ambitions, and landscape change demonstrated a high degree of uniformity. More than most groups, Mormons emphasized cooperation, community, and economic success as means to establish and solidify their occupation of a region.

Settlement sites were usually selected by leaders, and the settlers either volunteered or were called by church leaders to move to each new location. The Mormon landscape that evolved has a number of distinctive features, all intimately related to the religious identity. Speaking in 1874, church president George Smith observed: 'The first thing, in locating a town, was to build a dam and make a water ditch; the next thing to build a schoolhouse, and these schoolhouses generally answered the purpose of meeting houses. You may pass through all the settlements from north to south, and you will find the history of them to be just about the same.'

Francaviglia (1978) describes the Mormon landscape in detail, based on extensive travels within the region. Mormon towns were laid out in approximate accord with the City of Zion plan as detailed in 1833 by the church's founder, Joseph Smith: there was a regular grid pattern, with square blocks, wide streets, half-acre lots that included backyard gardens, brick or stone construction, and central areas for church and educational buildings. Towns laid out along these lines continue both to reflect and to enhance the Mormons' community focus, and represent a distinctive town landscape within the larger region of the mountain west.

The rural landscape of the Mormons is also distinctive, emphasizing arable agriculture rather than pastoral activities (which necessarily involve greater movement and lower population densities). Among the notable features of the farm landscape are networks of irrigation ditches, lombardy poplars, unpainted fences and barns, and hay derricks. Together, these features serve both to distinguish the Mormon landscape from the surrounding areas and to impose a sense of unity on it.

The wide main street in Manassa, Colorado; the central temple in Cardston, Alberta; and a hay derrick at Cove Fort, Utah (William Norton photos).

Two aspects of culture are particularly important in understanding our human world: language and religion. Not only are they important in themselves, but they are also good bases for delimiting cultures, and hence regions, and they affect both behaviour and landscape.

Language

Both language and religion have traditionally been of interest to cultural geographers, but neither has been recognized as a core concern. Park's observation (1994:1) that 'the study of geography and religion remains peripheral to modern academic geography' applies also to the study of geography and language.

Yet—as we will see in our discussions of ethnicity (Chapter 8) and political states (Chapter 9)—language and religion are often the fundamental characteristics by which groups distinguish themselves from other groups. In the current chapter we will look at both language and religion in terms of classification, origins, diffusion, regionalization, links to identity, and relations to landscapes. No attempt is made in either case to produce comprehensive human geographies. This is appropriate because, as noted, neither linguistic geography nor religious geography is a well-recognized subdiscipline of human geography; rather, both are cultural variables that relate to many different aspects of human geography.

A Cultural Variable

Language, probably the single most important human achievement, is of interest to human geographers for several reasons. *First*, language is a cultural variable, a learned behaviour that initially evolved so that humans could communicate in groups, probably for the purpose of organizing hunting activities. It is possible that early humans, concentrated in the area of origin, all spoke the same early language. According to McWhorter (2002), this brief phase of human unity occurred about 150,000 years ago in East Africa. However, as people moved across the earth, different languages developed in different areas offering new physical environmental experiences. From the beginnings of cultural evolution, therefore, language has been a potential source of group unity (and therefore of total-popula-

tion disunity)—a topic of great interest to human geographers.

A *second* important feature of language is its usefulness in delimiting groups and hence regions. Language is the means by which a culture ensures continuity through time—and the death of a language is often seen as the death of a culture. It is for this reason that such determined efforts have been made to ensure the continuity, or even revival, of traditional languages in places such as Wales and Quebec. As a fundamental building-block of nationhood, language is of particular interest to human geographers concerned with the plight of minority groups, and the relationships between language and nationalism.

Our *third* interest in language concerns its interactions with environment, both physical and human. Spatial variations in language are caused in part by variations in physical and human environments, while language itself is an effective moulder of the human environment. The importance of language to the symbolic landscape links it to group identity, returning us to our first interest.

How Many Languages?

Like many cultural variables—indeed, like culture itself—language began as a single entity (or at most a few different ones) and diversified into many. Of the roughly 7,000 distinct languages (not counting minor dialects) that existed four hundred years ago, approximately 1,000 have disappeared, leaving about 6,000 languages today. It is feared that many more may vanish over the next few hundred years.

Languages die for two general reasons. First, a language with few speakers tends to be associated with low social status and economic disadvantage; hence those who do speak it may not teach it to their children. Second, because globalization depends on communication between previously separate groups, it is becoming essential for more and more people to speak a major language such as English or Chinese. There are varying reactions to the idea that perhaps half of our present languages will disappear during the current century. For many observers, language loss represents culture loss and is just as serious a threat as loss of biodiversity. Others take a more positive view, seeing the loss of linguistic diversity as

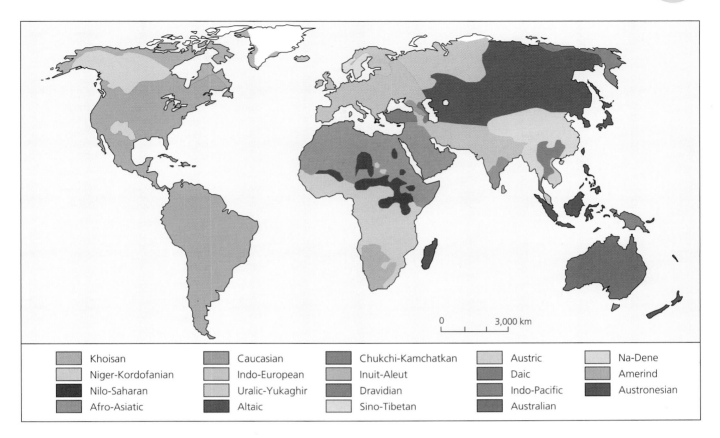

Khoisan

Niger-Kordofanian

Nilo-Saharan

Afro-Asiatic

Caucasian

Indo-European

Uralic-Yukaghir

Altaic

Chukchi-Kamchatkan

Inuit-Aleut

Dravidian

Sino-Tibetan

Austric

Daic

Indo-Pacific

Australian

Na-Dene

Amerind

Austronesian

a sign of increasing human unity, countering the divisive tendencies identified at the beginning of this chapter.

Classification and Regions

By studying language families—groups of closely related languages that suggest a common origin—scholars have found evidence of language evolution dating from perhaps 2.5 million years ago and evidence of language distribution patterns dating back some 50,000 years. Nevertheless, it appears that the distribution shown in Figure 7.8, just prior to the changes initiated by European overseas expansion, is a product of relatively recent migrations, during the last 5,000 years or so. Note that this map would be significantly different today, because of the expansion of several of the Indo-European languages since the sixteenth century.

As human groups moved, their language changed, with groups that became separated from others experiencing the greatest language change. Something similar has happened over the past several hundred years as Europeans especially have moved overseas. Although obvious differences in dialect have developed (compare the varieties of English

spoken in southern England, the southern United States, and South Africa, for example), no new languages have come into being as a result of these more recent movements, for two general reasons: because not enough time has elapsed for this to occur and because the groups that moved have remained in close contact with their original groups. This is not to say that languages cannot change significantly over a relatively short period of time, however: just compare the English of Shakespeare with that spoken today.

Table 7.3 provides a numerical summary of the number of speakers in each of the 19 language families. The largest family is the Indo-European, followed by the Sino-Tibetan. Table 7.4 identifies the eight languages spoken by the most people. Mandarin has more speakers than any other language, with English ranking second.

Interestingly, the Indo-European family is spatially dispersed today, while the Sino-Tibetan is essentially limited to one area. The difference reflects the fact that many Indo-European-speaking countries colonized other parts of the world, whereas Sino-Tibetan-speaking countries did not. Thus the numerical importance of Mandarin is attrib-

Figure 7.8 World distribution of language families before European expansion.
Source: M.C. Ruhlen, *A Guide to the World's Languages: Volume 1, Classification* (Stanford: Stanford University Press, 1987).

utable simply to the large population of China (similarly, the numerical importance of Hindi reflects the large population of India), whereas the numerical importance of English, Spanish, and Portuguese is attributable to, among other things, colonial activity.

Table 7.3 LANGUAGE FAMILIES

Language Family	Number of Languages	Number of Speakers (000)
Indo-European	144	2,300,000
Sino-Tibetan	258	1,200,000
Altaic	63	250,000
Niger-Kordofian	1,064	181,000
Austronesian	959	180,000
Afro-Asiatic	241	175,000
Dravidian	28	145,000
Daic	57	50,000
Uralic-Yukashir	24	22,000
Amerind	583	18,000
Nilo-Saharan	138	11,000
Austric	4	7,000
Caucasian	38	5,000
Indo-Pacific	731	2,735
Na-Dene	34	202
Khoisan	31	120
Inuit-Aleut	9	85
Australian	170	30
Chukchi-Kamchatkan	5	23

Source: Adapted and updated from M.C. Ruhlen, *A Guide to the World's Languages: Volume 1*, Classification (Stanford: Stanford University Press, 1987).

Table 7.4 NUMBERS OF SPEAKERS, MAJOR LANGUAGES

Language	Number of Speakers (millions)
Mandarin	1052
English	508
Hindi	487
Spanish	417
Russian	277
Arabic	246
Bengali	211
Portuguese	191

Source: *The World Almanac and Book of Facts*, 1999.

The Indo-European language family

Rather than review each of the language families in detail, we will investigate one family, Indo-European, and one component language, English. This choice is not arbitrary. English is the 'language of the planet, the first truly global language' (McCrum, Cran, and MacNeil 1986:19).

The parent of all Indo-European languages, what we might call proto-Indo-European, probably evolved as a distinct means of communication in either the Kurgan culture of the Russian steppe region, north of the Caspian Sea, or a farming culture in the Danube valley. The likely period of origin is between 6000 BCE and 4500 BCE. Our knowledge of these origins has been gleaned largely from reconstruction of culture on the basis of vocabulary. The evidence suggests a culture that was partially nomadic, that had domesticated various animals (including the horse), that used ploughs, and that grew cereals. The fact that contemporary Indo-European languages have similar words for snow but not for sea places the original language in a cold region some distance from the sea.

As this cultural group dispersed, using the horse and the wheel, the language moved and evolved (Figure 7.9). Different environments and group separation gave rise to many changes. The earliest Indo-Europeans to settle in what is now England were Gaelic-speaking people who arrived after 2500 BCE. They were followed by a series of Indo-European invaders: the Romans in 55 BCE and the Angles, Saxons, and Jutes in the fifth century CE (Figure 7.10). The latter three groups introduced what was to become the English language, forcing Gaelic-speakers to relocate in what is now the Celtic fringe. Surprisingly, little mixing took place between Gaelic and the incoming Germanic language. Old English, as it is called, was far from uniform, and was subsequently enriched with the introduction of Christianity (Latin) in 597, the invasions of Vikings between 750 and 1050, and the Norman (French) invasion of 1066. Each of these invasions led to a collision of languages, and English responded by diversifying. Today it has an estimated 500,000 words (excluding an equal number of technical terms); by contrast, German has 185,000 words and French 100,000.

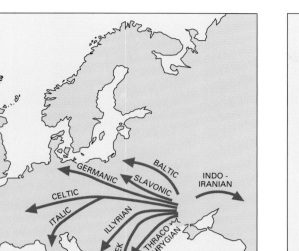

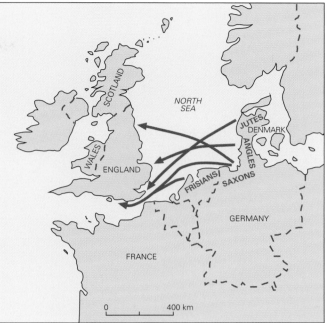

Figure 7.9 (left) Initial diffusion of Indo-European languages.

Figure 7.10 (right) Diffusion of Indo-European languages into England.

Beginning in the fifteenth century, the English language, along with other Indo-European languages, spread around the globe. This spread encouraged the diversification of English, including the development of many American varieties. Not only is English the first language of many countries, but it is the second language of many others. In India, for example, there are perhaps 70 million people who speak English as a second language; this is a vital unifying force in a country of many languages. English is indeed becoming the global language, entirely replacing some minority languages and becoming increasingly necessary as a second language in many parts of the world, including Europe.

Language and Identity

For many groups, language is the primary basis of identity—hence the close links between language and nationalism, the desire to preserve minority languages, and even the various efforts that have been made to create a universal language. A common language facilitates communication; different languages create barriers and frictions between groups, further dividing our divided world.

Language and nationalism

The link between language and **nationalism** is clear. Rarely, however, are the boundaries of language regions clearly defined—a fact that has resulted in many difficulties for those countries aspiring to declare their national territory on the basis of a common language. In medieval Wales there was a single word for 'language' and 'nation', and in present-day Ireland the survival of the Gaelic-speaking area is considered to be 'synonymous with retention of the distinctive Irish national character' (Kearns 1974:85). There are two reasons why language is often seen as the basis for delimiting a nation. First, a common language facilitates communication. Second, language is such a powerful symbol or emblem of 'groupness' that it serves to proclaim a national identity even where, as in Ireland, the language itself does not serve a significant communication function.

Before the nineteenth century, the boundaries between states and languages rarely coincided. France and the United Kingdom were single-language states, but most of Europe was politically divided on the basis of factors other than language. From the beginning, the rise of nationalism, in the nineteenth century, represented an effort to integrate language and state. As a result of this effort, the state of Italy was established in 1870 and the state of Germany in 1871.

Multilingual states

Other areas in Europe, however, did not follow this trend. Switzerland is a prime example of a viable political unit with several official languages: 70 per cent of the population speak German, 19 per cent French, 10 per cent

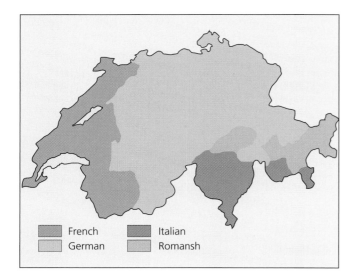

French
German

Italian
Romansh

Antwerpen
Brugge   Gent   ANTWERPEN   LIMBURG
WEST
FLANDERS   EAST
FLANDERS   Brussels   Hasselt

BRABANT   Liège
HAINAUT   LIÈGE

Mons   Namur
NAMUR

LUXEMBOURG

Arlon

Dutch
French
German

Figure 7.11 (left) Four official languages in Switzerland.

Figure 7.12 (right) Dutch, French, and German in Belgium.

Italian, and 1 per cent Romansh (Figure 7.11). The political stability for which Switzerland is known reflects several factors, including the pre-1500 evolution of the Swiss state, its long-standing practice of delegating much governmental activity to local regions, and its neutrality in major European conflicts.

More typical examples of **multilingual states** are Belgium and Canada. In both cases, the lack of a unifying language appears to have kept the country relatively weak. Despite conscious efforts to follow the Swiss example, Belgium is politically unstable. Created as late as 1830 as an artificial state—a move that pleased other European powers but not necessarily the people who became Belgians—Belgium has not managed to remain detached from European conflicts and today is divided between a Dutch-speaking north and French-speaking south (Figure 7.12). Although it is a bilingual state, the two areas are regionally unilingual. The fact that the capital, Brussels, is primarily French-speaking, even though it is located in the unilingual Dutch area, only aggravates a difficult situation: not only is Belgium not a bilingual country in practice, but it does not have a bilingual transition zone.

The situation is similar in Canada (Figure 7.13). There is no continuous language transition zone between English and French Canada—only a series of pockets, especially around the cities of Montreal and Ottawa and in Acadian areas of the Maritime provinces. This pattern suggests a parallel with Belgium (Box 7.6).

Minority languages

Any language is most likely to survive when it serves in an official capacity. **Minority languages** without official status typically experience a slow but inexorable demise. Thus minority-language speakers often strive for much more than the mere survival of their language: they may press for the creation of their own separate state, using the language issue as justification.

Examples of minority languages include Welsh and Irish in Britain, Spanish in the United States, French—despite its official status—in Canada, Basque in Spain, Hausa and other languages in Nigeria, and Cantonese in China. This is a highly condensed list; most countries in the contemporary world have at least one minority language. The consequences are varied. In Britain both Welsh and Irish continue to strive for survival and indeed independence (Box 7.7). In Spain the Basque language—one of very few languages in Europe that do not belong to the Indo-European family—is at the centre of a powerful and often violent Basque independence movement. In Nigeria the official language is English—a result of colonial activity and of the fact that, although Nigeria has several indigenous languages, no single one is predominant.

Communications between different language groups

Historically, some languages have played important roles even when they are not the first language of populations. In India, follow-

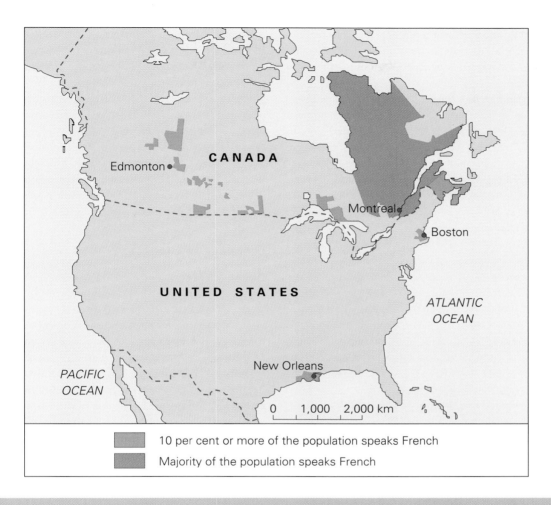

Figure 7.13 French and English in North America.

7.6

LINGUISTIC TERRITORIALIZATION IN BELGIUM AND CANADA

Although the specific causes are quite different, the Belgian and Canadian language situations are basically similar. Belgium is clearly divided into two linguistic territories, while Canada appears to be approaching the same situation.

Since about the fifth century, the area that is now Belgium has been divided into a northern area of Flemish speakers (Flemish is the ancestor of modern Dutch) and a southern area where Walloon (French) is spoken. Thus Belgium lies on both sides of the linguistic border separating the Germanic and Romance sub-families of the Indo-European family. To expect a viable state to emerge in this context as late as 1830 was quite ambitious, particularly when the French-speakers explicitly wished to join France. In fact, Belgium is an artificial state created especially by the British and Germans, neither of whom favoured territorial expansion for France. Thus it comprises two distinct language areas with a minimal transition zone; it has attempted to follow the example of Switzerland, but without success.

Canada, settled by both French and English during its formative years, is also a bilingual state. With Confederation in 1867, both French and English were adopted as official languages. The French formed a majority in Quebec and the English a majority in Ontario. Most of the remainder of Canada west of Ontario was settled either by English-speakers or by immigrants from other countries who settled in English-speaking areas; immigration of French-speakers has been very limited. Over centuries, interaction between French and English had given rise to distinct transition zones in western New Brunswick, southern Quebec, and eastern Ontario. However, analysis of local migration and interaction between the two groups (Cartwright 1988) suggests that the zone is disintegrating into a series of pockets, and that the linguistic territorialization evident in Belgium since 1830 is becoming a reality in Canada as a consequence of spatially delimited language differences.

ing independence, English joined Hindi as an official language. In parts of east Africa, Swahili is an official language; combining the local Bantu with imported Arabic, Swahili is an example of a *lingua franca,* a language developed to facilitate trade between different groups, in this case Africans and Arab traders. In some other colonial areas **pidgin** languages have developed as simplified ways of communicating between different language groups. Pidgins are common in southeast Asia, where they are usually based on English, with some Malay and some Chinese. A pidgin that becomes the first language ('mother tongue') of a generation of native speakers is known as a **creole**. Creoles are relatively common in the Caribbean region and vary according to whether the principal European component is English, Spanish, or French; the other components are the native Carib and imported Bantu.

A number of attempts have been made to promote the use of a single universal language; artificial languages have even been invented for this purpose. In 1887, the most popular such language, Esperanto, was introduced, but it failed to make a significant impact. Perhaps part of the reason was that, even though it was based on roots common

to several European languages, it had no ties to any specific tradition, culture, or environment. For the same reason, most human geographers reject the idea of a universal language—even though they recognize that, in principle, a universal language could promote communication and understanding between groups, thus minimizing division and friction. In any case, it seems unlikely that any artificial language will ever succeed in playing such a role. Today, the best hope for such a language rests with English.

Language in Landscape

Place names

Our discussion so far has emphasized the centrality of language to culture and group identity, and its effects on our partitioning of the earth. But language also plays a key role *in* landscape, in the form of place names or **toponyms**. We name places for at least two reasons. The first is in order to understand and give meaning to landscape. A landscape without names would be like a group of people without names; it would be difficult to distinguish one from another. Second, naming places probably serves an important psy-

7.7

THE CELTIC LANGUAGES

The Celts were one of the most important early groups to diffuse from the Indo-European core area. Beginning about 500 BCE, they spread across much of Europe. But after some 500 years of expansion, as other, more organized groups came into contact with them, they gradually retreated into some of the more inhospitable and isolated areas of western Europe. Today the remains of the once-large Celtic group live in four small pockets: western Wales, western Scotland, western Ireland, and northwest France. Each of these four areas still has some Celtic speakers: in Wales about 500,000 speak Welsh, in Scotland about 80,000 speak Gaelic, in Ireland about 70,000 speak Erse (Irish Gaelic), and in northwest France about 675,000 speak Breton.

In Britain the Celts were pushed to the western limits by the Anglo Saxon in-movement. Over time, various Celtic languages, such as Cornish (southwest England) and Manx (Isle of Man in the Irish Sea), disappeared, leaving the four pockets already noted.

The remaining Celtic languages are all in precarious positions, basically because the languages are not associated with political units. The Irish-language area, called the Gaeltacht, cov-

ered about 33 per cent of Ireland in 1850 and had perhaps 1.5 million speakers; today it covers about 6 per cent and has lost 95 per cent of its speakers. The first concerted efforts to save the Gaeltacht came in 1956. The basic assumption is that language is a key requisite for any group that aspires to retain a traditional culture. Consequently, the Irish government has, since 1956, actively encouraged retention of the language and cautious social and economic development for the rural Gaeltacht region. So far, the various government efforts to foster development have been quite successful and have not resulted, as some feared they might, in loss of the Irish language.

Welsh, a minority language intimately tied to traditional Welsh culture and a rural way of life in an increasingly Anglicized environment, is in a similar position. Welsh Wales is restricted to the extreme northwest and southwest, and the Welsh language is threatened, despite vigorous local efforts to preserve it. As a part of the United Kingdom, Wales is not in as strong a position as Ireland when it comes to implementing local development and language retention policies.

chological need—to name is to know and control, to remove uncertainty about the landscape. For these two reasons, humans impose names on all landscapes that they occupy and on many that they do not (the moon is a prime example).

Place names, then, are a significant feature of our human-made landscapes, often visible in the form of road signs and an integral component of maps—our models of the landscape. Many place names combine two parts, generic and specific. 'Newfoundland', for example, has *Newfound* as the specific component and *land*—the type of location being identified— as the generic component.

There are many ways of classifying place names. In a detailed analysis of Finnish settlements in Minnesota, Kaups (1966) applied a classification based on the mechanisms of naming. Possessive names indicate an association, possibly ownership, by an individual (personal) or a group (ethnic). Of 92 Finnish place names, 52 were possessive; 16 commemorated an important place or person; and 19 were descriptive, identifying an easily recognizable characteristic of the location, such as marshland. Only five place names could not be easily accommodated by this classification.

Analysis of place names can thus provide information about both the spatial and the social origins of settlers. In addition, names such as Toivola (Hopeville) in Minnesota or Paradise in California tell us something about their aspirations. Place names typically date from the first effective settlement in an area. Because the Finns in Minnesota were among the later European settlers, their opportunity to name places was restricted to the local scale, whereas other groups, most notably the French and Spanish, were able to place their languages on the landscape at the regional scale; witness the profusion of French names in Quebec and Louisiana and of Spanish names throughout the American southwest.

Place names are often an extremely valuable route to understanding the cultural history of an area. This is especially clear in areas where the first effective settlement was relatively recent, but it is also the case in older settled areas. In many parts of Europe, for example, former cultural boundaries can be identified through place-name analysis. Jordan (1988:98) provides an example from

before 800, when the Germanic–Slavic linguistic border roughly followed the line of the Elbe and Saale rivers in what is now Germany. Although German-speakers began moving east after 800, evidence of the earlier boundary remains: place names are German west of the line and Slavic east of it. Similarly, throughout much of Britain it is possible to identify areas settled by different language groups by studying place names; in northeast England, for example, place-name endings such as *–by* are evidence of Viking settlement.

The 'Great American Desert'

Clearly, then, language is everywhere in landscape. But language can also make landscape. A striking example is the nineteenth-century use of the term 'desert'—as in 'Great American Desert'—to describe much of what is now the Great Plains region. Most European Americans knew nothing of the plains until several expeditions in the early nineteenth century returned with reports of a 'desert' that would surely restrict settlement. This misconception arose because individual explorers recorded what impressed them most, because the small areas of sand desert were indeed impressive, and because the absence of trees did make the plains a desert in comparison with the heavily forested east. Among the consequences was delayed settlement as migrants bypassed the plains in favour of the west coast. Thus the language used to describe a specific environment affected human geographic changes in the landscape.

Landscape in Language

Not only is language in landscape, but landscape is in language. First, because physical barriers tend to limit movement, there is a general relationship between language distributions and physical regions (this is evident on a world scale in Figure 7.8). Second, the vocabulary of any language necessarily reflects the physical environment in and with which its speakers live; hence there are many Spanish words for features of desert landscapes and few comparable English words. Similarly, many words are available to discriminate between different types of snow in the Inuit language, or between the colourings of cattle in the Masai language. As languages move to new environments, it also becomes necessary to add new words; the

Figure 7.14 Hearth areas and
diffusion of four major religions.

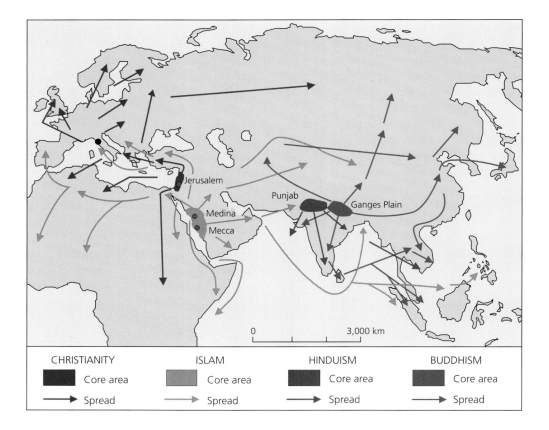

Figure 7.14 Hearth areas and diffusion of four major religions.

CHRISTIANITY	ISLAM	HINDUISM	BUDDHISM
■ Core area	■ Core area	■ Core area	■ Core area
→ Spread	→ Spread	→ Spread	→ Spread

word 'outback', for example, entered English only with the settlement of Australia. Finally, the human landscape is in language to the extent that language reflects class and gender; within a given language, word choices and pronunciations say a great deal about social origins. This area of research is now labelled sociolinguistics.

Religion

A second fundamental cultural element is religion. Although the specific origins of religious beliefs are no less difficult to trace than those of language, the universality of religion suggests that it serves a basic human need. Essentially, a religion consists of a set of beliefs and associated activities that are in some way designed to facilitate appreciation of our human place in the world. In many instances, religious beliefs generate sets of moral and ethical rules that can have a significant influence on many aspects of behaviour.

It is important to note that there are often major distinctions between women and men with regard to religious behaviour. It is usually women who make up the majority of the followers of a religion, and women play a critical role in teaching reli-

gion to their children. Further, most major religions—with the exception of Islam—have at least some female figures of worship (although the most important figures are generally male). Yet in many religions women are excluded from serving in a formal, structured role; in the Christian tradition, although women are now admitted to the ministry in many Protestant churches, they remain largely excluded from the Catholic and Orthodox traditions. Men have played the dominant official roles in Hinduism, Buddhism, Christianity, and Islam, to name only the four largest organized religions.

Classification and Regions

Figure 7.14 identifies the origin areas of the four largest religions, as measured by numbers of adherents—Hinduism, Buddhism, Christianity, and Islam—while Figure 7.15 maps the contemporary distribution of these and other religions. These maps make two things especially clear. First, there are two major religious 'hearth' areas: Indo-Gangetic and Semitic. Second, two religions, Christianity and Islam, have diffused over large areas, whereas Hinduism has not experienced sig-

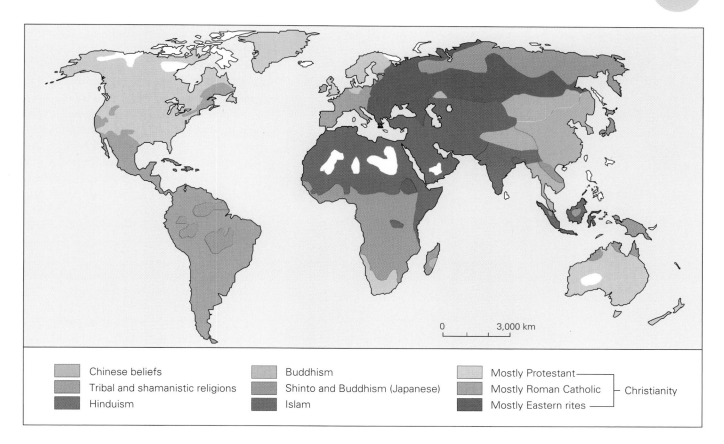

Chinese beliefs
Tribal and shamanistic religions
Hinduism

Buddhism
Shinto and Buddhism (Japanese)
Islam

Mostly Protestant
Mostly Roman Catholic — Christianity
Mostly Eastern rites

nificant spread and Buddhism has effectively relocated to China. These conclusions are confirmed by Table 7.5, which provides detailed data on the distribution of religious groups by selected major world regions. Christianity is the leading religion numeri-cally, with 33 per cent of the world popula-tion declaring themselves, at least nominally, Christian, followed by Muslims (followers of Islam) at 20 per cent, Hindus at 13 per cent, and Buddhists at 6 per cent. A high 13 per cent are classed as non-religious.

Figure 7.15 World distribution of major religions.

Table 7.5   MAJOR WORLD RELIGIONS: NUMBER OF ADHERENTS (THOUSANDS), 2001

Religion	Africa	Asia	Europe	Latin America	North America	Oceania	Total	%
Christianity	368,244	317,759	559.359	486,591	261,752	25,343	2,019,052	33.0
Islam	323,556	845,341	31,724	1,702	4,518	307	1,207,148	19.7
Hinduism	2,384	813,396	1,425	775	1,350	359	819,689	13.4
No religion	5,170	611,876	105,742	16,214	28,994	3,349	771,345	12.6
Chinese folk religions	33	385,758	258	197	857	64	387,167	6.3
Buddhism	139	356,533	1,570	660	2,777	307	361,985	5.9
Ethnic religions	97,762	129,005	1,258	1,288	446	267	230,026	3.8
Atheism	432	122,408	22,555	2,787	1,700	369	150,252	2.5
New religions	29	101,065	160	633	847	67	102,801	1.7
Sikhism	54	22,689	241	0	535	19	23,538	0.4
Judaism	215	4,476	2,506	1,145	6,045	98	14,484	0.2
Spiritism	3	2	134	12,169	152	7	12,466	0.2
Baha'i faith	1,779	3,538	132	893	799	113	7,254	0.1
Confucianism	0	6,277	11	0	0	24	6,313	0.1

Source: *2002 Encyclopedia Britannica Book of the Year* (Chicago: Encyclopedia Britannica, Inc., 2002).

Hindu devotees fill earthen pots with water from the holy river Ganges at Baidyabati, India, 3 August 2003. Pilgrims then walked 35 kilometres to take the water to the Shiva temple at Tarakeswar (AP photo/Bikas Das).

A useful classification of religions, favoured by geographers because it relates closely to spatial distributions, distinguishes between universalizing religions, which actively seek converts, and ethnic religions, which are closely identified with a specific cultural group. Although very useful, this twofold classification necessarily excludes hundreds of numerically small religions. For example, many cultures embraced a form of **animism**—in which a soul or spirit is attributed to various phenomena, including inanimate objects—especially before the diffusion of universalizing religions, notably Christianity and Islam.

Ethnic religions

The principal ethnic religions emerged earlier than their universalizing counterparts. They are associated with a particular group of people and do not actively seek to convert others. Of the several hundred religions in this category, the largest is Hinduism, which evolved in the Indo-Gangetic hearth—a lowland area of north India that drains into the Indus and Ganges rivers—about 2000 BCE. Hinduism was the first major religion to evolve in this area, from which it spread east down the Ganges and then south through India, eventually to dominate the entire region. As it diffused, it absorbed and blended with other religious beliefs. From India it was carried overseas, but it has not

retained significant numbers of adherents outside of India. Today, Hinduism is an Indian religion associated with a specific place and a specific cultural group. Hinduism has no dogma and only a loosely defined philosophy by religious standards. It is polytheistic (worshipping more than one god) and has close ties to the rigid social stratification of the caste system. Like many other religions, Hinduism has spawned numerous offshoots. Jainism, for example, rejects Hindu rituals but shares many basic tenets of Hinduism, including the belief in reincarnation and *ahimsa* (the ethical doctrine that humans ought to avoid hurting any living creature). A more recent offshoot is Sikhism, a hybrid of Hinduism and Islam that arose about 500 years ago in the Punjab region of India.

A second ethnic religion is Judaism. The first monotheistic religion (worshipping one single god), Judaism originated about 2000 BCE in the Near East. Following the Romans' destruction of Jerusalem in 70 CE the Jews were driven out of their homeland and eventually dispersed throughout Europe; the entire body of Jews living outside Israel, in Europe and elsewhere, is known as the Diaspora. Only in 1948 was the long-term goal of a Jewish homeland achieved with the creation of the state of Israel. Judaism contains significant internal divisions reflecting theological and ideological differences. Other ethnic religions include Shinto (the indigenous religion of Japan) and Taoism and Confucianism, both of which are primarily associated with China.

Universalizing religions

Both Hinduism and Judaism have given rise to major universalizing world religions. Buddhism was the first universalizing religion, an offshoot of Hinduism founded in the Indo-Gangetic hearth by Prince Gautama, who was born in 644 BCE. During his lifetime he preached in northern India, but after his death Buddhism spread into other parts of India. It was carried to China after about 100 BCE and then to Korea, Japan, southeast Asia, Tibet, and Mongolia. Today there are two principal versions of Buddhism: Theravada Buddhism in southeast Asia and Mahayana Buddhism in east Asia.

A second universalizing religion, Christianity, developed about 600 years after Buddhism

The seventeenth Karmapa (left), at the Gyuto Ramoche Monastery near Dharmsala, India, in February 2000. One of the highest lamas in Tibetan Buddhism, Urgyen Trinley Dorje had escaped from Chinese-occupied Tibet in January and reached Dharmsala—the home of the Tibetan government in exile—after a 1400-km trek over the Himalayas (AP photo/ John McConnico).

as an offshoot of Judaism in the Semitic hearth area. Christianity began when disciples of Jesus accepted that he was the expected Messiah. The religion spread slowly during his lifetime and more rapidly after his death as missionaries carried it initially to areas around Jerusalem and then through the Mediterranean to Cyprus, Turkey, Greece, and Rome. Christianity spread rapidly through the Roman empire. A major east–west division between the Roman Catholic and Eastern Orthodox forms of Christianity occurred in 1054, and the north–south line dividing these two remains the most basic religious boundary in contemporary Europe (see Figure 9.9). Following the Protestant Reformation of the 1500s—an effort to reform Roman Catholic dogma and teachings that led the Protestants to break away and establish a number of independent churches—southern Europe remained predominantly Catholic and northern Europe became largely (though by no means exclusively) Protestant. Both Catholic and Protestant versions of Christianity spread to other parts of the world in the course of European movement overseas. The importance of Christianity in the contemporary world reflects not only the large number of its adherents and their wide spatial spread, but also the fact that Christian thinking has been a cornerstone of western culture, affecting attitudes and behaviours at the level both of the group and of the individual.

The third major universalizing religion, Islam, also arose in the Semitic hearth area; it is related to both Judaism and Christianity but has additional Arabic characteristics. Founded by Muhammad, who was born in Mecca in 570, by the time of his death in 632 it had diffused throughout Arabia. Further diffusion was rapid as a result of Islamic political and military expansion. Arab Muslims created an empire that stretched west to include parts of the northern Mediterranean as far as Spain and much of north Africa, and east to include the areas of Iraq, Iran, Afghanistan, and Pakistan. Islam has also spread into much of southeast Asia. Today there are two principal branches of Islam. Sunnis, who represent about 90 per cent of the total, dominate in Arabic-speaking areas, as well as in Pakistan and Bangladesh; Shiites are a majority in Iran and Iraq. Although Islam is often considered a religion of the Middle East, the largest Muslim populations today are in Indonesia, India, Pakistan, and Bangladesh. Rather like Christianity, Islam is an all-encompassing worldview, shaping both group and individual attitudes and behaviours.

Religion, Identity, and Conflict

The importance of religion for individual and group identity varies considerably. In much of the more developed world, especially in urban areas, religion is not central to human activity, and many people consider it

Nearly two million Muslim pilgrims at Friday prayers inside and outside Islam's holiest shrine, the Great Mosque in Mecca, Saudi Arabia, 2 March 2001. The pilgrimage to Mecca is one of the five central duties of every Muslim (AP photo/Amr Nabil).

irrelevant or marginal. Some countries have actively rejected religion and replaced it with a political belief, such as communism, as a guide to beliefs and behaviour.

Despite these examples of the declining relevance of religion, however, in reality most humans are unable to separate themselves entirely from religion, some because they choose to be actively religious and others because religion is typically one part of state identity. Indeed, many people consider religion generally to be the basis of life, and regard their own as the only true religion. Not only do religions divide our world, they also often encourage people to engage in what may appear, from a detached perspective, to be inappropriate behaviour; recall James's comment quoted at the beginning of this chapter.

Our human tendency to identify with a specific religion has given rise to many military conflicts, from the medieval Crusades and the European religious wars of the sixteenth and seventeenth centuries to the recent and ongoing conflicts in Pakistan (Muslim–Hindu), Lebanon (Christian–Muslim), and Northern Ireland (Catholic–Protestant Christian). In some cases religious differences are used as excuses for aggression that actually has other motives. But there is no doubt that, in many instances, religion promotes mistrust of non-believers, an attitude that, combined with a general lack of understanding, leads to hostility

and conflict. As we will see in Chapter 9, some contemporary scholars point to hostility between Islam and Christianity in particular as a major cause of conflict in the contemporary world, although other scholars consider this a gross over-simplification. Not surprisingly, this debate has intensified following the terrorist attacks on New York and Washington in 2001 and the war in Iraq in 2003.

In some cases, religion serves as an even more potent force for group unification than language. Religion is even more capable than language of resisting external influences. The North American experience of immigration and settlement by people from many different regions of the world suggests that religion is often the most lasting feature of a culture, retained long after language has been lost. The roles played by both religion and language in the creation, continuation, and, sometimes, collapse of political states will be considered in more detail in Chapter 9.

Religions as civilizations

One of the more contentious debates in the human geographic study of religion concerns the question of links between religions and civilizations. As we saw in Box 7.3, the term 'civilization' is not restricted to one group of people or way of life. Rather, it is used to refer to a variety of technological and related changes occurring in particular regions at particular times. In the twentieth century, scholars began working to identify and label different civilizations as they are now distributed around the world; we referred earlier to the work of Toynbee and then extended our account using the term 'world cultural regions' in place of 'civilizations'.

Most recently, it has become common to define two civilizations in terms of their religious bases in Islam and Christianity, and to interpret belonging to one or the other as a fundamental basis for human identity. Not many other religions are typically regarded as defining civilizations.

Equating Islam and Christianity with civilizations can lead to some unfortunate consequences. One problem is that both are universalizing religions, which means that, strictly speaking, both have a duty to spread their message of truth to all people: when these messages differ, mutual respect—even tolerance—may be difficult. As we have seen, Toynbee identified three 'civilizations' associated with two different forms of Christianity. Obviously the civilization he called 'Western Christendom'—now perhaps more commonly known simply as 'the west'—includes a great many non-Christian elements. In the case of Islam, however, we have only the one word, 'Islam', to designate both the religion and the civilization. As a result, people in the west often fail to distinguish between the two. In recent years this tendency has been especially common in the context of terrorist activity. Thus terrorism is seen as representative of Islam itself—even if most Muslims would disclaim it as antithetical to their faith. The debate over the degree to which religion and civilization are connected is especially important in a world that is currently experiencing serious frictions between Christian and Islamic identities.

Religious Landscapes

Religion and landscape are often inextricably interwoven, for three principal reasons. First, beliefs about and attitudes towards both the physical environment—nature—and the human environment are integral parts of many religions. Second, religion often influences land use. Third, many religions explicitly choose to display their identity in landscape.

Religious perspectives on humans and nature

In many cases it is an important function of religion to serve as an intermediary between humans and nature—although the type of relationship favoured varies. Judaism and Christianity place God above humans and humans above nature. In other words, they incorporate an attitude of human dominance over the physical environment, an attitude that is reflected in numerous ways. Christians in particular see themselves as fulfilling an obligation to tame and control the land. Other religions, however, take a very different view. Eastern religions in general see humans as a part of, not apart from, nature, and both as having equivalent status under God. The result is a very different relationship between humans and land (Box 7.8).

Many religions ascribe a special status to certain features of the physical environment. Rivers and mountains may be sacred places (the Ganges for Hindus and the Jordan for Christians; Mount Fujiyama in Shintoism). Human environments may also achieve sacred status (Mecca for Islam and Lourdes for Catholicism). More generally, almost any religious addition to landscape—church, temple, mosque, cemetery, shrine—is sacred.

7.8

THE MITHILA CULTURAL REGION

An excellent discussion of the implications for landscape of the human–land relationship implicit in Hinduism is available in an article by Karan (1984). In the course of discussing the folk art of the Mithila region in north India, between the Ganges and the Himalayan foothills, Karan also describes the people and the place they have created.

The Mithila region is an alluvial plain occupied by some 15 million Vedic Hindus at the high density of 722 per square kilometre. Religion, emphasizing the unity of humans and all nature, is central to all life, and the Maithilis believe in awareness of life as manifested in all things. Plants, animals, clouds, the sky, and water share the life-force equally with humans. This concept of oneness affects attitudes both towards land and towards others. The key social unit is the extended family, and the individual's ultimate responsibility is to the group.

Cultural landscapes in the region reflect religious attitudes. Because buildings are of local clays, human impact is minimized. The rice fields in the alluvial plain blend imperceptibly with the physical environment. This is not a landscape made by humans, but rather one that humans share with all other aspects of nature. The folk art of the people is a profound expression of their being a part of nature; the colours used are made from natural materials, while the subjects of their paintings are largely religious. The vast scope and seeming timelessness of the forces of nature are also fundamental themes in the folk art of the region. In addition to providing an example of a human–land relationship that differs fundamentally from the Judaeo-Christian model, this brief discussion also highlights the relevance of folk art as a part of a group's cultural record. This is relatively new territory for the human geographer.

Roman Catholic pilgrims at the shrine in Lourdes, France. Believers attribute miraculous cures to waters from a spring said to have been revealed by the Virgin Mary to a local peasant girl in 1858 (Victor Last, Geographical Visual Aids).

Religion and use of land

Religious beliefs about human use of land, plants, and animals can have significant impacts on regional economies. For example, the pig is a common domesticated animal in Christian areas, but absent in Islamic and Jewish areas. It was the traditional Catholic avoidance of meat one day per week that prompted European fishermen to move to Newfoundland waters long before Europeans settled in the New World. And Hindus regard cows as sacred (Box 7.9). These are only the most familiar examples of how beliefs influence human behaviour.

Different religions incorporate different beliefs and attitudes, and what is important to one group may not be important to another. When different groups value similar places, as in the case of sites in Jerusalem regarded as sacred by Judaism, Islam, and Christianity, the result is often conflict. When religion affects the way we use land, it can be a powerful cultural factor operating against economic logic.

Religious symbols in landscape

Landscape is a natural repository for religious creations, a vehicle for displaying religion. Sacred structures, in particular, are a part of the visible landscape. Hindu temples are intended to house gods, not large num-

bers of people, and are designed accordingly; Buddhist temples serve a similar function. By contrast, Islamic temples (mosques) are built to accommodate large congregations of people, as are Christian churches and cathedrals. The size and decoration of religious buildings often reflect the prosperity of the local area at the time of construction.

In Calcutta, traditional Hindu temple architecture reflected the belief that temples are the homes of gods. Hence, because mountains are also traditional dwelling places of gods, temple towers were built to reflect mountain peaks, while small rooms inside the temples resembled caverns. In the late eighteenth century, these temples were replaced by flat-topped two- or three-storey buildings built next to the homes of the very rich. Finally, in this century, a series of new temple styles appeared as a result of the pressures of urbanization and related institutional processes. These styles are described in detail by Biswas (1984) and are a clear example of the effects of changing power relations.

Sacred structures are important religious and tourist locations. The Golden Temple at Amritsar, India, is sacred to Sikhs but also attracts many others, as does the Vatican in Rome or Westminster Abbey in London.

Places to house gods or to gather for worship are typical features of most landscapes. In many Christian communities, the church is a religious and social centre serving many extra-religious functions. Although some religious groups actively reject such external expressions of religion as churches and create landscapes devoid of religious expression (Box 7.10), such cases are unusual.

Other religious practices that create a distinctive landscape include the construction of roadside shrines and the use of land for burying the dead. Hindus and Buddhists cremate their dead, but Muslims and Christians typically bury them, a practice that requires considerable space.

These brief comments do little more than highlight some of the many ways in which religious beliefs may be expressed in landscape. The account of the Mormon landscape in Box 7.5, included in the earlier discussion of cultural landscapes, is relevant to this section as well, and additional examples of religion and landscape symbolism will be noted in the following chapter.

Cultural Globalization

The contents of this chapter reflect the importance to human geography both of regions—one of our three recurring themes —and of the culturally based divisions so aptly described by James at the beginning of the chapter. Today, however, it seems possible that that those divisions may be reduced in the process of cultural globalization.

Jerusalem: the Western or 'Wailing' Wall and Dome of the Rock mosque (AP photo/Eyal Warshavsky).

RELIGION AND 'IRRATIONAL' CHOICES

Why do Hindus refuse to eat beef even when it is the only 'food' available? The simple answer is that the cow is sacred to Hindus because it is a symbol of everything alive, the mother of life. But why is this so? Is there a logical reason behind the sacred status of the cow? Or is this an example of a mistaken belief that leads to irrational economic behaviour?

When the Indian constitution was drafted, it included a bill of rights for cows. As a result, stray animals invade private property, and government agencies provide homes for old cows. To western eyes, this situation is nonsensical.

A careful analysis of the larger economic system in India, however, leads to a different conclusion. As Harris (1974) has pointed out, there is a shortage of oxen (male cattle), which are used as draft animals. In the absence of tractors, cattle are essential to agriculture. Cows also provide dung used as fuel and as a household flooring material. In short, cows are an invaluable component in Indian agriculture and life, and the use made of them is highly efficient. The fact that it does not appear that way to the Christian economist in the developed world does not alter this truth. Hindus venerate cows because cows are so important. They do not kill them because that would, in the long run, disturb the overall pattern of Indian agriculture.

This is one example. Other seemingly perplexing customs can also be logically explained. Cultural riddles and irrational behaviours associated with particular religions do make sense once we see them in their larger context.

Five baptized Sikhs lead a procession at the Golden Temple in Amritsar, India, in 2003, marking the 337th anniversary of the birth of the tenth guru, Gobind Singh (AP photo/Aman Sharma).

One way to conceive of cultural globalization is to imagine a process that began before the rise of the nation state, when cultures and identities were essentially local. With the emergence of the nation state, a second option became available: membership in a national culture. In this sense, nation states can be seen as cultural integrators, bringing together various local identities in such a way that it became possible for individuals to understand themselves and their lives in both traditional local and newer national contexts—although (as we will see in Chapter 9) the transition from a singular to a dual cultural identity has not been easily accomplished in many parts of the world. Today, some suggest that we have reached a stage where a global cultural identity is developing.

Proponents of this argument see a homogeneous global culture replacing the multitude of local cultures that has been characteristic for most of human history (for an essentially Marxist interpretation of this phenomenon, recall Box 2.6, p. 55). The culture that is diffusing is western in character and largely derived from the United States. Among the mechanisms that allow this culture to spread spatially are various aspects of the mass media and consumer culture—newspapers, magazines, the Internet, music, television, films, videos, fast-food franchises, fashions. That these aspects of popular culture are being diffused around the world, and that

they are influencing aspects of nonmaterial culture such as religion and language, is undeniable and will be discussed in further detail in Chapter 8. However, assessing their significance is not as easy as it might seem.

In fact, it appears unlikely that globalization will erase the power of local places and local identities—what Vidal called *pays* and *genres de vie*. Physical environments vary throughout the world, and there is no denying that the connections between physical and human geographies are often intimate. Most places still look different from other places. In some ways, at least, their inhabitants still behave differently from other people, and they still hold some attitudes, feelings, and beliefs that differ from those of other people. Despite clear evidence that the number of the world's languages is decreasing, the roughly 6,000 languages that still exist today are compelling examples of cultural variation from place to place. We are all human, but there are still significant differences between places and groups of people, resulting primarily from differences in cultural identity. Certainly the world we live in is not homogeneous. Thus it may be that a conflict is developing between parochial ethnicity and global commerce.

With regard to globalization, the distinguished human geographer David Harvey wrote:

> The more global interrelations become, the more internationalized our dinner ingredients and our money flows, and the more spatial barriers disintegrate, so more rather than less of the world's population clings to place and neighborhood or to nation, region, ethnic grouping, or religious belief as specific marks of identity. . . . Who are we and to what space/place do we belong? Am I a citizen of the world, the nation, the locality? Not for the first time in capitalist history . . . the diminution of spatial barriers has provoked an increasing sense of nationalism and localism, and excessive geopolitical rivalries and tensions, precisely because of the reduction in the power of spatial barriers to separate and defend against others (Harvey 1990:427).

Harvey raises the challenging possibility that there is a second way of thinking about cultural globalization. Instead of making the world more and more homogeneous, perhaps it is reinforcing the distinctiveness of local places and identities. Certainly globalization does not appear to mean the end of diversity or the imposition of a single global culture. On the contrary, the discussion of political geography in Chapter 9 will provide considerable support for the argument that localism is increasing rather than decreasing.

Indeed, thinking about cultural globalization obliges us to acknowledge that there are no uncontested and unidirectional processes at play in the contemporary world. Globalization, in the sense of an ever-increasing connectedness of places and peoples, is a fact, but it is not the only important fact. Some regional economies remain distinctive, and it is possible that some regional differences are actually being enhanced with the rise of regional trading blocs, as we shall see in

Chapter 10. It is hard to believe that the cultural world is becoming uniform at a time when so many ethnic groups are reasserting their identities—at least partly in reaction against the declining importance of national political and cultural identities.

But there is a third, more emancipatory, way to think about the consequences of cultural globalization. If western values in particular are spreading around the world, then those values include the basic ideals of liberal democracy, including freedom of speech and freedom of cultural expression. If we are free to be what we wish to be, then perhaps we will not choose either to be the same as others or to separate ourselves from them. Rather, perhaps we will choose to value both our own personal rights and freedoms and those of others, including the inalienable right to improve their economic circumstances. Our future—wherever we are—may

RELIGIOUS LANDSCAPES: HUTTERITES AND DOUKHOBORS IN THE CANADIAN WEST

7.10

Most landscapes include evidence of religious occupation. Places of worship in particular are visible even in an increasingly secular developed world. Box 7.5 described the Mormon landscape, which not only includes religious features, such as ward chapels and temples, but was built in direct response to directions from church leaders and in accord with Mormon religious doctrine. This box considers the landscape imprint of two groups, Hutterites and Doukhobors, in parts of the Canadian west. Unlike the Mormons, these two groups have not settled across a broad region, and neither have they aspired to express themselves in landscape. Hence the impact of their religion on landscape is not quite so visible.

A number of Protestant Anabaptist groups emerged during the Reformation in Europe. All shared a belief in adult baptism and a literal interpretation of the Bible, but they varied in other areas of religious belief and practice. Three such groups, all of which emigrated to North America because of persecution in Europe, are the Hutterites, Mennonites, and Amish. Each expressed some aspects of their religious belief in landscape. Hutterites emigrated beginning after the First World War and continue to live in the Canadian prairies, creating a landscape that reflects their belief in community and their desire for isolation from the larger world (Simpson-Housley 1978). At the centre of each communal settlement are a kitchen complex and long houses, and around them

are buildings that are used for economic functions; the two types of building are painted different colours. Hutterite communal settlements do not include commercial stores or bars. The Hutterite farming landscape typically involves a greater diversity of activities than do neighbouring farms.

Doukhobors, like Hutterites, are a Christian sect. They broke from the Orthodox church in the eighteenth century and were banished to the Caucasus region, where they built a flourishing community. As a result of persecution, they imigrated to Canada beginning in 1898. Doukhobors reject the 'externalities' of religion; their settlements are thus without houses of worship and there are no crosses or spires in the landscape. In short, the Doukhobor areas lack any religious symbolism. Doukhobors also believe in the equality of life; hence their settlements are communal and all work and financial matters have a group rather than an individual focus. Communal living arrangements required the construction of distinct double houses that accommodate up to 100 people (Gale and Koroscil 1977).

Although most Christian landscapes show evidence of religious symbolism and beliefs, such as houses of worship, cemeteries, and roadside shrines, some areas, settled by distinctive community-based religious groups, reflect group organization even when they do not include obvious symbolic features. Both Hutterites and Doukhobors belong in the second category.

well be one of greater pluralism, more choices, and (most important) enhanced mutual respect. These comments—which are in close accord with the 1948 United Nations Universal Declaration of Human Rights—offer a more hopeful counterpart to the sombre scenario outlined at the beginning of this chapter.

CHAPTER SEVEN SUMMARY

Our divided world
The human world is divided both physically and, more important, culturally.

Culture and society
The terms 'culture' and 'society' are difficult to define. Neither has a single universally accepted meaning, and sometimes the two are even used interchangeably. For our purposes, 'culture' refers to humans' ability to knowingly change physical landscapes in directions suggested by experience, while 'society' refers to a cluster of institutionalized ways of doing things. Traditionally, North American geographers have used the term 'culture', whereas European geographers have used 'society'. This chapter and the one following reflect the importance of integrating the two concepts.

Human scale of analysis
Geographers can study humans at any scale, from all humans in the world to single individuals. Typically, however, cultural geography operates at the scale of a group of people with some recognizably common set of operating rules; hence geographers study language groups and religious groups.

Cultural evolution
Preagricultural groups, probably numbering between 10 and 30 individuals, used language, fire, and tools. Cultural variation prior to the development of agriculture was probably limited to features directly related to the physical environment. Agriculture and civilization were closely related; among the factors that may have contributed to the development of both are irrigation, social change, class conflict, and population pressure. Once in place, agriculture is typically permanent, whereas civilization has not proved to be. Most civilizations have collapsed as a result of either natural disaster or conflict (internal or external).

Cultural regions
Regions can be delimited on various scales. The world scale can provide a useful overview, but lacks precision. North America can be usefully regionalized using the concept of first effective settlement. Europe is usefully defined by reference to specific traits.

Cultural landscapes
Cultural regions have distinct cultural landscapes because of the impact of culture on land and the variations in human–land relationships. 'Cultural adaptation' is cultural change in response to environmental and cultural challenges. Geographers studying cultural adaptation have proposed a variety of explanations, including (in addition to first effective settlement) culturally habituated predisposition; cultural preadaptation; and the core, domain, and sphere model.

Languages
Probably the single most important human achievement, language is an essential key to understanding human groups, their attitudes, beliefs, and behaviours. Indeed, it is not far-fetched to assert that language is the single most appropriate indicator of culture.

The basis for group communication, language is the earliest source of group unity and the means by which cultures continue through time. For many groups, language *is* culture. Today there are perhaps 6,000 languages in the world. Individual languages can be grouped into families—languages that share a common origin. The language family with the most speakers is Indo-European. Mandarin has more speakers than any other single language, but English is the most widespread and the nearest to a world language.

Language as identity

Languages create barriers between groups and facilitate the development of group identity. For many groups, language is the principal basis for a national identity. Multilingual states are characteristically less stable than unilingual states.

Language and landscape

Place names—toponyms—are the clearest expression of language on landscape. Place-name studies help us understand early settlement. Language also helps to make landscape in the sense that a place may become what it is named—as in the case of the Great American Desert. Both physical and human landscapes are reflected in language.

Religions

For many people, religion is the basis of life. Thus religion is a useful variable for region-alizing and may be even more powerful than language in reinforcing group identity. Because different religions affect attitudes and behaviour in different ways, they help to distinguish one group from another and to promote group cohesiveness.

Religion and landscape

Many religions function as intermediaries between humans and nature. Often, specific physical environments are ascribed a special status. Some religious beliefs can affect regional economies. Landscape is also a nat-ural vehicle for religious expression.

Cultural globalization

There is increasing evidence to suggest that globalization is transforming our culturally divided world, mainly through the diffusion of western-derived attitudes, beliefs, and behaviours.

LINKS TO OTHER CHAPTERS

- Landscape:
    Chapter 1 (Humboldt, Ritter, Vidal, and Sauer).

- Globalization:
    Chapter 2 (concepts)
    Chapter 4 (ecosystems and global impacts)
    Chapter 5 (population growth, fertility decline)
    Chapter 6 (refugees, disease, more and less developed worlds)
    Chapter 8 (popular culture)
    Chapter 9 (political globalization)
    Chapter 10 (economic globalization)
    Chapter 11 (agriculture and the world economy)
    Chapter 12 (global cities)
    Chapter 13 (industrial restructuring)

- World cultural divisions:
    Chapters 8 and 9.

- 'Culture' and 'society' as terms in human geography:
    Chapter 1 (disciplinary evolution, especially Sauer and Vidal).

- Social scale:
    Chapter 2 (concept of scale).

- Beginnings of civilization:
    Chapter 2 (Marxism)
    Chapter 3 (races and racism)
    Chapter 4 (energy sources)
    Chapter 10 (origins of agriculture).

- World regions:
    Chapter 2 (concept of region)
    Chapter 3 (environmental regions)
    Chapter 5 (demographic regions)
    Chapter 6 (more and less developed regions)
    Chapter 9 (political and cultural regions)
    Chapter 10 (agricultural regions)
    Chapter 11 (urban regions)
    Chapter 12 (industrial regions).

- Cultural adaptation:
    Chapter 2 (environmental determinism possibilism);
    Chapter 4 (natural resources and human impacts in general).

- Language and identity; religion and identity:
  Chapter 8 (ethnicity)
  Chapter 9 (nationalism).

- Language and landscape; religion and landscape:
  Chapter 2 (humanism)
  Chapter 8 (symbolic landscape).

- Women and religion:
  Chapters 5, 6, 8, and 10 (gender).

- Religion and land use:
  Chapter 10 (agricultural land use).

FURTHER EXPLORATIONS

BRONOWSKI, J. 1973. *The Ascent of Man*. London: BBC.
An outstanding study of cultural evolution from prehistory to the present.

BUTLIN, R.A. 1993. *Historical Geography: Through the Gates of Space and Time*. New York: Arnold.
A survey of the many different approaches to reconstructing the geographies of the past.

EDWARDS, J. 1985. *Language, Society and Identity*. Oxford: Blackwell.
A non-geographic study of the role played by language in larger cultural issues.

FRANCAVIGLIA, R.V. 1978. *The Mormon Landscape*. New York: AMS Press.
An unusually detailed description of one of the most visually distinctive cultural landscapes in North America.

GADE, D.W. 1999. *Nature and Culture in the Andes*. Madison: University of Wisconsin Press.
A book made up of essentially discrete chapters on aspects of human–nature relationships in the Andean region; unified by commitment to the Sauer-inspired school of cultural geography and reflecting considerable fieldwork.

GASTIL, R.D. 1975. *Cultural Regions of the United States*. Seattle: University of Washington Press.
A thoughtful regionalization that can be interestingly compared with that of Zelinsky.

GROTH, P., AND T.W. BRESSI, eds. 1997. *Understanding Ordinary Landscapes*. New Haven: Yale University Press.
A clear summary of the approaches to cultural landscape study, followed by a series of useful analyses; also useful for Chapter 8.

HEAD, L. 2000. *Second Nature: The History and Implications of Australia as Aboriginal Landscape*. Syracuse: Syracuse University Press.
A well-researched and well-presented study of the contested understandings of cultural landscapes in postcolonial societies; effectively integrates physical and human geographic subject matter.

JORDAN-BYCHKOV, T.G., AND B. BYCHKOVA JORDAN. 2002. *The European Culture Area: A Systematic Geography*. 4th ed. London: Rowman and Butterfield.
An excellent book. One of the few cultural geographies of a major world region; includes a highly original synthesis and analysis of language, religion, and other cultural traits.

LEIGHLY, J. 1978. 'Town Names of Colonial New England in the West'. *Annals, Association of American Geographers* 68:233–48.
One example of a study of toponyms used to analyze settlement history.

MEINIG, D.W. 1969. *Imperial Texas: An Interpretive Essay in Cultural Geography*. Austin: University of Texas Press.
One of several fine regional studies by this author; includes several original approaches to region and landscape analysis.

PARK, C. 1994. *Sacred Worlds: An Introduction to Geography and Religion*. New York: Routledge.
Focuses on the spatial distribution of religion, the processes by which religion and religious ideas

spread through space and time, and the visible manifestations of religion in the cultural landscape.

ROONEY JR, J.R., W. ZELINSKY, and D.R. LOUDON, eds. 1982. *This Remarkable Continent: An Atlas of United States and Canadian Society and Cultures*. College Station, TX: Texas A and M University Press.

An original attempt to map and discuss a wide variety of cultural traits.

SALTER, C., ed. 1971. *The Cultural Landscape*. Belmont: Duxbury.

A collection of readings organized in a novel and stimulating fashion.

SHORTRIDGE, J.R. 1977. 'A New Regionalization of American Religion'. *Journal of the Scientific Study of Religion* 16:143–53.

A comprehensive statistical regionalization of some religious differences.

SOPHER, D.E. 1967. *Geography of Religions*. Englewood Cliffs: Prentice-Hall.

The first book-length study of this topic, full of useful ideas and facts.

SPENCER, J.E. 1978. 'The Growth of Cultural Geography'. *American Behavioral Scientist* 22:79–92.

One of many overviews, but especially useful to the new student of geography.

WAGNER, P.L. 1974. 'Cultural Landscapes and Regions: Aspects of Communication'. *Geoscience and Man* 5:133–42.

An early statement regarding the importance of communication to our understanding of culture.

ZELINSKY, W., and C.H. WILLIAMS. 1988. 'The Mapping of Language in North America and the British Isles'. *Progress in Human Geography* 12:337–68.

A detailed review article.

ON THE WEB

http://www.ethnologue.com
A website with information about languages of the world.

http://geonames.nrcan.gc.ca/
Published by Natural Resources Canada; provides detailed information on place names in Canada.

http://mapping.usgs.gov/www/gnis/
A source of information on physical and cultural geographic features in the United States.

http://www.unesco.org/
The UNESCO home page is a good starting point for seeking information on global cultural issues.

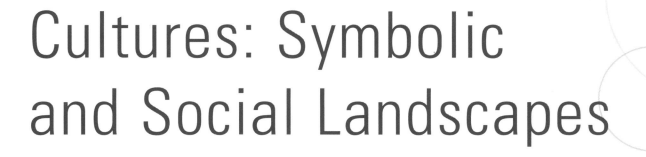

CHAPTER **8**

Cultures: Symbolic and Social Landscapes

This chapter continues the discussion of landscapes as they reflect relationships between humans and land: the first of our recurring themes. However, the emphasis here is rather different. Specifically, although some of the material in Chapter 7—notably the discussions of cultural adaptation and linguistic and religious identity—had conceptual, social, or symbolic overtones, the overall approach was empirical.

By contrast, the thrust in this chapter is strongly theoretical. Following a reintroduction to 'culture', which includes a symbolic interpretation and emphasizes the plurality of cultures, we outline three types of society: feudalism, capitalism, and socialism. Then, under the general heading of social theory, we offer some brief additional comments on Marxism and humanism and introduce three newer conceptual approaches—structuration, postmodernism, and feminism—that will reappear in later chapters, especially 12 and 13.

Informed by these new theoretical perspectives, we then return to landscape, this time as the spatial constitution of culture. Now a third cultural variable is introduced: like language and religion, ethnicity is often a fundamental factor in establishing and maintaining group identity. Following a discussion of landscapes and power, we consider folk and popular culture as they are reflected in landscape and as they relate to globalization processes, and conclude with a brief look at the idea of social engineering.

'Paris Las Vegas': this resort hotel and casino in the Nevada desert offers tourists a simulacrum of the quintessential symbol of France (Jochen Tack/DAS FOTOARCHIV).

Rethinking Culture

Chapter 7 restricted itself to a single view of culture—albeit an important one with strong roots in anthropology. In recent years, however, several other perspectives on culture have had a significant influence in human geography.

In the traditional interpretation, especially for Sauer and other members of the landscape school, culture was a given; accordingly, these geographers analyzed the impact of culture on landscape, especially as it was reflected in the material and visible landscape and the formation of regions. Sauer's view of culture as cause remained largely unquestioned in geography until about 1980, when debate about the superorganic content of this view surfaced (Box 7.2, p. 220). This debate, along with increasing interest in both Marxism and humanism, encouraged a substantial rethinking of the culture concept. The term 'new cultural geography' distinguished the revised concepts of culture and the analyses they prompted from the traditional landscape-school view introduced by Sauer.

A Symbolic View

With the rise of humanistic geography in the 1970s, a symbolic interpretation of culture became prominent. This view broadens the concept of culture to embrace more fully non-material culture and to emphasize that individuals create groups through communication. This broader view of culture allows human geographers to consider topics beyond landscape: specifically, topics that fall under the general heading of what we have called the spatial constitution of culture. As P. Jackson and S.J. Smith explain it:

> Culture, in the sense of a system of shared meanings, is dynamic and negotiable, not fixed or immutable. Moreover, the emergent qualities of culture often have a spatial character, not merely because proximity can encourage communication and the sharing of individual life worlds, but also because, from an interactionist perspective, social groups may actively create a sense of place, investing the material environment with symbolic qualities such that the very fabric of landscape is permeated by, and caught up in, the active social world (Jackson and Smith 1984:205).

This interpretation of culture had been anticipated a decade earlier by P.L. Wagner (1975:11): 'The fact is that culture has to be seen as carried in specific, located, purposeful, rule-following and rule-making groups of people communicating and interacting with one another.'

This shift in geographical thinking was in line with a major school of thought in sociology called **symbolic interactionism**. Essentially, this set of closely related theories, derived from the ideas of the American philosopher G.H. Mead (1934), argues that humans learn the meanings of things through social interactions, and that our behaviour in any given situation is the product of our response to the perceived environment. In other words, interactions with other people provide us with meanings for things that we are then able to use to understand those things. Once we have acquired such an understanding, we use it to define the situations that we encounter and then act accordingly.

New Cultural Geography

Closely related to the symbolic view of culture is the idea that in fact there is no single, fixed entity called culture, but rather a plurality of cultures, understood as those values that members of human groups share in *particular* places at *particular* times. Considered from this perspective, cultures are not objects but mediums or processes—what Jackson (1989:2) described as maps of meaning: the 'codes with which meaning is constructed, conveyed, and understood'.

This interpretation has led geographers to study a great many topics besides landscape: from previously ignored groups, new cultural forms, and 'otherness' to Eurocentrism and ideologies of domination and oppression. Most generally, human geographers now tend to base their questions about human identity on the logic of **constructionism** rather than the more traditional logic of **essentialism**. Whereas the essentialist view sees the characteristics that constitute identity as inherited and largely unchanging, the constructionist view stresses not only that those characteristics are socially made or acquired, but also that they are contested in the sense that there are no unequivocal meanings: different characteristics are important in different places and at different times. The distinction is important philosophically, since the essentialist view is typically associated with empiricism, while the constructionist view is closely related to feminism and postmodernism (to be discussed later in this chapter). This shift in perspective is reflected here: in Chapter 7 we discussed the identity characteristics of language and religion from a primarily essentialist perspective, while this chapter discusses place, ethnicity, and gender, and the following chapter discusses nationality, from a primarily constructionist perspective.

The rethinking of culture has opened a number of new directions for cultural geographers, some of whom have now rejected the landscape school's emphasis on the regional mapping of material and visible features of landscape on the grounds that such mapping assumes the existence of a single distinctive and unchanging cultural group constituting *a culture*. Favoured instead are approaches that acknowledge the existence of a plurality of *cultures*, located in specific times and places. Among the topics for study suggested by recognition of multiple cultures are cultural identities and their links to place, and issues of cultural dominance and subordination.

Along with these new views of culture has come a new understanding of landscape itself, including the natural landscape, as something that is socially constructed. Studies in this area focus on two aspects of landscape: symbolic (specifically, the meaning contained in landscape) and represented (landscapes as represented in literature and art, as well as the more usual visible and material landscapes). Such work often treats landscape as a text that is open to interpretation, and, recognizing the importance of images, includes among its research methods **iconography**: the description and interpretation of images to uncover their symbolic meanings (see Box 8.5, p. 265). The idea that nature is socially constructed—in effect, that it is part of culture—has several important implications. We need to recognize that our understanding of nature is filtered through human representations of it, and that these representations vary with time and place. We also need to be aware that the representation of nature is never neutral: any such representation—and there may be several of them for any part of the natural world—is ideologi-

cally loaded, and it is part of the geographer's task to interpret them.

With these rather different views of culture and landscape in mind, we can now turn to a brief discussion of different types of society, specifically feudalism, capitalism, and socialism, and then to an evaluation of various approaches to the subject matter of this chapter.

Types of Society

Our contemporary cultures reflect a long evolutionary process that has followed different routes in different parts of the world. This section summarizes the European transition from feudalism to capitalism and the experience of socialism. Using Marxist logic, each of these societal types is an example of a mode of production (refer to the discussion of Marxism in Chapter 2).

Feudalism

Feudalism was a non-centralized system of governance and social and economic organization that developed in northern Europe over the centuries following the collapse of the Roman empire (*c.*375). Under the feudal system, all land was owned by the king, who effectively delegated control of it to his warrior lords (vassals) in return for their military and political support. The 'direct producers' who worked the land were the peasants, who were permitted to live on the land in exchange for their labour and were subject to the legal and political control of their individual lords.

Feudalism has several important implications for both place and people. In a feudal society, people were defined by their social class and social mobility was extremely limited. Further, peasants were not free to choose whom they would work for; this was the situation that Marx would describe as exploitation.

Capitalism

A new type of social and economic organization began to emerge as early as the late sixteenth century, and was fully in place in many parts of Europe by the eighteenth century. The term used to describe this system, **capitalism**, was first popularized by Marxists in the late nineteenth century. Capitalism is characterized by the transformation of labour into a commodity that can be bought and sold and the separation of the producer (the worker) from the means of production, which are owned by the 'capitalist' class. Today, capitalism is the dominant form of economic and social organization in the western world.

Among the distinctive characteristics of capitalism, according to Marx, are its capacity for self-expansion through ceaseless centralization and concentration of capital; continual technological changes to the production process; the cyclical nature of the associated process of development; and the divisions it creates between classes, leading to class conflict. In Marx's view, capitalism, like feudalism, exploits the peasant or working class, and because people under capitalism are not fully free, they are alienated (Box 8.1).

Another important social theorist, Max Weber, saw capitalism in quite a different light, as an **ideal type**. He traced its origins to the religious ethic of Protestantism, the growth of cities, and the legal and political framework provided by the rise of a new type of nation state.

After the end of the Second World War, **competitive capitalism** (as it is now usually called) changed considerably as a result of the rapid growth of major (often transnational) corporations and increased involvement by the state in the economy (often through public ownership). This is known as **organized capitalism**. Most recently, there is evidence to suggest that a further transformation is now underway: the term **disorganized cap-**

italism refers to a new form that is characterized by a process of disorganization and industrial restructuring. The transition from organized to disorganized capitalism (also referred to as the transition from **Fordism** to **post-Fordism**) is in part a reflection of the malfuctioning of capitalism suggested by economic instability, social injustice, poverty, and unemployment.

The concept of **class** is often employed by social scientists in discussions of capitalism. Specifically, sociologists use the term either as a structural category with respect to status—typically distinguishing lower, middle, and upper class—or as an expression of self-identity. The first interpretation has proven of little interest to human geographers, at least partly because the categories have different meanings in different parts of the world. The second interpretation ought perhaps to be of interest, both because different classes tend to locate in different areas—this is most apparent in the internal geography of cities—and because questions of self-identity are of increasing concern in the new cultural geography. Despite these attractions, however, human geographers have used the concept infrequently and somewhat ineffectively. The principal exceptions are some Marxist geographers who consider the concept of class to be a more useful general concept than culture; for Blaut (1980), class also has the advantage of implying links between groups, with some classes exercising authority over other classes. It can be argued that cultures are discrete and separate entities, whereas classes are linked by power relations

8.1

THE CONCEPT OF ALIENATION

According to Marx, the capitalist system has a dehumanizing effect on individuals living within it, an effect that he termed **alienation**: because they sell their labour and do not control the means of production, they lack control over their own lives. At the same time, the capitalist state uses democracy and guarantees of individual human rights to legitimize the maldistribution of political and economic power. It has even been suggested that the state itself evolved to legitimize capitalism and prevent substantial popular opposition to the circumstances of alienation (Johnston 1986:176). As capitalism has spread across the globe, states at all levels of the world system—core, semiperiphery, and periphery—have experienced alienation. Although it is most detrimental in the periphery and least detrimental in the core (since living standards are highest in the latter), the basic effects are the same.

The concept of alienation is central to many discussions of global problems. For example, the alienated individual can no longer interact directly with the natural world; all our relations with that world are in some way organized by forces that are not part of us or of nature.

(although much of the new cultural geography explicitly interprets cultures too in terms of power relations). Viewed in this light, class involves patterns of dominance and subordination that are reflected in our attitudes and behaviours and in the landscapes we create. The links between this interpretation of the concept of class and the Marxist philosophy outlined in Chapter 2 are clear.

Socialism

In the twentieth century, a number of cultures rejected capitalism in favour of **socialism**, a form of social and economic organization based on common ownership of the means of production and distribution of products. The common feature of all socialist endeavours is the opposition to capitalist individualism; as the term itself suggests, socialism focuses on community, equality, the well-being of society as a whole, and the vision of a classless society.

Multiple versions of socialism were proposed in the nineteenth century, communism representing the most extreme, social democracy the least. It is notable, however, that communism as it was eventually implemented in Russia and elsewhere differed significantly from Marx's own vision. Moreover, the late-twentieth-century rejection of communism in the former USSR and throughout eastern Europe has resulted in an expansion of the capitalist world. Today, the prospect of a transition from capitalism to socialism seems unlikely. For many people 'socialism' now signifies nothing more radical than the provision of certain basic welfare measures (such as state-funded medical care) within a basically capitalist economy.

Social Theory

Diversity of Current Approaches

Social theory is not the property of any one discipline; all the social sciences ask questions about ways of life and human behaviour. During the middle part of the twentieth century, the social sciences were dominated by a positivistic approach based on ideas borrowed from the physical sciences and focused on the development of theory, the testing of hypotheses, and the creation of laws (see the discussion of positivism in Chapter 2). Questions of interpretation were not considered.

There have, however, been dramatic changes in our theoretical preferences since the late 1960s. The weakening of the influence exerted by positivism has permitted the flowering of a multitude of alternative approaches linked both by their rejection of two positivistic claims—that social science can be value-neutral, and that the theory/hypothesis/law approach necessarily produces the best explanation—and by their acceptance of the idea that social science is an interpretative endeavour: in other words, one in which questions of meaning and communication are relevant.

Several of these newer approaches trace their intellectual origins to earlier writers: for example, Marxism to Marx and humanism (phenomenology) to various nineteenth- and twentieth-century geographers. Others are of more recent origin, notably structuration and postmodernism.

The current proliferation of theoretical approaches has provoked two quite different responses among practising social scientists. Some see it as a problem: if social theorists are unable to agree among themselves, what possible use can social theory serve for those actively engaged in research? Others see it as an invaluable aid in helping scholars to avoid the dogmatism that results from a single dominant approach. This is an important question—and one for which there is not a single correct answer.

Humanism and Marxism Revisited

One topic of interest to contemporary human geographers is the social significance of space and place. Humanists are interested in the elements that combine to produce a sense of place, in the symbolism of landscape, and in iconography. Those with a more Marxist perspective are interested in social inequality with respect to such variables as language, religion, ethnicity, class, gender, and sexuality, especially as they are reflected in space. Both groups recognize the importance of space in the constitution of social life. The Marxist perspective also introduces the social constructionist idea of nature as a product of culture, the idea that in addition to the natural world as conventionally understood, apart from humans, there is a second nature that emerged with industrial landscapes—nature transformed as the product of human labour.

Three more recent theoretical approaches that are now making important contributions to human geography are feminism, structuration, and postmodernism.

Feminist Theory

There is no single body of feminist theory; rather, there are various schools of feminist thought associated with larger bodies of theory such as liberalism, Marxism, socialism, or postmodernism. Nevertheless, all schools of **feminism** are united in their commitment to improving the social status of women and securing equal rights with men.

Women are systematically disadvantaged in most areas of contemporary life. The fundamental reason for this inequality is **patriarchy**: a social system in which men dominate women. Under the traditional division of labour, women were economically dependent on male breadwinners, and the few women who did work outside the home were paid less than men received for equivalent work. *Culture* is seen as a key factor in the construction of **gender** differences through various socialization processes. *Sexuality* and *violence* are both seen as forms of social control over women. Finally, the *state* is seen as typically reinforcing traditional households and failing to intervene in cases of violence against women.

Of the several traditions of feminist thought and action, the oldest, dating back to the late eighteenth century, is *liberal feminism*, aimed at securing equal rights and opportunities for women. Of the more recent traditions, developed since the 1970s, two of the most important argue that the oppression of women cannot be corrected by superficial change because it is embedded in deep psychic and cultural processes that need to be fundamentally changed. *Radical feminism* contends that gender differentiation results from gender inequality, and that the subordination of women is separate from other forms of social inequality, such as those based on class. *Socialist feminism* similarly emphasizes gender inequality, but links that inequality to class; what this means is that both men and capital benefit from the subordination of women.

Structuration

The sociologist Anthony Giddens has developed a social theory that addresses the links between human agents (people) and the social structures within which they function. Developed out of numerous earlier contributions to social theory, including Marxism, **structuration** theory focuses on the capacities that permit people to institute, maintain, and alter social life (see, for example, Giddens 1984).

Structuration theory sees individuals as agents operating within local social systems, sometimes called **locales**, and the larger social structures of which they are a part; capitalism is one such structure—a set of rules created by humans to facilitate human survival. Perhaps the key element of structuration theory, however, is its identification of the dualities associated with social structure and human agency. First, social structure enables human behaviour, while at the same time behaviour can also influence and reconstitute culture. Second, the rules of any social structure are at once constraining, because they limit the actions available to individuals, and enabling, because they do not determine behaviour; there is an intriguing conceptual parallel here with environmental determinism.

Some human geographers have responded favourably to the fundamental logic of structuration (see Gregson 1986) because Giddens has explicitly incorporated human geographic ideas in the theory and because a number of influential human geographers have made significant theoretical and empirical contributions to it—for example, Gregory (1981). The long-term impact on human geography, however, remains to be seen. Box 8.2 provides one example of a structuration approach to a human geographic problem.

From the perspective of the beginning human geographer, one of the advantages of considering structuration theory is that it obliges us to think more explicitly about the relative value of focusing on structure and/or agency in human geographic analysis. The fact is that, traditionally, human geographers have emphasized structure over agency in that they have endeavoured to identify general processes responsible for the formation of the human geographic world. A specific example of this type of thinking is environmental determinism; more generally, both the positivist and Marxist approaches assume that it is possible and useful to identify general processes. The explicit focus on questions of agency in human geography is relatively

8.2

AN APPLICATION OF STRUCTURATION THEORY

The aims of the research summarized in this box are not significantly different from those of much previous work: to understand the changing landscape and character of a place, in this case the Swedish province of Skåne. What is different is that the basic approach is derived from structuration theory; this means that both the underlying concepts and the language used to express ideas are different from those associated with the more traditional empirical, descriptive approach.

The key idea is that place is a human product: 'it always involves an appropriation and transformation of space and nature that is inseparable from the reproduction and transformation of society in time and space' (Pred 1985:337). Applying structuration concepts to the analysis of Skåne allows the researcher to discern a multitude of meaningful local variations in the evolution of place resulting from local variations in, for example, the timing of land enclosure, agricultural production, diets, and language.

The central concern is with the 'becoming' of a place and with place as a 'historically contingent process'. These terms are most easily explained by reference to Figure 8.1, which shows that:

> any place or region expresses a process whereby the reproduction of social and cultural forms, the formation of biographies, and the transformation of nature and space ceaselessly become one another at the same time that power relations and time-specific path-project intersections continuously become one another in ways that are not subject to universal laws, but vary with historical circumstances (Pred 1985:344).

Seven general propositions further clarify how structuration theory is used (Pred 1985):

1. Structuration is a continuous process. There are in fact two interconnecting processes of social reproduction (structure) and individual socialization (agency).
2. What occurs in any given place and the meanings attached to that place are tied to the structuration process in point 1 above. Indeed, they are tied to the structuration process in that place and elsewhere.
3. Power relations are central to the social structure of a place as it is becoming. In accord with the idea of duality of structure, however, they can be transformed.
4. Power relations also influence what people know and perceive.
5. The manner in which power relations are tied to the becoming of a place depends on the degree to which local institutions are controlled locally or non-locally.
6. The becoming of a place may be dominated by institutional projects relating to production and distribution as these imply spatial and social divisions of labour.
7. The becoming of a place is also tied to individual biographies.

Notwithstanding the success of this and some other applications of structuration, relatively few authors have applied structuration as a research methodology, not least because the theory itself is complex. In recent years, Pred himself has favoured other conceptual schema, notably aspects of postmodernism—see, for example, Pred and Watts (1992).

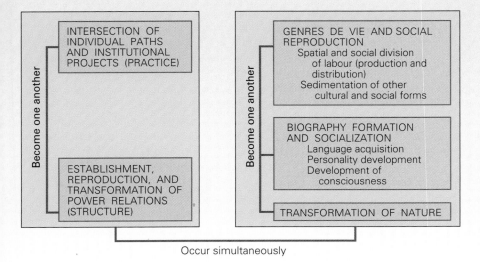

Figure 8.1 Components of place as a historically contingent process.

Source: A. Pred, 'The Social Becomes the Spatial, the Spatial Becomes the Social: Enclosures, Social Change and the Becoming of Place in the Swedish Province of Skåne', in *Social Relations and Spatial Structures,* edited by D. Gregory and J. Urry (London: Macmillan, 1985):336-75.

recent, and usually reflects a humanistic emphasis. As we saw in Chapter 2, humanism tends to focus on the intentional human actor as an active participant, rather than a passive being moulded by general processes.

Postmodernism

Most contemporary human geographers would agree that **postmodernism** is an especially difficult body of ideas to understand. At least one reason is that postmodernism is, by definition, neither structured nor unambiguous. Another is that there are many versions of postmodernist theory, most of them developed in disciplines far removed from human geography—most notably architecture, literature, and other expressions of culture. Despite these very real difficulties, postmodernism is playing an increasingly important role in contemporary human geographic research.

Modernism

As the name implies, postmodernism emerged as a reaction to **modernism**. In cultural history, 'modernism' is a general term used to refer to any number of movements that, beginning in the mid-nineteenth century, broke with earlier traditions. Modernism developed most fully in art and architecture, but is also full of implications for social science methodology, as evidenced by the positivism that first appeared in sociology in the nineteenth century. It assumes that reality can be studied objectively and can be validly represented by theories (as in positivism), and that scientific knowledge is practical and desirable. Modernism is also closely linked to the industrial revolution and the rise of capitalism, emphasizing such classic liberal themes as the rationality of humans, the privileged position of science, human control over the physical

8.3

A POSITIVIST RESPONSE TO POSTMODERNISM

It is hardly surprising that an approach such as postmodernism, which effectively undermines much previous academic work because of its insistence that there are multiple ways of reading any reality, will be subject to opposing views; nor is it surprising that human geographers who have worked primarily in a positivist tradition will be at the forefront of such criticism.

Brian Berry is one of North America's most distinguished human geographers and, according to the Social Science Citation Index, the world's most frequently cited geographer since the 1960s. In the course of reviewing a collection of essays edited by Abler, Marcus, and Olson (1992), *Geography's Inner Worlds: Pervasive Themes in Contemporary American Geography*, Berry (1992) noted the postmodern emphasis in several essays and took exception to the implications of such an emphasis for human geography. He expressed specific concern about the implications of two claims:

1. 'While scientific method remains accepted in physical geography, critical social theorists have raised challenges to the use of formal modeling in human geography. . . . Human geographers continue to be engaged in lively epistemological debate in which there is little consensus except for considerable negativism towards logical positivism' (quoted in Berry 1992:491–2).
2. 'Contemporary human geography reflects . . . a postparadigm condition in which disciplinary practices and concepts appear, for good or bad, to have broken loose from any notion of disciplinary closure and unity' (quoted in Berry 1992:492).

More important, Berry identified a fundamental contradiction between the editors' assertion that geography needs to be a 'coherent, synthetic, global discipline focused on human use of the earth' (quoted in Berry 1992:493) and their warm approval of the postmodern diversity represented in the book:

The success of *disciplines* is that they think in ways *disciplined* by theory; for them, paradigm shifts occur because older theories are found wanting and are pushed aside by theories that offer better explanation. But postmodernism carries with it a New Age social vision that embodies not only ideas of holism (the interdependence of all systems, spiritual as well as material) and earth awareness (the interdependence of all things on earth, including humankind) but also a particular view of human rights (the rights of all individuals to chase their *sadhana* and live transformative lives) that, in academic practice, translates into a concept of inner meaning (geographer's inner worlds) that is fundamentally antitheoretic. . . . If this be geography's *karma* so be it, but the implications are not pleasant. Detached from the world of science, a generation of geographers will, like the regionalists, have little accumulative wisdom to share with others, simply disparate fragments—some insightful, some beautiful, but none coming together into the whole cloth of theories that provide order to inquiry and help frame practice (Berry 1992:493).

environment, the inevitability of human progress, and a search for universal truths.

The postmodernist alternative

Postmodernism rejects all the assumptions of modernism. Reality cannot be studied objectively because it is based on language. Rather, it should be thought of as a **text** in which all aspects are related (it is 'intertextual'). Therefore reality cannot be accurately represented. Taken to the extreme, this means that truth is relative and, for practical purposes, non-existent. Causality does not exist, and theory construction has no meaning

So how does a postmodernist approach work? The emphasis is on the **deconstruction** of texts and the construction of narratives that do not make claims about truthfulness; such narratives tend to focus on differences, uniqueness, irrationality, and marginal populations. Deconstruction questions the established readings of a text and highlights alternative readings. Overall, postmodernism considers modernist claims to be arrogant, even authoritarian. 'Postmodernism and deconstruction question the implicit or explicit rationality of all academic discourse' (Dear 1988:271).

Among the principal attractions of postmodernism for contemporary human geography is its emphasis on cultural otherness, its openness to previously repressed experiences such as those of women, homosexuals,

and those lacking power and authority generally. Later sections of this chapter will reflect this emphasis on the diversity of human experience.

Diverse postmodernisms

Not all postmodernism is quite as described above. In fact, the concept varies considerably between disciplines and even within them. Some who embrace postmodernism nevertheless continue in the progressive directions suggested by modernism, becoming involved in social movements or working to break down the barriers between researchers and subjects so that people are allowed to speak for themselves. These versions are relatively close to some earlier concepts of culture, such as symbolic interactionism, and some other social philosophies, such as humanism. That there is no single unequivocal version of postmodernism in human geography is not surprising, given that the central message of postmodernism is the importance of diversity. For critical responses to the postmodern approach from two different perspectives, see Boxes 8.3 and 8.4.

Other human geographers acknowledge the strengths of the postmodern perspective, such as its emphasis on cultural otherness, but express concern about the postmodern tendency to focus on topics that could be seen as trivial. For example, Hamnett (2003:1)

8.4

A MARXIST RESPONSE TO POSTMODERNISM

Another established geographic approach that takes exception to postmodernism is Marxism. David Harvey, the most influential Marxist human geographer since the early 1970s, explained his reservations in his book *The Condition of Postmodernity* (1989). The greatest concern for Marxists is the postmodernist claim that Marxism is unable to explain the growth of disorganized capitalism (as opposed to organized capitalism). Harvey firmly believes in the power of Marxist theory to explain these processes and events.

From a Marxist perspective, Harvey argues that postmodernism reflects 'a particular kind of crisis within . . . [modernism]',

one that emphasizes the fragmentary, the ephemeral, and the chaotic side (that side which Marx so admirably dis-

sects as integral to the capitalist mode of production) while expressing a deep scepticism as to any particular prescriptions as to how the eternal and the immutable should be conceived of, represented, or expressed. But postmodernism, . . . with its concentration on the text rather than the work, its penchant for deconstruction bordering on nihilism, its preference for aesthetics over ethics, takes matters too far (Harvey 1989:116).

Most critically, Harvey (1989:117) sees the rhetoric of postmodernism as dangerous because it ignores 'the realities of political economy and the circumstances of global power'.

Although we obviously cannot resolve any of these debates about postmodernism, it is important to recognize some of the tensions that exist in contemporary human geography.

worries that human geography has become a 'theoretical playground where its practitioners stimulate or entertain themselves and a handful of readers, but have in the process become increasingly detached from contemporary social issues and concerns'.

The 'Cultural Turn'

Another way to think about recent additions to the conceptual arsenal available to human geographers is to suggest that a 'cultural turn' has taken placed. This phrase—which has been used in the context of recent changes in many of the social sciences and humanities, including the 'new cultural geography' noted earlier in this chapter—essentially refers to an increased appreciation of the importance of culture in understanding humans and their political and economic activities.

The impact of this cultural turn has been especially notable in studies of human identity and human difference and of the politics related to these matters. There is now a persistent questioning of traditional concepts, classifications, and categories such as those used in our discussions of language and religion in Chapter 7. Identity is increasingly being understood in terms of constructionism rather than essentialism, that is, as something that is socially created and therefore subject to ongoing change rather than something that is predetermined and fixed. This more sophisticated understanding of identity has prompted human geographers to examine the power relations between dominant groups and other groups, as well as the politics of difference.

Also associated with the cultural turn is the idea of **contextualism**: a recognition of the importance of acknowledging that a specific discourse (see glossary) is employed in all human geography. This implies that we need to be aware of, and sensitive to, precisely how knowledge is being constructed—in other words, readers need to know who is conducting the research and what their agenda is. It is for this reason that some current work in human geography explicitly acknowledges the positionality and situatedness of the author.

Landscape as Place

Landscapes are not simply locations; they are also *places* in the sense that they convey meaning. This idea is reflected in the increasing numbers and sophistication of iconographic analyses (Box 8.5). The meaning that a place has is a human meaning, dependent on social matters. There is an important circularity here. Humans, as members of groups, create places; in turn, each place created develops a character that affects human behaviour. This circularity is characteristic of all landscapes, but is perhaps most relevant in the extreme cases—landscapes of the advantaged and the disadvantaged, the privileged and the underprivileged, insiders and outsiders, rich and poor, and men and women.

Places as Social Creations

Much of this chapter deals with ideas that are relatively new to human geography and that are continually being tested in various contexts as they unfold. One distinguished human geographer (Johnston 1991:67–8) argues that our focus should be on 'regions' (also known as places, localities, or locales). The thrust of his argument as follows:

1. Region creation is a social act; regions differ because people made them that way. Thus differences in physical environment are relevant, but are not a cause.
2. Regions are self-reproducing entities because they are the contexts within which people learn. Thus people are made by the places that they create.
3. Regional cultures do not exist separately from the individual members of the culture.
4. Regions are not autonomous units; they interact with other regions.
5. Regions are often the deliberate creations of those in control.
6. Regions are potential sources of conflict.

Each of these six points will reappear in the discussions and examples presented later in this chapter.

Studying Places

The humanistic concept of place is central to this chapter, and some extensions of that concept will aid our understanding of the material that follows. *First*, our experience of place and the meaning we attach to place are not simply individual matters: they are intersubjective (shared). This idea is in accord with our symbolic rethinking of culture. The most compelling example of the intersubjective character of place is the idea of home. Whatever specifically the word 'home' refers to—a dwelling, other people, the earth itself—for most people, a home is shared

Second, our experience of any place may be characterized either by topophilia, a positive attitude towards place, or by topophobia, a negative attitude towards place (both terms were introduced in Chapter 2 and are included in the glossary). These terms refer to the emotional attachment that exists between person and place, so that the specific feelings of topophilia or topophobia are related to specific characteristics of both the individual and the environment in question. Individual characteristics include personal well-being

and familiarity with place. Environmental characteristics include the attractiveness of places as judged by individuals; for example, a pastoral setting may be attractive, while an urban slum may be unattractive.

A *third* extension of the place concept is the degree to which something in the environment exhibits the characteristics of place or placelessness (see the glossary). As we saw earlier in a Marxist context (Box 2.6), there is a tension present in all places, a tension between the local and the global. Those things in an environment that are local in origin, such as a festival or an indigenous building style, enhance place identity, whereas those things that are global in origin, such as chain department stores or an imported

8.5

THE ICONOGRAPHY OF LANDSCAPE

Iconography is the description and interpretation of visual images to uncover their symbolic meanings. To the extent that landscapes can be regarded as depositories of cultural meanings, it is possible to subject them to iconographic analysis. A successful iconographic analysis will reveal how a landscape is shaped by and at the same time shapes the regional culture of which it is a part. Two Canadian examples are briefly summarized here.

Osborne (1988) focused on the development of a distinctive national Canadian iconography both in artistic images of lands and peoples (notably the paintings of Tom Thomson and the Group of Seven) and in the various responses these images provoked. The declared aim of those artists was to assist in the development of a national identity, largely through their paintings of the Canadian north. During the first half of the twentieth century, they helped to create a distinctive image of Canada: 'rock, rolling topography, expansive skies, water in all its forms, trees and forests, and the symbolic white snow and ice of the "strong North"' (Osborne 1988:172). Although this image is clearly limited, and is now just one among a vast range of national images, it remains a compelling example of iconography.

A different application of iconographic logic can be seen in the examination, by Eyles and Peace (1990), of images of the city of Hamilton, Ontario, in their societal context, asking why these images appear as they do. Derived primarily from newspaper accounts, the images were varied, but they were dominated by a single—negative—theme: Hamilton as steeltown. This image most frequently arose in comparisons with neighbouring Toronto, in which Hamilton appeared as the blue-collar

city of production and Toronto as the city of consumption. Because Hamilton is an industrial city, it was seen as a polluted landscape representing the past rather than the future, as an area of economic decline.

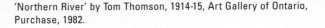

'Northern River' by Tom Thomson, 1914-15, Art Gallery of Ontario, Purchase, 1982.

building style, contribute to placelessness. Even if all placeless areas do not look alike, they lack the local characteristics that make a place distinctive, comfortable for insiders, and perhaps incomprehensible for outsiders. Global characteristics are those that make a place similar to other places and therefore relatively placeless, perhaps less comfortable for insiders but more comprehensible for outsiders (Entrikin 1991). The concept of placelessness is an important one because it suggests that the more places become placeless, the more similar human experience becomes. The idea that the modern world is losing environments with distinctive local features and gaining environments with global features has implications for the subjects discussed in the rest of this chapter—vernacular regions, ethnicity, landscapes and power, and popular and folk culture—as well as some discussed in later chapters, such as tourism (Chapter 13).

Vernacular Regions

The cultural regions discussed in Chapter 7 were formal regions; areas with one or more cultural traits in common, they were the geographic expressions of culture. A vernacular region may or may not be formally defined: what matters is that it is perceived to exist, by those living there and/or by people elsewhere. Such regions introduce the idea of the spatial constitution of culture. Clearly, a formal region may also be a vernacular region; the Mormon landscape, for instance, can be defined formally (Francaviglia 1978), but it is also generally perceived to exist as a distinct entity by Mormons and others.

Creating a Vernacular Region

In other areas the link is not so clear. The area of French Louisiana, which is the only surviving remnant of the vast French Mississippi valley empire, is perceived as a distinctive vernacular region in which the population is associated with one group identity, namely Cajun. According to Trépanier (1991), however, this perception is not an accurate reflection of cultural reality. There are in fact four important French subcultures in the area: White Creoles, Black Creoles, French-speaking Indians, and the descendants of the Acadians who fled Nova Scotia after 1755 (now known as Cajuns). Furthermore, until

recently, outsiders' perceptions of the Cajun group identity were largely negative, and it is only since the late 1960s that a process of image modification, which Trépanier (1991:161) called 'beautification', has been under way. The Creole identity, on the other hand, carried a positive image for both Black and White French-speakers. How are we to explain this anomaly?

The decision to identify the area with the Cajun group was made by the state government. In 1968, the Louisiana legislature created the Council for the Development of French in Louisiana (CODOFIL) to gain political benefits by cultivating a French image, but this did not garner any real popular support. Once created, CODOFIL was determined to unify French Louisiana using the then-negative Cajun label rather than the more positive Creole label. Trépanier (1991:164) interpreted this effort as a way to guarantee a White identity for the French-speaking area. Once this decision was made, it became necessary to improve the popular image of the Cajun identity, which was achieved through publicity campaigns, including the organization of Cajun festivals. The fact that a new governor was elected under the Cajun banner helped enormously.

For many, French Louisiana now exists as a vernacular region with a Cajun identity that disguises the variety of French-speaking people and the fact that both the Black Creoles and the French-speaking Indians reject the Cajun identity. As this example vividly illustrates, vernacular regions are in fact much more than areas perceived to possess regional characteristics; they are also regions to which specific meanings and values are attached—they have social and symbolic identity.

Delimiting Vernacular Regions

North American geographers have delimited many vernacular regions, typically by collecting information on individual perceptions. Probably the most elaborate survey was conducted by Hale (1984); it gathered 6,800 responses from such people as local newspaper editors. In order to identify true vernacular regions, any survey has to ensure that the responses received in some way represent those of average people.

Zelinsky (1980) avoided the survey problem by studying the frequency of a specific

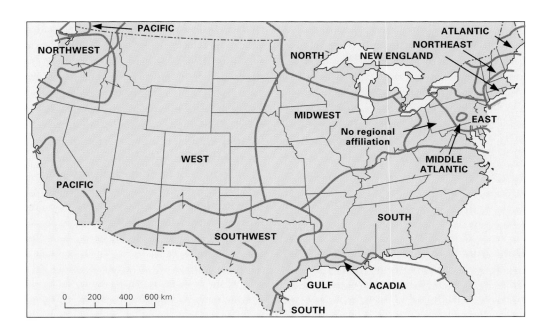

Figure 8.2 North American vernacular regions.

Source: W. Zelinsky, 'North America's Vernacular Regions', *Annals, Association of American Geographers* 70 (1980):14.

regional term and a more general national term in metropolitan business usage; for example, if the region in question is the American south, the incidence of the term 'Dixie' might be compared with that of the term 'American'. Figure 8.2 shows Zelinsky's findings: 14 vernacular regions, some areas that lack regional affiliation, and two areas that are included in either one or both of two regions. It is instructive to compare this map and Figure 7.5 (p. 228), the map of cultural regions based on the concept of first effective settlement. The fact that the south, middle west, and New England are similarly located on both maps clearly indicates that the culture region and the vernacular region are related to similar underlying factors. From a Canadian perspective, it is notable that the western region extends deep into the prairie provinces.

Vernacular Regions and Sense of Place

If an area has an identity specific enough to be named, this suggests that the identity is meaningful to those using the name. In other words, vernacular regions are not simply locations: they are places. The name of the region conveys a meaning—or possibly more than one. There are parts of the world that possess such a powerful regional identity that the mere mention of the name conjures up vivid mental images. Most North Americans, regardless of religious affiliation, recognize

that the term 'Holy Land' refers to the land bordering on the eastern Mediterranean. Other places have one meaning for one group and a quite different meaning for another. For country-music lovers, Nashville is likely to be most meaningful as the home of the Grand Ole Opry; for others it may be simply the capital of Tennessee.

Vernacular regions are often viewed more positively by those living within the region than by those outside. For many living in the American Bible belt, the name is a source of great pride; yet for some outsiders it is a term of derision. In other cases, if the name of the vernacular region has been imposed for purposes of tourism or commercial promotion, with no real roots in the region, it may mean little to either residents or outsiders.

Psychogeography

The term 'psychogeography' is used to refer to 'how people feel about, experience, paint themselves into the world and take that portrait back into themselves as literal parts of who they are and hence their well-being' (Stein and Thompson 1992:63). The aim in a psychogeographic study is to identify an internal self-image of a region, without reference to external perceptions (see Box 8.6).

Homelands

Another way to think about psychogeography—vernacular regions that are clearly perceived by those living within them—is in terms of a **homeland**. According to Nostrand and Estaville (1993:1), there are four basic ingredients to a homeland: people, place, sense of place, and control of place. 'Sense of place' refers to the emotional attachment that people have to a place, while 'control of place' refers to the requirement of a sufficient population to allow a group to claim an area as their homeland. Most homelands are associated with groups defined on the basis of common language, religion, or ethnicity.

Ethnicity

Definition

Ethnicity is often poorly defined; one researcher who examined 65 studies of ethnicity noted that 52 of them offered no explicit definition at all (Isajiw 1974). The geographer K.B. Raitz argued that **ethnic** should refer to any group that has a common cultural tradition, that identifies itself as a group, and that constitutes a minority in the society where it lives: 'ethnics are custodians of distinct cultural traditions. . . . the organization of social interaction is often based on ethnicity' (Raitz 1979:79). Thus the group may be delimited according to one or more cultural criteria, but it is necessary that the group not be living in their national territory: Swedes in Sweden do not constitute an ethnic group, whereas Swedes in the United States do, because they identify themselves as one.

Although this definition is clear and reasonable, it is not universally employed. Indeed, it is important to recognize that 'ethnicity' is generally regarded as one of the most confusing terms in social science. The greatest confusion occurs when terms such as 'race' or 'minority' are used interchangeably with 'ethnic'.

We learned in Box 3.4 that, biologically speaking, there are no such things as races (subspecies) within the human species, and Box 3.5 outlined the historical use of supposed racial differences as an excuse for identifying certain groups as inferior to others. Today the term 'race' is commonly used to

8.6   PSYCHOGEOGRAPHY: THE SENSE OF OKLAHOMANESS

Understanding a regional cultural landscape requires consideration of regional identity, which includes popular attitudes that have developed through long experience in a landscape. In a study of regional self-awareness, Stein and Thompson (1992:65) used the state of Oklahoma 'as an example of a community of meaning within a politically-defined territory'—what they call a cultural identity system. '"Oklahomaness" connotes what is distinctively Oklahoma; that is, the boundaries of the identity and the contents within it' so that 'Oklahoma is first and foremost a state of mind which springs from a common pool of self images, a community of meanings which lends the state much of its regional character' (Stein and Thompson 1992:66).

What are the self-images, the popular attitudes, that shape the identity of Oklahoma today? First, there are images associated with Native Americans: the Trail of Tears resulting from the forced migration of the Five Civilized Tribes from the southeastern United States to Oklahoma; reservation lands taken from them by the United States government to give to White Americans. Second, there are images associated with a famous series of land rushes to the 'unassigned' land taken from Native Americans between 1889 and 1901: images of unbridled enthusiasm and opportunism summed up in the labels 'Boomers' and 'Sooners'.

As a result of these land rushes, Oklahoma came to occupy a special place in the American psyche; it is no accident that the musical show (later a film) *Oklahoma!* was so successful on Broadway from 1943 to 1948, a time of exuberant American nationalism.

A third component of Oklahoma's regional identity is the tragic and haunting image of the 'Dust Bowl' of the 1930s, associated with the folk songs of Woody Guthrie and so movingly described in John Steinbeck's novel *The Grapes of Wrath*. Cowboys and open spaces combine to produce a fourth image, one that today implies an 'ideological statement of moral superiority over those with whom Oklahomans have compared themselves and felt inferior' (Stein and Thompson 1992:73). Both of these images are linked by Stein and Thompson to a fifth: the symbolism of land and sky.

These images, along with a number of others (such as the college football mentality) are essential to understanding the state of Oklahoma as a meaningful cultural identity system; in effect, they are boundaries delimiting an area that has real meaning for those inside. These images are not limited to a single group of people and are in fact closely tied to relations between those inside and those outside the region. The identity of Oklahoma is continually changing as relationships with others change.

set apart outsiders whose physical appearance does not accord with some generally accepted norm; the key divide is the most visible one: skin colour. It is important to recognize that the fact that races do not exist does not prevent the label from being applied

Some groups are generally regarded—either by themselves or by themselves and others—as minorities because they are in some way different from (and therefore excluded by) the majority. Common bases for delimiting a minority are language, religion, ethnicity, perceived racial identity, and recent immigrant status, or any combination of these.

Shared Identities: Changing Identities

The term 'ethnicity' usually suggests some enduring collective identity through a shared history—a shared time as well as a shared space. Most groups who identify themselves as ethnic base their ethnicity on one or both of the two principal cultural variables discussed in the previous chapter: language and religion. Like language and religion, ethnicity is both inclusionary and exclusionary. Some people are defined as insiders because they

share the common identity of the group, while others are seen as outsiders because they are different; thus the majority of people belong to at most one ethnic group. Yet it is quite possible for an ethnic group to change its identity and behaviour over time.

'Ethnic' is a convenient term, partly because it defies explicit definition. Thus human geographers often use it to identify and discuss cultural regions. Generally speaking, an ethnic region or neighbourhood is an area occupied by people of common cultural heritage who have voluntarily chosen to live in close spatial proximity.

Ethnic areas

Most immigrant ethnic groups, especially those moving into urban areas, experience an initial period of social and spatial isolation that may lead to low levels of well-being, relative deprivation, and the development of an ethnic colony, enclave, or **ghetto**. We can interpret these common initial experiences of deprivation and residential segregation as social and spatial expressions of outsider status. Often a local group identity is continually reinforced by **chain migration**, the process whereby migrants from a particular

Situated a little south of Madison, Wisconsin, in a region of rolling hills reminiscent of the Swiss alpine farmlands, the small town of New Glarus calls itself 'America's Little Switzerland'. The village was founded in 1845 by 108 Swiss immigrants, and succeeding generations have retained the community's Swiss-German language, vernacular architecture, and folk traditions. In this photo the Swiss architectural influence is evident in the church steeple as well as the chalet-style roof on the left. The decorative banners pair the symbolism of the Swiss flag with the American stars and stripes, and the 'Alpine' store on the right is clearly intended for the tourist market (Photo courtesy of the New Glarus Chamber of Commerce).

area follow the same paths as friends and relatives who migrated before them.

Despite often negative initial experiences, most new immigrant groups do eventually experience either **assimilation** or **acculturation**. Most European groups that move into cities in the United States steadily lose their ethnic traits and assimilate, eventually becoming part of the larger culture. Other groups, however, are more likely to acculturate: to function in the larger culture but retain a distinctive identity. In the United States, acculturation has been more common than assimilation.

One key factor determining whether or not assimilation occurs is the degree of residential propinquity: if group members live in close spatial proximity, then social interaction with the larger culture is limited and assimilation unlikely. Once again we see an intertwining of space and culture. A second factor is the impact of state policies. In Canada there have been policies actively encouraging **multiculturalism** since 1971. The political logic for the introduction of multiculturalism was succinctly stated by then Prime Minister P.E. Trudeau: 'Although there are two official languages, there is no official culture, nor does any ethnic group take precedence over any other' (quoted in Kobayashi 1993:205).

Immigrants moving into rural areas, especially if they do so as a group or in a chain migration, tend to retain aspects of their ethnic identity longer than do those moving into urban areas. Many regional landscapes, especially in areas of European expansion, are characterized by distinct ethnic imprints, primarily because of the relative recency of the occupation. This is especially likely to be the case if the group has a formal structure, as do the Mormons discussed in Chapter 7, but is typical regardless because of the relative isolation of rural areas. Historical and cultural geographers often study the extent of cultural change that accompanies and follows settlement and the creation of landscapes (see McQuillan 1993).

Landscapes and Power Relations

Our discussions of social theory and related contemporary concerns in cultural geography make it clear that peoples and places are now being considered in a variety of conceptually sophisticated ways. Human geographers are increasingly aware of the diversity of peoples and places as a result of the insights suggested by humanism and Marxism and, more recently, by structuration, postmodernism, and feminism. Together, these varied approaches invite us to explore more than cultures as ways of life and the visible landscapes related to them (the geographic expressions of culture); they invite us to explore culture as a process in which people are actively involved and landscapes as places constructed by people (the spatial constitution of culture). But this was not always the case.

Discovering Difference and Inequality

As we learned in Chapter 2, traditional regional geography was essentially empirical in emphasis, focusing on the description of people and places; it did not concern itself with uncovering the power differences that we now know to underlie human geographies. Similarly, as we also saw in Chapter 2, the spatial analysis school was explicitly positivist in philosophy, taking it for granted that such work was by nature objective, unaffected by the identity of the human geographer conducting the research. We now understand that such assumptions are often unwarranted. It was only about 1970, with the emergence of the new philosophical approaches based on humanism and, especially, Marxism that human geographers began to consider questions of meaningful human difference in areas such as access to resources and quality of life. In short, human geographers added a political dimension—a concern with power and inequality—to their studies.

A second reason why difference and inequality did not emerge as topics of interest before about 1970 was that until then the field of geography—like most other academic disciplines—was dominated by white, middle-class, heterosexual, able-bodied males. In the era when the dominance of patriarchal discourse (see glossary) was near-universal, questions of difference and inequality were rarely acknowledged.

About 1970, however, a new radical geography, sometimes called social geography, began to consider previously neglected aspects of human landscapes, such as the evidence of differences in power and related differences: in well-being, incidence of crime, health, and availability of health care. Over time, such studies have broadened to include analysis of different qualities of landscapes—what we call élitist landscapes and landscapes of stigma—and of gender as it relates to landscape and to economic development. Of central concern here is what we might call the quality of life. This is a very difficult concept to discuss objectively, for it can be interpreted only in some specific context. This section offers an overview of these important human geographic interests.

As was suggested above in the accounts of feminist and structuration theory, and as described in our discussion of types of society, there are significant social and spatial variations in the distribution of **power** and **authority**. Further, the logic of world systems theory (Chapter 6) relies on the argument that there is a dominant core and a subordinate periphery in the contemporary world. It is not surprising, then, that to understand peoples and places it is essential to recognize the importance of the distribution of power and authority.

Gender in the Landscape

Gender, like class, implies a distinction between power groups—in this case, dominant males and subordinate females. For feminist geographers, gender is a key category of analysis for two reasons: because it involves power relations and because it involves specific roles for males and females. That human geographers did not begin paying serious attention to gender differences until recently is regrettable, since the former insistence on discussing humans in general effectively meant ignoring all the ways in which, around the world, the lives and experiences of women and men are different. For example, it is still women who typically perform domestic work, which is unpaid, repetitious, and often boring, and the jobs available to women outside the home are often low-paid and low-skilled. In short, women and men tend to do different work, in different places, and to lead different lives, which include different visions of the world and of themselves. These differences are not explained in terms of biological differences (sex) but in terms of cultural differences (gender).

Interestingly, gender differences are not always as apparent in the landscape as class or ethnic differences. Yet landscapes are shaped by gender, and they do provide the contexts for the reproduction of gender roles and relationships (Monk 1992).

There are numerous examples of how landscapes, both visible and symbolic, reflect the power inequalities between women and men in their embodiment of patriarchal cultural values. In urban areas, *statues and monuments* typically reinforce the idea of male power by commemorating male military and political leaders: 'conveyed to us in the urban landscapes of Western societies is a heritage of masculine power, accomplishment, and heroism; women are largely invisible, present occasionally if they enter the male sphere of politics or militarism' (Monk 1992:126).

The design of *domestic space* is strongly influenced by ideas about gender roles. In many cases, including traditional Chinese and Islamic societies, areas for women and areas for men are separated, and those for women are often isolated from the larger world. In western societies, the favoured domestic design has centred on the home as the domain of women and as a retreat from the larger world.

This same patriarchal logic influenced the expansion and morphology of city **suburbs**: men were seen as commuters and women as home-makers/consumers requiring ready access to shops and schools. It is now argued that this arrangement further disadvantages women, who may become isolated because of distance from town or who may struggle to cope with work at home and limited opportunities for paid employment in the suburbs. The morphology of the city therefore works to reinforce traditional gender roles and identities and serves as an obstacle to change.

Gender and Human Development

There are many expressions of gender in landscape, expressions that typically disadvantage women, but perhaps the clearest evidence can be found in data on human development. Table 6.10 (p. 200) detailed the Human Development Index (HDI) for a number of countries, and Table 8.1 provides details of a more sophisticated, gender-sensitive HDI (GDI) that shows comparative data on female and male estimates of life expectancy, adult literacy, mean years of schooling, employment levels, and wage rates. The results are illuminating.

It is significant that several countries change position in the rankings when the gender-sensitive measure of human development (GDI) is used. The top three countries on the HDI list—Norway, Sweden, and Canada—each fall two places, to third, fourth, and fifth on the GDI, while Australia and Belgium (fifth and fourth on the HDI) rise to the top. Finland also rises, from tenth spot to eighth, while the US and Iceland remain in sixth and seventh position. It is interesting to note that, except for the three countries that rose in the ranking, the GDI figure is lower than the corresponding HDI: what this indicates is the existence of gender inequality. The same inequality is evident for most of the countries for which a GDI is calculated.

Landscapes of Resistance

A compelling example of the expression of power in landscape was the **apartheid** landscape of South Africa, formally set in place beginning in 1948 and finally dismantled in 1994 (Box 8.7 and Figure 8.3).

Two important general ideas emerge from the account of apartheid, and also from the account of gender in landscape. *First,* dominant groups have constructed landscapes that take certain characteristics—the characteristics of the dominant group—for granted. Thus urban areas in the western world have typically been constructed assuming heterosexual nuclear families, women dependent on men, and able-bodiedness. Because of these assumptions, those individuals and groups who do not conform to these societal expectations are perceived as different— and are often excluded or disadvantaged as members of society. When such people intrude into landscapes that were not constructed for them, the result is often controversial, both culturally and spatially. Indeed, a distinctive feature of the contemporary world is the unsettling effect that the expression of other identities and the crossing of spatial boundaries have on dominant groups.

The *second* general idea concerns the way some groups of people understand themselves in relation to others. In the case of apartheid, the dominant Afrikaners imposed their views of themselves and the indigenous Black population on the latter. Like all imperialists, they found their justification in a logic that in the context of Asia and the Middle East has come to be known as **orientalism**. According to this logic, which implies a relational concept of culture, the indigenous people of areas that Europeans wanted to colonize were '**others**'—not only different from but less than the Europeans, who as 'superior beings' had a natural right to use them and their lands as they chose.

The same logic has often been extended beyond the colonial context, as it is conventionally understood, to operate in many more locally defined situations. In many cases, those colonized have responded by creating landscapes of resistance.

It is not surprising that places often become sites of conflict between different groups of people, as in the example of the apartheid landscape. Throughout history, groups have endowed parts of the earth's surface with meaning: that is, they have created places. But the understanding of place held by one group may very well be different from that held by another group; hence the identity and ownership of places is often contested. Many examples of these landscapes of resistance are associated with new social movements—in support of the environment, social justice, ethnic separatism, and so on—and they are best interpreted as expressions of opposition to power. Other subcultures, especially those associated with youthful populations, express opposition to mainstream lifestyles and mainstream places.

Table 8.1	GENDER-RELATED HUMAN DEVELOPMENT, 2002			
Country	**HDI rank**	**HDI value**	**GDI rank**	**GDI value**
Norway	1	0.942	3	0.941
Sweden	2	0.941	4	0.940
Canada	3	0.940	5	0.938
Belgium	4	0.939	2	0.943
Australia	5	0.939	1	0.956
United States	6	0.939	6	0.937
Iceland	7	0.936	7	0.934
Netherlands	8	0.935	9	0.933
Japan	9	0.933	11	0.927
Finland	10	0.930	8	0.933

Source: United Nations Development Programme, *Human Development Report, 2002: Deepening Democracy in a Fragmented World*. (New York: Oxford University Press, 222).

THE ORIGINS OF APARTHEID

The first permanent European settlement in South Africa was established in 1652 to grow food to supply Dutch ships; population growth was slow until the nineteenth century. The colony became British in 1806, and from 1806 until 1948 South Africa was predominantly British in political character. From the election of the Nationalist party in 1948 until the political victory of the African National Congress in 1994, the dominant political influence was Afrikaner (Dutch). The Union of South Africa, founded in 1910, is now fully independent.

The history of South Africa since 1652 has been characterized by a series of conflicts between several relatively distinct societies. Conflicts between White and Black Africans in particular continue today. Segregation of different groups was characteristic of the South African landscape prior to 1948, but in that year legislation was passed that politically institutionalized the system known as apartheid, or separate development. Thus apartheid had roots in long traditions of both social conflict and relatively informal segregation. Understanding apartheid, however, clearly requires further consideration of the social group that formally institutionalized it: the Afrikaners.

The Afrikaners, as their name suggests, have separated themselves from their European heritage, although their language is derived from Dutch and traditionally they have belonged to the Dutch Reformed Church. By the twentieth century they identified themselves fully with Africa, often being called 'the White tribe'. In 1948 the Afrikaners gained political control of a state made up of Whites, Blacks (of various linguistic and tribal groups), Coloureds, and Asians. On gaining power, the Afrikaners instituted apartheid on the grounds that it was the best way to allow a socially fragmented country to evolve; their argument was that integration of different groups leads to moral decay and racial pollution, whereas segregation leads to political and economic independence for each group. Ten African homelands were created, four of which were theoretically independent (their independence was never recognized by the United Nations or by any government outside South Africa).

According to the Afrikaners, then, apartheid provided Blacks with independent states and rights comparable to those of Whites. In practice apartheid simply enabled the Whites to maintain their own cultural identity and political power while exploiting Black labour. Interestingly, the dismantling of the apartheid system, which began in the late 1980s and concluded with the 1994 election of the first democratic government, proceeded relatively peacefully.

During the apartheid era, desirable landscapes, such as this Indian Ocean beach north of Port Elizabeth, South Africa, were often designated for 'whites only'. Similar signs were placed in many urban parks, while official buildings, such as post offices, had separate entrances for whites and other groups (William Norton photo).

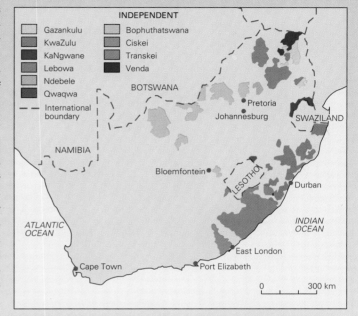

Figure 8.3 Apartheid on the national scale: the 'homelands' planned by the South African government in 1975.

Gays and lesbians parade in downtown Montreal at the close of week-long festivities in August 2000. Public celebrations of sexual identity are now regular events in many large urban centres, where they are becoming major tourist attractions (CP photo/ La Presse/Alain Roberge).

two suggests that the identical behaviour may be interpreted by the majority differently, depending on whether it represents a landscape challenge to the heterosexual majority (see Johnston 1997).

Ethnicity, gender, and sexuality are not the only grounds for exclusion from the landscape created by the dominant group. The homeless, the unemployed, the disabled, and the elderly all find themselves in some way incapable of fitting into the dominant landscape. Some of these groups are more visible than others, and some are certainly more controversial, but all are in some way disadvantaged because the landscapes created by dominant groups presume a set of identity characteristics that they do not possess. For example, dominant groups 'presume able-bodiedness, and by so doing, construct persons with disabilities as marginalized, oppressed, and largely invisible "others"' (Chouinard 1997:380).

Well-being

The term **well-being** is used to refer both to the overall condition of a group of people and to specific components of well-being— economic, social, psychological, and physical. Landscapes at various spatial and social scales differ all too widely in terms of the well-being of their occupants. On the global scale there are areas of feast and areas of famine; some children are born into poverty and others into affluence. On the regional scale, many landscapes in areas of European overseas expansion have continued to evolve in response to the needs of the imperial power, to the detriment of the well-being of most indigenous inhabitants.

On a more local scale, as we saw above, the landscapes of disadvantaged groups such as the homeless also reflect the fact that there are dominant and subordinate groups in any culture. Human geographers are paying increased attention to these landscapes, at a variety of scales, and to the social and power relations that lie behind them.

In the 1960s, concern about social problems in the United States prompted the development of spatial social indicators as measures of well-being. In a pioneering study, Smith (1973) identified seven sets of indicators representing seven different components of well-being:

Such groups may choose to display their difference through the adoption of alternative lifestyles, including musical and clothing preferences, and through their attachment to places such as inner city neighbourhoods.

Another area in which dominant groups construct landscapes that reflect their characteristics concerns sexuality. It is not uncommon, especially in large cities in the western world, for particular sexual identity groups to challenge the mainstream perception of space as heterosexual. An obvious example is the gay pride parade, march, or demonstration. In Auckland, New Zealand, the first two gay pride parades took place outside of the area understood to be gay, and hence were seen as challenges to the larger heterosexual community—as expressions of resistance in landscape. The third parade, in 1996, took place within an area widely perceived to be gay. The fact that the 1996 parade proved to be less controversial than the first

1. income, wealth, and employment;
2. the living environment, including housing;
3. physical and mental health;
4. education;
5. social order;
6. social belonging; and
7. recreation and leisure.

Examining the extent to which quality of life differs for different groups in different places, Smith found extreme inequalities at all spatial levels in the US.

All these works, however, raise a common question: do all people in fact have the same needs and the same ideas of what is good and what is bad? One attempt to conceptualize along these lines argues that all of us share a basic goal, to avoid harm (Doyal and Gough 1991). Achieving this basic goal requires satisfaction of two needs: physical health and the ability to make informed decisions concerning personal behaviour. These two in turn require that the following conditions be satisfied:

1. adequate supply of food and water;
2. availability of protective housing;
3. safe workplace;
4. safe physical environment;
5. necessary health care;
6. security while young;
7. relationships with others;
8. physical security;
9. economic security;
10. safe birth control and child-bearing; and
11. required education.

It is, of course, one thing to list such needs and another thing to measure them effectively in such a way that the integrity of different groups of people is not lost. As we have seen, the United Nations has led the way in the measurement of well-being, variously called quality of life or human development.

Much of the work done in these areas is labelled **welfare geography**; this is a general approach to a wide range of issues, but with an emphasis on questions of social justice and equality. Concern with the geography of well-being and welfare geography has encouraged research into such important issues as the geography of education (focusing on the often inequitable spatial distribution of services and facilities) and the geog-

raphy of justice (focusing on the variations in the spatial availability of social benefits).

An important issue in this work is the existence of spatial inequalities, something that most of us take for granted. Indeed, our capitalist society and economy are predicated on the presence of inequalities. Capitalism allows a few to be very wealthy, many to be comfortable, and large numbers to live in poverty. This does not need to be the case. It is possible to strive for greater spatial equality such that at least basic human needs are met for all people, in all places. Two important areas where spatial differences are clearly evident are crime and health and health care.

Crime

Human geographers have addressed the study of criminal behaviour and activities from a range of philosophical perspectives, but with the central concern of relating crime to relevant spatial and social contexts (Evans and Herbert 1989). Mapping the incidence of crime has helped identify places where criminal activity is most prevalent, such as inner cities and other areas of poverty, low-quality housing, high population mobility, and social heterogeneity. These essentially spatial, and often empiricist or positivistic, studies are accompanied by studies of the social environments of criminals, the victims of crime, and the geography of fear, all of which are more clearly Marxist or humanist in focus.

Two topics of particular concern at present are the relationship between areas of criminal activity and urban decline, and the ways in which our use of space is affected by our concerns about the likelihood of a criminal act (Box 8.8).

Health and Health Care

Susceptibility to disease, morbidity (illness), mortality, and health-care provision are all closely related to questions of well-being and the quality of life, and all have been analyzed by human geographers (see Chapters 5 and 6). Regardless of spatial scale, there has been a tendency to focus research on the less privileged members of society to demonstrate how health issues relate to a wide range of social and economic conditions. It is clear that health problems are related to the physical environment, to social factors such as housing and living conditions, and to access to health services. Several studies have emphasized inequalities in the availability of services (see Jones and Moon 1987).

Canada's Native peoples offer a compelling example, as both their health status and their access to care are much poorer than those of other Canadians. This situation is one aspect of a way of life that is characterized by poverty and low self-esteem. These result at least partly from a long period of oppression and, more specifically, a residential school system that removed children from their parents and imposed an alien way of life on them. Today, many Native communities have high rates both of physical and sexual abuse and of alcohol and substance abuse. As we saw in Box Intro.2, Native Canadians have higher rates of infant mortality and a lower life expectancy than do other Canadians. Major causes of death include suicide and tuberculosis (a poverty disease), while a common chronic condition is diabetes mellitus (indicative of poor diet).

Élitist Landscapes

Much of the discussion in this chapter, and elsewhere in this text, focuses on the different types of landscapes occupied by different types of people. As we have seen, there are landscapes for the privileged and the less privileged, the advantaged and the disadvantaged. These two sorts of landscapes are now generally labelled élitist landscapes and landscapes of stigma. There are obvious links between the theoretical questions of class and gender discussed above and these contrasting landscape types.

The existence of élitist landscapes is obvious, although they have seldom been analyzed. Geographers studying cities have traditionally recognized that all cities include a range of areas associated with different social classes; it is not unusual for the higher-class areas to be at a higher elevation than other areas, to be close to a lake, river, or ocean, or to be so located that they are not affected by the pollution generated by industry. In British cities, it is usual for higher-status areas

8.8

THE GEOGRAPHY OF FEAR

Geographers have made considerable advances in the analysis of social and spatial variations in the distribution of fear. There is compelling evidence to suggest that the fear of crime is a major problem affecting our perceptions of certain areas and our behaviours in them (Pain 1992; Smith 1987; Evans 2001). The problem is especially acute for women, the elderly, and other groups within society who are relatively powerless and vulnerable.

Fear of domestic violence, sexual harassment, and rape limits women's access to and control of space within and outside the home and also imposes limitations on social and economic activities. A study of the geography of rape and fear in Christchurch, New Zealand, by Pawson and Banks (1993), demonstrates the extent of the problem. Christchurch has a population of 292,000, of whom 91 per cent are of European origin.

Pawson and Banks had two goals: first, to determine the spatial and social incidence of rapes that were publicly reported in the press; second, to map the spatial and social distribution of the fear of crime. The results of their research, based on study of their data and some national statistics, showed a spatial distribution that emphasized the inner city area and a general cor-

relation with patterns of other criminal activity (such as burglary), areas of youthful population, and rental homes. They demonstrated that younger women under 25 years are especially at risk from rapists; that about 50 per cent of the rapes occurred in the victim's home, not a public place; that most occurred at night during or immediately after the hours of social contact and at a time of minimal community surveillance; and that there was a marked pattern of seasonality, with fewer rapes in the winter months.

To map the spatial and social distribution of the fear of crime, a survey of about 400 persons was conducted. The results showed that although fear was widespread, there was a distinct spatial pattern, with the low-income, least stable, and high-rental areas reporting the greatest levels of fear. Socially, women and the elderly displayed the greatest concern, and fear was much greater at night than during the day. The evidence clearly showed that women and the elderly are often reluctant to use either their neighbourhoods or the city centre at night.

The conclusions of this research are generally in accord with those of several other studies. As this book has already emphasized in other contexts, we live in an unequal world. For women, the city at night may well be a landscape dominated by males.

to be spatially segregated and to include such features as golf courses and specialty retail areas. In Melbourne, Australia, an élite residential area has been a consistent feature of the urban landscape, one that moved steadily south as the city grew. In an important sense, identity and landscape are very closely interwoven in these élite areas.

Élite regions can also be identified. In Britain a long-standing distinction between south and north clearly reflects a distinction between privilege and lack of privilege. Even more obvious examples of privileged landscapes are tourist areas, especially in less developed countries.

Landscapes of Stigma

Other landscapes do not have a good name. In recent years, human geographers have identified pariah landscapes, landscapes of despair, and landscapes of fear. Examples of pariah landscapes include many ghetto areas in cities and Aboriginal reserve lands in Canada and the United States. The experiences of discharged mental patients prompted introduction of the term 'landscapes of despair', a vivid descriptor that applies

equally well to pariah landscapes. One of the most publicized landscapes of despair in recent years is the inner city landscape of despair of homeless people. All such landscapes are expressions of exclusion, spatial reflections of social injustice (Box 8.9).

As already noted, violence is closely identified with specific locations. Many people must cope on a regular basis with landscapes of fear; elderly people and women are especially susceptible to violence, usually by males. Most people have mental maps identifying some areas as safe, others as unsafe—though certain 'safe' areas may actually be safe only for groups and/or during daylight hours. Large, open spaces, such as parks, are often perceived as unsafe, as are closed areas with limited exits, such as trains. The recent emergence of 'neighbourhood watch' schemes is a clear reflection of increased fear and decreased social interaction in many city areas. Landscapes of fear result in restricted use of public space.

Folk Culture and Popular Culture

The distinction between 'folk' and 'popular' culture adds another dimension to this account of the social and symbolic aspects of

The planned community of Seaside, Florida, is an élitist landscape. An early example of the movement known as the new urbanism, or neo-traditionalism, it was established in the early 1980s in a conscious effort to create a modern version of the idealized old-fashioned American town (© Steven Brooke).

Three homeless men lie on benches outside a church in downtown Toronto. Scenes such as this are increasingly common as many cities struggle to cope with growing numbers of people living on the street (Dick Hemingway).

landscape, one that is also relevant to our earlier discussion of the concepts of place and placelessness. Folk culture today survives mainly in rural areas, often among groups linked by a distinctive ethnic background and/or religion and language; it tends to resist change and remain attached to long-standing attitudes and behaviours. By contrast, popular culture is largely urban and tends to embrace change—even to the point of seeking change for its own sake, as in the case of fashion.

This is not to say that elements of folk culture cannot be found in urban areas. In large

8.9

CONSIGNED TO THE SHADOWS

The title 'Consigned to the shadows'—borrowed from an article in a geographical magazine (Evans 1989)—could refer to many disadvantaged groups. In fact it refers to people who are mentally ill. Those of us living in larger urban centres are probably accustomed to the occasional horrific news story about deplorable conditions in some home for the mentally ill. However, you may not have heard about the Greek island of Léros, where some 1,300 mentally ill people live with little food, minimal medical care, and inefficient sanitation; when the odour becomes too unpleasant, they are hosed down. People with various disorders are grouped together without proper care, let alone hope of recovery.

Conditions on Léros may appal us today, but the island is not an extreme case historically. Mentally ill people have typically been separated from the majority with no real attempt to cure them. Indeed, there is no real understanding of mental illness. In the United States, until recently, homosexuality was regarded as a mental disorder; in Japan, private mental hospitals now house about 350,000 people (one of the highest per capita rates in the world), at least partly because it is relatively easy to have a person so detained.

The problem of properly housing the mentally ill is increasing. The World Health Organization predicts a dramatic rise in the numbers of people with serious mental disorders in the less developed world. This is not surprising, given that such illnesses can be caused by a wide range of factors, including medical and dietary factors. A recent WHO estimate places the number of mentally ill people worldwide at 100 million.

multicultural cities, attitudes and behaviours specific to many different ethnic groups not only survive but flourish, and in some cases have even become features of the broader popular culture—prime examples include the Caribbean festivals held every year in Toronto and London, England, or the dragon boat races mounted by the Chinese communities in various Canadian cities. In fact, the boundary between 'folk' and 'popular' is far from clear-cut. Nevertheless, it is possible to identify some basic characteristics of each.

Folk Culture

In general, folk cultures are more traditional, less subject to change, and, in principle, more homogeneous than popular culture. (This third identifying characteristic is of limited value, however, since popular culture landscapes can also be homogeneous.) Folk cultures have preindustrial origins and bear relatively little relation to class. Religion and/or ethnicity are likely to be key unifying variables, and traditional family and other social traditions are paramount. In general, folk cultures are characterized by resistance to change and have a strong sense of place.

Domestic architecture, fence styles, and barn styles vary from one folk culture to another. However, the landscapes of individual folk culture groups, such as the Amish or Mennonites, are uniform and largely unchanging—a reflection both of central control and of individuals' desire to conform to group norms. Diet too reflects a preference for long-established habits (see Box 8.10); the mass-produced fast foods associated with popular culture are firmly rejected. The same is true of musical preferences: traditional styles associated with a particular region (e.g., western swing in north Texas and Oklahoma) are strongly preferred over the products of North American/global popular culture.

A member of Vancouver's Yau Kung Moon Athletic Association operates the head of a dragon at the opening ceremonies of a Dragon Boat Festival in Kelowna, BC, in 1999 (CP photo/Kelowna Daily Courier/Darren Handschuh).

8.10

GEOPHAGY

Geophagy, a folk custom that takes different forms in different places, is the intentional eating of dirt. Although it is most commonly practised in sub-Saharan Africa, it also occurs in other tropical areas, the southern United States, Saudi Arabia, Turkey, and India, and has been historically documented in Europe. Specific types of dirt are preferred: usually clays, but sometimes sand. There appear to be several reasons behind the custom, the dominant one being the belief that earth provides essential minerals, especially for pregnant women and the ill. In the southern United States, slaves with hookworm sometimes ate clay to reduce gastric pain. It is also possible that slaves may have practised geophagy in the hope that that it would enable them to return to their homeland after death.

In the southern United States, some young Black children and pregnant Black women continue the custom. Geophagy is a good example of a tradition condemned by many that nevertheless is still followed because it is a custom. It also appears to be a custom that has remained largely unaffected by cultural change. There are many other folk remedies and customs, but few of them are as unusual as geophagy.

The 'World Waterpark' in the West Edmonton Mall. Built in three phases from 1981 to 1986, the mall is a prime tourist destination—a massive complex of department stores, shops, restaurants, recreation areas, amusements, and services that also includes a luxury hotel (Courtesy West Edmonton Mall).

Popular Culture

Trends in popular culture, including new attitudes and behaviours, tend to diffuse rapidly, especially in more developed areas where people have the time, income, and inclination to take part. Today, the diffusion of popular culture is an important part of the cultural globalization introduced in Chapter 7.

Jackson (1989) describes the rise of popular culture in nineteenth-century Britain, showing how the social activities of the working-class populations were constrained by more powerful classes. New places of entertainment (such as music halls) became major issues of class struggle, as did activities such as prostitution. Many issues today similarly involve the control of space, and Jackson (1989:101) argues that the 'domain of popular culture is a key area in which subordinate groups can contest their domination'. One of the better-known instances of social inequality that has expression in landscape relates to matters of sexuality. For example, in most large cities there are gay/lesbian neighbourhoods that serve as powerful examples of the spatial constitution of culture.

Other geographic analyses of popular culture focus on specific landscapes (shopping malls, sports, gardens, urban commercial strips) and regions (musical regions, recreational regions). Shopping malls—artificial landscapes of consumption, located in most urban areas in the more developed world—provide a vivid illustration of the impact of popular culture on landscape. Usually enclosed, windowless and climate-controlled, catering to the homogenized consumer tastes of the moment, malls are characterized above all by their sameness. In fact, they exemplify the placelessness that, as we saw in Chapter 2, is one of the by-products of globalization (discussed in detail in Chapter 10). When completed in 1986, the West Edmonton Mall in Edmonton, Alberta, was the largest in the world, with 836 stores, 110 restaurants, 20 movie theatres, a hotel, and a host of recreational features (see Jackson and Johnson 1991). As products of popular culture, shopping malls encourage the further acceptance of popular culture.

Sport is another important part of popular culture. In fact, major organized sports, such as baseball and soccer, originated in the late nineteenth century in association with the social changes initiated by the industrial revolution. Regions can be delimited in terms of sporting preferences among spectators and participants alike. Rooney (1974) discovered clear regional variations in the United States: for example, professional American football players come primarily from a cluster of southern states (Texas, Louisiana, Mississippi, Alabama, and Georgia). Geographers often see links between such preferences and vernacular region (certainly the five states noted above can be identified as the American deep south).

Geographic studies of both folk and popular culture are plentiful in North America and Europe. At present folk studies still tend to be closely associated with the traditional landscape school of Sauer; however, the more theoretically informed methods outlined and employed in this chapter are likely to become increasingly prominent. Studies of popular culture are already being conducted in both the traditional landscape school and the newer arena of cultural analyses enriched by social theory.

Group Engineering in a Globalizing World?

This brief concluding section is a provocative postscript to both the present and the previous chapter. We can argue that the real significance of culture is that it conditions our attitudes and behaviour. Do we then want to engineer ourselves? Is there value in attempts to change our cultural beliefs and practices? These are difficult and sensitive questions that provide a challenging conclusion to our discussions of culture.

In this chapter we have highlighted the value of perceiving landscapes or regions as places with social and symbolic content. We have pursued this idea using a wide variety of theoretical underpinnings and focusing on how a number of important variables—such as gender and ethnicity—interact to form groups and related landscapes. A recurring feature has been the reality of inequality: unequal social groups living in unequal landscapes.

Groups in particular places have very different experiences; income levels vary, as do levels of health, education, and overall environmental quality. All human geographers are concerned about these issues, but the importance that each one attaches to them is clearly related to philosophical persuasion. The empiricist might emphasize the need to describe facts accurately; the positivist might focus on statistical precision and the development of theoretical explanations; the humanist might centre attention on the experiences of those living in such places; the Marxist might strive to explain issues, identify causes, and advocate change; the feminist might uncover the inequalities associated with the gendering of landscape; the structurationist might emphasize the process by which the place has 'become' through time in relation to the distribution of power; the postmodernist might focus on the writing of alternative geographies inspired by the experiences

of previously excluded groups. Each of these approaches has merit, and we may hope that, together, they will help us find solutions to the uneven distribution of well-being.

The previous chapter began with a series of powerful metaphors: 'mountains of hate, rivers of inflexible tradition, oceans of ignorance' (James 1964:2)—physical geographic terms used to describe aspects of our human condition. Language, religion, ethnicity, class, gender, and sexuality have all at some time in some place been used as justifications for war, cruelty, hypocrisy, or dogma. The goal of a single universal language remains elusive; the prospects for uniting diverse religious beliefs are negligible; ethnic groups continue to value traditions that in some cases may be inappropriate to modern societies; class distinctions may even be increasing; and the likelihood of eliminating the cultural implications of gender and sexuality seems slight. All of these variables (including gender-divided cultures) are human constructions, and they are inevitable consequences of cultural evolution. Not only does each variable involve divisions within the human population, but divisive attitudes and behaviours are actively encouraged by many cultural groups; for religious groups especially, dogma is fact.

Human diversity, then, has a dangerously divisive aspect. We are now at a point in human cultural evolution when we can give serious thought to social engineering: designing ourselves, and hence our future. It is therefore appropriate to ask whether or not we ought actively to attempt such a project—to devise new, overarching values and moral standards that might bridge the divisions between us. These are heady issues. Today, the more developed world is actively questioning traditional value systems. Only 100 years ago, racial hierarchies were proposed by serious scholars and accepted by other serious scholars; even more recently, most White people took the racial inferiority of others to be fact. Today, such theories are largely discredited and such attitudes are increasingly rare.

A world without human diversity at both group and individual levels may be unimaginable. Perhaps what we really need to achieve is a diversity that involves mutual respect between groups. Only then will our divided world be free of 'hate . . . inflexible tradition . . . [and] ignorance'.

CHAPTER EIGHT SUMMARY

Symbolic interactionism

Sees cultural groups as being created and maintained through communication; closely associated with humanistic concepts.

New cultural geography

A welcome addition to Sauerian cultural geography, based on a revised concept of culture, which introduces new approaches and much new subject matter, especially relating to issues of dominance and subordination. The new cultural geography often implies constructionist rather than essentialist interpretations of such topics as human identity.

Types of society

There are three types of society ('modes of production', in Marxist terminology): feudal, capitalist, and socialist. Both the feudal and capitalist types explicitly divide people into unequal groups or classes. Our contemporary social and economic world is dominated by capitalism.

Current social theory

There several social theories to which human geographers can turn to facilitate their analyses of people and place. In addition to the humanistic and Marxist approaches introduced in Chapter 2, these include the approaches known as feminism, structuration, and postmodernism. Human geographers are also actively using radical and socialist versions of feminist theory, and pursuing various aspects of postmodernism. Some of these additional conceptual inspirations reflect a 'cultural turn' in the social sciences and humanities.

Place

Today, landscapes are generally recognized as signifying systems that have meanings; we used the term 'place' (Johnston prefers 'region') in this context. In addition, they are interpreted as the continually changing product of ongoing struggles between groups and between different attitudes and behaviours. The distinction between 'place' and 'placelessness' is particularly significant, as is that between the related ideas of local and global environments.

Vernacular regions

Vernacular regions are perceived to exist by people living inside and/or outside them; they may be created institutionally and typically possess a strong sense of place. In North America, the map of vernacular regions closely resembles the map of regions delimited using the concept of first effective settlement. A related area of interest is known as psychogeography. Some vernacular regions may qualify as homelands if they are especially closely identified with a distinctive cultural group.

Ethnicity

Ethnic groups are often loosely defined; the key linking variable may be language, religion, or common ancestry. They are minorities and may be seen as set apart from the larger society, but over time many experience acculturation or assimilation. Canada is an officially multicultural society.

Landscapes and power relations

Human well-being is unequally distributed. Two particularly important variables in this context are gender and ethnicity.

Landscapes are gendered to reflect the dominance of patriarchal cultures: dominant men and subordinate women. Gender differences in well-being and the quality of life in general are especially evident at the aggregate scale using measures of human development. More generally, human geographers are increasingly concerned with landscapes of resistance (constructed by those who are excluded from the landscapes constructed by and for dominant groups); topics such as crime and health as they reflect differences from place to place; and what might be called élitist landscapes and landscapes of stigma.

Studies focusing on landscape and power relations mark a dramatic break with the traditional focus initiated by Vidal and Sauer especially. In brief, we are seeing increasing interest in Marxism, humanism, feminism, and postmodernism; decreasing concern with the landscape per se; and growing acknowledgement of the role of human agency in both social relations and power relations. These newer concerns first emerged in the 1970s.

Folk culture

Folk cultures prefer to retain long-standing attitudes and behaviours. A wide range of landscape features and folk activities are typically studied in the manner of the Sauer landscape school.

Popular culture

The term 'popular culture' is applied to groups whose attitudes and behaviours are constantly subject to change. Geographic studies of popular culture employ both traditional and more recent conceptual backgrounds and include analyses of landscapes and regions. Today, popular culture is usually discussed in the larger context of globalization processes.

Future cultural identity

Cultural variables are human constructions. They promote divisions between people and landscape that have often led to conflict between groups. Geographers now ask questions about our future cultural identity and the possibility of human engineering—deliberately creating new cultures.

LINKS TO OTHER CHAPTERS

- Globalization:
    Chapter 2 (concepts)
    Chapter 4 (ecosystems and global impacts)
    Chapter 5 (population growth, fertility decline)
    Chapter 6 (refugees, disease, more and less developed worlds)
    Chapter 7 (cultural globalization)
    Chapter 9 (political globalization)
    Chapter 10 (economic globalization)
    Chapter 11 (agriculture and the world economy)
    Chapter 12 (global cities)
    Chapter 13 (industrial restructuring)

- Rethinking culture:
    Chapter 7 (culture and society).

- Capitalism and socialism:
    Chapter 1 (disciplinary history)
    Chapter 2 (Marxism)
    Chapter 6 (world systems analysis).

- Humanism and Marxism:
    Chapter 2.

- Modernism:
    Chapter 4 (environmental ethics)
    Chapter 6 (global inequalities).

- Feminism:
    Chapter 5 (especially Box 5.8)
    Chapter 6 (gender).

- Concept of place:
    Chapter 2 (humanism).

- Vernacular regions:
    Chapter 2 (concepts)
    Chapter 7 (regions, including Mormon landscape).

- Ethnicity:
    Chapter 3 (races and racism)
    Chapter 7 (language and religion).

- Landscapes and power relations:
    Chapter 6 (global inequalities, especially indexes of human development)
    Chapter 9 (politics of protest).

- Folk culture:
    Chapter 7 (landscape).

- Popular culture:
    Chapter 12 (intraurban issues).

- Group engineering:
    Chapter 7 (divided world).

FURTHER EXPLORATIONS

AGNEW, J.A., and J.S. DUNCAN, eds. 1989. *The Power of Place: Bringing Together Sociological and Geographical Imaginations*. Boston: Unwin Hyman.

An edited volume, including both conceptual and empirical studies; focuses on the intellectual history of concepts of place and the linking of power and place.

ATKINS, P., AND I. BOWLER. 2001. *Food in Society: Economy, Culture, Geography*. New York: Oxford University Press.

Covers a wide range of topics, including food consumption patterns, regional differences in food preferences, and links between food and gender.

BALE, J. 2003. *Sports Geography*. New York: Routledge.

Drawing together material from a disparate literature. this book probably includes something to interest every geography student, regardless of philosophical persuasion or conceptual preference.

CARNEY, G.O., ed. 1995. *Fast Food, Stock Cars and Rock-n-Roll: Place and Space in American Pop Culture*. Lanham, MD: Rowman and Littlefield.

One of the few geography books on the topic of popular culture; includes discussions of music, food, clothing, religion, architecture, politics, and sports.

CATER, J., and T. JONES. 1989. *Social Geography: An Introduction to Contemporary Issues*. London: Arnold.

An excellent overview of a wide range of subject matter, including ethnicity, crime, and gender.

COSGROVE, D.E. 1997. *Social Formation and Symbolic Landscape*, 2nd ed. Madison: University of Wisconsin Press.

A text written at a relatively advanced level that created widespread interest when first published in 1989; this second edition has a very useful introduction placing the larger work in context.

CRANG, M. 1998. *Cultural Geography*. London: Routledge.

A textbook that is concerned with aspects of the new cultural geography from a British perspective; written at an accessible level for undergraduates.

DUNCAN, J., and D. LEY. 1993. *Place/Culture/Representation*. London: Routledge.

A series of essays aimed at representing cultural geography using diverse social theories; includes several examples with a Canadian focus.

GATRELL, A.C. 2002. *Geographies of Health: An Introduction*. Malden, MA: Blackwell.

A comprehensive textbook covering basic concepts, philosophies and related methods of explanation, techniques, and a wide range of empirical discussions; includes discussion of health issues related to human impacts on environment.

HAMNETT, C., ed. 1996. *Social Geography: A Reader*. New York: Arnold.

An edited collection of readings covering such topics as gender, class, race, and social justice.

HANSON, S., and G. PRATT. 1995. *Gender, Work, and Space*. London: Routledge.

An important book that explains how gender differences are constructed and how these differences are constituted in and through space.

KATZ, C., and J. MONK. 1993. *Full Circles: Geographies of Women over the Life Course*. New York: Routledge.

Analyzes the lives and expectations of women of all ages in both more and less developed countries with reference to class, ethnicity, national identity, and individual values.

KEITH, M., and S. PILE, eds. 1993. *Place and the Politics of Identity*. New York: Routledge.

A radical text that critically examines the traditional power relations of male/female, heterosexual/homosexual, and White/Black.

LEY, D. 1983. *A Social Geography of the City*. New York: Harper and Row.

An excellent text that includes discussion of ethnic neighbourhoods, ghettoes, and landscapes of stigma.

McDOWELL, L. 1999. *Gender, Identity, and Place*. Minneapolis: University of Minnesota Press.

Probably as good an introduction to feminism, feminist geography, and geographies of gender as is available; written in a very accessible style and with numerous references to earlier work.

———. 1989. 'Women, Gender and the Organization of Space'. In *Horizons in Human Geography*, edited by D. Gregory and R. Walford. London: Macmillan, 136–51.

An overview of the need for and implications of incorporating gender in human geography: 'add women and stir'?

MASSEY, D. 1993. 'Questions of Locality'. *Geography* 78:142–9.

An argument in favour of locality studies as the heart of geography.

MITCHELL, D. 2000. *Cultural Geography: A Critical Introduction*. Oxford: Blackwell.

A good overview of the new cultural geography with a thoughtful discussion of its origins.

Momsen, J., and V. Kinnaird, eds. 1993. *Different Places, Different Voices: Gender and Development in Africa, Asia and Latin America*. New York: Routledge.
   Twenty case-studies of the living conditions of women in the less developed world; a radical challenge to western feminist approaches.

Noble, A.G., ed. 1992. *To Build in a New Land: Ethnic Landscapes in North America*. Baltimore: Johns Hopkins University Press.
   An excellent first book-length study of North American ethnic groups and their landscapes; largely traditional cultural geography in content, but with some humanistic overtones.

Norton, W. 2000. *Cultural Geography: Themes, Concepts, and Analyses*. Toronto: Oxford University Press.
   A textbook that incorporates and attempts to integrate the traditional landscape school of cultural geography and the new cultural geography.

Nostrand, R.L. 1992. *The Hispano Homeland*. Norman: University of Oklahoma Press.
   A seminal contribution to homeland studies that identifies a distinctive group and their landscape through a reconstruction of their history.

Pile, S., and M. Keith, eds. 1997. *Geographies of Resistance*. New York: Routledge.
   This edited book includes a number of thought-provoking examples of terrains of resistance.

Rosenau, P.M. 1992. *Post-Modernism and the Social Sciences: Insights, Inroads and Intrusions*. Princeton: Princeton University Press.
   A clearly written account of a complex subject; distinguishes between two principal versions of post-modernism and defines the key terms and issues.

Shields, R. 1991. *Places on the Margin: Alternative Geographies of Modernity*. New York: Routledge.
   A series of four case studies, including Niagara Falls and the Canadian north, that emphasize the social construction of space.

Short, J.R. 1991. *Imagined Country: Environment, Culture and Society*. New York: Routledge.
   An exploration of the relationship between people and the physical world, focusing on how images are reflected in the North American cinema, the British novel, and Australian painting.

Shurmer-Smith, P., and K. Hannam. 1994. *Worlds of Desire, Realms of Power: A Cultural Geography*. New York: Arnold.
   A textbook on new cultural geography that eschews traditional interpretations of culture, emphasizing instead a postmodern perspective.

Spain, D. 1992. *Gendered Spaces*. London: University of North Carolina Press.
   A cross-cultural exploration of the spatial construction of gender; demonstrates that segregation, power, and gender are inseparable from one another in many cultures.

Stratford, E., ed. 1999. *Australian Cultural Geographies*. New York: Oxford University Press.
   A collection of writings on various relationships between people and place from the perspective of the new cultural geography; although the examples come from Australia, they are also relevant in a wider context.

Valentine, G. 1989. 'The Geography of Women's Fear'. *Area* 21:385–90.
   A brief factual account of the impact of violence on women's use of space.

Winchester, H. 1992. 'The Construction and Deconstruction of Women's Roles in the Urban Landscape'. In *Inventing Places: Studies in Cultural Geography*, edited by K. Anderson and F. Gale. Melbourne: Longman Cheshire, 139–56.
   All the contributions to this edited volume make worthwhile reading; Winchester's piece is especially helpful as an indication of how radical feminism can be combined with postmodernism in an empirical analysis.

On The Web

http://www.americanfolk.com/
A site concerned with all aspects of American folk and popular culture.

http://www.capitalism.org/
A site that argues the case for capitalism as a system based on the rights of the individual.

http://www.worldsocialism.org/
A site that argues the case for socialism as a system designed to make the world a better place for all.

CHAPTER 9

The Political World

One of the most basic divisions in the world today is the division into political states. Political states play an important role in individual and group behaviour. They also provide much of the data on which human geographers rely. A focus on political states follows logically from our discussions of language and religion (Chapter 7) and ethnicity (Chapter 8).

To explain how the world map came to look as it does today, we begin by exploring the subjects of nationalism, the nation state, and colonialism. A discussion of geopolitics introduces some of the general models that geographers have used to facilitate understanding of world affairs. Considering why some states are relatively stable and others subject to stress, this section helps to explain why human geographers can make important contributions to understanding many of the problems that states encounter. Subsequent sections address the role of the state in everyday life, the rise of new political movements, and the geography of elections.

We conclude with a discussion of the geography of peace and war and the possible waning of the nation state. Drawing on the understanding provided by Chapters 7, 8, and 9, this concluding section raises the big question—what will the future bring?—and outlines eight possible scenarios.

A British soldier on highway checkpoint patrol in southern Iraq, April 2003 (PA photo/Brice Adams/Daily Mail/MoD Pool).

If human geography is about the role played by space in the conduct of human affairs, politics is about struggles for power: specifically, the power to exercise control over people and the spaces they occupy. Like many other animal species, humans have an inherent need both to delimit territory—an area inside which we feel secure against outsiders—and to apportion space among different groups. The creation of specific territories is the basis for political organization and political action. The political partitioning of space creates the most fundamental of geographic divisions: the sovereign state. Most states are recognized as such by other states; their territorial rights are typically respected by others; they are governed by some recognized body; and they have an administration that runs the state.

State Creation

For most people, the most familiar map of the world is probably the one that shows it divided into states. Our familiarity with the concept of international boundaries reflects our recognition that security, territorial integrity, and political power lie enmeshed in this pattern of states. Each state, large or small, is a sovereign unit.

Defining the Nation State

'Nation' is not an easy word to define. As we have seen in Chapters 7 and 8, humans are divided into numerous cultures based on such variables as language, religion, and ethnicity; cultural affiliation is important not least because it provides people with an identity and a sense of community. The term **nation** is potentially applicable to all cultural groups. But not all cultural groups aspire to be nations, and a key question is why some cultures regard themselves as possessing national identity while others do not.

Humans are also divided into formally demarcated political units known as states. The **state** is a set of institutions, the most important being the potential means of violence and coercion; a state also makes the rules that govern life within its territory, encouraging cultural homogenization.

A **nation state** is a clearly defined cultural group (a nation) occupying a spatially defined territory (a state). Each nation state is, in principle, a political territory including all members of one national group and excluding members of other groups. In practice, few nation states fit that strict definition because the vast majority are not composed of just one national group. Nevertheless, the concept of the nation state is important to us for two reasons. First, because the phrase is part of a dominant discourse (see the glossary and our discussion of human geographic concepts in Chapter 2), commentators often use the phrase 'nation state' to refer to the political unit of a state. Second, the concept of the nation state has played, and continues to play, an important role in shaping the contemporary political world.

Predecessors of the Nation State

Our current political divisions are the products of a long evolutionary process that began with claims to territory and that is now characterized by independent states, most of which aspire to be nation states—that is, political groupings of people with a common identity. In Europe before c.1600, the sovereignty of most individual rulers depended on the allegiance of people rather than control over a clearly defined territory; however, the classical Greek and medieval Italian city states were exceptions to this generalization.

Classical Greece was not one political unit but an amalgam of city states. Both Plato and Aristotle gave thought to the size, location, and population of the ideal state. Plato favoured a self-sufficient and inward-looking state, while Aristotle favoured a self-sufficient but outward-looking state. The principal state in Greek times was the Persian empire, a large and loosely structured unit that was built through conquest and dependent on efficient leadership. One of Aristotle's pupils, Alexander the Great, actively pursued the expansionist (outward-looking) policies favoured by his teacher, destroying the Persian empire and creating an even larger empire, although it would be short-lived.

From the retreat of the Romans until the sixteenth century, Europe experienced a succession of empires whose major divisions were typically based on religious grounds (for example, between Christian and Islamic areas and between Christian and pagan areas). Most states in medieval Europe were feudal in structure (see Chapter 8). Their boundaries were uncertain and their spatial extent was constantly subject to change. Political control was usually exercised by a few individuals, and there was a clear social hierarchy. Gradually, these feudal states were replaced by absolutist states that concentrated power in a monarchy (as, for example, in France); starting about two hundred years ago, these states in turn were replaced by nation states.

Historically, then, the fact that a political unit exists does not mean that its boundaries delimit a national identity. Many states (such as city states or tribal units) have been too small to qualify as nations, while other states (such as empires) have been so large as to be multinational.

The Rise of the Nation State

Nationalism

To understand the rise of nation states, we need to refer again to the concept of nationalism. As we saw in Chapter 7, nationalism reflects the belief that a nation (cultural group) and a state (political unit) should be congruent; it assumes that the nation state is the natural political unit and that any other basis for state delimitation is inappropriate. Thus an aspiring nation state will usually argue that

- all members of the national group have the right to live within the borders of the state;
- it is not especially appropriate for members of other national groups to be resident in the state; and
- the government of the state must be in the hands of the dominant cultural group.

It is not difficult to understand how application of these arguments can lead to conflict, whether external (with other states) or internal (between cultural groups). Although this principle of nationalism is taken for granted in the modern world, as our brief historical account of states has indicated, it is a relatively recent idea.

Also important to our understanding of the rise of nation states is the concept of **sovereignty** as it became invested in states. The idea that every state is a sovereign unit became established following the 1648 Treaty of Westphalia, which concluded a long period of European conflict. At that time it was accepted that the ruler of a realm had the authority to determine the religion of its people. Eventually, the concept of sovereignty would contribute to the emergence of independent states.

Explaining nationalism

The map of Europe as it was redrawn in 1815, following the Napoleonic Wars, depicted states demarcated on the basis of dynastic or religious criteria; no real attempt was made to correlate the national identities of populations with their states or their rulers. Indeed, such an exercise would have proved difficult, because the cultural map of Europe was exceedingly complex and cultural groups included many social divisions. Such an exercise would also have been inappropriate, because nationality was not of great importance at the time; people did not question boundaries based on dynastic, religious, or community differences. This raises an important question: Why, despite obvious problems of definition, did national identity emerge as the standard criterion for state delimitation during the nineteenth century? Here are five common theories:

1. Nation states emerged in Europe in response to the rise of nationalist political philosophies during the eighteenth century.
2. Humans want to be close to people of similar cultural background.
3. The creation of nation states was a necessary and logical component of the transition from feudalism to capitalism, as those who controlled production benefited from the existence of a stable state (a Marxist argument).
4. Nationalism is a logical accompaniment of economic growth based on expanding technologies.
5. The principle of one state/one culture arises from the collapse of local communities and the need for effective communication within a larger unit.

In Europe, both Germany and Italy were created in the nineteenth century in response to centrally organized efforts at political union of a national group. The transition came earlier for Britain and France. Britain evolved gradually throughout the eighteenth century, while in France monarchical absolutism was intact until the 1789 revolution; in both of these cases there was an early general correspondence between culture and state.

Police keep separatists (right) and federalists (left) apart during a protest by English rights activists in Montreal demanding English signs at the Eaton's department store in 1998 (CP photo/Ryan Remiorz).

Nation states in the contemporary world

Despite the transition to nation states, our contemporary political map still includes many states that contain two or more nations. Among the multinational states are many African countries whose boundaries were drawn by Europeans without reference to African national identities. Although nationalistic ideology has been clearly articulated in the context of anticolonial movements, no substantial changes to the European-imposed boundaries have been achieved to date.

Many multinational states are politically unstable, prone to changes of government and/or expressions of 'minority nation' discontent. Canada and Belgium are examples of politically uncertain binational states: both include more than one language group and both are experiencing internal stresses related to the differing political aspirations of these groups. States that approach the nation-state ideal, such as Italy and Denmark, are found

Figure 9.1 The British empire in the late nineteenth century.

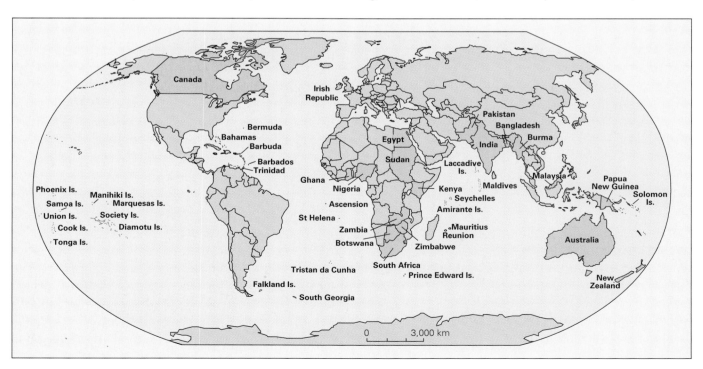

primarily in Europe. States that incorporate members of more than one national group are not necessarily unstable; the United States has largely succeeded in creating a single nation out of disparate groups, while Switzerland is probably the prime example of a stable and genuinely multinational state.

Exploration and Colonialism

Clearly, not all states in recent centuries have aspired to the nation-state model. European countries in particular attempted to expand their territories overseas to create empires similar to those that existed in earlier periods in Europe and elsewhere. Such empires are multinational by definition, and were typically regarded as additions to rather than replacements for nation states.

European countries began developing world empires about 1500, and most such empires achieved their maximum extent in the late nineteenth century. During the twentieth century, most of these empires have been dissolved. Figure 9.1 shows the British empire as it was in the late nineteenth century, when large areas of the world were under British control. Today the British empire is limited to a few islands, mostly in the Caribbean and south Atlantic (Box 9.1).

Exploration

Most empires began as a result of exploratory activity (see Box 1.3 on the contentious term 'exploration'). In general, geographers have paid little attention to exploration—defined as the expansion of knowledge that a given state has about the world—other than to list and describe events. One geographer, however, has derived a conceptual framework for the process of exploration, a framework that focuses on the links between events (Overton 1981). As depicted in Figure 9.2, the process begins with a demand for exploration and may, in due course, lead to the 'development' of the new area. Such 'development' is likely to mean exploitation of the explored area as a source of raw materials needed by the exploring country. Britain, for example, viewed Canada as a source of fish, fur, lumber, and wheat, and Australia as a source of wool, gold, and wheat.

9.1

REMNANTS OF EMPIRE

In recent years, two former colonies have changed their political status in accord with an agreed-upon timetable: Hong Kong was transferred from Britain to China in 1997, and Macao was transferred from Portugal to China in 1999. However, approximately 40 other political units, representing some 3 million people, are still best described as colonies. Although some colonies currently controlled by France, Denmark, the Netherlands, and New Zealand are likely to become independent as economic circumstances change, others are likely to remain politically dependent for some time to come, for two general reasons.

First, some areas are perceived as particularly valuable to the colonial power. Both France and the United States use islands in the Pacific for military purposes, and the US is known to be reluctant to relinquish control over Guam and Diego Garcia.

Second, it appears that some colonial areas would simply not be viable as independent states. A prime example is St Helena, an island of 122 km^2 (47 square miles) with a population of 5,700 located in the south Atlantic Ocean (see Figure 9.1).

First reached by the Portuguese in 1502, the island later became the object of competition between the Dutch and the British, and today it remains part of the much-reduced British empire. The competition reflected St Helena's location, which made it valuable as a stopover on the Europe–India route. Over time, movements of slaves from east Africa and indentured labour from China, as well as British and some other European settlement, created an ethnically diverse population who supported themselves by providing food and other supplies for ships. By the late nineteenth century, however, technological advances such as steamships and refrigeration, and the opening of a new route using the Suez Canal in 1869, had made St Helena unnecessary as a stopover, and the economic problems soon became evident. With a very limited resource base, St Helena suffers from its small size and small population. For much of the twentieth century the agricultural industry was dominated by flax; at times, half the working population was employed in flax production. Today, however, agriculture is limited to livestock and vegetables produced for subsistence purposes only.

The future of St Helena is uncertain. It is isolated, dependent on financial support from Britain, and unlikely to become a tourist attraction. From the British perspective, it is no longer a prestigious overseas possession but a financial liability (Royle 1991).

Figure 9.2 Principal elements in
the process of exploration.
Source: Adapted from J.D. Overton,
'A Theory of Exploration', *Journal of
Historical Geography* 7 (1981):57.

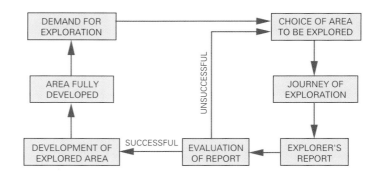

Colonialism

The way a new area was used by the explor-ing country was typically a function of per-ception and need. Economic, social, and political activity in explored areas that became colonies was typically determined by and for the exploring power. This was true of European empires and of the territories acquired by the United States from 1783 onwards (Figure 9.3). In most cases, Aboriginal populations were eliminated or moved if their numbers represented a mili-tary or spatial threat. The economic activities encouraged were the ones that benefited the central power. This is the process known as colonialism (see Chapter 6).

In recent European history, colonialism took the form of territorial conquest throughout the Americas, much of Asia (including Siberia), and most of Africa. Competition for colonies was a major feature of the world political scene from the fif-teenth to the early twentieth century. The earliest colonial powers were Spain and Portugal, followed by other west European states, the United States, and Russia.

Explaining colonialism

The reasons behind colonial expansion are complex; although often summarized as 'god, glory, and greed', they may also have includ-ed various changes in Europe involving the prevailing feudal system and economic com-petition at the time. It is worth recalling that, shortly before the beginning of European colonialism, China and the Islamic world were the two leading world areas; indeed, gunpowder, the mariner's compass, printing, paper, the horse harness, and the water-mill were all invented in China. The fact that Europe was the region that initiated global movements appears to be related to the

demands of the economic growth that began in the fifteenth century, as well as to internal social and political complexity, turmoil, and competition. Reasons for the colonial fever that swept Europe, especially in the nine-teenth century, included the ambitions of individual officials, special business interests, the value of territory for strategic reasons, and national prestige. But perhaps the most compelling motive was economic: colonial areas provided both the raw materials need-ed for domestic industries and additional markets for industrial products.

Colonialism effectively ended following the Second World War. The defeat of Japan and Italy immediately removed them as colo-nial powers, and most of the remaining parts of the British empire achieved independence over the next few years. Three European powers—France, the Netherlands, and Portugal—tried to retain their colonies, but all were unsuccessful.

Decolonization

In recent years the number of states in the world has increased, from 70 in 1938 to more than 190 in 2003. Most of the new states have achieved independence from a colonial power. Many of them reflect national groupings, but some are areas around which Europeans drew boundaries for their own reasons.

As we have seen, the idea of a national identity originated in Europe, but it has been welcomed by colonial areas that were dis-contented with colonial rule, aspired to inde-pendence, and suffered psychologically from foreign rule. The transition to independence has been violent in some cases and peaceful in others, depending in part on the way indi-vidual European powers responded as the independence movements in their colonies

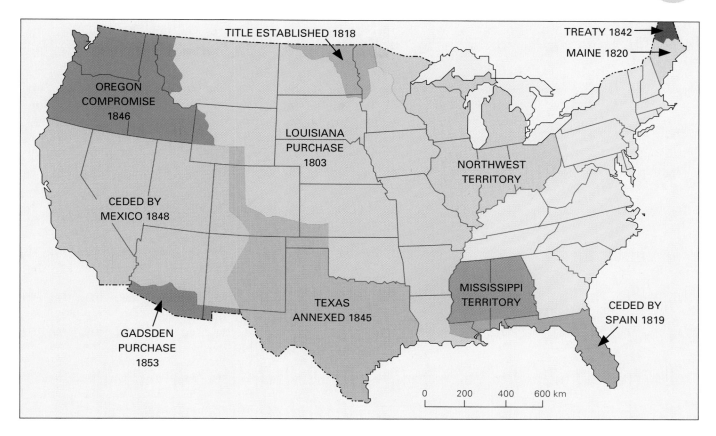

gathered momentum in the 1950s. Most of the former British colonies experienced a relatively smooth transition to independence, and almost all of them are now members of the Commonwealth, a voluntary organization of 54 nations. Some other former colonies, however, achieved independence only after long civil wars; examples include Algeria (France), Mozambique (Portugal), and the Congo region (Belgium).

Effects of colonialism

It appears that many colonies may have resulted in net losses to the national economies of the colonial powers, benefiting the stock exchange and a few individuals rather than state treasuries. But there were other benefits. For example, the vast expanses of Canada, Australia, and South Africa offered a way for Britain to reduce population pressure. The effects on the areas colonized are complex. As we saw in Chapters 5 and 6, many areas have experienced massive population growth and associated economic problems because of the colonial experience and related reductions in death rates. Many believe that **imperialism**—in which one powerful state or terri-

tory seeks to control another, weaker territory—has had extremely negative consequences for the less powerful area. This is one of the central arguments of the world systems theory introduced in Chapter 6. A related set of ideas, **dependency theory**, contends that African and Asian countries became poor as a result of their colonization, while the colonial powers advanced at their expense. Specifically, dependency theory claims that less developed countries depend for their survival on their links with dominant countries in the more developed world—links formed during the colonial period, when occupation by the imperialist powers led to the disintegration of indigenous cultures and the formation of the less developed world.

State Creation: Some Conceptual Discussions

Philosophical discussions of state creation and expansion can be traced as far back as Plato and Aristotle; Strabo took a favourable view of the Roman empire; ibn-Khaldun distinguished between the territories of nomadic and sedentary peoples; and Bodin and Montesquieu both favoured the emerg-

Figure 9.3 Territorial expansion of the United States. The phrase 'manifest destiny' reflected the belief that the new United States had an exclusive right to occupy North America. In 1803 President Jefferson purchased the Louisiana territory from France. Spain ceded the Florida peninsula in 1819, while war with Mexico resulted in the acquisition of Texas and the southeast. The Gadsden purchase of 1853 completed the southern boundary of the US. The northern boundary along the 49th parallel was established in 1848. Subsequently, American interests expanded north to Alaska and across the Pacific to include Hawaii.

ing nation-state model. Three more recent discussions are outlined below.

Ratzel

The great early nineteenth-century German geographer Ritter viewed states as organisms. He saw both cultures and states as progressing through a life cycle, an idea that Ratzel would elaborate on. Ratzel's seven laws concerning the spatial growth of states (Box 9.2) are not the product of rigorous scientific logic; rather, they are generalizations based on observations of a supposed ideal world. Central to Ratzel's thinking was the notion of the state as a living organism. Although this notion was criticized even in Ratzel's day on the grounds that it reified (assumed the independent existence of) something that was actually a human creation, Ratzel's ideas have been highly influential, both conceptually and practically.

Jones

A rather different view of states was developed by S.B. Jones (1954) on the basis of some ideas put forward earlier by Gottman, Hartshorne, and the political scientist Deutsch. Jones proposed a chain of events beginning with some political idea and concluding with the creation of a political area (Figure 9.4). In this way, the idea of state creation may lead in due course to actual state creation. Between idea and area are a positive decision to implement the initial idea; movement of some variety following a political decision; and a field of activity, created by that movement, which ultimately leads to the delimitation of a political area. Jones's application of these ideas to the creation of the state of Israel is outlined in Box 9.3: the political idea is Zionism; the political decision is the Balfour Declaration; the principal movement is the immigration of Jews; the resulting field is settlement and government activity; and the political area is Israel.

Figure 9.4 Jones field theory.

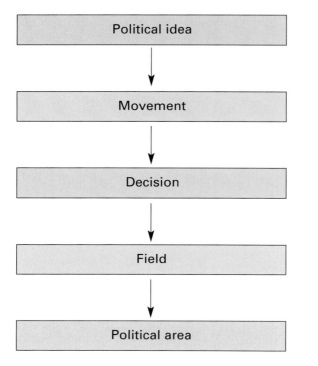

9.2

LAWS OF THE SPATIAL GROWTH OF STATES

Ratzel's 'laws', published in 1896, are in fact generalizations:

1. The size of a state increases as its culture develops.
2. The growth of a state is subsequent to other manifestations of the growth of a people.
3. States grow through a process of annexing smaller members. As this occurs, the human–land relationships become more intimate.
4. State boundaries are peripheral organs that take part in all transformations of the organism of the state.

5. As a state grows, it strives to occupy some politically valuable locations.
6. The initial stimulus for state growth is external.
7. States' tendency to grow continually increases in intensity.

The key idea is that a state grows as its level of civilization rises.

These laws present an interesting parallel with Ravenstein's laws regarding migration (Box 6.2). In neither case is the logic employed strictly deterministic.

THE JEWISH STATE

The Jews are a religious group with roots in the area that is now Israel. Jerusalem, in particular, is an important symbolic location for Jews—and for Christians and Muslims. During the time of Jesus, this area (known as Palestine) was under Roman rule. In 636 it was invaded by Muslims and from then until 1917 (with a brief Christian interlude during the Crusades), it was a Muslim country, specifically Turkish after 1517. Over the centuries, in a process known as the Diaspora, Jewish populations moved away to areas throughout much of Europe and overseas. At the end of the First World War (1914–18), Syria and Lebanon were taken over by France, and Palestine and Transjordan (now Jordan) were taken over by Britain, under League of Nations (the predecessor of the United Nations) mandates. The Palestine mandate—specifically the Balfour Declaration of 1917—allowed for the creation of a Jewish national home but without damaging Palestinian interests. An impossible goal was being set.

In 1920 Palestine included about 60,000 Jews and 600,000 Arabs. Most of the Jews had arrived since the 1890s. Jewish immigration increased under the mandate, especially as persecution of Jews in Hitler's Germany intensified. After the Second World War, Jewish survivors struggled to reach Palestine, which by 1947 included 600,000 Jews, 1.1 million Muslim Arabs, and about 150,000 Christians (mostly Arab). Arab–Jewish conflicts were by now commonplace, and Britain, unable to cope, announced that it intended to withdraw.

The United Nations then produced a plan to partition Palestine into a Jewish and an Arab state, with Jerusalem as an international city. The Arabs did not accept the plan, and the Jews declared the independent state of Israel (1948). Conflict immediately flared between Israel and nearby Arab states and between Jews and Palestinian Arabs, and has continued to the present. One longstanding issue of contention is Israel's occupation of the West Bank (of the Jordan River) and Gaza Strip, which it captured (from Jordan and Egypt respectively) during the Six Day War of 1967. Although Israel has established relations with Egypt (in 1979), the Palestine Liberation Organization (in 1993), and Jordan (in 1994), serious obstacles to peace remain. Among them are the absence of diplomatic relations between Israel and Syria, the uncertain status of Jerusalem, the continuing presence of Jewish settlements in the West Bank and Gaza, and, finally, ongoing terrorist actions against Israel by Palestinian militants and Israeli military actions against the Palestinians. Significantly, in 2002 the United States acknowledged the need for an independent, democratic, and viable Palestinian state existing in peace and security with Israel and its other neighbors.

Israeli police investigators search for evidence in front of a Jerusalem café following a suicide bombing in September 2003. The explosion killed at least four people and wounded approximately forty others (AP photo/Lefteris Pitarakis).

A Palestinian boy searches for his family's belongings in the rubble of a four-storey building at the Nusseirat refugee camp in the Gaza Strip, March 2003. The building was destroyed in a battle between Israeli troops firing from tanks and helicopters and dozens of Palestinian gunmen. Seven Palestinians, including a four-year-old child, were killed. Scenes like these two reflect the seemingly endless cycle of violence between Israel and the Palestinians (AP photo/Karel Prinsloo).

Deutsch

A third concept was proposed by Deutsch (see Kasperson and Minghi 1969:211–20), who describes a process of state creation that can involve as many as eight stages:

1. transition from subsistence to exchange economy;
2. increased mobility leading to formation of core areas;
3. development of urban centres;
4. growth of a network of communications;
5. spatial concentration of capital;
6. increasing group identity;
7. rise of national identity;
8. creation of a state.

Both Jones and Deutsch emphasize evolution and focus primarily on human actions. Neither regards states as organisms. As we have seen, nations do not exist as conveniently packaged bundles of people waiting for state boundaries to be drawn around them (Taylor 1989). Nations and states are both human creations; explaining the existence of nations and national identity requires consideration of ethnic distributions and the political construction of nations. An example is contemporary Quebec, where aspirations for a separate political identity appear on occasion to be politically orchestrated. Box 9.4 presents a discussion of the process by which the Canadian state came into being, a discussion that includes a demonstration of a thought-provoking—if intellectually risky—approach known as the **counterfactual** method.

These conceptual discussions bring us to one of the most exciting and provocative aspects of the study of our political world: geopolitics.

Geopolitics (and *Geopolitik*)

Geopolitics is the study of the relevance of space and distance to questions of international relations. Geopolitical discussions originated in the late nineteenth century, but manipulation of the concept by some German geographers and Nazi leaders in the 1930s and 1940s led others to reject it. Only since the 1970s has geopolitics revived and once again become a legitimate area of interest.

The intellectual origins of geopolitics lie in the work of Ratzel, specifically in his seven laws of state growth. The term 'geopolitics' was coined early in the twentieth century by a Swedish political scien-

9.4

A DIFFERENT CANADA?

The renowned historical geographer A.H. Clark has posed an intriguing question about Canada (Clark 1975): What would have happened if the French explorer Samuel de Champlain had not established a French empire on the St Lawrence in 1608?

His 1603 voyage up the St Lawrence as far as Montreal island left Champlain, an excellent geographer, unimpressed with the St Lawrence as an entry point to North America. Hence he spent the summers of 1604, 1605, and 1606 exploring the North American coastline south of his winter base in Port-Royal (Nova Scotia), seeking a better entry point. All three expeditions were characterized by misfortunes and delays. In each case, Champlain was unable to leave Port-Royal as early as planned because supply ships from France had not arrived and his men were still suffering the after-effects of disease-ridden winters. As a result of these delays, together with the meticulous care he took in surveying, he never travelled as far south as the mouth of the Hudson River (present-day New York City).

Clark argues—convincingly—that only a series of unfortunate circumstances prevented Champlain from reaching the Hudson, and that history would have been very different if he had done so.

First, Champlain would surely have recognized the value of the Hudson as an entry point and committed the French to that location rather than the St Lawrence. Now let fantasy rule for a few moments. Champlain could have travelled the Hudson to its confluence with the Mohawk (present-day Albany) and formed an alliance with the local Five Nations people. The result could have been a powerful French core area, a better environment for the seigneurial land-holding system, and a generally more attractive region for French settlement. Conceivably, a French empire based in what is now New York State might have been indestructible.

Canada would be a very different place; indeed, it would be *in* a different place. This intriguing argument should encourage us to appreciate the value (and risk!) of creative thinking. In human geography, the actual and the possible both merit consideration.

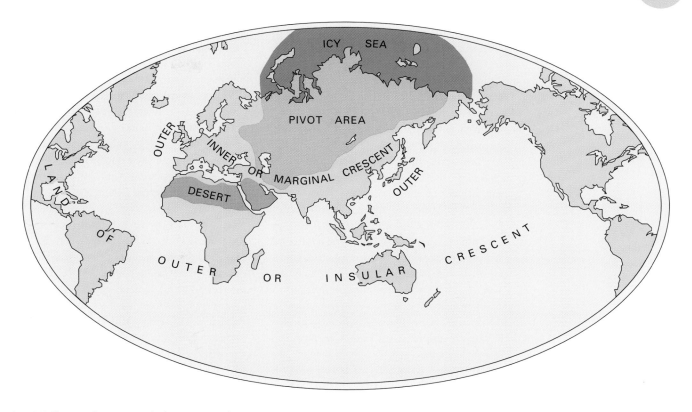

ICY   SEA

PIVOT   AREA

OUTER

INNER   OR   MARGINAL   CRESCENT

OUTER

LAND

DESERT

OF

OUTER   OR   INSULAR   CRESCENT

tist, Kjellen, who expanded on Ratzel to argue that territorial expansion was a legitimate state goal. Clearly, this argument is only one aspect of the larger field, the relevance of space and distance to questions of international relations. It was, however, this narrow interest that dominated geopolitical discussions until they were largely abandoned in the 1940s.

Geopolitical Theories

A leading British geographer, Mackinder, writing in 1904, was the first to formulate a geopolitical theory. Mackinder's **heartland theory** has strong environmental determinist overtones, reflecting British concerns about perceived Russian threats to British colonies in Asia, especially India. Mackinder contended that the Europe–Asia land mass was the 'world island' and that it comprised two regions: an interior 'heartland' (pivot area) and a surrounding 'inner or marginal crescent' (Figure 9.5):

> Who rules East Europe commands the Heartland;
> Who rules the Heartland commands the World Island;
> Who rules the World Island commands the World.
> (Mackinder 1919:150)

Mackinder was arguing that location and physical environment were key variables in any explanation of world power distribution. The theory is flawed because it overemphasizes one region (eastern Europe) and because Mackinder could not anticipate the rise of air power, but it did influence other writers in general and may well have exercised a continuing influence on the United States' policy in particular.

Certainly the earlier work of Ratzel, Kjellen, and Mackinder influenced the rise of *geopolitik* in the 1920s. **Geopolitik** is a specific interpretation of more general geopolitical ideas. It focuses on the state as an organism, on the subordinate role played by individual members of a state, and on the right of a state to expand to acquire sufficient *lebensraum* (living space). The individual most responsible for popularizing these ideas was Haushofer, a German geographer in Nazi Germany who was bitterly disappointed by the territorial losses Germany experienced as a result of the First World War. Haushofer's academic justification for German expansion coincided with Nazi ambitions. The extent of Haushofer's influence on actual events is uncertain, but it is the case that the German interest in *geopolitik* resulted in a general disillusionment with all geopolitical issues.

Figure 9.5 Mackinder heartland theory.

Despite this disillusionment, several scholars continued to present their views on the global distribution of power. Spykman, writing in the 1940s, built on Mackinder's 'inner or marginal crescent' (rimland) to develop a **rimland theory**, arguing that the power controlling the rimland could control all Europe and Asia and therefore the world. As an American, Spykman saw considerable advantage in a fragmented rimland—a view that has certainly influenced post-Second World War American foreign policy in Asia (Korea and Vietnam) and in eastern Europe. Also during the 1940s, de Seversky stressed that the US needed to be dominant in the air to ensure state security.

Since the resurrection of geopolitics in the 1970s, numerous discussions have centred on the 'new geopolitics' and a critical geopolitics that places emphasis on the social construction of national and other identities. The emphasis on a culturally informed geopolitics is especially interesting in light of the new cultural geography introduced in Chapter 8.

New World Order—or Disorder?

During the **Cold War** period, from the end of the Second World War until the early 1990s, the world was ideologically divided between the democratic capitalist states belonging to the US-dominated North Atlantic Treaty Organization (NATO) and the communist states belonging to the USSR-dominated Warsaw Pact. This bipolar division no longer exists: NATO has expanded from its original 12 member countries to a total of 26, and all of the new members are former communist states in Eastern Europe. Other new members are likely in the near future.

The first stage in this geopolitical transition was Poland's installation of a non-communist government—approved by the USSR—in 1989. Events followed with remarkable speed: the collapse of communist governments elsewhere in eastern Europe, the symbolic breaching of the Berlin Wall on 9 November 1989, the reunification of Germany in 1990, and the collapse of communism in the USSR in 1991. We are now living within a new and uncertain geopolitical world order.

What has followed the cold war? Some political geographers continue to focus on conventional geostrategic issues; deBlij (1992) is making renewed use of Mackinder's heartland model, while Cohen (1991) sees eastern Europe as a 'Gateway Region' between maritime and continental areas. Other political geographers, however, propose a geoeconomic view of the world (O'Loughlin 1992). They see the world as consisting of three major core regions— North America, western Europe, and east Asia—each of which is undergoing a process of economic integration: respectively, the North American Free Trade Association (NAFTA), the European Union (EU), and the Association of Southeast Asian Nations (ASEAN). At the same time, the fundamental divisions—between north and south, more developed and less developed—remain, with all the potential for conflict they imply; see Kaplan (1993, 1996) and Chapter 6. We will revisit the unfolding geopolitical structure of the world in the conclusion to this chapter.

The Stability of States

Maps of our political world are subject to frequent and often dramatic change, whether as a result of war or, less dramatically, as a result of economic agreements. Many factors may affect the stability of a given state.

Centrifugal and Centripetal Forces

The geographer Hartshorne (1950) distinguished between centrifugal and centripetal forces. **Centrifugal** forces tear a state apart; **centripetal** forces bind a state together. When the former exceed the latter, a state is unstable; when the latter exceed the former, a state is stable. The most common centrifugal forces are those involving internal divisions in language and religion that lead to a weak *raison d'être* or state identity. Other centrifugal forces include the lack of a long history in common (the case in many former colonies), and state boundaries that are subject to dispute. The most common centripetal force is the presence of a powerful *raison d'être*: a clear and widely accepted state identity. Other centripetal forces include a long state history and boundaries that are clearly delimited and accepted by others. The various centrifugal or centripetal forces are often closely related; for example, the presence of internal divisions generally indicates that groups do not share a long common history and that boundaries are not agreed upon. There are also close links between stability and peace and between instability and conflict within and between states.

Internal Divisions

Secessionist movements arise when nations within multinational states want to create their own separate states; examples include the Québécois in Canada, the Flemish and Walloons in Belgium, and the Welsh and Scottish nationalist movements in the United Kingdom. In other cases 'nations within' may want to link with members of the same nation in other states to create a new state;

the Basques in Spain and France and the Kurds in Iraq, Iran, Syria, Turkey, Armenia, and Azerbaijan are two examples (Box 9.5). **Irredentism** is the term used in a third situation, when one state seeks the return from another (usually neighbouring) state of people and/or territory formerly belonging to it; Somali irredentism is an example.

Nation and state discordance in Africa

Discordance between nation and state is especially evident in Africa (Figure 9.6). There are only a few instances where a distinct national

9.5

THE PLIGHT OF THE KURDS

According to some commentators, the biggest losers in the Middle East over the past 100 years have been the Kurds. Since the collapse of the Ottoman (Turkish) empire at the end of the First World War, other groups—Turks, Arabs, Jews, and Persians— have consolidated or even created their own states. Yet more than 20 million Kurds remain stateless, dispersed in Turkey (10.3 million), Iran (4.6 million), Iraq (3.6 million), Syria (1 million), and in smaller numbers in other neighbouring states. Their failure to achieve state identity is all the more surprising given that, in 1920, Britain, France, and the United States all acknowledged the need for a Kurdish state: 'No objection shall be raised by the main allied powers should the Kurds . . . seek to become citizens of the newly independent Kurdish state' (Article 64, Treaty of Sèvres, signed 20 August 1920, quoted in Evans 1991:34).

Why have the Kurds not succeeded in imposing their identity on a region to create a state? Part of the answer lies in the question of Kurdish identity itself. Most Kurds are Sunni Muslims, but not all; their society, located in a mountain environment, is typically poor and divided tribally; their dialects are varied; and there is no one political party to speak for all of them. In short, it could be argued that the Kurds are united more by persecution than by any sense of shared identity. A second part of the answer is the fact that a Kurdish state would weaken the authority of all the states in which Kurds now live. To discourage the development of Kurdish nationalism, Turkey banned the use of the Kurdish language in schools until 2002 (when it abandoned that policy as part of its unsuccessful effort to gain entry to the European Union), and since the 1990s has resisted any suggestion of independence for Iraqi Kurds living next to its own Kurdish population.

The Kurds' preferred location for their own state would be in the mountainous border regions of Turkey, Iraq, and Iran.

Members of a Kurdish family prepare to leave their home in the Kurdish town of Chamchamal, near Kirkuk in northern Iraq, in March 2003. Local residents feared a chemical attack by Saddam Hussein's forces in the event of a US-led war (AP photo/Newsha Tavakolian).

Unfortunately, this is an area valued by existing states for its resources, which include oil supplies at Kirkuk. Iraq has used chemical warfare against the Kurds and, following the end of the 1991 Gulf War, mounted an extensive campaign to slaughter the Kurds in Iraq, which caused massive population movements into Turkey. Ironically, Turkey is far from a safe haven: in 1979, the country's prime minister stated: 'the government will defeat the disease (of Kurdish separatism) and heads will be crushed' (quoted in Evans 1991:35). The future is not promising. As of 2003, the future remains uncertain. The 2003 war in Iraq may lead to some degree of Kurdish autonomy in that country, but the larger picture is unclear and not promising.

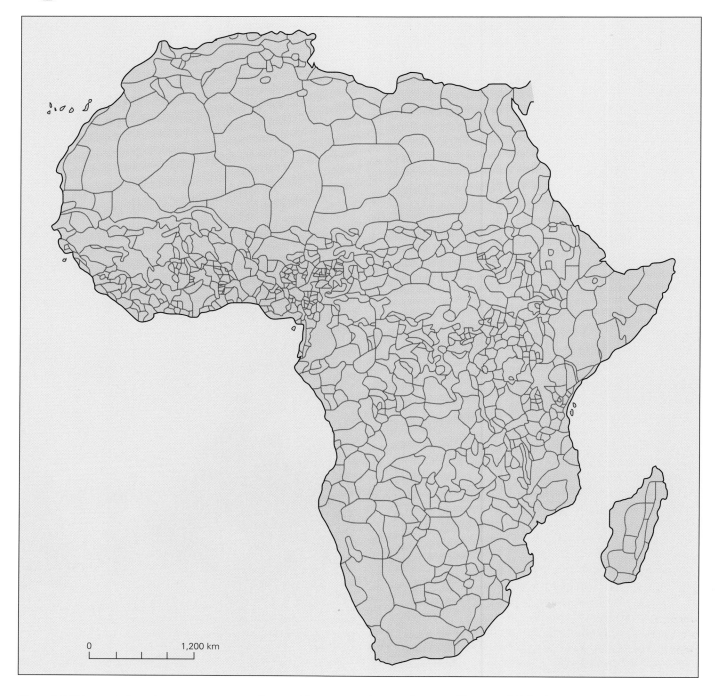

Figure 9.6 African ethnic regions.
Source: L.D. Stamp and W.T.W. Morgan, *Africa: A Study in Tropical Development*, 3rd ed. (New York: John Wiley & Sons, 1972):41. Reprinted by permission of John Wiley & Sons, Inc.

group correlates with a state; the southern African microstates of Lesotho and Swaziland are examples. Ironically, perhaps, in recent years it has been suggested that Lesotho might join South Africa (Lemon 1996). The principal argument in favour of Lesotho's sacrificing its identity in this way is that it is the only sovereign state in the world, apart from the microstates of San Marino and the Vatican, to be completely surrounded by a single neighbour. Among the many African nations that once had related states but no longer do are the

Hausa and Fulani nations in Nigeria, the Fon in Benin, and the Buganda in Uganda. The states of contemporary Africa are not products of a long African history, but creations of colonialism that subsequently achieved independence. State boundaries, shapes, and sizes are all colonial creations, reflecting past European rather than African interests. In consequence, Africa is characterized by political fragmentation. Some states have a high degree of contiguity (when there are many states, most states have many neighbours); some are very small in

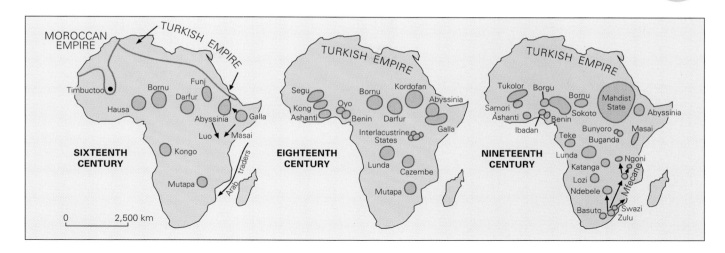

Figure 9.7 African political areas in the sixteenth, eighteenth, and nineteenth centuries.

size and/or population; some have awkward shapes and long, often environmentally difficult boundaries; and some lack access to the sea. Each of these difficulties is related to the colonial past, a past that ignored national identity in the process of creating states.

National identity was evident in many of the precolonial African states. Two early states in west Africa were Ghana and Mali (names later given to modern states covering differ-

ent areas); both of these states succumbed to Islamic incursions in the eleventh century. States continued to grow and decline. Figure 9.7 shows maps for the sixteenth, eighteenth, and nineteenth centuries. The colonial impact is especially clear when we compare Figures 9.6 and 9.7; European neglect of African national identities is a prime cause of African nationalism and contemporary instability in many African states (Box 9.6).

TRIBES, ETHNICITY, POLITICAL STATES, AND CONFLICT IN AFRICA

9.6

Culturally and politically, precolonial Africa was organized in tribes, defined by kinship links and shared cultural characteristics, rather than states as in Europe. Yet there is little evidence of precolonial tribal conflicts. By creating states based on factors other than tribal identity, the colonial powers disrupted African tribal identities and provided a basis for the many tribal or—perhaps more accurately—ethnic conflicts that are evident today.

At the root of recent and current African conflicts was the tendency of colonialism to privilege one group over others and thus to stimulate competition over resources. During the decolonization phase, it was not unusual for different groups to unite against a common enemy, the colonial power. But independence has in many cases exposed the fragility of such alliances; not surprisingly, African countries have not been able to construct nations out of different ethnic groups. Overall, African countries find it difficult to keep in place the many politically fragile states that Europeans created. Their task has not been facilitated by the many problems—including rapid population growth, inadequate food supplies, and declining prices for commodities—that they have encountered. Tragically, violence becomes widespread as ethnic groups compete for resources and political power.

In central Africa, conflict erupted in Rwanda in 1994, and a related conflict began in the Democratic Republic of Congo (formerly Zaire) in 1996. The two principal competing ethnic groups, Hutu and Tutsi, are not clearly defined; indeed, they speak the same language and share many cultural characteristics. Prior to the colonial period, they lived side by side without significant conflict. But European colonialism created four countries—Rwanda, Burundi, the Democratic Republic of Congo, and Tanzania—each containing both Hutu and Tutsi populations. Europeans then reinforced the ethnic division by favouring the Tutsis, whom they regarded as superior; hence today Hutus and Tutsis occupy different social positions. Differences between these two groups were not the cause of their recent conflicts. Rather, the causes included the lack of human development, high population densities, lack of agricultural land, and a collapse in the price of coffee, the region's most important commercial crop. The conflict assumed an ethnic character, but was not directly caused by ethnic hostility.

Unfortunately, it appears that once ethnic violence begins, it encourages additional and more widespread ethnic violence. The postcolonial era in central Africa, therefore, has witnessed a cycle of ethnic killings.

Figure 9.8 Areas with nationalist/separatist movements in Europe.

Shelled buildings stand empty and desolate in Bosnia (CIDA photo/Roger Lemoyne).

The future of African political identities remains uncertain. The Organization of African Unity was established in 1963 with the explicit goal of providing a common voice for the emerging independent African states. Sadly, it failed; most commentators found that in practice it did more to impede than to promote democracy, human rights, and social and economic growth. In 2002 the OAU was replaced by the African Union. There is real hope that this new organization will succeed in enhancing national identities, encouraging cooperation, and permitting Africa to speak with one voice in a global context.

Nations and states in Europe

In Africa, then, present-day postcolonial states rarely reflect nationalist aspirations. The situation is different in Europe, where present-day states correspond more closely to national identity. There are nevertheless many instances of internal division, most of which are based on language and/or religion. Rokkan (1980) identifies four functional prerequisites for the existence of a state—economy, political power, law, and culture—and argues that the tensions between European cores and their peripheries reflect the fact that peripheries have less political power, less developed legal structures, and less dominant cultures. He identifies two axes: a north–south cultural axis from the Baltic to Italy, and an east–west economic axis. The north–south cultural axis was Protestant (a religion that favoured national aspirations) in the north and Catholic (a religion that cut across national boundaries) in the south. Along the east–west economic axis, the west had key commercial centres (good partners in state formation), but in the east the urban centres were economically weak (and hence unable to offer the resource base needed for the building of states). Arguing that the success of efforts at nation-building depended on cultural conditions, whereas the success of efforts at state-building depended on economic conditions, Rokkan proposed that the north–south cultural axis determined the former, while the east–west axis determined the latter. Hence the most favourable conditions for building nation states existed in the northwest.

Contemporary nationalist movements in western Europe can thus be seen as relics of a Rokkan-type core/periphery cleavage in areas not yet homogenized into an industrial society with its related class divisions. In our present context, this view is certainly useful.

Separatist movements in Europe

Regionalism and associated efforts to assert separate identities remain strong in Europe (Box 9.7). Figure 9.8 shows several areas with political parties whose policies are explicitly regional and nationalist. Some favour separation; others, self-government within a federal state structure, perhaps by means of **devolution** (the transfer of power from central to regional or local levels of government). In most of these European cases, language is a key factor. However, it is worth noting that

CONFLICTS IN THE FORMER YUGOSLAVIA

From the mid-fifteenth to the mid-nineteenth century, the area that was to become the federal state of Yugoslavia was part of the Ottoman empire. By the beginning of the twentieth century, some parts—notably Serbia—were independent, while others belonged to the Austro-Hungarian empire. At the end of the First World War, Yugoslavia was founded as a union of south (Yugo) Slavic peoples, although the name was not adopted until 1929. Following the Second World War, a socialist regime was established under the leadership of Tito, who dominated the political scene until his death in 1980, after which a form of collective presidency was established. The ethnically diverse area of Yugoslavia was organized as six socialist republics and two autonomous provinces (Table 9.1).

The ethnic diversity evident in Table 9.1 is further complicated by the major European cultural division between eastern and western Christianity that runs through former Yugoslavia as well as through Romania, Ukraine, and Belarus (Figure 9.9). It is therefore not surprising that this federal state was dissolved in 1991, following Serbia's efforts in 1988 to exert a greater influence in the federation by assuming direct control of the two autonomous provinces (Kosovo and Vojvodina).

The process of dissolution began in 1991 with the secession of Slovenia, which was quickly followed by that of Croatia. In Bosnia bitter conflict between Croatian Muslims and Serbs introduced that terrible euphemism 'ethnic cleansing'. In 1999 violence flared in the southern province of Kosovo as Serbians and ethnic Albanians contested an area that has meaning to both groups; western powers intervened on behalf of the Albanian population. Perhaps the final chapter in this prolonged period of conflict and political change came in 2003, when the republics of Serbia and Montenegro decided to drop the name 'Yugoslavia' from the map. Following the collapse of the Ottoman, Austro-Hungarian, and Soviet empires in the twentieth century, the demise of Yugoslavia might be seen as marking a symbolic end to efforts to bind multiple ethnic groups into artificial single entities.

Identity—national, ethnic, linguistic, religious—has been a central feature of recent conflicts in the former Yugoslavia. It was with this thought in mind that S.P. Huntington described the conflicts as one example of a 'fault line' war:

Croats saw themselves as the gallant frontier guardians of the West against the onslaught of Orthodoxy and Islam. The Serbs defined their enemies not just as Bosnian Croats and Muslims but as 'the Vatican' and as 'Islamic fundamentalists' and 'infamous Turks' who have been threatening Christianity for centuries. . . . The Bosnian Muslims, in turn, identified themselves as victims of genocide, ignored by the West because of their religion, and hence deserving

of support from the Muslim world. All the parties to, and most outside observers of, the Yugoslav wars thus came to see them as religious or ethnoreligious wars (Huntington 1996:270–1).

One especially tragic component of the conflicts that erupted as Yugoslavia disintegrated was the practice of **genocide** ('ethnic cleansing'). The most notorious example of genocide in that war was carried out in 1995 at Srebenica—in a supposedly safe area under the control of UN peacekeeping forces— where 23,000 Bosnian Muslim women and children were expelled by Serbs and over 7,000 Muslim men and boys murdered, in some cases after being blindfolded and bound (Wood 2001).

Ethnic tensions have also surfaced within the larger region of the Balkans. Since the collapse of their communist regimes, Bulgaria has felt threatened by its Turkish population and Romania by its Hungarian population.

Table 9.1 ETHNIC GROUPS IN THE FORMER YUGOSLAVIA

Republic

Serbia	85% Serbs
Croatia	75% Croats, 12% Serbs
Slovenia	91% Slovenes
Bosnia-Herzegovina	40% Slavic Muslims, 32% Serbs, 18% Croats
Montenegro	69% Montenegrins, 13% Slavic Muslims, 6% Albanians
Macedonia	67% Macedonians, 20% Albanians, 5% Turks

Province

Kosovo	77% Albanians, 13% Serbs
Vojvodina	54% Serbs, 19% Hungarians

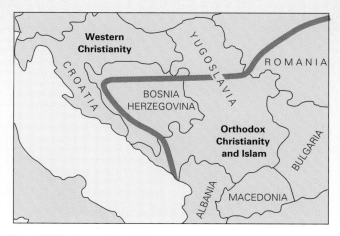

Figure 9.9 The former Yugoslavia.

Supporters of the Basque nationalist party Batasuna, out-lawed in Spain, march through Saint-Jean-Pied-de-Port in southwestern France, carrying portraits of Basque prisoners and a banner that reads: 'Basques are not legal / Let the people talk' in April 2003. Straddling the border between France and Spain, the Basque region has been the subject of a vocal and sometimes militant independence movement since the 1950s. The fact that their language does not belong to the Indo-European family is one indication of the Basque people's distinctiveness (AP photo/Bob Edme).

most of the areas asserting a distinct identity are peripheral and economically depressed:

Northern Ireland: Since partition in 1921, a large Catholic minority has sought to separate from the United Kingdom and integrate with predominantly Catholic Ireland. The most significant movement towards resolution of the conflict came in 1998, when the 'Good Friday' agreement established a new framework to help decide future constitutional change.

Wales: Plaid Cymru, the Welsh national party, was formed in 1925 to revive the Welsh language and culture; it favours self-government for Wales. The devolution referendum of 1978 saw 12 per cent in favour and 47 per cent against a devolved assembly. However, in a 1997 referendum Welsh voters approved proposals—introduced by the British government—for a devolved National Assembly. Since 1999, the National Assembly has been responsible for the development and implementation of policy in a wide range of areas including agriculture, industry, the environment, and tourism.

Scotland: The Scottish National Party was formed in 1934 and achieved notable political success in 1974; by 1978, 33 per cent of voters favoured a devolved assembly, with 31 per cent opposed. As part of the same initiative that led to the creation of the Welsh Assembly, proposals for a Scottish Parliament were introduced by the British government and a 1997 Scottish referendum approved its creation by a substan-

tial majority. The Parliament's responsibilities are similar to those of the Welsh assembly. Whether the new Scottish and Welsh assemblies will lead to further separation remains to be seen. Today, the Scottish National Party is focusing attention on representation in the European Community.

Flanders and Wallonia: The two linguistically distinct regions of Belgium were combined as a single state in 1830 and have never achieved national unity.

Brittany: Breton (a member of the Celtic family) is the original language of this region in northwestern France, and is still spoken by 10 per cent of the population. Although various groups agree that culture and language in the region need to be defended, separation is not typically favoured.

Alsace: A German-speaking area that is now part of France, Alsace changed repeatedly between German and French rule in 1871, 1918, 1940, and 1945.

Basque country: A distinctive language region divided by the French–Spanish border. One separatist organization practises terrorism, but others strive for separate identity through democratic means. A Basque assembly was convened in 1977.

Catalonia: The Catalan language is being revived, and today 17 per cent of the population speak it. A Catalan assembly was convened in 1977.

Corsica: One militant group seeks full independence from France, and today the French government perceives Corsica as a real threat to national unity. Schools are now permitted to teach Corsican.

Tyrol: An area of German-speakers in northern Italy; total autonomy has sometimes been sought.

Other areas in western Europe that have recently seen movements for autonomy or independence include Galicia (northwest Spain), Andalusia (southeast Spain), southeast Belgium (German-speakers), the Jura region (France–Switzerland), and Sardinia.

Thus, although language is the principal centripetal force in many western European states, there are numerous specific centrifugal forces. Much the same can be said of pre-1989 eastern Europe and the former USSR, where nationalist sentiments were rarely publicized. Events since 1989 have highlighted the national problems of states such as Romania, where a Hungarian minority has been repressed; Bulgaria, where the majority attempted to assimilate Turks in their efforts to create a united Bulgaria; and Yugoslavia, where Serbian nationalism has proved to be highly destructive (Box 9.7).

The former USSR

After 1988, the former USSR experienced considerable separatist pressure from the three Baltic republics—Latvia, Lithuania, and Estonia—and from southern republics with significant Islamic populations. In 1990 conflict flared once again between Christian Armenians and Muslim Azerbaijanis in what was then the Soviet Republic of Azerbaijan. It is important to note that the former USSR was not a typical state; rather, it was the last great world empire. In fact, in 1918 the three Baltic republics, Ukraine, Byelorussia, Georgia, Azerbaijan, and Armenia all fought for independence from the Russian empire; see Box 9.8.

With the collapse of the USSR the five central Asian republics of Kazakhstan, Uzbekistan, Tajikistan, Turkmenistan, and Kyrgyzstan emerged as independent countries. However, one closely related group with similar aspirations for independent state status seems unlikely to see those aspirations fulfilled in the immediate future. The Uighurs, who live in the northwestern Chinese province of Xinjiang, are part of the same larger group to which the people of the five new central Asian countries belong. This group conquered much of Asia and Europe before converting to Islam in the fourteenth century; at that time the larger region was one of great wealth and included such important centres as Tashkent and Samarkand. But by the mid-nineteenth century, the entire region had been conquered, either by Russia or, in the case of the Uighur group, by China. Today only the Uighurs have not achieved independence, and their future is uncertain; so many Han Chinese immigrants have moved into the region that the Uighurs now make up only about half of the total population of some 16 million.

South Asian conflicts

Ethnic and other tensions in South Asia are reflected in a highly unstable political landscape. Among the factors behind those tensions are a complex distribution of ethnic groups and associated religious rivalries (especially, but not exclusively, Hindu–Muslim); the **caste** system; and the 1947 departure of the British as colonial rulers (Zurick 1999).

South Asia had been the 'jewel in the crown' of the British Empire. With the withdrawal of the British in 1947, the two new states of India and Pakistan were born, and in 1948 Ceylon became independent (it would be renamed Sri Lanka in 1972). However, the goal of creating national identities has proved elusive. India was intended to be predominantly Hindu and Pakistan predominantly Muslim, with the boundaries to be imposed (on the basis of 1941 census data) where none had existed before, dividing several well-established regions—notably Kashmir and the Punjab in the northwest and Bengal in the northeast. One immediate result was the movement of about 7.4 million Hindus from Pakistan to India and 7.2 million Muslims from India to Pakistan. It is estimated that several million died in the chaotic redistribution of population that ensued. The new state of Pakistan was divided into two parts, West and East, with roughly equal numbers of people separated by 1,600 km (1,000 miles) of Indian territory. The capital was located in West Pakistan, first at Karachi, then at Islamabad, and—perhaps inevitably—relations between the two sections were poor. By 1971 a large part of the Pakistani army was in East Pakistan fighting Bengali guerillas. In that year the eastern section seceded, creating the independent state of Bangladesh.

Today conflict is endemic in much of South Asia (see Figure 9.11). Five areas of tension are located in the Himalayan Mountains alone. Most notably, the northwestern region of *Kashmir* is the subject of ongoing struggles between India, which officially controls the entire state, and Pakistan, which in practice administers the upper portion of it and also lays claim to the predominantly Muslim

9.8

THE COLLAPSE OF THE USSR

Figure 9.10 The former USSR.

The Union of Soviet Socialist Republics was never really a union; rather it was an empire tightly controlled by one of the republics, Russia (Figure 9.10). In this sense, it differed little from its predecessor, the pre-1917 Russian empire.

Russian expansion from the small core area around Moscow is typically explained in terms of a continuing search for good agricultural land, seaports, and borders that could be easily defended against regular invasions by Vikings, Poles, Germans, and Mongols. To the north, the Russians built the White Sea port of Archangel by 1584; to the west, they established St Petersburg in 1703. Meanwhile Russia was also expanding to the east, across Siberia; Russians crossed the Urals in 1582, reached the Sea of Okhotsk at the edge of the Pacific by 1639, claimed Alaska in 1741, and during the nineteenth century established fur-trade settlements as far south as California. To the south, Russia moved down the rivers Don, Dnieper, and Volga, acquiring Ukraine and gaining access to the Black Sea by 1800. During the nineteenth century, Russia moved into the area between the Black Sea and the Caspian Sea.

Russia retreated from North America in 1867, when it sold Alaska to the United States, but the empire that the new communist government inherited in 1917 was still massive. It expanded even farther with the suppression of local independence movements and the reannexation of several areas that had broken away from the Russian empire after the First World War, including Ukraine, Moldova, Estonia, Latvia, and Lithuania.

From 1917 until the collapse of the USSR in 1991, considerable numbers of ethnic Russians moved into other republics of the Union, which adopted policies favouring the Russian language and suppressing the practice of religion. The collapse of the USSR, as part of the sweeping changes that occurred in Europe in the late 1980s and early 1990s, left a difficult and uncertain legacy of ethnic tensions and often ethnic conflict. Table 9.2 provides data on ethnic populations in the former republics of the USSR, which are now 15 independent states, and highlights the high percentages of Russians in most of these states.

Table 9.2 ETHNIC GROUPS IN THE FORMER USSR

Country (former republic)	Population (millions)	Titular Nationality (%)	Principal Minorities
Russia	146.5	82	Tatars 4%
Ukraine	49.9	73	Russians 22%
Uzbekistan	24.4	71	Russians 8%
Kazakhstan	15.4	40	Russians 38%, Ukrainians 5%
Belarus	10.2	78	Russians 13%
Azerbaijan	7.7	83	Russians 6%, Armenians 6%
Tajikistan	6.2	62	Uzbeks 23%, Russians 7%
Georgia	5.4	70	Armenians 8%, Russians 6%, Azeris 6%
Moldova	4.3	65	Ukrainians 14%, Russians 13%
Turkmenistan	4.8	72	Russians 9%, Uzbeks 9%
Kyrgyzstan	4.7	52	Russians 22%, Uzbeks 12%
Armenia	3.8	93	Azeris 2%
Lithuania	3.7	80	Russians 9%, Poles 7%
Latvia	2.4	52	Russians 35%
Estonia	1.4	62	Russians 30%

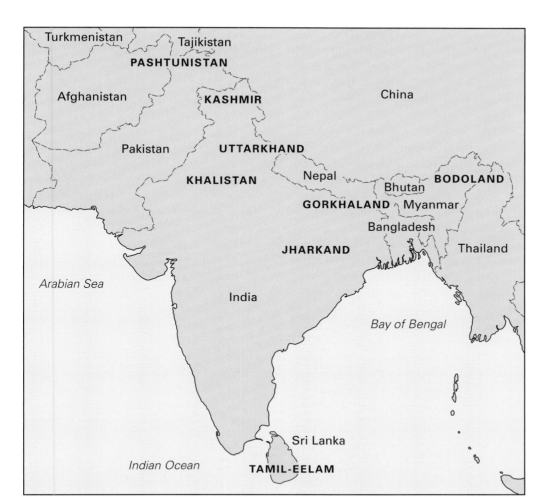

Figure 9.11 Some areas of conflict in Southeast Asia.

Kashmir Valley to the south; it is widely feared that these tensions could erupt into war. At the same time, multiple Kashmiri separatist movements are demanding independence from both India and Pakistan. Northwest of Kashmir in *Pashtunistan*, Pathan tribes are divided between Afghanistan and Pakistan but make strong claims for statehood. South of Kashmir, Pahari-speaking people favour creation of a separate state which they call Uttarkhand, while in the eastern Himalayan region of *Gorkhaland* some ten million ethnic Nepalese aspire to statehood and have been involved in several uprisings in recent years. Farther still to the east, in *Bodoland*—where India borders with Burma and China—a number of groups live independently both of one another and of central state control.

Meanwhile, south of the Himalayas in the Punjab, Sikhs aspire to independence from India and the creation of their own state, *Khalistan*. The struggle for control led to serious violence in the 1980s, but recent years have seen Sikhs turning to more peace-

ful political activity. Farther east, numerous groups collectively known as the adivasi (indigenous peoples) assert their right to a state known as *Jharkand*. Finally, Sri Lanka has for many years been in a state of civil war between the Buddhist Sinhalese majority, who run the government, and the Hindu Tamils, who want their own independent homeland, *Tamil-Eelam*. The first Tamils arrived on the island around the eleventh century, but a second group, known as Indian Tamils, were taken there by the British in the late 1800s to work on tea plantations. Conflict first flared in the early 1900s when Tamils moved onto Sinhalese land, prompting Sinhalese resentment and political and other discrimination. Sinhalese nationalism became more assertive over time, and Tamils have mounted many organized terror campaigns since the 1970s (see Box 6.9, p. 201).

The above account covers only a smattering of the conflicts that plague South Asia. In some cases ethnic groups war against the state; in others, ethnic groups war against one another; and in yet other cases the conflicts have international dimensions. Further complicating the scenario are ongoing Hindu–Muslim differences and the complexities of the caste system.

Internal divisions in Canada

Are there internal divisions in Canada? It is a huge country of 9.8 million km² (3.8 million square miles), 31.3 million people, and large unsettled areas. Much of the land is inhospitable to humans (Cartier described the southern coast of Labrador as the land that God gave to Cain). Further, Canada comprises ten provinces and three territories, with marked regional variations. Ways of life range from traditional to very modern. It is a plural society made up of Aboriginal peoples, the two founding European cultures (English and French), and a host of other national groups. Together, these physical and human factors have produced strong regional feelings—a major cause of instability. In recent years, the unity of Canada has been threatened not only by Quebec (French-language) separatism (which suffered a set-

back with the 2003 election of a Liberal provincial government in Quebec City), but by various western Canadian separatist movements as well (Box 9.9).

Cores and peripheries

One of the principal causes of social unrest in general is the **core–periphery** economic spatial structure—a common feature of modern states. Rich cores and poor peripheries often exacerbate separatist tendencies in the peripheral areas. Thus most of the areas of unrest shown in Figure 9.8 (p. 302) may feel disadvantaged not only in terms of national identity but also in terms of economic well-being. Many peripheral areas have a specialized or short-lived economic base that is particularly subject to economic problems such as unemployment.

Boundaries

Boundaries mark the limits of a state's sovereignty. They are 'lines' drawn where states meet or where states' territorial waters end. A state's stability often reflects the nature of its boundaries.

9.9

REGIONAL IDENTITIES AND POLITICAL ASPIRATIONS

Much of this chapter (and, to a lesser extent, Chapters 7 and 8) is concerned with regions and the grievances that many regional populations feel when they are part of a political unit with which they feel unable to identify. Certainly, the desire to be an independent political unit is at the head of the agenda for many groups. This desire is understandable, given the importance of states in the contemporary world: 'they behave very much like Greek gods, so much so that, given the great powers which they wield, they need to be taken very seriously' (East and Prescott 1975:1).

Is the existence of local and regional nationalisms a good thing or a bad thing? Of course there is no correct answer; but the question is still worth thinking about. The current world political map, as we know all too well, is not sacrosanct. Changes occur, and often for very good reasons. It may seem that regionalism is in fashion today. But is the trend towards the creation of new states, reflecting ever more localized identities, something that should be encouraged? The world as a whole is uncertain. On the one hand are the processes of globalization and movement towards the integration of some states; on the other are numerous groups

arguing that they are different from others and therefore should be separate from them. Most states in which certain groups are seeking some degree of independence react negatively to the possible loss of territory, population, and prestige.

Is it better for a group to look outwards and encourage cosmopolitanism or to look inwards and encourage ethnicity? Or is it possible to accomplish both these goals? Does support for a sub-state level of nationalism result in an ethnocentric world view and encourage lack of respect for others, or does it allow for increased self-respect and a better understanding of others? Although there are no agreed-upon answers to such difficult questions, the literature in social psychology suggests strongly that simply classifying people into groups prompts a bias in favour of one's own group (Messick and Mackie 1989). Such biases may arise even between groups that are interdependent and cooperative, as is the case with English- and French-speaking Canadians.

Canada is not the only country in which different groups hold different opinions of movements to achieve some degree of autonomy within a larger state.

A US border patrol agent drives along the US–Mexico border in Jacumba, California, as prospective immigrants wait to attempt an illegal crossing (CP/ AP/Susan Sterner).

The characteristics that give identity to a nation, such as language, are rarely as abruptly defined as its boundaries are. For this reason there are many countries that include at least one significant minority population. In principle, such situations are less common where the boundaries are *antecedent* (established before significant settlement began), because settlers moving into areas close to the boundary must acknowledge its presence. Boundaries of this kind are often geometric—as in the case of the US–Canada boundary west of the Great Lakes, which follows the 49°N parallel. The boundaries of Antarctica are a second example; in this case, they were actually defined before significant exploration began, to minimize conflict between competing states over such matters as mineral rights.

Other boundaries are *subsequent*: defined after an area has been settled and the basic form of the human landscape has been established. Such boundaries may attempt to reflect national identities (for example, the present boundary of France is an approximation of the *limites naturelles* of the French nation) or may totally ignore such distinctions (this is the case with most colonial boundaries). Many subsequent boundaries are continually subject to redefinition, as in western Europe.

All boundaries are artificial in the sense that what is meaningful in one context may be meaningless in another. Rivers are popular boundaries, since they are easily demarcated and surveyed, but they are generally areas of contact rather than separation, and so tend to make poor boundaries. Rivers may also be poor boundaries if they are wide and contain islands (the Mekong River in southeast Asia) or if they change course (the Rio Grande).

In some cases, groups have erected barriers not to demarcate boundaries but rather to serve as physical obstructions preventing others from entering. Well-known examples include the Great Wall of China, constructed in the third century BCE to prevent invasion from the north, and Hadrian's Wall, constructed across northern England in the second century CE to prevent invasion from Scotland. Along some parts of the US–Mexico border fences are in place to restrict illegal immigration. In one exceptional instance—the Berlin Wall—the aim was to prevent people from leaving the eastern sector for the west. In 2003 Israel began building an elaborate system of fences and military checkpoints along the border with the West Bank in the hope of reducing terrorist incursions.

Divided states

Some states have been divided into two or more separate parts; this situation increases the likelihood of boundary problems. A classic example, noted earlier, was the partition of India as part of the decolonization of Britain's Indian empire, beginning in 1947. In some cases of partition the boundaries between the newly formed states have been especially inappropriate. Examples include Germany (East and West from 1945 until 1990) and Korea (North and South since 1945). In both cases, one nation was divided into two states for reasons that had more to do with the political wishes of other states than with the wishes of the people themselves. There are sound reasons to argue that artificial, externally imposed boundaries such as these are rarely long-lasting.

Groupings of States

The contemporary world is characterized by two divergent trends. First, as we have already seen, many regions and peoples are actively seeking to create their own independent states; this is a continuation of the processes of nationalism and decolonialization (see Boxes 9.5–9.9). Second, some groups of states are actively choosing to unite, sometimes even at the cost of sacrificing aspects of their sovereignty.

European integration

The principal example of such a voluntary grouping was established following the end of the Second World War in 1945—the date that marked the end of the old Europe. Various moves towards European union resulted in the 1957 formation of the European Economic Community (EEC) by France, Belgium, the Netherlands, Luxembourg, Italy, and West Germany. These six states had already created a European Coal and Steel Community and a European Atomic Energy Community, both of which merged with the Economic Community in 1967 to form a single Commission of the European Communities. A common agricultural policy was adopted. Other European countries that did not favour such close integration formed the European Free Trade Association in 1960.

Gradually, the EEC, now the EU, assumed dominance, with Britain, Denmark, and Ireland joining in 1973, Greece in 1981,

Portugal and Spain in 1986, and Austria, Sweden, and Finland in 1995 (Norwegians voted against joining in 1994). The most dramatic expansion of the EU took place in 2004 with the addition of ten countries, bringing the total to 25. Eight of the new members—Estonia, Latvia, Lithuania, Poland, the Czech Republic, Slovakia, Hungary, Slovenia—are in eastern Europe. The remaining two are the Mediterranean islands of Malta and Cyprus. The latter poses a political problem because it was partitioned after an attempted Greek coup in 1974, and since then a buffer zone policed by the United Nations has divided the Greek-controlled Republic of Cyprus from the northern area, which is controlled by Turkey and not internationally recognized.

When these ten countries were accepted for membership, three others—Romania, Bulgaria, and Turkey—were considered not ready to join. Romania and Bulgaria will probably be admitted in the near future, but Turkey has a record of human-rights violations that makes its case problematic. It is also possible that the EU will expand into North Africa, which would raise complex questions about 'European' identity.

Why have so many sovereign states been willing to sacrifice some components of their independence? Probably the most important reason is the inherent appeal of a united Europe in a world dominated before 1991 by the US and USSR and now by the US alone. Today the 25-member EU has a population of 453 million—larger than the total population of the three countries (Canada, the US, and Mexico) linked by the North American Free Trade Agreement (NAFTA). During the formative years of the EU, the US was seen as a real economic threat and the former USSR as a real military threat. Thus it seemed that there was a place for Europe in the world—but not for some patchwork of European states. As the proponents of union saw the two superpowers, both were large; both occupied compact blocks of territory with a low ratio of frontier to total area; both had substantial east–west extent that increased the area of comparable environment. Both were mid-latitude countries with large populations and densely settled core areas, and both had a wide variety of natural resources. Individual European countries did not share

Figure 9.12 European ethnic regions.
Source: L. Kohr, *The Breakdown of Nations*
(Swansea: Christopher Davies, 1957).

any of those crucial characteristics—but a united Europe would. Unlike the US, however, a united Europe would be multinational, and a voluntary multinational state might be more stable than an involuntary one such as the empire of the former USSR.

A radically different view of Europe's future was proposed by L. Kohr (1957). Arguing that aggression is a result of great size and power, Kohr suggested that Europe would be a more peaceful place if, instead of uniting, it divided into many small ethnic

states as shown in Figure 9.12. Such an arrangement would minimize the problems associated with borders and national minorities and in effect return Europe to its medieval form. Although at present Europe appears to be moving in a different direction, in fact there is evidence of increasing regional authority within existing states.

Indeed, even as Europe is becoming increasingly integrated, the profusion of states and independence movements within it suggests that fragmentation might be seen as more characteristic of its component parts. Since the end of the Second World War, certainly, political tensions in Europe have centred on intranational rather than international disputes. We are thus confronted with a complex scenario in which individual states are integrating with one another at a time when many of those same states are experiencing serious internal stresses.

Other groupings of states

Apart from the EU, the principal groupings of states involved in establishing trading areas are the two noted earlier in this chapter, NAFTA and ASEAN; these are discussed further in Chapter 10. But there are other such groupings, among them the Commonwealth, the Economic Community of West African States, the Latin American Integration Association, the Organization of American States, the Organization of Petroleum Exporting Countries, the Organization of African Unity, the South East Asia Treaty Organization, and the Maghreb Union (Box 9.10). In eastern Europe an economic group named Comecon (the Council for Mutual Economic Assistance) began meeting secretly in 1949, but was terminated in 1991 following the collapse of the communist bloc. Finally, there is the prospect of a single world government. Since 1945, the United Nations has striven to assist member states in limiting conflict. Today most of the world's states are

members, and states in the developing world now make up the majority.

The Role of the State

Most of the world's people are citizens of a specific state and, as such, are subject to its laws; they have limited power—if any—to change the state significantly, and they are spatially tethered. Our everyday life is irrevocably involved with government, for all states 'are active elements within society, providing services which are consumed by the public' (Johnston 1982:5). Political units correspond spatially to cultural, social, and economic units, and human geographers often find it useful to define regions on a political basis. The state is a factor affecting human life; it has evolved with society and is a key element in the mode of production. Because of the state's crucial role in the lives of people and places, it is important to recognize that there are many different ways in which a state can be governed.

Forms of Government

The two principal political philosophies that are accepted today are capitalism and socialism, but a number of related or alternative ideas are also important to us. Fundamental to the capitalist form of government is **democracy**: rule by the people. Democracy implies five features: regular free and fair elections to the principal political offices;

9.10

THE MAGHREB UNION

'Maghreb'—an Arabic word meaning 'west'—has long been used to designate Africa west of Egypt and north of the Sahara. The Maghreb Union, an economic union made up of Morocco, Algeria, Tunisia, Libya, and Mauritania, was formally established in 1989. With the country of Western Sahara (not a member of the union) these Arab states comprise a geographic region in both physical and cultural terms.

Given the course of African history, this region has never been a single political unit and the contemporary boundaries are a consequence of European colonialism. Libya was colonized by Italy, and the other four states by France. Libya

achieved independence in 1951, Tunisia and Morocco in 1956, Mauritania in 1960, and Algeria (after eight years of war) in 1962.

The principal reason for the union appears to be a general assumption that integration will benefit the economies of all five states. However, there is also a link with recent developments in Europe, for when Spain and Portugal joined the European Economic Community in 1986, both Morocco and Tunisia lost major markets for their agricultural products. In any event, the union was formed at a time when differences between the member states appeared to be diminishing (Arkell and Davenport 1989).

universal suffrage; a government that is open and accountable to the public; freedom for state citizens to organize and communicate with each other; and a just society offering equal opportunity to all citizens. Democracy was important for a period in classical Greece, but only in the nineteenth century did it reappear as a major idea, and only in the twentieth century was it generally approved and commonly practised.

Monarchy is rule by a single person. Constitutional monarchies do survive in some countries, but typically without any real power. Britain's **monarchy** legitimizes a hierarchical social order; those of the Netherlands, Denmark, and Norway are more democratic. **Oligarchy** is rule by a few, usually those in possession of wealth. The term was introduced by Plato and Aristotle; the favoured term today is 'élite'.

Government by **dictatorship** implies an oppressive and arbitrary form of rule that is established and maintained by force and intimidation. Military dictatorships are common in the contemporary world; an example is North Korea. Fascism is an extreme form of nationalism that, especially in Europe between 1918 and 1945, provided the intellectual basis for the rise of political movements opposing governments whether they were capitalist or socialist. In Italy and Germany, fascist parties achieved power through a combination of legal means and violence.

The political philosophy of **anarchism** may emphasize either individualism or socialism and was advocated by two notable nineteenth-century geographers, Kropotkin and Réclus. It rejects the concept of the state and the associated division of society into rulers and ruled.

Socialist Less Developed States

'Socialism' is an imprecise term, but we can identify two general characteristics of socialist regimes:

1. They aim to remove any and all features of capitalism, especially private ownership of resources, resource allocation by the marketplace, and the class structure associated with them.
2. They have the power, in principle, to make substantial changes to society.

Recent experiences in the socialist developed world of eastern Europe have made it clear that neither of these characteristics has met with popular approval.

Nevertheless, socialism has had a major impact in the less developed world. Several African countries have attempted to combine tradition with socialist ideals; perhaps the most successful has been Tanzania. In Latin America, Cuba's socialist system has survived since 1959 in defiance of the dominant capitalist model, while socialist movements have played important roles in mobilizing the poor in various other countries. But it is in Asia that socialism has been able to exert its most significant and continuing influence (Hirsch 1993). Today there are socialist or communist governments in four major Asian countries: China, Laos, Vietnam, and North Korea. Of these, only North Korea still maintains a traditional hard-line, secretive, closed system. Interestingly, North Korea today is run by a communist royal family; before his death in 1994, Kim Il Sung groomed his son, Kim Jong Il, as heir. North Korea remains a closed country, although meaningful contact was made with democratic South Korea in 2000.

One particular version of socialism is based on the revolutionary thought and practice of Mao Zedong, the leader of the peasant revolution in China that led to the 1949 creation of the People's Republic of China. **Maoism** has two components: a strategy of protracted revolution to achieve power, and the practice of socialist policies after the revolution succeeds. Both aspects continue to be influential in other parts of the world.

In many less developed countries, socialism has a strong anticolonial, nationalist component. Unlike the former socialist states in the developed world, which are largely urban and industrial, most of these states are firmly rural in character. Although the details vary, it is fair to say that in these states, even more than in capitalist states, individual behaviour is often determined by larger state considerations. Fertility policies, for example, are likely to be rigorous, as in China. Perhaps most important, central planning— that is, planning by the central government—means considerable state involvement in people's everyday lives.

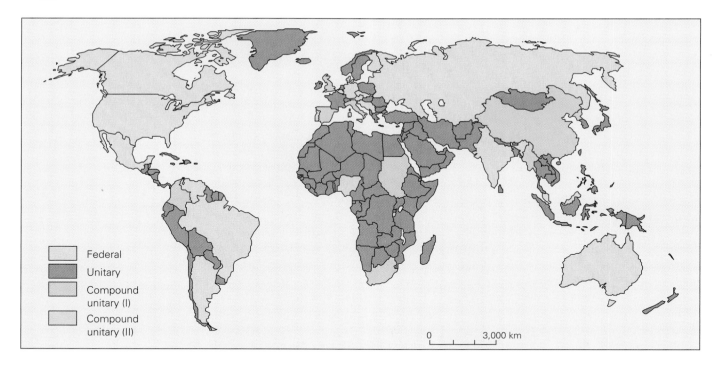

0 ⊢⊣⊢⊣ 3,000 km

Figure 9.13 World distribution of state types. This map reflects the distribution of states in the early 1980s, before the collapse of the Soviet Union.
Source: R. Paddison, ed., *The Fragmented State: The Political Geography of Power* (New York: St Martin's Press, 1983):32.

Substate Governments

In many states, political authority is not entirely centralized at the national level. The political geographer Paddison (1983) has distinguished three levels of centralization (Figure 9.13):

1. *Unitary*: The most centralized form of government; local governments are used by the central state to organize the political hinterland.
2. *Federal*: The least centralized; examples include Australia, Canada, and the United States. One purpose of **federalism** is to prevent one level of government from dictating to another.
3. *Compound unitary*: Midway between the federal and unitary types, these systems devolve substantial powers to subnational governments, but less power than in the federal case. There are two types: Type I has established regional governments, and Type II involves small (usually peripheral) areas that maintain some distinct identity while retaining ties to the state.

According to this classification, most of the large-population states (except China) are federal, although the most common type generally is unitary. Any understanding of how a particular state works must therefore consider all appropriate levels of government.

Australia is a state in which subnational units, called states, have considerable authority. The present states began as separate British colonies, and their boundaries reflect decisions of the British government as well as various intercolonial rivalries. One impact of the colonial boundaries was recorded by Thomas Knox, an American traveller of 1888:

> Beyond Goulbourn the railway carried our friends through the district of Riverina. . . . At Albury they crossed the Murray River and entered the colony of Victoria; a change of gauge rendered a change of train necessary, and Fred remarked that it seemed like crossing a frontier in Europe, the resemblance being increased by the presence of the custom-house officials, who seek to prevent the admission of foreign goods into Victoria until they had paid the duties assessed by law (in Larkins and Parish 1982:49).

Although the transition from colonies to states, in 1901, has reduced the importance of these subnational units, any discussion of Australian geography and landscape must consider the role played by the components of the federation, past and present.

More generally, an increasing trend towards decentralization means that various administrative, political, and financial responsibilities are being transferred from national to subnational governments. In South Africa, the government is hoping that decentralization will help to unify a country so long divided by apartheid legislation. In both Ethiopia and Bosnia–Herzegovina, decentralization has been introduced in order to accommodate various ethnic tensions. There is also evidence of increasing decentralization in the federal state of India, with several states devolving powers to local governments. Finally, even the unitary state of China is also moving in this direction.

Exercising State Power

In capitalist countries, state power is exercised through various institutions and organizations (Clark and Dear 1984); this state apparatus includes the political and legal systems, the military or police forces required to enforce the state's power, and mechanisms such as a central bank to regulate economic affairs. Typically, there are significant spatial variations in the management of this **state apparatus**, as well as in public-sector income and spending and the provision of **public goods**, including services such as health care and education. Inequalities in the distribution of public goods often reflect government efforts to influence an electorate prior to an election (a popular American term for this behaviour is 'pork barrelling'). Clearly the form of government and the particular political philosophy favoured directly affect the manner in which state power is exercised.

One of the most critical issues concerning the power exercised by individual states is the need for international cooperation in solving global environmental problems. As we saw in Box 4.2, 'The Tragedy of the Commons or Collective Responsibility?', state power is needed to ensure that private-sector industries do not harm the environment. In the same way, some international authority is needed to ensure that individual states behave responsibly. But as yet there is no such international power—only the occasional meeting of states convened to address particular problems.

There are at least two dilemmas here (Johnston 1993). *First*, governments in more developed countries may be unwilling to protect the global environment when such actions will result in losses of jobs and wealth (and therefore votes) in their own states.

Second, governments in less developed countries typically contend that they cannot afford to implement environmentally appropriate policies and practices, and that they are not the principal causes of environmental problems.

The politics of protest: Social movements and pressure groups

The relationship between a state and its population is always subject to change. In extreme cases, a substantial proportion of the people may reject the state's authority altogether—a situation that can lead to significant internal conflict. An example is Northern Ireland, where many citizens believe that the British state has no legitimate authority over them. In other cases, the people may accept the legitimacy of the state itself but oppose the specific governing body; in such cases, efforts to replace the people in charge with others deemed more appropriate may lead to civil war. Even in states that are relatively stable and peaceful, however, there are usually several groups striving to influence government policy in various areas.

In many parts of the more developed world today, it is not uncommon for state governments to face protest from groups with specific agendas; such 'beyond the state' social movements have been organized around a vast range of concerns, from racism and labour issues, to women's issues, environmental conservation, gay and lesbian rights, and nuclear weapons, to housing conditions and urban redevelopment. We have already encountered one such movement in Box 4.10, on the conflict surrounding logging in Clayoquot Sound.

Social movements are collective endeavours to instigate change, characterized by active participation on the part of their members and—unlike the state—a relatively informal structure (Carroll 1992; Wilkinson 1971). Specific details vary, of course, but it is not uncommon for social protests to begin following the failure, from the perspective of the interested group, of earlier attempts at persuasion and collaboration. Nevertheless, social protest does not always lead to social change.

Elections: Geography Matters

Much attention has been paid to the question of voting since E. Krebheil's (1916) analyses of geographic influences in British

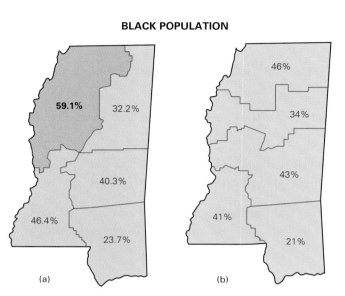

BLACK POPULATION

(a)   (b)

Figure 9.14 (above) The original 'gerrymander', intended to concentrate the vote for one party in a few districts. The term (introduced in the *Boston Gazette*, 26 March 1812) reflected the name of the governor who signed the law—Gerry—and the supposedly salamander-like shape of the new configuration.
Source: R. Silva, 'Reapportionment and Redistricting', *Scientific American* 213, no. 5 (1965):21.

Figure 9.15 (right) Congressional districts in 1960s Mississippi: (a) the pre-1966 districts; (b) the post-1966 districts that guaranteed a minority of Blacks in all districts—an example of deliberate gerrymandering.
Source: Adapted from J. O'Loughlin, 'The Identification and Evolution of Racial Gerrymandering', *Annals, Association of American Geographers* 72 (1982):180.

elections. Typically, the focus has been on spatial and temporal variations in voting patterns and on such causal variables as environment, economy, and society. In any analysis of elections, geography is an important factor: in the boundaries of voting districts, voting behaviour, government activity, and larger world issues.

Creating Electoral Bias

In 1812 the governor of Massachusetts, Elbridge Gerry, rearranged voting districts (Figure 9.14) to favour his party. Today, **gerrymandering**—a word coined by Gerry's political opponents—refers to any spatial reorganization designed to favour a particular party. It aims to produce electoral bias in one of two ways: either by concentrating supporters of the opposition party in one electoral district, or by scattering those supporters so that they cannot form a majority anywhere. Racial gerrymandering has been especially common. Figure 9.15 shows how congressional district boundaries in Mississippi were manipulated in the 1960s: prior to 1966, Blacks formed a majority in one of the five districts, but the redrawn boundaries ensured that Blacks were a minority in all districts. Gerrymandering was especially prevalent in the early 1990s, when several US states fiddled with their electoral borders to create some districts dominated by black voters and others dominated by whites.

In both cases the intention was to benefit the party in power (the Republicans) by reducing the impact of the black vote, which has traditionally gone mostly to the Democrats: in the first instance by ensuring that many black votes would be wasted (since the winning candidate would be elected by a huge majority) and in the second by ensuring that black voters would be too few to elect the candidate of their choice.

Gerrymandering is a deliberate effort to produce an electoral bias; yet it has only recently been deemed a violation of the Constitution in the United States. A second method, particularly useful to parties that appeal to rural voters, is **malapportionment**. This method reflects the fact that rural areas are typically much less densely populated than urban areas. In this case, boundaries are drawn so that rural electoral districts contain a relatively small number of voters, who are thus able to exert a much greater influence than the same number of opposition voters in a large urban district.

Clearly, then, electoral bias is not difficult to produce. But it can also be produced quite unintentionally. Under what circumstances can electoral boundaries be considered fair? Some observers argue that a district ought to be a meaningful spatial unit in a human geographic context; others argue that districts ought to be as diverse as possible. Probably all that can be agreed on is that the number of voters in each district should be as close to equal as possible, and that districts should be continuous.

Voting and Place

It is not uncommon to assert that class is a dominant influence on voting behaviour. Thus in Britain there is a distinction between the Labour party, traditionally representing the views of workers against those of the capitalist employers, and the Conservative party, representing the employers. A similar distinction can be made in the US between the Democrats (worker-based) and Republicans (employer-based). In Canada the situation is less clearly defined: although the New Democratic Party (NDP) and the Liberals are more oriented towards the working class than the Conservatives, ethnic and regional divisions can play a more important role than class in Canadian politics. Recent election results in Northern Ireland, Scotland, and Wales also suggest a tendency for voting to be linked to ethnic or nationalist interests.

But is class really a key explanation of voting behaviour? Do people from the same class but different regions vote similarly? We have already noted that nationalism can disturb this relationship. Place also matters. Geographic research has clearly demonstrated that where a voter lives is crucial. This is especially well documented by Johnston (1985), who used data from the 1983 British election to show that national trends cannot simply be transferred to the local scale. Four types of local influences on voting can be identified:

1. 'Sectional effects': 'a long-standing geographical element of voting whereby differences in local and regional political culture produce spatial variations in the support given to the various political parties' (Johnston 1985:279). Sectional effects have been clearly demonstrated in presidential elections in the US. Longstanding geographic cleavages require any successful candidate to build an appropriate geographic coalition; these cleavages are so factored into the electoral college system that, to be elected president, votes alone are not enough—they must be in the right place.
2. Environmental effects: for example, in the 1983 British election it was found that the

higher the level of unemployment in an area, the more successful the Labour party candidate was. Candidate incumbency is another environmental factor: candidates who already hold their seats usually attract more votes than challengers do.
3. Campaign effects: 'Vote-switching was more common in the safer seats and the longer-established parties (Conservative and Labour) did better in the marginal constituencies' (Johnston 1985:287).
4. Contextual effects: individuals may be influenced in their voting decisions (as they are in other behaviours) by their social contacts—friends, neighbours, relatives, and fellow employees with whom they are in contact. The emphasis on social contacts here is in general accord with the 'symbolic interaction' interpretation of culture noted in Chapter 8.

Voting, then, is influenced by both class and place. Thus any successful political party needs to develop a strong social and spatial base, and any meaningful analysis of elections needs to consider both factors.

The Geography of Peace and War

'In geographical terms, this planet is not too small for peace but it is too small for war' Bunge (1988). Bunge's comment underlines the fact that—unlike even the First and Second World Wars—any future world war would be truly global in impact.

Conflicts

The concepts of war and peace were not clearly distinguished in European thought until the eighteenth century, when capitalism and nation states emerged. It was in this period that war became a state activity with the creation of professional armies and the elimination of private armies, and war became a temporary state of affairs, with long periods of peace between conflicts. Between the French Revolution and the First World War wars were waged between nation states, but after 1918 conflicts became more ideological—variously between communism, fascism, and liberal democracy. Only since the end of the Cold War have non-western cultures begun to play a major role in global politics. Since 1945, most conflicts have been centred in the less developed world. Before

1945, most casualties were caused by global wars; after 1945, most casualties were caused by relatively local wars.

The contemporary world contains many traditional enmities (Box 9.11). Most of these relate to ethnic rivalries and/or competition for territory. Conflicts between peoples, a feature of human life for eons, have become increasingly formalized and structured over time as the number of independent states has increased. Indeed, some authors believe that aggression is a necessary human behaviour (see Box 9.12, p. 322). Since 1945, the United Nations has offered member states the opportunity to work together to avoid conflict, and in Korea many countries fought together under the UN flag to resist aggression. Table 9.3 lists a wide array of disputes and conflicts in which the UN has become involved.

Conflicts may be grouped in four categories: (1) traditional conflicts between states, (2) independence movements against foreign domination or occupation, (3) secession conflicts, and (4) civil wars that aim to change regimes.

Category 1 conflicts since 1945 include three India–Pakistan wars, four wars involving Arab states and Israel, and the Vietnam War. Category 2 conflicts have arisen primarily as a consequence of decolonization; examples include the conflicts in the Belgian Congo (1958–60), Mozambique (1964–74), and Indonesia (1946–9). Category 3 conflicts, over secession, include the struggles of Tibet (1955–9), Biafra (1967–70), and Philippine Muslims (1977–). Category 4 conflicts include the civil wars in China (1945–9), Cuba (1956–9), Bolivia (1967), and Iran (1978–9).

Conflicts between states (category 1) have the greatest potential to disrupt the human world. Geographers have focused on various theories of international relations to explain such conflicts. Among the principal variables in such theories are power, environment, and culture. Much of the related empirical work is strongly quantitative.

Although the disruption caused by category 4 conflicts—civil wars—may be somewhat less, there is good reason today to be especially concerned about them. Most wars now are civil wars—there were approximately 25 ongoing as of 2003—and they typically continue for years. Governments in the more developed world often assume that such wars are rooted in ancestral ethnic

9.11

SOME TRADITIONAL RIVALRIES

The contemporary world includes many examples of long-standing international rivalries. Although these rivalries may not be evident at any particular time, they point to potential conflict areas. The following highly selective list does not include rivalries involving minority groups within states.

In Europe there are several traditional rivalries:

Poland–Germany
Poland–Russia
Romania–Hungary
Turkey–Russia
Turkey–Greece
Turkey–Bulgaria

In the Middle East there is a traditional enmity between Turkey and Syria and a whole set of post-1948 enmities between Israel and Arab states.

In Asia there are traditional enmities involving:

Thailand–Cambodia
Thailand–Burma
China–Russia
China–Vietnam
China–Mongolia
India–Pakistan
Japan–Korea

In Africa there are numerous ethnic rivalries, but they are rarely expressed at the state level. In southern Africa there are rivalries between the Zulu and other groups in South Africa and between Shona and Ndebele in Zimbabwe. In eastern Africa there is an Ethiopia–Somalia rivalry, and throughout the Sahel region there are rivalries between Arabs to the north and Black Africans to the south.

In South America there are rivalries between Brazil and Argentina and, in general, between Indian and Hispanic groups.

and religious hatreds and therefore that little can be done to prevent them. Yet a 2003 World Bank study suggests that this assumption is not necessarily valid. Rather, the study found that the principal cause of civil war is lack of development. As countries develop economically, they become progressively less likely to suffer violent conflict, and this in turn makes further development easier to achieve. By contrast, when efforts at development fail, a country is usually at high risk of civil war, which will further damage the economy, increasing the risk of further war.

Civil wars cause population displacement (as discussed in Chapter 6), mortality, and poverty among local populations, and have spillover effects on neighbouring countries. But they also have global impacts. For example, in a country engaged in civil war, there may well be areas where it is impossible for the state to exercise the usual control over illegal activities such as drug production or trafficking—about 95 per cent of the world's hard drugs are produced in countries experiencing civil war. Tragically, civil wars may be prolonged because a few people benefit financially from such activities.

Recent events suggest that it may be possible to identify a fifth category of conflict: action taken against states that support terrorism. Both the 2001 invasion of Afghanistan in 2001 and the 2003 war on Iraq could be classified in that group. However, in both cases terrorism was probably only one of several considerations.

Two farmers harvest opium from their poppy plants near Kandahar, Afghanistan, in May 2002. According to the UN drug control agency, growers took advantage of a power vacuum created by the US-led war and the collapse of the Taliban to re-establish Afghanistan as the world's largest producer of opium
(AP photo/Victor R. Caivano).

Table 9.3   DISPUTES AND CONFLICTS INVOLVING THE UN, 1945–1990

Operations by UN Forces or Military Observer Groups

Greek frontiers, 1946–9
Indonesia, 1947–51
Palestine/Israel, Egypt, Lebanon, 1947–
Kashmir, 1948–
Korea, 1950–3
Congo/Zaire, 1960–4
West Irian, 1962–3
(North) Yemen, 1963–4
Cyprus, 1964–
Dominican Republic, 1965
India–Pakistan, 1965–6
Iran–Iraq, 1988–
Afghanistan, 1989–
Angola, 1989–
Namibia, 1989–90
Central America, 1989–
Haiti, 1990

Former UN Trust Territories (now independent or merged with a neighbouring state)

Togolands
Cameroons
Rwanda
Burundi
Tanganyika
Somalia
NE New Guinea
Nauru
Western Samoa

Other Disputes and Conflicts That Have Involved the UN

West Berlin, 1948–9
Taiwan, 1949–71
South Africa, 1952–
Hungary, 1956
Algeria, 1956–61
Indochina, 1958–66

Tibet, 1959–61
Tunisia, 1961
Kuwait, 1961–3
Mozambique, 1961–75
Cuba, 1962
Rhodesia/Zimbabwe, 1962–79
Borneo, 1963–5
Gibraltar, 1965–
Falklands, 1965–
Czechoslovakia, 1968
Western Sahara, 1974–
East Timor, 1975–
Cambodia, 1979–
Iran–United States 1979–80
Afghanistan-USSR, 1979–
Malta–Libya, 1980
Belize–Guatemala, 1981
Libya—United States, 1986
Venezuela–Guyana, 1986–7
Iraq–Kuwait, 1990–

Source: A. Boyd, *An Atlas of World Affairs*, 9th ed. (New York: Routledge, 1991):30–1.

The Costs of War

It is estimated that, between 1945 and 1990, the world spent perhaps $16 trillion for defensive or offensive reasons. Today the developed world spends twice as much as it did in 1960, and the less developed world spends six times as much (both figures allowing for inflation). Some 29 million people are employed by the world's armed forces and some 11 million in various weapons industries. Our capacity to destroy is greater than ever and always increasing. Possibly as many as 25 million people have died in conflict since 1945—as many as died in the Second World War. Although most of these deaths have been in the less developed world, most of the weapons used originated in the developed world. Military expenditures are a great drain on many economies: Israel spends some 27 per cent of its gross national product in this way, and many poor countries spend more than 10 per cent, compared to a world mean of 5.6 per cent.

The world may not be at war, but it is certainly not at peace with itself.

The Geography of Nuclear War

Geographers have contributed significantly to our understanding of the causes of conflict. Perhaps more important, they have helped to clarify the consequences of war, especially nuclear war. In Britain, Openshaw, Steadman, and Greene (1983) published an extensive series of estimates of probable casualties following any nuclear attack. Bunge (1988) published a provocative yet penetrating *Nuclear War Atlas*, and other geographers have explored the possible climatological effects of a nuclear exchange (Elsom 1985). Nuclear war would mean national suicide for any country involved, and a large-scale nuclear war would affect all environments. Large areas of the Northern hemisphere would likely to experience subzero temperatures for several months regardless of the season. Low temperatures and reductions in sunlight would adversely affect agricultural productivity. A nuclear winter would result in many deaths from hypothermia and starvation. It is not necessary to belabour these points. Let us consider instead the role that geographers can play in influencing public awareness and alerting the makers of public policy to the folly of nuclear war.

Geographers have two important responsibilities: to teach people to love the land and the peoples of their state, and to teach people not to hate and fear other states. Undoubtedly, human geographers are in an enviable position to achieve such goals. They have studied the environmental and human consequences of nuclear war. They have used cartography (long a highly effective propaganda tool) to teach, and so are able to identify any deliberate territorial biases. And their study of landscapes, places, and their interrelationships can help to develop the global understanding that is so desperately lacking in our contemporary world.

Our Geopolitical Future?

One of the most interesting, and most important, questions we might ask concerns the character of the geopolitical world as it is unfolding today. As we have seen throughout this chapter, that world is a complex mix of relative stability and great uncertainty. Some of the political changes described here have followed quite predictable directions, while others have seemed quite unexpected. It would not have been difficult to predict the eventual decolonization of Africa, for example, as it took place over the decades after 1945. On the other hand, most observers were not prepared for the mass expressions of discontent throughout eastern Europe that led to the end of the Cold War, or for the peaceful termination of apartheid in South Africa. So, with some trepidation, we ask: what might our geopolitical future have in store? Many answers to this question have been suggested; this section will discuss eight intertwined possibilities.

Perpetual Peace

Is it possible that we are approaching the beginning of what Kant called 'perpetual peace'? One school of thought believes that we have now seen the end of major wars between states, although local wars will continue. This 'end of history' thesis is based on the idea that, with the termination of the Cold War in the early 1990s, the principles of

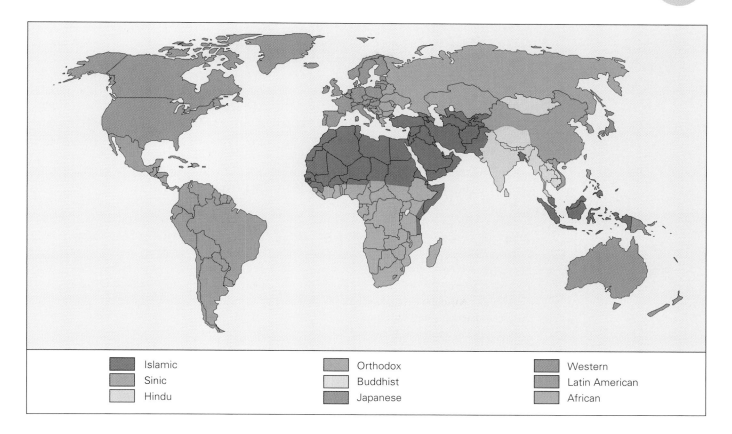

Islamic	Orthodox	Western
Sinic	Buddhist	Latin American
Hindu	Japanese	African

liberal democracy are increasingly accepted around the world, and that liberal democracies do not wage war against one another (Fukuyama 1992). This may be so, but the many local and regional conflicts that have erupted since the end of the Cold War make it difficult to see the phrase 'perpetual peace' as applicable to the contemporary world. Nationalist, populist, and fundamentalist movements have proliferated in recent years, with resulting conflicts in many areas, including the Persian Gulf region, the former Yugoslavia, Somalia, and central Africa.

From a western perspective, the most significant reasons to doubt the likelihood of perpetual peace were the 2001 terrorist attacks by Osama Bin Laden's al-Qaeda group that destroyed the twin towers of the World Trade Center in New York City and part of the Pentagon in Washington. The United States responded first with a military campaign in Afghanistan designed to replace the Taliban government, which was sympathetic to al-Qaeda, and then with an invasion of Iraq that toppled the regime of Saddam Hussein.

Clash of Civilizations

Another possibility is what Huntington describes as a clash of civilizations or cultures:

It is my hypothesis that the fundamental source of conflict in this new world will not be primarily ideological or primarily economic. The great divisions among humankind and the dominating source of conflict will be cultural. Nation states will remain the most powerful actors in world affairs, but the principal conflicts of global politics will occur between nations and groups of different civilizations. The clash of civilizations will dominate global politics. The fault lines between civilizations will be the battle lines of the future (Huntington 1993:22).

Huntington (1996) identifies nine major cultures: western, Sinic, Buddhist, Japanese, Islamic, Hindu, Orthodox, Latin American, and African (Figure 9.16). It is instructive to compare these with the regions proposed by Toynbee (p. 225) and the cultural regions depicted in Figure 7.2, and to reflect again

Figure 9.16 World civilizations. Source: S.P. Huntington, *The Clash of Civilizations and the Remaking of World Order* (New York: Simon and Schuster, 1996):26-7.

on the quotation from James that introduced Chapter 7. It can be argued that conflict will be based on cultural divisions for six basic reasons:

1. Cultural differences—of language, religion, tradition—are more fundamental than differences between political ideologies. They are basic differences that imply different views of the world, different relationships with a god or gods, different social relations, and different understandings of individual rights and responsibilities.
2. As the world becomes smaller, contacts will increase and awareness of cultural differences may intensify.
3. The ongoing processes of modernization and social change separate people from long-standing local identities and weaken the state as a source of identity. Increasingly, various fundamentalist religions are filling the gap, providing a basis for identity.
4. The less developed world is developing its own élites, with their own—non-western—ideas of how the world should be.
5. Cultural characteristics, especially religion, are difficult to change.
6. Economic regionalism is increasing and is most likely to be successful when rooted in a common culture.

The argument that cultural difference will be the principal basis for future global conflicts is strong. We have already discussed numerous regional and local conflicts that are rooted in cultural differences, especially of language and religion, in this chapter and the two preceding it. Humans may not be naturally aggressive (see Box 9.12), but conflict, much of which is not necessarily harmful, does seem to be fundamental to our existence at all social scales.

Terrorism

A third possibility is that the future will be significantly affected by the threat of terrorism. Much of the immediate reaction to the 2001 terrorist attacks was couched in the terms introduced by Huntington, with many commentators reflecting on the long history of conflict between Islam and Christianity and discussing the attacks in this context. Others, however, found that view superficial and argued that 'Islamic' terrorism is the work of individuals and small extremist groups—not representative of Islam itself (see Chapter 8). Certainly it is important to recognize that organizations like al-Qaeda do not reflect larger religious identities. Yet it is also the case that cultural and ethnic differences, often based in language or religion, have been at the root of many of the conflicts discussed elsewhere in this chapter.

Coping with terrorist organizations is especially difficult because—unlike states—

9.12

NATURALLY AGGRESSIVE?

A fundamental debate in social science concerns a particular human characteristic: aggression. Are we naturally (genetically) aggressive or do we learn (through culture) to be aggressive?

Certainly aggression is a normal part of the world we live in. It operates at many spatial and social scales and takes a variety of forms. But there are cultural variations in aggression: some cultures display considerably more aggression than others. So, are we naturally aggressive?

Desmond Morris believes we are. In his popular book *The Naked Ape* (1967), he argues that humans fight for three rea-

sons—to establish dominance, to defend territory, and to defend family—and hence that aggression is biologically programmed. In *On Aggression* (1967) another popular writer, Konrad Lorenz, presents a similar view, developed on the basis of a detailed study of animal behaviour. Ashley Montague, by contrast, does not see aggression as being genetically controlled. In *The Nature of Human Aggression* (1976), he argues that early human societies, far from being aggressive, were characterized by cooperation; it was with the advance of technology that humans became increasingly violent. For Montague, humans are neither 'naked apes' nor 'fallen angels'.

they are not spatially tethered components of the political landscape. This does not mean, however, that they are able to function in a geographic vacuum. Most terrorist groups are supported, at least informally, by states—hence the American invasions of Afghanistan and Iraq following the 9-11 attacks. While the latter were clearly of a different order from any previous terrorist activities, terrorism has been a threat in many countries for some time—England, Ireland, France, Spain, and Israel have been especially vulnerable. Cohen (2003: 90–1) identifies roughly a hundred states that have been exposed to terrorist activities since the end of the Second World War. It is interesting to note that 34 of them are Islamic states subject to attack from extremist Islamic groups—a fact that may raise questions about the logic of the 'clash of civilizations' scenario.

Conflict over Resources

As we saw in Chapter 4, human activities can have very serious implications for resource availability. Today, some commentators predict that resource shortages will play a key role in future conflicts (Klare 2001). Perhaps most obviously, there may be increased competition for access to oil and gas, water, and commodities such as timber, copper, gold, and precious stones. In such a scenario, the areas containing those resources—many of which are located in contested and unstable areas of the less developed world—would be major sites of conflict.

In the case of oil and gas, areas vulnerable to conflict include the Caspian Sea basin, the South China Sea, Algeria, Angola, Chad, Colombia, Indonesia, Nigeria, Sudan, and Venezuela, not to mention the Persian Gulf region (it has been widely suggested that one motive for the 2003 war in Iraq was to gain control of the region's oil reserves). In the case of water, the most vulnerable areas might be those where a major water body is shared between two or more states, as in the case of the Nile (Egypt, Ethiopia, Sudan, and others), Jordan (Israel, Jordan, Lebanon, Syria), Tigris and Euphrates (Iran, Iraq, Syria, Turkey), Indus (Afghanistan, India, Pakistan), and Amu Darya (Tajikistan, Turkmenistan, Uzbekistan). There may be competition for diamonds in Angola,

the Democratic Republic of Congo, and Sierra Leone, for emeralds in Colombia, for gold and copper in the DR of Congo, Indonesia, and Papua New Guinea, and for timber in many tropical countries.

A Single Superpower

Prior to the Second World War, world politics was dominated by several European states. Between the end of that war, in 1945, and the end of the Cold War, in the early 1990s, the Soviet Union and the United States competed for world supremacy. Since the collapse of the Soviet Union, the US has been the sole world superpower. Thus we have moved from a multipolar to a bipolar and finally a unipolar world. The implications of this shift are uncertain, although one indicator came in early 2003, when the US ignored opinion in much of the rest of the world and invaded Iraq. Many observers, concerned about this unprecedented unipolarity, see the United Nations as playing an essential role in world affairs now, since it represents most countries and therefore a broad range of interests. As of the early twenty-first century, it seems clear that the US will remain the principal power, although some foresee a stronger Europe and also a stronger Asian region, perhaps centred on China.

Officers of the Royal Ulster Constabulary inspect the debris left by a bomb that killed 28 people and injured 220 in the market town of Omagh, Northern Ireland, in August 1998. A telephone warning had encouraged police to move people closer to the blast area (AP photo/ Alastair Grant).

Some commentators suggest that the contemporary political world comprises three dominant regions: North America, Europe, and east and southeast Asia. Trade within these regions has increased more than trade between them, and there is evidence from Europe that economic integration might lead to political integration, at least at the regional scale.

Democracy and National Identity

As we noted above, regarding the possibility of perpetual peace, it seems likely that democracy will continue to spread around the world. Since 1980, 81 countries have taken significant steps towards democracy, and 47 of them have become fully democratic. Since 1990, the number of countries that have ratified the six main human rights conventions and covenants has risen from about 90 to about 150.

Around the world, citizens are increasingly unwilling to accept authoritarian rule: since 1980 alone, 33 military regimes have been replaced by civilian governments. Challenges to dictatorship may be internal, external, or both. But of course the spread of democracy does not preclude conflict. Democratic states frequently experience challenges from within, especially from ethnic or regional groups that feel marginalized in some way. Indeed, acceptance of democratic principles may encourage challenges to national identity. Identities—whether at the level of the nation itself or of the group within the nation—are not fixed but dynamic, continuously playing themselves out in a complex negotiation. True, for many states there is a fairly obvious basis in language and/or religion and also some sense of a shared history. But many individuals have several competing identities—for example, a sense of belonging at once to a local community, a larger region, and a state.

Political Globalization

In many respects, this seventh possible scenario is the most complex, overlapping as it does with several earlier scenarios. For many observers it also seems the most plausible. There is much evidence to support the claim that a process of political globalization has been underway for quite some time.

Political globalization is typically seen as one component of the overall globalization introduced in Chapter 2, which will be more fully examined in Chapter 10. Thus improvements in technology, specifically communications technology, and the spread of increasingly dominant transnational corporations, are producing an integrated world economy in which, as trade increases, consumer behaviour patterns become more and more similar. The idea of political globalization, however, also incorporates two explicitly political circumstances.

First, following the end of the Cold War, a new world political order emerged in which the United Nations was finally in a position to play the role for which it was created. Previously, it could be argued, both the US and the Soviet Union actively sought to prevent the UN from fulfilling its mandate.

Second, globalization not only relies on the spread of democracy but also tends to promote it. Because democratic states naturally tend to form closer links with one another than with other political forms, states that want to trade with democratic states will be persuaded to become more democratic themselves. Perhaps more important, globalization may also promote peace if the 'perpetual peace' scenario is correct in suggesting that democracies do not wage war with one another. It can be argued that the process of political globalization has been underway for a very long time: in 1500 BCE the world contained perhaps 600,000 autonomous polities, but by 2003 that number had been reduced to about 190. On the other hand, developments in the twentieth century suggest that, in the short term at least, the trend may be towards more rather than fewer political units. The end of the First World War brought the collapse of several empires and the rise of several new states. The end of the Second World War ushered in both a period of previously unheard-of cooperation between European states and a 'Cold War' that divided much of the world into two camps, American and Soviet. The decades since 1945 have seen some apparently anti-globalizing trends as colonial areas have achieved independence; approximately 95 new states have come into being this way, and together the collapse of the Soviet Union and the disintegration of Yugoslavia were responsible for the creation of about 19 more. Additional new states may be formed as a result of the trend towards decentralization discussed earlier in this chapter.

Decline of the state

Because political globalization challenges the integrity of the individual state, it entails a reduction in the importance of the role played by states. By the early 1990s, Ohmae (1993:78) was arguing that the nation state 'has become an unnatural, even dysfunctional unit for organizing human activity and managing economic endeavor in a borderless world', and suggesting that on 'the global economic map the lines that now matter are those defining what may be called "region states"'. A region state is a natural economic area. It may be part of a nation state, as in the case of north Italy, or it may cross boundaries, as in the case of Hong Kong and south China before 1997 (when China regained control of Hong Kong). Its principal economic links are global rather than national. In the Canadian context, it might be suggested that British Columbia forms a region state with the American west; other parts of Canada also have north–south relationships with areas of the US. In Canada, as elsewhere, the rise of region states has significant implications for national unity.

Certainly the role of the state is changing. But perhaps states themselves will be able to adjust to the changing circumstances resulting from political globalization. In the past, many states sought power as measured by territory and population; today, by contrast, states seek markets. The challenge for the state in the twenty-first century may well be to integrate with other states without sacrificing national identity. On the other hand, the nation state may prevail after all, since people tend to resist change as long as they are reasonably comfortable. Thus most Canadians have wanted Quebec to remain part of Canada, if only because of the insecurity that its secession would entail.

Tribalism: Not One World, but Many?

Our final scenario focuses on globalism's opposite number: tribalism, or the tendency for groups to assert their right to a state separate from the one they occupy. Those who see tribalism as gaining ground point to secession movements, civil war, terrorism, and cross-border ethnic conflict fed by the resurgence of local identities. Support for this scenario can be found in two other trends.

First, as we have seen in this chapter and the two before it, groups of people with some common identity are increasingly emphasizing their distinctiveness, often at the expense of the states to which they 'belong'. During the 1990s a number of new states were created as, for example, the former Czechoslovakia divided peacefully into the Czech Republic and Slovakia, and the former Yugoslavia disintegrated in bitter conflict (see Box 9.7, p. 303). At least some of the new states created in this way may turn out to be more meaningful, culturally, than their predecessors. In any event, the potential for the creation of additional new states is considerable.

Second, in some states, such as Canada and Australia, the principle of multiculturalism is widely accepted and cultural pluralism is valued. It is possible that acceptance of multiculturalism will lead to a weakening of central authority as 'tribal' identities become stronger.

Uncertain Times

This overview of possible scenarios suggests that future generations are likely to look back on the early twenty-first century as a period of uncertainty. We may hope that they will also see it as a transitional period in which—however gradually—conflict declined, democracy spread, and the chances improved for people around the world to live without being oppressed, discriminated against, or persecuted. In truly human geographic terms, we all have a right both to have a home and to feel at home in the world.

CHAPTER NINE SUMMARY

Dividing territory

Human beings partition space. The most fundamental of the divisions we create is the sovereign state. Almost all people today are subjects of a state. Loosely structured empires were typical of Europe before 1600. The link between sovereignty and territory emerged most clearly in Europe after 1600.

Nation states

In principle, a nation state is a political territory occupied by one national group. Despite the importance of nationalism and the territory-state-nation trilogy, there are many binational or multinational states today. Numerous scholars have attempted to explain how states are created: Ratzel viewed states as organisms; Jones developed a field theory involving a progression from idea to area, and Deutsch outlined an eight-stage sequence. Nationalism is the belief that each nation has a right to a state; only in the twentieth century has nationalism become a prevalent view.

Empires: Rise and fall

World history includes many examples of empire creation and collapse. European empires evolved after 1500 and typically disintegrated after 1900, greatly increasing the number of states in the world. Imperialism is the effort by one group to exert power, via their state, over another group located elsewhere. The colonies created in this way quickly became dependent on the imperial states. The most compelling motivation for colonialism was economic.

Geopolitics

Geopolitics is the study of the roles played by space and distance in international relations. Early geopolitical arguments by Ratzel and Kjellen included the idea that territorial expansion was a legitimate state goal. This idea, in the form of *geopolitik*, dominated geopolitics until it was discredited in the 1940s because it was too closely associated with Nazi expansionist ideology. Other important geopolitical theories include Mackinder's heartland theory and Spykman's rimland theory. In the early 1990s the cold war that had started in 1945 ended and a geopolitical transition began.

States: Internal divisions

The stability of any state is closely related to the relative strength of centrifugal and centripetal forces within it. A key centripetal force is the presence of a powerful sense of shared identity; a key centrifugal force is the lack of a strong shared identity, perhaps because of the presence of minority groups—which themselves may wish to secede. In Africa, discordance between nation and state is common where state boundaries imposed by former colonial powers without reference to national identities. European states typically have greater internal homogeneity, although many countries do experience conflict over minority issues, some of which might be explained by reference to core/periphery concepts.

The collapse of the former USSR, an ethnically diverse empire, is one example of instability; the disintegration of Yugoslavia, a country with several republics and a deep divide between Christians and Muslims, is another. Canada's instability is related to a wide variety of physical and human factors. Many of the areas asserting a distinct identity are peripherally located in their states and economically depressed.

Boundaries of states

Stability may be closely related to boundaries. State boundaries are lines that can be depicted on maps, whereas national boundaries are typically zones of transition. Antecedent boundaries precede settlement; subsequent boundaries succeed settlement.

Groupings of states

There are two opposing trends in our political world: minority groups increasingly aspire to create their own states, while states increasingly desire to group together, usually for economic purposes. The EU is the principal instance of the second trend. As of 2004, it includes 25 countries.

Substate governments

Political authority is decentralized in federal states. Many of the states with large populations are federations. Like state boundaries, substate boundaries often do not reflect physical or human differences.

The power of the state

There are numerous types of states and related political philosophies. States perform many important functions, both internally, with regard to the distribution of public goods, and externally, in their relationships with other states, especially regarding conflict and environmental concerns.

Elections

Geography is a central factor in elections. Some electoral boundaries are deliberately drawn to create an electoral bias using gerrymandering or malapportionment. The supporters of different political parties are often spatially segregated. Place has a significant influence on voting behaviour.

Peace and war in the twenty-first century

Peace and war, with their obvious implications for both people and place, are inherently geographic. Geographers can help us understand the causes and consequences of conflict, and can contribute to increased understanding between states. Even countries that are not at war are heavily committed to military and related expenditures. Although the geopolitical future remains uncertain, most scholars agree that conflict will continue.

LINKS TO OTHER CHAPTERS

- Globalization:
  - Chapter 2 (concepts)
  - Chapter 4 (ecosystems and global impacts)
  - Chapter 5 (population growth, fertility decline)
  - Chapter 6 (refugees, disease, more and less developed worlds)
  - Chapter 7 (cultural globalization)
  - Chapter 8 (popular culture)
  - Chapter 10 (economic globalization)
  - Chapter 11 (agriculture and the world economy)
  - Chapter 12 (global cities)
  - Chapter 13 (industrial restructuring)

- State creation, nation states, nationalism, state stability:
  - Chapter 7 (language and religion)
  - Chapter 8 (ethnicity).

- Group formation and group identity:
  - Chapters 7 and 8.

- Colonialism, dependency theory, socialism in less developed states:
  - Chapter 1 (history of geography)

  - Chapter 6 (the less developed world; world systems analysis).

- European colonial activity:
  - Chapter 3 (Figure 3.8, on global environment regions).

- Core and /periphery:
  - Chapter 13 (uneven development).

- State groupings (nafta, eu, asean); globalization:
  - Chapter 9 (trade).

- Nation–state discordance in Africa, including Box 9.6:
  - Chapter 6 (refugees).

- Politics of protest:
  - Chapter 4 (Box 4.9)
  - Chapter 8 (landscapes and power).

- Possible future conflicts:
  - Chapters 4 and 6 (future trends in population, food, and environment);
  - Chapters 7 and 10 (world regions).

FURTHER EXPLORATIONS

ABLER, R. 1987. 'What Shall We Say? To Whom Shall We Speak?' *Annals, Association of American Geographers* 77:511–24.

Focuses on the role of geography in our contemporary world, emphasizing the need to educate others on matters of key political concern.

AGNEW, J. *Geopolitics: Re-visioning World Politics*. 1998. New York: Routledge.

An overview of an ever-changing geopolitics arguing that political geography has always served the interests not just of scholars but of particular states; focuses on the aftermath of the cold war and impacts on geopolitics.

———. 2002. *Making Political Geography*. London: Arnold.

An informative basic political geography text, covering all the topics addressed in this chapter; especially useful on the reasons why combining geography and politics aids our understanding of the contemporary world. Includes both traditional content and appropriate postmodern perspectives.

ARCHER, J.C., et al. 1988. 'The Geography of US Presidential Elections'. *Scientific American* 259, no. 1:44–51.

An important article explaining that the US electorate is divided by enduring geographic cleavages such that any winning candidate must build a geographic coalition.

BELL-FIALKOFF, A. 1993. 'A Brief History of Ethnic Cleansing'. *Foreign Affairs* 72, no. 3:110–21.

A summary article that helps place the Bosnian conflict in a sound historical perspective; also explains the importance of homelands in a political context.

BOYD, A. 1991. *An Atlas of World Affairs*, 9th ed. London: Methuen.

An indispensable factual account of our contemporary political world; companion atlases deal with Africa, North America, and the European Union.

BUNGE, W. 1973. 'The Geography of Human Survival'. *Annals, Association of American Geographers* 63:275–95.

A thought-provoking and deeply passionate piece by a challenging, concerned, and often polemical writer.

COHEN, S.B. 2003. *Geopolitics of the World System*. Lanham: Rowman and Littlefield.

A very substantial volume that covers geopolitical issues from a pragmatic perspective, identifying a hierarchy in the world system (geostrategic realms, geopolitical regions, national states, quasi-states, and territorial subdivisions), along with other such features as shatterbelts, gateways, and compression zones; includes predictions for the future.

CUTTER, S.L. 1988. 'Geographers and Nuclear War: Why We Lack Influence in Public Policy'. *Annals, Association of American Geographers* 78:132–43.

An important statement addressing the actual and potential role of geography in peace education and nuclear awareness.

GOTTMAN, J. 1973. *The Significance of Territory*. Charlottesville: University Press of Virginia.

An original essay by a well-known political geographer; substantial conceptual and historical content.

GRAY, C.S., and G. SLOAN, eds. 1999. 'Special Issue on Geopolitics, Geography and Strategy', *The Journal of Strategic Studies* 22: numbers 2 and 3.

An interesting collection of often challenging revisionist interpretations of traditional geopolitical thinkers, such as Mackinder, Mahan, and Haushofer, along with a number of pioneering new analyses on such topics as geography in the space age and the impact of information technologies.

KASPERSON, R.E., and J.V. MINGHI, eds. 1969. *The Structure of Political Geography*. Chicago: Aldine.

A pioneering collection of forty articles; includes some works by Ratzel, Mackinder, Whittlesey, Hartshorne, and Wallerstein.

KLIOT, N., and S. WATERMAN, eds. 1983. *Pluralism and Political Geography: People, Territory and State*. London: Croom Helm.

A series of readings that includes good discussions of theory and important examples.

KNIGHT, D.B. 1982. 'Identity and Territory: Geographical Perspectives on Nationalism and Regionalism'. *Annals, Association of American Geographers* 72:514–31.

A clear discussion of some of the difficulties of defining nation and state, as well as the related problems of group delimitation.

LIND, M. 1994. 'In Defence of Liberal Nationalism'. *Foreign Affairs* 73, no. 3:87–99.
A succinct summary of the contemporary implications, especially in Europe, of the simple idea that every nation should have its own state.

MONMONIER, M. 2001. *Bushmanders and Bullwinkles: How Politicians Manipulate Electronic Maps and Census Data to Win Elections*. Chicago: University of Chicago Press.
An overview of how contemporary American politicians are using cartography to help them adjust electoral district boundaries to their advantage; modelled on the term 'gerrymander', 'Bushmander' refers to the boundary changes prompted by the US Justice Department under former President George H.W. Bush; 'Bullwinkles' refers to one New York Congressional District that was so misshapen that it resembled the antlers on Bullwinkle, the animated moose.

MUIR, R., and R. PADDISON. 1981. *Politics, Geography and Behavior*. New York: Methuen.
A different type of political geography text, focusing on how political factors and processes influence and interact with spatial behaviour; detailed examination of values, attitudes, images, and decision-making.

MURPHY, D.T. 1997. *The Heroic Earth: Geopolitical Thought in Weimar Germany, 1918–1933*. Kent, OH: Kent State University Press.
A revisionist study of *geopolitik* that claims these ideas had less influence on Hitler than is often supposed; emphasizes the *geopolitik* concern with space in contrast to the Nazi concern with race.

O'SULLIVAN, P. 1986. *Geopolitics*. London: Croom Helm.
Focuses on geographic interpretations of the relationships between states; includes substantive conceptual discussion.

O TUATHAIL, G., and S. DALBY. 1998. *Rethinking Geopolitics*. London: Routledge.
A study of critical geopolitics, focusing on approaches that are not state-centred and that are suitable in contemporary situations; includes some insightful discussions of the relevance of culture and identity as these are socially constructed.

PARKER, G. 1998. *Geopolitics: Past, Present, and Future*. London: Pinter,.
A study of the 'new geopolitics' that developed from the 1970s onwards, focusing on international relations from a geographic or spatial perspective.

———. 1983. *A Political Geography of Community Europe*. Toronto: Butterworths.
A detailed analysis aimed at determining whether the region is acquiring distinct geopolitical characteristics: that is, whether a 'new' region is emerging.

PEPPER, D., and A. JENKINS, eds. 1985. *The Geography of Peace and War*. Oxford: Blackwell.
An excellent set of readings focusing on such issues as the cold war, the arms race, nuclear war, and peace education.

ON THE WEB

http://www.europa.eu.int/
The home page of the European Union; a good source of news and detailed statistics.

http://www.nato.int/
The home page for the North Atlantic Treaty Organization; another excellent source of news and statistics.

http://www.chass.utoronto.ca/~dwelch/netlinks.html
A website containing links to information on world politics.

http://www.worldpolitical.com/
Part of the world news network, this site examines world politics today.

Interaction and Economic Globalization

This chapter builds on the distance concept introduced in Chapter 2. We begin with the challenging idea that distance is now less important, as a human geographic concept, than it was in the past. An outline of five kinds of distance (physical, time, economic, cognitive, and social) is followed by an overview of diffusion—a concept that has already played an important role in our discussions of migration, cultural regions, and colonial activity, and that will resurface with reference to the spread and growth of both agriculture and industry. Here we identify three approaches to diffusion research: cultural geography, spatial analysis, and political economy. The next section considers transportation, specifically the modes available for the movement of goods: by water, railway, road, and air. We then turn to the factors affecting trade; various theories of trade; and efforts at regional integration—a discussion that leads to our final two sections.

Many commentators believe that globalization is transforming the human world from a collection of connected but often very different places to a single entity in which differences are greatly reduced, if not eliminated. We will discuss this theory in a long-term context. Finally, we will take a detailed look at the processes involved in economic globalization, especially the rise of transnationals and the changing technologies of information transmission. This final discussion is an essential preface to our discussions of agriculture in Chapter 11, settlement in Chapter 12, and industry in Chapter 13.

The Information Centre at the Toronto Stock Exchange (Dick Hemingway).

Distance Friction and Frictionless Distance

Why are things located where they are? This is one of the key questions that human geography seeks to answer. The standard assumption has always been that location—whether of people, farms, towns, industries, or anything else—is not random, but follows a pattern. Sometimes the phenomena of human geography are close to one another and sometimes they are separated by great distances, but 'spatial regularity' means that we are able to make sense of those locations. For example, as we saw in Chapter 6, at the most basic level global population distribution can be explained by physical geography: some parts of the world have been better suited than others to support human life.

With this idea of spatial regularity in mind, one way to think about human geography is as 'a discipline in distance'—a phrase used by the distinguished British geographer Wreford Watson (1955) in explicit acknowledgement of the way humans use space and distribute their activities as a way of adjusting to distance.

Why is distance important in this way? Because overcoming it requires effort, and, as the American sociologist Zipf (1949:6), explained, 'an individual's entire behavior is subject to the minimizing of effort.' Applied to human movement, this **principle of least effort** suggests that location decisions are made to minimize the effort required to overcome what we call the **friction of distance**. Human geographers agree that many of our current distribution patterns reflect the friction of distance (Box 10.1). We saw in Chapter 7 how the first law of geography—that near things are more closely related than distant things—applied to cultural regions. Similarly, Chapters 11, 12, and 13 respectively will present many more examples of the relationships between one set of geographic facts and another: the choice of crops grown by farmers and access to market; the choice of residential location and workplace location; industrial sites and the locations of resources and markets. In every case, the friction of distance has played a central role.

Today, however, it seems that that in many instances this friction of distance is decreasing, perhaps even to the point where some distances, at least, will become frictionless. If that is true, is human geography still a discipline in distance? The answer is yes—and no.

On the one hand, even a cursory glance at the world today shows that distance still matters, since in most cases overcoming it still takes some effort. On the other hand, that same glance is likely to show that in many instances distance now matters somewhat less than it did in the past. The reason for the declining importance of distance as a factor determining where we locate things is that friction decreases as communication tech-

10.1

EKISTICS: A SCIENCE OF HUMAN SETTLEMENTS

In the late 1950s a Greek planner, Doxiadis, developed a science of human settlements, which he named *ekistics*. His central assertion was that all human settlements are part of a complex system that includes nature, humans, the built environment, and the networks that link humans. Within the context of this complex, all-encompassing system, Doxiadis argued that humans strive to:

1. maximize contacts with all parts of the system;
2. minimize the effort expended in achieving the first goal;
3. optimize use of space so as to maintain contacts but, at the same time, maintain an appropriate distance;

4. optimize the quality of all contacts; and
5. optimize the integration of the first four goals.

Doxiadis concluded that the location of settlements is determined with these goals in mind. Thus movement takes place along specific paths, which form the basic network of contacts. Settlement locations and transport routes are direct consequences of the behavioural imperatives underlying those five goals. This stimulating contribution is a useful reminder of the important role played by movement and contacts in human life. The principal source for information on this topic is the journal *Ekistics*, which has been published since 1957.

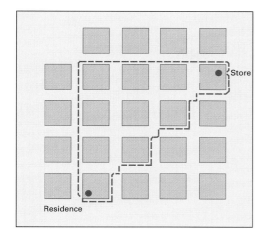

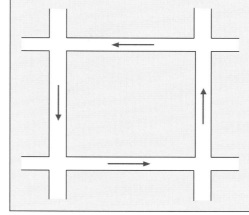

Figure 10.1 (left) A non-straight line, shortest distance route.

Figure 10.2 (right) A one-way system. To travel a block, the distance covered may be one block or three, depending on direction.

nologies improve: what is increasingly important today is the way locations are integrated with high-speed communication networks. For most of us, the most potent examples of an essentially frictionless distance are email and the Internet—technologies that make almost-instantaneous contact with faraway places a routine matter.

Highlighting the changing importance of distance, this chapter moves from the topic of diffusion to transportation, trade, and, finally, economic globalization. The friction associated with distance will become noticeably less as we progress through these topics. Nevertheless, for most of us, distance continues to play a central role in everyday life. You only have to think about your own daily life—travelling between home and school, visiting friends, choosing which restaurant or movie theatre to go to—to realize that the friction of distance and the principle of least effort are still important considerations in many respects.

Concepts of Distance and Space

Different interpretations of distance and space are appropriate for different activities. For example, in some cases distance may be best measured in terms of time or money, and by extension the same is true of sets of distances: spaces. Our discussions of distance and space are traditionally couched in terms of Euclidean geometry. This tradition has been described as a 'container view of space' (Harvey 1969:208), since it sees space as a framework in which geographic facts are located and geographic events occur. However, it is important to note that there are many different types of geometry. Riemann geometry, for example, is more

appropriate than Euclidean geometry when we are concerned with a spherical surface, such as the surface of the earth, because in Riemann space the shortest distance between two locations is a curved line, not a straight one. The characteristics of a particular geometry can be described in terms of the path of minimum distance between two points: geodesics. For example, when we say that economic and time distance are different, we are saying that economic space has a different geodesic than time space.

This chapter will consider five different but related concepts of distance and space: physical, time, economic, cognitive, and social.

Physical Distance

The spatial interval between points in space is the physical distance. It is often measured with reference to some standard system and is therefore a precise measurement. Sometimes, however, physical distance is measured in less precise terms. In ancient India, the standard unit of distance measurement was the *yojana*: the distance that the royal mode of transportation—an elephant—could travel from dawn to dusk (Lowe and Moryadas 1975:22). Even in advanced technologies, imprecise measurements may be used; in North American cities it is common to measure distance by reference to numbers of city blocks.

The shortest travel distance between points is often not a straight line. In a grid-pattern city, it is a series of differently oriented straight lines (Figure 10.1). In other instances, the physical distance between two points may be related to the direction of travel, as in the case of vehicle movement in a one-way street system (Figure 10.2).

Figure 10.3 (right) Time distance
in Edmonton.
Source: Adapted from J.C. Muller,
'The Mapping of Travel Time in Alberta,
Canada', *Canadian Geographer* 22
(1978):197-8.

Figure 10.4 (below) Toronto in
physical space and time space.
Source: G.O. Ewing and R. Wolfe,
'Surface Feature Interpolation on
Two-Dimensional Time-Space Maps',
Environment and Planning A9
(1977):430, 435.

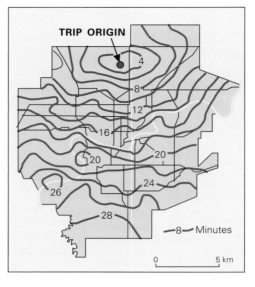

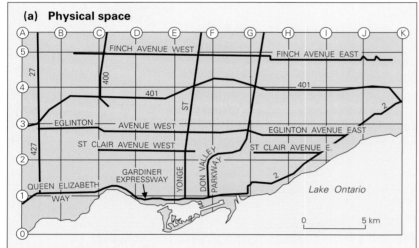

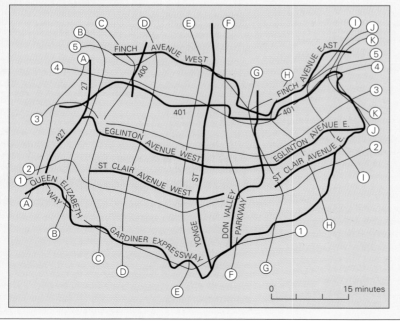

Time Distance

For some movements, especially those of
people as opposed to materials or products,
time is important; hence the preferred route
may be the quickest rather than the shortest.
Time distance is related to the mode of
movement, traffic densities, and various regulations regarding movement. Figure 10.3
depicts isochrones—lines joining points of
equal time distance from a single location—
for the city of Edmonton, Alberta. Clearly,
travel time is not directly proportional to
physical distance; if it were, the isochrones
would be equally spaced concentric circles.
Figure 10.4 presents an imaginative extension of these ideas depicting Toronto in (a)
physical space and (b) time space. The
time–space map shows space stretching in
the congested central area and shrinking in
the outlying areas—a direct consequence of
the greater time needed to travel a given distance on congested as opposed to freely
flowing routes. This pattern is not fixed, of
course. The extent of stretching and shrinking varies according to time of day and day
of the week (rush hour on a business day
generates the most stretching).

Converging locations

The idea that travel times typically decrease
with improvements in the technology of
transport has been labelled **time–space
convergence** (Janelle 1969). We can conceive of locations converging on each other;
these are locational changes in relative space
(recall that in Chapter 2 we described relative
space as subject to continuous change). It is
possible to calculate a convergence rate as the
average rate at which the time required to
move from one location to another decreases over time. For example, Janelle (1968) calculated the convergence rate between
London, England, and Edinburgh, Scotland,
from 1776 to 1966 as 29.3 minutes per year.
More generally, it took Magellan three years
(1519–22) to circumnavigate the globe.
Today we can fly around the world in less
than two days. With the laying of the first
telegraph cable across the North Atlantic
seabed in 1858, the 'distance' between
Europe and North America was reduced
from weeks to minutes, and now information
can be transmitted around the world almost
instantaneously.

Landing the telegraph cable at Heart's Content Bay, Newfoundland, from the *London Illustrated News*, 8 Sept. 1866 (National Archives of Canada, c–066507).

Economic Distance

Movement from one location in space to another usually entails an economic cost of one kind or another. Thus 'economic distance' can be defined as the cost incurred to overcome physical distance. As with time distance, there is not necessarily a direct relationship between physical distance and other measures. If we consider the movement of commodities, for example, it is not uncommon to find that costs increase in a step-like fashion and that the cost curve is convex (see Figure 13.5, p. 464). Similarly, in some cities taxi fares are determined not by physical distance but by the number of zones crossed.

For many industries and other businesses, cost distance is of paramount importance. There is considerable logic to the notion that economic activities should be mapped in economic space, not physical or container space.

In Britain in 1950 the average person travelled about 8 km (5 miles) a day, mostly on foot. Today this average distance has increased to about 45 km (28 miles), most of it not on foot. Similar changes have taken place throughout the more developed world. Much of the increase in distance travelled can be attributed to improvements in transportation networks, and the growth of suburbs that in many cases require commuters to cover greater distances. Distance travelled is closely linked to economic status—people who regularly overcome large distances for business and pleasure are sometimes described as 'hypermobile'.

Cognitive Distance

Spatial cognition is a matter of individuals' capacity to be precise in determining locations and distances. Each individual has a different interpretation of reality—that is, of actual locations and actual physical distances. Understanding individual interpretations may be essential if we are to understand human spatial behaviour. Indeed, cognitive distance may be a better explanation for human movement than physical distance—but how can it be measured? Physical, time, and economic distance can all be measured precisely, but cognitive distance cannot. Instead, cognitive distances are determined by asking respondents about the relative proximity of various locations to one given location. The importance of cognitive distance to behaviour is then assessed by analyzing the discrepancies between physical and cognitive distance. Also of interest are the reasons why cognitive distances differ from physical ones.

Most research has demonstrated that cognition affects both our measures of distance and our larger spatial comprehension. Cognition itself is affected by the individual and social characteristics of humans. Most of us are able to be relatively precise about distances and locations in the area that we know well, but as physical distance increases, cognitive distance becomes less precise. In addition, most of us judge favoured locations to be closer than locations we dislike. In North America, for example, many people feel that the west coast is an attractive place to live and

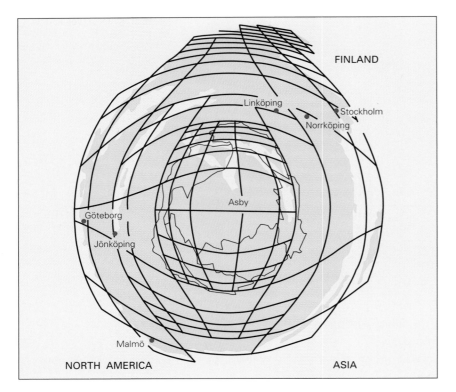

Figure 10.5 Logarithmic transformation of distances from Asby, Sweden.

hence may judge it to be closer than it actually is, whether they measure the distance in physical, temporal, or economic terms. (For more on mental maps, see Chapter 6, especially Figure 6.2, p. 185.)

Although it is clear that cognition affects behaviour in space, including movement, geographers are still uncertain whether any specific cause-and-effect relationships exist. It remains to be determined whether cognitive distances and spaces can be used to predict spatial behaviour as well as spatial preferences.

Social Distance

The concept of social distance is complex, and the sociological literature offers a variety of interpretations. **Proxemics** is the study of social distances as defined by individuals in particular circumstances (for example, social interactions such as conversation), and is typically a subject for social psychologists, not geographers (Hall 1966). At the group scale, the correspondence between social and physical distance has been studied with special reference to residential distributions. Following the lead of urban sociology, geographers have analyzed the links between social status (often interpreted as class) and residence. Decisions about residence result in the creation of relatively distinct social spaces. One way to meas-

ure social distance is by the amount of interaction between groups—for example, the degree of intermarriage. Social distance is most complete in cultures with rigid hierarchical **caste** systems.

Transformations of Distance

In geographic studies, the use of physical distance may often be inappropriate; the relevant measure may actually be temporal, economic, cognitive, or social. This is one reason it may be useful to transform physical distance data (others have to do with certain statistical procedures). As an example, we turn to the classic work of the Swedish geographer Hagerstrand (1967).

Recognizing that movement intensity decreased sharply with increasing distance from the focal point because potential movers were less familiar with distant places than with nearer ones, Hagerstrand mapped the logarithms of the physical distances (Figure 10.5). The focal point for his analysis was the small town of Asby, Sweden, which is located at the centre of a map that represents the information space of Asby residents. The cartographic result of this logarithmic transformation is that nearby distances appear exaggerated, while distances farther away seem to shrink (to appreciate the displacement of the other locations identified, compare this map of Sweden with a more conventional one). Many types of human interaction, as well as human movement, exhibit the same negative relationship to distance. Figure 10.5 may well be the correct cognitive map. From a practical cartographic viewpoint, a logarithmic-transformed map creates room for the most information where it is most needed, at the shorter distances.

A rather different type of transformation is the **topological map**. The correct spatial order of locations is maintained, but space itself is elastic because distance and direction can be altered. An excellent example of a topological transformation is a map of the London underground system (Figure 10.6). If we assume that most of us view distance not in precise quantitative terms but rather in relative terms (such as 'less than' or 'greater than'), some topological transformations may be highly relevant to the geography of movement—perhaps more relevant than the original untransformed physical-distance data.

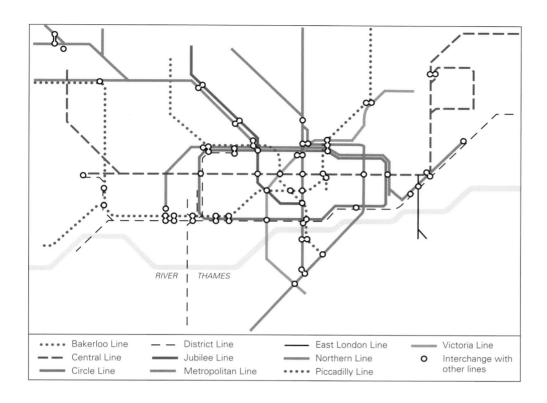

Figure 10.6 The London underground system.

Legend:
- ••••• Bakerloo Line
- – – Central Line
- —— Circle Line
- – – District Line
- ▅▅▅ Jubilee Line
- ▅▅▅ Metropolitan Line
- —— East London Line
- —— Northern Line
- ••••• Piccadilly Line
- —— Victoria Line
- ○ Interchange with other lines

RIVER | THAMES

Gravity and Potential Models

Many geographic studies of movement between locations have found a close link between the quantity of movement on the one hand and both the physical distance between locations and the sizes of locations on the other.

Gravity model

The gravity model describes the relationship between quantity of movement and distance as follows:

$$I_{ij} = \frac{M_i \times M_j}{D_{ij}^{\,b}}$$

where

I_{ij} is the interaction (movement) between locations i and j

M_i and M_j are the sizes (usually population sizes) of locations i and j

D_{ij} is the distance between locations i and j

b is a distance decay function

This formulation is a human equivalent to the gravity concept in physics. It proposes that the quantity of movement between two locations increases as their size increases: in other words, the larger $M_i \times M_j$ is, the greater the movement. It also proposes that the quantity of movement between two locations decreases as D_{ij}—the distance between the locations—increases. Empirical evidence suggests that the impact of distance is not uniform and that D_{ij} needs to be raised to some power: thus $D_{ij}^{\,b}$, where b varies, depending on circumstances, from as low as 0.5 to as high as 3.0. Note that a high value of b implies considerable distance friction, while a low value of b implies the reverse. As noted in the opening comments to this chapter, transportation and technological advances cause the value of b to decrease; that is, they decrease the frictional effects of distance (Box 10.2).

This valuable model was introduced by the nineteenth-century social scientist H.C. Carey. Carrothers quoted Carey as follows:

> Man, the molecule of society is the subject of Social Science. . . . The great law of *Molecular Gravitation* [is] the indispensable condition of the existence of the being known as man. . . . The greater the number collected in a given space the greater is the attractive space that is there exerted. . . . Gravitation is here, as everywhere, in the *direct* ratio of the mass, and the *inverse* one of distance (Carrothers 1956:94).

We have already encountered one of the first uses of this concept in our discussion of migration. In 1885, Ravenstein used a variation of the model when analyzing migration flows between English cities. Other applications appeared from the 1920s onwards and included a 'Law of Retail Gravitation' (Reilly 1931).

The gravity model assumes that interaction—in the form of the movement of people or goods, telephone communication, or whatever else—takes place between locations. It also assumes that the greater the size of the location (such as an urban centre), the greater the interaction will be; in so doing, it treats all individuals as the same. Of course this assumption is overly simplistic, since individuals are not equal in ability to generate interaction. Nevertheless, the gravity model is attractive because it is simple, it has parallels in physical science, and it has provided useful descriptions.

Potential model

The gravity model is concerned with relationships between pairs of points. The potential model is conceptually similar to the gravity model, but is concerned with the influence that many points have on one point. It can be expressed as follows:

$$V_i = \sum_{j=1}^{n} \frac{M_i}{D_{ij}^b}$$

where

V_i = potential at location i
M_i = size of location i
D_{ij} = distance between locations i and j
b = a distance decay function

Thus if a town, i, is located in a region with other towns, then town i has some potential for interaction with other towns. The potential at i is an aggregate measure of the influence of all other $j = 1$ to n places.

10.2

THE TYRANNY OF DISTANCE

The Tyranny of Distance is the title of a history of Australia by the eminent Australian historian Geoffrey Blainey. Blainey (1968:2) begins: 'in the eighteenth century the world was becoming one world but Australia was still a world of its own. It was untouched by Europe's customs and commerce. It was more isolated than the Himalayas or the heart of Siberia.' The twin ideas of distance and isolation frame the processes of change in Australia in the European era. The four principal 'distances' are:

1. distance between Australia and Europe;
2. distance between Australia and nearer lands;
3. distance between Australian ports and the interior; and
4. distance along the Australian coast.

The distance concept proposes new ways of explaining why Britain sent settlers to Australia in 1788 and why the colony was at first weak but gradually succeeded. Similarly, it aids understanding of Chinese immigration in the 1850s and later immigrations from southern Europe. Within Australia, the distances from coast to interior and along the coast are closely linked to economic and other change. The first Australian staple, wool, was effectively able to overcome the distance from pasture land to port and the distance from port to Europe. Wool opened up much of southeastern Australia and was followed briefly by gold and then wheat. These products, along with technological innovations, combined to overcome the problems of distance.

Blainey is acutely aware that distance was not the only relevant variable—climate, resources, European ideas, and events were also important. Similarly, he continually acknowledges that distance is not a simple concept: it is fluid, not static, and can be measured in many ways. In later writings, Blainey has noted that distance is also a key to understanding the Aboriginal history of Australia.

Sensibly used, the distance concept can help us to understand the unfolding of life in many parts of the world. Blainey's work is eminently geographical.

Although the distance concept has not been explicitly applied by Canadian historians, it is central to a leading thesis in Canadian history. This 'Laurentian' thesis, proposed by Harold Innis, places the Canadian historical experience in the context of staple exports and communications that—as in Australia—effectively overcame the tyranny of distance.

As in the gravity model, distances may be raised to some power to accommodate the friction of distance. This model too has been extensively used by human geographers.

Studies such as these, in which physics analogies are applied to social phenomena, are known as **social physics**. Not all human geographers are comfortable with an approach that treats humans as molecules and places as masses. Nevertheless, as Carrothers puts it, 'while it may not be possible to describe the actions and reactions of the individual human in mathematical terms, it is quite conceivable that interactions of groups of people may be described this way' (Carrothers 1956:201).

Extensions of the basic models

The gravity and potential models are simplifications of complex phenomena, and, quite appropriately, numerous variations and extensions of the basic models have been proposed. Two closely related sets of models are worth noting here. First are the entropy-maximizing models introduced by Wilson (see Wilson and Bennett 1986). The term **entropy**, which originated in thermodynamics, refers to the disorganization in a system: entropy-maximizing models identify the most likely state of a system— that is, how many geographic facts are located in certain areas, given certain constraints. They have been extensively used for modelling spatial interaction. Second are location-allocation models, which can be used to determine the optimal location—that is, the location that minimizes movement and other costs associated with facilities, such as hospitals.

Links to Regional Science

In the late 1940s, the hybrid discipline of regional science developed in response to dissatisfaction with the achievements of regional economic analysis. Largely the creation of the American economist Walter Isard, who wrote two major works (1956, 1960), it favoured the use of neoclassical economic theory and statistical techniques. This influential new discipline proved attractive to the many human geographers who favoured spatial analysis as the dominant approach to their discipline. Regional science is important in the present context

because it proved to be one of the principal sets of ideas through which social physics procedures were accepted and practised by geographers such as Warntz (for example, 1957). Since about 1970, in response to the humanist and Marxist critiques of positivistic spatial analysis, interest in regional science and social physics has waned considerably. As will become evident in Chapters 11, 12, and 13, contemporary geographers are more attracted to approaches such as those based on political economy.

Diffusion

Spatial patterns are rarely static. To understand present patterns, it is often helpful to analyze past patterns; similarly, understanding current circumstances may help us to predict the future. We have frequently had occasion to acknowledge (at least implicitly) this point— and the importance of time generally—in this book.

Diffusion is best interpreted as the process of spread in geographic space and growth through time. Migration—the movement of people—can be regarded as a form of diffusion, although the term is more often used in the context of some particular **innovation**, such as a new agricultural technique. Diffusion research has a rich heritage in geography, and has been associated with three approaches in particular: cultural geography, spatial analysis, and political economy.

Diffusion Research in Cultural Geography

Until the 1960s, most diffusion research took place under the general rubric of historical and cultural geography in efforts to understand cultural origins, cultural regions, and cultural landscapes. It is specifically associated with the landscape school initiated by Sauer. Typical studies focused on the diffusion of particular material landscape features, such as housing types, agricultural fairs, covered bridges, place names, or grid-pattern towns. The usual approach was to identify an origin and then describe and map diffusion outwards from it; the work of Kniffen (for instance, 1951) is a good example.

Various issues emerged from this research. In many cases, debate centred on the question of single or multiple invention, or the numbers and locations of hearth areas, or the problem of agricultural origin source areas and subsequent diffusion. The importance of ethnic or social characteristics in relation to particular groups' receptivity to innovations was frequently acknowledged. For example, a detailed study of cigar tobacco production in the United States that focused on the significance of ethnicity—'In each of the tobacco producing districts, tobacco culture came to be identified with hard work, clever farming techniques, economic independence, and with an ethnic group' (Raitz 1973:305)—found that in Wisconsin there was a very close relationship between people of Norwegian descent and tobacco production.

Diffusion continues to be a central concern in cultural geography.

Diffusion Research and Spatial Analysis

The character of diffusion research changed substantially following the work of Hagerstrand (1951, 1967), who pioneered the use of models and statistical procedures. This shift was associated in North America with the rise of spatial analysis as a major approach in human geography after about 1955. Although the specific interests of these researchers were quite different from those of cultural geographers, the central concern was unchanged: to study diffusion as a process effecting change in human landscapes. The importance of chance factors was explicitly acknowledged by Hagerstrand in his use of a procedure known as Monte Carlo **simulation**, which allowed for the likelihood of any given acceptance of an innovation to be interpreted as a probability. Spatial analysis-oriented diffusion research also introduced a number of themes that we might call empirical regularities, because they are consistently observable in geographic analyses. Four of these empirical regularities are outlined below.

Neighbourhood effect

An innovation such as a new farming technology will likely be adopted first close to its source and later at greater and greater distances. This is the neighbourhood effect—a term that describes the situation where diffusion is distance-biased. In general, the probability of new adoptions is higher for those who live near the existing adopters than it is for those who live farther away. The simplest description of this situation, of course, is a diagram showing a series of concentric circles of decreasing intensity with increasing distance—similar to the ripple effect produced by throwing a pebble into a lake.

The neighbourhood effect occurs in those circumstances where an individual's behaviour is strongly conditioned by the

10.3

CHOLERA DIFFUSION

In the nineteenth century, North America experienced three major cholera pandemics: in 1832, 1848, and 1866. Cholera is generally spread by water contaminated with human feces, but it can also be carried by flies and food. Climatically, its diffusion is promoted by warm, dry weather.

Each of the three pandemics diffused differently as the urban and communication systems changed. The 1832 outbreak spread in a neighbourhood fashion, the 1866 outbreak in a hierarchical fashion. The 1848 outbreak demonstrated aspects of both types of diffusion (Pyle 1969).

The 1832 pandemic occurred at a time when both urbanization and the communication system were limited. Urban centres were few and small, and movement was largely by water—coastal, river, and canal. Data on cholera spread show a neighbourhood process. After initial outbreaks near Montreal and New York City, the disease moved along distance-biased, not city-size-biased, paths. The 1866 pandemic occurred in the context of a very different urban and communication system: eastern North America was by then well served by rail, and an integrated urban system was developing. These changing circumstances allowed cholera to diffuse hierarchically. From New York City it travelled along major rail routes to distant second-order centres, as well as along a number of other minor paths. In many cases, cholera took much longer to reach areas close to the urban areas where it started than it did to reach more distant large cities.

This example is a useful reminder of the importance of economic and other infrastructures to any diffusion process. In general, a neighbourhood effect is more likely in circumstances of limited technology and on a local scale, while a hierarchical effect is more likely in a developed technological context and in a larger area.

local social environment. It is most common in small rural communities with limited mass media communication, where the most important influences on behaviour are personal relationships—in traditional or '*Gemeinschaft*' societies (see page 416). The logarithmic transformation illustrated in Figure 10.5 describes such circumstances. The neighbourhood effect is likely to be least evident in urban settings and in circumstances of improving transport and communication technologies that reduce distance friction.

Hierarchical effect

The hierarchical effect is evident when larger centres adopt first and subsequent diffusion spreads not only spatially but also vertically down the urban hierarchy. Thus the innovation jumps from town to town in a spatially selective fashion rather than spreading in the wave-like manner associated with the neighbourhood effect.

This effect is likely to occur in circumstances where the receptive population is urban rather than rural, and is increasingly evident as transport technology improves (Box 10.3).

Resistance

Of course, reception of an innovation does not guarantee acceptance. Resistance to innovation is a sociological phenomenon. The greater the resistance, the longer the time before adoption occurs. Resistance too shows a spatial pattern: urban dwellers are typically less conservative than their rural counterparts.

S-shaped curve

Probably the best-supported empirical regularity is the S-shaped curve. When the cumulative percentage of adopters of an innovation is plotted against time, the typical result is an S-shaped curve (see Figure 2.8, p. 62). This curve describes a process that begins gradually and then picks up pace, only to slow down again in the final stages.

A simple conceptual framework

A simple framework for conceptualizing the process of innovation diffusion is derived from a variety of sources, but especially from Rogers (1962).

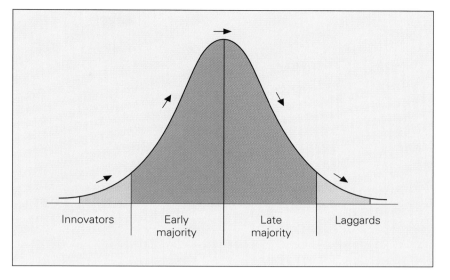

Figure 10.7 Distribution of innovativeness. As the innovation is adopted over time, an S-shaped curve emerges.

The process of innovation diffusion has four components: the innovation, a population of potential adopters, the communication process, and the adoption process. The *innovation* is assumed to be a desirable advance upon earlier procedures. For such an innovation to be diffused and either accepted or rejected, a *population* is necessary. At any given time during the process of diffusion, there are two classes of individuals in the population: adopters and non-adopters. There are three explanations for the presence of non-adopters: they have not heard of the innovation; they have heard and are in the process of making a decision; or they have heard and have made the decision to reject it. In addition, a third category of individuals, called change agents, is often relevant. These are people who promote the innovation, but who may be regarded as external to the system; for convenience, they cannot be either adopters or non-adopters themselves.

The *communication process* is the process by which an innovation, or knowledge relating to it, is diffused. It operates when either a change agent or an adopter communicates with a non-adopter. Communication may take two basic forms: pairwise (individual to individual) or mass (individual or organization to group). Adoption of an innovation is rarely immediate upon receipt of information; it takes time. The *adoption process* may be described as follows: awareness—interest—evaluation—trial—adoption. Individuals vary in their innovativeness and may be classified as innovators, early majority, late majority, and laggards (Figure 10.7).

The global diffusion of products such as Coca Cola is facilitated by the more general diffusion of American cultural and economic characteristics. These two images are from Guadaloupe and Malaysia (Victor Last, Geographical Visual Aids).

The pattern of innovation diffusion is assumed to be equivalent to the pattern of adopters, and the adoption process is assumed to be a learning process. The entire process, then, is envisaged as a stimulus–response situation, with receipt of information as the stimulus and the decision to adopt or reject as the response. Adoption or learning may occur as a result of reinforcement, insight, or both.

This framework recognizes processes of communication and adoption that allow an individual to be in one of three states:

1. *No knowledge*: the state prior to the successful completion of the communication process.
2. *Knowledge*: the state after receipt of the information, but prior to the decision-making.
3. *Adoption*: the state after the decision to adopt is made.

Like any conceptual framework, this one deliberately simplifies a complex reality, and is intended merely to aid in diffusion research.

Diffusion Research and Political Economy

For many researchers, the framework outlined in the preceding section represents yet another example of the dehumanizing approach characteristic of the spatial analysis school. Accordingly, after about 1970 a third approach was developed that focused not on the mechanics of the diffusion process but rather on the explicitly human consequences of diffusion. Although there are various aspects to

this third approach, it is perhaps most frequently associated with Marxist perspectives (discussed in more detail in Chapter 11).

Any diffusion process adds to existing technological capacity, but it also affects the use of resources in some way. For example, some innovations result in significant time savings and hence change the daily time budgets of individuals; others actually demand more time, as in the case of the introduction of formal schooling. Analyzing such innovations means analyzing cultural change.

Contemporary research into innovation diffusion also considers the extent to which there may be a spatial pattern of innovativeness related to a wide range of social variables such as relative wealth, level of education, gender, age, employment status, and physical ability. These variables go far beyond the broad spatial variables we have noted before now, such as rural–urban or ethnic differences. Not surprisingly, such research also tends to consider overarching social, economic, and political conditions over which most individuals exercise little or no control.

In a study of the diffusion of agricultural innovations in Kenya since the Second World War, Freeman (1985) discovered that the diffusion process could be greatly affected by the pre-emption of valuable innovations by early adopters. The difference between the two curves shown in Figure 10.8 reflects the influence of an entrenched élite who, as early adopters, saw advantages in limiting adoption by others. From their positions of social authority, members of the élite were able to take political action, lobbying for legislation to prevent further spread of an innovation or limiting access to essential agricultural processing facilities. This pattern was observed in Kenya for three agricultural innovations (coffee, pyrethrum, and processed dairy products).

This example makes it clear that the process by which an innovation spreads can be just as important as the innovation itself.

The Diffusion of Disease

Geographers concerned with mapping and modelling disease diffusion have used a variety of procedures reflecting diverse philosophical perspectives. The discussion of cholera diffusion in Box 10.3 is presented as a case-study using Hagerstrand-derived empirical regularities in the spatial analysis tradition to describe the diffusion process. In addition, research on the diffusion of infectious diseases has sometimes used mathematical modelling techniques (for example, Cliff, Haggett, and Ord 1986; Cliff and Smallman-Raynor 1992).

Mapping has been particularly useful with respect to the diffusion of AIDS; Gould (1993) demonstrates the value of maps in understanding the spread of the disease, and Smallman-Raynor, Cliff, and Haggett published an AIDS atlas (1992). Work such as this emphasizes the complexity of the diffusion process, identifying cultural, social, behavioural, economic, political, and transportation factors that have all played a part in past and present distribution patterns of AIDS. Thus physical and human geographers are contributing to our understanding of AIDS through their interest and expertise in space, movement, and diffusion.

Transportation

Humans have continually striven to facilitate movement across the surface of the earth by reducing distance friction and the costs of interaction. One result has been the evolution of transport systems. Complex components of the human landscape, transport systems use specific modes (such as road, rail, water, or air) to move people and materials, to allow for spatial interaction, and to link centres of supply and demand. The fact that in much of the less developed world transport systems are largely deficient has a direct economic effect, limiting activities such as commercial agriculture and mineral production.

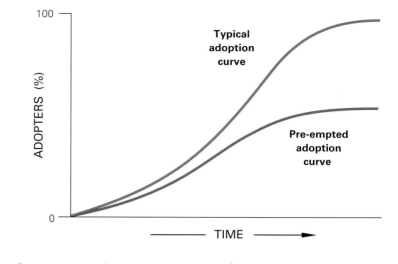

Figure 10.8 Effects of pre-emption on the adoption curve.

In some countries, a transport system has been constructed to encourage national unity and open up new settlement regions. It is not a coincidence that this has occurred in the two largest countries in the world, both with large areas of difficult environment: Canada and Russia. Canada's transcontinental railway, completed in 1886, served as a crucial centripetal factor and encouraged settlement of the prairie region. The trans-Siberian railway between Moscow and Vladivostok has been operating since the beginning of the twentieth century.

Transport Geography: An Overview

Geographic study of transport is not as well developed as the studies of agriculture, settlement, and industry discussed in the following chapters. Nineteenth-century geographers such as Ratzel and Hettner viewed transport routes as landscape features and as factors related to more general landscape changes, and the early twentieth-century French school of human geography regarded transport as a critical component of the geography of circulation. Nevertheless, transport geography remained largely unexplored until the 1950s, when several geographers published studies of particular modes of transport. The most significant developments came with the rise of spatial analysis; they involved quantitative and modelling studies, many of which had a planning orientation. Even today, transport geography has a notably positivistic flavour, and has been relatively unaffected by humanist, Marxist, and other types of social theory (Hoyle and Knowles 1992).

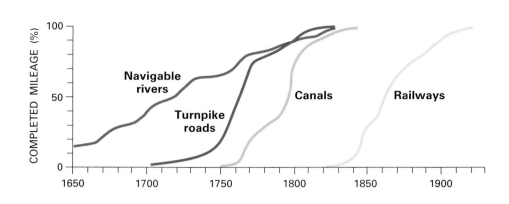

Evolution of Transport Systems

Three processes characterize the evolution of transport systems: intensification, or the filling of space; diffusion, or spread across space; and articulation, or the development of more efficient spatial structures. However, changes in transport systems do not occur at a steady pace. Just as agriculture, settlement, and industry have experienced times of revolutionary (as opposed to evolutionary) change, so has transportation. In fact, the changes that have occurred in transportation reflect many of the same basic forces that have prompted revolutionary change in other aspects of the human geographic landscape. In short, transport systems change in response to (1) advances in technology and (2) various social and political factors. A review of the evolution of transport networks in Britain will clarify these generalizations.

Britain

The first organized transport system in Britain was created by the Romans between 100 and 400 CE for political and military reasons. Unlike the earlier system of local routes intended simply to permit social interaction between settlement centres, the Roman system focused specifically on facilitating rapid movement between London and the key Roman centres. Once the Romans withdrew from Britain, their system fell into disuse because it did not serve the needs of the local population. Only in the seventeenth century did a new national system emerge, reflecting the emerging national interest; this system took the form of carriers and coach services. By the early eighteenth century, two innovations diffused rapidly across Britain: turnpike roads (an organizational innovation) and navigable waterways, especially canals (a techno-

logical innovation). In the nineteenth century, the technological innovation of railways appeared, and in the twentieth century, the road system responded to the new technology of automobiles. Finally, air transport added a new dimension to the overall transport system.

Turnpike roads, charging tolls, transferred the cost of road maintenance from local residents to actual road users. No new technology was involved, and thus, in principle, a turnpike system could have been set up at any time. Beginning about 1700, turnpikes diffused rapidly. During the early industrial period, *canals* played an important role in serving mines and ironworks. After 1790 they became the dominant transport mode, and until about 1830 they played an important role in industrial location decisions and related population distributions. Canals were tied to industry, since they were best suited for moving heavy raw materials, whereas turnpikes were better suited to the movement of people and information. Thus the two new systems were largely complementary rather than competitive.

Neither of these eighteenth-century developments compared to railways in long-term impact. *Railways* dominated British transport from about 1850 to about 1920. The first railway was completed in 1825. The early railway network was relatively local and related to industry, but after 1850 railways linked all urban centres. Railway construction in Britain was organized and financed by small groups of entrepreneurs with minimal government intervention. Elsewhere in Europe and overseas, governments typically played a larger role in railway construction.

Each of these three innovations—turnpike roads, canals, and railways—displays the characteristic S-shaped growth curve (Figure 10.9): a slow start, rapid expansion, and a slow completion period. The pattern by which transport systems evolved in Britain is fairly typical. Today, most parts of the developed world have transport systems that include navigable waterways, railways, roads, and air traffic, and are continually changing in response to technology and demand.

Theories of transport system evolution

Several human geographers have studied the evolution of transport systems from a theoretical perspective. For the less developed world, Taafe, Morrill, and Gould (1963) proposed four stages of development (Figure 10.10):

1. A series of ports are scattered along an entry zone (most likely a coastline); there is no real network (Fig. part a).
2. A few of the early ports prosper and develop networks inland (parts b, c).
3. A number of feeder routes and lateral connections arise (parts d, e).
4. High-priority linkages emerge (part f).

This generalized model usefully complements detailed empirical studies and has inspired a substantial body of research (Box 10.4). A second and rather different theoretical focus concerns the complex relationship between transport change and larger issues of economic change. Transport is both cause and effect and must be considered in analyses of, for example, agricultural, settlement, and industrial change (Box 10.5).

Networks

A network consists of specific locations linked together by routes to form some interconnected system. The locations that are linked by routes are usually referred to as *nodes*. Nodes are both sources and destinations of all types of movement; they may be individuals or aggregates of individuals in the form of specific businesses or settlements. In the latter two cases, nodes are spatially fixed and their distributions can be described as dense, sparse, clustered, dispersed, and so forth. Whereas central place theory is closely concerned with settlements as nodes and their relative locations, in this chapter we are more concerned with the routes that facilitate movement between nodes. *Routes*—the channels along which interaction occurs—are sometimes spatially fixed, as in the case of

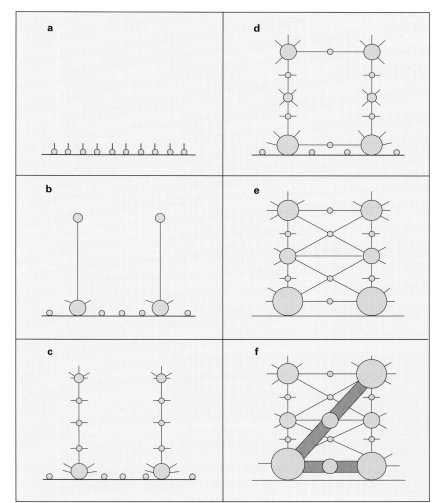

Figure 10.10 Evolution of a transport network.

roads and railways, and sometimes more flexible, as in the case of air and sea routes.

Geographers have long been interested in the characteristics of routes, especially their locations. Lines of communication are an essential factor in all economic geography: farmers and their products must travel to market and industrialists must move materials and products. In one sense, then, routes are the skeleton on which the flesh of human geography is built. 'Communications are themselves a part of the landscape' (Appleton 1962:xviii). As for geographers, we ask our usual questions about where and why.

Graph theory

Networks are most easily analyzed by using graph theory. Although they involve simplification, graphs make objective analysis possible. Geographers were introduced to graph theory by one of the pioneers of theoretical geography, William Garrison (1960). Lowe and Moryadas (1975:79) provide an example of the procedure. Figure 10.11a shows Martinique, a mountainous island with a circuitous road network; Figure 10.11b converts the network to a graph. The graph simplifies the network by excluding incidental characteristics but retaining the essential topology. The elements of the graph are nodes and routes—elements that can be used to generate a series of measures such as the number of nodes and routes. Kansky (1963) introduced a large number of other graph theoretic measures such as the beta index, the number of linkages per node:

$$beta = r/n$$

where

n = number of nodes
r = number of routes

A beta value greater than one suggests that alternative routings exist between some pairs of nodes. For the Martinique road network, beta = 1.49. Beta is generally higher in more developed countries than in less developed ones. A second measure introduced by Kansky (1963) is the gamma index: the ratio between the actual and the possible number of routes (minimum 0, maximum 1): gamma = $r/3(n - 2)$. For the Martinique road network, gamma equals 0.52. This measure is often expressed as a percentage; 0.52 is converted to 52 per cent, meaning that the network has a 52 per cent level of connectivity.

The beta and gamma measures indicate the utility of graph theory and are used extensively. Nevertheless, graph theory shares the drawbacks of all such simplified procedures; information is lost (deliberately), and highly variable nodes and routes are treated as though they were identical. Many networks have a hierarchical structure, in which the nodes that are close together are especially well connected.

Modes of Transport

In some cases, different modes of transport that may have evolved at different times, and under very different human geographic circumstances, combine to create an integrated system. In other cases, route duplication and a general lack of coordination may combine to create a system that as a whole is somewhat incoherent, although each specific mode may be a perfectly logical system in itself.

Each of the various transport modes has advantages and disadvantages. *Water* transportation is a particularly inexpensive means of moving people and goods over long distances because water offers minimal resistance to movement and because waterways are often available for use at no charge. But it is slow, and may be circuitous. In the case of inland waterways, local relief can be a major

10.4

PORT SYSTEM EVOLUTION

A study of the ports of Ghana concluded that the system developed from 'a highly unstable scattering of numerous primitive surf-ports with restricted hinterland links' to 'a gradual concentration of traffic and the emergence of a stable system with dependence on two efficient deep-water ports' (Hilling 1977: 104).

In this Ghanaian example, transport development and port development are linked:

1. First there was a long period (1482–1900) of primitive surf-port operations corresponding to the first scattered-ports stage of the Taafe, Morrill, and Gould (1963) model. Surf-ports—necessary because there were no good natural harbours—usually involved a pier extending beyond the surf zone.

2. Between about 1900 and 1928 the transport system penetrated inland and two ports, Sekondi and Accra, achieved dominance. Both of them acquired additional port facilities.

3. After 1928 the interior transport system developed a series of interconnections, while deep-water ports were constructed at the two most suitable sites, Tema (near Accra) and Takoradi (near Sekondi).

4. Finally, a high-priority route system developed, focusing on the two deep-water ports, and most of the smaller ports lost trade.

As a consequence of these changes, Ghana now has a contemporary port and a transport system that is cost-efficient.

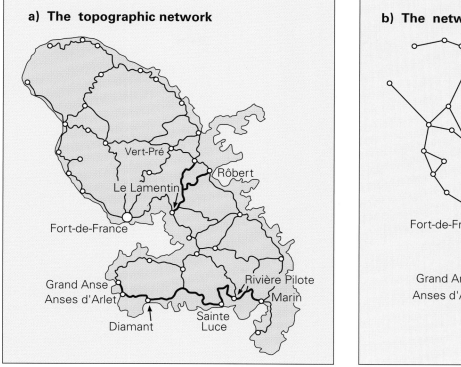

a) **The topographic network**

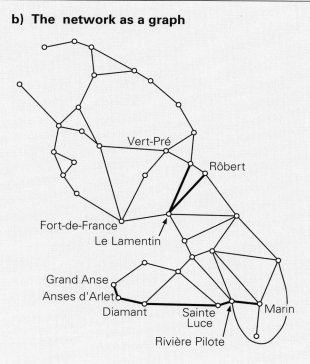

b) **The network as a graph**

Figure 10.11 The road network of Martinique.

Source: J.C. Lowe and S. Moryadas, *The Geography of Movement* (Boston: Houghton Mifflin, 1975):80.

obstacle. In some areas weather conditions—especially freezing and storms—may cause particular problems for water transport: in northern Manitoba, for example, the port of Churchill on Hudson Bay is open only from mid-July to mid-November. Canals continue to be a critical component of many transportation systems. In Europe, the Rhine–Main–Danube Canal was completed in 1992, allowing for a continuous water link between the North Sea and the Black Sea.

Land transportation includes railways and roads. *Railways* are the second least expensive form of transportation, after water, and are suitable for moving bulk materials when waterways are not available. However, in some areas local relief makes railway construction (and maintenance) expensive, if not impossible. The most significant railway construction in recent years has taken place in Europe. The 'Channel tunnel' between England and France, opened in 1994, is in

10.5

RAILWAYS AND ECONOMIC GROWTH

Traditionally, transport has been regarded as a primary causal variable in economic growth and given a pre-eminent role in several key theories of economic change. Rostow (1960), for example, saw railways as the initiators of economic change.

In the 1960s, this traditional view was seriously questioned by a group of 'new' economic historians who drew heavily on the concept of social savings: the difference between actual national income and national income without railways. The usual approach was to employ a counterfactual scenario (recall Box 9.4). Thus the following research question was posed: In a given area at a given date, what would have happened had there not

been a railway system? Central to any such analysis is the argument that we can understand the importance of railways to economic growth only if we describe the area with and without railways. Needless to say, any such work is subject to controversy over both the general approach adopted and the specific uses made of it. In any event, the general conclusion was that railways were not the primary cause of economic growth—the 'social savings' were minimal. In geographic terms, the related conclusion is that, without railways, the human landscape would have differed little in general content, although the details would, of course, have differed considerably.

Passengers arriving from London at the Gare du Nord in Paris. The train that makes the Channel run is called the Eurostar (Victor Last, Geographical Visual Aids).

that it is less likely to be hindered by human-created barriers. International trade is playing an increasingly important role in the world, as it has grown much more rapidly in recent years than international production. This confirms the apparent trend towards international integration in the global economy.

Factors Affecting Trade

The movement of goods from one location to another reflects spatial variations in resources, technology, and culture. The single most relevant variable related to trade is distance. Like all forms of movement, trade is highly vulnerable to the friction of distance—a friction that decreases as technology advances. Other relevant variables include:

1. the specific resource base of a given area (needed materials are imported and surplus materials exported);
2. the size and quality of the labour force (a country with a small labour force but plentiful resources is likely to produce and export raw materials); and
3. the amount of capital in a country (higher capital prompts export of high-quality, high-value goods).

All these variables are closely linked to level of development, and in fact trade between more and less developed countries is often an unequal exchange: the former export goods to the latter at prices above their value, while the latter export goods to the former at prices below their value. Moreover, the less developed countries are inclined to excessive specialization in a small number of primary products, or even a single staple. More developed countries may also specialize, of course, but they do so in manufactured goods. This difference reflects the technologically induced division of labour between more and less developed worlds. The fact that the less developed countries are typically dependent on the more developed countries as trading partners is central to the world systems theory discussed in Chapter 6. Today the majority of world trade moves between developed countries, although the rise of the newly industrializing countries (NICS; see Chapter 13) is affecting trade flows. For a simplified explanation of commodity flows, see Box 10.6.

fact three tunnels—two for trains and one for services. In the Scandinavian region a series of rail and road links (including tunnels and bridges) link Sweden and Norway through Denmark to Germany (Scanlink, completed in 2000). In recent years several train tunnel projects have been undertaken in the Alpine region, on the routes between Italy and Germany.

Roads accommodate a range of vehicles and are typically less expensive to build than railways; hence they are often favoured in regional planning and related schemes. Most of the internal movement within Europe is by road, with inland waterways second and railways third.

Transportation by *air* is the most expensive, but it is rapid and is usually favoured for small-bulk, high-value products. Air transport is particularly subject both to rapid technological change and to political influence. Even small countries tend to favour maintaining a national airline for reasons of status and tourist development. In many cases airport construction and maintenance are funded by governments.

Trade

Trade can take place if the difference between the cost of production in one area and the market price in another will at least cover the cost of movement.

In principle, domestic trade is identical to international trade. The major difference is

Trade Theories

It may be true that, as Isard (1956:107) wrote, 'trade and location are as the two sides of the same coin.' Yet from a geographic viewpoint, trade theory excludes the central concept of location theory, namely distance. Quite literally, trade theory typically considers countries as though there were no space between them. Integrating location theory and trade theory seems a difficult task, not least because trade theory often lacks empirical support.

Classical trade theory originated in the work of economists such as David Ricardo (1773–1823) and focuses on the natural or historically created differences between the countries specializing in production and related trade; its key variable is labour. Trade is explained by the *law of comparative advantage* as follows: In the absence of barriers to trade, a country will specialize in the production and export of those commodities that it can produce at a comparatively low cost and import those goods that can be produced at a comparatively lower cost in other countries.

Neoclassical trade theory extends classical theory by using geometric techniques and adding further concepts such as the notion of external economies, which helps to explain why trade can be conducted between areas that are similar in most respects. External economies are the benefits that an industry enjoys as a result of the expansion of the industry to include related activities.

Modern trade theory extends the range of variables considered. Whereas classical theory focused on labour, modern theory also includes such variables as production, capital, land, and entrepreneurship, comparing countries in terms of their overall endowment with these factors. Applied to a historical topic such as the growth of trade associated with the transition from feudalism to capitalism, modern theory might focus on new producers seeking ever larger markets as an example of entrepreneurship.

Regional Integration

The fact that trade occurs across international boundaries means that it can be regulated, and in fact most governments do actively intervene in the importing and exporting of goods. For example, they may set up trade barriers to protect domestic production against relatively inexpensive imports. The most popular form of regulation involves the imposition of a **tariff**.

Regional integration—the linking of separate states in some form of economic union—has become important since the

10.6

EXPLAINING COMMODITY FLOWS: ULLMAN

The study of commodity flows has long been a topic of concern to geographers. In North America, the most influential writer on this topic has been Edward Ullman. Ullman was especially interested in transportation, and even defined geography as the study of spatial interaction (apparently after hearing a sociologist define sociology as social interaction). On the basis of detailed studies, Ullman (1956) concluded that commodity flows could be reasonably explained by reference to three variables:

1. *Complementarity*. When a surplus of a certain commodity in one area is matched by a deficit of the same commodity in a second area, this complementarity permits the commodity to flow from the first area to the second. This logic applies at all spatial scales.

2. *Intervening opportunity*. Complementarity can result in product movement only if there is no other area between the complementary pair that is an area of either surplus or deficit. If such an intervening opportunity exists, it will either serve the deficit area or receive the surplus because it is closer to both areas than they are to each other.

3. *Transferability*. Commodity flows are affected by the ease with which a commodity can be transferred. A product that is difficult or expensive to move will be less mobile than one that is easier or less expensive.

Together, these three factors provide a simplified explanation of commodity flows at various times and at a multitude of spatial scales.

Second World War. The basic motive behind integration is the opportunity for increased trade and the expansion of potential markets. Until recently, most integration was quite limited, but today there is compelling evidence of a widespread desire among separate states to unite economically. In many cases economic integration also involves some degree of political integration. There are five stages in the process of integration (Conkling and Yeates 1976:237):

1. The loosest form of integration is the *free trade area*, consisting of a group of states that have agreed to remove artificial barriers, such as import and export duties, to allow movement and trade among themselves. Each state retains a separate policy on trade with other countries. Two major examples of free trade areas are the North American Free Trade Area—NAFTA, comprising Canada, the United States, and Mexico and established in 1994—and the ASEAN Free Trade Area, comprising Brunei, Indonesia, Malaysia, the Philippines, Singapore, and Thailand and established in 1967 ('ASEAN' is an acronym for *Association of Southeast Asian Nations*.)

2. The second stage in international economic integration is the creation of a *customs union*, in which member states not only remove trade barriers between themselves but impose a common tariff barrier. In this case there is free trade within the union and a common external tariff on goods from other states.

3. A *common market* has all the characteristics of a customs union, and in addition allows the factors of production, such as capital and labour, to flow freely among member states. Members also adopt a common trade policy towards non-member states.

4. There are two levels of integration beyond the common market that groups of states may achieve. An *economic union* is a form of international economic integration that includes a common market and harmonization of certain economic policies, such as currency controls and tax policies.

5. Finally, at the level of *economic integration* member states have common social policies, and some supranational body is established with authority over all. The extent to which individual states may be prepared to sacrifice national identity and independence to achieve full economic integration is not yet clear.

At present, the most developed example of regional economic integration is the European Union (EU), which established a single European market in 1992. The importance of transportation as a means of encouraging integration is evident in the several ambitious projects recently completed or underway (the Channel tunnel, transalpine routes, and Scanlink). As the process of regional integration continues, states left outside the major groupings may suffer economically (Cleary and Bedford 1993). This is one reason why many countries—including countries in Western Asia and North Africa that would generally be considered non-European—aspire to join the EU.

The Association of South East Asian Nations (ASEAN) consists of ten countries—Brunei, Cambodia, Indonesia, Laos, Malaysia, Myanmar, Philippines, Thailand, Singapore, and Vietnam—and encourages free trade between members. It is also moving towards free trade with China and strengthening trading connections with Japan and India. Economic progress in this area was significantly impeded by the Asian economic crash of 1997.

The South American economic grouping of Argentina, Brazil, Bolivia, Chile, Paraguay, and Uruguay, known as Mercosur, is a customs union that has experienced real difficulties because of financial crises in Argentina. Whether the region will follow the route taken by the EU remains to be seen—one potential stumbling block is the risk that further integration might imply some loss of national sovereignty. The formation of a Free Trade Area of the Americas (FTAA) has been proposed.

It seems likely that African countries also will move towards some form of integration in the near future, especially with the 2002 creation of the African Union (Chapter 9). Among the problems that those countries face are a generally low level of economic development, by global standards, and the fact that the areas in which African countries are most competitive—agriculture and textiles—are two of the major areas still subject to protective tariffs (see Box 10.8, p. 356) in the west. Integration of all areas into the emerging global economy is considered in the next section.

Towards One World

The rest of this chapter is devoted to globalization. We will begin by establishing the broad context necessary to place the phenomenon in perspective, before addressing some specific aspects of economic globalization. This is not, of course, our first discussion of globalization, nor will it be our last. But it is a pivotal one. Let's quickly review our previous discussions and identify those still to come.

In Chapter 4 we saw how decisions taken in one part of the world can have consequences elsewhere. For example, your decision to eat a hamburger contributes, however slightly, to the removal of the tropical rainforest, and your decision to drive a car contributes, however slightly, to global warming. In Chapter 5 we learned that increasing gender equality—specifically, empowerment of women through education, whether in school or through the mass media—is helping to reduce the numbers of births and slow the rate of world population growth. In Chapter 6, discussions of the less developed world emphasized the fact that it is not possible to consider regional problems without taking a global perspective that enables us to see the links between more and less developed areas. In Chapter 7 we reviewed some evidence suggesting that a process of cultural globalization is underway, and in Chapter 8 we identified links between such globalization and popular culture. Finally, in Chapter 9 we identified political globalization as one possible geopolitical scenario for the future.

For most observers, globalization is primarily economic. In this chapter we will examine economic globalization in more detail, and the next three chapters will build on this material in their discussions of agriculture, settlement, and industry. In each case we will see how the friction traditionally associated with distance has been reduced with the advent of globalization. First, however, to help us place globalization in the big picture, reinforce our current understanding, and shed some additional light on why our world is the way it is—and why it is constantly changing—let us take a brief journey through time.

Shaping the Contemporary World

Geography is ambitious, as befits a discipline with the fundamental goal of 'writing about the earth'. Accordingly, it is sometimes important to step back and acknowledge how many factors, past and present, have shaped the world we see today. What fundamental things do we need to know in order to understand contemporary human geographies?

Physical geographies (Chapter 3)

Regardless of our technological achievements, human and physical geographies continue to be closely related. This is true whether we are thinking at the scale of the world as a whole (e.g., determining global population distribution and density), or at the local, personal level (e.g., deciding what clothes to wear, depending on the season). In physical geographic terms, all parts of the world are not the same, and different environments are often associated with different human geographies. Human geographers may no longer subscribe to any rigid notion of environmental determinism, but it is still essential to take physical geographies into account in our analyses.

Life on earth (Chapter 3)

We are not alone on the earth. Humans are only one of many lifeforms, and we cannot consider ourselves in isolation. Just as we need to take into account basic physical geographies, so we need to consider our relationships with plants and other animals. The fundamental lesson of ecology is how completely we depend on other organisms—from the trees that produce oxygen to the food that we eat to the bacteria that enable us to digest it.

Human origins (Chapter 3)

Current evidence suggests that *homo sapiens sapiens* first appeared in Africa only about 100,000 years ago and eventually spread around the world in a process of diffusion closely related to physical geography. The fact that, biologically speaking, humans are a single species exposes the fallacy of the concept of race. Since the emergence of *homo sapiens sapiens*, human evolution has been cultural, not biological.

Culture and diversity (Chapters 7 and 8)

We may all belong to one biological species, but human culture is astonishingly diverse. Culturally, humans have evolved in a seemingly infinite number of directions over the past 100,000 years—as evidenced by our many languages, religions, and identities—and this is what makes our world such a divided place. Whether or not you agree with James's interpretation of cultural diversity (see p. 218), you will probably find at least some truth in the idea that cultures set up barriers that are often difficult to cross.

Origins of agriculture (Chapters 4 and 11)

The transition to agricultural life that began about 12,000 years ago, often known as the agricultural revolution, was a critical technological advance that permitted population growth, new forms of social organization, the beginnings of urbanization, and the development of specialized occupations. Because urbanization was largely dependent on agriculture, many of our cities are located in agricultural areas. During the long period since domestication began, agricultural technologies have diversified and advanced, allowing humans to settle new areas and achieve increasing yields.

European overseas movement (Chapters 1, 6, and 9)

For various reasons, Europeans moved themselves, their cultures, and their economies across large parts of the world beginning in the fifteenth century. This process was not really completed until the late nineteenth century, when European states carved up much of Africa between themselves. It was during this period of approximately 450 years that many of the characteristics of our contemporary world were put in place. Most notably, according to world systems theory (Chapter 6), European expansion was responsible for the emergence of the two worlds we know today: the less and the more developed. It also promoted the spread and growth of Christianity, of several European languages, and of Eurocentric racist logic.

Social and economic organization: Modes of production (Chapters 2 and 8)

Modes of production are important because they both enable and constrain human activities; in any particular mode, there are some things that can be done and some things that cannot. In the contemporary world the dominant mode of production is capitalism, which involves what is often described as the ceaseless accumulation of capital.

States (Chapter 9)

States are the basic building blocks that most of us have in mind when we think about world events. The nation state may be declining in importance, as we saw in Chapter 9, but it remains the case that most people in the world are citizens of a particular state and are tethered to it, with little opportunity to move permanently to another state.

The 'Industrial Revolution' (Chapter 13)

The term 'industrial revolution' refers to a series of technical changes between about 1750 and 1850 that involved the large-scale use of new energy sources, especially coal, and the introduction of new machines that necessitated the construction of factories. These factories generated urban growth, and industrial cities replaced earlier forms of settlement. New transport links—canals, railways, and roads—were essential for moving raw materials to factories and finished products to markets. Agriculture became increasingly mechanized, and the proportions of the total labour force that were engaged in agriculture declined rapidly. In much of Europe this was also a period of large-scale overseas migration.

The twentieth century

Many dramatic changes took place during the twentieth century. A basic account of these changes is included in Box 10.7.

Globalization

Because the processes of globalization are still unfolding, it is not necessarily easy to recognize all of them, let alone to understand all their implications. The rest of this chapter will address the most potent aspect of the phenomenon: economic globalization. This form of globalization involves both technological change, especially the instantaneous transmission of information, and organizational change, especially the rise of transnationals able to 'slice their way at will through national boundaries and

10.7

LOOKING BACK AT THE TWENTIETH CENTURY

The temptation to look back at the twentieth century and see those one hundred years as representing some meaningful period of time proved irresistible to many writers, scholarly and popular, in the late 1990s. Even though our division of time into centuries is clearly an arbitrary human choice, it is fair to say that centuries have a distinctive place in the popular imagination—long enough to show evidence of meaningful change, but still short enough to comprehend. This box provides an overview of the principal human geographic changes that took place during the twentieth century and that offer some insight into the twenty-first-century world. All the facts noted here are discussed in the appropriate larger context elsewhere in the book.

The fundamental difference between the world of the early twentieth and early twenty-first centuries is the increase in the number of people—from 1.6 billion to 6.1 billion in a mere hundred years. Furthermore, each of us is expected to live longer—in the more developed world life expectancy has increased from 45 to 75 years, and in the less developed world from 25 to 64 years. These changes reflect the fact that mortality rates, especially for infants, have declined throughout the world because of technological advances. In addition, the locations of the greatest concentrations of people have shifted, from Europe to East and South Asia, South America, and increasingly also to Africa. In 1900 the largest cities in the world were London, Paris, Berlin, New York, and Chicago, whereas the largest cities in 2000 included Tokyo, Bombay, Lagos, Dhaka, and São Paulo.

The way that we make our living has changed. In 1900, about 90 per cent of the world's people practised agriculture: this figure has fallen to about 50 per cent, and approximately 45 per cent of us now live in cities. Urban living encourages smaller families and new conceptions of appropriate family structures. Other social changes of concern to human geographers include a multitude of improvements in quality of life as measured in terms of access to health services and educational facilities. There has also been a decline in organized religion, although in some instances religious fundamentalism has increased.

Forms of discrimination, especially racism and sexism, have diminished in both intensity and impact. The old European assumption that white-skinned, usually Christian people were inherently superior to others has been effectively repudiated—although not without the horrific experiences of the Holocaust and apartheid. In much of the world, women are no longer excluded from public arenas such as politics and business. Thus the authority both of white-skinned people and of men, especially heterosexual men, has been successfully challenged, although it is also clear that the transition towards equality and social justice remains incomplete. Another important attitudinal change is evident with respect to our understanding of the relationship between humans and nature. We appear to be undergoing a transition from an ethnocentric to an ecocentric viewpoint, and developing a corresponding appreciation that our world is an ecosystem—in other words, that actions in one place have impacts in other places.

Technological advances have brought marked changes in our ability to move from place to place and to transmit information. Significant improvements in the technology of information transmission were evident before 1900 with the invention and widespread adoption of the telephone and the electric telegraph. During the first half of the twentieth century, communications improved markedly with the advent of automobiles and air travel, and the world shrank accordingly, but it was the period after the Second World War that saw the introduction of computers and the virtually instantaneous transmission of information that they permit.

The political geographic world experienced dramatic shifts during the twentieth century. The world of 1900 was dominated by colonialism, notably European, but changes over the century ultimately resulted in a world without empires. The vast overseas possessions of such countries as Britain and France have disappeared, as have the land-based empires of Austria–Hungary, Russia, and the Ottoman world. One consequence of decolonization has been a remarkable increase in the number of states in the world, from about 50 in 1900 to about 190 today. The transition might be characterized as a shift from a small number of great powers dominating the world to a multitude of ethnically defined states. Further, most contemporary states are democratic, accepting the basic principle that authority is legitimated from below and not from above.

The economic organization of the world has changed markedly. In 1900, economic power was concentrated where political power was concentrated, and the major players practised protectionism. Today a transition is underway towards a global economy and a corresponding emphasis on free trade. Nevertheless, we continue to live in a world characterized by gross inequalities: much evidence confirms that the rich are becoming richer and the poor poorer.

One broad conclusion to be drawn from this hurried survey is that we are now able to talk intelligibly on a global scale as well as in local and regional terms—an idea that lies at the heart of many accounts of our contemporary world. Of course, globalization has not made local and regional scales meaningless, but it has certainly become increasingly important to take global considerations into account in mentally structuring our lived worlds. Indeed, it is often asserted—by human geographers and others—that globalization, understood broadly as an ever-increasing connectedness of both people and places, is the greatest challenge facing humans at the present time.

render all state policy-makers redundant' (Dicken 1992:48).

Economic Globalization

To begin, it is important to point out that economic globalization is not the same thing as economic internationalization. The internationalization of economic activity began with the development first of empires and then of economic links—especially through the movement of capital and goods—between colonial powers and their colonies. But this process of internationalization did not extend to production, which continued to take place in a single national territory. Trade between countries is not globalization. Only when production is no longer contained by national boundaries, and this change has been brought about primarily by the rise of transnational corporations, can we properly speak of globalization. As Figure 10.12 shows, the increasing economic importance of transnationals means that the production process is now organized *across* rather than *within* national boundaries.

The Global Economic System

To understand how transnational corporations have been able to emerge, it may be helpful to review recent changes in the global economic system. Certainly, economic cooperation between states has ebbed and flowed over the past century. In much of the western world, including Canada, the 1920s and 1930s were a period of trade barriers and limits on foreign investment.

The beginnings of what we now think of as economic globalization came at the end of the Second World War, when the United States became the world's principal economic power and was able to initiate the Bretton Woods agreement of 1944. This agreement led to the creation of two important economic institutions: the International Monetary Fund (IMF) and the World Bank (initially called the International Bank for Reconstruction and Development). The IMF was to provide short-term assistance to countries whose currencies were tied either to gold or to the US dollar, and the World Bank was to provide development assistance to Europe after the war. An important component of the economic order introduced at Bretton Woods concerned international capital movement, which was to be regulated primarily through national systems of exchange controls, with the US dollar playing the key role.

Another feature of the post-war economic system was free trade. This was the concern of a third international institution, created in 1947: the General Agreement on Tariffs and Trade (GATT), intended to reduce barriers to trade between countries. Between 1947 and 1994 there were eight rounds of GATT trade talks; the most recent ('Uruguay') round, which ended in 1994, accomplished the most sweeping liberalization of trade in history (see Table 10.1).

However, the economic order introduced in the 1940s under the leadership of the United States has changed significantly in two ways. First, the movement of capital is now virtually immediate (because it is achieved electronically), and almost unregulated. This represents a dramatic change from the Bretton Woods system. The principal geographic expression of this new form of capital movement is the rise of global cities. In additional changes from the Bretton Woods system, the roles of the IMF and World Bank have been expanded, and in 1995 the GATT was replaced by the even more powerful World Trade Organization (WTO) (Box 10.8).

Second, as we saw above, there are now several close-knit regional organizations that are essentially discriminatory and protectionist in their approach to trade. Some

Figure 10.12. Conventional international trade and intra-firm trade within a transnational corporation.
Source: P. Dicken. 1992. *Global Shift: The Internationalization of Economic Activity.* London: Paul Chapman, 49.

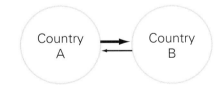

(a) Conventional 'arm's length' international trade between two countries

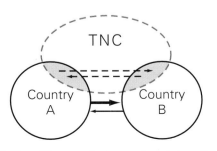

(b) The difference between 'arm's length' trade and intra-firm trade

commentators even suggest that we are moving towards a world composed of three regional trading blocs that will function not only as economic unions (Figure 10.13) but as political ones as well, eventually replacing states. The latter remains to be seen. In any event, in addition to the EU, NAFTA, and ASEAN, the list of regional organizations today includes the Caribbean Community (CARICOM), Central American Common Market (CACM), Commonwealth of Independent States (CIS), Economic Community of West Africa (ECOWAS), Central African Customs and Economic Union (UDEAC), the Maghreb

Oxfam aid workers wearing fibreglass heads of the G8 world leaders, outside the World Trade Organization meeting in Cancun, Mexico, in September 2003. To symbolize the waste of European and American agricultural subsidies, the 'Big Head' leaders sat down for a breakfast of 'dumped' cereal, sugar, and milk (AP photo/Ginnette Riquelme).

Table 10.1 FROM PROTECTIONISM TO FREE TRADE: A CHRONOLOGY

Date	Events
1947	GATT founded in Geneva with 23 members; some tariff cuts.
1948	First GATT meeting, held in Havana, Cuba.
1949	Second round of talks, held in Annecy, France; 500 tariff cuts; ten new countries admitted.
1950–1	Third round, held in Torquay, England; 8,700 trade concessions; four new countries admitted.
1956	Fourth round, held in Geneva, Switzerland; extensive tariff cuts.
1957	The Haberler Report outlines new guidelines for GATT; problems facing less developed countries considered; impact of new European Economic Community considered.
1960–2	'Dillon round' (named after US Under-Secretary of State) held in Geneva; 4,400 tariff concessions.
1964–7	'Kennedy round' negotiations, also in Geneva, depart from the product-by-product approach used in previous talks and adopt an across-the-board method of cutting tariffs for industrial goods; 50% tariff cut in many areas.
1973–9	'Tokyo round', involving 99 countries; reduction of import duties and other trade barriers by industrial countries on tropical products exported by developing countries; average tariff on manufactured goods produced by nine largest markets cut from 7% to 4.7%.
1982	Ministerial meeting in Geneva reaffirms validity of GATT rules for the conduct of international trade and commits to combating protectionist pressures.
1986–93	'Uruguay round'; cuts in industrial tariffs, export subsidies, licensing and customs valuation; first agreements on trade in services and intellectual property.
1995	WTO established as a replacement for GATT .
1996	First WTO ministerial conference, in Singapore.
1998	WTO ministerial conference in Geneva.
1999	WTO ministerial conference in Seattle, WA; major anti-globalization protests by a wide variety of interest groups.
2001	WTO ministerial conference at Doha, Qatar, provided mandate for negotiations on a range of subjects, and other work including issues concerning the implementation of the present agreement.
2003	WTO ministerial conference (Cancun, Mexico); there are now 144 members, including China, with Russia expected to join shortly. In September an alliance of poorer countries, led by China, India, and Brazil, challenge the right of the US and EU to determine the details of trade deals; when talks collapse, each side blames the other.

Figure 10.13 The contemporary geo-economy. Dicken emphasizes that the power structure of the contemporary world economy is based on three regions: North America, the European Union, and East and Southeast Asia. Each contains one of the three global cities identified in Table 12.5 (p. 445). Clearly a key consideration in the creation of these trading blocs is spatial proximity. This situation represents a significant change from the mid-twentieth century, when European colonial empires were the crucial regions and the international division of labour was a simple core–periphery structure.
Source: P. Dicken, *Global Shift: The Internationalization of Economic Activity* (London: Paul Chapman, 1992):45.

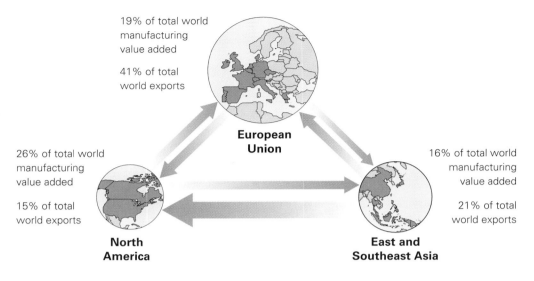

19% of total world manufacturing value added

41% of total world exports

European Union

26% of total world manufacturing value added

15% of total world exports

North America

16% of total world manufacturing value added

21% of total world exports

East and Southeast Asia

10.8

THE WORLD TRADE ORGANIZATION

Formed in 1995 to replace the GATT as the guardian of international trade, the WTO has become closely associated with globalization. In 2003 it had 144 member countries, and at least 30 more were hoping to join (see Table 10.1). The WTO's mandate includes facilitating smooth trade flows, resolving disputes between countries, and organizing negotiations; it is the only international body with such authority (Hoad 2002). WTO decisions are absolute and binding on member countries. The GATT and the WTO have been spectacularly successful: world trade is increasing rapidly, and tariffs are on average about one-tenth what they were in 1947. Nevertheless, protectionism remains significant in three important areas: agriculture, clothing, and textiles.

In late November–early December 1999, a WTO meeting of member countries' trade ministers in Seattle was overwhelmed by well-organized protest demonstrations in which labour organizations, environmental activists, consumer groups, and human rights advocates shared their concerns about the circumstances and presumed consequences of economic globalization.

The concerns are many and varied. Some argue that the WTO is too secretive, and that its discussions should be held in a more public context, involving a much broader range of participants. Some accuse the organization of being a front for corporate interests, especially influential transnationals. Some see it as benefiting the wealthy, more developed countries and neglecting the poor of the less developed world (certainly the most powerful players are the European Union, Japan, and the US). Some point out that the WTO is inconsistent in promoting unrestricted movement for some goods while permitting restrictions on the movement of others. In particular, these critics object to the continuation of agricultural subsidies in more developed countries that allow those countries to export artificially low-priced produce to the less developed world while preventing the importation of produce from less developed countries. Perhaps most fundamentally, some consider it ironic that the WTO's wealthiest member states continue to resist the unrestricted movement of labour in the form of international migration. With this contradiction in mind, Massey (2002) points out that many politicians and business leaders in the more developed world take different stances depending on the issue under discussion: if that issue is free trade, then they support the idea of a world without borders, whereas if the issue is international migration, they support the nationalist idea of separate states with defensible borders.

The WTO rejects most of these criticisms, arguing that the best way to address global inequality is through trade liberalization, so that free trade becomes fair trade. In many respects the WTO is the most democratic of international organizations, having a one member—one vote system, and serving as a forum for discussion between governments, the vast majority of which are democratically elected; Rigg (2001) notes that many of the WTO's opponents are more secretive, less accountable, and less democratic than the organization itself. There is probably some truth on both sides of the debate, with the WTO acknowledging the need for some internal reforms and many critics accepting that the WTO is necessary in some form.

Union (see Box 9.10), the Southern African Development Coordination Conference (SADCC), and Mercosur. On the other hand, the Asia Pacific Economic Forum (APEC), comprising 18 countries (including the three largest economies in the world—the United States, Japan, and China) is attempting to move even further along the road of trade liberalization.

There were also political changes in the late 1980s and early 1990s that had significant economic repercussions. With the collapse of communism in the former Soviet Union and eastern Europe, and the end of apartheid in South Africa, several new participants entered the world trade arena. And the country with the largest population in the world, China, is adopting increasingly capitalist economic policies.

In any event, these political changes are not directly related to the economic globalization we are discussing here. In that context, the principal organizational change is the rise of transnational corporations (corporations that operate in two or more countries), while the principal technological changes are taking place in the realm of communications and information flow.

Transnational Corporations

Table 10.2 provides sales data for the leading transnational corporations, with GDP data from selected countries for comparative purposes. Because transnationals are able to adjust their activities not just within countries but between them, they are able to take advantage of variations in factors such as land costs and labour costs at both scales. They undermine national economies because they are able to organize movements of information, technology, and capital between countries.

Different transnationals carry out different proportions of their production at home and abroad. In 1997, for example, foreign employment accounted for 53 per cent of the total employment at the Ford Motor Co., 40 per cent at Toyota Motor, and 31 per cent at General Motors. For the Swiss-based Nestlé, however, the foreign-employment figure was 96 per cent, and for the UK-based BAT Industries it was 86 per cent

Most of the transnationals are based in the US, Europe (especially the UK and

Germany), or Japan. But others have their home bases in the newly industrializing countries in Asia, or even in countries (such as Canada) that are more usually regarded as host countries for **foreign direct investment**—a term that refers to the investment of capital into manufacturing plants located in countries other than the one where the company is based. In general, countries that are major sources of foreign direct investment are also major hosts. Indeed, one way to identify economic globalization is to focus on the amount and location of international direct investment, which has grown even more rapidly than international trade since the Second World War. In most cases both the investors and the targets of their investment are located in the more developed countries (Table 10.3).

Typically, a transnational will have its headquarters in the more developed world, but locate its manufacturing activities in the less developed world, where wages are lower. There is, then, a division of production that is referred to as the **international division of labour** (see Chapter 13, especially Figures

| Table 10.2 | SALES DATA FOR TRANSNATIONALS COMPARED TO GDP FOR SELECTED COUNTRIES, 1997 |

Transnational or Country	Total Sales or GDP (US$ billions)
General Motors	164
Thailand	154
Norway	153
Ford Motor	147
Mitsui and Co.	145
Saudi Arabia	140
Mitsubishi	140
Poland	136
Itochu	136
South Africa	129
Royal Dutch/Shell Group	128
Marubeni	124
Greece	123
Sunitomo	119
Exxon	117
Toyota Motor	109
Wal Mart Stores	105

Note: A rather different database, reported in the New York Times, 26 December 1999, compared the GDP of selected countries with the then market value—an amount that changes daily—of various companies. At US$593 billion, Microsoft's value was similar to the GDP of Spain, while, at US$456 billion, General Electric's value was similar to the GDP of Thailand.

Source: Calculated from data in *Forbes Magazine*, vol. 162, no. 2 (27 July 1998), 116–54.

Activists in Hong Kong called for a boycott of McDonald's in August 2000 after it was reported that in Shenzen, China, children as young as 14 were working 16-hour days for just 1.50 yuan (less than 30 cents) an hour making promotional toys for the fast-food chain. The characters under the 'golden arches' read 'exploiting workers' (Dickson Lee, *South China Morning Post*).

13.11 and 13.12, which show the percentages of national labour forces in industrial and service employment respectively).

Since 1945, trade has increased much more rapidly than production—a clear indication of how much more interconnected the world has become. Most of the products dealt in by transnationals fall into two groups: either they are technologically sophisticated, or they involve mass production and mass marketing. In addition, much of the recent growth in trade reflects the increasing role of services—commercial, financial, and business activities—conducted by transnationals.

Technological Changes

Of all the space-shrinking and time-compressing technologies that have been developed in recent years, the most important to economic globalization are those that facilitate the almost instantaneous transmission of information regardless of distance. These include advances in the mass communications media, and in the use of cables, faxes, and satellites.

Other technological changes relate especially to the production process, the principal change being from Fordism to post-Fordist 'flexible production'. As we will see (in Chapter 12), 'flexible production' means that manufacturers are able to produce specific goods inexpensively in small quantities, rather than having to mass-produce as in Fordism.

Globalization: Good or Bad?

You may not be surprised to learn that this textbook will not answer the question in the heading above. Rather, it will provide some food for thought, in the hope that you will

Table 10.3 — NET FOREIGN DIRECT INVESTMENT FLOWS: SELECTED COUNTRIES AND GROUPINGS OF COUNTRIES, 1985 AND 1997

Country	Net Foreign Direct Investment Flows (US$ millions)	
	1985	1997
Canada	1,357	8,246
United States	20,010	90,748
United Kingdom	5,480	36,897
All 45 High HDI countries	44,388	266,255
Mexico	1,984	12,101
Malaysia	695	3,754
Turkey	99	606
All 94 Medium HDI Countries	10,311	126,766
Bangladesh	1	145
Zambia	52	70
Mali	3	15
All 35 Low HDI Countries	980	2,449

Note: The countries selected for inclusion are representative of their groups. The three groups of countries are those defined by UNDP on the basis of the HDI statistic (see Table 6.10 and related discussion). The totals for the three groups provide striking evidence of the concentration of foreign direct investment in the more developed world.

Source: United Nations Development Programme, *Human Development Report*, 1999 (New York: Oxford University Press, 49–52).

use it to clarify your own thinking. As with environmental issues (Chapter 4) and the state of the world generally (Chapter 6), many people express strong opinions on globalization; and, as with those other topics, opinions often reflect ideological positions—personal views of the world and how it ought to be. Let us begin with the arguments put forward by those who oppose globalization.

Opposing globalization

Why is opposition to globalization so widespread? There are several interrelated reasons.

In the broadest terms, globalization is seen as benefiting the more developed world at the expense of the less developed. More specifically, it is seen as benefiting those few countries within the more developed world where transnational corporations are based, at the expense of the many (specifically working) people, especially those in the less developed world. There are also more specific criticisms, however, many of which are directed at the World Trade Organization in its capacity as guardian of world trade. For example, when the members of the WTO meet, they tend to insist on closed discussions. By excluding other interested parties, such as labour unions and environmental and anti-poverty groups, the WTO gives the impression that it is undemocratic and working to benefit only its members—or indeed (since the WTO now includes less as well as more developed countries) only a select few of those. Certainly the movement towards free trade—or what might perhaps be better described as selective free trade—can be interpreted as serving the interests of the major economic powers, such as the US, Japan, and the European Union, at the expense of the less developed countries that are only now becoming industrialized. It is worth noting that when the major powers themselves were industrializing, in the nineteenth and early twentieth centuries, they all enjoyed the benefits of protectionist trade policies, government intervention, and subsidies designed to encourage domestic industrial growth.

It is often charged that globalization is contributing to the ever-widening gap between the rich and the poor. Certainly it is reasonable to suspect that if those who are already rich are in a position to dictate the rules of the game, they may become richer at the expense of the poor. On the other hand, the gap between rich and poor is not a new phenomenon. Since the beginning of European overseas expansion and early capitalism, those countries and groups of people with the power to exercise control over others have benefited at the latter's expense; this is the basic logic of the world systems and dependency theories (Chapter 6).

For some critics, the most serious problem associated with globalization is that it gives priority to export-centred economies, with the result that the merits of locally sustainable economies are downplayed. Exploitation of resources in the less developed world to satisfy demand in the more developed world is likely to have a negative impact on local economies. There are related worries about the environmental consequences of globalization, particularly with respect to resource extraction. Increasingly, local landscapes are being treated as global commodities, and local communities are losing their traditional resource bases.

In addition to these primarily economic issues, specific concerns have been expressed about the rise of supranational organizations that could replace nation states as the basic units of the political world; increasing cultural homogenization and the loss of local cultural identity; and social consequences in areas such as gender equality.

Finally, the criticism that may be the most profound is also the most uncertain. It has been suggested that globalization is unnatural in that many of the processes associated with it do not reflect any natural pattern of economic change—rather, they are the products of carefully designed economic strategies. Of course, similar criticism could be directed at the earlier protectionist phase. The extent to which the current situation represents a departure from the past can be debated, but public opinion in general does appear to be suspicious of globalization.

Supporting globalization

But are the reasons for opposing globalization the only ones worth thinking about? Of course not. Most of those who support globalization do so for business-oriented reasons. However, Johan Norberg defends globalization for reasons that are essentially moral. In his book *In Defence of Global*

A Nike factory in Indonesia
(Irene Slegt/Panos Pictures).

Capitalism (2001), Norberg argues in favour of capitalistic economic globalization on the grounds that it represents the best hope for eliminating poverty and fulfilling human potential. It is his moral perspective that makes Norberg's argument compelling for many people. Using data from many countries, including the two largest, China and India, he argues that lowering trade barriers and investing capital in less developed countries helps to reduce global poverty, and hence is not only defensible but morally imperative. As an example of the way transnational investment in less developed countries increases wages, Norberg points to a Nike shoe manufacturing plant in Vietnam: when it first opened, in the early 1990s, workers walked to the factory; three years later they used bicycles; three years after that they used scooters; and finally they are beginning to use cars. Although Norberg differs from many other proponents of globalization in favouring the free movement of people as well as goods (see Box 10.8, above), in general he maintains that improvements in food supply, education, equality, human rights, and gender issues show that globalization is making the world a better place for most people.

To the extent that global capitalism increases opportunities and wealth for the world's poor, obviously it is a positive thing. The problem, of course, is that globalization is far too complex a set of processes for us to be sure that all its consequences will be so benign. As is so often the case, reasonable people may well take very different positions, and many will come down somewhere in the middle, recognizing the benefits that globalization can bring, but also aware of the constant effort that will be required to prevent or mitigate adverse consequences.

A Global Village?

Today space is shrinking, time is being compressed, and national borders are becoming less important than they used to be. These transformations are particularly evident in four recent developments:

1. New markets have arisen: global markets in services; global consumer markets with global brands; deregulated, globally linked financial markets.
2. New participants have emerged: most important among these are transnationals, the WTO, and the several regional trading blocs; there have also been numerous corporate mergers and acquisitions.
3. Faster and less costly means of communication have developed: the Internet, cellular phones, and fax machines facilitate the transmission of information, and there are also ongoing technological advances in the movement of goods by air, rail, and road.
4. New policies and practices are being put into place concerning matters such as human rights and environmental issues.

But do all these ongoing changes mean that the world we live in, the lives we live, and the landscapes that we occupy are changing? Specifically, do we now all live in one global village? Certainly many people have suggested that this is the case, but it is important not to exaggerate the point. The reality is that, for most of the people in the world, it is still life at the local, regional, and national levels that matters. For example, use of the Internet is high in the more developed world—a 2000 survey found that more than 50 per cent of Canadians had used the Net in the preceding year—but it is much lower in the less developed world. Similarly, crossing national borders is much more difficult for unskilled than for skilled workers. *Globalization is a reality, but by no means an uncontested one; we may think of the world as a global village, but the truth is that most of its people are not yet villagers.*

CHAPTER TEN SUMMARY

A discipline in distance

The spatial location of geographic facts is not random: humans typically choose to minimize movement in order to minimize the frictional effects of distance.

Types of distance

Physical distance is the spatial interval between two points that are usually (though not necessarily) measured with reference to some standard system. Time distance is often more important than physical distance, especially where movement of people is concerned. The valuable concepts of plastic space and time–space convergence reflect the fact that time distance between any two locations is a function of changing circumstances. 'Economic distance' is the cost incurred in overcoming physical distance; it is often the paramount consideration for industries and other businesses. 'Cognitive distance' reflects individual perceptions of distance and is especially difficult to measure. 'Social distance' is particularly relevant to residential distributions—and the creation of distinctive social spaces.

Gravity and potential models

Movement between locations is affected by both the distance to be covered and the population size of the locations—a human equivalent to the gravity concept in physics. The 'potential model' is similarly derived from physics and expresses the potential that a particular location has for interaction with other locations. Both models are attractive simplifications of complex real-life situations.

The diffusion process

Following the pioneering work of Hagerstrand, diffusion research in geography changed from a study of landscape features to a study of process. Four empirical regularities are the neighbourhood effect, the hierarchical effect, resistance, and the S-shaped curve. Most recently, diffusion studies have focused less on the process and more on the human consequences of diffusion, especially its potential to cause spatial inequalities.

Transport system evolution

Transport systems are both cause and effect of other economic aspects of landscape. Major changes are associated with political expansion and technological advances. In the case of Britain, the most dramatic changes came with the industrial revolution: turnpike roads, navigable waterways, railways, and finally an improved road network. Theoretical approaches to the evolution of transport systems focus on the identification of stages that are related to larger geographic issues.

Networks and modes

Transport networks are made up of nodes and routes and are often analyzed in a simplified form using graph theory. Each of the principal modes of transport—water, land, and air—has some advantages and some disadvantages.

Trade

Although trade is clearly related to the distance between locations, the three principal branches of trade theory all exclude any reference to distance. Trade is also related to resource base, labour force, and capital. Today, much world trade is regulated by various agreements between countries concerning economic integration.

Shaping our world

To understand our contemporary world we need at least a basic understanding of the forces that have shaped it, many of which are touched on elsewhere in this book. Among them are physical geography, human dependence on other life forms, human origins and early movements, cultural evolution, agricultural technologies, European overseas movement, diverse modes of production, the rise of states, industrial technologies, twentieth-century changes, and current globalization trends.

Economic globalization

Although the reality of globalization seems clear, the details of its impact are contested. For example, on the one hand, a new global economy is emerging in response to the internationalization of capital, production, and services; on the other hand, there are several major examples of regional integration and associated regional protectionism. One thing that is certain is that globalization is meeting considerable opposition from a variety of groups.

LINKS TO OTHER CHAPTERS

- The study of distance is geography par excellence; hence this chapter links with all other chapters in this book.

- Globalization:
  Chapter 2 (concepts)
  Chapter 4 (ecosystems and global impacts)
  Chapter 5 (population growth, fertility decline)
  Chapter 6 (refugees, disease, more and less developed worlds)
  Chapter 7 (cultural globalization)
  Chapter 8 (popular culture)
  Chapter 9 (political globalization)
  Chapter 11 (agriculture and the world economy)
  Chapter 12 (global cities)

- Five concepts of distance and space:
  Chapter 2 (concepts)

  Chapters 7 and 8 (cognitive and social distance in relation to cultural regions and landscapes)
  Chapters 11, 12, and 13 (time and economic distance in relation to land- use theories).

- Diffusion research in cultural geography:
  Chapter 7 (creation of cultural landscapes).

- 'Neighbourhood effect':
  Chapter 11 (von Thünen theory).

- Diffusion and political economy:
  Chapter 11 (political ecology and agriculture).

- Regional integration and globalization:
  Chapter 9 (economic groupings)
  Chapters 11, 12, 13 (economic restructuring).

FURTHER EXPLORATIONS

BERRY, B.J.L., E.C. CONKLING, and D.M. RAY. 1997. *The Global Economy in Transition*, 2nd ed. Upper Saddle River, NJ: Prentice-Hall.
> An up-to-date and readable economic geography textbook that includes chapters on the forces promoting globalization and the factors reinforcing regionalization, along with a detailed account of trade.

CHAPMAN, K. 1979. *People, Pattern and Process: An Introduction to Human Geography*. London: Arnold.
> An introductory text centred on spatial interaction and spatial diffusion; provides a detailed account of geography as a discipline in distance.

CHORLEY, R.J., and P. HAGGETT. 1970. *Network Analysis in Geography*. New York: St Martin's Press.
> A detailed account of many of the topics raised in this chapter, with a consistent focus on spatial analysis.

CLARK, G.L., M.P. FELDMAN, AND M.S. GERTLER, eds. 2000. *The Oxford Handbook of Economic Geography*. Toronto: Oxford University Press.
> A valuable resource covering an unexpectedly wide range of topics in economic geography; includes excellent overviews of global economic integration and global changes.

DONERT, K. 2000. 'Virtually Geography: Aspects of the Changing Geography of Information and Communications'. *Geography* 85: 37–45.
> A discussion of the impact of new technologies and the geography of the Internet; outlines a new discipline of cybergeography.

FULLERTON, B. 1975. *The Development of British Transport Networks*. London: Oxford University Press.
> A brief account of transport change from 1750 to 1970, covering waterways, railways, and roads.

GATRELL, A.C. 1983. *Distance and Space: A Geographical Perspective*. Oxford: Clarendon.

> A clear account of distance and spatial concepts with many examples of appropriate analyses.

HAGERSTRAND, T. 1967. *Innovation Diffusion as a Spatial Process*, translated by A. Pred. Chicago: University of Chicago Press.

> A seminal work integrating the cultural and spatial analytic approaches to diffusion; the postscript, by Pred, is an excellent summary of diffusion research before 1967.

HAGGETT, P. 2000. *The Geographical Structure of Epidemics*. Oxford: Clarendon.

> An informative book by one of the best known and most original of twentieth-century geographers; examines epidemics using diffusion concepts, and provides valuable insights into the infection patterns several contemporary diseases, including AIDS.

HANSON, S., ed. 1986. *The Geography of Urban Transportation*. New York: Guilford Press.

> An edited volume of well-written articles on topics such as urban transport planning, transportation and energy, and the environmental impacts of transportation.

HAY, A.M. 1973. *Transport for the Space Economy: A Geographical Study*. London: Macmillan.

> A useful, though somewhat dated, study of transport issues, including the analysis of networks.

LEINBACH, T.R. 1983. 'Transport Evaluation in Rural Development: An Indonesian Case Study'. *Third World Planning Review* 5:23–35.

> An article covering a range of issues, from the role of transport in development to urban transport problems.

MORRILL, R.L., and J.M. DORMITZER. 1979. *The Spatial Order: An Introduction to Modern Geography*. North Scituate, Mass.: Duxbury.

> An introductory textbook with a novel organization and a strong theoretical content that is especially useful for discussion of the whole range of movement-related topics.

PAWSON, E. 1977. *Transport and Economy: The Turnpike Roads of Eighteenth-Century Britain*. London: Academic Press.

> A scholarly work that combines detailed analysis with sound generalizations related to diffusion and economic change.

RIDDELL, J.B. 1985. 'Urban Bias in Underdevelopment: Appropriation from the Countryside in Post-Colonial Sierra Leone'. *Tijdschrifte voor Economische en Sociale Geografie* 76:374–83.

> A detailed analysis that explains the rejection of intensive rice cultivation by peasants in Sierra Leone in terms of urban bias.

VANCE, JR, J.E. 1986. *Capturing the Horizon: The Historical Geography of Transportation*. New York: Harper and Row.

> A major conceptual and factual work dealing with the growth of transport networks.

WALLACE, I. 1990. *The Global Economic System*. London: Unwin Hyman.

> A clearly written discussion of broad spatial economic issues.

ON THE WEB

http://globalization.about.com/mbody.htm
Looks at the various issues surrounding globalization.

http://www.ifg.org/
The International Forum on Globalization brings together leading activists concerned with the consequences of globalization processes.

http://www.ph.ucla.edu/epi/snow.html
A site hosted by the UCLA Department of Epidemi-

ology and devoted to the work of John Snow, one of the first medical geographers; includes a link to Snow's famous London cholera map.

http://www.wto.org/
The home page of the World Trade Organization; contains a wealth of information on the organization itself as well as trade and related issues.

Agriculture

Agricultural geography—the identification, measurement and explanation of spatial variations in agricultural activities—is generally considered to be a branch of economic geography. The other principal branches (settlement and industrial geography) are discussed in Chapters 12 and 13. Thus it may be helpful to think of these three chapters as forming a unit. Economic geography is obviously affected by the economic globalization discussed in the preceding chapter. However, cultural and political factors also play very important roles. In the case of agriculture, for example, cultural factors such as religious beliefs and ethnicity may influence the varieties of crops or animals produced in certain areas, and state intervention in the form of quotas and subsidies also has a direct effect.

This chapter begins with an examination of the way the discipline of economic geography has changed in recent years as it has incorporated various concepts from social science. Drawing on the spatial analysis tradition—the third of our three recurring themes—we then examine why agricultural activities are located where they are, and discuss in detail a classic theoretical argument that many geographers have found useful.

The next major section explains the origins and evolution of major agricultural activities and identifies nine principal agricultural regions around the world. This is followed by discussions of the political economy and political ecology of agriculture. The chapter concludes with a brief look at recent changes in food consumption patterns and preferences.

Grain elevator, Moosomin, Saskatchewan (Victor Last, Geographical Visual Aids).

Geography is an eclectic and fashion-prone discipline. The attention span . . . for major theoretical or methodological perspectives is rather short-lived. For many a geographer, it is very hard to keep up with the endless re-formulations of spatial or geographical perspectives and theoretical influences (Swyngedouw, 2000:41).

Inventing, and Reinventing, Economic Geography

The subjects of this chapter and the two that follow—respectively agriculture, settlement, and industry—represent the core of what is usually labelled 'economic geography'. Although that label remains appropri-ate, it may be helpful to begin with an overview of the way our understanding of these traditionally 'economic' subjects is increasingly being informed by cultural, social, and political perspectives. Such an overview may be especially helpful in light of the quotation above, which succinctly identifies a feature of human geography that is one of its great strengths but also an occa-sional weakness.

Human geography is indeed an eclectic discipline that welcomes new approaches. In Chapter 2 alone, we saw how it incorpo-rated four very different philosophical approaches:

11.1

HUMAN GEOGRAPHY AND ECONOMICS

To understand the relationship between economics and human geography, it is helpful to be aware of the main schools of economic thought:

- *Classical economics:* This school, which dominated from the mid-eighteenth century until about 1870, was founded by Adam Smith and exemplified by the work of David Ricardo. It explained the price of goods by reference to the labour theory of value, which sees price as reflecting the labour time that goes into the production of the good.
- *Marxian economics:* This school shares with classical eco-nomics an interest in the labour theory of value. Part of Marx's larger political economy, it applies his theories of value and exploitation to price theory. It also exposes the class relations that explain the inequalities of capitalist economies.
- *Neoclassical economics:* This 'new' school replaced the dominant classical view in the later nineteenth century, introducing the concept of use value, or utility, and empha-sizing that price depends on supply and demand—what people are willing to pay. Neoclassical economics is closely associated with capitalism, and has dominated the disci-pline until the present.
- *Political economy:* Today, many economists have a strong political and related policy orientation. The political economy approach is derived in part from Marxian economics and rejects the basic market-based logic of neoclassical economics.
- *Neo-Ricardian economics:* Neo-Ricardians—an important school since about 1960—use the labour theory of value and develop aspects of Marxian theory in the context of the problems of late twentieth-century capitalism.

What is the significance of these developments for human geography? Following the institutionalization of geography in 1874, a new branch of the discipline emerged. Usually called **commercial geography**, it focused on world and regional production and links to the physical environment, and its content was based explicitly on environmental determinism. This early economic geography had close ties to economics and, from the economics perspective, could be usefully described as 'spatial economics'. By the 1950s, however, in the broad field of economic geography there was increasing dissatisfaction with environmental determin-ism, and the regional focus, as we have seen, was in decline. A fresh approach was needed.

Economics, especially in the then-dominant—and still very important—form of neoclassical economics, offered that fresh approach. Geographers turned to positivism, the philosophy favoured by neoclassical economics, and to various econom-ic theories, to test and further develop economic concepts in a spatial framework. Problems in economic geography related to agriculture, industry, settlement, planning, and transportation thus occupied an especially prominent place on the research agenda of human geographers from the early 1960s to some time in the 1970s. More recently, human geog-raphers have also turned to both Marxian economics (in the form of the political economy approach) and neo-Ricardian economics to support specific research endeavours. The remaining three substantive chapters of this book reflect the influence of the last two approaches as well as that of the dominant neoclassical school.

1. empiricism (1900–mid-1950s);
2. positivism (mid-1950s–c.1970);
3. humanism (c.1970–); and
4. Marxism (also c. 1970–).

In Chapter 8 we identified three more approaches:

1. feminism (1970s–)
2. structuration theory 1980s; and, finally,
3. postmodernism (late 1980s–).

Reflecting the 'cultural turn' of recent years (see Chapter 8), most of our newer approaches have built on versions of humanist, Marxist, feminist, or postmodern ideas.

With this chronological summary of approaches in mind, let us now review what we might call the invention of economic geography—and its ongoing reinvention.

Descriptive Regional Economic Geography

As the name suggests, 'descriptive regional economic geography' is the traditional empirical approach outlined in Chapter 1. Until about the mid-1950s, most economic geography focused on economic activities within a regional framework. Geographers mapped and described land use, identifying regions on the basis of their major economic activities. Some of this work was general in character, focusing on a dairy belt or a manufacturing region, for example, while in other cases the ultimate goal was to delimit appropriate regional boundaries. Such studies remain a central part of geography, as evidenced by the accounts of world agricultural regions in this chapter and world industrial regions in Chapter 13.

Spatial Analysis

From the vantage point of the early twenty-first century, spatial analysis dominated economic geography for only a short time, but it represented a conceptual, theoretical, and quantitative revolution. Abandoning traditional regional geography, with its focus on the empirical and idiographic (unique and particular) for the spatial analysis school, with its nomothetic approach (focusing on universal laws), geographers in the mid-1950s were inspired by developments in other social sciences, especially economics.

As a discipline, economics is concerned with many issues relevant to human geography: from the allocation of limited resources (such as the land and soil needed for agriculture or the minerals and power supplies needed for industry), to exchange transactions and the accumulation and distribution of wealth (issues central to human geographic analysis settlement and transportation). Box 11.1 outlines some of the links between human geography and economics. Especially relevant are the brief sketches of the classical and neoclassical traditions in economics, since these were the ones that interested geographers of the spatial analysis school as they shifted their attention from region to location and interaction.

Of course this revolutionary shift in focus owed much to economics, but it also made a valuable contribution to economics in the form of a new hybrid discipline called regional science. For geographers, one obvious shortcoming of economics was that many of its theories seemed to have been constructed 'in a wonderland of no spatial dimensions' (Isard 1956:25). As they set out to rewrite economics in spatial terms, human geographers not surprisingly turned first to theories that already included a spatial dimension.

Positivist theory

Much of the rest of this book is concerned with location—specifically, the location of human activities and the interactions between locations. Inspired by classical and neoclassical economics, **location theory** explains geographic patterns at any given place and time, while **interaction theory** explains movements between locations.

As we saw in Chapter 2, theories are sets of interrelated statements that explain reality—for example, patterns of economic activity. Theories are usually generated inductively, from facts; then hypotheses are deduced from them and tested in the real world (see Figure 2.1, p. 48). Perhaps the greatest virtue of theories developed following the scientific method is their rigour: once a theory is generated, the process of hypothesis deduction and verification will not (in principle) vary with the researcher's abilities. Verified hypotheses eventually assume the status of laws, which can later become initial statements in other theories.

By definition, a theory is merely a simplification of complex reality, and may even be untrue. Yet even a theory that proves not to be true can still offer insight and facilitate further work. Indeed, advocates of positivism argue that any science—including human geography—depends equally on fact (subject matter) and theory (the means of explaining the subject matter).

We will examine three important examples of positivist theory in this chapter and the two that follow. All three are closely linked to the classical and/or neoclassical economic traditions. The *first* example, addressing the question of why agricultural activities are located where they are, was developed by an economist, Johann Heinrich von Thünen, but uses distance as the key explanatory variable. The *second*, in Chapter 12, seeks to explain why urban settlements are located where they are; this theory was developed by a geographer, Walter Christaller, inspired by the von Thünen theory. *Finally*, in Chapter 13 we will encounter a theory explaining why industrial activities are located where they are; like von Thünen's theory, this one was developed by an economist (Alfred Weber), but uses distance as the key explanatory variable.

Incidentally, when positivist-type theory was introduced to human geography, some critics argued that theories had no place in geography because all locations, indeed all geographic facts, are unique. While this statement is literally correct, it is highly misleading: however unique individual locations may be, it is always possible to recognize general types of location.

Theories explaining economic location have typically been **normative** in nature—concerned with what ought to be. By definition, then, they do not purport to explain reality itself, but rather what reality would be like in some hypothetical ideal situation. For example, our principal agricultural location theories are based on the concept of the 'economic operator' (Box 11.2).

Evaluating spatial analysis

Research in the area of spatial analysis was innovative and challenging. For the first time, human geographers became excited about conceptual advances, constructing theory, and the possibility of testing theoretical predictions quantitatively. By the early 1970s, as we have seen, such work was widely criticized as a version of human geography that lacked the human element. As a result, spatial analysis and the regional science associated with it were very influential for only a brief period. Nevertheless, these approaches left an important legacy, and they still play an important role in some contemporary work.

Reinventing Economic Geography

Since the early 1970s, economic geography, like the larger discipline of human geography, has not been dominated by any single

---

11.2

THE ECONOMIC OPERATOR CONCEPT

The **economic operator** concept (also known as **rational choice theory**) is a normative concept according to which each economic operator minimizes costs and maximizes profits as a result of perfect knowledge and a perfect ability to use such knowledge in a rational fashion. This theory has its roots in classical economics, but is one of the basic propositions of the neoclassical school. There is, of course, no such person as an economic operator in the real world—none of us is blessed with the required 'omniscient powers of perception' and 'perfect predictive abilities' (Wolpert 1964:537). What makes this unrealistic concept useful in economics and geography is that it allows us to generate hypotheses that are not encumbered by the complexities of human behaviour. In fact, the economic operator represents one extreme on a continuum of possibilities that range from optimization to minimal adaptation. We know that in reality most human behaviour falls between the two extremes and can be described as **satisficing**—aimed at satisfying the individual rather than optimizing the situation. The concept of satisficing behaviour is a basic alternative to rational choice theory.

approach. However, most work in the field has been linked in some way to one or more of the post-positivist philosophical developments, especially variants of Marxism, feminism, and postmodernism.

Marxism

The emergence of Marxist perspectives in human geography was linked to social and economic issues including the civil rights and anti-Vietnam war movements, increasing unemployment and inflation, and the declining health of many industrial areas. Together, these developments encouraged an interest in various forms of radical geography. The pioneering statement was a book by Harvey (1973; see Box 12.7, p. 438) that served to introduce geographers to the Marxist political economy described in Box 11.3.

Those working from a Marxist perspective strenuously reject the supposed value neutrality of spatial analysis, arguing that such an approach can only perpetuate the status quo. Rather, they say that the forces of capital accumulation and related social structures are responsible for the creation and ongoing recreation of the economic geographic landscape. The issues on which Marxist-inspired geographers focus—poverty, unemployment, industrial decline, uneven regional development—had been largely ignored by spatial analysts concerned primarily with explaining locations.

By the 1980s, however, much Marxist research was being criticized for focusing on the geographic outcomes of the large-scale processes of capital accumulation and not taking enough interest in political contexts and human intentionality. Some criticisms built on structuration theory (see p. 260) to highlight the importance of localities as the appropriate scale for analyzing the space economy of capitalism. Others reflected humanist and especially feminist perspectives.

Feminism

The central argument behind feminist critiques of Marxist work is simply that gender is just as important as class, if not more so. Notably, many feminists contend that gender analysis does much more than add an extra set of questions to the discussion of a subject like economic geography: it quite literally transforms it. McDowell (2000) suggests three examples:

1. Feminist economists and economic geographers have shown that the work done by women is often ignored. In particular, domestic labour—including caring for children or elderly family members—is unpaid and widely unacknowledged as work. Further, national economic statistics do not take women into account: for example, data on income and health typically reflect the overall circumstances of

---

11.3

MARXIST POLITICAL ECONOMY

According to Marx, capitalism is contradictory, characterized by social tensions and conflicts. The essential argument is as follows. A capitalist society is based on the circulation of capital—through production, exchange, and consumption—and this circulation leads to economic growth, or what is often called the ceaseless accumulation of capital. Because capital circulation involves the movement of money, goods, and labour, it creates geographies of production, consumption, and interaction. The requirement that circulation lead to economic growth implies that lack of growth, or decline, is untenable. Indeed, for the capitalist system to work, labour must

produce more value in the production process than it receives in the form of wages, so that business owners can appropriate that 'surplus value' as their profit. For Marx, the accumulation of capital by owners at the expense of workers inevitably leads to class conflict.

The capitalist system as Marx describes it has three key geographic implications: social and spatial divisions of labour (e.g., along class and gender lines); competition between owners for space, resources, and economic infrastructure; and, most generally, inherent instability (e.g., periods of depression and inflation).

the household and do not capture differences between household members—in the allocation of resources, for example.

2. In the labour market, assumptions about women tend to limit them to a relatively narrow range of occupations that are coded as feminine and typically seen as inferior.

3. Finally, it is possible to interpret features of the economic landscape—for example, workplace practices—not in terms of Marxist political economy but rather in terms of gender identities.

The body of research into these and related issues is substantial and growing. In addition, some feminist geographers draw on postmodern theory, notably when they focus on the social construction of gendered identities.

Postmodernism

On the one hand, postmodernism rejects the traditional empiricist faith that human geographers can provide factually reliable accounts of the world. On the other, it rejects the assumption—central to both positivism and Marxism—that theory can be relied on to explain reality. In both cases, postmodernism has had an enormous influence on the work of geographers and other academics and practitioners, for it has demanded new ways both of viewing and of representing the world. We have already noted some of the implications of these ideas in Chapter 8, especially the concern with listening to previously repressed voices. The sheer diversity of the postmodern endeavour appeals to many geographers. Postmodern theory has suggested new directions for research—for example, focusing on consumption rather than production, or the commodification of people and place. The interest in research directions such as these is one component of 'the cultural turn' as it has affected economic geography. Today it is widely acknowledged that cultures, at all scales, affect economic activity, and that economic activity in turn is one aspect of culture and of cultural change.

In conclusion, it is particularly important to note that postmodernists see geographic theory and practice as social activities, inevitably affected by the geographer doing the work. Thus they demand that geographers situate themselves, stating their identity and circumstances as these relate to their work. Barnes (2001: 557) aptly describes this demand as reflecting the reality that all geographic writing is necessarily a 'view from *somewhere*'—an observation that may recall the humanistic critique of positivism that we noted in Chapter 2.

Summary

This account of changing approaches to the subdiscipline of economic geography mirrors the larger history of human geography from 1900 onwards outlined in Chapter 1. That history is further developed here because it is economic geography that most clearly illustrates these changing approaches—which in turn are mirrored in the contents of this chapter and the two that follow it:

1. Any introductory account of economic geography needs to provide basic factual information. Thus the empiricist regional approach continues to be an essential starting point.

2. Spatial analysis remains important, at least partly because such research addresses the fundamental geographic question: Why are things—farms, settlements, industries, or anything else—located where they are?

3. Marxist ideas have inspired a variety of important questions and related approaches, with particular emphasis on the role of class within a capitalist mode of production.

4. Feminist ideas have introduced important challenges to economic geography, questioning the validity of many basic assumptions and suggesting new directions for research that explicitly acknowledges gender differences.

5. Postmodernism (like other ideas associated with 'the cultural turn'), adds to the diversity and complexity of economic geography, enriching the theoretical background and opening the way to new and original directions in research. In particular, postmodernism has suggested a new focus on consumption and commodification.

One way to summarize the evolution of economic geography over the past century is to note how its focus has broadened, from the strictly economic (in the empiricist and spatial analysis traditions), to the economic and political (in the Marxist tradition),to the economic, political, and cultural (in the feminist and postmodern traditions).

This detailed analysis of the connections between apparently separate types of human geography may seem an unnecessary complication in an introductory textbook, but it reflects the world we live in, a world where economic, political, and cultural matters are irrevocably intertwined. One of our challenges in the rest of this book will be to analyze the three principal topics of traditional economic geography—agriculture, settlement, and industry—from multiple perspectives, taking advantage of the diverse approaches now at our disposal.

The Agricultural Location Problem

We begin our account of agriculture with a basic geographic question. Why are specific agricultural activities located where they are? Our answer will centre on economic issues, but first it will take into consideration several other factors—physical, cultural, and political—all of which are interrelated. Many geographers believe that the best way to approach agricultural issues is to think in terms of ecosystems. Low-technology agricultural systems are structured along the same lines as natural ecosystems, while higher-technology systems are so complex, involving so many different types of circumstances, that to attempt to isolate one variable would be seriously misleading.

Some Physical Factors

The agricultural landscape is typically made up of individual farms, and the details of the landscape reflect the decisions of farm owners, managers, and workers. Such decisions may lie anywhere along the continuum identified in Box 11.2—from optimization through satisficing behaviour to minimum adaptation—and reflect a variety of factors.

Physical factors are important influences in agricultural decisions and hence in the creation of agricultural landscapes. Animals and plants are living things and require appropriate physical environments to function efficiently. Farmers have two options: they can either ensure that there is a match between animal and plant requirements and the physical environment, or they can create artificial physical environments by, for example, prac-

tising irrigation or building greenhouses. Some farmers are prepared to create artificial environments for specific agricultural purposes, but others choose not to engage in activities that are environmentally inappropriate. Because many environments are suitable for more than one crop, the agricultural decision can be based on other factors; similarly, those other factors can play a role in decision-making when the demand for a product is lower than the quantity that can be produced.

Climate

Climatic factors are the main physical variables affecting agriculture. Plants have specific temperature and moisture requirements. Optimum temperatures vary according to the activity, but growing-season temperatures of between 18°C (64°F) and 25°C (77°F) are often required. Low temperatures result in slow plant growth, while short growing seasons may prevent a plant from reaching maturity. In many areas, frost may result in plant damage. Moisture is crucial to plant growth; too little or too much can damage plants. In some areas, especially semiarid zones, rainfall variability is a problem. Many of the world's major wheat-growing areas (such as the Canadian prairies) are highly vulnerable in this respect; Figure 11.1 indicates the relationship between wheat yield and mean annual rainfall at a time before fertilizer was widely used. Animals also have water requirements, especially dairy cattle; sheep are much more adaptable to water limitations.

Soils and relief

Two other important physical variables are soils and relief. Soil depth, texture, acidity,

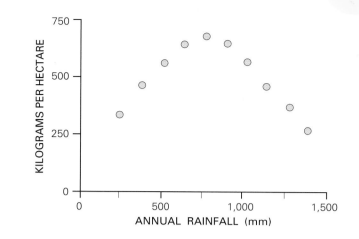

Figure 11.1 Relationship between mean annual rainfall and wheat yield in the US, 1909.
Source: Adapted from O.E. Baker, 'The Potential Supply of Wheat', *Economic Geography* 1 (1925):39.

and nutrient composition all need to be considered. Shallow soils typically inhibit root development; the ideal texture is one that is not dominated by either large particles (sand) or small particles (clay). Most crops require neutral or slightly acidic soils. The most crucial nutrients are nitrogen, phosphorus, and potassium. Soil fertility needs to be maintained if regular cropping is practised; methods include fallowing, manuring, crop rotation, and use of chemical fertilizers.

Relief (the shape of the land) also affects agriculture, specifically through slope and altitude. The angle, direction, and related insolation (exposure to sun) of slopes determine both the probability of soil erosion and the use of machinery. Generally, the flatter the land the more suitable it is for agriculture. Altitude affects temperatures: in temperate areas, the mean annual temperature falls 6°C (11°F) for each 1,000 m (3,280 ft) above sea level.

Some Cultural and Political Factors

In the past, agricultural location was generally considered to be determined by physical conditions:

> The Corn Belt is a gift of the gods—the rain god, the sun god, the ice god, and the gods of geology. In the middle of the North American continent the gods of geology made a wide expanse of land where the rock layers are nearly horizontal. The ice gods leveled the surface with glaciers, making it ready for the plow and also making it rich. The rain god gives summer showers. The sun god gives summer heat. All this is nature's conspiracy to make man grow corn (Smith 1925:290).

Today we have advanced well beyond environmental determinism, and we recognize the important roles that cultural and political factors can play. One important cultural factor is noted in Box 11.2: that farmers are not profit maximizers. Farmers in the more developed world typically favour security and a relatively constant income over a life dedicated to the unlikely goal of profit maximization, and poor farmers, in the more developed and less developed worlds alike, typically maximize product output for subsistence rather than profit. Even these preferences can be interpreted as consequences of cultural attitudes towards an essentially economic issue. The influence of culture is more obvious, however, in the case of religion and ethnicity.

Religion and ethnicity

Group religious beliefs may favour specific agricultural activities because of the value placed on either the activity or the product of the activity. Christianity, for example, values wine, which is used in the sacrament of Holy Communion. The church's demand for wine encouraged the spread and growth of viticulture, including its introduction to California by early Christian missionaries. Other agricultural activities are negatively affected by religious beliefs. Pigs are taboo in Islamic areas, while Hindus and Buddhists believe it is wrong to kill animals, especially cattle.

Immigrants in a new land often continue to adhere to the agricultural practices of their homelands. Thus an area settled by a variety of ethnic groups often presents a patchwork appearance. North American examples abound. One geographer has explained the location of cigar tobacco production in the United States by reference to ethnic variables (Raitz 1973). Another described two communities in south-central Illinois, one of German-Catholic ancestry and one of non-Catholic British ancestry (Salamon 1985). Although these communities are only 32 km (20 miles) apart and have similar soils, they differ substantially in farm size and organization. The German farms are small and diversified, including dairy, hogs, beef, and grain, whereas the British farms are monocultural grain. Arguing that these differences in farming practices stem from different attributes, Saloman contrasts the 'yeoman' Germans and the 'entrepreneurial' British (Table 11.1).

As Table 11.1 suggests, a variable such as land ownership can affect farmers' behaviour, especially with respect to innovation. Farms operated by their owners may be handled quite differently from rented farms or farms in socialist economies where decision-making is centralized.

The state

Political decisions affect much human behaviour, including the behaviour of farmers. Governments may have many reasons for deciding to influence farmers' behaviour, but in the contemporary developed world the most common motive is the desire to support an activity that, as measured by income, is in decline. In such cases government intervenes by fixing product prices and providing financial support for specific farm improvements such as land clearance (Box 11.4 and Figure 11.2). In the less developed world, governments provide assistance that enables farmers to adopt new methods and products.

The contemporary agricultural system is affected not only by state policies but by national and international trade legislation. Canada, the United States, Japan, and Europe all have marketing boards, quota requirements, government credit policies, and extension services to provide information to

A vineyard in Umbria, central Italy (Andrew Leyerle).

Table 11.1 CONTRASTING FARMING TYPES IN ILLINOIS

Yeoman	Entrepreneur
Goals	
Reproduce a viable farm and at least one farmer in each generation returns	Manage a well-run business that optimizes short-run financial returns
Strategy	
Ownership of land farmed preferred	Ownership plus rental land to best utilize equipment
Expansion limited to family capabilities	Ambitious expansion limited by available capital
Diversify to use land and family most creatively	Manage the most efficient operation possible
Farming Organization	
Smaller than average operations	Larger than average operations
Animals plus grain, crop variety	Monoculture cash grain
Land fragmentation	Land consolidation
Landowners often operators	Landowners frequently absentee
Expansion of community territory	Community territory stable
Family Characteristics	
Intergenerational cooperation	Intergenerational competition
Parents responsible for setting up son/heir	Incumbent upon son/heir to set up self
Many children, nonfarmers, live nearby	Often all children leave farming
Parents responsible for intergenerational transfer	Heirs responsible for intergenerational transfer
Early retirement geared to succession by children	Retirement geared to personal desires
Community Structure	
Village central focus of community	Village declining
Community loyalty	Weak community attachment
Population relatively stable	Population diminishing
Strong church attachment	Church consolidations
Farmers involved in village	Farmers uninvolved in village

Source: S. Salamon, 'Ethnic Communities and the Structure of Agriculture', *Rural Sociology* 50 (1985):326.

11.4

GOVERNMENT AND THE AGRICULTURAL LANDSCAPE

The Canada–United States border region between the Great Lakes and the Rocky Mountains provides an excellent example of how agricultural landscapes evolve in response to government policies. Around 1960, despite similar climatic, soil, relief, and drainage conditions, the landscapes differed considerably because the US government's agricultural policy was interventionist and its Canadian counterpart was much less so. By the 1970s, however, American intervention had decreased, while Canadian intervention had grown.

Figure 11.2 shows dramatic differences in crop and livestock combinations around 1960. The wheat allotment program in the United States meant that much wheat land was converted to raise barley, while the National Wool Act of 1954 encouraged sheep raising through a guaranteed price.

By the 1970s, however, the United States had moved away from support programs towards relatively market-oriented policies. By contrast, Canada had come to favour greater intervention to ensure adequate incomes for farmers. As a result of these changes, the differences between the two countries were reduced and the landscapes changed accordingly. The transboundary differences are significantly fewer today than in the 1950s and 1960s.

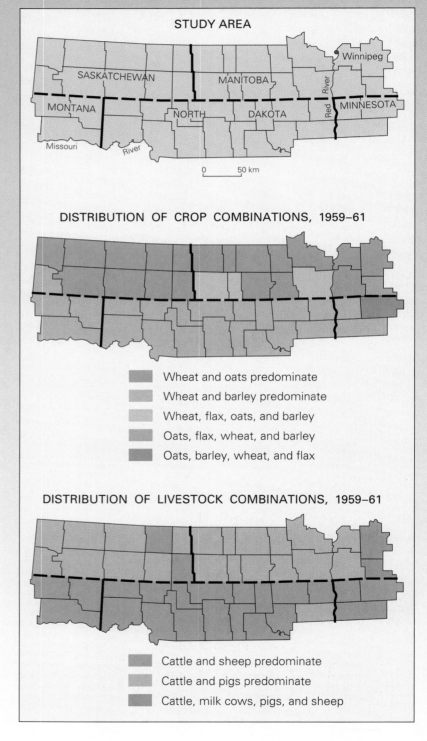

Figure 11.2 Crop and livestock associations along the US–Canada border.

Source: Adapted from H.J. Reitsma, 'Crop and Livestock Production in the Vicinity of the United States-Canada Border', *Professional Geographer* 23 (1971):217, 219, 221.

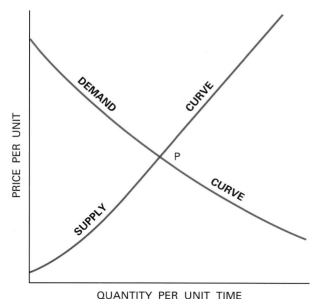

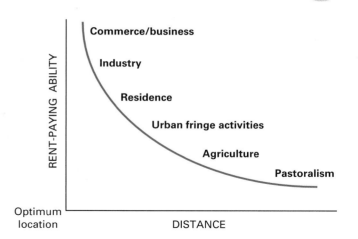

Figure 11.3 (left) Supply and demand curves.

Figure 11.4 (above) Rent-paying abilities of selected land uses.

the farm community, all of which affect the agricultural landscape. Perhaps the best-known examples of interventionist policies come from the European Union, where in some cases farmers have been paid not to grow crops (Bowler 1985).

Some Basic Economic Factors

The spatial patterns of agricultural activities are the end product of complex physical, cultural, and political variables. But human geographers agree that the most important variables generally—not necessarily in a specific location—are those labelled economic. Agricultural products are produced in response to market demand for them; thus farming is subject to the basic laws of supply and demand. Figure 11.3 depicts characteristic supply and demand curves. Since supply increases and demand decreases when the price received increases, an equilibrium price (P) can be identified at the intersection of the two curves. The ideal economic world of **commercial agriculture** is occupied by profit maximizers (economic operators) who respond immediately to any price changes.

But this type of response is neither feasible nor desirable for farmers in either the more or the less developed world. In the more developed world, farmers also value stability and independence, and may be willing to sacrifice profits accordingly; they may also choose to respond to decreasing prices by increasing rather than decreasing supplies. Farmers in the less developed world also

tend not to be profit maximizers, and their behaviour may well be rational. The aim of **subsistence agriculture** is to produce the amount of product required to meet family needs; profit maximization has no meaning for such farmers. The number of subsistence farmers worldwide has decreased significantly since about 1800, and it continues to decrease. Today, human geographers studying agricultural location focus instead on commercial farmers.

Competition for Land

Although several theories have been developed to explain observed variations in the spatial patterns of farming activities, they all focus on the fact that different activities compete for use of any location. Such competition arises because it is not possible for all activities to be carried out at their economically optimal location, and different activities may have identical optimal locations. The obvious question is how to determine the most appropriate use of a particular piece of land. Conventionally, land is assigned to the use that generates the greatest profits; thus, in a general sense, we can determine a hierarchy of land uses based on relative profits. But there is another way of viewing this situation. The greater the profits a particular use generates, the more that use can afford to pay for the land. The maximum amount that a given use can pay is called the **ceiling rent**. Figure 11.4 indicates the relative rent-paying abilities for a variety of land uses.

We can now identify the basic premise of much location theory: *the competition among land uses, a competition fought according to rent-paying abilities, results in a spatial patterning of those land uses.*

The Concept of Economic Rent

Agricultural location theorists use the above premise in combination with the concept of **economic rent**, which explains why land is or is not used for production. Land is used for production if a given land use has an economic rent above zero (Box 11.5). This concept can be related to one or more variables, but especially to measures of land use by the economist who formulated the economic rent concept, David Ricardo (1772–1823); however, the economist Johann Heinrich von Thünen (1783–1850) regarded distance as more important.

Von Thünen's Agricultural Location Theory

Von Thünen was a German economist and landowner who published two major works.

His treatise *The Frontier Wage* identified appropriate wages for agricultural workers. *The Isolated State*, published in 1826 (Hall 1966), reflected von Thünen's interest in economics (he was inspired by the work of Adam Smith) and his 40 years as manager of an estate on the north German plain.

The problem

To tackle the complex question of which agricultural activities should be practised where, von Thünen used a highly original method of analysis. He deliberately excluded several factors that are known to be relevant and proceeded on the basis of a framework that he called 'the isolated state':

> Assume a very large city in the middle of a fertile plain which is not crossed by a navigable river or canal. The soil of the plain is uniformly fertile and everywhere cultivable. At a great distance from the city the plain shall end in an uncultivated wilderness by which the

---

11.5

CALCULATING ECONOMIC RENT

It is important to clarify some details of the economic rent concept, notably how it is calculated and its spatial implications. Economic rent can be calculated as follows:

$$R = E (p - a) - Efk$$

where

- R = rent per unit of land (dependent variable)
- k = distance from market (independent variable)
- E = output per unit of land
- p = market price per unit of commodity
- a = production cost per unit of commodity parameters
- f = transport rate per unit of distance per unit of commodity

This equation is a straightforward linear relationship between rent and distance. Rent is a function of distance with the details of the relationship determined by $E (p - a)$, which gives the intercept, and Ef, which determines the slope of the line.

What are the spatial implications of economic rent as calculated above? Because economic rent declines with increasing distance from market, it eventually becomes zero, and when more than one agricultural activity is practised, a series of economic rent lines will appear, as in Figure 11.5. Where lines

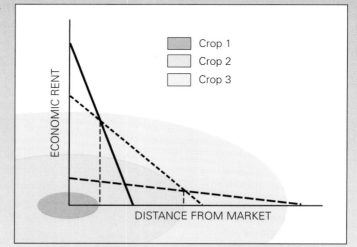

Figure 11.5 Economic rent lines for three crops and related zones of land use.

cross, one activity replaces another as the more profitable. The result is that agricultural activities are zoned around the central market and a series of concentric rings emerges.

The simple concept of economic rent as a function of distance results in a situation of spatial zonation.

state is separate from the rest of the world (Johnson 1962:214).

The assumptions

The isolated state model allowed for greatly simplified descriptions. Its assumptions may be elaborated as follows:

1. There is only one city, that is, one central market.
2. All farmers sell their products in this central market.
3. All farmers are profit maximizers—economic operators.
4. The agricultural land around the market is of uniform productive capacity.
5. There is only one mode of transportation by which farmers can transport products to market.

In effect, von Thünen was not so much excluding as holding constant such key variables as physical environment, humans, and transport. Only one key variable—distance from market—was allowed to vary. This was a distinctive and highly original contribution, and was, of course, in accord with his definition of economic rent, which centres on distance from market.

The answer

Given the above assumptions, von Thünen asked, 'How will agriculture develop under such conditions?' (Johnson 1962:214). The conditions are those of a controlled experiment isolating a single causal variable, distance from market. The answer is straightforward. Agricultural activities are located in a series of concentric rings around the central market, one zone for each product for which there is a market demand. To determine the location and size of each product zone, von Thünen used data from his own estate on production costs, market prices, transport costs, and so on. *The principle of concentric zones emerges from the concept of the isolated state, while the number, size, and content of zones is a function of particular places and times.* Combining the isolated state concept, his own data, and the economic rent concept, von Thünen came to two conclusions: (1) zones of land devoted to specific uses develop around the market, and (2) the intensity of each specific land use decreases with increasing distance from the market.

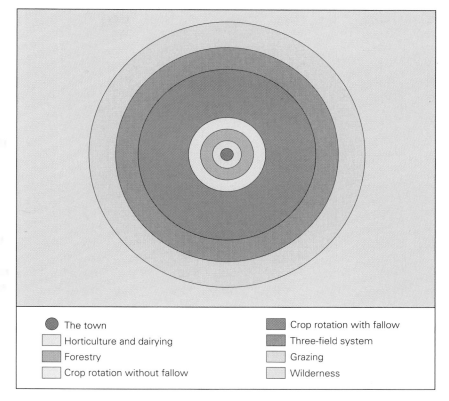

The town   Crop rotation with fallow
Horticulture and dairying   Three-field system
Forestry   Grazing
Crop rotation without fallow   Wilderness

Figure 11.6 Agricultural land use in the isolated state.

The crop theory

The first conclusion, the crop theory, is summarized in Figure 11.6, which illustrates how product location is affected by perishability and weight as they affect transport cost. Remember that while the specific activities depicted in this figure reflect north Germany in the early nineteenth century, the principle of zones applies generally. In zone 1 are market-gardening and milk production. Both fresh vegetables and dairy products are perishable, give high returns, and have high transport costs. Accordingly, they have a steep economic rent line reflecting a high Ef value (see Box 11.5), and a high intercept on the rent axis reflecting a high E (p − a) value. In zone 2 are forestry products (used for fuel and building). Forestry could command a location close to market because of the high transport costs involved in moving such bulky products. Rye, the principal commercial crop, is located in zones 3, 4, and 5. The differences between the zones reflect differences in the intensity of rye cultivation: zone 3 uses a six-year rotation, zone 4 uses a seven-year rotation, and zone 5 uses a three-field system. Livestock ranching—producing butter, cheese, and live animals—is located in zone 6. Beyond zone 6 is wilderness that can

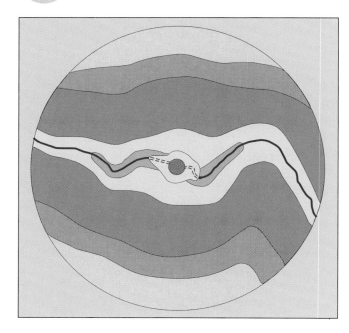

Figure 11.7 (left) Relaxing a von Thünen assumption: a navigable waterway.

Figure 11.8 (right) Relaxing two von Thünen assumptions: multiple markets and a transport network.

be brought into production if spatial expansion is needed in the future to meet increased product demand.

The intensity theory

Von Thünen's second conclusion is known as the intensity theory. Here von Thünen's own production data indicated that, for any given product, the intensity of production decreases with increasing distance from market. The easiest example to identify is the cultivation of rye in zones 3, 4, and 5, but the principle also applies inside specific zones.

Von Thünen's Contribution

In light of our discussion of theory and scientific method in Chapter 2, we need to ask to what extent von Thünen's work represents a true theoretical contribution. Although it does begin with a series of assumptions, they are not valid assumptions, or axioms, from which we can deduce hypotheses with confidence. Rather, they are simplifying assumptions that reduce the complexity of the problem. Hence we know from the outset that von Thünen's hypotheses—the crop and intensity theories—cannot be literally correct. On the other hand, we also know that they serve the purpose intended: that is, they describe an ideal state (interestingly, the original title of the work). The two hypotheses describe a state of affairs against which we can compare real-

world agricultural patterns. They describe what ought to be—a normative description. As in many economic models, a complex world is simplified to aid understanding; the spatial hypotheses produced are examples of ideal types (see glossary).

Labelling von Thünen's work as a theory is also complicated by the inclusion of real-world data in the procedures. It is nevertheless an excellent example of a type of theory to which geographers have been greatly attracted.

Some Modifications and Extensions

Von Thünen was well aware that his work was a simplification. Accordingly, he modified several assumptions when elaborating on the basic hypotheses. The assumption of a single mode of transportation to market was relaxed with the introduction of a navigable waterway serving to elongate zones (Figure 11.7). Relaxing the assumption of a single market and allowing a transportation network including roads resulted in a more complex pattern of zones (Figure 11.8). von Thünen also acknowledged that differential land quality would mean intensified use of the better areas.

Despite its age and some obvious weaknesses, von Thünen's work remains the fundamental answer to the question of agricultural location for at least two reasons. First, von Thünen's decision to simplify the issue by excluding certain variables was a highly original contribution; this basic method has been used by many other location theorists, as we will see in Chapters 12 and 13 especially. Second, in both of von Thünen's concepts, economic rent and agricultural location, the key geographic variable is distance as measured in terms of transport cost.

Human landscapes in general are clearly related to distance, regardless of how we choose to measure it—straight-line distance, transport-cost distance, time, or any other format. Relationships between distance and spatial patterns, such that some spatial regularity is evident, lie at the heart of much theory and analysis.

Distance, Land Value, and Land Use

Numerous geographers tested the von Thünen hypotheses or the general relations with distance in the context of agricultural patterns. The seminal work in this area is Chisholm's (1962), a remarkably thorough pioneering study of the scale ramifications of the concepts and an excellent survey of world evidence pertaining to the general issues.

Historical Examples

Von Thünen's own work is a historical empirical study, and it is hardly surprising that it is applicable in a general historical context. In fact, he was not the first to identify the existence of patterns in land-use zones. There are references to land use in Kent, England, regarding access to London, including a 1576 description and an 1811 account that identified four zones: clay pits, cattle pastures, market gardening, and hay.

Many analyses of agriculture in nineteenth-century North America highlight the von Thünen principles in a spatial and temporal framework. For example, one study of an area of Wisconsin centred on Madison found that the area close to the expanding centre was dominated by wheat production from 1835 to 1870, by market gardening from 1870 to 1880, and by dairy farming after 1880, demonstrating that zonal change for any one period becomes temporal change as the urban centre increases in size (Conzen 1971). A study of western New York State focused on the related idea that a prolonged period of falling transport costs is associated with increasing regional specialization of

agriculture; Leaman and Conkling (1975) showed that the expanding transport network of roads and canals allowed a frontier area, initially akin to von Thünen's wilderness, to rapidly become first an area of wheat cultivation and then an area of regional specialization related to access to market.

The Southern Ontario Frontier

Both of the American studies noted above found that areas experiencing increased settlement are transformed from non-agricultural areas into areas of subsistence agriculture and finally into a series of commercial agricultural types. Similar conclusions were reached in a study of nineteenth-century southern Ontario that focused explicitly on testing the von Thünen hypotheses (Norton and Conkling 1974).

The area analyzed lay north of the then-emerging centre of Toronto and can be regarded as one segment of the market area of that centre. Using data for 1861, the relationship between distance and land value is measured at $r = -0.6502$, producing an r^2 of 0.4228 and suggesting that 42 per cent of the spatial variation in agricultural land values is explained by distance to Toronto (Box 11.6 and Figure 11.9). The authors then follow the theoretical logic one step further by relating land values to a series of variables: distance to Toronto, distance to smaller regional market centres, distance to major lines of communication, and a measure of land capability for agriculture. The statistical result is $R = 0.7367$, producing an R^2 of 0.5427, which means that 54 per cent of the spatial variation in land values is explained by the set of independent variables.

Study of the 1861 pattern of land use showed an area divided into two basic zones: (1) close to Toronto a zone of fall wheat, peas, and oats, with production oriented for the Toronto market, and (2) an outer zone with much unoccupied land, used for spring wheat, potatoes, and turnips—an area anticipating commercial development. The incipient zonation in place by 1861 was to become fully realized in the later nineteenth century. This Ontario study also analyzed the intensity hypothesis and showed that the intensity of the key commercial crop (fall wheat) decreased with increasing distance from Toronto.

Examples in the Less Developed World

More surprising than these historical examples is considerable evidence suggesting von Thünen-type spatial patterns in less developed countries, where much agricultural activity is subsistence, not commercial, in orientation. A regional pattern of land use is often in fact a mix resulting from the different agricultural decisions made by relatively well-off commercial farmers and relatively less well-off subsistence farmers—groups that necessarily have different economic outlooks. The commercial farmers are concerned with profits and hence with rationalizing land use according to economic con-

straints, whereas the subsistence farmers are limited in their activities by lack of capital and inadequate size of holdings. Despite these complications, von Thünen-type patterns may emerge. Horvath (1969) analyzed land use *c.*1964 in the vicinity of Addis Ababa, Ethiopia. Although this highland area is more varied, physically and culturally, than the von Thünen model assumes, basic von Thünen patterns were evident: an area of eucalyptus forest around the city that was used for fuel and building material (an orientation outwards along roads was evident); vegetable cultivation within that forest, rather than in a separate zone; and a zone of mixed farming that included commercial and sub-

---

11.6

CORRELATION AND REGRESSION ANALYSIS

'A continuous theme in geographic research is that of analyzing the degree and direction of correspondence among two or more spatial patterns or locational arrangements' (King 1969:117). A traditional approach to the problem is to use map overlays—a procedure that is facilitated today by geographic information systems. Another approach is to apply a set of quantitative procedures known as correlation and regression analysis.

By simple correlation and regression, we may determine the degree of association between two variables. The correlation coefficient, r, is a statistical measure of the relationship, where r may vary between −1 (a perfect inverse relationship) and +1 (a perfect direct relationship). Squaring the r value gives a coefficient of determination, r^2, which may be expressed as a percentage and interpreted as the percentage variation in one variable explained by the other variable. A related set of procedures, known as multiple correlation and regression analysis, allows one variable to be related to two or more variables. In this case, the correlation coefficient is identified as R.

The procedure by which these results are obtained is as follows. The linear equation, $y = a + bx$, is used where

y = dependent variable
x = independent variable
a = y intercept
b = slope of the line

The economic rent equation is of this form $[R = E (p − a) − Efk]$. This linear equation is the simplest equation for predictive purposes. A method known as least squares is used to calculate the best-fit line on the graph of x plotted against y. Once the best-fit line is calculated, we can measure how well it fits. If all points fall on the line, it is a perfect fit and $r = −1$ or +1; more typically points are not located on the line and r lies between −1 and +1. In cases where y does not vary with changes in x, r is close to 0 (Figure 11.9).

Procedures are also available to test the significance of r. The results reported in the text are all statistically significant.

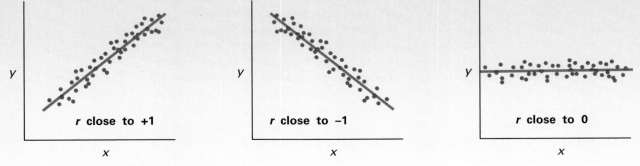

Figure 11.9 Scatter graphs, best-fit lines, and r values.

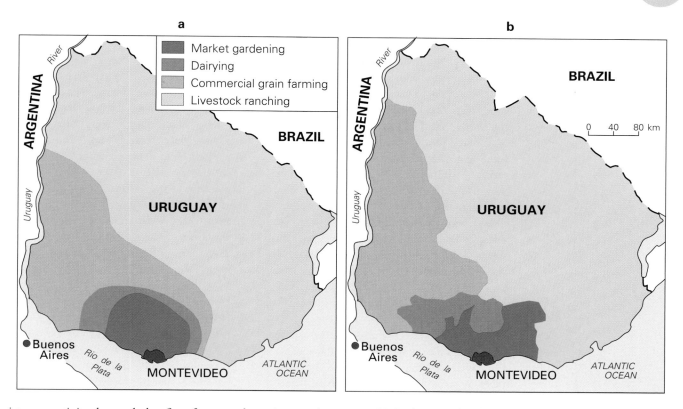

a

Market gardening
Dairying
Commercial grain farming
Livestock ranching

ARGENTINA
River
Uruguay
BRAZIL
URUGUAY
Buenos Aires
Rio de la Plata
MONTEVIDEO
ATLANTIC OCEAN

b

ARGENTINA
River
Uruguay
BRAZIL
URUGUAY
0   40   80 km
Buenos Aires
Rio de la Plata
MONTEVIDEO
ATLANTIC OCEAN

Figure 11.10 Agricultural land use in Uruguay: (a) as predicted by Thünen theory; (b) actual.

sistence activity beyond the first forest and vegetable zone. Horvath concluded that this was an example of incipient zonation, similar to the southern Ontario example above. Unfortunately, political circumstances in contemporary Ethiopia have made continuation of this analysis difficult.

A Contemporary Example

Griffin's (1973) study of Uruguay is a contemporary application of von Thünen theory. Uruguay possesses many of the essential characteristics of the isolated state, and Montevideo has the key attributes of the theoretical central city. Thus actual land use there (Figure 11.10) closely resembles the von Thünen model of ideal land use. The first zone is used for horticulture and truck farming, and followed by a zone of dairying, and finally a cereal zone, despite significant variations between model and reality in variables such as soil fertility, transport efficiency, and ethnicity.

Examples on the Continental and World Scales

Von Thünen's logic is relevant even at the continental and world scales. Several geographers have proposed concentric patterns of land use for North America, Europe, and the world. When Muller (1973) analyzed the nineteenth-century United States he found an expanding system of concentric rings. Jonasson (1925) described European agriculture in terms of an inner ring of horticulture and dairying, followed by zones of less and less market-oriented activity. On the world scale, Schlebecker (1960) traced the evolution of a world city from Athens to western Europe to western Europe–eastern US.

The growth of such a world market and the related expansion of the supply area can be analyzed by reference to import data. Over time, the mean distances over which agricultural imports have been moved have increased (Table 11.2). This spatial expansion of the export system has not been haphazard: it is clearly zonal in character. The changing details of the import system reflect a modified von Thünen model. Peet (1969) has detailed this situation specifically for wheat, showing that, in many instances, the arrival of wheat was preceded by extensive animal rearing (the Canadian prairies are an exception). In all cases, wheat cultivation was associated with major population growth, transport expansion, and the rise of an urban network. Spatially, wheat imports to Britain came first from Britain itself, then the Baltic region, followed by the United States, western Canada, and Australia.

Examples on the Local Scale

Von Thünen's theory has also been tested on the scale of individual villages and farms. In such analyses, the concept of distance is equated less with transport cost than with minimization of movement in relation to time expended. Various investigations of villages have tested modified versions of the crop and intensity theories and confirmed their relevance. Von Thünen-type patterns are also evident in the immediate vicinity of large cities. According to Sinclair (1967), even though the considerations affecting land use close to cities are not those identified by von Thünen, the spatial consequences in terms of zonation are very similar. One such consideration is the prospect of urban expansion, specifically the anticipation of such expansion and hence of profits to landowners. The closer the land is to the city, the greater the anticipation of urban expansion and hence the lower the incentive for capital investment in the land. This argument helps us understand why so many areas on the outskirts of cities are either vacant or put to some clearly temporary use.

Overall, the message is clear. The methods pioneered by von Thünen—specifically, the deliberate simplifications producing an ideal world against which reality can be assessed—are of enormous value to agricultural geographers and, as we shall see in later chapters, to industrial and urban geographers.

World Agriculture: Origins and Evolution

A discussion of agriculture on the world scale could focus on any number of topics: the influence of physical environment; the effects of regional population change; distances from markets; and so on. Perhaps the most useful approach, however, begins with the acknowledgement that the contemporary pattern of world agriculture is the still-changing product of a long history: 'The imprint of the past is still clearly to be seen in the world pattern of agriculture. To understand the present, it is essential to know something of the evolution of the modern types of agriculture' (Grigg 1974:1). To put it another way—as we did in the introduction to this book—it is largely human decision-making, past and present, in diverse cultural contexts, that has shaped and continues to shape our agricultural world.

Origins

The most fateful change in the human career can be said to have occurred . . . when the transition from foraging to farming began. During the preceding 150,000 years anatomically modern humans had successfully colonized almost all habitable and accessible areas and in so doing had learned to subsist, as 'hunter gatherers', on a great diversity of plant and animal foods. The most fundamental and far reaching consequence of the 'agricultural revolution' in the early Holocene was that it enabled more food to be obtained, and more people to be supported, per unit area of exploited land. It thus facilitated long-term sedentary settlement and the maintenance of larger and more complex social groups,

Table 11.2	AVERAGE DISTANCES FROM LONDON, ENGLAND, TO REGIONS OF IMPORT DERIVATION (MILES)				
	1831–35	1856–60	1871–75	1891–95	1909–13
Fruit and vegetables	0	521	861	1,850	3,025
Live animals	0	1,014	1,400	5,680	7,241
Butter, cheese, eggs	422	853	2,156	2,590	5,020
Feed grains	1,384	3,266	3,910	5,213	7,771
Flax and seed	2,446	5,229	4,457	6,565	6,275
Meat and tallow	3,218	4,666	6,018	8,093	10,056
Wheat and flour	3,910	3,492	6,758	8,286	9,574
Wool and hides	4,071	14,207	16,090	17,811	17,538

Source: After S. Leonard, 'Von Thünen in British Agriculture', *South Hampshire Geographer* 8 (1976):28.

which in turn enabled urban society to develop (Harris 1996:ix).

Agriculture originated in the **domestication** of plants and then animals. A domesticated plant is deliberately planted, raised, and harvested by humans; a domesticated animal is dependent on humans for food and shelter. As a consequence, domesticated plants and animals differ from their non-domesticated counterparts. From the human perspective, the domesticates are superior in that, for example, they bear more fruit or provide more milk; in fact, they have been deliberately engineered for these characteristics through selective breeding over long periods of time.

What was the first species to be domesticated? Where and when? How and why? In Chapter 7 we saw that, although no one knows the precise answers, in all likelihood agriculture began some 12,000 years ago. From an initial centre (or more probably centres), agriculture diffused to other areas, gradually replacing the main preagricultural economic activities, hunting and gathering. Although it was traditionally thought that agriculture first emerged in a few Asian centres, most notably southwest Asia (present-day Iraq), more recently it has been argued that it evolved independently in several centres in Asia (east, southeast, and southwest), Africa (northeast and south of the Sahara), the North American midwest, Central America, western South America, and southern Europe. It seems likely that, in each place, people began domesticating plants and animals in response to specific circumstances.

Early domestication

We are learning more and more about plant and animal domestication with advances in genetic research. Just as the study of genes is uncovering answers to specific questions about human origins, so it is illuminating many details of the domestication process. Most notably, such research is providing support for the idea that many plant and animal species were domesticated more than once, in different places at different times. Cows, pigs, sheep, goats, yaks, and buffalo were each domesticated at least twice, dogs at least four times, and horses on even more occasions. What this means is

Disappearing farmland in Richmond Hill, Ontario, north of Toronto. Scenes like this are increasingly common around the world as residential developments encroach on valuable agricultural land (Dick Hemingway).

that many different groups of people came up with the idea of domesticating other animals. It also suggests that the reason some animal species are not domesticated is not that no one tried, but rather that they proved unsuitable candidates (typically, the animals most susceptible to domestication already lived in groups and had some form of hierarchical organization).

The process of domestication

There is nothing difficult or complicated about the process of domesticating plants and animals; no knowledge of genetics is needed. It is simply a process that makes use of artificial selection as opposed to natural selection. Natural selection results in the survival and reproduction of those plants and animals that are best able to cope in a particular environment. Artificial selection involves humans allowing certain plants and animals to survive and breed because they possess features judged desirable by humans—for example, plants with bigger seeds or animals that are non-aggressive. The individual members of plant and animal species that humans favoured would reproduce and pass on the favoured characteristics, while the individual members with less desirable traits would be gradually eliminated.

Possible causes of domestication

Various explanations have been proposed to explain how humans became food producers. According to Sauer (1952), casual experimentation with plant and animal breeding probably began in a well-endowed environment that permitted a more sedentary way of life and a certain amount of leisure time. In his view, southeast Asia, especially wooded hilly areas away from possible floods, was the most likely hearth area.

Perhaps the most satisfactory explanation combines a number of relevant variables: demographic, environmental, and cultural. The key idea, not yet proven, is that population pressure might have prompted a search for new supplies. This hypothesis is based on the idea that although the process of domestication is not especially difficult, it can require a great deal of work and therefore is not especially attractive. Most societies did not need to increase food supplies because they had strategies to keep the population below carrying capacity. However, a change in climate at the end of the Pleistocene might have caused certain areas to become unusually rich in food resources, which would have made a more sedentary

way of life possible, and hence permitted an increase in population. Eventually, a larger population would have necessitated movement into marginal areas and competition for space, prompting the development of new strategies for food supply: specifically, domestication. Boserup (1965) and Binford (1968) are among the authors who have proposed such a scenario.

Whatever the details, by c.500 BCE agriculture had become a major economic activity. Grigg (1974:21–3) has identified five core areas and agricultural types evident by this date (Box 11.7).

Evolution

Our current agricultural types and regions are the outcome of a long evolutionary process. Since domestication began, plants and animals have diffused widely, far beyond the few areas where the wild varieties of the original species evolved. Several of the agricultural regional types to be discussed emerged in response to the European over-

---

11.7

AGRICULTURAL CORE AREAS, c.500 BCE

1. *Southwest Asia*
Basic crops: Wheat, barley, flax, lentils, peas, beans, vetch.
Basic animals: Sheep, cattle, pigs, goats.
Subtypes: Irrigated farming in the Nile Valley, Tigris and Euphrates, Turkestan, and Indus Valley; dry farming (without irrigation) was the basic southwest Asian complex mix of agricultural activities; Mediterranean agriculture was a mix of cereals and tree crops such as figs, olives, and grapes; in northern Europe, oats and rye were added to basic crops.

2. *Southeast Asia*
Two types originated on the southeast Asian mainland. Tropical vegeculture involved growing taro, greater yam, bananas, and coconuts, and raising pigs and poultry; a form of shifting agriculture. ('Shifting agriculture', practised in tropical forest areas, involves regular movement from one cut and burned area to another because of the rapidly declining soil fertility resulting from crop cultivation; also known as swidden agriculture, or slash-and-burn agriculture.) Wet rice cultivation gradually displaced the first type.

3. *Northern China*
Based on local domesticates (e.g., pigs, foxtail millet, soy beans, and mulberry) and imported domesticates (e.g., wheat, barley, sheep, goats, and cattle).

4. *Africa*
Two agricultural types, both shifting: tropical vegeculture (based on yams), and cereal cultivation (millets and sorghum).

5. *America*
Two agricultural types, both shifting: root crops and corn/squash/beans complex. No livestock.

Since c.500 BCE, links have been established between areas and types of agriculture. Of particular relevance are those agricultural changes prompted by European expansion from the fifteenth century onwards.

seas movement and/or to the demands of new population concentrations. In Europe, for example, some agricultural crops and methods prior to 1500 were the result of diffusion from the southwest Asian hearth areas; after 1500, two American crops, potatoes and corn, were widely adopted, and other agricultural changes reflected increased market size; these included intensification, expansion of cultivated area, and increasing commercialization. Commercialization was facilitated by technical advances in transportation, particularly after the mid-nineteenth century. Agricultural commercialization (Boxes 11.8 and 11.9) at that time was merely one component of substantial growth and change in the larger world economy.

A second agricultural revolution: Engand after 1750

The most radical changes in agricultural activities since the beginning of domestication have occurred in the larger European world, helping to create what is today 'the great gulf in productivity between the present agricultures of Asia, Africa and much of Latin America, and western Europe, North America and Australia' (Grigg 1974:53). In most cases, those changes did not begin until the seventeenth century, and it was only in the mid-nineteenth century that commercial agriculture became established on a large scale and the present pattern of agricultural

activity began to evolve. The link between industrial progress and agricultural progress seems clear.

A period of significant agricultural change—significant especially because of its links with other technological, social, and economic changes—began in England about 1750. Sometimes described as a second agricultural revolution, it involved a series of changes:

1. development of new farming techniques, including introduction of fodder crops and new crop rotations;
2. increases in crop output because of improvements in productivity;
3. introduction of labour-saving machinery; and
4. the ability to feed a growing population.

The first three of these represented changes in what Marx called the forces of production. But the new 'revolution' also involved changes in the relations of production associated with the decline of the feudal system and the emergence of capitalism. Among the institutional changes were the imposition of 'enclosure': the subdivision of large, open arable fields and large areas of pasture or wasteland into small fields divided by hedgerows, fences, or walls. Enclosure was part of a major social transformation from communal to private property rights.

11.8

FRONTIER AGRICULTURE: SUBSISTENCE OR COMMERCE?

To what extent were the early European settlers' agricultural activities oriented towards subsistence or commerce?

Many studies of frontier agriculture in North America have identified a brief subsistence phase followed by an increasing concern with commercial activities, which eventually became dominant. Evidence of commercialization is provided by the settlers' demands for improved communications with available markets and by a rapid transition from product diversification (characteristic of a subsistence economy) to product specialization (characteristic of a commercial economy). In South Africa, however, there is evidence to suggest that the subsis-

tence phase, during which farmers had limited contacts with markets, was considerably longer. What might explain this difference? One answer might be that whereas the motivation for frontier settlement in North America was largely economic, in South Africa it was largely social and political, reflecting the Dutch settlers' desire to escape the rule of the British in the established areas.

The message is clear: in seeking to understand agricultural activities, it is necessary to consider a wide range of factors. As the discussions in Chapters 11, 12, and 13 show, the reasons behind economic activities are often non-economic.

A second and related institutional change was the creation of farms that were typically rented from large landholders by capitalist farmers who employed labourers to work the land.

This second agricultural revolution—perhaps better described as an agrarian revolution—was one of the principal means by which England was able to begin supporting a growing population. 'The transformation of output and land productivity enabled the country to break out of a "Malthusian trap", allowing the population to exceed the barrier of 5.5 million people for the first time. Rising labour productivity ensured that extra output could be produced with proportionately fewer workers, so making the industrial revolution possible' (Overton 1996:206).

Nitrogen fertilizers

Of course, the changes initiated in English agriculture around 1750 are not the end of the story. Since about 1900, the more developed world has benefited from numerous institutional and technological changes that have facilitated massive increases in crop yield. Somewhat later, from the 1960s onwards, many of these benefits have become available to much of the less developed world. Together, these processes have transformed world food production.

Perhaps the most important development was the introduction of nitrogen fertilizers. Nitrogen is an essential nutrient for the cereal crops that have been staples for most people in most parts of the world since the beginnings of agriculture, and is added naturally to soil through rainfall. But rainfall alone does not add enough nitrogen, and it was soon discovered that growing a cereal crop on the same land year after year impoverished the soil and ultimately reduced yields. The favoured solution was to plant legumes along with the cereal crops, to replenish the soil's nitrogen content: peas and lentils in the Middle East, beans and maize in the Americas, soy and mung in Asia, and peanuts in parts of sub-Saharan Africa. Other strategies to prevent soil impoverishment included crop rotation and leaving land fallow for a season.

By the nineteenth century, the demand for cereals was increasing so rapidly that concerted efforts were made to find new sources of nitrogen. One source that was readily available in much of Europe was

11.9

THE PASTORAL FRONTIER

The pastoral frontier was a new landscape that emerged in those temperate grassland areas that experienced European overseas expansion, including parts of North and South America, South Africa, Australia, and New Zealand. The industrial revolution and related population increases in the eighteenth and nineteenth centuries created huge demands for wool, hides, and meat. The newly settled areas' responses to these market demands were largely dependent on social and political considerations.

In South America, pastoral activities were viewed favourably by governments that saw no alternative uses for the land, and as a result individual landholders were permitted to control large areas. In the United States, by contrast, government policies encouraged settlement by family farmers, and as a result the large areas necessary for grazing were in relatively limited supply. This difference in attitude is particularly well illustrated in the history of southeastern Australia.

Between about 1820 and 1860, much of Australia's southeastern interior was settled by sheep farmers without the benefit of roads or security of tenure, and in circumstances that demanded self-sufficiency. State governments were unable to control this economically motivated settlement until the 1860s, when other prospective settlers began looking for arable land. Now sheep farmers came to be seen as occupying inappropriately large areas of land that could be put to better use. Hence, a series of land laws were passed to allow arable farmers onto land previously held by sheep farmers. The aim of the land laws was to create stable and prosperous agricultural societies. Sheep farmers used various tactics to prevent arable farmers from laying claim to land; some even built huts on wheels so that they could be moved rapidly from site to site, to indicate to inspectors that the land was already occupied. Legal conflicts were common, and the transition from pasture to arable was not really complete until about 1900.

Members of an agricultural committee in Las Lajitas, Honduras, watch as a farmer spreads dried maize leaves on a terrace where coffee trees will be grown. The leaves will preserve soil humidity and, when they rot, provide organic fertilizer (International Fund for Agricultural Development photo/ Franco Mattioli).

horse manure. In Asia human waste was used. Another source was discovered about 1850 off the coast of Peru, on a few arid Pacific islands with whole cliffs made of guano (seabird excreta). It is estimated that some 20 million tons of guano were transported by ship from these islands to Europe between about 1850 and 1870. Once they had depleted this source, traders began exporting a fossil nitrate called caliche, found in desert areas in Chile, for use as a source of nitrogen. Even with the help of these organic fertilizers, however, crop yields were still too limited to permit human populations to grow beyond a maximum density of about 5 per hectare. Indeed, one of the principal constraints on population growth during the nineteenth century was the lack of a reliable source of nitrogen sufficient to enable farmers to increase their cereal yields significantly. This need was filled by two chemists who recognized that the vast ocean of air that surrounds us contains large quantities of nitrogen (Smil 2001).

The solution was ammonia synthesis—a process that converts atmospheric nitrogen to ammonia, which is then used to produce synthetic nitrogen fertilizers. This 'Haber–Bosch process' was discovered in the early twentieth century, and nitrogen fertilizer production began on a commercial scale in 1912. Nitrogen fertilizers were used widely by the 1950s, and today this process produces about 2 million tons of ammonia each week, providing almost 100 per cent of the inorganic nitrogen used in agriculture. China is the single largest producer. In an important sense, humanity today is dependent on nitrogen fertilizers, and it is this technological breakthrough, more than any other, that permitted population to rise from 1.6 to 6.1 billion in the twentieth century.

At the same time, nitrogen fertilizers have been associated with environmental damage—including soil and water contamination, increasing soil acidity, and the release of nitrous oxide (a potent greenhouse gas) into the atmosphere—as well as increased risk for some cancers. Concerns such as these have prompted many to argue for a return to organic farming (see Box 11.10).

The 'green revolution'

The second major agricultural advance in the twentieth century was what came to be known as the 'green revolution'—the rapid development of improved plant and animal strains and their introduction to the economies of the less developed world. As we have seen, such improvements had been ongoing in the western world since the eighteenth century, as one aspect of larger economic and technological changes. Until the 1960s, however, the less developed world did not benefit from these changes, and there were many instances, especially in Asia, of widespread hunger, malnutrition, and dependence on food aid. The initial break-through came as a result of scientific work in the more developed world.

In the 1960s, the Rockefeller and Ford foundations cooperated in the creation of an agricultural research system designed to transfer technologies to the less developed world. This system produced genetically improved strains of two key cereal crops, rice and wheat—often labelled High Yielding Varieties, or HYVs—and eventually other crops such as

11.10

ORGANIC FARMING

Critics of contemporary conventional farming strategies argue that they should be replaced by more sustainable and ecologically more appropriate methods. Collectively, the alternative methods they propose differ from conventional farming in three ways (Atkins and Bowler 2001:68):

1. Alternative methods are less centralized: there are more farms, and marketing is local and regional rather than national and international.
2. Alternative methods emphasize the community rather than the individual: the focus is on cooperative activity, using labour rather than technology when appropriate, and giving due consideration to all costs, material and non-material, as well as moral values.
3. Alternative methods are less specialized: farming is seen as a system of related activities rather than a series of individual components.

In short, alternative farming involves a completely different value system, in which farming is regarded as a way of life and not simply a matter of producing a product for the market.

Organic farming is the best known of these methods. Although it remains a minor component of the agricultural sector in all countries—in Canada only 1 per cent of farms were described as organic in 2001—it is viewed favourably by many (including the Prince of Wales) and is becoming increasingly popular in many western countries. Organic farmers may spend somewhat less than regular farmers on purchased inputs such as pesticides and fertilizers, but their production levels are significantly lower, and therefore their prices are higher. The farmers who pioneered organic production, and the consumers who supported them by purchasing their more expensive products, were motivated by real environmental and ethical concerns about conventional farming. As the demand for 'healthier' food increases, however, it seems clear that a significant part of organic sector today is motivated primarily by profit. With increasing numbers of organic producers turning to more conventional marketing strategies—including supplying national or even international markets—organic agriculture is not always what consumers believe it to be. In effect, producers who adopt the 'organic' label without the philosophy are hijacking it.

Employees of High Ground Organics in Watsonville, California, pack boxes with nine different kinds of organic fruits, vegetables, and herbs. In 2002, with its partner Mariquita Farm, the company was supplying approximately 700 subscribers with a box of fresh produce every week in exchange for an investment in the farm. (AP photo/Julie Jacobson).

Meanwhile, some observers are questioning whether foods produced using organic methods really are better for consumers than more conventional foods. Some say that what the organic sector offers is simply a lifestyle package designed to make consumers feel good about paying higher prices for their food.

A recent development with some similarities to the original organic farming movement is the movement for 'fair trade'. Still in its infancy, this is a grassroots social movement that appeals directly to people's sense of justice. The idea behind fair trade is that consumers pay a guaranteed price plus a small premium to groups of small producers supplying commodity goods such as coffee, tea, chocolate, and fruit. Supporters of this movement hope that fair trade will counter one of the evils often attributed to globalization: the world price of many commodities has barely risen in twenty years, with the result that many small producers in the less developed world operate at a loss. It will be interesting to see whether the fair trade movement is able to make a significant difference to those producers.

beans, cassava, maize, millet, and sorghum were also improved. Although the new strains tend to have lower protein content than their predecessors, they produce higher yields, respond well to fertilizer, are more disease- and pest-resistant, and require a shorter growing season. Besides genetic improvements, the 'green revolution' involved expanded use of fertilizers, other chemical inputs, and irrigation. In many cases, the adoption of HYVs and other technologies quickly doubled crop yields—hence the name 'green revolution'.

The new strains and technologies allowed farmers to grow enough not only to subsist but to market surplus production, raising farm incomes and thus stimulating the rural non-farm economy as well. Increasing incomes and lower prices led to better nutrition, with higher calorie consumption and more diversified diets. These positive impacts have been most evident in southeast Asia and the Indian subcontinent (Box 11.11).

Yet despite undoubted successes, the results have not always met expectations. As with efforts to introduce population control policies, full acceptance has sometimes been prevented, or at least delayed, by cultural and economic factors. Peasant farmers are usually conservative and reluctant to abandon traditional practices. To persuade them to accept new plant strains, governments have sometimes found it necessary to offer both subsidies and guaranteed prices. Accordingly, perhaps inevitably, the agricultural changes associated with the 'green revolution' have had some unfortunate consequences, and criticisms are plentiful.

One set of criticisms focuses on economic circumstances. New strains have been most readily adopted by those farmers who are better-off, who have some capital and relatively large holdings—with the result that in some cases improvements have led to the displacement of poor tenant farmers. Further, many of the poorest tenant farmers have been prevented from adopting improved strains because they cannot afford the fertilizer and pesticides required to grow them: the result is increased inequality and exacerbated poverty. A related criticism is that new strains and techniques have been accompanied by unnecessary mechaniza-

11.11

THE 'GREEN REVOLUTION' IN INDIA

For about twenty years after gaining independence in 1947, India depended on food aid, but 'green revolution' technologies have permitted steady increases in food production—although rapid population growth has offset some of the impact of these increases. You will recall from Chapter 5 that the Indian population is now over 1 billion and that India will shortly overtake China as the largest population country in the world. Part of the success of the 'green revolution' in India can be attributed to the presence of the Indian Agricultural Research Foundation (IARF), a government-funded organization that was established in 1905. With the adoption of the first new strain of rice in the late 1960s, funding to the IARF was increased, and together new technologies and a supportive infrastructure put India at the forefront of 'green revolution' advances.

The IARF is headquartered in Delhi and has nine regional research centres, a large library, and a modern mechanized farm. From the late 1960s onwards, its principal goals have been to facilitate the spread and growth of 'green revolution' technologies developed elsewhere while conducting its own research into the development of higher-yielding and more disease-resistant varieties, genetic enhancement, breeding for multiple cropping and intercropping, analyzing the links between photosynthesis and productivity, and improving the nutritional value of crops. This work has resulted in the development of 64 new crop varieties since the 1960s. In addition the IARF has worked on related projects including irrigation strategies, research on fertilizers and pesticides, and the use of manure.

Because India is a large country, diverse both physically and culturally, different strains and technologies are needed in different areas. For example, different physical settings need different environmentally friendly fertilizers. Efforts to develop monsoon-dependent crops continue, but so far without success. Overcoming cultural resistance is another problem, especially in areas of subsistence farming. Feeding the refugees who continue to move into some of the densely populated areas—including Bangladeshis escaping floods and Afghans fleeing political uncertainty in their homeland—is an additional challenge for Indian agriculture (Slatford and Fishpool 2002).

tion, which has reduced rural wages and increased unemployment.

A second set of criticisms focuses on undesirable environmental consequences. Excessive and sometimes inappropriate use of fertilizers and pesticides has resulted in water pollution, unwanted damage to insect and other wildlife populations, water shortages caused by increased irrigation, and some serious health problems.

Finally, a third type of criticism focuses on the broader social and political implications of the 'green revolution'. Specifically, there is a serious risk that imported knowledge, presumed to be expert, will be accepted at the expense of the local knowledge that has been acquired and successfully applied over centuries. This downgrading of local knowledge is one component of **neocolonialism** —which in turn may be seen as one aspect of globalization. On balance, it is clear that the 'green revolution' has helped to prevent possibly catastrophic food shortages. However, it is also clear that—ironically— adoption of new crop strains sometimes increases the dependency of the less developed country, as use of new technology increases the need to import fuel, fertilizer, and pesticides, and exposes the adopters to new risks, from environmental damage to rising oil prices. These problems notwithstanding, many observers see a need for additional 'green revolution' strains and technologies in sub-Saharan Africa.

Biotechnology: another agricultural evolution?

As a result of the green revolution, world grain production has increased spectacularly—about 45 per cent between 1970 and 1990. In the future, however, it seems likely that the most significant gains will be made through developments in the area of biotechnology.

Tissue culturing and DNA sequencing are two of the biotechnological methods that have helped to develop improved plant and animal varieties. Perhaps the most important—and controversial—aspect of biotechnology, however, is the alteration of the genetic composition of organisms, including food crops. Large areas of grain cultivation have already switched to genetically modified varieties, and in 1999 about 40 million hectares of land were used for genetically modified crops—a 44 per cent increase over 1998. As much as 33 per cent of US corn, 55 per cent of US cotton, and 90 per cent of Argentine soybeans come from genetically modified varieties. General Foods has made a caffeine-free coffee bean, and other products are expected to become available soon.

From the perspective of food production and the world food problem, the greatest benefits are likely to come with the creation of crops resistant to virus infections. Some observers also claim that foods will be produced that offer enhanced nutritional and possibly even medicinal qualities. So far, though, the main emphasis has been on engineering varieties that can resist specific herbicides (such as Monsanto's Roundup)— a development that is of less interest to consumers than to herbicide producers and farmers.

Some commentators argue that the real problems of food production are not to be solved by technological advances but rather by solutions addressing issues of spatial and human inequality (Chapter 6). Increasing numbers of interest groups, especially environmental groups and some consumers, are expressing concern about genetically modified crops and the impacts of farming on the environment. As we have seen, some favour a return to organic farming, which rejects the use of synthetic fertilizers and pesticides and promotes closer links between the producer and the consumer. Especially in Europe and Japan, consumer demand for organic produce is increasing. The European Union currently restricts the importation of many genetically modified products on the grounds that the consequences for both consumers and the environment are uncertain. Canada, the United States, and Argentina, on the other hand, argue that genetic modification poses no risks to either consumers or the environment. In January 2000, following a debate in Montreal, a global agreement was reached concerning safety rules for genetically modified products. This Biosafety Protocol to the UN Convention on Biodiversity includes rules designed to protect the environment from damage.

Nevertheless, it may not be an exaggeration to suggest that developments in biotechnology could prove to be the most significant changes in agricultural production since the first animals and plants were domesticated.

World Agriculture: Types and Regions

The noted American geographer Whittlesey identified nine major regional types of agriculture, and Figure 11.11 shows the global distribution of these nine types as elaborated upon by Grigg (1974); it includes a twofold division of one of the types, and identifies those areas with little or no agriculture.

Even a cursory review of Figure 11.11 highlights the relationships between agriculture and climate (Figure 3.7), between agriculture and generalized global environments (Figure 3.8), and between agriculture and population distribution and density (Figure 6.1). The figure also suggests the complex links between areas of production and areas of consumption.

Primitive Subsistence Agriculture

One of the earliest agricultural systems, 'primitive subsistence' or 'shifting' agriculture, is today practised almost exclusively in tropical areas. It involves selecting a location, removing vegetation, and sowing crops on the cleared land. Land preparation is minimal, and little care is given to the crops. Typically, agricultural implements are limited and livestock are not normally part of the system. After a few years, the land is abandoned and a new location is sought. In this farming system, land is not owned by an individual or family but rather by some larger social unit such as the village or tribe. Shifting agriculture has traditionally been for subsistence purposes, that is, for local consumption. Today, however, in addition to their subsistence crops many such cultivators also produce a different crop for sale, so that they have at least some market orientation.

Superficially, shifting agriculture appears to be wasteful and indicative of a low technology. Yet in fact there is evidence to suggest that it is a highly appropriate. Not only is it an effective way of maintaining soil fertility in humid tropical areas, and well-suited to circumstances of low population density and ample land, but it provides adequate returns for minimal capital and labour inputs. Shifting agriculture, then, can be explained by reference to a number of variables beyond economic considerations, including environment and population numbers.

A protestor dressed as a tomato crossed with a fish leads a demonstration in January 2000, calling on Ottawa to require labelling of genetically modified foods (CP photo/ Fred Chartrand).

Wet Rice Farming

Most of the rural population of east Asia is supported by wet rice farming, an intensive type of agriculture that requires only a small portion of the total land area—farms are small, perhaps only 1–2 ha (2.5–5 acres) and subdivided into fields—but large amounts of human labour. There are many versions of wet rice farming, but in all instances the crop is submerged under slowly moving water for much of the growing period. Wet rice farming is restricted to specific local environments, most notably flat land adjacent to rivers. Low walls delineate the fields and hold the water. Because this technology minimizes soil depletion, continuous cropping is often possible, producing multiple harvests each year. Cultivation techniques changed little until the 1960s, when some areas began using new rice varieties, fertilizers, and pesticides—innovations collectively known as the green revolution (see p. 387 above). There are close links between wet

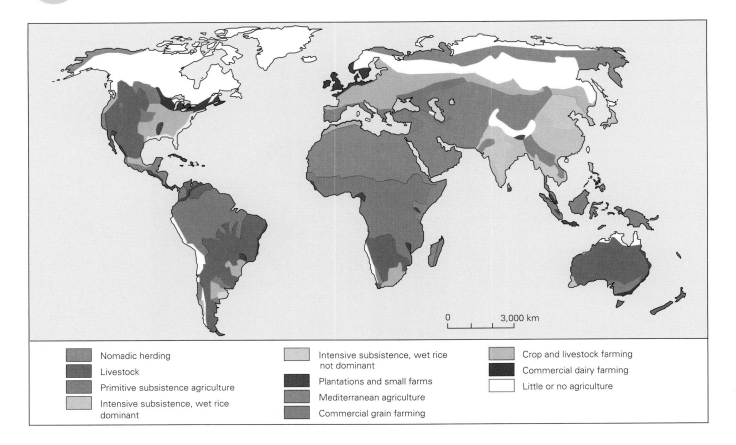

Figure 11.11 World agricultural regions.
Source: D.B. Grigg, *The Agricultural Systems of the World: An Evolutionary Approach* (New York: Cambridge University Press, 1974):4.

rice farming, high population densities, and flatland environments (wet rice fields on terraced hillsides are characteristic of south China and some areas of southeast Asia).

Pastoral Nomadism

Traditionally practised in the hot/dry and cold/dry areas of Africa, Arabia, and Asia, pastoral nomadism is declining today as a result of new technologies and changing social and economic circumstances; however, it remains an important type of agriculture in Somalia, Iran, and Afghanistan. Pastoral nomads are subsistence-oriented and rely on their herds for milk and wool. Meat is rarely eaten, and the livestock—cattle, sheep, camels, or goats—is rarely sold. Usually one animal dominates, although mixed herds are not unknown. Pastoral nomadism likely evolved as an offshoot of sedentary agriculture in areas of climatic extremes where regular cropping was difficult. Continually moving in search of suitable pastures, these groups achieved considerable military importance until the rise of strong central governments brought them under control (Box 11.12).

Mediterranean Agriculture

Mediterranean agriculture can be designated as a type because it is associated with a particular climate (mild, wet winters and hot, dry summers) and because it has played a major role in the spread and growth of agriculture in the western world. Traditionally, it has three components: wheat and barley, vine and tree crops (grape, olive, fig), and grazing land for sheep and goats. Whereas both wheat cultivation and grazing are extensive land uses, and low in productivity, vine and tree crops are intensive. Typically, all three activities are practised on all farms. Spatially, this type of agriculture evolved in the eastern Mediterranean. It had spread throughout the larger region by classical times, and then was exported to environmentally suitable areas overseas, such as California, Chile, southern South Africa, and southern South Australia. Population growth and technological changes have led to several variations, with a general increase in irrigation and a decline in extensive wheat cultivation.

Mixed Farming

The four agricultural types discussed so far are generally subsistence in orientation (Mediterranean agriculture might be described as mixed subsistence/commercial); the remaining five are commercial. Mixed farming prevails throughout Europe, in much of eastern North America, and in other temperate areas of European overseas expansion. It is clearly a variant of the earliest farming in southwest Asia, involving both crops and livestock. The transition to modern mixed farming involved a series of changes, such as the adoption of heavier ploughs and a three-field system (in the early Middle Ages), the reduction of fallow and the use of root crops and grasses for animal feed (*c*. the seventeenth century), and general intensification (mid-nineteenth century). Most of these changes have been related to population and market pressures.

Contemporary mixed farming is intensive and commercial (being closely associated with large urban areas), and integrates crops and livestock. The principal cereal crop varies according to climate and soil—it may be corn, wheat, rye, or barley—while root crops are grown for animal and human consumption. Crop rotation is standard, because it has both environmental and economic advantages, aiding soil fertility and minimizing the impact of price changes. As with the other farming types identified, there are many versions of mixed farming. Differences are particularly significant between western Europe and the American midwest, which is characterized by larger farms, especially high productivity, and advanced technology.

11.12

PASTORAL NOMADISM IN THE SAHARA AND MONGOLIA

Pastoral nomads in the Atlantic Sahara region of northwest Africa are declining in numbers for a variety of interrelated reasons, including government persuasion, economic change, drought, and war (Arkell 1991). For more than 700 years, the Sahrawis survived in a difficult desert environment by migrating vast distances to locate water and pasture for their herds of camels and goats. Even the imposition of artificial colonial boundaries in northwest Africa, dividing the traditional territory of the Sahrawis, did not prevent them from continuing their nomadic way of life. Only in the 1960s did economic changes, in the form of phosphate exploitation and related urbanization, prompt many of the herders to seek wage employment in urban areas.

More damaging changes occurred in the 1970s as a consequence of drought and war. The war began when Spain ceded its colonial interests in Western Sahara in 1975, leaving Morocco and Mauritania to compete for the territory, and continued after a guerrilla group (backed by Algeria) claimed to represent the interests of the Sahrawis in Western Sahara. Many Sahrawis fled into Algeria, and cross-border movements between the latter and those remaining in Western Sahara are now virtually nil, as Morocco has constructed a series of defensive walls (sand and rubble parapets) around much of Western Sahara. The 1989 creation of the five-country Maghreb Union (Box 9.10) has not yet helped to resolve the conflict, and the future of the nomadic pastoralist Sahrawis remains uncertain.

In the popular imagination, perhaps the most famous pastoral nomads were those from Mongolia who struck fear into

the fifth-century Roman empire and who, from the thirteenth century onwards, founded dynasties in China, Persia (Iran), and India. The physical environment that supported those people is the Mongolian steppe, a fertile grazing region resulting from grasses' ability to turn to biomass the energy of the summer sun. It is likely that several animal species, including sheep, goat, camel, and horse, were domesticated on the steppe. The present political state of Mongolia owes its existence to the conflict between pastoralists and the expansionist Chinese agrarian society. Part of the Chinese empire for two centuries, Mongolia achieved independence in 1911 and then became a communist state under Soviet tutelage in 1924. Communism was abandoned in 1990, and today Mongolia is a liberal democracy. The pastoral economy remains, but is subject to many stresses.

Perhaps the greatest stress is the contradiction between a free market economy and public ownership of land in the pastoral areas. The current trend is towards private ownership, and a 2002 law giving Mongolians the right to own plots of land in urban areas seems likely to be extended to the rural pastoral landscape. On the surface this seems a positive step, for, as our discussion of the 'tragedy of the commons' in Box 4.2 demonstrated, public ownership can lead to environmental degradation. On the other hand, private ownership could be a disaster, undermining the 'best practices' developed over centuries. Mongolia, like any other part of the world, is a changing environment and the consequences of change are necessarily uncertain.

Del Monte banana plantation, Costa Rica (Victor Last, Geographical Visual Aids).

oil-palm, cacao, coconuts, bananas, jute, sisal, hemp, rubber, tobacco, groundnuts, sugar cane, and cotton for export to Europe and North America primarily. This type of agriculture is extremely intensive, operating on a large scale using local labour (usually under European supervision) and producing only one crop on each plantation.

Because plantation agriculture evolved within the larger context of colonialism, it has profound social and economic implications and has been seen as exploiting both land and labour. However, the fault here lies with colonialism, not plantation agriculture in itself. Because the plantation economy was and is so labour-intensive, the first plantations, established in the fifteenth century by the Portuguese in Brazil to produce sugar cane, often used slave labour imported from west Africa; then, following the abolition of slavery, indentured labourers were brought in from India and China (see Box 6.3). As a consequence of the worldwide search for labour, alien peoples were introduced to many areas, and cultural conflict continues today.

Because plantations are designed to serve the needs of distant areas, plantation companies are often multinational: that is, they operate in more than one country. Many such companies are also characterized by what is known as vertical integration: the same company that produces the crop also refines, processes, packages, and sells the by-products. Thus, although political colonialism is largely gone, it has in many instances been replaced by economic colonialism.

Dairying

Specialization in dairying is closely related to urban market advances in transportation and other technological changes that began in the nineteenth century. It is particularly associated with Europe and areas of European overseas expansion. Farms are relatively small and are capital-intensive. The sources of the dairy products marketed in a given region often reflect distance from market, with those farms close to market specializing in fluid milk and those farther away producing butter, cheese, and processed milk. On the world scale, those dairy areas close to population concentrations—in western Europe and North America—specialize in fluid milk, while relatively isolated dairy areas—New Zealand, for example—focus on less perishable products. Fluid milk producers are making increasing use of the feedlot system, in which cattle feed is purchased rather than grown on the farm. In such cases dairying operations begin to look more like factory systems than the traditional image of an agricultural way of life. This development is most evident in North America.

Plantation Agriculture

A number of crops required for food and industrial uses can be efficiently produced only in tropical and subtropical areas. For this reason, as Europeans moved into the non-temperate world, they developed what became known as plantation agriculture. Plantations produce crops such as coffee, tea,

Ranching

Commercial grazing—ranching—is generally limited to areas of European overseas expansion and is again closely related to the needs of urban populations. Cattle and sheep are the major ranch animals, since beef and wool are the products most in demand. Much of the European expansion into temperate overseas areas in North and South America, Australia, New Zealand, and South

Africa was motivated by the desire to expand a grazing economy. Ranching is usually a large-scale operation because of the low productivity involved, and hence is associated with areas of low population density. In some areas the ranching economy has evolved into a livestock-fattening economy; the most notable example is the American corn belt, where corn and soybeans are produced for animal consumption. This trend appears likely to continue; ranchers, especially in semi-arid areas of the temperate world, are experiencing difficult economic times today.

Large-Scale Grain Production

In large-scale grain production the dominant crop is wheat grown for sale, often for export. Major grain producers are the United States, Canada, and Ukraine. Farms are large and highly mechanized. This type of production evolved in the nineteenth century to supply the growing urban markets of western Europe and eastern North America, where in many cases it displaced a ranching economy. In some areas, such as Ontario in the mid-nineteenth century, wheat became the staple crop:

> A more classic case of a staple product would be difficult to imagine. More specialized in wheat production than the farmers of present-day Saskatchewan, Ontario farmers of the mid-nineteenth century exported at least four-fifths of their marketable surplus. Close to three-quarters of the cash income of Ontario farmers was derived from wheat and wheat and flour made up well over half of all exports from Ontario (McCallum 1980:4).

Wheat and other grain staples were the engines of economic growth in other areas too. Today, however, most of those areas are more diversified, and some now rely more on mixed farming than grain production. On the Canadian prairies the transition from a predominantly wheat economy to a more diversified farming economy was evident even in the 1980s, reflecting both global trends and national economic strategies (Carlyle 1994; Seaborne 2001).

Globally, as we saw in Chapter 10, world commerce is increasingly liberalized and competitive, and as a result governments are

Harvesting canola near Ponoka, Alberta (south of Edmonton), August 2002. A severe drought that summer reduced crop yields by two-thirds (CP photo/Adrian Wyld).

being forced to reconsider their interventions into the agricultural sector. Particularly important for farmers on the Canadian prairies was the 'Crow rate', named for the Crow's Nest Pass Agreement of 1897, under which the Canadian Pacific Railway had guaranteed western grain farmers a special low rate 'in perpetuity'. In 1984 the terms of this agreement were changed and the 'rate' became a direct federal subsidy, which in 1995 was eliminated altogether. In recent decades non-grain crops have become increasingly important, among them oil seeds, especially canola and flax, and such specialty crops as peas, lentils, mustard seed, and canary seed. The hog industry is also growing rapidly: in Saskatchewan, hog numbers increased by almost 50 per cent between 1996 and 2001.

Overall, traditional areas of large-scale grain production are undergoing significant changes as the need to adjust to changing global, national, and regional circumstances becomes evident. In turn, changes in production lead to a multitude of other changes (Box 11.13).

A Marxist Perspective on Agriculture

In recent years, Marxist political economy (see Box 11.3) has provided a valuable perspective on agricultural change, especially in the context of an increasingly globalized world and food system. Marxist discourse has two important advantages for contemporary agricultural geography: it blurs the lines between agricultural and other economic activities, especially industry, and it integrates the agricultural experiences of the more and less developed worlds.

The Marxist perspective gained adherents when it became clear that neither the empiricist nor the positivist approach offered a meaningful account of the agricultural changes that began in the 1970s. Neither description nor location theory offered any conceptual insight into issues such as the industrialization of agriculture, the rise of agribusiness, state intervention, or the social impacts of technological change. A key insight revealed by the Marxist perspective is that capitalist development in the broad agricultural and food system has increased dramatically as non-farm interests such as banks and the food processing industry have gained increasing control of the agricultural production process.

Agriculture and the World Economy

For almost all of human history, food was consumed in the same location where it was produced. The fact that, today, producers and consumers are no longer in the same location is one reason the Marxist approach is so useful today. In the contemporary world, agriculture is just one component of larger systems of production and consumption that are often affected by regulations that have a global impact (e.g., trade agreements) and by the activities of transnational corporations.

The relative importance of agriculture in an economy decreases with economic growth (Grigg 1992). This can be demonstrated as follows:

1. For preindustrial economies, data for the World Bank category of low-income countries suggest that today about 32 per cent of their gross domestic product (GDP) is derived from agriculture, while about 68 per cent of their population is engaged in agriculture; similarly, the limited data available for contemporary industrial countries prior to the industrial revolution indicate that agriculture accounted for about 40 per cent of the GDP and about 70 per cent of the workforce.

11.13

CANADIAN FARMERS: FEWER AND OLDER

The trend towards fewer and older farmers in most parts of the more developed world is well established, and is one aspect of larger economic changes related to the changing nature of employment (a topic discussed more generally in Chapter 13). In Canada, the total number of farm operators declined by 10 per cent in the five years between 1996 and 2001. Provincially, the greatest declines were in Newfoundland, Prince Edward Island, and Manitoba. Although the decline in numbers of farmers has not reduced food production—with larger farms, those that remain are producing more than ever before—it is having a significant impact on rural life and the health and perhaps even the survival of many small towns.

The 1996–2001 decline was neither new nor unexpected. On the prairies, rural populations grew rapidly until the 1930s, when the numbers began the decline that has continued into the twenty-first century. Meanwhile, as the farm landscape is gradually depopulated, the infrastructure of railway lines, grain elevators, small towns, schools, and so forth is being dismantled, and the large amounts of capital required to survive in an increasingly high-risk environment mean that more and more agricultural land is now in the hands of corporations and huge farm conglomerates.

In short, it appears that the small family farm is a thing of the past. Most small farmers now require some off-farm income in order to support themselves, and the median age of farmers is steadily increasing across Canada; in Manitoba it rose from 46 to 48 in five years (1996–2001). To survive, some small farmers are moving into specialized areas—in Manitoba, some focus exclusively on raspberries or other soft fruits—or activities such as organic farming (see Box 11.10).

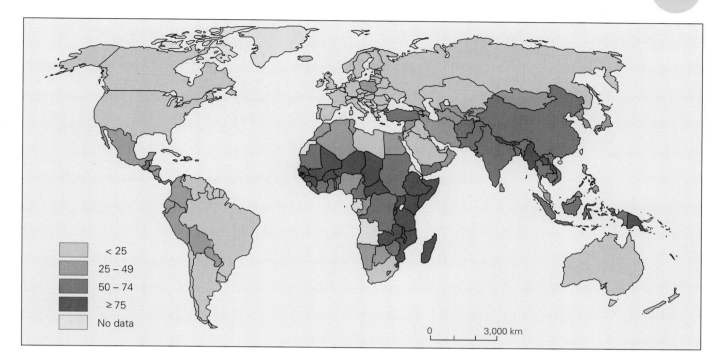

2. For industrial economies today, data for the World Bank category of high-income countries suggest that only 2.8 per cent of their GDP is derived from agriculture and that only 5 per cent of their population is actively engaged in agriculture.

These figures exaggerate the extent of the relative decline of agriculture because the agricultural data for contemporary industrial countries do not include food-processing and related activities, but the general picture is nevertheless clear: industrial and service activities have replaced agriculture as the most important sectors of national economies (Figure 11.12). Furthermore, agricultural incomes are lower on average than non-agricultural incomes, and this gap is widening all the time.

There are considerable spatial variations in the timing of the relative decline of the agricultural sector. The decline began in Britain, the home of the industrial revolution, where the agricultural percentage of the total labour force fell below 50 as early as the 1730s; in other European countries the decline to below 50 per cent had occurred by the 1840s; and in much of Africa and Asia it has not yet occurred.

Global trade in food

Two important causes of current agricultural change are the ever-increasing demand for food and the concentration of people in large urban centres. Given the differences in population growth rates between the more and less developed worlds, movements of food products from areas of surplus to areas of deficit, through both sales and aid, are bound to increase.

The global trade in food is a relatively recent phenomenon, associated with the industrial revolution and, in particular, technological advances in ocean transportation and refrigeration. Most of this trade involves imports to the more developed world of products from the less developed world, notably coffee, tea, and cocoa.

At least half of the world's people rely on one of three cereal crops: wheat, corn, and rice. In many cases these crops are produced for local consumption, but there is also a substantial global trade in them. Indeed, most of the foods exported from more developed countries are cereals; food exports are dominated by five countries—first and foremost the United States, but also Canada, Argentina, France, and Australia—while distribution of the surpluses is concentrated in five private corporations. All five of these corporations are international giants with diverse economic interests. The principal importer of grain in recent years has been the former USSR, but several Asian countries are increasing their imports, notably China, South Korea, Indonesia, the Philippines, and Japan.

Figure 11.12 Percentage of labour force in agriculture by country, 1996. This map provides a good indication of the importance of agricultural activity globally. The range is enormous. In many of the more developed countries less than 25 per cent of the labour force works in agriculture; in both Canada and the United States only 3 per cent do so. In some less developed countries the proportion is more than 75 per cent; in Niger and Mali, the value is 94 per cent. The correlation between development level and agriculture is far from perfect, however; several countries in Latin America have values below 25 per cent. The world average is 49 per cent.

Source: United Nations Development Programme, *Human Development Report* (New York: Oxford University Press, 1996):tables 16 and 32.

The case of China

The preceding discussions of plantation agriculture, the 'green revolution', and agribusiness all relate directly to the less developed world. Each of these agricultural developments is subject to criticism. Indeed, throughout much of the less developed world, uncritical acceptance of 'green revolution' innovations has led to problems, some of which are outlined in Box 11.14.

In China, economic liberalization began in 1978, when the country was first opened to international trade and foreign investment. Agriculture was the first sector of the economy to be reformed and modernized, in the 1980s (Morrish 1994). Since the collectiviza-

tion of agriculture following the revolution of 1949, all farm operations had been run by groups rather than individuals, to allow the communist government to exert maximum control over production. Some 60,000 rural communes, organized by officials, were the key social and economic units before 1979. Under the less rigorous system in place since 1979, families work together on a voluntary basis; farmers contract to use government-owned land, agreeing to deliver a specified amount of produce to the government, and are allowed to sell any surplus. The result of this reform was a steady increase in both yields (about 6 per cent per year) and incomes. Many peasant farmers are now opting not to

11.14

AGRICULTURAL CHANGE IN BHUTAN

Bhutan is a mountainous land-locked kingdom between China and India that is inhabited by people of Tibetan or Hindu Nepalese origin and is currently striving to emerge from its geographic isolation. Only 46,500 km² (17,800 square miles) in size, in 1999 it had a population of 0.9 million, only 15 per cent of which was urban; a CBR of 40; a CDR of 9; and an RNI of 3.1 per cent.

Agriculture, focusing on rice and livestock, is limited to about 3 per cent of the land, but is nevertheless the most important sector of the economy. Not surprisingly, farming operations are small in scale and oriented towards subsistence, sustainability, and household self-sufficiency. It appears that this system works relatively well, for although Bhutan is a poor country, poverty at the individual or household level is rare.

Recently, however, the government, with support from elsewhere, has been attempting to introduce changes in Bhutan's agricultural economy—specifically, to increase exports by using imported 'green revolution' improvements, such as higher-yielding varieties of rice and wheat.

Unfortunately, it appears that the crop varieties imported by Bhutan need pesticides and fertilizers and may not be well suited to the local environment—yet they are displacing cereal varieties that, while lower-yielding, are much more appropriate. Similarly, the government has adopted new policies to promote the production of two cash crops, cardamom (a spice) and potato, that appear to be environmentally and culturally inappropriate. There is an important moral to this discussion. Blind, or at least short-sighted, acceptance of new technologies, without careful evaluation of their consequences, may be disastrous. Any change is best preceded by a trial period and a careful appraisal of the broader cultural implications (Young 1991).

Farmers from Wangdi, Bhutan, threshing rice in the traditional way (International Fund for Agricultural Development photo/Anwar Hossain).

produce staple crops but are diversifying into new specialty crops.

Much remains to be achieved, however: although approximately 65 per cent of the labour force work in agriculture, this sector of the economy contributes only about 32 per cent of China's GDP.

Agribusiness

The term 'agribusiness' refers to the activities of transnational corporations. Plantation agriculture was the first example of agribusiness, and even today most such businesses concentrate on production in less developed areas for markets in more developed areas. Table 11.3 suggests the extent to which agriculture in the less developed world is controlled by transnationals operating from and marketing their products in the more developed world.

Agribusiness has been widely criticized, most often on the grounds that, as Table 11.3 clearly shows, it is a form of economic colonialism, with little concern for the social, economic, and environmental consequences of activities in the producing area. More specifically, agribusiness often contributes to the creation of an agricultural economy vastly different from the local producing economy; benefits foreign investors and local élites rather than peasant groups; occupies prime agricultural land; imports labour that may lead to local ethnic conflict (see Box 6.3); uses casual, part-time, and seasonal labour; and is reluctant to take responsibility for negative impacts on the physical environment.

State intervention

Much contemporary agricultural activity can be understood only in the larger context of state agricultural policies and negotiated trading agreements. Why has the state become increasingly involved in agriculture? The basic answer is that improvements in productivity have led to the oversupply of domestic markets, reducing both product prices and farm income. Inadequate income for farmers is the principal reason behind government intervention in the form of income support.

In the more developed world, state policies aim to ensure the security of food supplies, price and income stability, protection of consumer interests, and regional

development. Such policy objectives may be achieved through guaranteed prices, import controls, export subsidies, and various other amendments to the capitalist market system, such as marketing boards. Canada, for example, has price-support policies, income-support policies, and supply-management policies—all of which are subject to change as trading agreements become increasingly important. In the less developed world, most state policies are aimed at increasing productivity and achieving higher farm incomes and dietary standards.

As we saw in Chapter 10, international trading agreements, including the General Agreement on Tariffs and Trade (GATT; replaced by the World Trade Organization in 1995) and the Common Agricultural Policy (CAP) of the EU, have emphasized the need to reduce state support for agriculture.

Table 11.3 MULTINATIONALS AND CROPS IN THE LESS DEVELOPED WORLD

Crop	Companies	Observations
Cocoa	Cadbury-Schweppes, Gill and Duffus, Rowntree, Nestlé	These companies control 60–80% of world cocoa sales
Tea	Brooke Bond, Unilever, Cadbury-Schweppes, Allied Lyons, Nestlé, Standard Brands, Kellogg, Coca Cola	Combined, hold about 90% of tea marketed in western Europe and North America
Coffee	Nestlé, General Foods	Control 20% of world market
Sugar	Tate and Lyle	Buys about 95% of cane sugar imported into the EU
Molasses	Tate and Lyle	Controls 40% of world trade
Tobacco	BAT, R.J. Reynolds, Philip Morris, Imperial Group, American Brands, Rothmans	Combined, control over 90% of world leaf tobacco trade

Source: Adapted from C. Dixon, *Rural Development in the Third World* (London: Routledge, 1990):18.

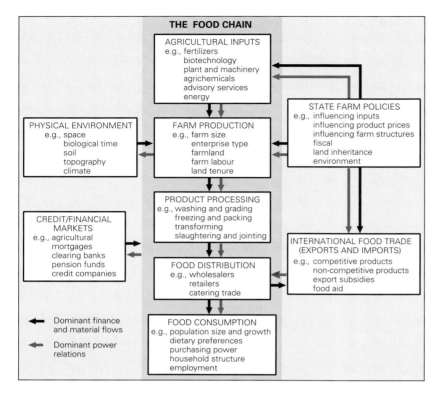

THE FOOD CHAIN

AGRICULTURAL INPUTS
e.g., fertilizers
      biotechnology
      plant and machinery
      agrichemicals
      advisory services
      energy

PHYSICAL ENVIRONMENT
e.g., space
      biological time
      soil
      topography
      climate

FARM PRODUCTION
e.g., farm size
      enterprise type
      farmland
      farm labour
      land tenure

STATE FARM POLICIES
e.g., influencing inputs
      influencing product prices
      influencing farm structures
      fiscal
      land inheritance
      environment

PRODUCT PROCESSING
e.g., washing and grading
      freezing and packing
      transforming
      slaughtering and jointing

CREDIT/FINANCIAL MARKETS
e.g., agricultural
      mortgages
      clearing banks
      pension funds
      credit companies

INTERNATIONAL FOOD TRADE (EXPORTS AND IMPORTS)
e.g., competitive products
      non-competitive products
      export subsidies
      food aid

FOOD DISTRIBUTION
e.g., wholesalers
      retailers
      catering trade

Dominant finance and material flows

Dominant power relations

FOOD CONSUMPTION
e.g., population size and growth
      dietary preferences
      purchasing power
      household structure
      employment

Figure 11.13 The food supply system.
Source: I.R. Bowler, 'The Industrialization of Agriculture', in *The Geography of Agriculture in Developed Market Economies*, edited by I.R. Bowler (London: Routledge, 1992):12.

The Industrialization of Agriculture

For many human geographers, the most fascinating—most 'geographical'—aspect of agriculture is that it is an activity carried out at the interface between humans and the physical environment. Agriculture, unlike most other economic activities, is constrained by physical geography, and its success, as measured by productivity or return on investment, is beyond the control of the farmer. But these physical constraints do not prevent what is now labelled the industrialization of agriculture. Rather, they guide that industrialization in particular directions.

Today, agriculture in the more developed world can be justifiably regarded as an industry, one in which the farmer is no longer the sole or even the primary decision-maker. According to Friedmann (1991:65), 'the distinction between agriculture and industry is no longer viable, and should be replaced by the conception of an agri-food sector central to capital accumulation in the world economy.' The industrialization of agriculture has become especially evident since the end of the Second World War. Both in the more developed world and in most parts of the less developed world, agriculture has become more intensive and more specialized. Today agriculture is one of the biggest

industries in the world, employing about 1.3 billion people, although the numbers employed in the more developed world are decreasing. Agriculture produces about US$1.3 trillion worth of goods per year, and world output of food continues to increase. Nevertheless, farmers are always obliged to strive to be more competitive, more concerned with the environment, and more aware of changing consumer tastes. Some farmers are even entering the information age—for example, using GPSs (see Chapter 2) to monitor soil and moisture conditions on a two-hectare square of land.

Changes to the food supply system

The food supply system is becoming increasingly complicated. Whereas the von Thünen theory focuses on horizontal relations on one level only (that of farmers), today there are not only horizontal relations on several levels but also vertical relations between different levels. As Figure 11.13 shows, the food supply system now integrates activities at five levels—inputs, farmers, processors, distributors, and consumers—and is increasingly globalized. In recent years, the greatest change in this system has been the expansion of the purchasing and selling power of the processing and distributing sectors.

Agricultural change in the more developed world has gone through four distinct phases since the Second World War: mechanization, chemical farming, food manufacturing, and biotechnology. The 1950s brought increasing mechanization, which raised yields and reduced labour requirements, and the 1960s saw the advent of large-scale chemical farming with nitrogenous fertilizers, herbicides, fungicides, and pesticides. These first two stages modernized agriculture. The 1970s witnessed significant growth in the manufacture-of-purchased-inputs sector and in the processing and distributing sectors (Figure 11.13), while the 1980s brought biotechnological advances in plant varieties and livestock breeds. The last two stages industrialized agriculture.

Smith (1984:362) demonstrates that these changes have had a 'marked imprint on the landscape' in the province of Quebec. With increasing vertical integration, retailers are assuming control of some processing, and processors are influencing farm operations;

thus farmers themselves are losing some of their capacity for independent decision-making. At the processing level, plants are now fewer and larger, and there is greater locational concentration. Similarly at the retailing level, a few large organizations tend to dominate the market.

Agricultural restructuring

A process central to many of these changes in the agricultural industry (and other industries, as we will see in Chapter 12) is **restructuring**. In a capitalist economy, this term can refer to changes in either the type of capital invested in (for example, from human labour to machines) or the movement of capital resulting from changes in technology or labour relations. In agriculture, restructuring may involve spatial changes in agricultural activities, movements of capital from one level to another, or changes in the organization of production. Restructuring can be seen as a process of ongoing adjustment to changing circumstances.

Political Ecology and Agriculture

Political ecology merges two important areas of contemporary human geography: traditional human ecology and political economy. This approach can be summarized as follows:

1. At the regional level, human geographers attempt to integrate the multiple social and ecological relations involved in the organization of agriculture and land use.
2. This approach permits a focus on the politics of place, especially regarding differences in power between groups and places (as discussed in a different context in Chapter 8).
3. It has clear implications for the design and implementation of rural development policies.
4. There is recognition of the need to involve local populations as well as governments and organizations in the planning process.
5. The interplay of social practices (agency) and political–economic conditions (structure) is often considered (as outlined in the account of structuration theory in Chapter 8).
6. In many cases the environmental impacts of agricultural activity can be addressed in political terms.

In this way a political ecology perspective can add significant new dimensions to more traditional human ecological approaches. Box 11.15 discusses an application of this type of analysis that provides a good indication of its procedures and merits.

Other examples include studies by Zimmerer (1991), integrating regional political ecology concepts, structuration theory, a politics of place, and production ecology in an analysis of agricultural change in highland Peru, and Grossman (1993:346), who used a political ecology perspective to 'highlight not only the impact of political–economic relationships on resource-use patterns but also the significance of environmental variables and how their interaction with political–economic forces influences human–environment relations'.

Of course, the political ecology viewpoint has much to say also about the environmental impacts of agricultural activities (see Chapter 4), especially with regard to the unequal distribution of those impacts. For example, the hurricane that devastated parts of Central America in October 1998 did not affect all agricultural operations equally. In fact, those farms using traditional methods suffered much less soil erosion and crop damage than did those using modern chemical-intensive methods. More generally, there is good evidence that although modern agricultural methods are usually more productive than traditional methods, environmentally they are less friendly. The four most common types of environmental damage caused by agriculture are particularly dangerous for the less developed world: soil degradation, soil and water pollution, water scarcity, and reduction of genetic diversity.

Gender Relations and Agricultural Restructuring

The significance of gender relations in the agricultural restructuring process is increasingly apparent, especially to those employing a political economy approach to the study of agriculture. In the more developed world, there is interest in the growing discrepancy between traditional gender relations and identities in farming communities and those in society as a whole. The trend for women to leave farming is generally considered a reflection of the conservative gender relations associated with farming communities.

In the less developed world, there is increasing recognition of the roles played by women, past and present. In the west African country of Gambia, for instance, women have played important roles in the traditional subsistence cultivation of rice and in recent initiatives in horticultural production. The smallest state in mainland Africa (11,300 km²/4,360 square miles), Gambia is a narrow strip of land bordering the Gambia River and encircled by Senegal except for a short Atlantic coast. In 1999 it had a population of 1.3 million, a CBR of 43, a CDR of 19, and an RNI of 2.4 per cent. The climate is subtropical, and wooded savanna covers most of the area. There are five principal ethnic groups, and some 85 per cent of the population are engaged in agriculture. Groundnuts are the one cash crop and account for about 90 per cent of exports.

Women play an important role in the subsistence cultivation of rice, but this activity does not generate any income. They have also been enthusiastically involved in recent government initiatives in dry-season horticultural production (Barrett and Browne 1991). If such initiatives increase women's incomes, they are clearly of great value; in fact, several of them yield well and generate the needed income. Typically, however, such efforts face two problems.

11.15

PEASANT–HERDER CONFLICT IN IVORY COAST

A study of conflict between two different ethnic and agricultural groups in the west African state of Ivory Coast provides a good example of how a political ecology approach can be applied to a traditional land-use problem (Bassett 1988).

In the northern Ivory Coast region, beginning about 1970, uncompensated crop damage by cattle in peasant fields gave rise to ethnic and economic tensions between Senufo farmers and Fulani herders, who accounted for 33 per cent of the country's beef production. By the mid-1980s, relations between the two groups had deteriorated to the point of open hostilities: 80 herders were killed in 1986. Some political candidates inflamed the situation by making election promises to banish the Fulani from Ivory Coast. However, the official government policy is to encourage herders so as to reduce dependence on imported beef.

What is the basic cause of this land-use conflict? Tradition-ally, human geographers would have pointed to a declining resource base relative to population. But this answer does not fully explain why some people lose while others gain land; nor does it explain why conflicts occur in such land-abundant areas as the northern Ivory Coast. A more helpful answer would focus on the politics of land use and the human ecology of agricultural systems.

In this case, the key is to determine which of the two groups is able to exert greater political power. The answer to this question will likely influence future land-use patterns. The Fulani are new to the area, while the Senufo are indigenous, but current trends give rights to minorities as well as established majorities (this was not the case in colonial times). The basic determinants of the conflict are outlined in Table 11.4. Note that the table includes both political ecological variables and traditional human ecological variables.

Table 11.4 DETERMINANTS OF PEASANT-HERDER CONFLICTS IN NORTHERN IVORY COAST

Ultimate Causes	Proximate Causes	Stressors	Counter-risks

Ivorian development model:
• surplus appropriation by
  – foreign agribusiness
  – the state
• livestock development policies
	Low incomes and beef consumption
	
	Insecure land rights
		Uncompensated crop damage
		
		Political campaigns
			Compensation
			
			Fulani expansion
			
			Crop and cattle surveillance

Savanna ecology Fulani immigration	Intersection of Senufo agriculture and Fulani semi-transhumant pastoralism	Theft of village cattle	Corralling animals at night

Source: T.J. Bassett, 'The Political Ecology of Peasant-Herder Conflicts in the Northern Ivory Coast', *Annals*, Association of American Geographers 78 (1988):456.

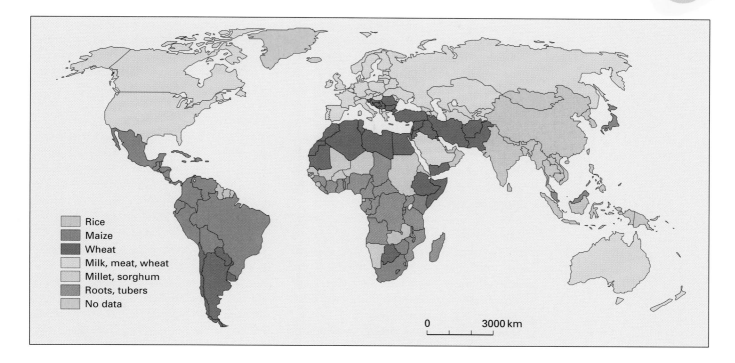

Figure 11.14 Global dietary patterns.

First, the new schemes are based not on the local knowledge of horticulture but on imported (or at least non-local) knowledge. Second, they are not environmentally or economically sustainable. Once again, there is evidence of government efforts to improve agricultural circumstances, but such efforts are probably misguided. In Bhutan and Gambia, long-established agricultural practices are ignored and new imported practices are favoured. It is not difficult to understand why governments are tempted by the new, but it is also important that the new learn from the old. The agricultural practices that have survived, perhaps little changed, for generations are usually grounded in economic and cultural realities; they may need to be improved, but this does not mean that they need to be replaced.

Consuming Food

Much of this chapter has emphasized the political component of the traditionally economic study of agriculture. In this final section, by contrast, we address the cultural component that has become such an important aspect of human geography in recent years. In the context of agriculture, the 'cultural turn' introduced in Chapter 8 is reflected in an interest not only in production but in consumption.

You Are What You Eat—and Where

Historically—as geographers such as Vidal recognized in the early twentieth century—the kinds of food consumed in various parts of the world usually reflected the available resources. Thus the food that people consumed represented one of the closest links between them and their environment, as well as one of the principal means by which groups of people were distinguished one from another. Today, despite the undeniable impact of processes of interaction and globalization, this is still the case in many parts of the world, as Figure 11.14 shows.

The big picture

Considering the history of food in terms of consumption and the consequences of production practices (rather those practices in themselves), Fernandez-Armesto (2201) identified eight key stages:

- Sometime before 150,000 years ago, humans began to use fire to cook food. This new technique for changing food—existing techniques included burying, drying, masticating, and rotting—had significant social consequences, providing a new focus for human interaction.
- In the *second* stage, groups began to see food as representing more than just some-

thing to eat. Table rituals and dietary taboos began to appear in recognition of the other dimensions associated with food, especially the moral and religious dimensions.

- In the *third* stage food became associated with social status: people in positions of authority were able to eat not only more food than others but also different kinds.
- The *fourth* and *fifth* stages together constitute what we earlier described as the first agricultural revolution: the domestication of plants and animals. These two stages—sometimes separate and sometimes related—represented a fundamental change, as humans began producing food, as opposed to collecting it. Plant domestication opened the way to new forms of social organization, greater population density, and the rise of urban centres.
- The *sixth* stage was a global movement of cultures in which food played several important parts. First, food has been one of the most important motivations for both overland and overseas movement over the past 1500 years. Indeed, demand for salt and spices played a major role in the development of transport routes and cultural contact. The wholesale movements of people over the following centuries involved widespread diffusion of animals and crops in all directions.
- The *seventh stage* is perhaps best known under the label 'McDonaldization,' although there is much more involved than the spread of American fast food to other parts of the world. In North America, for example, 'ethnic' foods—from countries such as Italy, Mexico, China, India and even New Zealand (kiwi and pavlova)—are increasingly popular.
- The *eighth* and most recent stage is the current trend towards fast food that is the first real shift away from the social interaction that was initiated in the first stage. To the extent that fast food makes it unnecessary for members of a family to sit down together, it may well contribute to the fragmentation of the family. It is also possible that fast food is contributing to a general decline in health, not only because its nutritional value may be poor but because, without the discipline and cama-

raderie associated with the shared table, the temptation to overeat can be hard to resist. Certainly the current increase in obesity seems to be closely related to the increasing popularity of fast food.

Consumption and place

Today, food consumption habits remain key indicators of human identity, reflecting regional characteristics as well as individual and group tastes., Building on the ideas promoted by Sauer, much traditional cultural geography was concerned with regional and local foodways, particularly what were called 'ethnic preferences'—that is, the preferences of the specific group under study. By contrast, much contemporary cultural geography addresses the complex intertwining of consumption and identity, taking a particular interest in the political implications of consuming 'ethnic foods'—that is, foods associated with groups other than the one under study. The popularity of 'ethnic food' in the latter sense may well reflect a modern version of the historic association between food and social status. In Britain in 2001, the Foreign Secretary reported that the new national dish was neither fish and chips nor roast beef, but chicken tikka masala, a dish that originated in India. Some commentators have suggested that this cross-cultural phenomenon perpetuates the former colonial relationship—although one could equally well interpret it as representing a reversal of that relationship.

Another consumption-related topic is the increasing popularity of specialty foods and drinks that are explicitly identified with a particular place (e.g., maple syrup as a distinctive product of eastern Canada). Through advertising, often directed at potential tourists, a conscious attempt may be made to construct images of product quality and the related place. This development is perhaps best seen as an extension of the longstanding association between products such as wine and cheese and particular places.

Eating away from home is increasingly popular, especially in North America. The remarkable variety of restaurants available, especially in urban areas, makes it possible not only to eat dozens of different sorts of food but to experience dozens of different cultures.

CHAPTER ELEVEN SUMMARY

Economic geography

Much of our human landscape-creating behaviour has strong economic motivations. Economics inspired much of the new human geography (often called spatial analysis) that came to the fore in the 1960s. Contemporary studies of agriculture, industry, settlement, and planning take various approaches and may incorporate economic, cultural, and political content. Cultural and political content is increasingly being emphasized.

The role of theory

Location theory explains geographic patterns, while interaction theory centres on movements between locations. Theories in general are valuable because of their rigour and simplicity. Many of the theories used by geographers are normative, describing what ought to be, rather than what is. The concept of the economic operator assumes that the goal of any economic activity is the maximization of profit in the context of perfect knowledge and a perfect ability to use such knowledge in a rational fashion.

The location of agricultural activities

Many factors can influence the location of agricultural activities. For some geographers, an ecological approach, considering all factors, is the most appropriate. Other geographers focus on particular factors, depending on the specific issue in question. Physical factors include climate, soils, and relief. As living things, animals and plants require appropriate physical environments, either natural or human-constructed. Cultural, social, and political factors also play a role. A farmer's religious and ethnic background affects his or her decisions, as do the policies of various levels of government. Economic factors include the basic principles of supply and demand.

Economic rent

The concept of economic rent is central to agricultural location theory. Land uses compete for locations because it is not possible for all activities to occupy their economical-ly optimal locations. Instead, land is devoted to particular uses according to a spatial pattern that reflects their differing economic rents (itself a specific measure of rent-paying ability). Economic rent can be based on one or more variables; Ricardo based such values on land fertility, while von Thünen used distance from market.

Von Thünen theory

In 1826 von Thünen, a German economist, published a landmark theory of agricultural location. A normative theory that holds all variables constant except distance from a central market, it generates two hypotheses: (1) zones of land use develop around the market; and (2) the intensity of each specific use decreases with increasing distance from the market. Von Thünen proposed particular contents of the zones for his time and place (early nineteenth-century north Germany). This theory and its variations have stimulated much geographic research.

Empirical analyses

Studies of the relationships (1) between land value and distance and (2) between land use and distance can be conveniently subdivided into those with a historical focus, those in less developed areas of the world, and those on various scales ranging from the world to individual farms. Overall, these analyses confirm the value of the von Thünen theory as a basis for investigating agricultural patterns.

Agricultural origins and change

Agriculture originated in the domestication of plants and animals, a long process that began about 12,000 years ago, probably in several different areas, and gradually diffused. Today preagricultural activities, such as hunting and gathering, are marginal. Major technological changes in the agricultural enterprise include a series of advances in eighteenth-century England, the twentieth-century development of nitrogen fertilizers, and the green revolution in the less developed world. Most recently, changing the genetic composition of food crops may be the greatest change in agricultural production since domestication began.

Regions and types

Today there are nine principal types of agriculture, each occupying a particular region or regions. Three types are classified as 'subsistence' (shifting agriculture, wet rice farming, pastoral nomadism); a fourth type, Mediterranean agriculture, is 'mixed subsistence/commercial'; and the remaining five are 'commercial' (mixed farming, dairying, plantations, ranching, large-scale grain production). All but one type can be traced back thousands of years; the exception is plantation agriculture, which developed as a result of European overseas expansion. The modern versions of dairying, ranching, and large-scale grain production are very different from their antecedents.

Agriculture in the world economy

Relative to other economic activities such as industry, agriculture has experienced a continuous decline in importance through time as measured by labour force and contributions to GDP.

Trade in food

Food is an important component of world trade. Much food is sent as aid to poor countries. The former USSR and several countries in Asia are major importers of grain. The United States is the principal exporter.

Agriculture in the less developed world

China is currently experiencing major changes in the organization of agricultural activities as part of a transition from communism. In some other areas, such as Bhutan, green revolution technologies have caused environmental problems. Recently, human geographers have employed a political ecology approach, combining traditional human ecology with political economy to address specific problems such as land-use change, land-use conflict, and the involvement of women in agriculture.

Agriculture as an industry

In many respects agriculture today functions as an industrial activity with several linked levels: input producers, farmers, processors, wholesalers and retailers, and consumers. National and international policies exert considerable influence on agricultural activities at all levels. Partly because of this organizational complexity, traditional von Thünen theory is inadequate to explain the spatial distribution of agricultural activities today.

Political ecology

There is much interest in the politics of place and gender relations in the context of agricultural activities. One critical comment on the green revolution relates to the need to involve local populations in the planning process.

Consuming food

Historically, the kinds of food consumed in various parts of the world primarily reflected available resources. Today much interest is focused on trends in consumption: the act of consuming is seen as a way of distinguishing oneself from others, and perhaps even of establishing superiority.

LINKS TO OTHER CHAPTERS

- Globalization:
  Chapter 2 (concepts)
  Chapter 4 (ecosystems and global impacts)
  Chapter 5 (population growth, fertility decline)
  Chapter 6 (refugees, disease, more and less developed worlds)
  Chapter 7 (cultural globalization)
  Chapter 8 (popular culture)
  Chapter 9 (political globalization)
  Chapter 10 (economic globalization)
  Chapter 12 (global cities)
  Chapter 13 (industrial restructuring)

- Cultural and political considerations:
  Chapter 7 (religion)
  Chapter 8 (ethnicity)
  Chapter 9 (role of the state).

- Location and interaction theories; the economic operator concept; von Thünen theory:
  Chapter 2 (positivism; concepts of space, location, and distance).

- Correlation and regression analysis:
  Chapter 2 (quantitative techniques).

- Origins of agriculture:
  Chapter 2 (environmental determinism; possibilism)
  Chapter 4 (energy and technology)

Chapter 5 (population growth through time)
Chapter 7 (beginnings of civilization).

- Second agricultural revolution:
  Chapter 2 (Marxist terminology)
  Chapter 5 (Malthusian concepts)
  Chapters 4 and 12 (industrial revolution).

- Nine agricultural regions:
  Chapter 2 (environmental determinism; possibilism)
  Chapter 3, especially Figure 3.8 (global environments).

- Plantation agriculture; green revolution; agriculture in the less developed world:
  Chapter 5 (fertility transition)
  Chapter 6 (migration, especially Box 6.3; population and food; world systems analysis).

- Industrialization of agriculture; global trade in food:
  Chapter 13 (restructuring)
  Chapter 9 (globalization).

- Political ecology of agriculture:
  Chapter 2 (Marxism)
  Chapter 4 (environmental ethics)
  Chapters 6 and 8 (gender)
  Chapter 8 (structuration theory).

FURTHER EXPLORATIONS

ATKINS, P., AND I. BOWLER. 2001. *Food in Society: Economy, Culture, Geography.* New York: Oxford University Press.
> Covers a wide range of topics, including food consumption preferences.

BAKER, O.E. 1925. 'Geography and Wheat Production'. *Economic Geography* 4:389–434.
> One of a series of articles by this early agricultural geographer, many of which exemplify a regional approach; see issues of *Economic Geography* between 1925 and 1933 for other examples.

BARRETT, H., and A. BROWNE. 1996. 'Export Horticultural Production in Sub-Saharan Africa: The Incorporation of The Gambia'. *Geography* 81:47–56.
> Details the recent involvement of sub-Saharan Africa in the international trade in fresh fruit, flowers, and vegetables; includes discussion of some social, including gender, implications.

BHATIA, S.S. 1965. 'Patterns of Crop Concentration and Diversification in India'. *Economic Geography* 41: 39–56.
> An early attempt to develop statistical measures for describing agricultural landscapes.

CONKLING, E.C., and M.H. YEATES. 1976. *Man's Economic Environment*. Toronto: McGraw-Hill.

> An excellent text that discusses agricultural theory and landscapes within a larger economic context.

CONZEN, M.P. 1971. *Frontier Farming in an Urban Shadow*. Madison: State Historical Society of Wisconsin.

> A detailed analysis of agriculture in an area experiencing the effects of rapid settlement and commercialization.

FOUND, W.C. 1971. *A Theoretical Approach to Agricultural Land Use Patterns*. London: Arnold.

> A sound overview of theoretical aspects of agricultural land use.

GILSON, J.C. 1989. *World Agricultural Changes: Implications for Canada*. Toronto: C.D. Howe Institute, Policy Study 7.

> A valuable survey of current changes in world agriculture focusing on government policies and needed reforms.

GRIGG, D.B. 1984. *An Introduction to Agricultural Geography*. London: Hutchinson.

> An excellent textbook dealing with factors affecting agricultural activities.

GROTEWOLD, A. 1959. 'Von Thünen in Retrospect'. *Economic Geography* 35:346–55.

> A dated but insightful commentary that is easy to comprehend.

HARRIS, D.R., ed. 1996. *The Origins and Spread of Agriculture and Pastoralism in Eurasia*. Washington, DC: Smithsonian Institution Press, ix-xi.

> An informative edited volume comprising a series of case-studies; comprehensive and balanced coverage reflecting current understanding.

ILBERY, B.W. 1985. *Agriculture: A Social and Economic Analysis*. Toronto: Oxford University Press.

> A well-written basic textbook.

JAKLE, J.A., AND SCULLE, A. 1999. *Fast Food: Roadside Restaurants in the Automobile Age*. Baltimore: Johns Hopkins University Press.

> An entertaining scholarly overview of the proliferation of roadside eateries in the twentieth-century United States. Authored by a geographer and a historian, this work makes explicit use of relevant geographic concepts; a mine of useful and insightful information.

LEAMAN, J.H., and E.C. CONKLING. 1975. 'Transport Change and Agricultural Specialization'. *Annals, Association of American Geographers* 65:425–37.

> A statistical analysis of agriculture in nineteenth-century upstate New York.

OVERTON, M. 1996. *Agricultural Revolution in England: The Transformation of the Agrarian Economy, 1500–1850*. Cambridge: Cambridge University Press.

> An advanced-level book, but clearly written and helpful in understanding agricultural change outside of England; especially useful for the links made to the development of capitalism.

PACIONE, M., ed. 1986. *Progress in Agricultural Geography*. London: Croom Helm.

> Covers a wide range of topics such as theory, classification, diffusion, government policies, and marketing.

SAUER, C.O. 1952. *Agricultural Origins and Dispersals*. New York: American Geographical Society.

> A classic study by the most prominent twentieth-century cultural geographer.

SHEPHARD, E., and BARNES, T.J., eds. 2000. *A Companion to Economic Geography*. Malden, Mass: Blackwell.

> A comprehensive overview of economic geography; includes accounts of history and diverse approaches, along with a focus on production, resources, social issues, and circulation.

SIMPSON, E.S. 1990. 'Plantations: Benefit or Burden?' *Geographical Magazine* 42, no. 1 (special supplement): 103.

> A brief but thought-provoking account of plantations as a component of colonialism.

WEBBER, P. 1996. 'Agrarian Change in Ghana'. *Geography Review* 9, no. 3:25–30.

> A good factual account of agrarian change in one village in northeast Ghana from 1977 to 1993, emphasizing the prevalence of poverty and the related need for more sustainable agricultural practices to reduce declines in soil fertility.

ON THE WEB

http://www.fao.org/
The site of the Food and Agriculture Organization; offers relevant facts and figures and topical discussions on all matters related to agricultural activity and related food supply; also very useful for Chapter 6.

http://www.ifad.org/
The International Fund for Agricultural Development, a specialized agency of the United Nations, was established as an international financial institution in 1977 following the food crises of the early 1970s that primarily affected the Sahelian countries of Africa. One important insight is that the causes of food insecurity and famine were not so much failures in food production, but structural problems relating to poverty and to the fact that the majority of the developing world's poor populations were concentrated in rural areas.

http://www.ifpri.org/
The mission of the International Food Policy Research Institute is to identify and analyze policies for sustainably meeting the food needs of the developing world.

http://www.wri.org/
The home page of the World Resources Institute; includes information on the full range of environmental issues, including climate change, biodiversity, ecosystem changes, and resource use and abuse, as well as regional analyses.

Settlement

This chapter focuses on two fundamental geographic issues: reasons for location and reasons for way of life. These issues are addressed separately for rural and urban settlements, with particular emphasis on the latter—a fair reflection of the interests of human geographers.

Around the world, the most common rural settlement pattern is nucleated, although dispersion is common in North America and parts of western Europe. Following an outline of two theories of rural settlement location, we look at how countrysides and the rural–urban fringe are changing.

A discussion of the origin and growth of cities, focusing on preindustrial cities, industrial cities, and urban planning, is followed by an examination of why urban settlements are located where they are. Several historical explanations are noted, but the main emphasis is on Christaller's central place theory, which offers clear conceptual parallels with von Thünen theory.

The next two sections examine the very different circumstances of contemporary cities in the more and the less developed worlds. For the former we look at the internal structure of cities; retailing, industrial, and residential land uses; some current postindustrial trends, including the evidence for the rise of global cities; the image of the city, and the city as image; and some of the problems faced by these centres of civilization. Discussion of cities in the less developed world focuses on the heritage of colonialism, 'exploding cities', and the many challenges that such cities face simply to provide the necessities of life.

Urban sprawl on the west side of Calgary, 2003 (CP photo/Jeff McIntosh).

Permanent settlements began to form with the introduction of agriculture. Today, only a minority of humans engage in hunting and gathering or agricultural activities that involve regular movement from place to place. Permanent settlements are associated with specific economic activities, initially producing food for subsistence, later distributing surpluses, and eventually developing into trade and industrial centres. Today, settlements that are principally agricultural are usually called rural; those that are principally non-agricultural are called urban.

The distinction between rural and urban is not, however, quite so simple. Different countries use different criteria to determine whether a settlement is rural or urban; Table 12.1 presents several examples. Most of the settled surface of the earth is rural, although the world is rapidly becoming increasingly urbanized (Figure 12.1). Estimates for 2002 defined 47 per cent of the world population as urban (according to the definitions used by each country); in the more developed world the figure was 75 per cent, and in the less developed world it was 40 per cent. It is likely that by 2015 a majority of the world's population will be living in urban areas.

The division between rural and urban reflects a long-standing geographic tradition. Rural settlement geography is often thought to have originated with the early twentieth-century studies by Vidal and other European geographers that focused on settlement patterns and their links to agriculture. Urban settlement geography is more recent in origin; the first English-language texts were published in the 1940s.

The following discussions of locational issues—rural, urban, and interurban—include both conceptual accounts based on historical experiences and theoretical accounts in the von Thünen tradition. Our discussions of way of life, both rural and urban, are conceptually more diverse; they reflect a wide range of economic and social theory, including Marxism, humanism, and postmodernism. Throughout our accounts of the urban way of life, two points are evident. First, cities are highly complex organizational and institutional forms, serving as centres for social interaction, cultural activities, economic development, political activity, and technological change. Second, cities are currently experiencing many changes related to growth and the economic globalization processes discussed in Chapter 10.

Rural Settlement

Before about 1950, rural settlement issues were at the core of human geography, but today urban issues receive most of the attention, perhaps because the urban experience is the dominant one in the more developed world and is increasingly important in the less developed world. In this section we will focus on rural settlement patterns, especially as they relate to agriculture; rural settlement

Table 12.1   SOME DEFINITIONS OF URBAN CENTRES

France	Communes containing an agglomeration of more than 2,000 inhabitants living in contiguous houses or with not more than 200 m between houses, and communes of which the major part of the population is part of a multicommunal agglomeration of this nature.
Portugal	Agglomerations of 10,000 or more inhabitants.
Norway	Localities of 200 or more inhabitants.
Israel	All settlements of more than 2,000 inhabitants, except those where at least one-third of the heads of households, participating in the civilian labour force, earn their living from agriculture.
Canada	(a) 1976: Incorporated cities, towns, and villages of 1,000 or more inhabitants and their urbanized fringes; unincorporated places of 1,000 or more inhabitants having a population density of at least 390 per square kilometre and their urbanized fringes.
	(b) 1981: Places of 1,000 or more inhabitants having a population density of 400 or more per square kilometre.
United States	Places of 2,500 or more inhabitants and urbanized areas.
Botswana	Agglomerations of 5,000 or more inhabitants where 75% of the economic activity is of the non-agricultural type.
Ethiopia	Localities of 2,000 or more inhabitants.
Mexico	Localities of 2,500 or more inhabitants.
Argentina	Populated centres with 2,000 or more inhabitants.
Japan	Cities (shi) having 50,000 or more inhabitants with 60% or more of the houses located in the main built-up areas and 60% or more of the population (including their dependants) engaged in manufacturing, trade, or other urban types of business. Alternatively, a shi having urban facilities and conditions as defined by the prefectural order is considered urban.

Note: This table confirms that there is no simple and agreed-upon definition of urban centre on the basis of population size, and emphasizes the danger of comparing data for different countries.

Source: H. Carter, *The Study of Urban Geography*, 4th ed. (London: Arnold, 1995):10–12.

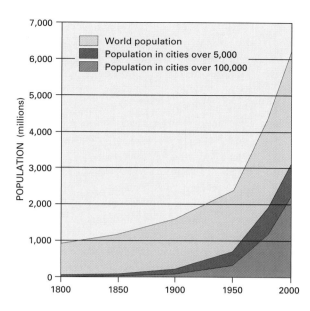

theory; changes in the rural way of life; and land-use transition from rural to urban.

Rural Settlement Patterns

Patterns of rural settlement range widely between the extremes of dispersion (random or uniform) and nucleation (clustering; see Fig. 2.4), depending on physical environment, culture, social organization, political influences, and economic activities. An important component of rural settlement landscapes is the pattern of fields, a landscape feature that often contributes greatly to the character of a place.

Dispersion

Humans are social animals, and throughout the world people have usually lived in close proximity to one another. Why, then, are many parts of western Europe today characterized by dispersed rather than nucleated rural settlements? The answer lies primarily in the economic and social changes associated with the eighteenth-century transition from feudalism to capitalism. Not only was dispersion thought to be more efficient than nucleation, but this pattern reflected the new emphasis on individual ownership of land. In England, for example, between about 1750 and 1850 the rural landscape was dramatically transformed by the process of enclosure (see Chapter 11), which included land consolidation, new field boundaries, and the construction of dispersed dwellings. Government played a leading role in this transformation. Like many other aspects

of the human landscape, field patterns can be seen as spatial expressions of power relations: the open pre-enclosure pattern reflected feudal social relations, while the enclosed pattern reflected individual ownership under a capitalist system.

In North America, dispersion was established as the norm from the outset of European settlement because in most cases the government surveyed the land and divided it into discrete lots before building even began. As a result, geometric patterns soon became standard. Perhaps the most familiar example of a government survey comes from New France, where the French government laid out long, narrow lots at right angles to the St Lawrence that gave all settlers access to water. However, the nucleated settlement pattern that this arrangement produced was an exception.

More commonly, in English-speaking North America the government survey systems encouraged dispersed settlement; typically, each settler was granted a one-quarter section and required to construct a dwelling on it. This system began in 1785, when the United States land survey laid out a baseline west to the Pacific from the point where the Pennsylvania border meets the Ohio River. For the government, the advantage of this type of survey was that it provided an orderly way to allocate land. A similar system was followed in the Canadian prairies, although in this area the landscape included some nucleated settlements as well (Box 12.1 and Figure 12.2).

Figure 12.1 (left) Growth of urban population relative to growth of world population, 1800–2000.

(above) The long-lot settlement pattern of the St Lawrence lowlands was established in the French colonial period (Victor Last, Geographical Visual Aids).

Nucleation

Historically, nucleated settlements dominated the rural landscape. At least part of the reason was the basic human need to communicate and cooperate with others; being with other people is important for security, social life, religious activities, and the regular exchange of goods and services. Other reasons for the development of nucleated settlements include a scarcity of good building land (for example, in areas that experience regular flooding), a need to defend the group against others, the need for group labour to construct and maintain a particular agricultural feature (such as terraces on steep slopes, or irrigation systems), and political or religious imperatives (as in the case of the Israeli kibbutz or the Mormon village).

Even today, nucleation has been favoured over dispersion throughout much of the rural world. In China, the 2002 rural population was estimated at 70 per cent of the total. Most of those people lived in nucleated village settlements. Although many of these villages have populations of several thousand, they are best described as rural settlements, since the inhabitants are farmers. Often these settlements are surrounded by very small fields.

Nucleated settlements are not uncommon in some parts of North America. The New England landscape reflects early settlers' desire to live close together, to reinforce group cohesion and provide some protection against attack. Similarly, the landscape of the Mormon cultural region in Utah and parts of adjacent states is characterized by nucleation

12.1

RURAL SETTLEMENT IN THE CANADIAN PRAIRIES

The rural landscape of the Canadian prairie region includes both dispersed and nucleated farm settlements. This pattern reflects both the policies of the central government and the wishes of some major ethnic groups.

The settlement that began in the 1870s was preceded by a survey that allowed for sectional settlement and dispersed farmsteads (Figure 12.2). Similar arrangements prevailed in many other areas of European overseas expansion, such as southern Ontario, much of the United States, and much of Australia. The Canadian government allocated some areas for settlement and others for grants to companies (especially railway companies) so that they could generate income by selling to settlers. Dispersed settlement was the norm, whether the land was obtained directly from the government or indirectly from a company. Most immigrants to the prairies accepted this arrangement—and the isolation that came with it—even though many had been accustomed to village life.

The principal exceptions to the dispersed settlement pattern were the farm villages established by specific ethnic groups. Doukhobors and Mennonites from Europe, Mormons from the western US, German Catholics, and Jews chose nucleation for religious and/or social reasons, to maintain a unified group. In these cases, the government rules requiring sectional dispersed settlement were waived on request. These nucleated settlements are an infrequent but distinctive feature of prairie rural settlement. Most of the farm villages were of the elongated-street variety (see Figure 12.3), although the Mormons favoured a square grid plan.

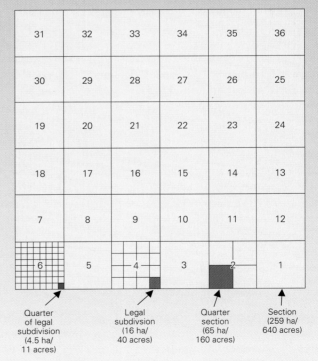

Figure 12.2 Township survey in the Canadian prairies. A typical township of 36 square miles (93 km²) is divided into 36 sections of 1 square mile (640 acres/259 ha) each, and subdivided into four quarter sections of 160 acres (64 ha) each. Settlers were usually assigned a quarter section. Close to urban centres, quarter sections might be divided into four legal subdivisions of 40 acres (16 ha) each.

as a consequence of the central planning practised by the Mormon Church (see p. 231).

Nucleated rural settlement offers many varieties of physical layout; see Figure 12.3 for examples. An irregularly clustered settlement suggests unplanned growth over time. Examples of regular settlements that may or may not result from some form of planning include elongated-street, green, and checkerboard settlements. Nucleated rural settlements range in size from a few dwellings (a hamlet) through to villages with populations of perhaps 25,000 in some parts of the world (as already noted, the distinction between rural and urban varies from country to country).

Rural Settlement Theory

Most studies of rural settlement focus on particular settlements or regions and include substantial descriptive content; this tradition was begun by the noted German researcher August Meitzen (1822–1910). Beginning in the 1960s, however, increasing attention has been given to more general questions, such as why rural settlements are located where they are.

Bylund theory

A Swedish geographer, E. Bylund (1960), developed deterministic theoretical formulations to explain rural settlements' expansion into previously unsettled territory. These formulations were based on two principal variables: distance from the parent settlement and relative attractiveness of the land (especially soils and climate). For an area in northern Sweden, Bylund assigned each possible location an attraction value and constructed a series of six settlement patterns. Like von Thünen theory, the Bylund formulations allow for simplified descriptions. When compared with maps of actual settlement in

Nucleated long-lot settlement, west of Quebec City (Victor Last, Geographical Visual Aids).

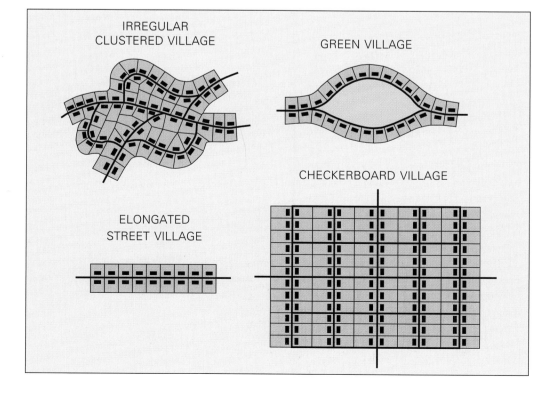

Figure 12.3 Examples of nucleated rural settlement patterns.

northern Sweden, the simplified settlement patterns proved quite similar, replicating the essential features of the actual settlement. Thus the importance of Bylund's two variables—distance from point of origin and land attractiveness—was confirmed.

Hudson theory

J.C. Hudson (1969) offers a deductive approach to rural settlement theory, based on other geographic theory and ecological distribution concepts concerning the dispersal of seeds from parent plants. Hudson proposed three stages for the rural settlement of an area. *Colonization*, the first stage, is the movement of people into a hitherto unoccupied area and their settlement in the landscape. *Spread*, the second stage, is the subsequent filling up of the area by the offspring of the first colonizers; this second stage leads to the establishment of some nucleated rural settlements. The third stage is *competition* between the newly created nucleations as they vie to attract support for their various activities (such as retail establishments) from surrounding dispersed rural settlements. Hudson associates each of his three stages with a characteristic pattern. The colonization pattern is one of concentric rings, the spread pattern is nebula-like with several distinct clusters, and the competition pattern is a regular lattice.

Some criticisms of rural settlement theory

Theoretical discussions of rural settlement can be of great value, but—like all geographic theories—they should be handled with caution. Probably the single most important criticism of Bylund and Hudson is that their ideas fail—intentionally, of course—to incorporate the multitude of social, cultural, political, and other human variables. In addition, Grossman (1971) has argued that biologically derived principles such as those used by Hudson are not applicable to rural settlements. Rather, Grossman argues that the process of settlement is largely dependent on human factors, and that human behaviour is not unvarying; for example, interference in the form of central planning will play an important role in some areas and not in others.

Rural settlement theory is not as well advanced as many other areas of geographic concern. These theories are, however, useful reminders of the value of investigating geographic topics in a conceptual manner. They can also greatly enhance our understanding of specific cases.

Rural Settlements in Transition

Gemeinschaft and Gesellschaft

Numerous studies in diverse parts of the world have established that there are important differences between rural and urban ways of life. In 1887, for example, the sociologist Ferdinand Tönnies (see Cater and Jones 1989: 170–1) distinguished between **Gemeinschaft** (community), human association based on close personal contact, as in village settings, and **Gesellschaft** (mass society), human association that is depersonalized, as in urban settings. It is true that, historically, rural settlements have been characterized by spatial isolation and limited spatial mobility—circumstances that may well tend to bring their inhabitants into close contact with one another. However, the notion that members of rural communities therefore enjoy close informal social relations may be misleading, since the typical social framework in a rural setting was (and in many areas still is) a rigid social hierarchy.

The seminal work on differences between rural and urban areas was accomplished by C. Wirth (1938), who contended that increases in the size and density of populations result in increased anonymity, further division of labour, social heterogeneity, and enhanced mobility. Wirth thus argued that urban areas were distinguished by a distinctive way of life. The lifestyle differences between rural and urban are generally well documented, although R.E. Pahl (1966) has argued that, in some cases, the social differences between rural and urban areas are unclear.

Contemporary rural society

In most countries in the more developed world, 'rural society has been subject to a process of cultural colonization in that the dominant images of rural life have been formulated by (middle class) urbanites and projected on to the countryside' (Cater and Jones 1989:194). The results—notwithstanding nostalgic myths of a rural idyll and better yesterdays—are urban dominance and rural dependence.

Some human geographic research into rural areas (and, to a lesser extent, related agricultural landscapes) in the more developed world is being conducted under the general rubric of the 'cultural turn' outlined in Chapter 8. Of particular concern is the widely held image of the 'rural idyll'. In the case of England, it might be argued that the 'idyll' image was constructed by white, middle-class males as an integral part of the larger English identity. Interestingly, in the ongoing debate about fox hunting—a pastime closely associated with 'Englishness'—most rural dwellers and people who associate rurality with an 'English' way of life remain sympathetic to the hunt despite widespread public criticism. More generally, the traditional image of rurality excludes those others, such as people of colour or disabled people, who do not conform to the imagined rural norm. The basic logic of this argument is that, like all social constructions, rurality necessarily has exclusionary qualities. It is in this context that some human geographers are now studying the experiences of 'others' in rural areas.

The reality is that rural areas have been substantially modified by industrialization. Rural–urban differences, especially in the more developed world, are decreasing as a result of improved rural services and the proximity of urban employment opportunities for many rural dwellers. Even in those rural areas that are far from urban centres, access to similar mass media ensures a certain uniformity in lifestyle.

Depopulation or repopulation?

A significant trend in rural areas today is depopulation as a result of the spatial concentration of economic activities in urban areas. The evidence for a shift in the national economies of more developed countries from the agricultural to the urban sector is compelling (this topic is discussed in greater detail in Chapter 13; see Table 13.5 for some relevant global data). A compelling example of regional depopulation is the prairie province of Saskatchewan (see Box 11.13). People are abandoning the small towns and villages because the numbers working in agriculture are shrinking and, without farm families and the community support and purchasing power they represent, the *raison d'être* for many small settlements no longer

exists. This decline has been evident for some time, the number of small towns and villages dropping from 317 in 1961 to 86 in 2002.

At the beginning of the twentieth century, settlers were moving to the prairies and nucleated centres were developing around the grain elevators established at 16-km (10-mile) intervals along railway lines. By the mid-1930s, however, the settlement process had effectively ended. The attractions of rural life gradually diminished as agricultural technology changed after the Second World War, becoming capital-intensive and requiring less labour, and as the recession of the 1980s reduced farm incomes. Today, small settlements in Saskatchewan are becoming less and less viable, both economically and socially. With declining farm populations, towns lose people, schools close, and businesses leave. A little northeast of Saskatoon, the former community of Smuts is a ghost town. The small town of Mendham (northwest of Swift Current), with a population of 40, received what might be seen as the final blow in 2001, when the Roman Catholic church that had served the community for 87 years was no longer able to function. Residents had earlier fought and lost battles to keep schools open. Allan, a little south and east of Saskatoon, was founded by German families in 1903 and grew impressively at first, but it too has faltered. There is still a school and a grain elevator, but continuing decline seems inevitable. These changes may well be normal consequences of larger global economic processes and national policies, but we may still mourn the loss to Canadian iconography that the disappearance of these towns represents.

Not all of Saskatchewan's small towns are losing population. Those that are thriving, however, usually have some economic base other than agriculture: Eastend, in the southeast of the province, has benefited from local dinosaur remains; Leader, near Mendham, is developing ecotourist facilities; and several towns remain viable because they are close to potash mines.

More generally, for Canada as a whole, a detailed analysis by Keddie and Joseph (1991) suggests that the period 1981–6 was one of sustained urban growth comparable to that of the 1960s. During this five-year period, the rate of urban growth in Canada was 5 per cent, compared to 0.8 per cent for rural growth. Table 12.2 summarizes the percentage change in regional rural population for the period 1981–6. Note that the figures vary substantially between regions: whereas the prairies experienced a rural population loss of 3.1 per cent, the Atlantic region's rural population grew by 3.5 per cent—a good example of the dangers of using data for large areas.

Urban growth is proceeding apace throughout the world. Or is it? At least until about 1970, the data were clear: people moved from less to more densely populated areas. Yet in the 1970s some parts of the more developed world saw a decline in this movement—and possibly even a countermovement. In some areas there was evidence of rural repopulation, a trend that gave rise to the term **counterurbanization** (Hall et al. 1973). According to some researchers, this trend is the result of industrial changes in a less spatially concentrated economy. Other possible explanations include an increasing appreciation for rural lifestyles, especially among the growing numbers of seniors, and the rapid proliferation of new communication systems that reduce the need to be spatially close to services. But there is a simpler explanation: birth rates in the more developed world are low, and hence there are fewer people to move from the less densely populated areas.

Significantly, the Canadian study that produced the data in Table 12.2 was prompted by evidence of some rural population growth in the 1970s, but no such trend was evident for the first half of the 1980s. Any rural repopulation in Canada during the 1970s was a 'fleeting and regionally specific phenomenon' (Keddie and Joseph 1991:379).

Unequal places

To put it bluntly, many rural communities are no longer the supportive, self-sufficient, and stable places we imagine they were in earlier times. Areas that have experienced depopulation have lost basic services as their market bases have declined, and typically any repopulation that may have occurred has not reversed this trend. The prevailing ethic in the more developed world is that of the market economy, and this ethic is not conducive to the establishment of services in rural areas.

Furthermore, those people who have moved to rural areas in recent years have generally not been seeking the same rural lifestyle; rather, repopulation involves commuters, retirees, and people seeking alternatives to the perceived unpleasantness of urban life. These new rural dwellers come from a wide range of social backgrounds and economic circumstances, and they have a wide range of attitudes. Unlike earlier settlers, they do not form a single cohesive group with shared aspirations and goals for their rural way of life.

The Rural–Urban Fringe

Transition zones between rural and urban areas often have multiple land uses and varied social and demographic characteristics. Indeed, the absence of clear boundaries between the two areas is a distinctive feature of contemporary land use and lifestyle. One way to look at this situation is to focus on the centrifugal forces that encourage urban land uses to expand into the fringe belt—a process that has come to be known as deconcentration, in contrast to the concentration more traditionally associated with urban development. The term **urban sprawl** is often used to describe the deconcentration that involves low-density expansion of urban land uses into surrounding rural areas (Box 12.2).

The Niagara peninsula

In much of the world, the transition from rural to urban land uses can be a cause for concern. One particularly sensitive example is the Niagara Peninsula region in southern Ontario (Figure 12.4). This region is one of

Table 12.2	PERCENTAGE CHANGE IN RURAL POPULATION BY REGION, CANADA, 1981–1986

Region	Aggregate % Change
Atlantic	3.5
Quebec	– 0.1
Ontario	3.4
Prairies	– 3.1
British Columbia	– 1.1
Canada	0.8

Source: Adapted from P.D. Keddie and A.E. Joseph, 'The Turnaround of the Turnaround? Rural Population Change in Canada, 1976 to 1986', *Canadian Geographer* 35 (1991):371.

only two important soft-fruit-producing areas in Canada, but it is close to large urban areas and itself contains a number of urban centres. Although concerns about urban encroachment were first raised in the 1950s, few efforts were made to limit the growth of urban areas until the 1970s, because growth was generally considered an unmixed good. As a result of community pressure and subsequent government action, rulings in the 1980s identified substantial areas that could be used only for agriculture or related purposes. These rulings represented a significant reversal of attitude.

Origins and Growth of Cities

Since about 1750, the world has experienced the rise of capitalist societies and economies, a rapidly increasing population, a proliferation of states, many new technologies, and growth in both the numbers and the size of urban

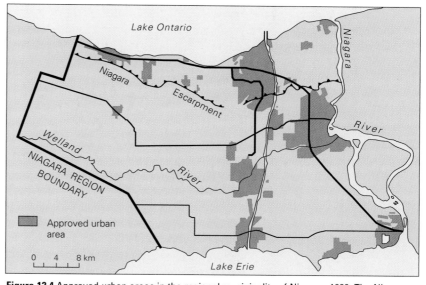

Figure 12.4 Approved urban areas in the regional municipality of Niagara, 1989. The Niagara region (population 370,000) is at the western end of an urban region containing 5 million people. North of the Niagara escarpment, a fertile plain—the Niagara fruit belt—edges Lake Ontario.
Source: Adapted from H.J. Gayler, 'Changing Aspects of Urban Containment in Canada: The Niagara Case in the 1980s and Beyond', *Urban Geography* 11 (1990):377.

12.2

EXURBANIZATION IN SOUTHERN ONTARIO

In many areas, deconcentration is taking the form of residential movement to the rural–urban fringe. Is such movement simply a matter of the extension of existing suburbs? Or does it represent a form of counterurbanization, or rural repopulation?

For Woodstock in southwestern Ontario, Davies and Yeates (1991) show that both perspectives are relevant. Focusing on **exurbanization**—the movement of households from urban areas to locations outside the urban area but within the commuting field—they distinguish between exurbanite households located in dispersed rural settlements (54 per cent of the total in the area around Woodstock) and exurbanite households located in villages (46 per cent of the total in the area around Woodstock). As shown in Table 12.3, 'the exurban-rural group is part of the new urban-to-rural migration pattern subsumed within the idea of counter-urbanization, while the exurban village group is part of the long-standing trend of suburbanization related to a search for more housing space and privacy at lower prices' (Davies and Yeates 1991:186). This conclusion is based on the relatively high valuation that the exurban-village group places on lower taxes and lower house prices and on the relatively high valuation that the exurban-rural group places on the amenity attractions of rural areas. Both groups place a high value on privacy, larger houses,

more land, the attractive rural landscape, a better quality of life, and less crime.

Another way to think about these trends is in terms of the push and pull factors introduced in the discussion of migration in Chapter 6.

Table 12.3 FACTORS INFLUENTIAL IN THE DECISION TO MOVE TO EXURBAN LOCATIONS AROUND WOODSTOCK, SOUTHWESTERN ONTARIO

Factor	Exurban-Rural % Important	Exurban-Village % Important
Better schools	34.4	42.4
Lower taxes	33.3	53.8
Less commuting	26.2	24.7
Friends and relatives	25.1	31.0
More privacy	95.1	82.3
Lower house prices	43.7	65.2
Larger house	63.9	58.9
Better services	15.3	19.0
More land	80.9	75.9
Proximity to urban area	48.6	51.3
Attractive landscape	79.2	72.8
Better quality of life	85.8	86.0
Less crime	56.8	72.2

Source: S. Davies and M.H. Yeates, 'Exurbanization as a Component of Migration: A Case Study in Oxford County, Ontario', *Canadian Geographer* 35 (1991):177–86.

centres. By 1850, the major world cities were concentrated in the newly industrializing countries, a pattern that continued well into the twentieth century. Most of these major cities were either European capitals or located on the eastern seaboard of the United States.

In recent decades, however, the majority of the world's cities (particularly the largest ones) have been located in former colonies in the less developed world. Before 1950, cities in the less developed world were typically transportation centres or colonial government centres that formally excluded most indigenous people. With independence, cities in the less developed countries became population magnets, offering employment and wealth to some, but only minimal benefits to the majority. In 1800 perhaps 3 per cent of the world population lived in cities; by 1900 the figure was 13 per cent; and by 2002 it was 47 per cent. Increasingly, cities are the focal points of our modern civilizations.

Today, the majority of large cities are in the less developed world. In both more and less developed worlds, such cities tend to occupy coastal locations. It is also notable that many capitals are in the large-city class.

Urban Origins

The city–civilization link is an ancient one; the two words share the Latin root *civitas*. Civilizations create cities, but cities mould civilizations. James Vance (1990:4) quotes Winston Churchill: 'We shape our houses but then they shape us.' The earliest cities probably date from about 3500 BCE and developed out of large agricultural villages. It is no accident that the emergence of cities coincided with major cultural advances, such as the invention of writing. From the beginning, cities have been 'the chief repositories of social tradition, the points of contact between cultures, and the fountainheads of inspiration' (Smailes 1957:8).

The origins of **urbanization** and the urban way of life, **urbanism**, are debatable. It seems likely that most cities originated in one of four ways. First, we know that cities were first established in agricultural regions. Certainly city life did not become possible until the progress of agriculture freed some group members from the need to be producers of food. Thus in some areas the first cities probably reflected the production of an *agricultural surplus*, possibly as a result of irrigation schemes. The earliest cities of this type were located in the Tigris–Euphrates region (present-day Iraq), the Nile Valley (Egypt), the Indus Valley (Pakistan), the Huang Valley (China), Mexico, and Peru, and were associated with the rise of agriculture in these regions (see Figure 7.1). These cities were the residential areas for those not directly involved in agriculture. They were small in both population and area. Populations generally ranged from 2,000 to 20,000, although Ur on the Euphrates and Thebes on the Nile may have numbered as many as 200,000. In 2001 a significant discovery was made on the coast of Peru. The city of Caral, located about 200 km north of Lima, has been dated as early as 2627 BCE—about 1,000 years before than any other city in the Americas. The inhabitants of Caral appear to have supported themselves by practising irrigation, growing squash, beans, and cotton, but (interestingly) not corn.

A second group of cities were probably established as *market places* for the exchange of local products. Some early cities were located on navigable waterways; others were established on long-distance trading routes: Venice was a prime example of a port city, while Baghdad was a port in the desert.

Third, some cities may have started as *military*, *defence*, or *administrative centres*. The Greeks built cities in their colonial areas as centres from which to exercise control; in many cases, they planned the city in advance as a grid pattern with a central space and a series of routes intersecting at 90°. The Romans, who knew that urbanization was the key to controlling conquered areas, followed the pattern established by the Greeks as their own empire expanded. Many north African, Middle Eastern, and European cities can trace their origins to either Greek or Roman times.

Fourth, it has been suggested that cities arose as *ceremonial centres* for religious activity. Chinese urban locations were selected using geomancy, an art of divination through signs derived from the earth; once an auspicious site was located, the city was laid out in geometric fashion. Chinese cities were square in shape, reflecting two fundamental beliefs: that the earth was square and that humans should be part of nature rather than dominate it (Wheatley 1971).

Preindustrial Cities

The term 'preindustrial city' is misleading to the extent that it suggests that all cities were the same before the industrial revolution; in fact there were notable differences between cities in different parts of the world, reflecting a wealth of cultural differences. It is a useful term, however, because it emphasizes the significance of the changes that accompanied industrialization. Before the advent of capitalism and the industrial revolution, most cities were concerned with marketing, commercial activities, and craft industries, and also served as religious and administrative centres. As new features in the human landscape, cities developed a new way of life—a new social system and a more diverse division of labour (Wirth 1938). Political, cultural, and social changes followed the economic change from preagricultural to agricultural societies. Rural and urban settlements were complementary, the former producing a food surplus and the latter performing a series of new functions. From the beginning, then, cities have been functionally different from surrounding rural areas.

Inside the preindustrial city

In modern industrial cities, different areas serve different functions. By contrast, in the preindustrial city homes, workshops, markets, and other functions were located in a relatively haphazard fashion. The principal evidence of any planning inside the city was a basic division between élite and other areas. According to Sjoberg (1960), in all preindustrial cities, regardless of time and place, the élite occupied the central core, which was also the economic, cultural, and political focal point. Vance (1971) argued that in Europe the guild system was more crucial to urban differentiation than the feudal system; thus, there was land-use zoning in the sense that workers in a particular craft were spatially concentrated.

Preindustrial urban growth

In Europe the process of urbanization faltered after the collapse of the Roman empire in the fifth century, as population movement and trade declined, and it did not resume for some 600 years. Finally, in the eleventh century, urbanization began to revive as commerce picked up, political units gained increasing power, population increased, and agricultural technology advanced. Old cities were revived and new ones were established throughout Europe, especially in areas well located for trade, on the Baltic, the North Sea coast, and the Mediterranean. Urbanization was encouraged in the seventeenth century by the development of **mercantilism**: an economic philosophy arguing that governments should play an active role in economic activities in order to help states attain their maximum economic potential. Thus mercantilism encouraged colonial expansion and city growth.

Meanwhile, throughout the centuries of Europe's decline, stagnation, and rebirth, in other areas—notably China and the Islamic world—the earliest cities flourished. Traditional cities outside Europe began to experience substantial change only with the arrival of European colonial activity.

Preindustrial cities, then, followed the development of agriculture and soon assumed distinctive functions and characters. The typical city functioned both as support for an agricultural population and as a trading centre—a function that was greatly enhanced by the acceptance of mercantilism. All this would change dramatically with the advent of industrialization.

Industrial Cities

The industrial revolution began in England about 1760 and combined a series of changes that altered human landscapes. The principal economic change was related to the emergence of a new merchant and entrepreneurial middle class who opposed the state's mercantilist-inspired interference and the limits it put on their own profits. With the rise of the merchant class, mercantilism faded and the final remnants of feudal society and economy disappeared. The new economic force of capitalism emphasized growth, profits, and a market economy. Capitalism, combined with a series of technological advances, particularly in the area of new energy sources, led to the industrial revolution and the emergence of the industrial city.

Capitalism and urban growth

Much recent city growth is related to the processes of economic globalization introduced in Chapter 10. Each of the three phases of capitalism outlined in Chapter 8

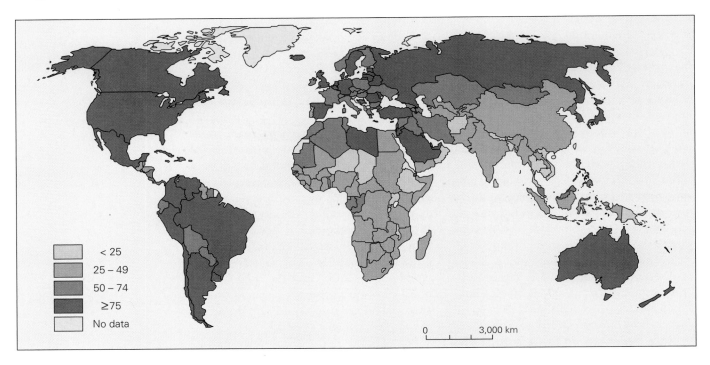

Figure 12.5 Percentage urban
population by country, 1996.
Source: United Nations Development
Programme, *Human Development
Report* (New York: Oxford University
Press, 1996):tables 20 and 40.

created new urban industrial geographies
because each involved changes in what was
produced, how, and where:

1. The first phase, **competitive capitalism**,
   may have begun as early as the late sixteenth century and extended to the middle of the twentieth. It was characterized
   by free-market competition between
   essentially local business activities, with little government regulation or interference.
   As this phase evolved, the scale of business
   activity increased, markets became more
   regional, and labour markets became more
   organized. From the eighteenth century
   onwards, competitive capitalism contributed to the growth of towns as the
   centres of industrial and related economic
   activity, and by the early twentieth century, it became closely associated with
   Fordism (see glossary). Together, mass production and mass consumption further
   contributed to the process of ongoing
   urban growth.
2. The second phase, **organized capitalism**, is usually considered to have begun
   after the Second World War and was associated with the development of larger
   companies, many of which eventually
   came to function as transnationals. Notable
   features of the latter part of this phase are
   the economic globalization processes out-

lined in Chapter 10, the rise of service
activities outlined in Chapter 13, and the
increasing emphasis on consumption
rather than production.

3. The third phase is variously known as
   advanced, late, or **disorganized capitalism**. With the success of organized capitalism and the emergence of new consumer
   preferences, producers found it necessary
   to seek out smaller niche markets for more
   unusual products, which then required
   special retail outlets and the flexible production systems associated with **post-
   Fordism**. More generally, the continuing
   emphasis on consumption means continuing change for urban areas as new shopping areas and specialty retail outlets are
   established.

Industrial urban growth

Urbanization has been one of the key phenomena of the industrial age, involving the
spatial movement of large numbers of people
and major changes in social life. Industrial
cities grew with remarkable rapidity in key
resource locations, especially on coalfields.
One of the first industrial areas in Britain was
Coalbrookdale, northwest of Birmingham,
where Abraham Darby first smelted iron ore
with coke instead of charcoal in 1709. After
1760, the steam engine became available and
old industrial activities, especially textile and

metal production, were transformed. The importance of coal for power and transportation resulted in major new concentrations of activity, machinery, capital, and labour in the new industrial cities.

Urban growth proceeded apace in the new industrial areas—in Britain from 1760 onwards and then in western Europe, the United States, parts of southern Europe, Russia, and Japan by the end of the nineteenth century. The rapidity of this growth is illustrated by the United Kingdom, where the urban population increased from 24 per cent in 1800 to 90 per cent by 2002. Since the Second World War, many parts of the less developed world have also experienced industrialization. Today in many states most of the population is located in cities, although most states continue to have a rural majority (Figure 12.5). Even in states that still specialize in agriculture, however, cities have grown as a result of the increasing importance of service activities.

Economic Base Theory

The changes described earlier have both economic and social causes. Generally, the economic causes are considered to be the financial benefits that accrue from city growth, namely reductions in the costs of assembly, production, and distribution. According to this view, cities are the most economic settlements. **Economic base theory** reduces urban economies to two interdependent sectors, basic and non-basic. The basic sector comprises all those activities that produce goods and services for sale outside the city; the non-basic sector, all those activities that produce goods and services for sale inside the city. This distinction leads to two important conclusions: (1) the larger the city, the less dependent it is on the basic sector, and (2) the larger the city, the more it is able to grow.

Explanations for city growth based on social causes revolve around the fact that groups offer security and that cities develop once the necessary social structures are in place. It seems reasonable to conclude that both economic and social causes play a role.

Urban Planning

Urban planning has a long history, but until recently the growth of most cities was unregulated and uncontrolled. Generally, urban planning results from an awareness of problems. It implies a loss of individual rights and a search for the common good. Although in some early cities the core areas were planned, resulting in central squares and geometric street patterns, other cities clearly grew haphazardly.

Origins

Modern urban planning originated in Europe and the United States only in the late nineteenth century. Interestingly, the origins of the planning movement had links with anarchism, through such notable figures as Patrick Geddes and the geographers Peter Kropotkin and Elisée Réclus, and utopian socialism, through the Scottish philanthropist Robert Owen. In 1898, Ebenezer Howard introduced the idea of the **garden city**—a city built according to a master plan to provide a spacious and high-quality environment for working and living. Each city was intended to house 32,000 people in an area of 2,429 ha (6,002 acres) and was characterized by a concentric pattern of land use, wide streets, low-density housing, public open spaces, and a **green belt.** Several settlements have been built following a modified version of the garden city plan, and the basic concepts have been highly influential in urban planning generally.

Le Corbusier and Wright

Perhaps the best-known urban planner was the Swiss architect Le Corbusier, who designed totally new cities, but who saw very few of his designs converted to reality. Le Corbusier advocated increasing open space and population density, and facilitating the movement of people, by rebuilding city centres. In the 1920s, these were revolutionary ideas. The new planned city was to have an organized spatial structure with residences located according to social class. Only two cities have been built on these lines: Chandigarh, the new capital of the Punjab, India, and Brasília, the new capital of Brazil. In North America, the urban designs of the celebrated architect Frank Lloyd Wright were never fulfilled, although his impact on architectural style has been substantial. Le Corbusier and Wright reflect the modernist (see 'modernism' in the glossary) taste for purity and simplicity in design.

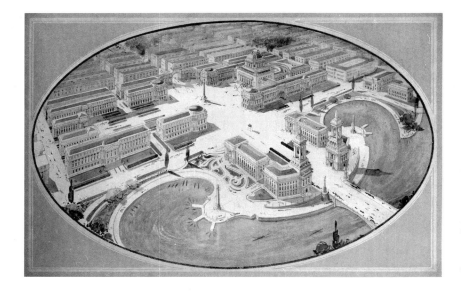

'The Civic Centre of Calgary as It May Appear Many Years Hence': this frontispiece of a plan by Thomas Mawson shows the influence of the 'city beautiful' movement, which favoured impressive malls with sweeping views.
(City of Calgary Archives, Town Planning Commission, File 6.)

Applications

The landscapes of both Britain and the United States have been affected by urban planning philosophies and practices with roots in various movements for social reform. Reformers believed that social conditions reflected the built environment, and that improving the quality of urban life was an essential part of the battle against crime and disease. In the early twentieth century, the 'city beautiful' movement was a brief but influential effort to impose a social order on urban areas by, for example, requiring uniform heights for buildings. In retrospect, the movement might be seen as a modernistic attempt to control the often chaotic growth of large cities.

Since the Second World War, planning has centred on physical design. In Britain, three key acts—the Distribution of Industry Act (1945), the Town and Country Planning Act (1947), and the New Towns Act (1958)—set the scene. Since 1946, 28 new towns have been created (Figure 12.6). In some cases the aim was to disperse population from overcrowded cities; in others, to support specific regional industrial policies. Probably the most significant consequence of Britain's planning legislation is that the growth of large urban areas has been limited. The United States has seen much less government intervention, and lacks a national framework. Typically, planning at the national scale has been limited to public housing and slum clearance. Zoning regulations are used to ensure that urban land is appropriately used.

Urban Locations I: Historical Explanations

Why are urban centres located where they are? Various answers to this question have already been noted. In most cases location reflects function: provision of services to rural populations, military strategy, or trade. Occasionally, the location of a set of centres can be explained by reference to government planning: for example, it has been suggested that the settlement pattern on the Nile *c.*1317–1070 BCE was determined by the state in order to maximize control of populations (Church and Bell 1988). Such situations are not common.

The Mercantile Model: Vance

According to Vance's mercantile model (1970), the initial growth of an urban centre is the product of external factors such as long-distance trade. The case of North America illustrates that model:

Stage I   Initial search of the North American coasts to gather information relevant to European interests.

Stage II   Initial testing of the quality of resources discovered and, in some cases, use of specific resources as staple products.

Stage III   Settlement at key points, both coastal and inland, to facilitate export of products.

Stage IV   Expansion of the economic region by establishing urban locations, usually at key transport sites on waterways.

Stage V   Emergence of the key trading locations as cities.

Close links between the mercantile model and what is known as the *staple model* of economic growth indicate that the former is effectively an offshoot of the latter. The staple model is often used by economic historians, especially for areas such as Canada and Australia, to describe situations where the exploitation of one or more key resources lays the groundwork for further economic growth. Canada's staples, in approximately chronological sequence, have been fish, fur, lumber, wheat, and minerals. For Australia, the sequence was wool, gold, and wheat.

The use of staple theory and related trading to explain the evolution of urban networks is novel, and reactions to Vance (1970) have been mixed. Sargent's (1975) study of Arizona towns confirmed the model, Earle (1977) found it inadequate to explain the pattern of early colonial towns in North America, and Meyer (1980) also considered it inadequate.

Metropolitan Evolution: Borchert

A second historical model, again applied to North America, identifies four technological stages and relates urban growth to each phase. Borchert (1967) proposed the following:

Stage I   Sail wagon, 1790–1830. During this stage, urban locations were primarily related to availability of water transportation.

Stage II   Iron horse, 1830–70. Rail networks focused on ports, but also encouraged urban growth along railway lines in the new hinterlands. Many new urban centres were established during this stage.

Stage III Steel rail, 1870–1920. Expansion into the large western region of North America and development of existing centres as industrial cities.

Stage IV   Auto–air–amenity, 1920 onwards. Creation of a dense road network, suburban sprawl, and new cities in the south.

This model resembles the mercantile model, but puts less emphasis on trade and more on the technology of transportation.

Selective Urban Growth: Muller

The model devised by Muller (1976) to explain North American towns is based on the mercantile model. Three stages are envisaged:

Stage I   Pioneer periphery. Initial location of urban centres related to early agriculture, few trading contacts, little commercial agriculture.

Stage II   Specialized periphery. Dominant urban centres emerge at points of regional contact.

Stage III Transitional periphery. The region is now fully linked to national networks, and the urban pattern is a continuation of past patterns, with exports of agricultural products and the establishment of new locations as manufacturing becomes important.

This third model implies that, after the first stage, some sites will lose their original advantages, especially with changes in transportation. The argument that towns were established from the outset of settlement was advanced earlier by Wade (1959:1): 'The towns were the spearheads of the frontier. Planted far in advance of the line of settlement, they held the west for the approaching population.'

Other Conceptual Contributions

Thus the mercantile model and related contributions do appear relevant to nineteenth-century North America. However, because

Figure 12.6 British new towns. Most of these towns were established between 1947 and 1950. In the south of England, most were located so as to relieve pressure on London; others were new industrial towns.

they are linked to the mercantilist economic philosophy that played such an important role between the feudal period in Europe and later capitalism, they may be less appropriate for other times and places

Attempts to explain the Australian and South African urban experiences have also used versions of the mercantile model. Conceptual contributions have focused on the importance of trade, related port development, and the role of distance. In addition, the core–periphery concept has been applied to the Australian and South African cases. Cores or central areas dominated by incoming Europeans included ports, capitals, and mining centres. Peripheral areas occupied primarily by native people experienced less urbanization and largely remained outside the growing economic region.

Rozman's model (1978) identified seven stages of premodern development (Table 12.4):

1. A preurban stage with unspecialized settlements;
2. Cities locate separately from one another and have only weak links to rural areas;
3. An administrative hierarchy develops;
4. High levels of centralization based on administration prevail;
5. Commercial centralization emerges, as do some periodic markets;
6. Increased commercial activity results in five or six levels of cities; and
7. A national marketing system emerges.

This model identifies the number of settlement levels—groupings by population size—associated with each stage. For England, the second stage evolved in the second century BCE, the third stage was bypassed, stage four evolved in the first century CE, stage five in the tenth century, stage six in the twelfth century, and stage seven in the sixteenth century. Although this model is extremely broad, it has the clear advantage of being less time- and place-specific than the earlier formulations.

Comparison of these various urban-location models with von Thünen's theory of agricultural location (Chapter 11) reveals substantial differences. Although von Thünen's work was stimulated by a specific set of facts, it largely succeeded in explaining agricultural location in general. By contrast, historical explanations of urban location are clearly limited to individual cases, and do not offer any general explanations. Fortunately, however, there is one urban-location theory that, like von Thünen's, has general applicability: central place theory

Urban Locations II: Central Place Theory

Von Thünen's most important contribution to geography is the concept of the isolated state: the use of a grossly simplified ideal situation to observe the effects of a particular variable. In von Thünen's study of agricultural location, the variable highlighted was distance expressed as transport cost. The German geographer Walter Christaller (1893–1969) used a similar 'isolated state' model to observe the role played by distance in the location of urban centres.

Although the influence of von Thünen is clear, Christaller's seminal work was perhaps even more directly influenced by Alfred Weber, a German economist who introduced theory to industrial geography (see Chapter 13) at a time when it was entirely descriptive. Like von Thünen and Weber, Christaller was a pioneer. The English title of his work, originally published in 1933 but not translated until 1966, is *Central Places in Southern Germany* (Christaller 1966). At the time of publication there was little interest in the book, but since 1945 it has proven highly influential in theoretical and empirical research, especially in Scandinavia and the United States. During the 1960s, the work

Table 12.4	SEVEN STAGES OF PREMODERN URBAN DEVELOPMENT

Stage	Number of levels	Characteristic
1	0	Preurban
2	1	Tribute city
3	2	State city
4	2, 3, or 4	Imperial city
5	4 or 5	Standard marketing
6	5 or 6	Intermediate marketing
7	7	National marketing

Source: Adapted from G. Rozman, 'Urban Networks and Historical Stages', *Journal of Interdisciplinary History* 9 (1978):79.

came to the fore as one component of the spatial analysis movement that swept English-language geography at that time.

In 1940 a second German academic, the economist August Lösch (1906–45), published *The Economics of Location*, a book that included many of Christaller's ideas. Although the present discussion of central place theory draws on the work of both theorists, it is largely couched in the language introduced by Christaller. Box 12.3 provides an interesting insight into the approach employed by this major theorist.

Assumptions

Theories begin with assumptions that simplify the key issue under investigation. The assumptions of Christaller's central place theory are similar to those utilized by von Thünen:

12.3

THEORY CONSTRUCTION BY CHRISTALLER

In a brief article originally published in German, Walter Christaller explained how he developed central place theory. The explicit references to von Thünen and Weber are illuminating. Christaller wrote:

Besides geography and statistics, I was also interested in sociology, which at that time (in 1913), when I was beginning my studies, had begun to exist as a new scholarly discipline. . . . At that time I witnessed the efforts of my teacher Alfred Weber in Heidelberg to create an industrial location theory.

. . . I continued my games with maps: I connected cities of equal size by straight lines, first of all, in order to determine if certain rules were recognizable in the railroad and road network, whether regular traffic networks existed, and, second of all, in order to measure the distances between cities of equal size.

Thereby, the maps became filled with triangles, often equilateral triangles (the distances of cities of equal size from each other were thus approximately equal), which then crystallized as six-sided figures (hexagons). Furthermore, I determined that, in South Germany, the small rural towns very frequently and very precisely were 21 kilometres apart from each other. This fact has been recognized earlier, but had been explained as being due to the fact that these cities were stopover places for long distance trade traffic, and that in the Middle Ages the distance daily by a cart was about 20 kilometres.

My goal was staked out for me: to find laws, according to which number, size and distribution of cities were determined.

. . . It was clear to me from the very beginning that I had to develop a theoretical schema for my regional investigation—a schema which, as is customary in national economics, is set up by isolating the essential and operative factors. It was, thus, as in the case of von Thünen's Isolated State,

abstracted from all natural and geographic factors, and, also, partly to be abstracted from all human geographic factors. It had to be accepted as a symmetrical plain, without obstructions such as rivers or mountain ranges, with a uniformly distributed population, in order to then determine where, under such conditions, the site of a central city or market could form. I thus followed exactly the opposite procedure that von Thünen did: he accepted the central city as already having been furnished, and asked how the agricultural land was utilized in the surrounding area, whereas I accepted the inhabited area as already having been furnished, and subsequently asked where the city must be situated, or, more correctly, where should the cities be situated. Thus, I first of all, as is said today, developed an abstract economic model. This model is 'correct' in itself, even if it is never to be found in the reality of settlement landscape in pure form: mountain ranges, variable ground, but also variable density of population, variable income ratios and sociological structure of the population, historical developments and political realities bring about deviations from the pure model. In the theoretical portion of my investigation, I thus did not satisfy myself with setting up a model for an invariable and constant economic landscape (thus, for a static condition)—but instead I also tried to show how the number, size and distribution of the central places change, when the economic factors change: the number, the distribution and the structure of the population, the demands for central goods and services, the production costs, the technical progress, the traffic services, etc.

. . . I was able to find surprising concurrences between geographical reality and the abstract schema of the central places (the theoretical model) especially in the predominantly agrarian areas of North and South Bavaria.

Source: W. Christaller, 'How I Discovered the Theory of Central Places: A Report About the Origin of Central Places', in *Man, Space and Environment*, edited by P.W. English and R.C. Mayfield (New York: Oxford University Press, 1972):601–10.

1. The land surface is flat and never-ending. Christaller introduces the assumption of flatness to avoid the complications implied by a physically variable landscape; similarly, the plain is never-ending in order to avoid any complications associated with boundaries.
2. This uniform plain has a uniform distribution of rural population; furthermore, all members of this population have identical behaviour and purchasing power. Christaller introduces this assumption in order to avoid the complications of human variations.
3. There is a homogeneous transport surface that allows equal ease of movement for all members of the population in all directions. Christaller introduces this assumption to avoid the complications of a network of roads and other lines of communication.
4. Finally, it is assumed that the above three conditions together describe some preliminary stage in landscape evolution, and that all subsequent evolution is related to the growth of cities as service centres.

Four Key Concepts

Christaller developed his theory using four key concepts:

1. The *central place*: an urban centre that performs a series of functions for the surrounding rural population. In fact, a central place originates in response to rural demands for functions.
2. The *range* of a good or service: the maximum distance that people are prepared to travel to obtain a particular good or service.
3. The *threshold*: a measure of the minimum number of people required to support the existence of a particular function.
4. *Spatial competition*: the idea that central places compete with each other for customers.

These four concepts apply at various scales and can be most easily illustrated by reference to a single business, such as a bakery. What conditions need to be satisfied for a bakery to operate profitably? First, there must be enough people wishing to purchase bakery products. For every central function, there is some minimum level of demand, which varies depending on the specific function (relatively low for a bakery, relatively high for a dentist); this is the threshold concept. The second condition to be satisfied relates to the distance people are prepared to travel to shop at the bakery, given certain prices. Third, there must be people located within the range of the bakery who are willing to purchase at the prices offered. Finally,

12.4

NEAREST NEIGHBOUR ANALYSIS

Geographers may describe the spatial pattern of urban centres (or of any other phenomena represented on a dot map) using a statistical procedure called nearest neighbour analysis. With this method a point pattern may be described as clustered, random, or uniform (see Figure 2.4).

Nearest neighbour analysis indicates the degree to which an observed distribution of points deviates from what might be expected if the points were randomly distributed in the same area. A single statistic, R_n, is calculated.

R_n equals distance observed (average distance between each point and its nearest neighbour) divided by distance expected:

$$\frac{1}{2\sqrt{\text{density of pattern}}}$$

If $R_n = 1.0$, the two distances are identical and the pattern is random. If $R_n < 1.0$, the observed distance is less than the expected distance, indicating that the points are relatively clustered. If $R_n > 1.0$ (to a maximum of 2.15), the points are relatively equally spaced—as proposed by central place theory.

Although a useful statistical procedure, nearest neighbour analysis needs to be employed at an approximate spatial scale (recall Figures 2.6 and 2.7). When only the first nearest neighbour of each point is used, a repeated pattern of two or more closely spaced points, occurring at large spatial intervals, gives a low value of R_n, although the pattern may be dispersed. Thus it is not uncommon to incorporate second-order and third-order neighbours or to confirm the results with an alternative procedure.

the people within the range who represent the demand must not have access to other competitive bakeries. All four of these conditions apply equally to individual business and the collection of businesses clustered in an urban centre.

Hexagons

In principle, to ensure the most efficient use of space, central places should be located as shown in Figure 12.7. Central places that compete with each other for the purchasing power of rural populations will tend to locate so as to produce a regular arrangement (Box 12.4). This triangular lattice represents the most economical use of space. Within a central place system, the ideal shape of each market area should be a circle; however, as Figure 12.8 shows, a set of circles will mean either that some areas are not served or that some areas are served twice. To completely cover an area without any overlap, the ideal shape is the hexagon; the nearest geometric figure to a circle, the hexagon has the greatest number of sides and provides total coverage without duplication.

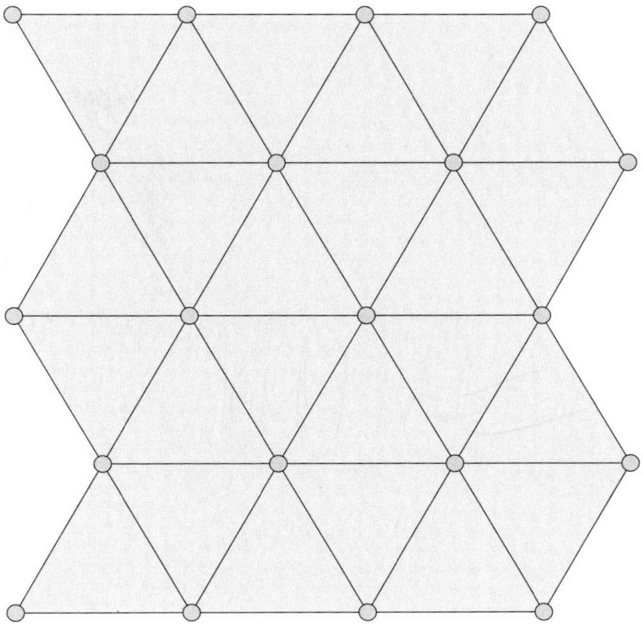

Hierarchies

Central places, of course, contain more than one central function. Bakery products are not the only ones purchased by rural populations. Furthermore, different central functions have different threshold populations and different ranges. We find, then, that some central places have many functions and are correspondingly large, while other central places have relatively few functions and are correspondingly small. Christaller identified seven such levels of settlement and showed that the highest-order settlements (the largest) will be few in number, while the lowest-order settlements (the smallest) will be more numerous. But the spatial regularity is not disturbed by these differences because for any given order, settlements are equidistant from one another. Figure 12.9 shows a two-order situation. Thus central place theory suggests a hierarchy of settlements and a nested hierarchy of trading areas.

Three Principles

We have just outlined what Christaller called the marketing principle. He also proposed the transportation and administrative princi-

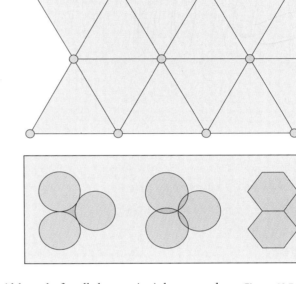

ples. Although, for all three principles, central places are located equidistant from each other, in any given region one of the three may dominate or, more probably, the effects of all three will be roughly equal.

The *marketing* principle invokes the principle of least effort: people will always go to the nearest centre that provides the required service. Christaller called this a $k=3$ arrangement, with k as the number of settlements at a given level in the hierarchy served by a central place at the next highest level. Figure 12.9 shows that each high-order centre serves three low-order centres (that is, one-third of each of the low-order centres that surround it, plus itself as a low-order centre). Note that higher-order centres include all the functions of lower-order centres.

The *transportation* principle allows for as many large centres as possible to be located on communication lines between centres. (Notice that we are now amending one of the initial assumptions.) The result is shown in Figure 12.10; there is a $k=4$ (the centre itself plus one-half of each of six lower-order centres) system in place.

Figure 12.7 (top) A triangular lattice.

Figure 12.8 (bottom) Theoretical trading areas.

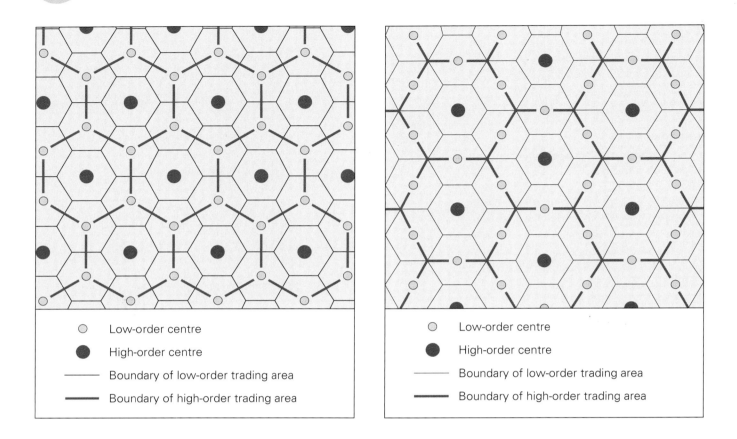

Low-order centre

● High-order centre

—— Boundary of low-order trading area

— Boundary of high-order trading area

Low-order centre

● High-order centre

—— Boundary of low-order trading area

— Boundary of high-order trading area

Figure 12.9 (left) A central place system: the marketing principle.

Figure 12.10 (right) A central place system: the transportation principle. In this instance, the low-order centres are located at the middle of each of the six sides and not at the corners of the hexagons of the high-order centres; hence each low-order centre is linked to two, not three, high-order centres.

The *administration* principle assumes that a central government would not risk subdividing trading areas or regions. Thus the hierarchy is built up through the addition of whole regions. Figure 12.11 depicts the arrangement of centres and shows that a $k=7$ situation applies: that is, a centre serves itself plus six lower-order centres. Box 12.5 offers some evidence to support the general hypotheses proposed in central place theory.

Other authors, especially Lösch (1954), have contended that Christaller's three principles are simply special cases of a more general situation. Extending this argument produces spatial arrangements that are much more complex than those proposed by Christaller.

The Rank Size Rule

Closely linked to central place theory, though separate in origin, is the rank size rule. Originally devised in 1913, the rule simply establishes a numerical size relationship between centres in a region. The rule is as follows:

$$P_r = P_1/R$$

where

P_r = population of centre r
P_1 = population of largest centre
R = rank size of centre r

Thus the population of a centre is inversely proportional to the rank of that centre. The largest centre in a region is named the **primate city**.

Geographers have noted that some regions conform to a rank size distribution pattern, in which the second largest city is one-half the size of the largest, and so forth, while other regions conform to a primate distribution pattern, in which one centre is more than twice the size of all other centres. This difference can be explained in various ways. For example, rank size distributions tend to occur in large countries, countries with a long history of urbanization, and countries that are economically and politically complex. Primate distributions tend to be found in small countries, those with a short history of urbanization, those with simple economic and political structures, and those at the centre of colonial empires.

Cities in the More Developed World

Perhaps no topic reflects the changing nature of geographic research quite so well as the geography of urban areas. In the 1960s, researchers in the spatial analysis school used simple models to focus on the internal structure of cities. Since then, the focus has shifted to the way the city is perceived by its inhabitants and as a symbolic landscape, the quality of life in the city, and cities as places. In addition, attention has been directed to the structure of the city, including the rise of postsuburbia and the postindustrial city. Conceptually, the focus has moved from positivism to humanism and various radical approaches, and most recently to other social theory. This transition in conceptual approaches to the contemporary urban area in the more developed world is outlined in Box 12.6.

Models of Internal Structure

A simple urban land-use model operates on the same principles as von Thünen theory, with different potential land uses having different economic rent lines. Thus land uses

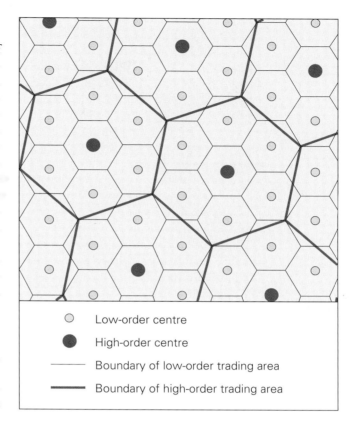

Figure 12.11
A central place system: the administration principle.

○ Low-order centre

● High-order centre

─── Boundary of low-order trading area

━━━ Boundary of high-order trading area

12.5

TESTING CENTRAL PLACE THEORY

The essential hypotheses of central place theory—relating to the spacing and size of centres and to the shape and size of market areas—have been evaluated many times and for many different regions. Together, these studies tend to confirm the basic hypotheses. Many of the early classic studies are described in detail by Berry (1967).

In evaluating central place theory, it is important to remember that it considers urban centres purely as service centres. It does not take into account urban centres established in response to, for example, some specific transportation or resource need. Thus Cape Town in South Africa began as a Dutch settlement on the route between Europe and Asia; the location decision had nothing to do with a surrounding rural population's demand for services. Similarly in northern Canada, many towns such as Sudbury, Ontario, are located where they are in order to exploit a specific resource—again without any stimulus from a surrounding rural population. Nevertheless, analyses show that central place networks allow for such factors as they evolve; a modified version of a central place network can evolve from non-central place beginnings.

One of the most interesting criticisms of central place theory concerns the extent to which a recognition of hierarchies is legitimate or whether it results from the application of convenient but non-specific terms, such as 'hamlet', 'village', and 'town'. There is no doubt that the relationship between number of functions and urban size varies, but it may be that there is a continuous functional relationship rather than a series of discrete stages.

A good example of a formal test of the hypothesis relating to urban centre spacing and size is Brush (1953). Brush identified three levels of settlement in southwest Wisconsin: hamlets, villages, and towns. There were 142 hamlets, 73 villages, and 19 towns; hamlets averaged two functions, villages 18, and towns 42; for hamlets, the average distance to the next centre was 8.8 km (5.5 miles); for villages, 15.8 km (10 miles); and for towns 33.9 km (21 miles). This research strongly supports Christaller's work.

It is difficult to exaggerate the importance of central place theory in the geography of urban centres and the spatial analysis school. Not only is it a major conceptual contribution, but it has helped to elevate urban geography in importance and stimulated further developments in location theory.

with the greatest values of economic rent at the city centre will have steep rent lines and occupy land adjacent to the centre. In principle, a city is made up of a series of concentric zones. A typical arrangement might have businesses at the centre with industrial and then residential uses at increasing distances from the centre (see Figure 11.4). The value of land decreases with increasing distance from the centre. A popular diagram of the urban land value surface that allows for a transportation network is shown in Figure 12.12. The three basic models of the internal structure of urban areas are shown in Figure 12.13.

The *concentric zone* model (a) was proposed by the pioneering Chicago sociologists (introduced in Box 12.6) who applied ecological concepts from biology and botany to urban areas, most notably Park, Burgess, and McKenzie (Park and Burgess 1921). This von Thünen-type model is an extension of economic rent logic. The central business district is at the city centre; the second zone is a transitional area comprising original industries and older houses; and the third,

fourth, and fifth zones are all residential, becoming increasingly affluent with increasing distance. This model, like von Thünen's, is a combination of theoretical logic and first-hand experience of the real world. As a descriptive device, it applies best to North American cities.

The *sector* model of urban land use (b) assumes that internal structure is largely conditioned by the location of routes that radiate outwards from the centre. This model was developed by Hoyt (1939) and, like the concentric zone model, is a descriptive formulation. It improves on that model, however, because it incorporates distance and direction. Part of the reason sectors develop is that, once a specific use is attracted to a particular line of communication, other uses locate elsewhere; industrial uses, for example, tend to repel residential uses.

The *multiple nuclei* model (c) incorporates an additional variable: the presence of several discrete centres in the urban area. Developed in 1945 by Harris and Ullman (1945), this model can take many forms in reality—the

12.6

MODERN AND POSTMODERN URBAN THEORY

Twentieth-century urban theory has undergone a series of changes. The first important theoretical focus developed in the 1920s. The Chicago school of *urban ecology* regarded the city as a social organism and interpreted change by drawing ecological analogies. For example, the most important process was the competition for urban space, which resulted in spatial segregation of groups. Other explanations of the internal structure of the city also centred on a group's dominance of an area and the invasion of an area by a group that then succeeded some other group in that area. A related school of thought, which played a leading role in the 1960s, centred on the concept of *neoclassical urban land rent*; these von Thünen-type theorists interpreted urban areas in terms of the rational 'economic operator' (see glossary).

Both of these approaches were criticized from various perspectives for their failure to explain many of the changes taking place in the modern city, especially those associated with the larger processes of transition from Fordist to post-Fordist modes of production (Chapter 8) or the related processes of economic and social restructuring (Chapter 10). *Political economists* pay

special attention to the conflicting interests of, on the one hand, financiers concerned with capital accumulation and, on the other hand, inner-city residential and small-business interests. In human geography, this Marxist approach is closely associated with the work of David Harvey (see Box 12.7).

Another focus, sometimes labelled *urban managerialism*, is derived from the work of the sociologist Max Weber and places particular emphasis on the state and other institutions. *Humanistic* concepts have also been applied to the study of urban areas, especially in the context of interpretive social research.

Most recently, there has been a movement towards some form of *postmodern* account of the city. A basic contention of postmodernism, as discussed in Chapter 8, is that all theories are to some degree repressive, because they privilege one explanation above all others. Postmodernists therefore question the legitimacy of all modern urban theories; see the discussion of postmodern urban landscapes later in this chapter.

Two excellent accounts of urban theory that include appropriate examples of all the above types are those of Bassett and Short (1989) and Cooke (1990).

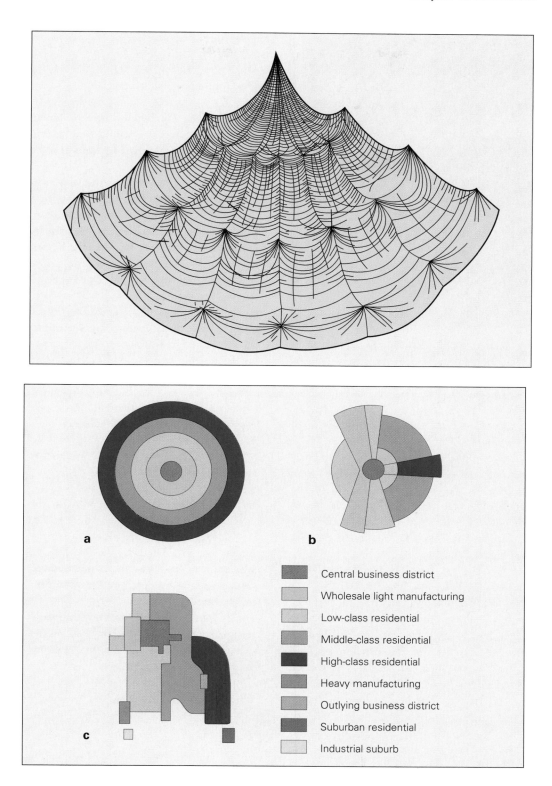

Figure 12.12 Urban land values: high-value areas are commercial or industrial locations on transport routes; low-value areas are residential or vacant.

Figure 12.13 Three models of the internal structure of urban areas.

Central business district

Wholesale light manufacturing

Low-class residential

Middle-class residential

High-class residential

Heavy manufacturing

Outlying business district

Suburban residential

Industrial suburb

figure shows only one example. The number of nuclei varies according to city size and details of development. For North American cities, five general land uses were proposed: central business district, wholesaling and light manufacturing area, heavy industrial area, residential areas, and suburbs.

All three models are generalizations; none of them purports to be universally correct, and for the most part they apply only to North America since the early twentieth century. To the extent that they simplify a complex reality, they are all essentially positivistic. There is a general failure to answer

The location of this Toronto Chinatown—in and around the city's long-time garment manufacturing district—is typical of many such ethnic enclaves
(Dick Hemingway).

oped world. There are seven components to his model:

1. The core area, which continues to function as the heart of the city, is the site of government, financial, and business offices. There is less retailing than previously. This zone grows upwards rather than outwards.
2. Surrounding the core is a zone that was previously light industrial and warehousing (as in the Park, Burgess, McKenzie model). In some cities this second zone is now stagnant, but in others it has benefited from business and residential investment.
3. Most cities include several areas of low-quality housing, characterized by poverty and often occupied by minority ethnic groups. Most of these areas are adjacent to the second zone, although some are located elsewhere.
4. Much of the rest of the city consists of middle-class residences, often divided into relatively distinct neighbourhoods. Although this is largest area of the city, it is not continuous; rather, it is interrupted here and there by the three remaining components.
5. Scattered throughout the middle-class area are élite residential enclaves. As with the poor areas, some of these are close to the city centre and some are suburban.
6. Also scattered throughout the middle-class area are various institutional and business centres such as hospitals, malls, and industrial parks.
7. Finally, most cities are growing outwards along major roads, and peripheral centres may develop that will be largely independent of the original city form.

This model is helpful because it provides a set of generalizations but is more flexible than the three classic models of internal structure.

Retailing

Much research on urban retailing has emphasized the important role played by the city centre region—the central business district (CBD) of the three models of internal structure. During the early 1950s especially, there was an almost exclusive concern with defining and delimiting the CBD. Then

(or even to ask) questions about processes, about the causes of particular spatial configurations. Thus although they are valuable as descriptions against which real-world structures can be assessed, these models do not adequately explain the internal structure of urban areas.

Interestingly, the distribution of population densities is remarkably consistent for all urban areas and bears little relation to the details of internal structure. Typically, densities decline as a negative exponential function of distance from the centre of the city (see Figure 2.5).

Inside the City

To provide empirical evidence to complement the conceptual work outlined above and reach a better understanding of why land uses and activities are located where they are inside the city, human geographers have done considerable research on the location of specific activities in urban areas, notably retailing, industry, and residences.

Before examining their findings, however, let us consider one more model. To reflect the recent changes affecting the internal structure of the city—changes such as deindustrialization and the rise of a service economy, decentralization of retailing activities, increased government intervention, increased automobile use and related suburban expansion—White (1987) revised the classic models to suggest the general form of an early twenty-first-century city in the more devel-

Murphy, Vance, and Epstein (1955) changed the terms of the discussion by drawing attention to the internal diversity of CBDs and their constantly changing boundaries. Subsequent studies have also emphasized change in land uses, as well as the fact that not all CBDs are the same. Today, city centres are subject to rapid change, often in response to changing tastes, and they are put to a multitude of uses; moreover, in the largest cities the central areas often contain several spatially distinct land uses, and in these cases the CBD concept itself has little value.

The above remarks notwithstanding, retailing obviously does play a dominant role in most city centres—although it also locates outside them.

Why do retailing activities locate where they do? One answer identifies eight factors that may influence such decisions (Nelson 1958):

1. trading area potential—an above-threshold population is needed;
2. access to trading area—retailers must compete for peak land-value locations;
3. growth potential—close to growing populations and rising incomes;
4. business interception—between employment and residential locations;
5. cumulative attraction—adjacent to other retail outlets;
6. compatibility—in an area of compatible uses;
7. minimizing competitive hazard—away from direct competitors; and
8. site economics—ease of local access.

Clearly, not all of these factors will be relevant in any one instance; for example, striving for both cumulative attraction and compatibility is likely to maximize rather than minimize the risks of competition.

In recent decades, urban retailing activity has undergone a number of organizational and related locational changes in response to broader postindustrial trends. There are three principal *organizational* changes. First, the numbers of independent traders have declined sharply as a consequence of competition between large business enterprises. Second, the largest retailers, of both goods and services, are now operating as transnationals, opening stores in more than one

A classic 'big box' store, in Scarborough, Ontario (Dick Hemingway).

country; major examples include Wal-Mart (American) and Marks & Spencer (British). Third, new communications technology has permitted the creation and rapid growth of television shopping networks.

The most evident *locational* change is a decentralization of shopping centres, many of which are now located in out-of-town areas accessible only by automobile. Such centres are controversial in many areas because they may compete with small rural retailers, replace attractive green sites, bring significant increases in traffic, and require large parking areas. In addition, a relatively small number of superstores are emerging, often located on previously vacant areas inside city boundaries; a specific form of this trend is the automobile mall consisting of several dealerships in close proximity. These locational changes challenge the viability of older retailing areas, especially in downtowns.

Industry

In the early twentieth century, it was usual for many industrial activities to be located adjacent to the CBD (a situation explicitly reflected in the concentric zone model) and along railway routes. These locations reflected the rapid industrial growth of the nineteenth century. In the contemporary city, there are many more types of industry, and they favour a much broader range of locations.

Several attempts have been made to categorize different industries and related favoured locations in and near contemporary urban areas. Castells and Hall (1994) noted five location types:

1. *Industrial complexes of high-technology firms* include linked research and development and manufacturing plants. The best-known example is Silicon Valley, south of San Francisco.
2. *Science cities* focus solely on research and development. A well-known example is Tsukuba, close to Tokyo: 'a national research centre funded totally by central government; its laboratories, engaged on basic research, are government laboratories' (Castells and Hall 1994:67).
3. *Technology parks* are intended to attract high-technology firms and thus generate industrial growth. These are increasingly common and can be found on the outskirts of most major cities.
4. *Technopoles* are designed as part of a larger program of regional planning aimed at upgrading depressed areas; they involve a variety of technologies and both public and private funding. Sendai, 300 km (186 miles) north of Tokyo, is one example.
5. *Regenerated older areas*. It is not uncommon to find these close to or inside major cities.

Thus both retailing and industry have moved outwards from inner city areas. To a significant degree, their place has been taken by service activities, office buildings, and public institutions. One of the principal functions of major cities today is to provide access to providers of advanced business services.

Residential areas

Human geographic analyses of residential areas in cities have built on the early concentric zone model to examine the spatial segregation of groups according to social status and ability to pay for desirable locations. Spatial segregation reflects consumer demand, the ability and willingness of suppliers to provide different quality areas, and government intervention. More generally, spatial residential segregation in the western world reflects a general acceptance of the idea that cities are not communities but rather institutions serving competing individual interests. In the 1960s, much of the research done on residential social areas (like the research into CBDs in the 1950s) took a descriptive approach based on statistical analysis. Another branch of research studied the residential location decision, which was widely regarded as a matter of free choice. More recent studies, however, have shifted focus and begun to analyze the multitude of constraints that restrict individual choices.

Among the most important constraints on individual residential location decisions are builders and developers, private and public; together, these make up what is often called the housing market. Analyses of the housing market closely reflect the political stance of the researcher. In the simplest terms, those with a socialist perspective believe that housing ought to be provided by the state as a social service like education or health care, while those with a capitalist perspective believe that housing is a commodity to be bought and sold without any state intervention. In most countries, attitudes towards housing lie somewhere between those two extremes. The details of residential location, and therefore of social areas, are closely related to changing political attitudes over time.

Urban sprawl

Perhaps the single most obvious development affecting retailing, industry, and residential areas is the ubiquitous problem of urban sprawl. Sprawl is especially prevalent where land is readily available, planning regulations are weak, populations are wealthy and can afford large homes, and levels of physical mobility are high. Sprawl can lead to the formation of **conurbations**—continuously built-up areas formed by the coalescence of several expanding cities that were originally separate—that are perhaps better described as city-regions. Such regions grow at the expense of the inner city, leading to reductions in the city's tax base, urban unemployment, underutilization of urban infrastructure, and lower property values inside the traditional city. Sprawl is also blamed for destroying farmland and increasing commuting times, which in turn may contribute to the deterioration of family life.

Producing the built environment

Human geographers analyze capitalist development in urban areas in terms of the major agents involved. The process by which rural land becomes urban may involve many participants: governments, rural landowners, speculators, architects, developers, builders, sub-contractors, real-estate agents, urban homeowners, and various financial and legal consultants. In order to understand the urban development process, it is essential to recognize that although most of these participants stand to benefit in one way or another from a change in land use, their motives vary widely. Governments at various levels establish planning policies that constrain or facilitate the process of land use change; rural landowners are primarily concerned with the agricultural productivity of land; speculators aim to buy land at a low price and sell at a high one; architects, developers, builders, and subcontractors together provide basic infrastructure and a finished dwelling and are motivated by the need to make a profit; real-estate agents mediate between sellers and buyers and have a vested interest in promoting an active property market; financial and legal consultants, including mortgage providers, earn fees for their services.

Neighbourhoods as places

Cities are internally complex, with different land uses, different experiences for different people, and different districts—some of high social status and others of low social status.

Perhaps the best way to understand the urban environment is to consider the many and complex ways in which social structures and individual actions relate to one another. As human geographers we understand group identity to be the mainspring of human society. Members of a single group have a shared understanding of place, and as a result there is no conflict over land use. In any urban area, however, there are many groups, each of which may have a different understanding of a particular place; locations that are used by many groups, such as roads, can take on a multitude of meanings and may become areas of conflict. Thus any urban area has two types of space: locations occupied by groups whose members share a common understanding, and locations used by diverse groups and belonging to none. The architect and planner Greenbie (1981) distinguished between proxemic spaces, such as neighbourhoods, and distemic spaces, such as roads. Proxemic spaces reflect group identity, while distemic spaces have various meanings,

Land for sale in Brisbane, Australia
(Victor Last, Geographical Visual Aids).

depending on the social background of the individual. Traditionally, the most successful and stable distemic space is the market.

Social theory that focuses on individuals as members of groups encourages us to acknowledge the meanings that different people attach to different locations. Such theory helps us to understand the values that people ascribe to a place and the conflicts that can emerge over land use.

Urban areas are characterized not only by varied land uses but also by distinctive residential areas. Typically, these areas are distinguished on the basis of class, ethnicity, or some other cultural variable. Especially in Europe, prior to the industrial revolution, the most distinctive urban residential district was the Jewish district, usually labelled the 'ghetto' (see glossary). An area apart from (as opposed to a part of) the larger city, a ghetto is held together by the internal cohesion of the group and the desire of non-group members to resist spatial expansion of the group. Box 12.7 looks at such conflicts from a Marxist perspective.

During the industrial revolution, class divisions became more evident and were expressed in spatial terms. By the beginning of the nineteenth century, British commentators were acutely aware of the distinction between the working-class districts close to the factories and the middle-class districts located elsewhere, especially on the outskirts of the urban area. Distinct residential areas emerged most obviously in large immigrant-receiving North American cities in the nineteenth century, with distinctions based on ethnicity as well as class. Once a particular ethnic group was large enough, it settled as a group, usually in an inexpensive area close to employment opportunities.

Elsewhere in the world, such segregation might be based on religion, as in Belfast, Northern Ireland. In South Africa, even before the formal institutionalization of apartheid, cities were clearly divided on explicitly ethnic lines. In other cases residential variation may reflect lifestyle preferences or the proximity of major employers, such as universities and hospitals. There is no doubt that society and space are irrevocably entwined.

The link between society and space is perhaps clearest in the case of neighbourhoods. Although the concept of the neighbourhood has not been formally defined, the term is generally understood to designate a district that reflects certain shared social values. Some geographers—for example, Smith (1985)—have suggested that the best indicator of neighbourhood identity is the presence of

neighbourhood activism. However, this argument is weak on at least two counts. First, where neighbourhood activism does exist, its purpose is not to establish or preserve identity, but to generate change (see Box 12.8). Second, some neighbourhoods have a clearly defined identity that is based not on activism but on some *negative* feature, such as a high crime rate, or the fact that it is dominated by members of some disadvantaged group—for example, substance abusers, dependent elderly, or people who are physically disabled. In their challenging work *Landscapes of Despair*, Dear and Wolch (1987) describe areas dominated by another such group: the mentally ill. Acknowledging that human landscapes reflect social processes, geographers recognize the need to eliminate such ghettos by eliminating their causes.

A Postsuburban Trend

One of the most distinctive features of contemporary cities in the more developed world is the emergence of 'postsuburbia' (Box 12.9) and, more broadly, what has been called the postindustrial city. Such cities have

---

12.7   A MARXIST INTERPRETATION OF THE URBAN EXPERIENCE IN A CAPITALIST WORLD

David Harvey is a human geographer who, after producing a major methodological work on positivism in 1969, has focused primarily on arguing the merits of a Marxist philosophy for understanding the urban way of life (Harvey 1969, 1973, 1982, 1989, 1996, 2000). According to Harvey, Marxism offers a broad theoretical perspective (a metatheory or higher-level theory) that allows geographers to approach such diverse issues as the built environment, urban economy, and local urban culture.

Using Marxist theory and adding spatial factors, Harvey (1982) argued that uneven spatial development is a necessary accompaniment of capitalism. This argument applies on various scales, including that of the urban area. Thus urbanization is seen as one more aspect of the uneven spatial development that occurs in a capitalist context. The spatial generality of these ideas justifies use of the term 'metatheory'.

Understanding urban places requires appreciation of both the flows of capital and surplus value through systems of cities and the class relations characteristic of a capitalist society. Explaining residential differentiation, for example, requires recognizing

that the more powerful classes have resources that allow them to acquire land of their choosing. In Marxist terms, the explanation lies in the superstructure. It is not appropriate, from this perspective, to see residential differentiation as resulting from the preferences of all people; it should be seen as resulting from the preferences of a powerful few.

More specifically, this Marxist perspective identifies the city as the site of three circuits of capital. The primary circuit concerns the structure of relations in the production process, such as the manufacture of goods for sale; the secondary circuit concerns investments in fixed capital, such as property development; and the tertiary circuit concerns investment in science and technology that helps to increase production. According to this theory, the transfer of capital from the primary to the secondary circuit is one cause of suburban expansion: thus in effect restructuring capital can lead to spatial restructuring. For some human geographers, Marxist theory provides critical insights into the forces behind the production of the built environment of the city; see also Box 11.3).

limited connections with industry; instead of offering manufacturing jobs, they offer work in the professions, management, administration, and skilled labour. Physically, such cities include office buildings and public institutions, but few factories. Cities that serve as government and business centres, such as Toronto, are prime examples, as are some major tourist and retirement centres. In many such cities, especially those that developed out of former industrial centres (as opposed to the all-new postsuburban cities) there is pronounced social polarization between the new middle classes and the working poor or unemployed (Ley 1993).

Gentrification

The decades since the Second World War have seen a partial reversal of the tendency for high-status populations to live away from the city centre. Not only is life in the inner city more convenient for those who work in the city centre, but it offers a number of less tangible 'quality of life' benefits. The process of transforming a formerly derelict or low-quality housing area is known as **gentrification** (from 'gentry', a term that originally referred to people of high social standing). Today most cities, large and small, in the more developed world are undergoing gentrification (Ley 1992).

According to the Chicago school's ecological model, neighbourhoods decline after reaching maturity, but clearly this is not always the case: gentrification is redevelopment prior to complete neighbourhood decline. This distinctive feature of some urban areas appears to be another consequence of

12.8

SLUMS AND THE CYCLE OF POVERTY

The rapid expansion of urban industrial areas in the nineteenth century resulted in high-density, poorly serviced housing areas of poor quality. These areas contrast markedly both with the neighbourhoods of preindustrial cities and with rural villages, where the separation of the poor and the rich was not nearly so marked. In 1845, the slums of Nottingham, England, were described as follows:

> . . . nowhere else shall we find so large a mass of inhabitants crowded into courts, alleys, and lanes as in Nottingham, and those, too, of the worst possible construction. Here they are so clustered upon each other; court within court, yard within yard, and lane within lane, in a manner to defy description. Some parts of Nottingham [are] so very bad as hardly to be surpassed in misery by anything to be found within the entire range of our manufacturing cities (Hoskins 1955:218).

As these nineteenth-century slums were cleared, they were replaced by equally poor forms of low-cost, often high-rise, housing. Both types of slum have been associated not only with poverty but also with high levels of crime, vandalism, and substance abuse. Many commentators have proposed that there is a **cycle of poverty** that perpetuates slums as a way of life and limits the prospects for integration with the larger urban area. Slums are often associated with distinct ethnic groups, although the terms 'ghetto' and 'slum' are not synonyms.

The problem of poverty, of which slums are only one component, is perhaps the biggest urban problem. Knox writes:

> . . . poverty comes as a package. Low income is at the core, but it is inextricably bound up with poor diets, poor physical environments, poor physical health, the psychological stress of continuously having to make ends meet, and the economic, social, and political disadvantages of being stigmatized as a 'loser' in a highly competitive society (Knox 1994:299).

Other human geographers approach urban social problems from different theoretical perspectives. Not surprisingly, as Box 12.7 suggested, Marxism provides one popular explanation, emphasizing the fact that the capitalist system necessarily both concentrates power and wealth in the hands of a few and also requires a majority who are relatively dispossessed both materially and socially. Some theories stress the consequences of economic change, such as the recent deindustrialization that results in increased levels of unemployment. Others, as suggested in Box 12.6, emphasize the roles played by those who have the power to control access to urban resources, such as real estate agents and mortgage companies (sometimes called 'gatekeepers'). The range of theoretical options is stimulating, but it cannot be denied that the issues are far from being resolved; most current evidence suggests that urban problems are increasing rather than decreasing.

the economic and social restructuring that has contributed to the rise of the postindustrial city. Like many changes, gentrification has both positive and negative aspects. While it represents a benefit for middle-class people who want to live close to the city centre, for the poor—especially renters—it represents a threat, since rising house prices may well result in their displacement. The issue of displacement raises an important question: should governments intervene to protect the more vulnerable residents?

The Image of the City

The following discussion relies heavily on a seminal work by Lynch (1960), in which the visual quality of urban areas was illustrated by case-studies of Boston, Jersey City, and Los Angeles. Visual quality is the ease with which residents are able to identify parts of the city and integrate these parts into the whole. What images do residents have of their cities? What are their mental maps? These questions are important both practically (for getting around) and emotionally (for making sense of where we live).

To determine how residents of different areas perceived their city, people were asked to identify features of the urban landscape. Five kinds of features were identified:

1. *Paths* are various routes through the city, from which many parts of the city can be seen.
2. *Edges* are boundaries between perceived districts.
3. *Districts* are parts of the city that are in some way perceived as different from surrounding areas; they are equivalent to geographic regions.
4. *Nodes* are key points in the city such as major intersections.
5. Finally, *landmarks* are distinctive physical objects such as buildings, statues, or hills. The results of the Los Angeles study are shown in Figure 12.14.

Such research is intriguing and appears to offer a valuable approach to urban planning, but so far it has not generated a substantial body of subsequent work.

The City as Image

Urban landscapes, in common with other human landscapes, can be interpreted symbolically—that is, as physical representations of cultural identities. In a series of iconographic studies, Cosgrove (for example, 1984) has analyzed the cities of Renaissance Italy in the context of humanistic culture, which was expressed in an urban landscape that displayed wealth through monuments and patronage of the arts.

12.9

SUBURBIA AND POSTSUBURBIA

Suburbs—areas outside the central core of a city and set apart from industrial areas—comprise housing and basic services designed for the working class and/or middle class. Although suburbs did exist in preindustrial times, large-scale suburbanization was prompted by industrial urbanization and the associated development of transportation. Suburbs represented a compromise between the need to live close to employment in the central city or industrial areas and the desire to be close to open spaces and rural areas. Many twentieth-century suburbs, especially in Europe, were designed for the working class. Incorporating large quantities of public housing, they have often become areas of urban deprivation, social anonymity, residential congestion, and pollution. The fact that, in many cities, public transportation systems have not been adequately modernized adds yet another set of difficulties to working-class suburban life.

In recent years, as suburban areas have matured, many of them have begun to acquire features formerly associated with city centres: residences, offices, and light industry are all included in the 'edge cities' of what is sometimes called postsuburbia. This trend involves the increasing decentralization of offices and factories, the decline of manufacturing employment, and the processes of economic restructuring. According to Knox (1992:26), 'Economic decentralisation has resulted in so much commercial and industrial development in the outer reaches of large US metropolitan areas that central cities and downtown Central Business Districts no longer dominate them the way they used to.' An excellent example of 'postsuburbia' is Tyson's Corner, Virginia: what was a rural corner in the mid-1960s now includes the ninth largest concentration of retail space in the United States.

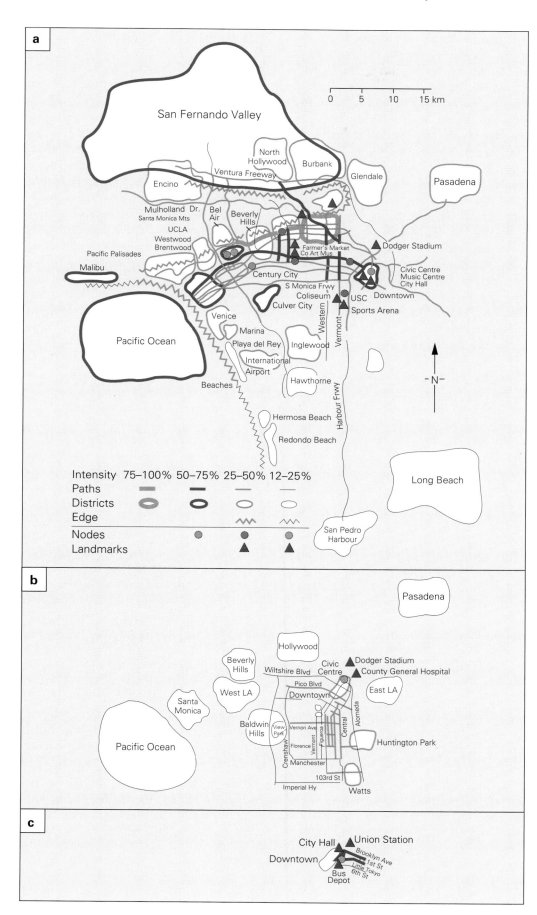

Figure 12.14 Three images of Los Angeles. These figures are frequently cited to illustrate the different images of a single urban area that different groups of people may hold. Figure 11.15a is the image of Los Angeles held by residents of Westwood, a predominantly White upper-class neighbourhood in the north of the city, close to Beverly Hills; 11.15b is that held by residents of Avalon, a predominantly Black area south of the city, close to Watts; 11.15c is that held by residents of Boyle Heights, a Spanish-speaking area close to the central industrial area. The message of these maps is painfully clear. Ethnicity, socio-economic status, and mobility all play a role in the process of image creation. Cities are different places for different groups.
Source: Los Angeles Department of City Planning, *The Visual Environment of Los Angeles* (Los Angeles: Department of City Planning, 1971).

Urban landscapes are constructed to represent and legitimate the power that the élite of the population are able to exercise (Duncan 1990). For example, cities that were planned to serve as capitals were usually designed to proclaim power and wealth, with grand avenues, large squares, impressive public buildings and monuments—all symbols of mastery. But reality often falls short of grand plans. Two relatively recent examples are Washington, DC, and Brasilia, the capital of Brazil: both designed as capital cities, both have developed massive problems of poverty, crime, and homelessness.

Like the rest of the human world of which they are a part, cities seem to have brought out the best and the worst in humans. Many cities that are centres of civilization for some of their inhabitants are nightmare landscapes for others.

Emerging Urban Problems

Cities throughout the more developed world typically experienced a prolonged period of substantial growth, beginning with a process of concentration that created the industrial cities of the nineteenth century followed by a process of decentralization that created the conurbations and urban sprawl of the twentieth century. Cities have always experienced problems of congestion and poverty (see Box 12.8). But today these urban areas are experiencing a number of new problems related to changes in economic base (and associated lack of investment) and decaying infrastructures.

Urban unemployment

For most cities, the transition to the current postindustrial phase has resulted in major losses of employment opportunities, and the deindustrialized areas have not received any new investment. The principal exceptions to this negative trend are the global or world cities discussed earlier in this chapter. Elsewhere, the new employment opportunities in the high-technology, service, and tourist industries are rarely open to former members of the industrial labour force. Increasing levels of unemployment lead to problems of social-service supply and, often, increasing social tension and criminal activity, not to mention personal difficulties.

Aging physical infrastructure

Another set of problems relates to the aging physical infrastructure of most cities in the more developed world today. Because most lack the financial base to upgrade, problems such as delays in road repairs and reductions in basic services are often evident throughout the city. But it is inner city areas, the oldest parts of the city, that tend to suffer the most and that are especially vulnerable because of the concentrations of disadvantaged people who generally live in these areas. In addition to homelessness and related problems (often linked to the deinstitutionalization of mental patients), many cities are experiencing increases in crime related to organized gang activity and violence associated with ethnic tensions. One common response, on the part of rich residents, is to widen even farther the gap between themselves and the poor; the differences between a downtown slum and a suburban gated residential community speak directly to the failure of many large cities to be communities for all residents.

Cities: Centres of civilization?

An example of an urban area that is at one and the same time a centre of civilization and a centre of social and environmental chaos is Los Angeles. Socially, Los Angeles has been described by Davis (1992) as 'fortress L.A.'. Environmentally, it is dependent on water imported several hundreds of miles via aqueducts—but it also has an expensive system of channels and dams to handle flooding caused by the annual snow-melt. In addition it experiences serious ozone pollution, produced when hydrocarbons and nitrous oxides (emitted by automobile exhaust) react in oxygen. Health warnings are issued when the ozone level exceeds a certain figure: in 1987 warnings were issued on 36 days; at Glendora, about 30 km (19 miles) downwind from LA, warnings were issued on 135 days.

An especially disturbing trend is the juxtaposition of privilege and deprivation. New Haven, Connecticut, is the home of Yale University, one of the wealthiest and most prestigious schools in the United States. In 2002 a news report described a tent city adjacent to the university. Established after authorities closed down an overflow homeless shelter, the tent city in turn was quickly cleared by police. The ironies are frightening. Proportionally, Connecticut has more millionaires than any other American state, and yet New Haven has an infant mortality rate comparable to that of Malaysia. It has suffered a devastating number of deaths associated with AIDS, and according to a 2002 report, almost 70 per cent of the city's children are without health insurance. Critics suggest that New Haven is a metaphor for the United States in early twenty-first century.

It may be that New Haven's tent city is an instance of what has been called 'the new poverty management' (Wolch and DeVerteuil 2001). This term refers to the organized efforts undertaken by the state and élite groups to maintain social control over the poor and underprivileged. In recent years, homeless people especially have been subject to involuntary moves between residential units, shelters, prisons, sober-living homes, hospitals, and the street. This 'circulatory' approach may have more to do with removing the poor from view than with finding a lasting solution to a social crisis.

Postmodern Urban Landscapes

Contemporary cities in the more developed world typically reflect twentieth-century growth and planning practices, industrialization, and modernity in general. Today a number of new and different processes are operating—what might be described as postmodern processes. The impact of postmodernism is particularly evident in architectural efforts to reproduce local vernacular building styles or to incorporate portions of old buildings in new developments. Similarly, there are more suburban office clusters being built than modernist office towers.

Some human geographers have suggested that what urbanism is experiencing today may be not just an evolution but a transformation within the context of a postmodern break from earlier cultural, economic, and political trends. Dear and Flusty (1998:50) pose the following questions: 'Have we arrived at a radical break in the way cities are developing? Is there something called a *postmodern urbanism*, which presumes that we can identify some form of template that defines its critical dimensions?' Their answers are in the affirmative. Arguing that most

previous research into the urban landscape has been based on the Chicago school of urban ecology, with its emphasis on concentric zones of land use, they propose a new research model based on a southern California-style urbanism and emphasizing global-local linkages, a reversal of the traditional relationship between hinterland and centre, and a ubiquitous social polarization.

This proposed Los Angeles model depicts globalization and restructuring processes as the driving forces behind the new urbanism. Because the postmodern city includes populations that are not only culturally heterogeneous but also politically and economically polarized, it is characterized by great disparities in wealth and styles of living. There are theme park areas, service-oriented edge cities, private housing areas based in common-interest developments, fortified zones, and mean streets dominated by gangs. A distinctive feature of this landscape is its fragmented character (Figure 12.15). It is notable that this new body of ideas, like much other postmodern work, requires a new vocabulary in order to express its ideas.

The idea of a postmodern urbanism is not without critics. Some have suggested that such ideas stress style over substance, that the basic premise of a transition from a modern to a postmodern epoch is questionable, and that the case of Los Angeles represents an exception. Certainly urban landscapes today reflect social changes; gentrification and the emergence of gated communities, designed to maintain privilege and exclude disadvantaged others, are only two examples. Yet at the same time many features suggest that there is at least as much continuity as there is change in the urban landscape today. In any event, for the beginning human geographer these ideas are an indication of the need for constructive re-evaluation of earlier ideas—and of the challenges to be faced in any attempt to make sense of contemporary urbanism.

Global Cities

As we noted at the beginning of this chapter, urbanization is affecting all parts of the world (see Figure 12.5). But increases in the numbers of urban dwellers and related increases in the numbers of very large cities (Figure 12.16) are not the only changes occurring.

Figure 12.15 A model of post-modern urban structure. The style and content of this diagram reflect the consequences of what is called keno capitalism (in a game of keno there is a game card with a numbered grid; during the course of the game, some squares are marked and others are not). Particular urban developments occur as capital is placed in particular locations in the city in an almost random manner. There is no apparent relationship between what is happening in one place and in any other place. The places, or interdictory spaces, are affected by the disinformation superhighway. 'Conventional city form, Chicago-style, is sacrificed in favor of a non-contiguous collage of parcelized, consumption-oriented land-scapes devoid of conventional centers yet wired into electronic propinquity and nominally unified by the mythologies of the dis-information superhighway' (Dear and Flusty 1998:66).
Source: After M. Dear and S. Flusty, 'Postmodern Urbanism', *Annals of the Association of American Geographers* 88 (1998):66.

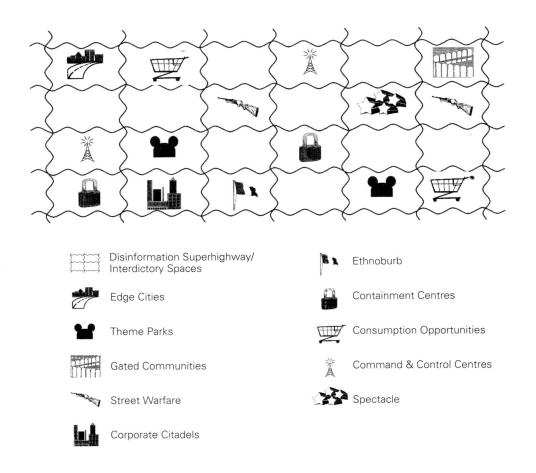

Disinformation Superhighway/ Interdictory Spaces

Edge Cities

Theme Parks

Gated Communities

Street Warfare

Corporate Citadels

Ethnoburb

Containment Centres

Consumption Opportunities

Command & Control Centres

Spectacle

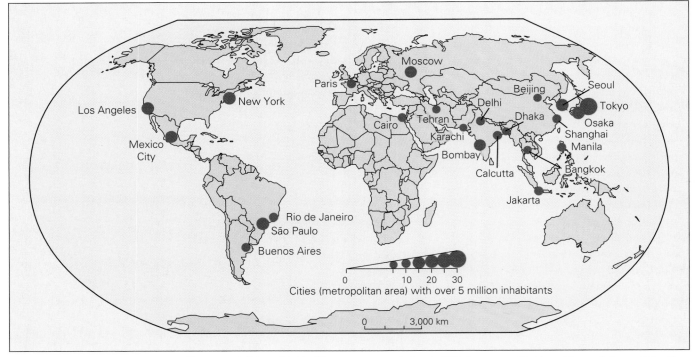

Figure 12.16 Cities with 10 million or more population. In 2000 there were 22 cities with populations of 10 million or more, most of them in the less developed world. Only New York, Los Angeles, Paris, Moscow, Tokyo, and Osaka are in the more developed world. But this fact is misleading in that it is the large cities in the more developed world that control the world economy. A large city is not necessarily a powerful city in global terms.

Many cities are undergoing changes that can be traced to larger changes associated with globalization, in particular the declining role of the state (discussed in Chapters 9 and 10) and the declining friction of distance (Chapter 10). Perhaps the most significant change of all is that a select few cities are now functioning as dominant players in global economic activity. These 'global cities' are not only outdistancing other cities but even replacing the nation state in this respect. The global city is a direct consequence of economic globalization.

Most commentators identify New York, London, and Tokyo as the three truly global cities today, not because of their size— although they are all large—but rather because of their role as financial control centres in a globalizing world (Sassen 1991). Building on the discussion in Chapter 10, we may identify several components of globalization that are especially relevant to cities: the restructuring of the international financial system, the globalization of property markets, investments by transnationals that prompt industrial dispersion, the rise of service activities, the 'stretching' of social relations as the links between widely separated locations intensify, and increased travelling and networking (Olds 2001).

The concentration of these components of globalization in a few specific cities enables them to function in a global capacity. For example, the more capital investment they receive, the more capital they attract. Transnational corporations and financial service groups locate their headquarters in these cities, which then are able to play determining roles in the location of much foreign direct investment, shifting the balance of investment and market power.

As noted, New York, London, and Tokyo are generally regarded as the three truly global cities (Table 12.5). Figure 12.17 suggests general spheres of influence for each of these three, as well as some regional centres. It is notable that two of these cities are not located in the regions for which they serve as centres (Miami serves Latin America, while London serves Africa and the Middle East), and that one of those two is not a first-level global city. Indeed, several other cities also approach world city status; Table 12.6 identifies five tiers.

Table 12.5   THREE GLOBAL CITIES

	Assets	Capital	Net Income
100 Largest Banks			
Tokyo	36.5	45.6	29.0
New York	8.6	8.8	20.8
London	4.2	5.7	13.2
All three	49.3	60.1	62.3
25 Largest Security Firms			
Tokyo	29.6	42.9	72.6
New York	58.6	50.0	22.0
London	11.1	4.9	2.8
All three	99.3	97.8	97.5

Note: This table shows the share in 1988 of the world's 100 largest banks and 25 largest security firms that are located in Tokyo, New York, and London, and confirms their status as the three pre-eminent global cities. These three cities also rank 1, 2, and 3 when we consider the locations of headquarters of major multinational firms (New York has 59, London has 37, Tokyo has 34); the next most important is Paris with 26.

Source: S. Sassen, *The Global City: London, New York, Tokyo* (Princeton: Princeton University Press, 1991):178–9

Table 12.6   FIVE TIERS OF GLOBAL CITIES

Tier 1	**Truly global cities.**
	There are only three: Tokyo, New York, London
Tier 2	**Cities with influence over large regions of the world system**
	Examples: Amsterdam, Frankfurt, Los Angeles, Miami, Singapore
Tier 3	**Important international cities with more limited or specialized international functions**
	Examples: Madrid, Mexico City, Paris, Seoul, Sydney, Zurich
Tier 4	**Cities of national importance with some transnational functions**
	Examples: Houston, Milan, Munich, Osaka, San Francisco
Tier 5	**Cities with a local leadership that has found a niche in the global marketplace**
	Examples: Atlanta (Georgia), Rochester (New York), Columbus (Ohio), Charlotte (North Carolina)

Figure 12.17. World cities and spheres of influence. This diagram is derived from a detailed analysis of office geography data that identifies 55 world cities of varying degrees of importance, and that questions many of the general assumptions about world cities. The patterns shown in this figure are quite different from some of the complex hierarchies described in other studies.
Source: P.J. Taylor, 'World Cities and Territorial States Under Conditions of Contemporary Globalization', *Political Geography* 19 (2000):25.

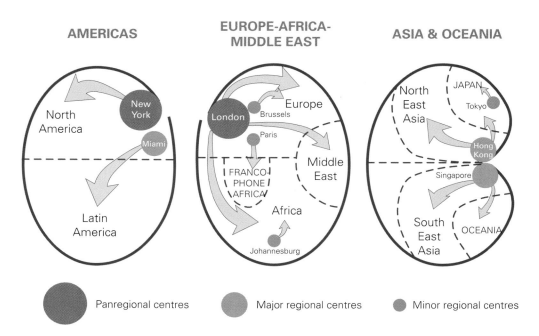

12.6 identifies five tiers.

The rise of globalized property markets has implications for other urban areas. World money and physically distant commodity markets are now influencing developments in many smaller communities and even neighbourhoods, making the task of understanding local change increasingly complex. It is at least partly in acknowledgment of this new reality that the contemporary city in the more developed world is often described as being postindustrial or postmodern. As the 'post' prefix suggests, human geographers know that new city forms are emerging, but so far they have no clear idea of what they will look like.

Cities in the Less Developed World

Colonialism and the Colonial Heritage

Many (though by no means all) of the cities in the less developed world have a colonial heritage. There are also many former colonial cities that are not in the less developed world, such as those in Australia and some of those in the United States and Canada.

Many of the cities in Africa and Latin America, and some in Asia, were essentially creations of European powers developed to serve European needs, a fact that in many cases is reflected in their pre-independence names,

their architecture, their locations, and their internal structure. In most instances, the pre-independence name has been changed for powerful symbolic reasons, although colonial architecture has generally remained. Cities were located in places that suited their role as export centres, but that often proved inappropriate for independent countries with radically different economic goals. With regard to internal structure, it was not uncommon for colonial powers to implement some degree of formal spatial segregation on the basis of ethnic identity, a situation that reached its extreme in South Africa before 1994 (see Box 8.7).

It is important to recognize such European-created cities as centres of exploitation: their functions were primarily administrative and military, and they controlled the export of valuable primary products. They failed to generate local growth, and little of the income generated benefited indigenous populations. As a consequence of the importation of indentured labour (see Box 6.3), many of these cities became culturally pluralistic, and as a result some have experienced group frictions. Because the colonial powers failed to initiate local industrial development, the rapidly growing cities of the post-independence period have had no pre-existing industrial base to build upon. For many of the cities in the less developed world, a number of the issues discussed in this section, especially the problems, had their origins in

Growth

Rapid city growth in the less developed world is a relatively recent phenomenon. In 1950, only 17 per cent of the population in the less developed world lived in cities; today the percentage is about 40 (41 per cent if China is excluded). This growth was the result of large movements of people from rural to urban areas, movements prompted by three general considerations:

1. First, as we saw in Chapter 5, falling death rates and high birth rates in the less developed world meant high rates of natural increase, and rapid growth of rural populations placed great stress on local rural resource bases.
2. Second, many workers were prompted to move by economic changes, such as the mechanization and commercialization of agriculture (which reduced agricultural labour requirements) and importation of manufactured goods (which undermined traditional peasant economies); urban areas offered both greater employment opportunities and higher wage levels.
3. Third, there was a general perception—not always correct—that the quality of life with respect to, for example, provision of basic services was better in cities than in rural areas.

One common feature in rural to urban migration has been a tendency to gravitate towards a single well-known city (see Box 6.2); this tendency has been especially strong in less developed countries, where a distinctive feature of urbanization has been the concentration of population in one city, usually the capital. Figure 12.18 maps cities in the less developed world with populations greater than 2 million. Most countries have a single city that is at least twice the size of the second largest city; this indicates a high level

The town of Ipoh was founded by Malaysia's British rulers in the 1890s. The colonial influence is apparent in the imposing architecture of its railway station (Victor Last, Geographical Visual Aids).

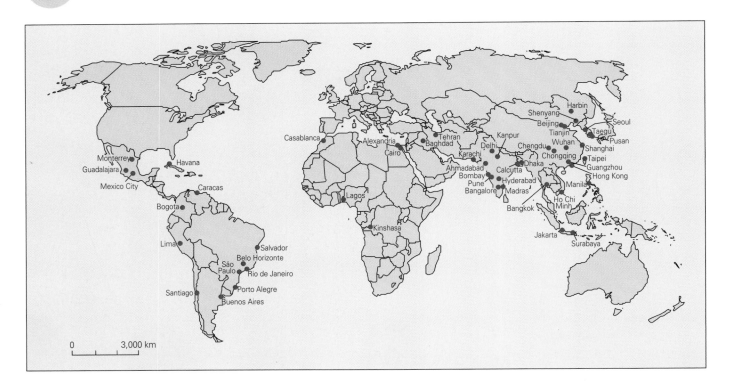

Figure 12.18 Cities in the less developed world with more than 2 million population.

of primacy (recall the earlier discussion of the rank size rule). In marked contrast to countries in the more developed world, it is not uncommon for the primate city in a less developed country to have between 20 per cent and 30 per cent of the total population of the country. It can be argued that these exceptionally large cities are too large, given the economic circumstances of the country; hence the large numbers of unemployed and the rise of the **informal sector**.

There are indications that urban growth in some parts of the less developed world is beginning to slow down. In both Latin America and much of Asia, declining fertility rates mean that the pool of potential migrants is being reduced; however, the situation in Africa and some parts of south Asia continues as before. It is expected that the Indian subcontinent will soon have six megacities: Bombay, Calcutta, Delhi, Dhaka, Karachi, and Madras.

There are, of course, many differences between the cities of the more and less developed worlds. Today many cities in the more developed world are experiencing suburbanization, counterurbanization, and exurbanization. Their counterparts in the less developed world, on the other hand, are probably best described as experiencing urbanization—that is, continued growth throughout the city.

Chinese Cities

'Chinese cities are different' (Macmillan 1996:20). They are unlike cities not only in the more developed world, but also in most parts of the less developed world. The rapid, uncontrollable growth so evident in most of the less developed world has not occurred in communist China, which has proved able to control population movements in a way that is not possible in free-market economies. Of course, as home to one-fifth of the world's population, China does have many very large cities (Figure 12.19).

Two major differences are especially apparent. First, unlike most countries, China has a relatively dispersed pattern of industrial location and therefore its cities are less specialized. In other words, most are relatively self-sufficient—not dependent, as many cities in the more developed world are, on trade in manufactured items. Basic goods such as cloth, bicycles, and watches are produced in all cities, not just a few. There are no single-industry towns.

Second, this industrial dispersion is somewhat countered by investment policies that favour the largest cities. Thus while most cities have a mixture of industrial activities, industrial activity overall is concentrated in a few relatively large cities.

Urban Problems

With reference to the less developed world, one commentator wonders: 'Does the growth of megacities portend an apocalypse of global epidemics and pollution? Or will the remarkable stirrings of self-reliance that can be found in some of them point the way to their salvation?' (Linden 1993:32). The answer is not obvious, although most current evidence suggests that the problems are many and becoming worse.

Cities that expand so rapidly that their growth is not controlled may be described as undergoing 'premature urbanization' and always experience problems. The slums of English industrial cities in the mid-nineteenth century were just as bad as those in many cities in the less developed world today, and people moved to them for essentially the same reason: no matter how difficult life there might be, for many people the opportunities outweigh the problems.

Squatter settlements

Much of the rapid urban growth in the less developed world has involved the growth of **squatter settlements**: areas of uncontrolled expansion on the periphery of a city, made up of low-quality, poorly serviced temporary dwellings. Squatter settlements have serious health, social, and economic problems. As they age, squatter settlements may become integral parts of the city, reducing pressure on rental areas inside the city and offering a base for new migrants. In many instances, squatter settlements have achieved a genuine integration with the larger city as unplanned but necessary growth (Box 12.10).

In general, cities in the less developed world are experiencing intense pressure to provide additional housing, but are incapable of responding adequately. Further, there are clear disparities in the ability of different groups in the city to gain access to improved services—disparities that are a direct reflection of relative power (Lowder 1994).

Given that squatter settlement growth seems inevitable, how can it best be handled? Two answers are suggested. First, internation-

al aid might be directed from rural to urban areas. Western governments and international agencies have long focused their attention on rural development, neglecting urban areas. It is true, of course, that rural areas need aid, but the reality is that the need for flush toilets is far greater in a congested city than in a relatively sparsely populated rural area. Second, as we have had occasion to note elsewhere (recall the Chapter 11 account of the green revolution, for example), it is not appropriate to simply impose change from outside. Improvements in urban areas need to actively involve local residents as participants. From a western perspective, it is easy to imagine the urban poor in the less developed world as some anonymous mass, incapable of handling their own affairs, but this is not the case. If the people living in squatter settle-

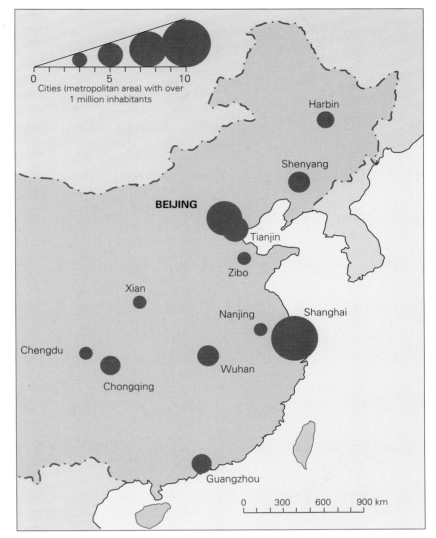

Figure 12.19 Cities in China.

ments had some security of tenure, they would likely be more willing to work to improve their homes (see Box 12.10).

Some examples

Regardless of location, many cities in the less developed world suffer from overcrowding, crime, poverty, disease, limited provision of services, traffic congestion, unemployment, damaged environments, and ethnic conflicts.

Consider *Mexico City*. This is a major world city, with a population of about 27 million, that has grown and continues to grow at an alarming rate. It is engulfed by shanty towns. Furthermore, it is located in an unfavourable environment that continually suffers from flooding, subsidence, and shortages of fresh water. The urban explosion began about 1900 as a result of development policies favouring industrialization. The city has doubled in size since 1970 and

may double again by about 2020. It is not an exaggeration to describe Mexico City as a social and environmental disaster—a disaster that is constantly worsening because of continuing growth. According to the United Nations Environment Program and the World Health Organization, it is the most polluted of large world cities: levels of sulphur dioxide, suspended particulate matter, carbon monoxide, and ozone exceed health guidelines by a factor of two or more, and levels of lead and nitrogen oxide are almost as bad. The principal source of these pollutants, which affect the health of anyone who breathes the city's air, is the burning of fossil fuels (Middleton 1994).

Consider *Calcutta*. The 'city of joy' is the largest in India, with a population of 15 million. Physically, it is located in a low-lying area that requires pumping of storm and sewage water. Some 40 per cent of the

URBANIZATION IN THE LESS DEVELOPED WORLD: BRAZIL

12.10

Brazil is typical of the less developed world in that it has experienced explosive urban growth since 1945, especially in recent years. Recent urban growth has accompanied an economic transition from agriculture to industry.

The earliest urban centres in Brazil were founded under Portuguese rule and were located on the coast to facilitate the export of sugar and other products. São Paulo, founded in 1554 by Jesuits, remained relatively unimportant until it became the key port for the export of coffee in the late nineteenth century. Growth has been accompanied by distinct spatial expressions of class. Working-class areas arose close to industrial sites, while élite areas developed on higher ground; in some cases, these areas are enclosed by security fences and patrolled by guards. Shanty towns, *favelas*, are common on the city's outskirts. Many new immigrants to cities such as São Paulo subsist by becoming involved in what is called the *informal sector*, selling goods in the street, or providing services such as car-washing. Such activities are necessarily low-paying, but are regarded as temporary. For migrants, the attraction to cities is not the shanty town or the possibility of informal-sector employment, but the perception that the city offers better schooling, better medical services, a better water supply, and the prospect of permanent employment. As shanty towns grow, enormous pressure is put on urban governments to provide the needed servic-

es, but then, inevitably, extended urban services attract even more migrants. This may be unfortunate for the migrants, because shanty towns are usually decrepit and lack basic amenities, but it can be viewed in a more positive light: shanty towns can be seen as reception areas for migrants, providing a transition between rural and urban ways of life.

The contemporary Brazilian city is thus characterized by marked class segregation and decentralization. Urban problems abound and appear to be far from resolution (Godfrey 1991). It is, however, instructive to relate the text comments on counterurbanization and the discussion on declining fertility in Box 5.3 to this issue. If birth rates decline substantially in the less developed world, then at least the problems of shanty towns will be decreased somewhat.

Perhaps the most hopeful sign, however, is that urban authorities are making efforts to formally integrate *favelas* into the city proper. Beginning in 1994, a plan was developed for Rio de Janeiro that anticipated the integration of the about 600 *favelas* into the city and that provided assistance for people to improve their housing. Although there have been many setbacks, this plan appears to be improving conditions in many areas. Of course, such ventures are not without critics; some Marxists argue that they serve to rationalize and perpetuate poverty while relieving governments of responsibility for the welfare of people.

population either live in slums or squatter settlements, or are homeless. Most homes lack proper sanitation, and garbage disposal is woefully inadequate. A 1991 survey identified an IMR of 123 (for comparison purposes, the 1991 IMR for the more developed world was 16, for the less developed world 81, and for India 95). The principal reasons for Calcutta's exceptionally high IMR are poor sanitation and overcrowding; specific causes of death include such avoidable problems as diarrhea, diphtheria, tetanus, and measles. Other problems reflect recent increases in industrial activity; it has been estimated that 60 per cent of the population suffer from respiratory tract problems because of poor air quality.

Consider *Kinshasa*. The Democratic Republic of Congo has large mineral reserves, extensive forests, and some good agricultural land, and has received huge amounts of development aid. The capital, Kinshasa, was a relatively efficient and thriving city until about 1990—but after that date the country and the city alike were plunged into chaos under a corrupt government that was not overthrown until 1997. Much of the aid money was diverted, ethnic tensions developed—partly as a consequence of the huge influx of Rwandan refugees in the early 1990s—and the capital city effectively ceased to function. In Kinshasa rioting began in 1991; both the economy and the urban infrastructure collapsed (the formal economy shrank by 40 per cent, inflation was 8,500 per cent in 1993 and 6,000 per cent in 1994); and diseases such as AIDS, tuberculosis, malaria, and cholera are spreading. A recent population estimate is 4.3 million.

Consider *Lagos*. The capital of Nigeria, with about 12 million people, combines serious physical and human problems; much of the site consists of swamp, and the land is unstable. Drainage and sewage problems worsen daily. With a population that is increasing rapidly because of larger economic issues prompting rural out-migration, the city is disease-ridden, congested, and virtually unlivable.

Consider *Cairo*. The largest city in the Middle Eastern region, with about 12 million people, treats less than half of its sewage: the remainder ends up in the Nile or in local

Poverty in Calcutta
(CIDA photo/Nancy Durrell McKenna).

lakes. Diarrhea and dysentery are common, and there is a real danger of typhoid and cholera outbreaks.

Consider *Bangui*. This city in the Central African Republic has a current population of about 500,000—yet its sewage system was built for a population of 26,000, and has never been extended or improved.

There is little to be gained by extending this catalogue of problems. However, it is clear that our earlier discussion of urban problems in the more developed world pales by comparison. The typically rapid and unplanned growth of less developed world cities leads to serious problems of poverty, housing supply, basic service provision, and criminal activity for much of the city area, and in most cases these difficulties are worsening. This point—made in a broader context in both Chapters 4 and 6 with reference to environmental issues and the prospects for improvements in global life—leads us directly into the 'brown agenda'.

The brown agenda

The issue of environmental conditions in the cities of the less developed world has been labelled the 'brown agenda' (Cohen 1993). This term is an explicit acknowledgement of the links between urban and economic growth and environmental deterioration. There are two principal concerns: conventional environmental health problems, such as limited land for housing and lack of services, and problems related to rapid industrialization, such as waste disposal and pollution. The significance of this twofold agenda is that cities are being challenged to address the second set of problems, which remain unsolved even by many cities in the more developed world, at a time when—as explicitly noted in the previous section—cities in the less developed world have not been able to solve the first set of much more fundamental concerns. Calcutta, as noted above, is a prime example.

The preceding section's outline of the problems facing a few cities gives substance to this brief, abstract sketch of the 'brown agenda'. To conclude these comments, it should be stressed that in all these problem situations it is the poor and underprivileged who suffer the most. There are close links between poverty, shelter, and the environment.

CHAPTER TWELVE SUMMARY

Why settle?

Permanent settlements were initially associated with economic changes, specifically the production of an agricultural surplus; they also reflect social needs.

Rural or urban?

This distinction is not a clear one. For 2002, 47 per cent of the world was classed as urban —75 per cent in the more developed world and 40 per cent in the less developed world.

Two key issues

Geographers concerned with settlement focus on why people settle where they do and on their way of life.

Patterns of rural settlement

Rural settlement patterns range from dispersed to nucleated. Nucleation, the oldest and most common pattern, may be regular or irregular. Dispersion is less common; in Europe it dates from the early eighteenth century, and it is typical of many temperate areas of European overseas expansion.

Rural settlement theory

The theoretical contributions of Bylund and Hudson provide simplified descriptions. Bylund explains locations in terms of distance and land attractiveness. Hudson compares settlements to seeds from plants and proposes a three-stage process, in which each stage prompts a specific pattern. Both theoretical contributions fail to consider a wide range of cultural and political variables.

Rural–urban differences and population movements

Especially in the more developed world, differences between rural and urban ways of life (*Gemeinschaft* and *Gesellschaft*) are decreasing. A transition from rural to urban is commonplace (especially close to urban areas) as urban residents move to rural areas in a process known as counterurbanization. Urban sprawl is also occurring. Today, there is evidence of both rural depopulation and repopulation.

Urbanization and urbanism

The world experienced a series of profound changes beginning about 1750: the rise of capitalism, great population increases, the emergence of nationalism, technological advances, and increases in both the size and the numbers of cities. In 1800, 3 per cent of the world's population lived in cities; by 1900 the figure had risen to 13 per cent, and in 2002 it was 47 per cent. This trend is increasing. The earliest cities developed about 3500 BCE as successors to large agricultural villages. The earliest cities (established before about 1760) and many cities in less developed areas can be conveniently labelled preindustrial: they function primarily as marketing and commercial centres. Urbanization was stimulated by the mercantilist philosophy that flourished in Europe from the beginning of European overseas expansion until the emergence of capitalism. With the rise of industrialization and capitalist philosophy, cities increased in number and size at a rapid pace. The locations of the industrial cities that emerged at this time were often determined by the location of the resources they depended on. For their inhabitants, ways of life in these cities represented a marked departure from earlier experiences.

Economic base theory

The basic sector of an urban economy consists of the multitude of activities that produce goods and services for sale outside the urban area; the non-basic sector consists of all those activities that produce goods and services for sale inside the city.

Urban planning

Most cities, especially in the more developed world, show the effects of some planning. The modern planning movement originated in Britain and the United States in the late nineteenth century. Major modernist figures include Le Corbusier and Frank Lloyd Wright.

Explaining urban locations

Many attempts have been made to explain why cities are located where they are. Traditionally, attention was focused on specific cities, but increasingly it is shifting to networks of cities. Most of the historical explanations that have been offered are most applicable to areas of European overseas expansion and identify a series of stages; examples include studies by Vance, Borchert, and Muller. The most influential and persuasive explanation, however, is the central place theory expounded by Christaller and Lösch. This theory, conceptually similar to the work of von Thünen, was published in 1933. A series of simplifying assumptions, combined with four key concepts, generates hypotheses concerning the size and spacing of cities. Many tests of these hypotheses have been conducted.

The rank size rule

This rule states that, in a given country, the population size of a city is inversely proportional to the rank of that city where the largest city is rank 1.

Changing urban theory

Major twentieth-century studies of the internal structure and identity of cities reflect a wide range of perspectives, including human ecology, neoclassical, Marxist political economy, Weberian, and humanistic. Most recently, postmodernists have questioned all such attempts at explanation on the grounds that any theory privileges one explanation above all others.

Inside more developed world cities

There are three basic models that purport to explain the internal structure of cities: the concentric zone, sector, and multiple nuclei models. Each appears to be valid to some degree, but all are of decreasing relevance to contemporary cities. Most recently, geographers have identified an emerging postindustrial city distinguished by such features as suburban office areas and gentrified inner areas.

Urban images and urban places

What image do residents have of the cities they inhabit? Lynch's work suggests that the typical resident recognizes five key features: paths, edges, districts, nodes, and landmarks. Most cities can be easily divided into areas of distinctive land use and even distinctive residential use. Residential areas that are different from other areas are most common in large immigrant-receiving cities; such differences reflect a number of cultural, social, and economic variables. Sometimes, parts of cities can be clearly demarcated as neighbourhoods. Much contemporary research recognizes the value of integrating spatial and social issues and is enriched by consideration of various social theories. Human geographers also recognize that cities are symbolic features that are often created to reflect power relations; this is especially true of capital cities.

A postmodern urban landscape

It may be that the traditional Chicago school model of urbanism, emphasizing structure and regularity, is less relevant to the contemporary city than a new Los Angeles–style model, emphasizing an almost random pattern of land uses. If such a transition in the urban landscape is occurring, it may be one component in a larger shift from modernism to postmodernism.

Global cities?

It is possible that two global processes—deconcentration of manufacturing and concentration of economic control—are contributing to the creation of global cities that are increasingly similar to one another, such as London, New York, and Tokyo.

Cities in the less developed world

The colonial heritage of many cities in the less developed world left them ill prepared to cope with the population explosions that occurred after about 1950. Many of these cities are experiencing social and environmental strains far more severe than those affecting cities in the more developed world.

LINKS TO OTHER CHAPTERS

- Globalization:
    Chapter 2 (concepts)
    Chapter 4 (ecosystems and global impacts)
    Chapter 5 (population growth, fertility decline)
    Chapter 6 (refugees, disease, more and less developed worlds)
    Chapter 7 (cultural globalization)
    Chapter 8 (popular culture)
    Chapter 9 (political globalization)
    Chapter 10 (economic globalization)
    Chapter 11 (agriculture and the world economy)
    Chapter 13 (industrial restructuring)

- Rural settlement location:
    Chapter 11 (second agricultural revolution world agricultural regions).

- Canadian prairie settlement:
    Chapter 7 (language)
    Chapter 8 (ethnicity).

- Rural settlement location theories; central place theory; intraurban theories:
    Chapter 2 (positivism; concepts of space, location, and distance)
    Chapter 11 (agricultural location theory)
    Chapter 13 (industrial location theory)
    Chapter 9 (distance).

- Rural depopulation:

    Chapter 6 (free migration)
    Chapter 11 (sectoral shift from agricultural to urban economies).

- Retailing and industry; postindustrial trends; impacts of industrial change on urban environment:
    Chapter 11 (industrial change).

- Postmodern processes:
    Chapter 8 (postmodernism).

- Images of the city; the city as image:
    Chapter 2 (humanism)
    Chapter 8 (landscapes as places).

- Colonial heritage in the less developed world:
    Chapter 6 (world systems analysis)
    Chapter 9 (colonial activity and the colonial experience).

- Urban growth in the less developed world:
    Chapter 5 (population growth).

- Chinese cities:
    Chapter 13 (industrialization in the less developed world).

- Urban problems in the less developed world:
    Chapter 4 (human impacts)
    Chapter 6 (less developed world).

FURTHER EXPLORATIONS

BERRY, B.J.L. 1981. *Comparative Urbanization: Divergent Paths in the Twentieth Century*. London: Macmillan.

An insightful overview of numerous urban issues by a leading geographer.

BOURNE, L.S., and D.F. LEY, eds. 1993. *The Changing Social Geography of Canadian Cities*. Montreal and Kingston: McGill-Queen's University Press.

An edited collection that addresses a range of social urban topics in the Canadian context; clear emphasis on contemporary social theory.

BRUNN, S.D., and J.F. WILLIAMS. 1977. *Cities of the World: World Regional Urban Development*. New York: Harper and Row.

A study in comparative world urban development that includes discussions of city growth, specific cities, problems of city living, and possible solutions.

BRYANT, C.R., and J.R.R. JOHNSTON. 1992. *Agriculture in the City's Countryside*. London: Frances Pinter.

A detailed survey that provides numerous insights into the changing land uses of the rural–urban fringe.

BUNTING, T., and P. FILION, eds. 1991. *Canadian Cities in Transition*. Toronto: Oxford University Press.

A substantial collection that considers Canadian cities from a wide range of urban geographic perspectives; includes evolutionary analyses, central place analyses, links with agriculture, industry, and transport, and a wide range of philosophical viewpoints.

CARTER, H. 1983. *An Introduction to Urban Historical Geography*. London: Arnold.

A thorough review of urbanism and urbanization through time; varied international content.

———. 1995. *The Study of Urban Geography*, 4th ed. New York: Arnold.

This fourth edition of a successful textbook maintains a traditional focus, beginning with the 'facts' of urbanism; it also reflects much of the work accomplished by urban geographers.

CLARK, D. 1982. *Urban Geography*. London: Croom Helm.

A well-written textbook covering all aspects of urban geography.

CLOKE, P., and J. LITTLE, eds. 1997. *Contested Countryside Cultures: Otherness, Marginalization, and Rurality*. London: Routledge.

An interesting collection of readings focusing on the social construction of rurality and the related exclusion of particular identities, 'others', from that social construction; examples include new-age travellers in Britain and lesbian groups in the United States.

CLOUT, H.D. 1972. *Rural Geography: An Introductory Survey*. Toronto: Pergamon.

An introductory survey focusing on Europe that highlights the diversity of rural geographic research topics.

DAVIES, W.K.D., and D.T. HERBERT. 1993. *Communities within Cities: An Urban Social Geography*. London: Belhaven Press.

A good urban geography textbook centred on the concept of community; includes an innovative account of community development.

GAD, G. 1991. 'Toronto's Financial District'. *Canadian Geographer* 35:203–7.

The first in a series on Canadian urban landscapes in this journal; later examples include the Toronto downtown, gentrification in Vancouver, markets in Ottawa, wholesaling in Edmonton, working-class suburbs in Hamilton, and neighbourhoods in Quebec City and Winnipeg.

GAYLER, H.J. 1990. 'Changing Aspects of Urban Containment in Canada: The Niagara Case in the 1980s and Beyond'. *Urban Geography* 11:373–97.

A detailed article focusing on a specific case-study and the wider implications of urban expansion.

HALL, P. 1996. *Cities of Tomorrow: An Intellectual History of Urban Planning and Design in the Twentieth Century,* 2nd ed. Oxford: Blackwell.

A scholarly and entertaining history of urban planning.

JAKLE, J.A. 2001. *City Lights: Illuminating the American Night*. Baltimore: Johns Hopkins University Press.

A very readable account of the history and consequences of public lighting in American cities; Jakle explains how lighting contributed to the modernization of cities, and stresses the place-enhancing qualities of lighting, especially with the development of a road network.

KNOX, P. 1994. *Urbanization: An Introduction to Urban Geography*. Englewood Cliffs: Prentice-Hall.

A substantial textbook that covers all aspects of urban geography; emphasizes contemporary changes in urban areas that involve decentralization and recentralization.

KNOX, P.L., and P.J. TAYLOR, eds. 1995. *World Cities in a World System*. Cambridge: Cambridge University Press.

An excellent book that provides a thorough and often innovative account of the contemporary urban world, setting the discussion of world cities in the Wallerstein world systems framework.

LEY, D. 1983. *A Social Geography of the City*. New York: Harper and Row.

An excellent textbook with full discussions of such issues as neighbourhoods and urban social change.

MARSHALL, J.U. 1989. *The Structure of Urban Systems*. Toronto: University of Toronto Press.

The most significant work to appear on the subject of central place theory since the 1960s; includes some clearly described and informative Canadian examples.

MITCHELL, R.D., and W.R. HOFSTRA. 1995. 'How Do Settlement Systems Evolve? The Virginia Backcountry during the Eighteenth Century'. *Journal of Historical Geography* 21:123–47.

A detailed research article that outlines three stages in the transition from dispersed agrarian communities to town-based settlement systems; relates these stages both to the change from subsistence to commercial production and to the desire of local settlers for town life.

SHORT, J. 1996. *The Urban Order: An Introduction to Urban Geography*. Oxford: Blackwell.

A useful textbook that addresses the various concerns raised in this chapter in much greater detail.

SMITH, R.H.T., E.J. TAAFE, and L.J. KING, eds. 1968. *Readings in Economic Geography: The Location of Economic Activity*. Chicago: Rand McNally.

Although dated, this volume includes excellent statements of location theory and a selection of important empirical studies; relevant to Chapters 10, 11, 12, and 13.

STELTER, G.A., ed. 1990. *Cities and Urbanization: Canadian Historical Perspectives*. Toronto: Copp Clark Pitman.

A useful collection of essays, including a good discussion of settlement system evolution.

WILLIAMS, S.W. 1997. 'The Brown Agenda'. *Geography* 82:17–37.

An excellent article that provides a succinct account of the links between poverty, shelter, and environment in the cities of the less developed world: includes a detailed consideration of Calcutta.

On The Web

http://www.bestpractices.org/

This United Nations site contains over 1,100 proven solutions, from more than 120 countries, to the common social, economic and environmental problems of an urbanizing world.

http://www.canurb.com/

This home page of the Canadian Urban Institute includes general information on the Canadian urban situation.

http://www.urban.org/

This is the site of a 'non-partisan economic and social policy research organization' that includes discussions of urban problems and issues.

CHAPTER 13

Industry and Regional Development

Why are industries located where they are? In this chapter, the answer once again takes the form of a neoclassical economic theory that displays all the advantages and disadvantages of that approach; the conceptual similarities with both agricultural and urban location theories are clear.

Following a look at the origins and evolution of twentieth-century industrial land-scapes—landscapes that (like so many other phenomena of interest to human geographers) owe their basic character to the industrial revolution—we turn to the world industrial map, emphasizing sources of energy, the differences between the more and less developed worlds, and the dramatic rise of Japan and the newly industrializing countries. Next, our discussion of the industrial restructuring that has been apparent since the 1970s (and was introduced in the previous chapter) emphasizes the growth and increasing concentration of service activities—a process that requires consideration of the transition from Fordism to post-Fordism and the deindustrialization evident in the more developed world, as well as a variety of theoretical perspectives, especially forms of Marxism. Following is an account of tourism and recreation that highlights commodification and trends in the area of consumption (as opposed to production). The chapter concludes with the geography of uneven development and the efforts that states have made to plan industrial growth so that it favours certain regions at the expense of others.

Two Tibetan girls in northwest China's Qinghai province stop to let tourists take pictures of them with their lambs, August 2000 (AP photo/Xinhua, Tang Zhaoming).

In the study of spatial variations and changes in economic activity, it has been customary for human geographers to distinguish three levels of economic activity. *Primary* production—activities such as farming, fishing, forestry, and mining—occurs where appropriate resources are available. *Secondary* production converts primary products into other items of greater value. This process is called manufacturing (literally, making by hand) and requires three things: a labour force, an energy supply, and a market. Most manufacturing is done in factories, and the result is a uniform product. *Tertiary* economic activities are those involved in moving, selling, and trading the goods produced at the first two levels, as well as activities such as professional and financial services. Recently, it has been helpful to identify a fourth level: *quaternary* activity is economic activity that specializes in assembling, transmitting, and processing information and controlling other business enterprises. It includes professional and intellectual services, such as those provided by management consultants and educators. Quaternary-sector employment is the favoured growth area in the postindustrial city discussed in Chapter 12.

Like any classification, this one is not ideal, and the divisions between the four levels are not absolute; but it is still useful. For our purposes, it is convenient to include all four kinds of economic activity in this chapter on industry, with two exceptions: transport-related activities and farming (discussed in Chapters 10 and 11 respectively).

The Industrial Location Problem

Locational questions and theories are central issues for geography, and they are just as important for industry as for agricultural and settlement issues. Once again, theories need to be considered in the context of particular social and economic systems, especially when we realize that industries strive to minimize costs, including labour costs, and maximize profits. Many different types of economic organization play roles in the industrial process. Even the household plays a role in production and reproduction, although that role is less important today than it was in the past; firms, which make up the commercial sector, are a second type of economic organization. Industrial location theory explains why firms locate their factories where they do (Box 13.1). Some firms are owned by one person; others are partnerships, cooperatives, or corporations. Corporations play the most important role in a country's economy.

Early Location Theory

For economists such as Adam Smith, J.S. Mill, and David Ricardo, industrial location was related to the location of agricultural food surpluses that could be used to feed industrial workers. Probably more important was Adam Smith's realization that under the new doctrine of capitalism, the central consideration in location decisions was purely financial. There is little doubt that this incentive remains paramount today. In some form, it has been at the heart of most industrial location theory since the writings of Adam Smith. The only major exception to this generalization is Marx, who saw industrial location as one of many issues that were best explained in terms of political inequalities. The classic industrial location theory, however, is the least-cost theory of the economist Alfred Weber.

Least-Cost Theory

The first attempt to develop a general theory of the location of industry was that of the German economist Alfred Weber (1868–1958), brother of the sociologist Max Weber. This work, published in 1909 and translated into English in 1929 (Friedrich 1929), is another example of a theory that does not purport to summarize reality. Rather, it aims to prescribe where industrial activities ought to be located; it is a normative model.

Weber follows the von Thünen tradition of setting up a number of simplifying assumptions:

1. Some raw materials are ubiquitous; that is, they are found everywhere. Examples suggested by Weber are water, air, and sand.
2. Most raw materials are localized; that is, they are found only in certain locations. Sources of energy fall into the localized category.
3. Labour is available only in certain locations; it is not mobile.
4. Markets are fixed locations, not continuous areas.

5. The cost of transporting raw material, energy, or the finished product is a direct function of weight and distance. Thus the greater the weight, the greater the cost; similarly, the greater the distance, the greater the cost.

6. Perfect economic competition exists. This means that the industry consists of many buyers and sellers, and no single participant can affect product price.

7. Industrialists are economic operators (see glossary) who are interested in minimizing costs and maximizing sales.

8. It is assumed that both physical geography (climate and relief) and human geography (cultural and political systems) are uniform.

These assumptions are clearly unrealistic, but they are necessary for Weber to arrive at a least-cost location format. Given these assumptions, industrialists will locate indus-tries at least-cost locations in response to four general factors: transport and labour (interregional factors) and agglomeration and deglomeration (intraregional factors).

Transport costs

Weber's first concern was transport cost. To find the point of least transport cost, two factors must be assessed: distance and the weight being transported. According to Weber, the point of least transport cost is the location where the combined weight movements involved in assembling finished products from their sources and distributing finished products to markets are at the minimum.

Weber also introduced the **material index**: the weight of localized material inputs divided by the finished product weight. If an industry has a material index greater than 1, then the point of least transport cost, and therefore the appropriate location, will be

---

13.1

FACTORS RELATED TO INDUSTRIAL LOCATION

Traditionally, a number of factors have been considered in the typical industrial location decision.

Various measures of *distance* have been very important. Distance from sources of raw material has become less important today as a result of improvements in transport and changing industrial circumstances, but it used to be a major consideration because such materials are unevenly distributed and vary in quality, quantity, cost of production, and perishability. Similarly, distance from an energy supply is now declining in importance: during the nineteenth century, it was essential to locate factories near a source of coal, but since then, the emergence of new energy sources such as electricity has freed industry from this locational constraint. Distance from market is another factor. Like the other distance variables, it tends to decrease in relevance with improvements in transportation.

Availability of *labour* also shows spatial variations in quality, quantity, and cost. Although theoretically perfectly mobile, labour is often limited to specific locations for social reasons or because of political restrictions on movement. *Capital*, in the form of tangible assets (such as machinery) or intangible assets (such as money), can be a key locational consideration; capital may be mobile. Finally, *transport* was traditionally a key factor, linked to several of the variables already noted.

Each of these factors plays a role in the typical location deci-sion, although the proportions of each required for a particular production process may vary; in some cases, for example, capital might be substituted for labour. When an industrialist has many options, the number of feasible locations is increased.

Also important for understanding industrial location are factors related to the nature of the product, as well as various internal and external economies. With regard to *internal economies*, each manufacturing activity has an optimum size and benefits may be gained from either horizontal integration (enlarging facilities) or vertical integration (involvement in related activities at one location). *External economies* often result when similar industries are located in close proximity.

Finally, industrial location decisions often reflect various human and institutional considerations. The most important human factor is *uncertainty*. Simply put, decision-makers do not have adequate information about all the relevant factors. They may wish to minimize their costs, but they are unlikely to be able to do so. We have already distinguished between optimiz-ing and satisficing behaviour, and in the current context this dis-tinction is crucial. Location decisions, then, often reflect uncer-tainty, individual judgements, and chance factors. The central institutional factor is the role played by *government*. There is often a conflict between economic logic and what governments see as socially desirable.

Figure 13.1 A locational triangle. The position of the least-cost location (*L*) will depend on the relative attractiveness of each of *R₁*, *R₂*, and *M*.

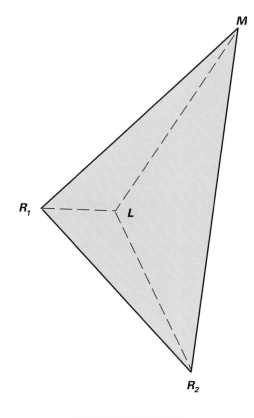

two raw material sources (R_1 and R_2) and a market (M), and the industry locates according to the relative attraction of each of the three. If only ubiquities are used, then the locational figure is reduced to one point, the market. If only one raw material is used and that material loses no weight in the production process, then the industry can locate at the material source, at the market, or anywhere on a straight line between the two. Clearly, the locational figure is needed only in those cases where weight-losing materials are used —but this is the most common type of manufacturing. Thus the locational figure will usually have more than three sides, generating a complex problem. Geographers usually solve such locational problems algebraically rather than by referring to diagrams.

nearer the localized material sources; if an industry has a material index less than 1, then the point of least transport cost, and therefore the appropriate location, will be nearer to the market. The historical importance of orientation to material resources has declined because today fewer industries use heavy, bulky materials and because transportation has improved.

In general, Weber's predictions are correct. Webber (1984:57) gives the example of the soft drink industry where 1 ounce of syrup, a ubiquitous material (water), and a 4-ounce can combine to produce 12 ounces of soft drink and a 4-ounce can. Thus the material index is $1 + 4/12 + 4 = 0.31$. This material index is less than 1, and indeed the soft drink industry is typically market-oriented. By contrast, industries such as the smelting of primary metals remain predominantly material-oriented, as do many agricultural processing industries, because in them the cost of materials still forms a substantial share of the total cost.

Locational figures

To tackle such problems, Weber used a graphic approach. A simple version of his figure is shown in Figure 13.1. In this case there are

Labour costs

Although transport costs are the key to Weber's work, he also acknowledged the importance of labour costs. Such costs represent a first distortion of the basic situations already described and can be approached by mapping the spatial pattern of transport costs and then comparing this pattern with that of the relevant labour costs.

To achieve this comparison, Weber introduced the concepts of **isotims** (lines of equal transport costs around material sources and markets) and **isodapanes** (lines of equal additional transport cost drawn around the point of minimum transport cost). Maps of isotims and isodapanes are examples of cost surfaces. Figure 13.2 shows isotims around two material sources and one market, while Figure 13.3 superimposes isodapanes on the isotims. Weber identified a critical isodapane, outside of which industrial location would not occur—industries can be attracted to low-labour-cost locations only if they lie inside the critical isodapane. A low-labour-cost location will attract industrial activity only if the savings in labour cost exceed the additional transportation costs of moving the raw materials to, and the finished products from, the low-cost labour site.

Agglomeration and deglomeration

Transport and labour are both interregional considerations. Agglomeration and deglomeration (see glossary) are intraregional con-

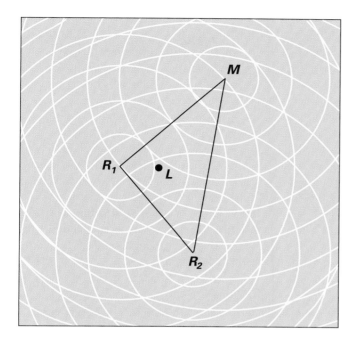

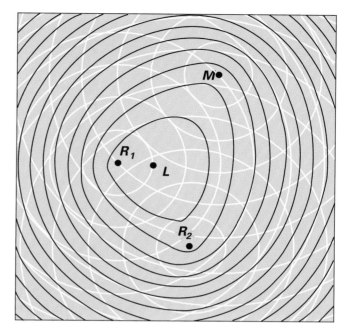

siderations that, like labour, can potentially cause deviation from the least-transport-cost location. *Agglomeration* economies result from locating a production facility close to similar industrial plants, allowing plants to share equipment and services, and generating large market-areas that aid the circulation of capital, commodities, labour, and information. *Deglomeration* economies are the reverse of agglomeration economies and result from location away from congested and high-rent areas.

Least-cost theory: An evaluation

Weber's theory is normative and does not attempt to describe or summarize reality. The major criticisms of it are directed at its simplifying assumptions. Markets, for example, are not the simple fixed points that Weber chose to assume. Labour markets are characterized by discontinuities associated with age, gender, ethnicity, and skill. Nevertheless, Weber's basic logic is simple and sound (Box 13.2).

Hoover (1948) continued Weber's focus on transport costs and added substantially to it. The Weberian locational figure changes when transport costs are not directly proportional to distance. The cost per unit of distance is less for long hauls than for short hauls (Figure 13.4), and a step structure is often used (Figure 13.5).

Market-Area Analysis

According to proponents of market-area analysis, Weber's fundamental error is his assumption that industrialists seek the lowest-cost location. Rather, for the market-area analyst, industrialists are profit maximizers. Least-cost theory is a form of variable cost analysis (concerned with spatial variations in production costs), while the market-area argument is a form of variable revenue analysis (concerned with spatial variations in revenue).

In market-area analysis, the location that will result in the greatest profit can be determined by identifying production costs at various locations and then taking into account the size of the market area that each location is able to control. The argument is that industries will attempt to monopolize as many consumers as possible—they seek a **spatial monopoly** and, in doing so, exhibit **locational interdependence**. A classic example of this argument is the Hotelling (1929) model, which proposes that the agglomeration of industry is a logical outcome of certain demand conditions (Box 13.3).

This argument about seeking a spatial monopoly is relevant, but if least-cost theory tends to de-emphasize demand, then market-area analysis tends to de-emphasize factors other than demand. The most important market-area theorist is Lösch, who (like

Figure 13.2 (left) A simple isotim map. The diagram shows isotims around the two raw-material locations *(R₁* and *R₂)* and the market *(M)*.

Figure 13.3 (above) An isodapane map. Superimposing isodapanes on an isotim map indicates how far from *L* an industry may locate to take advantage of a particular low-cost labour location.

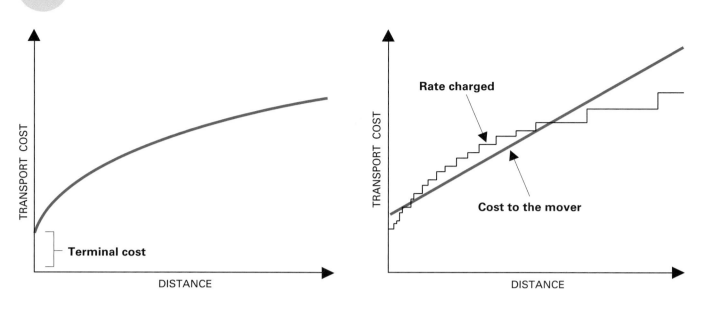

Figure 13.4 (above) Transport cost and distance. Note that the cost does not begin at $0, because there is a basic terminal charge.

Figure 13.5 (right) Stepped transport costs. Here the rate charged does not relate directly to distance because of the rate structures used by the movers.

Christaller) determined that the ideal market area is hexagonal. Like least-cost theory, market-area analysis is a normative approach.

Behavioural Approaches

Normative approaches allow us to identify rational or best locations and understand aspects of actual locations through comparison. But is such an approach satisfactory? Many geographers think not, and prefer to focus on explicit analyses of why things are where they are rather than on abstract analyses of where they ought to be, given certain assumptions. Normative approaches consider why things are where they are only in an indirect fashion. A set of approaches often labelled behavioural attempt to correct this situation.

In brief, behavioural approaches centre on the subjective views of individuals. The concept of the economic operator is rejected and the concept of satisficing behaviour (see glossary) is paramount. A simple example demonstrates the relevance of this approach. The beginning industrialist does not really debate where to locate; rather, the selected

---

13.2

TESTING WEBERIAN THEORY

Least-cost industrial location theory has been tested many times. This box discusses a classic example: the application of Weberian concepts to the particular industrial locations of the Mexican steel industry in the 1950s (Kennelly 1968). The choice of an example dating back half a century is deliberate, since least-cost theory was much more relevant then than it is today. The choice of the iron and steel industry is also deliberate, because least-cost theory can still be appropriately applied to this type of heavy industrial activity.

In 1950 the Mexican steel industry consisted of two main steel plants and a number of other plants. Production covered a range of products and accounted for about 50 per cent of Mexican requirements. Location of plants, particularly the major ones, was closely related to the availability of raw material resources, such as iron ore, coke, oil, and market scrap. The location of labour appeared to be of minimal importance, as unskilled labour was quite mobile. Access to market was important, however, with major markets in Monterey and Mexico City. After determining least-cost locations, Kennelly (1968) reported that, overall, the 1950 Mexican steel industry was in accord with basic Weberian principles.

Weberian theory is useful in explaining the locations of industries of this type, but its logic weakens with technological change and is less relevant to most contemporary location decisions than to those of the past. Other (especially Marxist) approaches offer insights into industrial location decisions that the traditional theories do not attempt to consider.

location is the existing location of the industrialist (or the nearest feasible location). Oxford, England, became the home of Morris, a major automobile manufacturer, for this non-least-cost and non-market-area reason. The location selected was the home area of the industrialist William Morris. Admittedly, had that home area been a remote Scottish island, then some other location would have been selected. Specific locations are often chosen for subjective reasons.

Contemporary industrial firms, especially large corporations, are complex decision-making entities. Location decisions made by small, often single-owner, firms are less likely to be economically rational than those made by large firms, but other considerations compete with least-cost and maximum profit logic for all firms. It is also clear that decision-makers do not typically have access to all relevant information and, furthermore, that such information is subject to various interpretations. Geographers are well aware of the less than perfect mental maps (see glossary) held by decision-makers. Although behavioural issues are not as easily theorized or quantified as the more directly economic issues, they are always factors to be considered in any analysis of industrial location.

The Industrial Revolution

Before the industrial revolution, industrial activity was important but limited. Every major civilization was involved in a few basic industrial activities such as bread-making, brick-making, pottery, and cloth manufacture. These activities were located in the principal urban centres and close to raw materials. In addition, each household was an industrial organization that produced clothing, furniture, and shelter. All industries involved minimal capital and equipment, were structured simply, and were small in size and output. Energy sources (such as wood) and raw materials (such as agricultural products and sand) were relatively ubiquitous, not localized. Given these characteristics, industries could be dispersed, and as a result, there was little evidence of an industrial landscape. Transport systems were limited and markets for products were local rather than regional or national. Everything changed, however, as a result of the developments that we call the industrial revolution (Box 13.4).

13.3

LOCATIONAL INTERDEPENDENCE: THE HOTELLING MODEL

In 1929 the American economist Harold Hotelling considered the idea that industrial firms producing similar products might be locationally interdependent because of shared demand. Assuming that any individual location decision must be tied to those of rival suppliers, Hotelling (1929) developed a model to show how one firm is attracted to the location of another firm. The classic simple example is that of two ice-cream sellers on a beach. In this situation, two producers are competing to supply identical goods to consumers evenly spread along a linear market, and there is the assumption that production costs are uniform, ice-cream selection is uniform, and demand is uniform. If the two vendors agree to locate back to back at the centre of the beach, each will be able to serve half of the total market; such locational decisions allow each producer to supply consumers at the extremities of the market (the two ends of the beach) without any loss of marketing advantage.

How does this locational pattern come about? Initially each vendor may locate at the centre of one half of the beach. Such a pattern is socially optimum, as it minimizes the average distance that consumers have to travel to purchase ice-cream and it also benefits each vendor equally, allowing each to serve half the beach. But if one of the two vendors moves towards the centre, then that vendor will be able to sell not only to her original half, but to a part of the other half as well; as a result, the other vendor will also attempt to move towards the centre. The eventual result is an equilibrium pattern with both vendors back to back at the centre of the beach.

The fundamental advantage of this model is that it provides an alternative to Weber's explanation of agglomeration. It is, of course, a highly simplified situation, based on identical products, identical production costs, the price of the product covering the cost of transport for the consumer, a linear market, and an unchanging demand. Furthermore, it does not allow for locational inertia—not many industrial firms are as footloose as ice-cream vendors!

Origins

Between 1760 and 1860, the industrial geography of England was dramatically transformed. In fact, England was the first industrial area. Elsewhere in western Europe the transformation did not begin until the nineteenth century, but was accomplished over shorter periods of time.

In general, the term 'industrial revolution' is apt, for it was a period of rapid and cumulative change unlike anything that had occurred previously: machines replaced hands, and inanimate energy sources were efficiently harnessed. However, the term may be misleading to the extent that it suggests that the process of industrialization took the same form in every case. In truth, the process was uneven, and other countries did not simply imitate the English example. It is also misleading to tie down the beginning of the industrial revolution too specifically. The English beginnings date back to the late sixteenth century, and traditional craftsmen were still very important as late as the mid-nineteenth century. Belgium followed the English example most closely, but—unlike England—emphasized the metallurgical industries far more than the textile industries.

13.4

THE PERIOD OF THE INDUSTRIAL REVOLUTION

Earlier in this book, we have referred to the industrial revolution even in contexts without any direct connection to industry. The reason is that the industrial revolution involved much more than industrial change. This box summarizes some of the principal changes associated with that revolution, which originated in Europe around the middle of the eighteenth century and continued until roughly the middle of the nineteenth. Collectively, these changes have contributed much to our contemporary world.

'Industrial revolution' is the name given to a series of technical changes that involved the large-scale use of new energy sources, especially coal, through inanimate converters. After the first successful steam engine was built in 1712, other new machines were able to use steam to improve industrial output. The introduction of new machines led to the construction of factories close to the necessary energy sources, raw materials, or transport routes. The resulting concentrations of factories promoted urban growth, including the emergence of distinct industrial landscapes and working-class residential areas, since workers needed to live near their places of employment. Industrial cities replaced the earlier preindustrial forms of settlement. As we saw in Chapter 10, to move raw materials to the factories and finished products to the markets, new transport links were essential; in England they initially took the form of improved roads, then canals, and, most critically, railways. Agriculture became increasingly mechanized, and there was a demand for new raw materials, especially wool and cotton, from overseas areas. Overall, the percentage of the total labour force engaged in agriculture declined rapidly.

Technical changes were accompanied by significant changes in demographic characteristics. As described in the demographic transition model, the first phase of the industrial revolution involved a dramatic reduction in death rates, followed by reductions in birth rates. The interval between these two drops was a period of rapid population growth: the world population was estimated at 500 million in 1650 and at 1,600 million by 1900. Industrialization and the associated population increases contributed to new areas of high population density, rural-to-urban migration, rapid urban growth, and considerable European movement overseas to the new colonial territories.

A major social and economic change was the collapse of feudal societies and the rise of capitalism. As labour was transformed into a commodity to be sold, the producer became separated from the means of production. Politically, this period witnessed the rise of nationalism and the emergence of the modern nation state, as well as the expansion of several countries around the globe to create empires. Although these empires were short-lived, they contributed enormously to the contemporary political map—and to the division between more and less developed worlds. The nation state, as we have already seen, remains one of the most potent forces in the contemporary world. The rise of modernism was linked to the industrial revolution through a common emphasis on the practicality and desirability of scientific knowledge. The transition to an industrial way of life was also a transition from tradition to modernity, from *Gemeinschaft* to *Gesellschaft*. Even though new postindustrial forces have emerged since about 1970, the world in which we now live is unmistakably a product of the industrial revolution.

There were also important advances in the organization of academic knowledge during this period; indeed, the social sciences are themselves related to these various changes. Geography was one of these disciplines. Major contributions to the advancement of knowledge included the works of Marx and Darwin.

The most important component of the 'revolution' was the rise of large-scale factory production. Developed out of earlier small establishments and domestic household activities, factories were both much larger and more mechanized; they also required more capital. Further, because they relied on new, localized energy sources, they tended to agglomerate. The key energy source was coal, and the first truly industrial landscapes arose in the coalfield areas of England. But these developments were only one part of the industrial revolution.

The late eighteenth century witnessed the rise of a new economic system that emphasized individual success and profits: capitalism. The new capitalists seized the opportunities provided by a whole series of technological developments in the metals and clothing industries: a furnace that burned coal to smelt iron (first used in 1709); a spinning jenny for multiple-thread spinning (1767); a steam engine used as an energy source (1769); an energy loom for weaving (1785). These are only a few examples of the new inventions, most of which were developed in England.

Recall our definition and discussion of energy in Chapter 4: energy is the capacity to do work, and the more successfully we can use forms of energy other than our own, the more successfully we can achieve our goals. Also, recall our definition of technology as the ability to convert energy into forms that are useful to us. The fact that the first agglomerations developed on the coalfields of northern and central England and in the traditional areas of cloth manufacture in northern England reflected the importance of coal for energy and the focus on the metal and clothing industries. These industrial agglomerations caused rapid increases in urban populations as workers moved closer to places of employment.

The basic logic of the Weberian least-cost theory is helpful in understanding these developments. Using localized energy sources meant that industries were located at those sources and, if possible, at the sources of raw materials. The need to reduce costs meant that the industrial revolution was paralleled by a transport revolution. In England, first the road system was improved, then the canal system, and finally railways were built.

To clarify some of the generalizations above, let us take a closer look at examples of nineteenth-century industrialization, especially the early British iron, steel, and wool industries.

Early Industrial Geography

Analyses of these early industrial landscapes include traditional descriptive accounts based on Weberian theory and other theoretical accounts that acknowledge the role played by economic power and organizations. Box 13.5 examines the Tyneside coalfield in northeast England in the light of concepts of power and organization.

Iron and steel

Prior to the industrial revolution, Britain had two principal centres of iron production— one in central England and one in southern England. Both used local ores and relied on wood (charcoal) as their energy source. The iron industry expanded rapidly after coal was substituted for charcoal from 1709 onward and an efficient steam engine was developed. As iron production increased, new uses for it were found: the first cast-iron bridge was built in 1779 in central England, and most new industrial machinery was constructed of iron.

By the early nineteenth century, new iron areas dominated as technological advances continued to favour coalfield locations such as south Wales and north-central England. An important new factor emerged after 1825 when the first steam-powered train was constructed in northeast England. The railway boom reached its height in the 1840s and gradually allowed some movement away from coalfields. Railways, together with the discovery of new iron ore sources, allowed a new industrial centre to develop in northeast England. Other major mid-nineteenth-century iron and steel areas were located in the English Midlands, south Wales, and central Scotland.

Textiles

Before the industrial revolution, the woollen industry was specialized and spatially concentrated in three areas: southwest England, east-central England, and northern England. The manufacturing process had remained unchanged from the fourteenth century to the late eighteenth century, when various inventions that were part of the larger industrial revolution prompted radical changes. Of the two major textile industries, cotton was the first to experience change, since most of the new procedures were more easily applied to cotton than wool. In both industries, the major changes followed the same pattern: a decline in domestic production and a rise in factory production; use of steam engines; rapid expansion in northern England, reflecting the availability of coal; and associated declines in the other two areas. In the case of the textile industry, the principal technological breakthroughs occurred in villages and small towns in northern England, as mechanized factories were constructed near suitable sources of water.

In the case of the cotton industry, raw material came initially from the East Indies, but supplies were irregular and a new source had to be found. The Caribbean and southern United States were suitable source areas, but the task of separating cotton fibres from seeds was so labour-intensive that, even with a slave economy, supplies were limited. The invention of the cotton-gin in 1793 allowed American plantations to flourish and English factories to be adequately supplied. The cotton mills of northern England became so

dependent on imports from the United States that mill workers were laid off when supplies were reduced during the American Civil War. In the case of wool, the new area of supply was southeastern Australia.

Industrial landscapes

The industrial revolution and related rapid population and urban growth combined to create new landscapes. Factory towns arose, especially in northern England, and migrants from the south and rural areas poured into the new employment centres. In environmental and social terms, the results were often disastrous.

Before the onset of industrial change, the landscape was agricultural and predominantly rural. Even before the introduction of steam energy, however, some industrial production was shifting from the household to factories and the landscape was becoming more congested. Probably the first real factory in Britain was a silk mill in north-central England that was completed in 1722; five to six storeys high, it employed 300 people and used water energy. This example shows that coal was not essential for industrialization;

13.5 GEOGRAPHICAL CHANGE AND INDUSTRIAL GROWTH: THE TYNESIDE COALFIELD

Understanding the changing geography of the Tyneside coalfield requires not only a grasp of traditional industrial location concepts but also an awareness of the larger social, political, and economic framework.

In the second half of the eighteenth century, the Tyneside coalfield in northeast England dominated the British coal industry. Capital and control in the coal industry were concentrated in a merchant guild, known as the Company of Hostmen, that had a monopoly on coal trading and was able to use it to dominate local government. Between 1700 and 1726, the coalfield underwent significant spatial expansion as the shallow, easily worked seams became exhausted; sinking deeper pits was not a technological option at the time. Much of this expansion was accomplished by capitalists who realized the value of controlling the various factors of production, especially land. Those responsible for expansion formed a group known as the Grand Allies in 1726. By 1750 the Grand Allies controlled production in the region and were regularly in conflict with other capitalists in

their efforts to introduce a price- and output-fixing arrangement.

The power exerted by the Grand Allies declined after 1770, by which time they controlled just under half the collieries. Their decline in power was in part the result of their failure to anticipate the development of new technology after 1750 that would permit expansion into new coal areas; instead, they had invested heavily in coal areas that were best exploited using older technology. As a result of technological change, therefore, the Grand Allies found themselves owning inappropriate coalfield areas. At the same time, changes in the economic environment, especially the banking system, helped to destroy their monopoly. The Grand Allies' place was quickly filled, however, by another capitalist organization, a form of cartel known as the 'Limitation of the Vend'.

As this brief summary makes clear, the Tyneside coal industry is best understood by reference not only to Weberian and market area arguments but also to social, political, and economic forces (Cromar 1979).

but it certainly accelerated the process of change. The first factory was followed by many others as the century progressed, but they produced neither smoke nor dirt. It was steam energy that resulted in major landscape change. More factories meant more workers crowded into small areas, and the use of coal quickly polluted the landscape. Land was expensive, and the areas of small terraced houses where workers lived quickly deteriorated into slums. These industrial cities were subject to heavy smoke pollution—a situation that in many cases was not addressed until the 1960s.

Diffusion of industrialization

After about 1825, the technological advances developed in Britain diffused rapidly to mainland Europe (especially Belgium, Germany, and France) and North America. Britain did not welcome the spread of its industrial innovations, and even tried to keep some advances secret; but it did not succeed. In the United States, Pennsylvania and Ohio became the early industrial leaders because of their high-quality coal sources, while in the Russian empire, Ukraine became the driving force when coal was discovered there. Japan began to industrialize after establishing cultural contact with the United States and rejecting feudalism in 1854.

In Europe the Ruhr region offered not only coal and small local iron ore deposits, but an excellent location for water transport. As the nineteenth century progressed, the demand for labour became so great that immigrants moved to the Ruhr from elsewhere in Europe, giving the region an unusually diverse ethnic character.

World Industrial Patterns

Sources of Energy

As we saw in Chapter 4, most energy today comes from fossil fuels: oil, natural gas, and coal. Since the beginning of the industrial revolution, coal had dominated. In the 1960s, however, oil replaced coal as the most important energy source globally. The peak period for oil was in the early 1970s, when it provided almost 50 per cent of the world's energy; since then, oil use has declined to about 40 per cent because of high prices, occasionally erratic supplies, more efficient use, and

The Tin Fouye Tabankort gas processing plant is the high-tech showpiece of Algeria's natural gas industry (AP photo/Bruce Stanley).

increasing reliance on other energy sources. However, oil continues to be the chief source of industrial energy, and the geographies of oil reserves, production, and consumption are important aspects of contemporary life.

A few countries tend to dominate world production of oil. In 1960 the major sources were the United States, Russia, and Venezuela; only about 15 per cent came from the Middle East and north Africa. Today the Middle East and north Africa dominate oil production, with about 40 per cent of the global total. Similarly, a small number of countries dominate in the production of natural gas: together the US (22.5 per cent), Russia (22.0 per cent), and Canada (7.0 per cent) produce more than 50 per cent of the global total; no other country produces more than about 4 per cent. Five countries are now responsible for about 70 per cent of world coal production: the US (26.3 per cent), China (24.4 per cent), India (7.2 per cent), South Africa (5.6 per cent), and Russia (5.4 per cent).

The International Energy Agency expects that by 2020 total world energy consumption will increase by about 50 per cent from current levels. Table 13.1 shows projected demand for the major energy sources for the years 2010 and 2020: oil is expected to fall from the current 40 per cent to about 37 per cent, natural gas to rise from the current 22 to 24 per cent, and coal to remain steady at about 23 per cent. In percentage terms, other energy sources such as nuclear and hydro are not expected to change significantly. Most of the demand growth will be in the less developed world because of population increase and industrialization. Table 13.2 provides statistical summaries of energy production, with reference to both types of fuel and global regions, comparing 1973 and 2000.

OPEC

The most important player to be identified in any account of global energy is the Organization of Petroleum Exporting Countries (OPEC), a cartel founded in 1960 by Iran, Iraq, Kuwait, Saudi Arabia, and Venezuela, which were later joined by Algeria, Gabon, Indonesia, Nigeria, Qatar, and the United Arab Emirates. OPEC's principal purpose is to set levels of production in order to determine the price of oil. Thus OPEC is able to wield enormous influence on the economies of more developed countries; a price increase in 1973, for example, was a leading cause of a widespread recession. Today OPEC works to keep the price of oil around US$22–28 per barrel. Box 13.6 offers capsule comments on the ten countries with the largest known reserves of oil.

So far, there are few substitutes for gasoline. Countries such as Brazil and Canada are converting agricultural products to alcohol that can be blended with gasoline, while some areas, such as Switzerland, are increasingly experimenting with electric cars. Different countries generate electricity in different ways. Nuclear reactors are heavily used in France, Belgium, and South Korea; Denmark is increasingly experimenting with wind; and many countries, such as Norway, India, New Zealand, and Canada, as well as countries in South America and Africa, rely primarily on hydroelectric power (Box 13.7).

Table 13.1   **OUTLOOK FOR WORLD TOTAL PRIMARY ENERGY SUPPLY, 2010 AND 2020 (%)**

Fuel	2010	2020
Oil	36.9	37.0
Coal	22.7	22.6
Natural gas	21.9	23.9
Nuclear	5.6	4.2
Hydro	2.3	2.3
Other	10.6	10.0

Note: Total primary energy supply (TPES) is made up of indigenous production, plus imports, and minus exports.

Source: BP Statistical Review of World Energy, 2002.

Table 13.2   **REGIONAL SHARES OF ENERGY PRODUCTION BY TYPE, 1973 AND 2000 (%)**

Region	Oil		Natural gas		Hard coal		Nuclear		Hydro	
	1973	2000	1973	2000	1973	2000	1973	2000	1973	2000
Middle East	36.8	29.7	2.1	9.1					0.3	0.5
OECD countries	23.7	28.3	71.5	43.4	49.7	39.7	92.8	86.6	71.3	51.2
Former USSR	15.0	11.9	19.7	27.8	22.8	8.4	5.9	8.4	9.4	8.5
Africa	10.1	10.5	0.8	5.3	3.0	6.0			2.2	2.7
Latin America	8.6	9.8	2.0	4.2	0.3	1.5			7.2	20.4
Asia	3.2	5.0	1.0	8.3	5.1	12.4	1.3	2.2	4.6	6.6
China	1.9	4.6	0.4	1.3	18.6	33.8			2.9	8.2
Non-OECD Europe	0.7	0.2	2.5	0.6	0.4	0.1		1.1	2.1	1.9

Notes: 1. The OECD (Organization for Economic Cooperation and Development) is an international organization devoted to helping governments tackle the economic, social, and governance challenges of a globalizing economy.

    2. This table separates China from the rest of Asia and treats it as a region on its own.

Source: International Energy Authority, 2003. http://www.iea.org/statist/keyworld2002/key2002/keystats.htm

13.6

OIL-PRODUCING COUNTRIES

Ranked by quantity of known reserves, the ten most important oil-producing countries are as follows (note that the first five countries are all in the Middle East):

- *Saudi Arabia.* The world's largest producer of oil (8.8 million barrels per day), with by far the largest known reserves (261.8 billion barrels—about 25 per cent of the global total), Saudi Arabia is nevertheless economically and politically dependent on the United States.
- *Iraq.* Iraq has the second-largest proven reserves, 112.5 billion barrels (less than half the Saudi total and almost 11 per cent of the global total), and produces about 2.4 million barrels per day, but production can be erratic. Even before the American-initiated war of 2003, Iraq was suffering from the war with Iran in 1980–88, the Gulf War in 1991, and more than a decade of sanctions.
- *United Arab Emirates.* There are 97.8 billion barrels of known reserves (a little over 9 per cent of the global total), and 2.4 million barrels are produced per day. The city of Dubai is the trading and financial centre for much of the larger Middle East region. The UAE is closely linked to the neighbouring state of Qatar, which has known reserves of only 15.2 billion barrels but a relatively high daily output.
- *Kuwait.* Kuwait's known reserves of 96.5 billion barrels are similar to those of the UAE, but its daily production is higher (3.7 million barrels). Kuwait maintains close ties with the west.
- *Iran.* Iran has known reserves of 89.7 billion barrels (about 8.5 per cent of the global total) and is OPEC's second largest daily producer (3.7 million barrels).
- *Venezuela.* The largest oil exporter in the Americas, Venezuela is unsympathetic to the United States. It has known reserves of 77.7 billion barrels and produces 3.4 million barrels per day.
- *Russia.* With known reserves of 48.6 billion barrels, Russia produces 7.1 million barrels per day. Several other states that were part of the former USSR also have oil reserves, notably Azerbaijan and Kazakhstan with 7 and 8 billion barrels of known reserves respectively. Traditionally, Russia has had close links with OPEC, especially Iraq.
- *United States.* The US is second only to Saudi Arabia as a producer, with 7.7 million barrels per day, and has known reserves of 30.4 billion barrels. But the US is also the world's largest

importer of oil, with about one-third of its imports coming from OPEC countries. Controversial exploratory activity is now underway in the Gulf of Mexico and in Alaska.
- *Libya.* Libya's oil production was curtailed by sanctions throughout much of the 1990s. Now it produces 1.4 million barrels per day and has known reserves of 29.5 billion barrels.
- *Mexico.* There are known reserves of 26.9 billion barrels, and the production rate is 3.6 million barrels per day.

Other important oil-producing countries and their known reserves include Nigeria (24.0 billion barrels), Algeria (9.2 billion), and Angola (5.4 billion), all in Africa; Brazil (8.5 billion barrels) in South America; Canada (6.6 billion barrels) in North America; the North Sea countries of Norway and the United Kingdom (with a combined total of 12.3 billion barrels); and Indonesia (5.0 billion barrels) and China (24.0 billion barrels), both in Asia.

About half of the world's oil consumption takes place in the United States and Canada (about 28 per cent) and Europe (about 22 per cent). More generally, the OECD countries together account for more than 62 per cent of all oil consumption. Japan and Korea are also major users of oil. Most of these consuming countries are likely to become increasingly dependent on imports. In short, there is a growing gap between the places where oil is produced and where it is consumed. Figure 13.6 shows the major patterns of trade in oil.

Note that the data in this box cannot give us any indication of when the world will run out of oil. The proven oil reserves in the Middle East and north Africa increased by almost 80 per cent in the period between 1973 and 2000, and other reserves continue to be discovered.

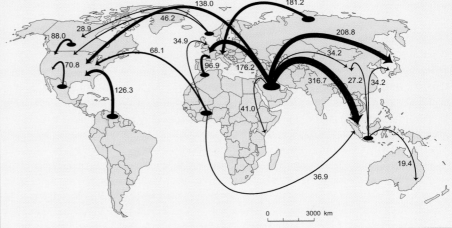

Figure 13.6 Major oil trade movements (million tonnes).

Energy supplies in Britain

In Britain the transition from coal to oil began in the 1960s. As we have seen, the industrial revolution started in Britain and relied heavily on coal. As local coal sources were depleted, imports became necessary, but they proved very expensive. As early as 1945, Britain began to import oil from the Middle East. Cost and politically related supply uncertainties prompted a search for oil and gas in the North Sea. Gas was discovered in 1965 and oil in 1969, and both were soon determined to be available in commercial quantities. By 1980 production was equal to domestic demand, with 36 oilfields and 25 gas fields developed. What does the future hold? The current supply areas are being depleted, and the key question is the extent to which declines can be countered by new discoveries. Some predictions suggest that 100 to 300 new fields will be developed by about 2020 and that the overall level of production will be maintained (Band 1991).

Industry in the More Developed World

The more developed world is responsible for a disproportionate share of manufacturing employment and output. Most of this industrial activity continues to be framed within a capitalist social and economic system. Thus decisions that affect the geography of industry and employment opportunities are made by industrial firms rather than government. While small firms must cope with larger economic changes, large firms often have the power to influence the economic environment within which they operate. Nevertheless, all firms must take international trends into account. In some cases, firms may find their home markets threatened by imports; in others, firms may rely heavily on export markets. The global manufacturing system is dominated by the more developed world, especially North America, western Europe, and the west Pacific (notably Japan). These three regions control the export and import of manufactured goods. Many industries are part of this global manufacturing system through trading activities. Other industries are transnational, owning or controlling production in more than one country. Transnational corporations and other large firms tend to be less committed to specific industrial locations than small firms because of the spatial separation of production from organization (Dicken 1992).

Industrial firms today are still concerned with costs and profits, and hence the basic logic of least-cost theory and market-area analysis continues to be relevant. While there is no doubt that many location (and other industrial) decisions are not optimal, this may not be crucial in the long term. Thus firms continue to make location decisions on the

---

13.7

AN ENERGY 'CRISIS': NEW ZEALAND IN 1992

Most of New Zealand's electricity is hydro-generated and thus dependent on adequate supplies of water. The principal source area is South Island, which has two mountains over 3,000 m (9,800 ft) and high levels of precipitation. Electricorp, a company in which the state is a major shareholder, is responsible for generating most of the needed electricity and selling to regional boards that in turn sell to consumers. Because the government requires Electricorp to make a profit, the company emphasized hydro generation, which is relatively inexpensive, closing down a coal-based generating station in 1990 and an oil-based station in early 1992. Unfortunately, hydro generation is dependent on weather conditions, and a drought that began in late 1991 posed a real threat to the generation of electricity.

By early 1992, lake levels were significantly lower than normal and a period of very cold weather increased demand for electricity. Electricorp was obliged to ask consumers to reduce consumption, to pay compensation to industrial users who reduced their use, to reopen the oil-fired station that had recently been closed, and to lower the water levels in storage lakes below the normally permitted minimum. The problem was resolved in mid-1992 when the drought ended, but the consequences have been considerable. At the time, it was estimated that the cost to Electricorp would be some NZ $300 million, and that the 1992–3 gross domestic product would be reduced by about 0.5 per cent because of the reduced manufacturing output (Murgatroyd 1993).

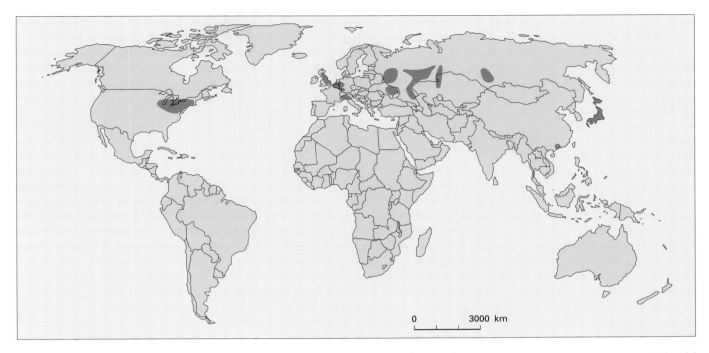

basis of the products, production technology, labour costs, sources of raw material and energy, capital availability, markets, and land costs—all considerations implicitly or explicitly contained within the traditional theories. Another key consideration, however, is the political context (Box 13.8).

Major industrial regions

As Figure 13.7 shows, there are five principal centres of industrial activity in the world and a large number of secondary centres. The five principal centres—eastern North America, western Europe, western Russia and Ukraine, Japan, and the Pearl River delta in southern China—account for over 80 per cent of global industrial production.

Eastern North America is the most productive industrial region in the world. Following European settlement, which began in the seventeenth century, it developed an industrial geography based on close

Figure 13.7 Major world industrial regions.

13.8

EXPLAINING LOCAL INDUSTRIAL CHANGE

To understand local industrial change (for example, the closure of a particular steel plant), it is necessary to consider global as well as local factors. There is no direct and predetermined link between local events and global circumstances, but global circumstances mediated through various levels of decision-making are often important.

In the case of the steel industry, major changes began in the early 1970s as the market for steel ceased to expand and production increasingly took place in newly industrializing countries such as Brazil, South Korea, and Japan. Together, these changes put pressure on plants in North America and Europe to cut production.

These global changes were then mediated through to the local scale. The key mediation occurred at the level of the state, which effectively exploited divisions in the labour force by, for example, offering redundancy pay to older workers and alternative employment to younger workers. The state also weakened unified opposition to proposed plant closures by identifying specific plants for closure rather than reducing production in all plants in a relatively uniform manner. Hudson and Sadler (1989) detail specific examples of such actions in the United States, France, West Germany, and Britain. Their analysis, like that presented in Box 13.5, is a good example of Marxist thought. In both cases, the frameworks considered most relevant are social and political rather than explicitly economic.

ties to Europe, the proximity of raw materials such as coal, iron ore, and limestone, availability of labour, rapid local urbanization and market growth, easy movement of materials and finished goods along the natural waterways, and the building of canals and railways. There are numerous local industrial areas within this region, including southern Ontario and southern Quebec, southern New England, the Mohawk Valley, the southern Lake Erie shore, and the western Great Lakes. Box 13.9 outlines the industrial geography of Canada.

Western Europe is second to North America in manufacturing output and, like the North American region, consists of numerous local industrial areas, the most important of which are central and northern Britain, the Ruhr and mid-Rhine valleys in

Germany, and northern Italy. Although Britain had the initial advantage of the earliest industrial start, by the mid-twentieth century it was having difficulty keeping up with more recent industrial developments. Much the same is true of major European coalfield areas. However, the Ruhr Valley in northwestern Germany, which contains the most important coalfield in Europe, developed a large iron and steel industry and has been much more successful in coping with industrial changes. In northern Italy, rapid diversification after the Second World War transformed a textile region into a textile, engineering, chemical, and iron and steel region using local gas and imported oil.

Until the dramatic political upheavals of the late 1980s and early 1990s, industrial location and production in the former USSR and

13.9

INDUSTRY IN CANADA

Plentiful natural resources mean that Canada has a well-developed primary (resource extraction) industrial sector. Its immense geographic extent and (consequently) dispersed national market pose a challenge to many firms because of high transportation costs. In addition, the primary sector is scattered throughout Canada, while the secondary sector is concentrated in the St Lawrence lowlands—southern Ontario and southern Quebec. Proximity to the United States has resulted in a manufacturing sector that includes many branch plants of American parent companies.

Following the initial in-movement of Europeans, a series of staple economies developed: fish, fur, lumber, wheat, and minerals. By the mid-nineteenth century Canada was a major exporter of primary (resource) products and an importer of secondary (manufactured) products. Manufacturing in Canada at that time was mostly limited to the processing of agricultural and other primary products for both domestic and export markets. Following Confederation in 1867, the Canadian government pursued two policies with direct relevance to the infant industrial geography: subsidizing railway building and introducing tariff protection for manufactured products. The immediate beneficiaries were the established areas of the St Lawrence lowlands, and a factory system was soon established. Montreal and Toronto in particular became industrial centres.

The capital needed to develop manufacturing industry initially came from Britain, but as British investment dropped off, a

branch-plant economy controlled by Americans developed. American firms were thus able to bypass the protective tariffs in place, while Canada benefited by the introduction of financial and other capital and technology. After the Second World War, the US became less globally competitive (especially compared to Japan), and as Canada became increasingly nationalistic, the branch-plant economy became less and less attractive for industrialists from the US.

Today the Canadian manufacturing industry is highly regionalized, and many geographers see Canada in terms of heartland and hinterland, or core and periphery. The hinterland produces the resources on which the manufacturing industries of the heartland depend. Toronto and Montreal have fabrication economies, while all other areas of Canada have resource-transforming economies. The two types of economy face different sets of problems. The industrial economies of Toronto and Montreal have suffered because of imports, especially from Japan and other Asian countries. The economy of the rest of Canada, which is largely dependent on raw materials, is subject to fluctuations in world price and demand over which Canadians have no control. At the national level, because Canada has many resources, economies facing problems are usually buffered by the successes of other sectors. At the local level, however, many areas are subject to booms and busts. Single-resource towns, of which there are many, are the most vulnerable (Britton 1996).

the eastern European countries were determined by central planning agencies. Recent changes have not, of course, significantly altered the spatial pattern of industrial activities. In the former USSR there are five major industrial areas, four of which are now in Russia. Those based in Moscow and Ukraine were established in the nineteenth century. The Moscow area is market-oriented and began its industrial development specializing in textiles; today it is diversified, with a wide variety of metal and chemical industries using oil and gas. In Ukraine industry is centred on a coalfield but also has access to supplies of iron, manganese, salt, and gas; Ukraine is a major iron and steel and chemical industrial area. The other three regions were developed by the USSR government after the 1917 revolution and are located in southern Russia. The Volga area to the east of Ukraine has local oil and gas sources and was a relatively secure location during the Second World War; its major industries are machinery, chemicals, and food processing. Farther east, the Urals area is both a source of many raw materials, especially minerals, and a major producer of iron and steel and chemicals; industrial growth in the Urals was promoted by the former USSR government because of its great distance from the western frontier. Farthest east is the Kuznetsk area, which is similarly endowed with minerals as well as a major coalfield.

Japan is relatively small in size; with few industrial raw materials of its own, it is forced to import almost everything it needs; and it is a substantial distance from major world markets. But Japan has overcome these obstacles by making use of its large population. Since the Second World War, Japan has become a major exporter of industrial goods by keeping labour costs low, and by 1970 its highly skilled labour force was increasingly focused on the production of high-quality goods. Japan is now a leading producer of computers and electronic equipment (Box 13.10). However, the world's most dynamic industrial region today is the Pearl River delta region of southern China, just north of Hong Kong (see p. 478).

Newly Industrializing Countries

Japan is no longer alone. The success story outlined in Box 13.10 has encouraged many imitators in other Asian countries. During the 1970s, the economies of South Korea, Taiwan, Hong Kong, and Singapore all accelerated rapidly, and those of Malaysia, Thailand, Indonesia, and the Philippines are accelerating today. These eight are the leaders among what have come to be known as the newly industrializing countries, or NICs (others include Brazil, Mexico, Greece, Spain, and Portugal). Industries in the Pacific Rim follow the Japanese example of low labour costs

13.10

INDUSTRY IN JAPAN

Japan comprises numerous islands scattered over an area roughly 2,500 km (1,500 miles) from north to south. Although cultural contact with Europe and North America did not begin until after 1854, with the end of feudalism, by 1939 Japan had expanded territorially and developed into an industrial power comparable to those in the west. Industrialization continued after the Second World War despite the loss of Korea and considerable bomb damage. During the 1950s and 1960s Japan excelled in heavy industry, especially shipbuilding, but by the late 1960s the emphasis had shifted to automobiles and electronic products, and most recently Japan has focused on computers and biotechnology. During each of these three phases since the Second World War, Japan has been a world industrial power.

The principal factors behind these remarkable industrial successes are low labour costs, high levels of productivity, an emphasis on technical education, minimal defence expenditures, aid from the United States (motivated by the perception that Japan served as a bulwark against Chinese communism), and a distinctive industrial structure in which small specialized firms are linked to corporate giants such as Nissan and Sony. This structure facilitates rapid acceptance of various new technologies.

Another aspect of this success story is Japanese business's considerable investment overseas. There are many Japanese companies operating in North America and Europe especially, and Japanese banks dominate international finance (Table 13.3).

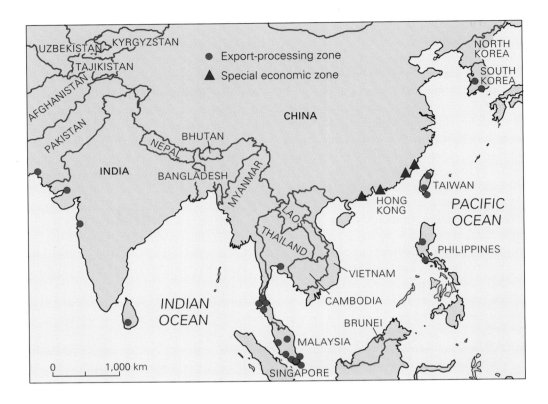

and high productivity. All eight countries are high-growth-rate areas experiencing the same labour shifts, from agriculture to industry, that much of Europe experienced in the nineteenth century.

The most successful of the NICs is South Korea. Until 1950 South Korea was a poor, less developed country characterized by subsistence rice production; today it is an industrial giant that began with heavy industry, then focused on automobiles, and now specializes in high-technology products. Its experience is the Japanese transformation repeated, and over a period of less than forty years.

Current evidence suggests that these trends will continue. One motivation for recent North American free trade agreements and continuing European integration is awareness of the need to respond to the dramatic industrial growth in much of Asia.

To attract transnational corporations, several NICs—including South Korea, Singapore, Taiwan, Hong Kong, China, the Philippines, and Mexico—and some other less developed countries have set up export-processing zones (Figure 13.8): manufacturing areas that export both raw materials and finished products. Industries are attracted to these zones for three general reasons: (1) inexpensive land, buildings,

energy, water, and transport; (2) a range of financial concessions in such areas as import and export duties; and (3) low workplace health and safety standards and an inexpensive labour force made up largely of young women, who are commonly regarded as less likely than men to be disruptive, and more willing to accept low wages and difficult working conditions. There are few advantages other than waged employment for the processing country, and even this advantage varies depending on larger international economic circumstances. The parallel between these export-processing zones and the agricultural plantations discussed in Chapter 11 is compelling.

There are especially close links between these export-processing zones and high-technology companies; in Mexico about 850 zones, known as *maquiladoras*, are set up within easy reach of the high-technology suppliers in such areas as the Santa Clara Valley in California (Silicon Valley) and the Dallas-Fort Worth area in Texas (Silicon Prairie). The high-technology materials, such as silicon chips, are manufactured in the United States and transported to the *maquiladoras* for the simple but labour-intensive process of assembly; then the finished product is transported to the US for sale.

Industry in the Less Developed World

The distinction between more and less developed is never straightforward. Because of their recent industrial success, the NICs are frequently referred to as economic miracles—and yet they are still regarded as belonging to the less developed world. More typical less developed countries have been much less successful industrially, and no miracles are expected of them.

One obvious distinction between the more and the less developed worlds is the degree of industrialization: many countries in the latter group have not yet experienced anything like an industrial revolution. It is widely hoped that industrialization might lead to employment for the unemployed or underemployed rural poor. In addition, many in the less developed world believe that industrialization would demonstrate economic independence, encourage urbanization, help to build a better economic infrastructure, and reduce dependence on overseas markets for primary products.

Problems

Potentially, industrialization has many advantages. Why is it so difficult for many countries to achieve it? A major obstacle is the legacy of former colonial rule. Historically, European colonial countries saw their colonies both as producers of the raw materials needed for their domestic manufacturing industries and as markets for those industries (13.4). This is a difficult legacy to move beyond. Most countries in the less developed world have neither the necessary infrastructure nor the capital required to develop it. Furthermore, fundamental social problems resulting from limited educational facilities do not encourage the rise of either a skilled labour force or a domestic entrepreneurial class. Finally, the domestic markets in many of these countries lack the spending capacity to make industrial production economically feasible.

Prospects

Despite these difficulties, considerable industrialization has taken place in many less developed countries since the Second World War. Major successes have been achieved in import substitution—manufacturing goods

Workers at the TRW maquiladora plant in Reynosa, Mexico, package seatbelt components for export to the US in 1996. At the time, Reynosa led all of Mexico in maquiladora construction, employment, and production, directly affecting the economy on the Texas side of the border (AP photo/L.M. Otero).

that were previously imported, often with the help of protective tariffs. Examples include light industries that are not technologically advanced, such as food processing and textiles. Heavy industry has been more difficult to develop, even though many of the raw materials used in the more developed world come from the less developed world. In general, the less developed world contradicts the standard primary–secondary–tertiary sequence. The more typical sequence is primary–basic tertiary–secondary; perhaps a quaternary sector will develop in due course.

China

As we noted in our discussion of agriculture in Chapter 11, liberalization of China's communist economy began in 1978 with the opening up of the country to international trade and foreign investment. Since the 1980s, both the agricultural and industrial economies have undergone reform and modernization. These important changes were initiated during the 18-year leadership of Deng Xiaoping (died 1997), who effectively determined that economic growth was more important than continued class struggle.

During the 1950s, industrial development was based on a combination of large technology-intensive, state-funded factories and small labour-intensive, locally organized units (Morrish 1994, 1997). The large state-funded units made up about 100,000 of the total of about 8 million units and produced 55 per cent of the national industrial output. Despite this high percentage, the strategy

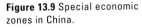

Figure 13.9 Special economic zones in China.

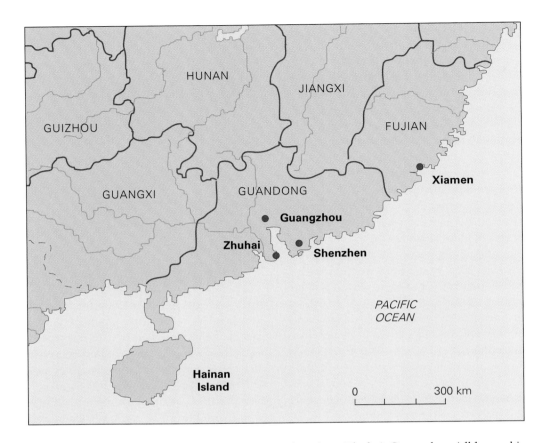

proved unsuccessful because the state-owned factories were expensive to operate and highly inefficient. Shortages of consumer goods were commonplace.

A new regional development policy was introduced in 1980 with the establishment of several special economic zones (similar to export-processing zones) that enjoy low taxes and other financial concessions to attract foreign investment. A reform program was initiated in 1984 that involved less state control, decreased subsidies, and increased response to market forces, and that encouraged local collective and private industrial enterprises. One result was that China's industrial production outstripped that of South Korea, increasing at 12 per cent per annum between 1980 and 1990. In 1994 the state relaxed its control even further and, to encourage collective and private enterprise, introduced a new program requiring urban industrial enterprises to respond to market forces directly, without state involvement.

The special economic zones have played a major role in the recent expansion of Chinese industrial activity (Figure 13.9). Four zones were established in 1980 at Shenzhen, Zhuhai, Guangzhou (all located in Guangdong province north of Hong Kong), and Xiamen (in neighbouring Fujian province). These zones have grown dramatically with the introduction of 'labour-intensive factories that manufacture everything from computer keyboards and dishwashers to leather coats and Mighty Morphin Power Rangers' (Edwards 1997:12). Most of the foreign investment has come from Hong Kong, attracted by cheap labour and land. Together, the three zones north of Hong Kong, all located in the Pearl River delta, now form what many consider to be the world's most dynamic industrial region, sufficiently large that it is affecting global trading patterns and investment flows. This delta area attracts about one quarter of China's foreign direct investment and generates about one-third of its exports. In a sense, it is the early twenty-first century equivalent of mid-nineteenth century industrial northern England —a workshop for the world.

In 1984, 14 locations were designated 'open coastal cities', free to trade outside China, and an additional three larger open zones were created one year later. A fifth, Hainan Island, was established in 1988. By the early 1990s it was clear that the coastal economy was growing too quickly, and the Chinese government began to focus attention on some inland areas. In fact, China's industrial growth has been so dramatic that it is a cause of some concern in other areas. The world's biggest net recipient of foreign investment, China now manufactures 60 per cent of the world's bicycles, and 50 per cent of the world's shoes. It is also by far the largest garment exporter, with 20 per cent of the global total today—a figure that is expected to reach 50 per cent by 2010. Wages are extremely low: garment workers are paid about US 40 cents an hour—less than a third what their counterparts in Mexico receive. More generally, China's economy is enormous, with the sixth largest GDP in the world. In terms of purchasing-power parity, its economy is second only to the United States. In addition to dominating production, China is increasingly important as a trading nation.

India

The example of India highlights many of the general points noted earlier. Following independence in 1947, India was a producer of agricultural products, but today it has a much more diversified industrial structure. Industrial landscapes have been created, and the government has attempted to reduce regional disparities. Since 1951, India has used a series of five-year plans to guide development. Initially, the emphasis was on heavy industry, but later plans have stressed self-reliance and social justice. Given that its industrial transformation did not begin until 1951, India has achieved remarkable success; the main reasons are its substantial market, available resources, adequate labour, and relatively sound government planning. Indian industry will likely continue to grow rapidly in the immediate future; indeed, many observers expect it to emulate China.

Globalization and Industrial Geographies

Globalization processes are playing a key role in contemporary industrial geography, affecting location, organization, and activity. Although these processes have already been discussed, especially in Chapter 10, it is important to review two key issues related to industrial change in both more and less

developed worlds. Industrial restructuring today amounts to a global shift in industrial investment and activity. This process was well underway by the 1980s, along with the other classic component of globalization—namely, a decline in the friction of distance that is a consequence of new communication technologies. Together these two processes have produced the new and dynamic industrial geography of today. Related to the two globalization processes is the transition from **Fordism** to **post-Fordism**.

Fordism to Post-Fordism

To understand the current 'post-Fordist' phase of industrial activity in the more developed world, it is important to understand the Fordism that preceded it. The term 'Fordism' refers to the methods of mass production first introduced by Henry Ford in 1920s America, particularly fragmentation of production—best illustrated by the use of assembly lines that reduced labour time and related costs. In sharp contrast to most nineteenth-century industrial activity, the Fordist system gave workers the necessary income and leisure time to become consumers of the many new mass-produced goods. Economic policies based on the theories of John Maynard Keynes, a British economist, were widely implemented, resulting in rising living standards throughout the 1950s and 60s. These two decades of sustained economic growth witnessed the rise of the first transnationals, among them Ford itself, based in the United States (automobiles), Nestlé, based in Switzerland (foodstuffs), and Imperial Chemicals, based in the United Kingdom (chemicals). These early transnationals were limited in their ability to invest outside their home countries, but they were able to take advantage of technological advances in transportation.

This economic boom slowed down in the early 1970s for two principal reasons. First was the termination of the provision in the Bretton Woods Agreement (see Chapter 10) that allowed the United States to convert overseas holding of US dollars to gold at a fixed rate. When the US ended this practice because of the expenses incurred in the Cold War and the resulting budget deficit, other currencies fluctuated, resulting in price fluctuations and some business losses. The second reason was the 1973 OPEC decision to raise the price of oil—a commodity essential to industrial production.

One important outcome of the recession of the early 1970s was industrial restructuring in the more developed world. Basically this involved deindustrialization, especially in such traditional industrial activities as textiles and shipbuilding and in automobile manufacturing, and corresponding reindustrialization as firms established branch plants overseas, either in NICs or in less developed countries (Box 13.11).

Industrial Restructuring

As Box 13.11 suggests, the transition to post-Fordism entails major changes in industrial geographies, resulting essentially from technological advances and globalization processes. Three technological changes are especially significant:

1. Production technologies, such as electronically controlled assembly lines and automated tools, are increasing the separability and flexibility of the production process.
2. Transaction technologies, such as computer-based, just-in-time inventory control systems, also increase locational and organizational flexibility.
3. Circulation technologies, such as satellites and fibre optic networks, facilitate the exchange of information and increase market size.

Together, these three changes represent a transition to **flexible accumulation**, making it easier for companies to take advantage of spatial variations in land and labour costs and to serve larger markets. The subsequent industrial restructuring takes three principal forms:

13.11

DEINDUSTRIALIZATION AND REINDUSTRIALIZATION

Deindustrialization and reindustrialization are spatial trends caused by economic restructuring: specifically, shifts in investment between the manufacturing and service sectors—or, to put it another way, shifts between production and consumption. Both can be seen as components of the transition to a postindustrial society. **Deindustrialization** is a reduction in manufacturing in more developed countries that is usually most easily measured by reference to employment data. It is most evident in the older industrial areas and in such industries as iron and steel, textiles, engineering, and shipbuilding. It is not so common in high-tech industries such as pharmaceuticals and electronics. The major social consequence is unacceptably high unemployment.

In the United States and the United Kingdom the percentage of workers employed in manufacturing fell from about 40 to about 20 per cent between 1900 and 2000. There are two general causes for this deindustrialization. First, as a result of globalization processes, including the rise of transnationals and improvements in communication technologies, much of the more developed world's manufacturing is being transferred to less developed countries—especially China. (This decline of manufacturing in the more developed world mirrors the earlier decline of agriculture.) In its place, service-sector activities have become increasingly important.

Deindustrialization is usually seen as a negative development. Because the decline takes place in areas where industry used to be highly concentrated, it initiates a larger economic and social decline. The fact that it tends to occur rapidly, often during periods of economic recession, makes regional and local adjustment all the more difficult.

Reindustrialization at least partially counters industrial decline. This process may take various forms. First, there is an increasing tendency for small and/or new firms to be more competitive. This is most likely to occur outside the traditional industrial areas, and reflects the information exchange made possible by electronic means. Second, high-tech industrial activities, especially microelectronics, are expanding rapidly in output, if not also in employment. Such industries are locating in environmentally attractive areas where skilled workers choose to live. A third aspect of reindustrialization is the expanding service industry. Rising incomes and changing lifestyles are bringing significant growth in tourism and recreation industries, with accompanying impacts on environmentally attractive areas. Other service industries, especially banking and information services, are also expanding in major urban centres.

In the more developed world the new industrial landscape is thus quite different from the nineteenth-century version. Not only is the landscape itself visually different with the decline of smoke-producing heavy industry, but both the work experience and the location of the landscape have changed. The need for coherent regional policies is obvious.

1. The relationship between corporate capital and labour is changing as machines replace people, manufacturing industry declines, and transnationals seek low-labour-cost locations (as noted in the account of export-processing zones above).
2. Both the state and the public sector are playing new roles with the shift away from **collective consumption** in areas such as education and health care to joint public–private projects and deregulation (discussed in Chapter 12).
3. There is a new division of labour at various spatial scales as the new technologies allow corporations to respond rapidly to variations in labour costs (as in the account of export-processing zones above as well as discussions in Chapter 12).

Locational Considerations

As the preceding discussion implies, decision-making, especially concerning location, by industrial firms today often differs markedly from the model proposed by Weber. Increasing emphasis on technology and decreasing emphasis on materials and localized energy sources means that, for many industries, transport cost is no longer the main criterion. Rather, many contemporary firms are trading off between two types of cost: the labour and land costs considered by Weber and a whole set of new costs associated with exchanging information between firms.

Thus a major debate today in studies of industrial activity, especially location decisions, concerns the implications of labour (and, to a lesser extent, land) costs on the one hand and the costs of information exchange on the other hand. For many industrial activities, information can be rapidly exchanged at low cost by electronic means, but there is a danger in such impersonal exchanges if the information is lacking in clarity. In principle, firms that make location decisions on the assumption that they will be able to exchange information successfully are able to seek out locations with low labour costs. The result is a decentralized (deglomerated) industrial pattern. A prime example is the successful expansion of many high-tech Japanese firms to other countries (Table 13.3). In situations where firms lack confidence in their ability to exchange clear and unambiguous information, they will prefer to locate in close proximity (to agglomerate) in order to facilitate personal, face-to-face exchange of information.

Acknowledging the contemporary relevance of information exchange introduces two possible patterns of industrial location. Decentralization occurs if long-distance electronic information exchange is feasible, but centralization occurs if the information is lacking in clarity and requires personal contact to be effective. Decentralization is typically associated with firms (often transnationals) that mass-produce a standard product and thus do not need to worry about ambiguities in the exchange of information, while centralization occurs when functionally related firms need to exchange information on a person-to-person basis because their products are not identical.

Service Industries

It is generally assumed that as an economy progresses, it undergoes a transition from primary (or extractive) to secondary (or manufacturing) to tertiary (or service) activities. Figure 13.10 provides a simple illustration of the relative importance, in terms of employment, of the different sectors of industrial activity. Indeed, some geographers contend that, especially in the more developed world, the emerging postindustrial society is increasingly service-oriented. This may be so, but service industries have existed for a very long time.

Table 13.3	**LOCATIONS OF R&D FACILITIES AND PLANTS OF NINE JAPANESE ELECTRONICS FIRMS, 1975 AND 1991**

Location	R&D Facilities		Production Plants	
	1975	1991	1975	1991
Japan	22	95	207	341
North America	0	13	7	69
EU	0	7	6	50
Asia	0	2	40	123
Other	0	6	20	34

Note: The nine electronics firms are Hitachi, Toshiba, Mitsubishi, Matsushita, Sony, Sanyo, Sharp, NEC, and Fujitsu.

Source: Adapted from Y.-M. Yeung and F.-C. Lo, 'Global Restructuring and Emerging Urban Corridors in Pacific Asia', in *Emerging World Cities in Pacific Asia*, edited by F.-C. Lo and Y.-M. Yeung (New York: United Nations University Press, 1996):37

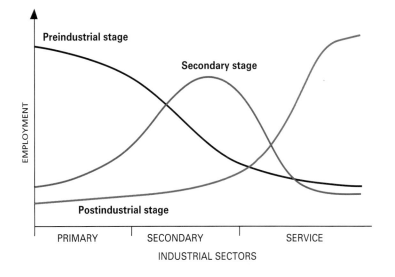

PRIMARY   SECONDARY   SERVICE

INDUSTRIAL SECTORS

Figure 13.10 A simplified diagram of the relationship between economic growth and distribution of employment.

According to Daniels (1985:1), a service 'is probably most easily expressed as the exchange of a commodity, which may either be marketable or provided by public agencies, and which often does not have a tangible form'. Clearly, service industries are diverse—so diverse that they encompass not only a tertiary sector but also quaternary and quinary sectors: transportation and utilities are tertiary; insurance and real estate are quaternary; while education, health, and government are quinary. Whatever their sector, the service industries are crucial components of any economy and merit the attention of geographic researchers.

The service industry grew along with manufacturing during the industrial revolu-tion, but it has achieved its most rapid expansion since the Second World War. In the US, for example, manufacturing employment decreased 3 per cent between 1980 and 1990, while service employment increased 20 per cent. In the less developed world, retailing and distribution are among the dominant service activities, while in the more developed world more specialized services such as banking and advertising are also important.

The global variations in employment in industry and in services are indicated in Figures 13.11 and 13.12 respectively, and some summary data are presented in Table 13.4. It is useful to compare the two figures with Figure 11.12, which maps the percent-age of the labour force employed in agri-culture. Together, the figures and table pro-vide some compelling additional evidence of the differences between the more and less developed worlds, as well as the changing structure of employment since 1960.

Geographic attempts to explain the loca-tion of service industries have focused on the central place model described in Chapter 12. This work has made it clear that service loca-tions can be determined by considering such standard causal variables as transport costs, market location, and economies of scale. In addition, services have an especially strong tendency towards agglomeration. Their loca-tions depend on population, and they tend to cluster. Furthermore, location decisions in the

New customer service agents learn US geography at the 24/7 Customer.com training centre in Bangalore, southern India, in 2001. Their training included coaching in American or British accents, depending on the clients they were to represent, as well as appropriate slang terms and cultural references from sports and television (AP photo/Namas Bhojani).

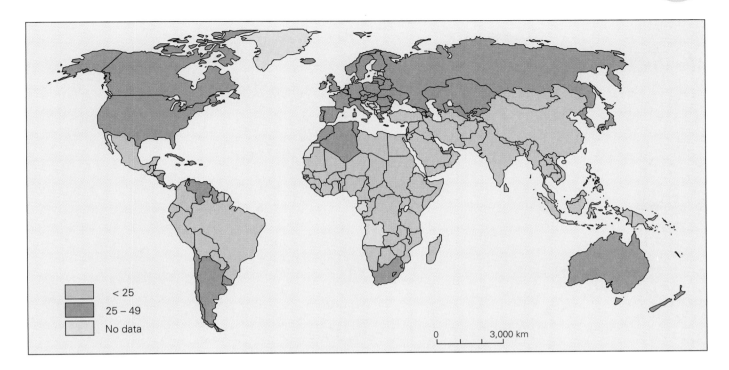

service industry appear to be especially vulnerable to behavioural variables such as the availability of information and the interpretation and use of information. A consideration of behavioural variables helps to explain the clustering tendencies. Information diffusion takes place most rapidly and effectively in local networks. Indeed, some argue that recent changes in the technology of information dif-

fusion have led not only to the clustering of certain services but also to spatial changes in the roles played by urban centres. Others, however, contend that new information technologies are leading to increased decentralization, possibly even (in some cases) home-based service work. Regardless of where service industries locate, there is little doubt that they are able to contribute significantly to

Figure 13.11 Percentage of labour force in industry by country, 1996. Figure 10.12, mapping percentages of the labour force in agriculture, showed not only enormous variations between countries but a basic distinction between the more developed and less developed worlds. Although this map of percentages in the industrial labour force shows less variety it also indicates a distinction between more developed and less developed worlds. In general, the percentages are between 25 and 50 in the more developed world and under 25 in the less developed world. The world average is 20 per cent. Source: United Nations Development Programme, *Human Development Report* (New York: Oxford University Press, 1996):tables 16 and 32.

Table 13.4	MINERAL PRODUCTION IN THE LESS DEVELOPED WORLD		
Mineral	**Country**	**% of World Production**	**Rank in World**
Bauxite	Guinea	16.5	2
	Jamaica	8.7	3
	Brazil	8.4	4
Iron	Brazil	18.5	2
Phosphate	Morocco	11.6	3
Silver	Mexico	16.0	1
	Peru	11.7	3

Note: The less developed world continues to be an important supplier of minerals to the more developed world—one more indicator of the relationship between these two regions. Other important minerals produced in the less developed world are copper, tin, nickel, cobalt, and fluorspar.

Source: J. Dickenson et al., *A Geography of the Third World* (New York: Routledge, 1996):159.

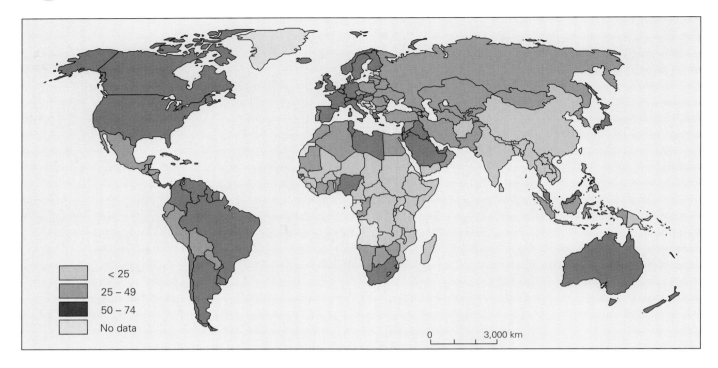

Figure 13.12 Percentage of labour force in services by country, 1996. This map shows the highest values (between 50 and 75 per cent) occurring in the more developed world and a few other countries (especially in Latin America). In general, the less developed world has lower proportions of service workers, often below 25 per cent. The world average is 31 per cent.
Source: United Nations Development Programme, *Human Development Report* (New York: Oxford University Press, 1996):tables 16 and 32.

Map legend: < 25; 25 – 49; 50 – 74; No data. Scale: 0 — 3,000 km

increases in productivity and living standards. In Canada, for example, service-sector growth is a crucial component of larger economic prosperity (Grubel and Walker 1989).

A geography of service industries must also consider the availability of those services that are central to human well-being, among them health care and educational facilities. In fact, any study of service industries is a virtual microcosm of contemporary geography, because services are linked to so many other geographic topics and because they are so important in contemporary economies and societies.

Industry and Society

The industrial geography discussed so far has said little about either humans or environment. Our concerns have been essentially spatial and economic—a fair reflection of the dominant traditions. Increasingly, however, as we have seen, human geographers are turning their attention to issues related to political and social circumstances. Since about 1970, many researchers have realized that much industrial geography can be best understood by reference to social theory. They argue that industrial activity does not take place in a politically and socially neutral setting, according to the principles of neo-classical economics: rather, to gain real insights into the geography of industrial cap-

italism it is necessary to turn to contemporary social theory. Unfortunately for beginners, the range of that theory is considerable, but most of it consists of variations on the Marxist themes discussed in Chapter 11. Massey (1984), for instance, produced a Marxist theoretical analysis of the changing industrial geography of the United Kingdom, while Graham et al. (1988) outlined a framework based on Marxism for investigating structural change in industries. Such industrial geographies represent significant departures from the usual Weberian economic theory; instead they give priority to relating spatial and social issues by reference to such topics as class and gender as they have evolved in a capitalist framework.

For example, the fact that transnationals actively seek low-wage locations for manufacturing activities has attracted charges of labour exploitation. High-profile companies such as Nike have been especially subject to criticism. There is also evidence that some transnationals locate plants in countries where the costs of production are kept low not only because the wages are low but also because there are fewer requirements concerning matters such as pollution control.

It is now commonplace to argue that no hypothesis about industrial (or other) locations can be complete unless it is placed in some appropriate social, political, and institutional context. Massey (1984) argued that places vary in many respects, from workplace social relations and the division of labour to local political circumstances, and each of these must be considered in any analysis of industrial location and the changing distribution of employment. Similarly, a focus on workplace social relations often requires consideration of gender differences.

Changing local labour markets

Traditionally, as we saw in Box 13.1, industrial geography treated labour as either a location factor or a commodity. Reacting against these interpretations, subsequent Marxist theory focused on the spatial division of labour—the idea that regional development and employment distribution are spatial expressions of the labour process, subject to control and manipulation. More recently, industrial geographers have focused on the geographies of labour markets, especially as these are affected by the transition from Fordism to post-Fordism. Table 13.6 summarizes the differences between these two phases.

Gendered employment

In most of the more developed world today, a majority of women of working age are in the labour force. This marks a significant change: in Canada, for example, female participation rates in the labour force more than doubled between 1951 and 1981. Yet for the most part the type of work done by females has not changed: a majority are still employed in clerical, service, and low-skill jobs. Why is this so? One clear reason is discrimination by employers. In an account of the steel industry in Hamilton, Ontario, Pollard (1989) cited evidence of sexist hiring practices by the principal steel company, Stelco. Between 1961 and 1978, Stelco received between 10,000 and 30,000 applications from women for production jobs—none of whom was hired. In the same time period, about 33,000 men were hired for production jobs. Following considerable publicity and intervention by the union and the Human Rights Commission, some women were hired. But the evidence suggests that the underlying social problem is much larger. While on the job, women were made to feel uncomfortable not only by men in the workplace but by members of

Hundreds of members of an advocacy group called 'Global Exchange' demonstrated outside a new 'NikeTown' store in San Francisco in February 1997. Protesting against 'sweatshop labour', they claimed that the company was paying its workers in Indonesia a mere 29 cents per hour (AP photo/Ben Margot).

the community, especially wives of male employees. Clearly, larger social traditions are at stake in any attempt to change the gender division of labour.

Environmental considerations

Industrial impacts on the environment were discussed in some detail in Chapter 4. Here we are interested in the locational implications of the environmental consequences of industrial activity. In many countries, industries are facing new rules designed to control pollution by regulating the quantities of waste produced and the disposal of waste products. In principle, environmental considerations add a new cost factor that may affect both location and production decisions. Interestingly, in a study of the effects of the

Table 13.5	**SUMMARY OF WORLD EMPLOYMENT**	
	1960	**1990**
Labour force in agriculture (%)	61	49
Labour force in industry (%)	17	20
Labour force in services (%)	22	31

Note: The changing character of the global economy is effectively summarized in this table. In percentage terms, the agricultural labour force is declining significantly, the industrial labour force shows a small increase, and the services labour force is increasing significantly.

Source: United Nations Development Programme, *Human Development Report* (New York: Oxford University Press, 1996):Table 16.

National Environmental Policy Act in the United States, Stafford (1985) concluded that the policy would not lead to major locational changes. As regulations become increasingly more rigorous, however, locational effects will likely become apparent.

Recreation and Tourism

Recreation, including time spent with family and friends, reading, and participating in or watching sports, is now a major part of most people's lives. Recreational activities include tourism, an industry that is generated by the more developed world, but that operates in both the more and the less developed worlds. The tourism industry has experienced dramatic growth since about 1960; according to the World Travel and Tourism Corporation, it is now the largest industry in the world and is growing at a phenomenal rate—23 per cent faster than the overall world economy. Globally, the number of tourists is estimated at about 700 million annually, and that number is expected to increase to some 1.5 billion by 2020, with an annual value of perhaps US$2 trillion.

But the importance of this industry cannot be measured simply in numerical terms, for at least two closely related reasons. First, 'tourism is a significant means by which modern people assess their world, defining their own sense of identity in the process'

Table 13.6	**LABOUR MARKETS: FROM FORDISM TO POST-FORDISM**	
	Fordist labour markets	**Post-Fordist labour markets**
Labour process	Mass production involving large workforces within firms. Productivity achieved by division of labour into detailed standardized tasks.	Flexible, specialized mass production. Productivity achieved by division of labour within firms into core and peripheral categories, with functional flexibility of core workers and numerical flexibility of peripheral workers.
Employment	Substantial proportion in manufacturing. Predominantly male. Most jobs full-time. High degree of job security. Generally full employment.	Small proportion in manufacturing, most in services. Increased female participation in labour force. Many jobs part-time and temporary. Increasing job insecurity. High unemployment.
Wages	Most set by collective bargaining (by industry-wide or nation-wide unions). Income inequalities stable or falling.	De-collectivization and localization of wage determination. Increasing polarization between high-paid full-time workers and low-paid part-time workers.
Labour relations	Highly formalized and confrontational. Labour organized into strong unions. Collective strike action by workers common and protected in law.	Much more individualistic and cooperative. Unions are in decline and solidaristic labour cultures in retreat.
Labour market regulation	Macroeconomic demand management to maintain full employment. Welfare-oriented. Benefits unconditional and funded by progressive taxation system. Extensive employment protection and workplace regulation.	Abandonment of full employment policies. Shift from welfare to workfare and training-fare. Benefits increasingly conditional. Deregulation of employment and workplaces to promote labour flexibility.
Spatial features	Relatively self-contained local labour markets. Distinct local employment structures and labour cultures. Typified by centres of mass production industry. Local disparities in unemployment are minimal.	Local labour markets are still important, but segmentation is greater within than between them. Typified by new local centres of flexible production and service activity. Substantial local disparities in unemployment and growth of localized concentrations of social exclusion.

Source: Adapted from R.L. Martin, 'Local Labour Markets: Their Nature, Performance, and Regulation', in *Oxford Handbook of Economic Geography*, edited by G.L. Clark, M.P. Feldman, and M.S. Gertler (New York: Oxford University Press, 2000):Table 23.1, 459.

(Jakle, 1985:11). Second, because it is 'one of the most penetrating, pervasive and visible activities of consumptive capitalism, world tourism both reflects and accentuates economic disparities, and is marked by fundamental imbalances in power' (Robinson 1999:25). In other words, tourism is a means by which both tourist and host communities create their respective identities and emphasize their difference from one another. It is because tourism offers significant opportunities for cultural contact that it creates opportunities for both cultural understanding and cultural conflict. In short, tourism is one of the means by which human geographies are created and recreated.

Tourist Attractions

It is in the more developed world that recreational activities, including tourism, are important features of modern life. In seventeenth-century Europe, the élite visited spas for medical purposes; later, the supposed health benefits of seaside resorts attracted a broader clientele. The idea that travelling for pleasure broadens the traveller's horizons first emerged in Europe during the eighteenth century. Mass tourist travel, however, depended on the introduction of the annual vacation (as opposed to holidays for religious observance) negotiated between employer and workforce—and that was a product of the industrial revolution. By the late nineteenth century the vacation proper was established and seaside resorts in Europe, especially Britain, were becoming playgrounds for the working classes. More recently, changes in employment patterns that allow for more leisure time, additional discretionary income for many people, decreasing travel costs, and increasing numbers of retirees have helped to diversify tourism and make it a year-round industry. There appear to be six principal attractions for contemporary tourists:

1. good weather (usually meaning warm and dry);
2. attractive scenery (coastal locations are especially favoured);
3. amenities for such activities as swimming, boating, and general amusement;
4. historical and cultural features (old buildings, symbolic sites, or birthplaces of important people);
5. accessibility (increasingly, tourists travel by air; in general, costs increase with the distance travelled); and
6. accommodation.

Locations with an appropriate mix of these features tend to be major tourist areas. In the more developed world, such areas are typically urban or coastal. In Europe, many of the most popular coasts for tourism are peripheral (for example in Spain and Greece), although this is not the case in countries such as the United States and Australia.

Falsifying place and time

Among the most distinctive tourist attractions today are **spectacles** such as sporting events or world fairs; theme parks such as Disney World (Ley and Olds, 1988), and large shopping centres such as the West Edmonton Mall (Shields, 1989; Hopkins, 1990; Jackson and Johnson, 1991). Many such manifestations of popular culture (see Chapter 8) are specifically created by developers to appeal to tourists as consumers, and lend themselves to a wide range of human geographic interpretations.

Attractions created where previously there was nothing to draw visitors are completely artificial. It is probably correct to say that what matters in this kind of venture is innovation, creating something that people want—sometimes even creating a demand for something that no one had thought of wanting before, and then making it available. Two important attractions of artificial sites are that they offer a controlled and safe environment while at the same time purporting to be exotic, providing a glimpse into other cultures and places.

In many parts of North America, tourist attractions such as gambling facilities capitalize on a previously unsatisfied demand. In many respects, Las Vegas is the pioneer in this area, and it regularly reinvents itself in order to continue as a prime tourist destination. Most of the early hotels in Las Vegas offered little more than basic accommodation for gamblers, but in 1966 the first of the large theme hotels, Caesars Palace, opened, setting the pattern for subsequent development. Today the city attracts many families and business conferences in addition to gamblers.

'Fantasyland' castle, Disneyland, California (Victor Last, Geographical Visual Aids).

One of the more interesting aspects of these essentially artificial tourist sites is that the experiences they offer are inauthentic—more about myths and fantasies than reality. Most observers see this form of tourism as bound up with postmodernism, the current emphasis on consumption rather than production, and the commodification of people, place, and time.

Even many supposedly real tourist sites may have more to do with myth than reality. Heritage and other historical sites, for example, may be so thoroughly reconstructed and packaged that it is difficult for tourist consumers to know how true to the past their experience is. Colonial Williamsburg in Virginia is often criticized on these grounds, as are many battlefields and heritage sites, at least partly because the meanings attached to the sites themselves are necessarily plural and contested. For some critics, our understanding of ourselves, our places, and our pasts is increasingly created by the tourist industry.

Mass tourism

Understood as part of a larger Fordist economy, mass tourism is a form of mass consumption: it involves the purchase of commodities produced under conditions of mass production; the industry is dominated by a few producers; new attractions are regularly developed; and different sites are essentially similar. Together, these characteristics make the tourist experience much the same everywhere, regardless of specific site. Further, as with other cases of mass production and con-

sumption, it is the producers of tourist sites that effectively determine the consumers' options, and thus have the power to direct what is sometimes called the 'tourist gaze'.

While commercial producers create attractions such as coastal resorts, governments too can direct the tourist gaze by creating parks and other favoured locations, although the tourist element in such cases is sometimes relatively small. For example, the ten new National Parks that Ottawa has created in the Canadian north since 1976 were intended primarily to preserve habitats and landscapes.

Tourism in the Less Developed World

Many of the favoured tourist destinations today are in the less developed world—in Mexico, numerous Caribbean and Pacific islands, and some Asian and African countries. Some of these places do offer an appropriate mix of the six attractions noted above, but they may also be attractive for other reasons. For example, low labour costs usually mean that these areas can offer competitive rates, and in addition some may represent—at least for Europeans or North Americans—an 'exotic' cultural experience. Certainly tourism is growing in the less developed world, especially in coastal areas, in response to increasing demand in the more developed world.

This trend poses several problems for the less developed countries, however. Most important from the economic perspective is that dependence on tourism makes a country vulnerable to the changing strategies of the tour companies in more developed countries. Many destinations offer sun, sand, and surf, and those that have no distinctive attractions must keep their rates low if they are to draw clients away from their competitors. Even so, most less developed countries appreciate the benefits that tourism can provide: local employment, stimulation for local economies, foreign exchange, improvements in services such as communications. In short, in less developed countries the tourist areas are generally much better off than the larger economy. Today, tourism is a major growth industry in the tertiary sectors of several less developed countries, and current evidence suggests continued growth (Box 13.12).

Alternative Tourism

The impact of tourism on local ways of life is contradictory. On the one hand, tourism encourages local crafts and ceremonies; on the other hand, it can destroy local cultures and dramatically change local environments. One commentator suggested that 'we are more prone to vilify or characterise mass tourism as a beast; a monstrosity which has few redeeming qualities for the destination region, their people and the natural resource base' (Fennell 1999:7). Increasing awareness of the problems associated with conventional mass tourism has stimulated development of various 'alternative' types of tourism centred on unspoiled environments and the needs of local people. Table 13.7 lists some of the advantages of such forms.

Recent years have seen a striking upsurge of interest in 'environmentally aware' or 'sustainable' tourism, sometimes called **ecotourism**. Belize, in Central America, is one of the best-known ecotourist destinations, and has hosted two major conferences on the topic. Tourism accounts for 26 per cent of Belize's gross national product and is becoming more important as prices for traditional cash crops, especially sugar cane, drop. Awareness of the fragility of coastal ecosystems led the government of Belize to encourage ecotourism; however, several observers have seen little difference between the developments designed for ecotourists in Belize and other, more traditional developments (Wheat 1994:18). In fact, specialized alternatives like ecotourism are not likely

TOURISM IN SRI LANKA

13.12

The Sinhalese ruler Parakramabahu (1153-86) built the city of Polonnaruwa, the site of this 14-m-long figure depicting the buddha entering nirvana (Euan White).

Sri Lanka is typical of tourist destinations in the less developed world in that its tourist industry is problematic and uncertain. Long described as a paradise by Europeans, it is a logical area, from a European perspective, to be marketed as a tourist destination. There are scenic landscapes, including beaches, tropical lowlands, and mountains, as well as a rich and diverse cultural heritage. During the 1960s, the Sri Lankan government created the necessary infrastructure of hotels, roads, and airline facilities, and improvements and additions to this infrastructure have continued. The industry experienced spectacular growth,

and by 1982 tourism was second only to tea as an earner of foreign exchange.

A period of decline between 1983 and 1989 reflected the ethnic conflict described in Box 6.9. Tourist numbers dropped by more than 50 per cent; northern and eastern Sri Lanka experienced the greatest losses in tourist numbers and hence the greatest economic damage. After 1989, the industry began to recover and attract visitors from elsewhere in Asia. As of the mid-1990s, the industry seemed more secure because of strong government support and careful regulation (O'Hare and Barrett 1993), but the future is still uncertain.

ever to attract the market that mass tourism does, although they may be able to make some contribution to local economies.

Creating Places and Peoples for Tourists

Culturally, the tourist industry can be seen as one more example of the dominance of the more developed world over the less developed. It can be argued that the activities of tourists both reflect and reinforce, perhaps even legitimize, existing patterns of inequality and the dominance of some groups over others. Increasingly, one reason for people to visit 'exotic' peoples and places is to experience cultural difference. Yet—not surprisingly—those peoples and places are rarely authen-

tic; in many cases they have been constructed specifically to satisfy what some critics refer to as 'the tourist gaze'. Travel companies often play on the desire for novelty by using 'exotic' images and descriptions in their brochures. In fact, for many tourists, the places and peoples they visit are commodities, and—as with many other commodities—the advertising used to present them to prospective consumers relies more on illusion than on factual information.

Human geographers are now analyzing images and representations such as those used in tourist brochures in order to unpack and understand the meanings they convey, and in many cases they are finding that such advertising stresses the power of the tourist in some

Variable	Conventional Mass Tourism	Alternative tourism
Accommodations		
spatial pattern	high density/concentrated	low density/dispersed
size	large-scale	small (local)-scale
impact	obtrusive (modifies local landscape)	unobtrusive (blends into local landscape)
ownership	non-local/big business	local/small business
Attractions		
emphasis	commercialized cultural/natural	preserved cultural/natural
character	generic, contrived	local, authentic
orientation	tourists only	tourists and locals
Market		
volume	high volume	low volume
frequency	seasonal (cyclical)	year-round (balanced)
segment	mass tourist	niche tourist
origin	a few dominant markets	diverse origins
Economic impact		
status	dominant sector	complementary sector
linkages	non-local	local
leakages	profit expatriation, high import level	low import level
multiplier	low multiplier	high multiplier
Regulation		
control	mainly non-local	mainly local
amount	minimal	extensive
basis	self-regulating free market forces	public sector emphasis
priority	develop, then plan	plan, then develop
motives (holistic/integrated)	economic (sectoral)	economic, social, environmental
timeframe	short-term	long term
development and tourist ceilings	no ceilings	imposed ceilings which recognize carrying capacities

Table 13.7 CHARACTERISTIC TENDENCIES: CONVENTIONAL MASS TOURISM VS. ALTERNATIVE TOURISM

Source: D.B. Weaver, 'Ecotourism in the Small Island Caribbean', *Geojournal* 31 (1993):458

dependent, often former colonial, area. Specifically, using the terminology introduced in Chapter 8, it is clear that 'tourist' places and peoples are being socially constructed—in terms of ethnicity, gender, food and drink—in order to stress the difference between 'them' and 'us' and enhance the elements of mystery and exoticism that attract tourists. Figure 13.13 shows one way of understanding the relationship between the places produced by the tourist industry and the places consumed by the tourist, suggesting that the social construction of tourist places comprises two sub-systems: place production by the tourist industry and place consumption by the tourists. Where these constructions are in agreement there is a zone of convergence.

As this discussion suggests, recreation and tourism are increasingly attracting the attention of human geographers interested in the new cultural geography introduced in Chapter 8. In particular there is an increasing interest in tourism as a process of consumption, a product of place construction, and a reflection of difference.

The Geography of Uneven Development

Much of our discussion has hinted at the relationship between industry, larger issues of economic change, regional disparities, and the possible value of regional planning. It is immediately evident that spatial inequalities exist at a variety of scales with respect to many factors that affect quality of life. How do we explain these differences and deal with them?

Explaining Variations in Regional Economic Growth

Explaining spatial variations in economic development and quality of life is not easy. Geographers are aware that disparities are common at all spatial scales, from the basic global division into two worlds all the way to local differences between, for example, residential areas in a city. Following are five of the many possible explanations for these disparities.

Physical geography

Knowing that human geographers no longer take simplistic environmental determinism seriously, you may be surprised by our first proposed explanation. Nevertheless, as we

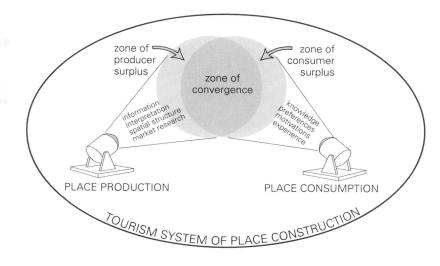

Figure 13.13 The tourism system of place construction
Source: After M. Young, 'The Social Construction of Tourist Places', *Australian Geographer* 30 (1999):386.

have seen on several occasions in this book, there is often a correlation between physical geography and a particular human pattern, such as the pattern of population distribution and density. On the global scale, there are obviously relationships between economic development and two aspects of physical geography: climate and proximity to a coastline. This is not to suggest that physical geography causes development, but clearly it can play an enabling or constraining role.

A GIS database of information on aspects of physical and human geography for 152 countries (accounting for 99.7 per cent of the world's population) shows that climate and proximity to coast are two essential considerations in a discussion of economic development (Mellinger, Sachs, and Gallup 2000). Temperate areas that are within 100 km of sea-navigable waterways are responsible for more than 50 per cent of global economic output, even though they account for only 8 per cent of the inhabited area of the globe. Findings such as these suggest that other areas are unsuited to long-term economic development: interior regions, for example, have high transportation costs, while tropical regions are more subject than others to infectious diseases and the productive quality of their soils is low.

Developmental stages

Societies have often been seen as travelling along a sequential path of development. Marx, for example, envisaged society passing from primitive culture to feudalism, to capitalism, to communism. Rostow (1960) saw a transition through five stages, from traditional society to mass-consumption society (Box 13.13). Others have seen society as experiencing shifts in the dominant occupational category, from primary to secondary to tertiary. Geographers have considered these various generalizations and contributed to them. In general, the sequence is as follows:

1. In the early stages of development, economy and society are fragmented, trade is limited, and primary activities dominate.
2. Subsequently, as transport improves and regions become more specialized and less subsistence-oriented, manufacturing develops and trading becomes important.
3. Finally, the service sector develops and regions become highly specialized.

This idea of developmental stages has also been subject to considerable criticism, however—recall the discussion of world systems theory in Chapter 6.

Staple theory

An alternative view is the staple theory of economic growth. A staple is a primary industrial product that can be extracted at low cost and for which there is a market demand. Originally developed by Canadian historians, the staple theory appears particularly applicable to areas of European overseas expansion. Staple success means economic growth both in the areas of extraction and in the export centres; hence staple production has direct impacts on regional growth. Indeed, other scholars have built models describing this process. Pred (1966), for example, focused on the staple's multiplier effects.

Core and periphery

A fourth general approach to regional differences in economic growth, proposed by Friedmann (1972), is based on the core–periphery concept. A core region is a dominant urban area with potential for further growth. Peripheries include areas of old established settlement characterized by stagnant, perhaps declining, economies, some of which may be former staple production areas; these are called downward transition areas. There are two other types of peripheral regions: upward transition regions are linked to cores and continue to be important resource areas, and resource frontier regions are peripheral new settlement areas. 'Peripheral regions can be identified by their relations of dependency to a core area' (Friedmann 1972:93).

13.13

THE ROSTOW MODEL OF ECONOMIC GROWTH

The Rostow model is a classic example of generalization about the economic development of capitalist states. On the basis of the European experience, Rostow proposed five stages:

1. *Traditional society:* Subsistence agriculture, domestic industry, and a hierarchical social system; a stable population/ resource balance.
2. *Preconditions for take-off:* Localized resource development because of colonialism or activities of a multinational corporation; export-based economy; often a dual economy (the Canadian Shield and the Canadian north might be considered to be in this stage).
3. *Take-off to sustained growth:* Exploitation of major resource; possibly radical and rapid political change (the NICs may be in this stage or the next).
4. *Drive to maturity:* Creation of a diverse industrial base and increased trade.

5. *Age of high mass consumption:* Advanced development of an industrial economy; evident in the more developed world when the model was developed in the 1960s.

This model is useful in an introductory human geographic context because it links easily to the demographic transition model (Chapter 5), the mobility transition model (Chapter 6), and the global diffusion of western cultures (Chapter 9 especially). As with these other models, however, the simplifications in the Rostow model can be risky: what happened in Europe and North America may not happen elsewhere (as Box 5.3 details with regard to fertility). It is also worth remembering that economic growth and development are not necessarily the same thing; recall the discussion of the human development index in Chapter 6.

Growth poles

A fifth approach argues that growth does not occur everywhere at the same time; rather, it manifests itself in points of growth. Thus a new area of resource exploitation will induce economic growth, but not necessarily at the source. Urban centres often serve as growth poles. Although a substantial literature has developed around the idea of growth poles, the concept is still not entirely clear.

Each of these five approaches has been widely applied, but each has serious deficiencies. As generalizations, they provide insights, but not explanations. Geographers investigating particular spatial and social issues tend to use the most appropriate model for their specific purposes. How, then, have geographers tackled the problems of unequal regional growth?

Developing the less developed world

Much of Chapter 6 was devoted to the disparities in quality of life between more and less developed countries. However, that chapter did not address the issue of significant regional differences within less developed countries. Governments and development agencies in most poor countries need to address these internal differences.

Since about 1980 many of these countries have been under sustained pressure from both the IMF and the World Bank to adopt top-down free-market economic policies that will result in structural adjustment. Specifically, governments are encouraged to open their economies to increase international trade, privatize previously state-owned enterprises, and reduce government spending. The logic is that such policies will generate wealth and development that will trickle down through all sectors of the economy to benefit all the people. This approach has been criticized on the grounds that in fact most of the wealth stays with the elite and any trickle-down benefits are minimal.

We saw in Chapter 11 that some failings of the green revolution have been blamed on the preponderance of top-down strategies. An alternative approach favoured by many national development agencies and by most NGOs working in poor countries, such as Oxfam, is to initiate development from the bottom, building on local community strengths and being sensitive to particular

A slum neighbourhood in Malabo, the capital of Equatorial Guinea, in August 2002. Despite an oil boom that by 2003 had attracted more than $3 billion in direct investment from US businesses, the country's ordinary people continue to be among the poorest on earth, and human rights groups have described the regime of Teodor Obiang as one of the most repressive in the world today
(AP photo/Christine Nesbitt).

local circumstances. Many poor countries are currently employing both approaches to regional economic and social development, and, and it may be that both are necessary to achieve meaningful growth.

Correcting Regional Economic Inequalities

Spatial variations in economic development and quality of life are normal outcomes of any process of change. Central or core areas will usually experience innovations before outer or peripheral areas, and the uneven distribution of resources prompts spatial variations. Unfortunately, in many parts of the world these natural spatial variations have resulted in differences that society considers inappropriate. Hence most countries have regional development policies.

The Canadian example

Governments have long intervened in the processes of economic change. In Canada, for example, companies were granted monopolies, the progress of settlement was dictated, and tariff barriers were installed—all before 1900. In the twentieth century, however, such intervention began to play a major role in redressing spatial imbalances. Typically, regional development policies have been designed to encourage growth in depressed, usually peripheral, areas.

A number of regions in Canada qualify as relatively underdeveloped. Indeed, some would contend that almost all of Canada outside the St Lawrence lowlands qualifies for this description. In the case of the Atlantic provinces—Newfoundland, New Brunswick, Nova Scotia, and Prince Edward Island—economic and related social problems stem from a peripheral location some 1,600 km (1,000 miles) from the major centres, a poor resource base, national tariff policies, and increasingly centralized transport and production systems. Inequality is evidenced by low wages and high unemployment. The Canadian government has attempted to address these issues by stimulating growth through low-interest loans to farmers and others and tax subsidies for industrialists. Some results have been achieved, but the basic problems have not been solved. The basic goals are to foster a self-sustaining entrepreneurial climate, more successful medium and small business enterprises, lasting employment opportunities, and an expanding competitive economy. These are not simple goals in a globalizing world that tends to reinforce the dominance of already successful regions rather than spread investment and opportunity to peripheral locations.

In the eyes of many Canadians, the Canadian Shield and the North—which together make up some 75 per cent of Canada—are frontiers that produce primary products for export that benefit primarily the Canadian core area; in many respects, then, these resource extraction areas are part of the less developed world. Others, notably Native populations, see these areas as a homeland. These very different images are not easy to reconcile (see Bone 1992).

The European example

The European Union is a second example of a political unit that contains core and peripheral regions and, as a result, regional inequities. As of 2004 there are 25 members. Recent moves towards enlargement and inte-gration have been taking place within a context of significant regional disparities. The number of peripheral regions with low incomes and high unemployment rates increases with each expansion. Several countries that joined the EU in the 1980s, especially Greece (1981), Spain (1986), and Portugal (1986) brought serious regional development problems, while the 2004 addition of ten members, eight of them from eastern Europe (Estonia, Latvia, Lithuania, Poland, the Czech Republic, Slovakia, Hungary, Slovenia) increases the breadth of regional variations. There are, therefore, very strong arguments for a European regional policy. To date, however, the principal policy has simply been to provide funds to depressed regions. An effective regional policy is clearly needed.

Decaying industrial areas

In core and peripheral regions alike, many areas are depressed as a result of outdated industrial infrastructures (see Box 13.11). A useful way to consider industrial change is to recognize four stages:

1. Infancy: initial primary activities and domestic manufacturing.
2. Growth: the beginnings of a factory system.
3. Maturity: full-scale development of manufacturing and related infrastructure.
4. Old age: decline and inappropriate industrial activity.

Today, the typical depressed region is located on a coalfield, lacks diversification, and is overly reliant on heavy industry. Unemployment is high and out-migration is normal. Depressed regions are similar in principle to peripheral regions that were formerly dependent on a staple, as both are unable to diversify when circumstances change.

The United Kingdom, which has both peripheral and depressed areas, has experienced considerable government intervention since the 1930s. Areas are variously designated as special development areas, development areas, and intermediate areas—designations that change according to political circumstances. Government intervention in the United Kingdom takes the form of development controls, financial incentives, and creation of industrial estates.

Planning in socialist countries

The clearest examples of regional development policies involve socialist-planned economies. By definition, socialist states do not rely on the capitalistic dynamics of private ownership and entrepreneurship. Rather, they depend on the rationalization of industry and the implementation of comprehensive planning. The majority of socialist states have been in the less developed world. Regional policies have included efforts to industrialize large cities (as in China during the Great Leap Forward [1958–9], when perhaps 20 million rural residents moved into cities) and to limit large city growth. The seventh five-year plan in China (1986–90), which focused on uneven development and rapid urbanization, was a direct reflection of the fact that earlier plans had largely failed to address two issues: income differences between urban and rural areas, and the domination of industry by urban areas located on the country's coast. The constants in Chinese planning are state direction, collective ownership of the means of production, and control over population movements—all features rare in capitalist countries.

We end this chapter with an observation about developmental inequality from Deng Xiaoping, the political leader of China for more than two decades before his death in 1997. Apparently Deng once remarked that it was perfectly in order for some parts of China to become rich before other parts; in fact, this was the way it had to be (Freeberne 1993:420).

CHAPTER THIRTEEN SUMMARY

Levels of production

Traditionally, three levels are recognized: primary (extractive), secondary (manufacturing), and tertiary (service). A fourth level (quaternary) involves the transfer of information.

Levels of economic organization

Two levels are recognized. Households produce and reproduce; as economic organizations they are of decreasing importance. Firms make up the commercial sector; manufacturing firms operate in factories.

Factors related to location

Industrial location theory explains why factories are located where they are. Traditionally, the relevant factors have included transport; distance from raw material sources, energy supplies, and the market; availability of labour and capital; the nature of the industrial product; internal and external economies; entrepreneurial uncertainty; and governmental considerations.

Least-cost theory

The typical location decision aims to maximize profits by minimizing costs or maximizing sales (or both). Least-cost theory, developed by Weber and published in 1909, is a normative theory that identifies where industries ought to be located. Following a series of simplifying assumptions, Weber concluded that industries locate at least-cost sites determined by transport costs, labour costs, and agglomeration–deglomeration benefits. Solving locational problems required defining concepts such as material index, locational figure, isodapane, critical isodapane, and isotim. Other researchers, especially Hoover, have improved Weber's formulation of the transport-cost variable.

Market-area analysis

Market-area analysis, the second major theoretical approach, centres on profit maximization rather than cost minimization. Lösch is the principal theorist.

Behavioural approaches

Behavioural approaches to the industrial location problem focus on the subjective views of the people involved and the common preference for satisficing rather than optimizing behaviour. Behavioural issues are important, albeit more difficult to quantify or theorize than strictly economic considerations. Contemporary industrial firms are, typically, complex decision-making entities.

Industrial revolution

Prior to the industrial revolution, industrial activity was domestic, small in scale, and dispersed. A revolution in industrial activity and related landscapes occurred between 1760 and 1860, beginning in Britain. Factories replaced households as production sites, mechanization increased, and localized energy sources were used. All these changes were related to the rise of capitalism, which emphasized individual initiative and profits. New industrial landscapes appeared on the coalfields of Britain and were accompanied by city growth, rural to urban migration, and a series of transport innovations. Inner-city slum landscapes developed in many of the new agglomerations on coalfields and in the traditional textile areas of northern England.

Energy

Coal was the main fuel during the industrial revolution, but was replaced by oil in the 1960s. OPEC is a major player in international affairs because of the importance of oil in the contemporary world. Some countries, such as New Zealand, are dependent on hydro-generated electricity supplies, which may be disrupted by droughts.

Industry in the more developed world

Industry is responsible for a disproportionate share of manufacturing employment and output. Most of this activity continues to take place in a capitalist economic and social framework. The global manufacturing system is dominated by North America, western Europe, and the west Pacific. Multinational corporations are increasing in importance.

Canadian industry

The dominant characteristics of Canada's industrial geography include abundant natural resources, a well-developed extraction industry, a branch-plant economy reflecting the country's proximity to the United States, a dispersed national market, a dispersed primary sector, and highly concentrated secondary and tertiary sectors.

Newly industrializing countries

Japan was the first of the NICs and has been followed by other Asian countries (notably South Korea), some southern European countries, and Brazil and Mexico. Export-processing zones using inexpensive and relatively unskilled labour have been established in some countries (for example, the *maquiladoras* in Mexico).

Industry in the less developed world

Industrial growth is highly desired by many less developed countries to bolster weak economies, provide employment, demonstrate economic independence, encourage urbanization, help create a better economic infrastructure, and reduce dependence on overseas markets for their primary products. But industrialization is difficult to achieve, not least because of the legacy of colonialism. Many countries have experienced their greatest industrial successes in import substitution. India has thrived since 1951, while China opened the doors to international trade and foreign investment and also reduced state control of industries during the 1980s. The most dynamic industrial region in the world today is the Pearl River delta in south China, just north of Hong Kong.

Industrial restructuring

Technological changes in production, transaction, and circulation are currently prompting some major changes in the global economy and in national and regional industrial geographies.

Information exchange

Industrial location decisions used to be made largely on the basis of transport costs, but today the costs of information exchange and labour are much more important. Efficient electronic transfer of reliable information allows firms to decentralize by seeking out areas of low labour cost. Where information exchange requires personal contact, location patterns tend to be centralized.

Service industries

As part of the process of economic restructuring and the transition from Fordism to post-Fordism, industrial societies are moving to a postindustrial service stage. Because the service industry is so diverse, a wide range of factors may determine locations. Central place theory offers the most important set of explanatory concepts, but behavioural approaches also seem relevant. New information technologies are undoubtedly affecting the location of services.

The tourism industry

The largest industry in the world, tourism is generated by the more developed world, but many favoured destinations are in the less developed world. Spectacles and theme parks are a distinctive form of tourist attraction. Many less developed countries take advantage of attractive climates and landscapes to cultivate a tourist industry, although their success can be affected by volatile political circumstances, as in the case of Sri Lanka. Tourist areas may benefit economically, but may also experience negative cultural and environmental consequences. Some countries, such as Belize, are encouraging ecotourism. Human geographers are increasingly interested in the ways in which the places and peoples consumed by tourists are represented.

Some social issues

Industrial geography, like other branches of economic geography, has neglected social issues. Now that industrial geography is becoming more concerned with issues such as the gender division of labour, it is increasingly turning to various social theories.

Spatial inequalities

Correcting spatial inequalities is seen as a government responsibility. Governments in capitalist countries frequently implement policies to improve the economic status of regions that are less developed, either because they are peripheral or because their industrial infrastructure is outdated. The most obvious instances of planning the spatial economy take place in socialist countries. Economic restructuring continues to contribute to uneven development.

Links to Other Chapters

- Globalization:
    - Chapter 2 (concepts)
    - Chapter 4 (ecosystems and global impacts)
    - Chapter 5 (population growth, fertility decline)
    - Chapter 6 (refugees, disease, more and less developed worlds)
    - Chapter 7 (cultural globalization)
    - Chapter 8 (popular culture)
    - Chapter 9 (political globalization)
    - Chapter 10 (economic globalization)
    - Chapter 11 (agriculture and the world economy)
    - Chapter 12 (global cities)
    - Chapter 13 (industrial restructuring)

- Industrial location theory:
    - Chapter 11 (agricultural location)
    - Chapter 12 (settlement location)
    - Chapter 2 (positivism).

- Industrial revolution:
    - All chapters.

- Energy:
    - Chapter 4 (changing energy sources).

- Differences in industrial geography of more and less developed worlds:
    - Chapter 6 (inequalities)
    - Chapter 11 (differences in agriculture)
    - Chapter 12 (differences in settlement).

- Industry in Japan; NICs; industrial geography of China:
    - Chapters 11 and 12 (multinationals).

- Tourist attractions:
    - Chapter 8 (popular culture).

- Ecotourism:
    - Chapter 4 (sustainable development)
    - Chapter 6 (less developed world).

- Representations of peoples and places in the tourist industry:
    - Chapter 8 (introduction to concepts).

- Industrial restructuring (especially in the service industry):
    Chapter 12 (industry and services in cities; global cities).

- Industry and society:
    Chapter 2 (Marxism)
    Chapter 8 (Marxism, class, gender).

- Environmental considerations in industry:
    Chapter 4 (environmental issues).

- Uneven development:
    Chapter 11 (core and periphery)
    Chapter 9 (trade and trade blocs)

FURTHER EXPLORATIONS

BALE, J. 1981. *The Location of Manufacturing Activity*. Edinburgh: Oliver and Boyd.
    An easy-to-read basic text that is full of examples of industrial location issues; focuses on British examples.

BATHELT, H., and A. HECHT. 1990. 'Key Technology Industries in the Waterloo Region: Canada's Technology Triangle (CTT)'. *Canadian Geographer* 34:225–34.
    A detailed analysis of thirty-three high-technology firms showing that the smaller firms are the most research-intensive, that all the firms are strongly linked to local economic activities, and that the strongest locational factors are the availability of skilled labour and local residence or education.

BLACKBOURN, A., and R.G. PUTNAM. 1985. *The Industrial Geography of Canada*. London: Croom Helm.
    A very readable account that includes discussions of each Canadian region and raises many of the issues that face manufacturing industries in the more developed world.

BRITTON, J. 1996. *Canada and the Global Economy: The Geography of Structural and Technological Change*. Montreal: McGill-Queen's University Press.
    An excellent book that helps to explain the contemporary global economy and Canada's role in it; focuses on the importance, for Canada, of openness to external influences, regional variations in resource bases and urbanization, rapid changes in technology and markets, and government strategies.

BRITTON, S., R. LE HERON, and E. PAWSON, eds. 1992. *Changing Places in New Zealand: A Geography of Restructuring*. Christchurch: New Zealand Geographical Society.

An account of the experiences of New Zealand following ideologically driven government decisions after 1984 to favour economic deregulation as a way of coping with changes in the international economy.

BURTON, R. 1995. *Travel Geography*, 2nd ed. London: Pitman.
    A factually detailed overview of both the world's geographical resource base for tourism and the spatial patterns of world tourist activity.

DANIELS, P. 1993. *Services in the World Economy*. Oxford: Blackwell.
    A thorough account of the service component of industrial activity, with detailed discussion of examples.

DICKEN, P. 1993. 'The Changing Organization of the Global Economy'. In *The Challenge for Geography: A Changing World, A Changing Discipline*, edited by R.J. Johnston. Oxford: Blackwell, 31–53.
    An eminently readable account of contemporary trends in the global economy, with details of some changes in industrial geography.

EDGINGTON, D.W. 1993. 'The New Wave: Patterns of Japanese Direct Foreign Investment in Canada during the 1980's'. *Canadian Geographer* 38:28–36.
    An analysis of how Japanese investment during the 1980s helped diversify and strengthen the economic importance of the Canadian heartland in southern Ontario.

ENSTE, H., and C. JAEGER, eds. 1989. *Information Society and Spatial Structure*. London: Belhaven Press.
    A series of readings with a focus on new information technologies and their relation to service industries.

HAMILTON, F.E.I., ed. 1974. *Spatial Perspectives on Industrial Organization and Decision Making*. London: Wiley.

An explicit reaction to neoclassical orthodoxy; an early indication of behaviourists' concern with how industrial location decisions are actually made, as opposed to how they ought to be made.

HARRINGTON, J.W., and B. WARF. 1995. *Industrial Location: Principles and Practices*. New York: Routledge.

A valuable book that covers a wide range of material; includes both conceptual and empirical analyses.

HAY, A.M. 1976. 'A Simple Location Theory for Mining Activity'. *Geography* 61:65–76.

A useful contribution to a largely neglected area; illustrated by an educational game.

JONES, K. 1989. 'Editorial'. *The Operational Geographer* 7, no. 2:2.

An introduction to a series of useful articles on services and marketing activities in Canada, especially Ontario.

KEEBLE, D. 1989. 'Core-Periphery Disparities, Recession and New Regional Dynamisms in the European Community'. *Geography* 74:1–11.

A detailed account of regional issues with some clear discussions of deindustrialization, reindustrialization, and tertiarization.

KRUGMAN, P. 1995. *Development, Geography, and Economic Theory*. Cambridge, Mass: MIT Press.

A thoughtful and thought-provoking book by an important economist that reflects some non-traditional views; regrets the rarity of 'spatial' considerations in economics; useful also for Chapters 10 and 11.

MASSEY, D. 1995. *Spatial Divisions of Labour*, 2nd ed. New York: Routledge.

Now considered a classic; a valuable introduction to the type of industrial geography that emerged during the 1980s, emphasizing labour and employing a Marxist approach (as opposed to the previous neoclassical focus on raw materials, energy sources, and transport).

MATHER, C. 1993. 'Flexible Technology in the Clothing Industry: Some Evidence from Vancouver'. *Canadian Geographer* 37:40–7.

An account of the impact of flexible technology on the Vancouver textile industry; shows that some old technologies are still important, and that impacts on labour are largely negative.

MORGAN, N., AND A. PRITCHARD. 1998. *Tourist Promotion and Power: Creating Images, Creating Identities*. Toronto: Wiley.

A thoughtful and challenging overview of images in the tourist industry; draws on ideas from both the cultural studies and human geographic traditions; includes numerous short case studies.

NORCLIFFE, G. 1993. 'Regional Labour Market Adjustments in a Period of Structural Transformation: An Assessment of the Canadian Case'. *Canadian Geographer* 38:2–17.

A detailed account of changes associated with the transformation from mass production to flexible production; demonstrates that the Canadian experience is distinctive because the transformation has been mediated by a specific staple regime of accumulation.

SANDBERG, L.A. 1989. 'Geographers' Perception of Canada in the World Economic Order'. *Progress in Human Geography* 13:157–75.

An interesting account of recent and current perceptions of Canada in the global economic context that aids understanding of regional disparities.

SMITH, D.M. 1981. *Industrial Location Theory: An Economic Geographical Analysis*, 2nd ed. New York: Wiley.

A comprehensive account of all aspects of industrial location theory; one of the best books available on the general theme of the location of economic activity; integrates cost models with demand factors over time.

WATTS, H.D. 1987. *Industrial Geography*. London: Longman.

An excellent survey of the field with a distinctive focus on the human consequences of contemporary industrial change.

WEAVER, D.B. 1998. *Ecotourism in the Less Developed World*. New York: CAB International.

Places ecotourism in a larger context, as one component of alternative tourism; includes chapters on Costa Rica, Kenya, Nepal, Thailand, and the Caribbean and South Pacific.

WEBBER, M.J. 1972. *Impact of Uncertainty on Location*. Cambridge, Mass: MIT Press.

An excellent if rather advanced statement concerning decision-making in industrial and other location decisions.

On The Web

http://www.bp.com/centres/energy2002/ downloads/ index.asp
The annual report produced by BP provides a wealth of data and analysis concerning energy.

http://www.iea.org/statist/keyworld2002/key2002/ keystats.htm
The International Energy Authority, established in November 1974, has gained recognition as one of the world's most authoritative sources for energy statistics. Its massive annual studies of oil, natural gas, coal and electricity are indispensable tools for scholars as well as energy-policy makers and companies involved in the energy field.

http://info.ic.gc.ca/
Industry Canada publishes this site, which mostly covers news items and governmental matters.

http://www.oecd.org/EN/home/0,EN-home-0-nodirectorate-no-no-no-0,00.html
Homepage of the OECD.

http://parkscanada.gc.ca/
Parks Canada publishes this site, which includes links to numerous discussions of issues related to tourism and recreation.

http://www.world-tourism.org/
The World Tourism Organization is the leading international organization in the field of travel and tourism. It serves as a global forum for tourism policy issues and a source of practical information and statistics.

The Human Geography of the Future . . . and the Future of Human Geography

The introduction to this book identified three recurring themes: relations between humans and land, regional studies, and spatial analysis. By now you should have a clearer idea of why these themes are so central to human geography.

The humans-and-land theme is the underlying focus of Chapter 4, on human impacts on land, but it also played an important part in the discussions of cultural regions and landscapes in Chapters 7 and 8. Philosophically, discussions of topics such as these are generally possibilist in character. The empirically oriented regional studies theme frequently reappears—not surprisingly, given that the aim of regionalizing is to classify locations. Chapter 3 describes a series of physical geographic regions; the population geography introduced in Chapter 5 anticipates the global regionalization into more and less developed worlds that is detailed in Chapter 6; Chapters 7 and 8 discuss cultural regions, including the vernacular variety, and the distributions of languages and religions. The regional theme is also important in the political geography introduced in Chapter 9, especially in the context of ethnic identities, and in the outline of world agriculture in Chapter 11. Finally, the spatial analysis theme and the positivist approach associated with it

are especially well represented in the four chapters (10, 11, 12, and 13) concerned with economic geography, particularly with respect to location theories.

The introduction also pointed out that one goal of human geography—and this book in particular—is to help us understand our changing world as it is today, and how it came to be so. Human geography is a discipline of tremendous breadth and diversity, with especially close links to the natural sciences, particularly physical geography, and a healthy academic pedigree. A human geographic perspective—a geographical imagination—is an invaluable asset. The editor of a major world newspaper, *The Times*, has written of school geography:

> It seemed a discipline that took a child out of the classroom, into the street and the park and said: 'Look what's here! How did it come to be here? What is it made of?' On such empirical enquiry rests all learning. Geography offered such empiricism in the most comprehensible and immediate form. It also offered the basis for argument, for disagreement, for controversy, for the spirit of dialectic that makes for true understanding, not rote learning (Jenkins 1992:193–2).

If it is not possible to do full justice to the title of this final chapter, we must at least attempt to comment on current changes both in our subject matter itself—human behaviour as it affects the earth's surface—and in our methods of studying it.

Changing Human Landscapes

Human geography is an exciting field. If the first chapters of this book suggested to you that human geographers today must look back nostalgically to the golden days when every journey meant new facts to be discovered, catalogued, and understood, by now you probably realize that the end of the exploration era did not by any means signify an end to discovery. Humans are constantly changing landscapes in a myriad of ways—cultural, social, political, and economic. There may not be a wealth of new facts over the horizon, waiting to be discovered, but the facts within the horizon are constantly changing, as is our understanding of those facts. Indeed, the facts are changing at an ever-increasing pace: because there are so many of us to cause change, because technology is always improving, and because the world we live in is shrinking as a result of advances in communications and other forms of movement. Three examples of changing human landscapes—the human geography of the future—will clarify these generalizations.

People and Resources

Population and society

Human geography is about human beings, and probably the single most important change related to human beings is the continuing increase in our numbers. As we saw in Chapter 5, the rate of natural increase in the world population peaked in the 1960s and is now declining (the fertility transition), but the total number of people continues to increase. The predicted world population for 2050 is 9.1 billion. Most of the current growth is taking place in the less developed world; many countries in the more developed world have natural increase rates close to zero (see Table 5.4). The composition of the world population is also changing as the numbers of elderly people increase, and important social transformations are under

way as gender roles and family patterns change (recall the Box 5.9 discussion of the conferences held in Cairo in 1994 and Beijing in 1995).

With population growth below replacement levels, population aging, and corresponding declines in the labour force, immigration is likely to become an increasingly important—and divisive—issue in the more developed world. Meanwhile, the shift will continue away from industrial activity and towards 'knowledge' or 'information' as the economic engine of society. Since knowledge travels freely, without regard for physical borders, the prospects for upward social mobility through higher education will increase—but so will the competition that individuals face.

As recently as about 1750, most people lived in rural areas and derived their living from agriculture. It was in these conditions of relative isolation from others that many of the different human group identities apparent today developed—before the massive changes initiated by the industrial revolution. These differences of language, religion, and ethnicity have left a legacy of cultural and social diversity that continues to shape the contemporary world (recall the discussion of regional identities and political aspirations in Box 9.9).

Urbanization

Today an ever-increasing percentage of the world's population lives in urbanized areas, and this trend is likely to continue (see Figure 12.1); the number of very large cities will also continue to increase, especially in the more developed world (see Figures 12.14 and 12.16). These ongoing increases will bring further changes in the urban way of life, and if the problems evident in urban areas of the less developed world are any indication, many of these changes will be in an undesirable direction.

The experience of urban life is also changing in the more developed world, as the example of Tokyo vividly illustrates. With up to 60,000 persons per square kilometre, Tokyo is the most crowded city in the world. There is a constant demand for more space (Greenlee 1988). Initially, this demand expressed itself in urban sprawl, but more recently it has involved internal reorganization. The population of 28 million-plus in the larger urban region is now making ingenious use of the urban area by building upwards, downwards, and into the sea. Building upwards is a relatively recent alternative for a city prone to earth tremors. In addition to skyscrapers, there are elevated highways and railways and even multi-storey golf driving ranges. Building downwards is also popular; some buildings use at least four below-ground levels for retail and office purposes. Perhaps most dramatically, Tokyo is reclaiming land from the sea to permit more expansion. It has even been suggested that the city might extend out onto a floating platform in Tokyo Bay. Tokyo is an example of an ever-changing urban area with some unusual innovations.

Human impacts

More people mean more demands both on resources and on agricultural and industrial outputs. Currently, energy use per capita is much higher in the more developed world than it is in the less developed world. Necessarily, then, energy demands in the less developed world are increasing as expectations rise and industrialization proceeds. The basic conclusion is unavoidable: the impacts on our fragile home, as discussed in Chapter 4, will also increase. Although it appears that awareness of human impacts is encouraging a shift, at least in the more developed countries, from an anthropocentric to an ecocentric view of the world—an idea discussed in the context of sustainable development in Chapter 4—there is still a long way to go. To put the issue in slightly different terms, many human geographers now challenge the basic premise on which industrial societies have been established: the idea that progress depends on conquering nature and increasing production. This idea, centred on what has been called the 'treadmill of production' (Schnaiberg and Gould 1994:v), can be seen as the cause both of environmental degradation and of inequalities in the social well-being of different groups. Problems of the global environment and social justice are not separate issues.

A Changing Political World

Every day, the media report on events of geographic interest. To understand the Iraq war of 2003 and its aftermath, for example,

Reclaimed land in Tokyo harbour; in the background is Mount Fuji. The two windmills are part of an effort to harness wind power (AP photo/Koji Sasahara).

we need to know something about at least five thoroughly geographic subjects: the way the states in the area were created; the spatial distribution of various ethnic groups (Shiite Muslims in the south, Sunni Muslims in the centre, and Kurds in the north); the importance of oil and natural gas to the economies both of producing countries and of their customers in the more developed world; the relations of dependency that exist between more and less developed worlds; and the geopolitical perception of the United States following the 2001 terrorist attacks.

The same is true of the terrible conflict unleashed by the political collapse of the former Yugoslavia in the early 1990s as different linguistic, religious, and ethnic groups appealed to perceived ethnic identities and historical injustices. As we saw in Box 9.7, understanding the Yugoslavian conflict demands recognition of the region's complex cultural and political geography and the history of its domination by other empires.

In 1997, internal conflict erupted in the neighbouring state of Albania—sometimes described as the only Third World country in Europe. Among the many complex reasons were the difficulties experienced in the transition from a planned to an open market economy, following the 1991 collapse of communism, and the perception among those in the south that they were less well off than those in other regions of the country.

Then in 1999 a new round of violence erupted in Kosovo, a southern province of Serbia inhabited mostly by ethnic Albanians. Many commentators attributed that conflict to the particular effect of Kosovo on the Serbian national psyche, for it was in Kosovo that the Serb Prince Lazar was killed by Turks in 1389—a defeat that would lead to Serbia's annexation by the Ottoman Empire in 1459. Nevertheless, as a result of Prince Lazar's reported heroism in the face of adversity, many Serbs—or at least Serb leaders striving to reinforce Serbian identity—came to regard Kosovo as a sacred space. Moreover, according to the Serbs, the ethnic Albanian population of the region had sided with the Turks. Not surprisingly, the Albanian people of Kosovo see things differently: according to their traditions, Albanians were living in the region before the Serbs, and fought with the Serbs against the Turks. When, in 1999, Serb forces entered Kosovo, purportedly to quell an armed independence struggle by the Kosovo Liberation Army, the conflict led to several instances of ethnic cleansing and massive refugee movements of ethnic Albanians. These conflicts appeared to finally play themselves out in 2003, when the republics of Serbia and Montenegro agreed to drop the name of Yugoslavia.

Similarly, the ethnic conflict in the African state of Rwanda during the mid-1990s can be better understood if one has an appreciation of the human geography of the region (see Box 9.6). Rwanda is situated in one of the most densely populated regions in Africa—the highland region surrounding Lake Victoria. Part of German East Africa in the late nineteenth century, both Rwanda and neighbouring Burundi were placed under Belgian administration at the end of the First World War and became independent, as separate states, in 1962. About 90 per cent of the population in Rwanda are Hutu, Bantu-speaking herders and farmers (82 per cent in Burundi). These people were formerly subordinate to the minority Tutsi, who are pastoralists and probably of Hamitic descent. The ethnic violence that erupted between these two groups in the 1990s was only the latest instance of a longstanding conflict.

Another state in central Africa, the Democratic Republic of Congo (formerly Zaire), descended into chaos and then experienced a change of government in early 1997. As we saw in Chapter 12, in a brief reference to the capital city, Kinshasa, the country has large mineral reserves, extensive forests, and some good agricultural land, and has received huge amounts of development aid. However, between about 1990 and 1997, conditions declined rapidly because of a corrupt government, ethnic tensions that developed partly as a consequence of the huge influx of Rwandan refugees in the early 1990s, and the larger context of a poor population experiencing rapid increase and, as a result, increasing pressure on the country's resource base. According to many current commentators, similar scenarios may well be played out in other African states. The formation of the new African Union in 2002 may offer some hope for the future.

It also seems highly probable that there will be further assertions of autonomy by groups who perceive themselves as different from the majority groups in their states. A fairly recent example comes from China, where the Uighurs (see Chapter 9) in the northwestern province of Xinjiang, also known as Turkmenistan, launched a new round of independence claims in 1997, following the death of longtime Chinese leader Deng Xiaoping. Building on earlier discussions of group identity in Chapters 7 and 8, the autonomy theme was central to much of Chapter 9.

Changing People in a Changing World

Recent years have seen the collapse of many spatial barriers, and each such collapse has many impacts. Yet as we saw in Chapter 10, the collapse of barriers in the context of globalization is only one trend—if perhaps the most obvious—among several: specifically, regional integration, the staking of

Rwandan refugees arriving
in Tanzania (CIDA photo/
Roger Lemoyne).

territorial claims by particular ethnic groups, and a possible weakening of the role played by the nation state.

The uniqueness of human beings resides in both our biological attributes and our ability to adapt culturally. We humans have managed to avoid extinction because of our culture— essentially, our ability to develop ideas from experience and to act on the basis of them. Once established, however, culture can actually be an impediment to change. The fact that most people are most comfortable with familiar patterns goes a long way towards explaining our desire to identify with place in a world where, for many of us, places are changing at a bewildering pace—constantly being created and destroyed (recall the discussion in Chapter 8 of the importance of place). In his book *A Question of Place*, Johnston (1991) argues that geography needs to study what he calls 'milieux': places within which ways of life are constructed and reconstructed—in short, places in their entirety.

Disappearing peoples

One effect of the collapse of spatial barriers is that the numbers of people who live in relatively traditional ways, detached from the larger global context, are rapidly decreasing. Tribal peoples use deceptively simple technologies to live in close harmony with their physical environment. All too often, however, they are in the way of what

most of us consider to be progress. Such 'progress' has destroyed numerous tribal peoples perceived as backward and ignorant by the prejudiced, culturally arrogant advocates of 'progress'. Although we are becoming increasingly aware of the negative consequences of such 'progress', little is being done to protect the few tribal groups that remain. It is estimated that as many as ten such groups disappear each year. In 1989 Survival International—a group formed in 1969 to help tribal peoples protect their rights—estimated that only 29 such peoples remained in existence (Handbury-Tenison and Shankey 1989) (see Table Conclu.1).

The Batwa are members of the Pygmy group included in Table Conclu. 1. Once they roamed the forest area of southwest Uganda, but today they are destitute. Although they had lived in the region for over 50,000 years as foragers and were generally recognized as belonging to the high forest area, in the nineteenth century they were displaced by other groups who began using much of the area for agriculture. Today the Batwa are socially subordinate to the newcomers, for whom they work, and are subject to widespread discrimination. Ironically, in 1991 the Batwa were even forced out of some of their traditional territory when it was set aside as a refuge for the rare mountain gorilla. Today only about 1,850 Batwa survive; making up about half of the Ugandan pygmy population, they are mostly squatters or

The Iban people of Sarawak were once known as Borneo's most fearsome headhunters. Today they make up about a third of the population of Sarawak. The state's official tourism website describes them as a 'generous, hospitable and placid people' who have adopted 'a peaceful agrarian lifestyle' (Andrew Leyerle).

tenants on land owned by others, and subsist largely by begging. Their culture is lost and their numbers are decreasing.

In the case of the Batwa, it may be that the people themselves are disappearing along with their culture, but in other cases a people may simply abandon their traditional culture in the process of modernization. There are at least two ways of thinking about the disappearance of a traditional way of life. Most scholars in the field would probably regret the loss. Others, however, might argue that such regret suggests a preference for keeping primitive people primitive—in effect, reducing them to the equivalent of zoo animals.

Changing Human Geography

Chapters 1 and 2 of this book outlined the evolution of human geography and identified key philosophies, concepts, and techniques of analysis. One purpose of these accounts was to demonstrate how our discipline is always

Table Concl.1 DISAPPEARING PEOPLES

Group	Location	Number Remaining 1989	Principal Threats to Survival
Inuit	Alaska/Canada	100,000	Loss of identity
Iroquois	St Lawrence region	22,600	Loss of identity
Hopi	Arizona	10,000	Forced relocation
Hawaiians	Hawaii	9,000	Loss of identity, tourism
Lacandon	Mexico	300	Deforestation, settlement
Kuna	Panama/Colombia	50,000	Deforestation, tourism
Kogi	Colombia	2,000	Settlement, tourism
Yanomami	Venezuela/Brazil	21,000	Mining, epidemics
Emerillion	French Guyana	200	Loss of identity
Waorani	Equador	80	Oil exploration, settlement
Nambikwara	Brazil	300	Settlement, epidemics
Caraja	Brazil	800	Industrialization
Jurana	Brazil	50	Physical extinction
Caingua	Paraguay/Argentina	3,000	Destruction of environment
Taureg	Sahara Desert region	300,000	Loss of identity
Pygmy	Zaire/Congo/Gabon	200,000	Deforestation, settlement
Bushmen	Kalahari Desert region	60,000	Loss of identity
Sammi	Norway/Sweden/Finland/USSR	58,000	Loss of identity
Gond	India	3,000,000	Industrial activities
Vedda	Sri Lanka	2,000	Destruction of environment
Onge	Andaman Islands (India)	100	Physical extinction
Semang	Malaysia	2,500	Deforestation, forced conversion to Islam
Penan	Sarawak region	10,000	Logging, lack of land rights
Akha	China/Burma	50,000	Lack of land rights
Ainu	Japan	3,500	Loss of identity, tourism
Tasaday	Mindanao (Philippines)	30	Physical extinction
Papuans	West Papua (Indonesia)	1,000,000	Settlement, oil exploration
Aborigines	Australia	250,000	Loss of identity
Maoris	North Island (New Zealand)	300,000	Loss of identity

Source: Adapted from R. Handbury-Tenison and C. Shankey, 'An Uphill Struggle for Survival', *Geographical Magazine* 61 (1989):29–33.

changing, largely in response to the changing needs of society. Human geography, like other academic disciplines, exists to serve society. There is, then, every reason to believe that the discipline described in this book will be a different discipline tomorrow. Attempting to predict the details of such change is hazardous, but we may at least identify what seem to be the most pressing societal demands on human geography today.

Space and Place

First, the need for an academic discipline focusing on space and place seems self-evident. Space is important because we need to understand location, distances between locations, and movements between locations if we are to make sense of the world. It is for this reason that the single most important geographic contribution to human life is the map. Maps are the means by which geographers make sense of the world; maps allow us to understand and explain why things are where they are, and to express the closely related concepts of space, location, region, and distance. It is because maps are so important that geographers have taken advantage of recent technological advances to develop the more sophisticated tool of geographic information systems (GIS).

Place is important because understanding the meaning that people assign to locations is crucial if we are to comprehend their attitudes and behaviours. All people have an innate sense of place, but human geographers are able to refine this innate sense into a knowledge that facilitates understanding of the changing human world. Our recurring discussions of globalization, regional character, and local change have continually reinforced these points.

Together, space and place as they relate to human feelings, beliefs, attitudes, and behaviour lie at the core of human geography.

A Synthesizing Discipline

The second reason the discipline of geography is important to society is that it offers the best framework for investigating human impacts on global and local environments. Most ecological research benefits immeasurably from some understanding of the links between human and physical geography. For example, many disciplines may try to address

the consequences of global deforestation, but only geographic responses have the benefit of a solid background in both physical and human sciences.

The nineteenth-century American geographer and congressman George Perkins Marsh wrote in 1864:

> . . . the earth is fast becoming an unfit home for its noblest inhabitant, and another era of equal human crime and human improvidence, and of like duration . . . would reduce it to such a condition of impoverished productiveness, of shattered surface, of climatic excess, as to threaten the depravation, barbarism, and perhaps even extinction of the species (Marsh 1965:43).

Focused, like Marsh, on the relationship between humans and land, geographers continue to play an important role in both environmental analysis and environmental education.

Handling Data

A third way in which both human and physical geography are serving society today is through the development and application of remote sensing and electronic data processing. In an increasingly complex world with ever larger data sets, remote sensing and global positioning systems allow us to acquire data rapidly and repeatedly, while electronic data processing permits the conversion of data into maps and facilitates data analysis. Perhaps the best-known recent advance is the development of geographic information systems—integrated computer systems for the input, storage, analysis, and output of spatial data.

These new ways of handling data are much more than mere extensions of earlier procedures. In an important sense, they are changing the nature of geographic information and the role that such information plays in society. A traditional map is a single product that is manufactured for users, whereas a GIS can be manipulated by the user: data can be added or removed depending on specific interests, scales can be changed (the user can 'zoom in' on a particular area, for example), and the data can be updated as new information becomes available. Remember, new tools prompt new thoughts.

Understanding and Solving Problems

Fourth, human geography facilitates understanding of people and place by explaining how different peoples see and organize space differently. Hence geographers are acutely aware that there are no simple or universal solutions to problems. Effective changes can be made only following full analysis of spatial variations. Sensitive to spatial variations and the significance of place, human geographers know that solutions to a problem in one area may not be solutions to similar problems in another area. This sensitivity is evident, albeit in different ways, in all four of our major philosophical traditions—empiricist, positivist, humanist, and Marxist. Understanding space and place is important in itself and because it allows us to propose solutions to social and environmental problems. This focus is particularly evident in the new and vital subdiscipline of cultural geography introduced in Chapters 7 and 8. This progressive subdiscipline is increasingly concerned with highlighting regional and global inequalities and working towards solutions.

Values

What underlying values direct human geographic research? The general answer is that human geographic values are socially specific, reflecting the values that legitimate the societies in which we live. As we saw in Chapter 2, whereas the positivist school of geography purported to be value-free, Marxist and humanistic geography contributed the important insight that values are socially constructed and therefore subject to change. This general answer does not, of course, address specific questions about values, such as the relative merits of environmental preservation on the one hand and economic gain on the other hand (Buttimer 1974).

Being Human Geographers: Where We Began

By now the advantages of a training in human geography should be clear; the advantages of being human geographers follow logically. A training in human geography allows us to understand our world; being human geographers allows us to put that training into practice. Most students of human geography do not formally become human geographers; rather, they find employment in a wide range of professions. A training in human geography encourages understanding of the importance of spatial scale, global and local alike. It emphasizes interactions between humans and land as well as emotional attachments to place; and it develops an appreciation of the major global changes in population numbers, culture, political identity, and land use. This understanding is combined with an enviable set of skills that includes not only traditional quantitative and qualitative research procedures, but also the distinctive tools of cartography, remote sensing, and geographic information systems. It is not surprising that human geographers find employment in such diverse arenas as teaching, business, industry, government, consulting, urban and regional planning, conservation and historical preservation, and libraries and archives.

Most important, any training in human geography provides life-long benefits—benefits that we might summarize under the heading of 'geographical imagination'.

Geographical Imagination

What is geographical imagination? Simply put, it is appreciation of the relevance of space and place to all aspects of human endeavour. Peattie (1940:33) recognized this more than half a century ago when he wrote that geography 'explains ways of living in all their myriad diversity'. More recently, Gregory (1994) offered a detailed commentary on geographical imagination as sensitivity to space and place in the effort to understand ways of life.

Geographical imagination is independent of philosophical preference:

- From an essentially *positivistic* viewpoint, Morrill (1987:535) observed, 'If there is not a convincing theory of why and how humans create places and imbue them with meaning, then it is time to develop that theory.'
- From an essentially *humanistic* viewpoint, Prince (1961:231) asserted, 'Good geographical description demands not only respect for truth, but also inspiration and direction from a creative imagination. . . . it is the imagination that gives [the facts] meaning and purpose through the exercise of judgement and insight.'

- Finally, from a *Marxist* viewpoint, Harvey (1973:24) observed, 'This imagination enables the individual to recognize the role of space and place in his own biography, to relate to the spaces he sees around him, and to recognize how transactions between individuals and between organizations are affected by the space that separates them.'

Three different geographers, three different theoretical perspectives, one fundamental conclusion. The concept of geographical imagination is an appropriate note on which to conclude this introductory text. I hope that as students of human geography you have gained an appreciation of and a sensitivity to the importance of space and place in understanding the world we call home.

CONCLUSION SUMMARY

Changing human landscapes

Although the days of discovering new worlds are over, human geography is by no means fixed or static. As old facts disappear and new ones appear, our world is constantly changing. Among the most important areas where changes are occurring are population numbers, urban growth, and resource use. There are also changing political, economic, and cultural circumstances at the global scale, the precise consequences of which remain uncertain, although one clear impact is increasing pressure on tribal peoples, who are disappearing at an alarming rate.

Changing human geography

Human geography is an academic discipline that changes in response to the changing requirements of the society of which it is a part. Space and place are constantly being reinforced as our key concepts; our links with physical geography are a major strength; new technical approaches offer ever-improving ways to handle data, and our increasing understanding of issues is leading to more and better solutions.

Being a human geographer

Develop and use your geographical imagination.

FURTHER EXPLORATIONS

All of the following offer excellent, if sometimes idiosyncratic, commentaries on human geography. All are by distinguished geographers and all are well written and highly entertaining. Read and enjoy!

EYLES, J., ed. 1988. *Research in Human Geography: Introduction and Investigations*. Oxford: Blackwell.

GOULD, P. 1985. *The Geographer at Work*. London: Routledge and Kegan Paul.

HAGGETT, P. 1990. *The Geographer's Art*. Oxford: Blackwell.

———. 1996. 'Geographical Futures'. In *Companion Encyclopedia of Geography: The Environment of Humankind*, edited by I. Douglas, R. Huggett, and M. Robinson, 965–73. London: Routledge.

JOHNSTON, R.J. 1985. 'To the Ends of the Earth'. In *The Future of Geography*, edited by R.J. Johnston, 3–26. London: Methuen.

STODDART, D.R. 1986. *On Geography*. Oxford: Blackwell.

———. 1987. 'To Claim the High Ground: Geography for the End of the Century'. *Transactions, Institute of British Geographers* n.s. 12:327–36.

Because there is such a close relationship between geography and many world events—both human and physical—it is essential for geographers to keep abreast of current developments. Two excellent factual sources, both published monthly, are available in major reference libraries: *Keesing's Record of World Events* and *Current History*. Finally, one of the most useful ways to explore any aspect of geography in greater detail is to refer to the following publications:

DOUGLAS, I., R. HUGGETT, and M. ROBINSON, eds. 1996. *Companion Encyclopedia of Geography: The Environment of Humankind*. London: Routledge.

GAILE, G.L., and C.J. WILLMOTT, eds. 1989. *Geography in America*. Toronto: Merrill.

NATIONAL RESEARCH COUNCIL. 1997. *Rediscovering Geography: New Relevance for Science and Society*. Washington, DC: National Academy Press.

ON THE WEB

http://www.facingthefuture.org/trends/trends-index.htm
A discussion of some global problems, focusing on scarce resources, poverty, conflict, and the environment.

Also, now is a good time to revisit the sites listed at the end of the Introduction.

REFERENCES

INTRODUCTION
BONE, R.M. 1992. *The Geography of the Canadian North*. Toronto: Oxford University Press.
FAIRGRIEVE, J. 1926. *Geography in School*. London: University of London Press.
GRITZNER, C.F. 2002. 'What Is Where, Why There, and Why Care?' *Journal of Geography* 101, 38–40.
JOHNSTON, R.J. 1985. 'To the Ends of the Earth'. In *The Future of Geography*, edited by R.J. Johnston, 326–38. New York: Methuen.

CHAPTER 1
HART, J.F. 1982. 'The Highest Form of the Geographer's Art'. *Annals, Association of American Geographers* 72:1–29.
HARTSHORNE, R. 1939. *The Nature of Geography: A Critical Survey of Current Thought in the Light of the Past*. Lancaster: Association of American Geographers.
MAY, J.A. 1970. *Kant's Concept of Geography and Its Relation to Recent Geographical Thought*. Toronto: University of Toronto Press.
MENZIES, G. 2002. *1421: The Year China Discovered the World*. London: Bantam Press.

CHAPTER 2
BLAINEY, G. 1968. *The Tyranny of Distance*. Melbourne: Macmillan.
BUNGE, W. 1968. 'Fred K. Schaefer and the Science of Geography'. *Harvard Papers in Theoretical Geography*, Special Papers Series, Paper A (mimeographed).
———. 1979. 'Fred K. Schaefer and the Science of Geography'. *Annals, Association of American Geographers* 69:128–32 (condensed version of Bunge 1968).
DOHRS, F.E., and L.M. SOMMERS, eds. 1967. *Introduction to Geography: Selected Readings*. New York: Crowell.
FEBVRE, L. 1925. *A Geographical Introduction to History*. London: Routledge and Kegan Paul.
HARTSHORNE, R. 1955. '"Exceptionalism in Geography" Re-examined'. *Annals, Association of American Geographers* 45:205–44.
———. 1959. *Perspective on the Nature of Geography*. Chicago: Rand McNally.
HOLLAND, P., et al. 1991. 'Qualitative Resources in Geography'. *New Zealand Journal of Geography* 92:1–28.
HUNTINGTON, E. 1927. *The Human Habitat*. New York: Van Nostrand.
JACKSON, R.H., and R. HENRIE. 1983. 'Perceptions of Sacred Space'. *Journal of Cultural Geography* 3, no. 2:94–107.
LEWTHWAITE, G.R. 1966. 'Environmentalism and Determinism: A Search for Clarification'. *Annals, Association of American Geographers* 56:1–23.
MARTIN, G.J. 1989. 'The Nature of Geography and the Schaefer-Hartshorne Debate'. In *Reflections on Richard Hartshorne's The Nature of Geography*, edited by J.N. Entrikin and S.D. Brunn. Association of American Geographers, Occasional Publication:69–88.
MINSHULL, R. 1967. *Regional Geography: Theory and Practice*. London: Hutchinson.
MONMONIER, M. 1991. *How to Lie with Maps*. Chicago: University of Chicago Press.

PEET, R. 1989. 'World Capitalism and the Destruction of Regional Cultures'. In *A World in Crisis: Geographical Perspectives*, 2nd ed., edited by R.J. Johnston and P.J. Taylor, 175–99. Oxford: Blackwell.
RAPER, J.F., and N.P.A. GREEN. 1989. 'The Development of a Tutor for Geographic Information Systems'. *British Journal of Educational Technology* 20:164–72.
RELPH, E. 1976. *Place and Placelessness*. London: Pion.
SAARINEN, T.F. 1974. 'Environmental Perception'. In *Perspectives on Environment*, edited by I.R. Manners and M.W. Mikesell, 252–89. Washington: Association of American Geographers, Commission on College Geography Publication 13.
SCHAEFER, F. 1953. 'Exceptionalism in Geography: A Methodological Examination'. *Annals, Association of American Geographers* 43:226–49.
SEMPLE, E. 1911. *Influences of Geographic Environment*. New York: Henry Holt.
SPATE, O.H.K. 1952. 'Toynbee and Huntington: A Study in Determinism'. *Geographical Journal* 118:406–28.
TOBLER, W. 1970. 'A Computer Movie'. *Economic Geography* 46:234–40.
TUAN, Yi-Fu. 1979. *Landscapes of Fear*. Oxford: Blackwell.
———. 1982. *Segmented Worlds and Self: Group Life and Individual Consciousness*. Minneapolis: University of Minnesota Press.
———. 1983. 'Geographical Theory: Queries from a Cultural Geographer'. *Geographical Analysis* 15:69–72.
———. 1984. *Dominance and Affection*. New Haven: Yale University Press.
WHITTLESEY, D. 1954. 'The Regional Concept and the Regional Method'. In *American Geography: Inventory and Prospect*, edited by P.E. James and C.F. James, 19–68. Syracuse: Syracuse University Press.
WOOD, D. 1993. 'The Power of Maps'. *Scientific American* 268, no. 5:88–93.
YEATES, M.H. 1968. *An Introduction to Quantitative Analysis in Economic Geography*. Toronto: McGraw Hill.

CHAPTER 3
ARKELL, T.C. 1991. 'The Earth's Fault'. *Geographical Magazine* 63, no. 8:19–21.
DEAR, M., and J. WOLCH. 1989. 'How Territory Shapes Social Life'. In *The Power of Geography: How Territory Shapes Social Life*, edited by J. Wolch and M. Dear, 3–18. Boston: Unwin Hyman.
FEDER, K.L., and M.A. PARK. 1997. *Human Antiquity: An Introduction to Physical Anthropology and Archaeology*, 3rd ed. Toronto: Mayfield.
GOULD, S.J. 1981. *The Mismeasure of Man*. New York: Norton.
———. 1985. 'Human Equality Is a Contingent Fact of History'. In *The Flamingo's Smile*, edited by S.J. Gould, 185–98. New York: Norton.
———. 1987. 'Bushes All the Way Down'. *Natural History* 96, no. 6:12–19.
KENNEDY, K.A.R. 1976. *Human Variation in Space and Time*. Dubuque: Brown.
KIRCHNER, J.W. 1989. 'The Gaia Hypothesis: Can It Be Tested?' *Reviews of Geophysics* 27, no. 2:223–35.

LEWIN, R. 1982. *Thread of Life: The Smithsonian Looks at Evolution.* Washington, DC: Smithsonian Books.

LOVELOCK, J. 1979. *Gaia: A New Look at Life on Earth.* Toronto: Oxford University Press.

———. 1988. *The Ages of Gaia.* New York: W.W. Norton.

MONTAGUE, A., ed. 1964. *The Concept of Race.* New York: Collier.

TRENHAILE, A.S. 1990. *The Geomorphology of Canada.* Toronto: Oxford University Press.

CHAPTER 4

AGNEW, C. 1990. 'Green Belt Around the Sahara'. *Geographical Magazine* 62, no. 4:26–30.

ANDERSON, D. 1999. 'Abrupt Climatic Change'. *Geography Review* 13,1:2–6.

BAHN, P., and J. FLENLEY. 1992. *Easter Island: Earth Island.* New York: Thames and Hudson.

CARSON, R. 1962. *Silent Spring.* Boston: Houghton Mifflin.

CHAMBERS, F. 1998. '"Global Warming": New Perspectives from Palaeoecology and Solar Science'. *Geography* 88:266–277.

DREGNE, H.E. 1977. 'Desertification of Arid Lands'. *Economic Geography* 53:322–31.

GALE, R.J.P. 1992. 'Environment and Economy: The Policy Models of Development'. *Environment and Behavior* 24:723–37.

GLANCE, N.S., and B.A. HUBERMAN. 1994. 'The Dynamics of Social Dilemmas'. *Scientific American* 270, no. 3:76–81.

GOUDIE, A. 1981. *The Human Impact: Man's Role in Environmental Change.* Oxford: Blackwell.

GROVE, R.H. 1992. 'Origins of Western Environmentalism'. *Scientific American* 267, no. 1:42–7.

HARDIN, G. 1968. 'The Tragedy of the Commons'. *Science* 162:1243–8.

JOHNSTON, R.J. 1992. 'Laws, States and Superstates: International Law and the Environment'. *Applied Geography* 12:211–28.

———, and P.J. TAYLOR. 1986. 'Introduction: A World in Crisis?' In *A World in Crisis: Geographical Perspectives,* edited by R.J. Johnston and P.J. Taylor, 1–11. Oxford: Blackwell.

KAPLAN, R.D. 1994. 'The Coming Anarchy'. *Atlantic Monthly* 273, no. 2:44–76.

———. 1996. *The Ends of the Earth: A Journey at the Dawn of the 21st Century.* New York: Random House.

KELLY, K. 1974. 'The Changing Attitudes of Farmers to Forest in Nineteenth Century Ontario'. *Ontario Geography* 8:67–77.

LOMBORG, B. 2001. *The Skeptical Environmentalist: Measuring the Real State of the World.* New York: Cambridge University Press.

MARSH, G.P. [1864] 1965. *Man and Nature, or Physical Geography as Modified by Human Action,* edited by D. Lowenthal. Cambridge, Mass.: Harvard University Press.

PONTING, C. 1991. *A Green History of the World.* New York: St Martin's Press.

POWELL, J.M. 1976. *Environmental Management in Australia, 1788–1814.* New York: Oxford University Press.

REPETTO, R. 1990. 'Deforestation in the Tropics'. *Scientific American* 260:36–42.

RUCKELSHAUS, W.N. 1989. 'Toward a Sustainable World'. *Scientific American* 261, no. 3:166–75.

SIMON, J., and H. KAHN, eds. 1984. *The Resourceful Earth.* Oxford: Blackwell.

SMIL, V. 1987. *Energy, Food and Environment.* New York: Oxford University Press.

———. 1989. 'Our Changing Environment'. *Current History* 88:9–12, 46–8.

———. 1993. *Global Ecology: Environmental Changes and Social Flexibility.* New York: Routledge.

SUPPLY and SERVICES CANADA. 1991. *The State of Canada's Environment.* Ottawa: Supply and Services Canada.

THOMAS, W.L., et al., eds. 1956. *Man's Role in Changing the Face of the Earth.* Chicago: University of Chicago Press.

TUAN, Y.F. 1971. *Man and Nature.* Washington, DC: Association of American Geographers, Commission on College Geography, Resource Paper No. 10.

WILSON, E.O. 1989. 'Threats to Biodiversity'. *Scientific American* 261, no. 3:108–16.

WORLD COMMISSION on ENVIRONMENT and DEVELOPMENT. 1987. *Our Common Future.* Oxford: Oxford University Press.

CHAPTER 5

BOSERUP, E. 1965. *The Conditions of Agricultural Change.* London: Allen and Unwin.

DANIEL, M.L. 2000. 'The Demographic Impact of HIV/AIDS in Sub-Saharan Africa'. *Geography* 85:46–55.

DEMENY, P. 1974. 'The Populations of the Underdeveloped Countries'. In *The Human Population,* 105–15. San Francisco: W.H. Freeman.

DOENGES, C.E., and J.L. NEWMAN. 1989. 'Impaired Fertility in Tropical Africa'. *Geographical Review* 79:101–11.

DWYER, D.J. 1987. 'New Population Policies in Malaysia and Singapore'. *Geography* 72:248–50.

EBERSTADT, N. 2002. 'The Future of AIDS'. *Foreign Affairs* 81, no. 6:22–45.

ECONOMIST, THE. 1993. 'Eastern Germany: Living and Dying in a Barren Land'. *The Economist* 331, no. 23 (April):54.

ECONOMIST, THE. 2002. 'The Next Wave'. 19 October:75–6.

EHRLICH, P. 1968. *The Population Bomb.* New York: Ballantine Books.

HALL, R. 1993. 'Europe's Changing Population'. *Geography* 78:3–15.

JOWETT, J. 1993. 'China's Population: 1,133,709,738 and Still Counting'. *Geography* 78:401–19.

KING, R. 1993. 'Italy Reaches Zero Population Growth'. *Geography* 78:63–9.

MEADOWS, D.H., et al. 1972. *The Limits to Growth.* New York: Universe Books.

PETERSON, P.G. 1999. 'Gray Dawn: The Global Aging Crisis'. *Foreign Affairs* 78:42–55.

RIDKER, R.G., and E.W. CECELSKI. 1979. 'Resources, Environment and Population: The Nature of Future Limits'. *Population Bulletin* 34, no. 3:3–4.

ROBEY, B., S.O. RUTSTEIN, and L. MORRIS. 1993. 'The Fertility Decline in Developing Countries'. *Scientific American* 269, no. 6:60–7.

TOOLIS, K. 2000. 'While the World Looks Away'. *The Guardian Weekend:* 2 December.

UNITED NATIONS. 2002. *World Population Aging: 1950–2050.* New York: United Nations, Department of Economic and Social; Affairs, Population Division.

VANDERPOST, C. 1992. 'Regional Patterns of Fertility Transition in Botswana'. *Geography* 77:109–22.

CHAPTER 6

BARBERIS, M. 1994. 'Haiti'. *Population Today* 22, no. 1:7.

BOGUE, D.J. 1969. *Principles of Demography.* New York: Wiley.

BONGAARTS, J. 1994. 'Can the Growing Human Population Feed Itself'. *Scientific American* 271, no. 3:36–42.

BRANDT, W. 1980. *North-South: A Programme for Survival.* London: Pan.

DAVIS, M. 2001. *Late Victorian Holocausts: El Nino, Famines, and the Making of the Third World.* New York: Verso.

DEG, M. 1992. Natural Disasters: Recent Trends and Future Prospects'. *Geography* 77:198–209.

DICKENSON, J., et al. (1996). *A Geography of the Third World*, 2nd ed. New York: Routledge.

GEE, M. 1994. 'Apocalypse Deferred'. *The Globe and Mail*, 9 April:D1 and D3.

GOULD, P., and R. WHITE. 1986. *Mental Maps*, 2nd ed. Boston: Allen and Unwin.

GRIGG, D 1977. 'E.G. Ravenstein and the "Laws" of Migration'. *Journal of Historical Geography* 3:41–54.

HALFACREE, K., and P.J. BOYLE. 1993. 'The Challenge Facing Migration Research: The Case for a Biographical Approach'. *Progress in Human Geography* 17:333–48.

IGNATIEFF, M. 1996. Review of 'The Ends of the Earth', by R.D. Kaplan. *The New York Times Book Review*, 31 March.

KAPLAN, R.D. 1994. 'The Coming Anarchy'. *Atlantic Monthly* 273, no. 2:44–76.

———. 1996. *The Ends of the Earth: A Journey at the Dawn of the 21st Century*. New York: Random House.

KNOX, P., and J. AGNEW. 1994. *The Geography of the World Economy*, 2nd ed. London: Arnold.

LEE, E.S. 1966. 'A Theory of Migration'. *Demography* 3, no. 1:47–57.

MAHMUD, A. 1989. 'Grameen Bank Bangladesh: A Workable Solution'. *Geographical Magazine* 61, no. 10:14–16.

MOON, B. 1995. 'Paradigms in Migration Research: Exploring "Moorings" as a Schema'. *Progress in Human Geography* 19:504–24.

OSTERGREN, R.C. 1988. *A Community Transplanted: The Trans-Atlantic Experience of a Swedish Immigrant Settlement in the Upper Middle West, 1835–1915*. Madison: University of Wisconsin Press.

PETERSEN, W. 1958. 'A General Typology of Migration'. *American Sociological Review* 23:256–65.

PORTER, G. 1992. 'The Nigerian Census Surprise'. *Geography* 77:371–4.

RAVENSTEIN, E.G. 1876. 'Census of the British Isles, 1871; Birthplaces and Migration'. *Geographical Magazine* 3:173–7, 201–6, 229–33.

———. 1885. 'The Laws of Migration'. *Journal of the Statistical Society* 48:167–227.

———. 1889. 'The Laws of Migration'. *Journal of the Statistical Society* 52:214–301.

SIMPSON, E.S. 1989. 'An Island Divided'. *Geographical Journal* 61, no. 10:1–5.

SMIL, V. 1987. *Energy, Food, Environment: Realities, Myths and Options*. Oxford: Clarendon.

———. 2000. *Feeding the World: A Challenge for the Twenty-First Century*. Cambridge, MA: MIT Press.

SOWDEN, C. 1993. 'Debt Swaps—For or Against Development'. *Geographical Magazine* 25, no. 12:56–9.

TAYLOR, P.J. 1989. 'The Error of Developmentalism in Human Geography'. In *Horizons in Human Geography*, edited by D. Gregory and R. Walford, 303–19. London: Macmillan.

TINKER, H. 1974. *A New System of Slavery: The Export of Indian Labour Overseas 1830–1920*. London: Oxford University Press.

TODD, H. 1996. *Women at the Center: Grameen Bank Borrowers After One Decade*. Boulder: Westview.

UNITED NATIONS DEVELOPMENT PROGRAMME. 1996. *Human Development Report, 1996*. New York: Oxford University Press.

——— (2002). *Human Development Report 2002: Deepening Democracy in a Fragmented World*. New York: Oxford University Press.

WALLERSTEIN, I. 1979. *The Capitalist World Economy*. Cambridge: Cambridge University Press.

WOLPERT, J. 1965. 'Behavioural Aspects of the Decision to Migrate'. *Papers of the Regional Science Association* 15:159–69.

WORLD BANK. 1998. *Global Development Finance, 1998: Analysis and Summary Tables*. Washington: The World Bank.

YOUNG, L. 1996. 'World Hunger: A Framework for Analysis'. *Geography* 81:97–110.

ZELINSKY, W. 1971. 'The Hypothesis of the Mobility Transition'. *Geographical Review* 61:219–49.

CHAPTER 7

BISWAS, L. 1984. 'Evolution of Hindu Temples in Calcutta'. *Journal of Cultural Geography* 4:73–84.

CARNEIRO, R. 1970. 'A Theory of the Origin of the State'. *Science* 169:733–8.

CARTWRIGHT, D. 1988. 'Linguistic Territorialization: Is Canada Approaching the Belgian Model?' *Journal of Cultural Geography* 8:115–34.

CHILDE, V.G. 1951. *Man Makes Himself*. New York: Mentor Books.

FEDER, K.L., and M.A. PARK. 1997. *Human Antiquity: An Introduction to Physical Anthropology and Archaeology*, 3rd ed. Toronto: Mayfield.

FRANCAVIGLIA, R.V. 1978. *The Mormon Landscape*. New York: AMS Press.

FRIEDL, J., and J. PFEIFFER. 1977. *Anthropology: The Study of People*. New York: Harper and Row.

GALE, D.T., and P.M. KOROSCIL. 1977. 'Doukhobor Settlements: Experiments in Idealism'. *Canadian Ethnic Studies* 9:53–71.

GIDDENS, A. 1991. *Introduction to Sociology*. New York: Norton.

HARRIS, M. 1974. *Cows, Pigs, Wars and Witches: The Riddles of Cultures*. New York: Vintage Books.

HARVEY, D.W. 1990. 'Between Space and Time: Reflections on the Geographical Imagination'. *Annals, Association of American Geographers*. 80:418–34.

HUNTINGTON, E. 1924. *Civilization and Climate*. New Haven: Yale University Press.

HUXLEY, J.S. 1966. 'Evolution, Cultural and Biological'. *Current Anthropology* 7:16–20.

JAMES, P.E. 1964. *One World Divided*, 2nd ed. Toronto: Xerox College Publishing.

JORDAN, T.G. 1988. *The European Culture Area: A Systematic Geography*, 2nd ed. New York: Harper and Row.

———, and M. KAUPS. 1989. *The American Backwoods Frontier: An Ethnic and Ecological Interpretation*. Baltimore: Johns Hopkins University Press.

KARAN, P. 1984. 'Landscape, Religion and Folk Art in Mithila: An Indian Cultural Region'. *Journal of Cultural Geography* 5:85–102.

KAUPS, M. 1966. 'Finnish Place Names in Minnesota: A Study in Cultural Transfer'. *Geographical Review* 56:377–97.

KEARNS, K.C. 1974. 'Resuscitation of the Irish Gaeltacht'. *Geographical Review* 64:82–110.

KROEBER, A.L., and T. PARSONS. 1958. 'The Concepts of Culture and of Social System'. *American Sociological Review* 23:582–3.

LEWIS, B. 2003. *The Crisis of Islam: Holy War and Unholy Terror*. New York: Weidenfeld and Nicholson.

MCCRUM, R., W. CRAN, and R. MACNEIL, eds. 1986. *The Story of English*. London: BBC.

MCWHORTER, R. 2002. *The Power of Babel: A Natural History of Language*. New York: W. H. Freeman.

MEINIG, D.W. 1965. 'The Mormon Culture Region: Strategies and Patterns in the Geography of the American West, 1847–1964'. *Annals, Association of American Geographers* 55:191–220.

MITCHELL, R.D. 1978. 'The Formation of Early American Cultural Regions'. In *European Settlement and Development in North*

America, edited by J.R. Gibson, 66–90. Toronto: University of Toronto Press.

PARK, C. 1994. *Sacred Worlds: An Introduction to Geography and Religion*. New York: Routledge.

RUSSELL, R.J., and F.B. KNIFFEN. 1951. *Culture Worlds*. New York: Macmillan.

SAUER, C.O. 1925. 'The Morphology of Landscape'. *University of California Publications in Geography* 2:19–53.

SIMPSON-HOUSLEY, P. 1978. 'Hutterian Religious Ideology, Environmental Perception, and Attitudes Towards Agriculture'. *Journal of Geography* 77:145–8.

SMIL, V. 1987. *Energy, Food and Environment*. New York: Oxford University Press.

SPENCER, J.E., and R.J. HORVATH. 1963. 'How Does an Agricultural Region Originate?' *Annals, Association of American Geographers* 53:74–92.

TOYNBEE, A.J. 1935–61. *A Study of History*, 12 volumes. New York: Oxford University Press.

WITTFOGEL, K. 1957. *Oriental Despotism: A Comparative Study of Total Power*. New Haven: Yale University Press.

ZELINSKY, W. 1973. *The Cultural Geography of the United States*. Englewood Cliffs: Prentice-Hall.

CHAPTER 8

ABLER, R.F., M.G. MARCUS, and J.M. OLSON, eds. 1992. *Geography's Inner Worlds: Pervasive Themes in Contemporary American Geography*. New Brunswick: Rutgers University Press.

BERRY, B.J.L. 1992. Review of *Geography's Inner Worlds: Pervasive Themes in Contemporary American Geography*. *Urban Geography* 13:490–4.

BLAUT, J.M. 1980. 'A Radical Critique of Cultural Geography'. *Antipode* 12:25–9.

DEAR, M.J. 1988. 'The Postmodern Challenge: Reconstructing Human Geography'. *Transactions, Institute of British Geographers* NS 13:262–74.

DOYAL, L., and I. GOUGH. 1991. *A Theory of Human Need*. London: Macmillan.

ENTRIKIN, N. 1991. *The Betweenness of Place*. Baltimore: Johns Hopkins University Press.

EVANS, D. 2001. 'Spatial Analyses of Crime'. *Geography* 86:211–23.

EVANS, D.J., and D.T. HERBERT. 1989. *The Geography of Crime*. London: Routledge.

EVANS, R. 1989. 'Consigned to the Shadows'. *Geographical Magazine* 61, no. 12:23–5.

EYLES, J., and W. PEACE. 1990. 'Signs and Symbols in Hamilton: An Iconology of Steeltown'. *Geografiska Annaler* 72B:73–88.

FRANCAVIGLIA, R.V. 1978. *The Mormon Landscape*. New York: AMS Press.

GIDDENS, A. 1984. *The Constitution of Society*. Cambridge: Polity Press.

GREGORY, D. 1981. 'Human Agency and Human Geography'. *Transactions, Institute of British Geographers* NS 6:1–18.

GREGSON, N. 1986. 'On Duality and Dualism: The Case of Time Geography and Structuration'. *Progress in Human Geography* 10:184–205.

HALE, R.F. 1984. 'Vernacular Regions of America'. *Journal of Cultural Geography* 5:131–40.

HAMNETT, C. 2003. 'Editorial. Contemporary Human Geography: Fiddling While Rome Burns?' *Geoforum* 34:1–3.

HARVEY, D.W. 1989. *The Condition of Postmodernity*. Oxford: Blackwell.

HERBERT, D.T., and D.M. SMITH, eds. 1989. *Social Problems and the City: New Perspectives*. Oxford: Oxford University Press.

ISAJIW, W. 1974. 'Definitions of Ethnicity'. *Ethnicity* 1:111–24.

JACKSON, E.L., and D.B. JOHNSON. 1991. 'Geographic Implications of Mega-Malls, with Special Reference to West Edmonton Mall'. *Canadian Geographer* 35:226–32.

JACKSON, P. 1989. *Maps of Meaning: An Introduction to Cultural Geography*. London: Unwin Hyman.

———, and S.J. Smith. 1984. *Exploring Social Geography*. London: Allen and Unwin.

JAMES, P.E. 1974. *One World Divided*, 2nd ed. Toronto: Xerox College Publishing.

JOHNSTON, R.J. 1986. 'Individual Freedom and the World Economy'. In *A World in Crisis: Geographical Perspectives*, edited by R.J. Johnston and P.J. Taylor, 173–95. Oxford: Blackwell.

———. 1991. *A Question of Place: Exploring the Practice of Human Geography*. Oxford: Blackwell.

JONES, K., and G. MOON. 1987. *Health, Disease and Society: An Introduction to Medical Geography*. London: Routledge.

KNOX, P.L. 1975. *Social Well-Being: A Spatial Perspective*. Oxford: Clarendon Press.

KOBAYASHI, A. 1993. 'Multiculturalism: Representing a Canadian Institution'. In *Place/Culture/Representation*, edited by J. Duncan and D. Ley, 205–31. London: Routledge.

McQUILLAN, A. 1993. 'Historical Geography and Ethnic Communities in North America'. *Progress in Human Geography* 17:355–66.

MEAD, G.H. 1934. *Mind, Self and Society*. Chicago: University of Chicago Press.

MONK, J. 1992. 'Gender in the Landscape: Expressions of Power and Meaning'. In *Inventing Places: Studies in Cultural Geography*, edited by K. Anderson and F. Gale, 123–38. Melbourne: Longman Cheshire.

NOSTRAND, R.L., and L.E. ESTAVILLE, Jr. 1993. 'Introduction: The Homeland Concept'. *Journal of Cultural Geography* 13, no. 2:1–4.

OSBORNE, B.S. 1988. 'The Iconography of Nationhood in Canadian Art'. In *The Iconography of Landscape: Essays on the Symbolic Representation, Design and Use of Past Environments*, 162–78. New York: Cambridge University Press.

PAIN, R. 1992. 'Space, Sexual Violence and Social Control: Integrating Geographical and Feminist Analyses of Women's Fear of Crime'. *Progress in Human Geography* 15:415–31.

PAWSON, E., and G. BANKS. 1993. 'Rape and Fear in a New Zealand City'. *Area* 25:55–63.

PRATT, G. 1989. 'Reproduction, Class and the Spatial Structure of the City'. In *New Models in Geography, Volume 2*, edited by N. Thrift and R. Peet, 84–108. London: Unwin Hyman.

PRED, A. 1985. 'The Social Becomes the Spatial, the Spatial Becomes the Social: Enclosures, Social Change and the Becoming of Place in the Swedish Province of Skåne'. In *Social Relations and Spatial Structures*, edited by D. Gregory and J. Urry, 336–75. London: Macmillan.

———, and M.J. Watts. 1992. *Reworking Modernity: Capitalisms and Symbolic Discontent*. New Brunswick: Rutgers University Press.

RAITZ, K.B. 1979. 'Themes in the Cultural Geography of European Ethnic Groups in the United States'. *Geographical Review* 69:79–94.

ROONEY, J.F. Jr. 1974. *A Geography of American Sport*. Reading, Mass.: Addison Wesley.

SMITH, D.M. 1973. *A Geography of Social Well-Being in the United States*. New York: McGraw-Hill.

SMITH, S.J. 1987. 'Fear of Crime: Beyond a Geography of Deviance'. *Progress in Human Geography* 11:1–23.

STEIN, H.F., and G.L. THOMPSON. 1992. 'The Sense of Oklahomaness: Contributions of Psychogeography to the Study of American Culture'. *Journal of Cultural Geography* 11, no. 2:63–91.

TRÉPANIER, C. 1991. 'The Cajunization of French Louisiana: Forging a Regional Identity'. *Geographical Journal* 157:161–71.

UNITED NATIONS DEVELOPMENT PROGRAMME. 1996. *Human Development Report, 1996.* New York: Oxford University Press.

WAGNER, P.L. 1975. 'The Themes of Cultural Geography Rethought'. *Yearbook: Association of Pacific Coast Geographers* 37:7–14.

ZELINSKY, W. 1980. 'North America's Vernacular Regions'. *Annals, Association of American Geographers* 70:1–16.

CHAPTER 9

ARKELL, T., and M. DAVENPORT. 1989. 'Africa's Fractious Five'. *Geographical Magazine* 61, no. 12:16–27.

BUNGE, W. 1988. *Nuclear War Atlas.* Oxford: Blackwell.

CARROLL, W.K. 1992. *Organizing Dissent: Contemporary Social Movements in Theory and Practice, Studies in the Politics of Counter-Hegemony.* Toronto: Garamond Press.

CHOUINARD, V. 1997. 'Guest Editorial. Making Space for Disabling Differences: Challenging Ableist Geographies'. *Environment and Planning D: Society and Space* 15:379–387.

CLARK, A.H. 1975. 'The Conceptions of Empires of the St Lawrence and the Mississippi: An Historico-Geographical View with Some Quizzical Comments on Environmental Determinism'. *American Review of Canadian Studies* 5:4–27.

CLARK, G.L., and M.J. DEAR. 1984. *State Apparatus: Structures and Language of Legitimacy.* Boston: Allen and Unwin.

COHEN, S.B. 1991. 'Global Geopolitical Change in the Post Cold War Era'. *Annals, Association of American Geographers* 81:551–80.

———. 2003. *Geopolitics of the World System.* Lanham: Rowman and Littlefield.

DEBLIJ, H. 1992. 'Political Geography of the Post Cold War'. *Professional Geographer* 44:16–19.

DOUGLAS, J.N.H. 1985. 'Conflict Between States'. In *Progress in Political Geography*, edited by M. Pacione. London: Croom Helm.

EAST, W.G., and J.R.V. PRESCOTT. 1975. *Our Fragmented World: Introduction to Political Geography.* London: Macmillan.

ELSOM, D. 1985. 'Climatological Effects of a Nuclear Exchange: A Review'. In *The Geography of Peace and War*, edited by D. Pepper and A. Jenkins, 126–47. Oxford: Blackwell.

EVANS, R. 1991. 'Legacy of Woe'. *Geographical Magazine* 63, no. 6:34–8.

FUKUYAMA, F. 1992. *The End of History and the Last Man.* New York: Free Press.

HARTSHORNE, R. 1950. 'The Functional Approach in Political Geography'. *Annals, Association of American Geographers* 40:95–130.

HIRSCH, P. 1993. 'The Socialist Developing World in the 1990s'. *Geography Review* 7, no. 2:35–7.

HUNTINGTON, S.P. 1993. 'The Clash of Civilizations'. *Foreign Affairs* 72, no. 3:22–49.

———. 1996. *The Clash of Civilizations and the Remaking of World Order.* New York: Simon and Schuster.

JOHNSTON, L. 1997. 'Queen(s') Street or Ponsonby Poofters? Embodied HERO Parade Sites'. *New Zealand Geographer* 53, 2:29–33.

JOHNSTON, R.J. 1982. *Geography and the State: An Essay in Political Geography.* New York: St Martin's Press.

———. 1985. *The Geography of English Politics.* London: Croom Helm.

———. 1993. 'Tackling Global Environmental Problems'. *Geography Review* 6, no. 2:27–30.

JONES, S.B. 1954. 'A Unified Field Theory of Political Geography'. *Annals, Association of American Geographers* 44:111–23.

KAPLAN, R.D. 1994. 'The Coming Anarchy'. *Atlantic Monthly* 273, no. 2:44–76.

———. 1996. *The Ends of the Earth: A Journey at the Dawn of the 21st Century.* New York: Random House.

KLARE, M.T. 2001. 'The New Geography of Conflict'. *Foreign Affairs* 80:49–61.

KOHR, L. 1957. *The Breakdown of Nations.* Swansea: Christopher Davies.

KREBHEIL, E. 1916. 'Geographic Influences in British Elections'. *Geographical Review* 2:419–32.

LARKINS, J., and S. PARISH. 1982. *Australia's Greatest River.* Melbourne: Rigby.

LEMON, A. 1996. 'Lesotho and the New South Africa: The Question of Incorporation'. *Geographical Journal* 162:263–72.

LORENZ, K. 1967. *On Aggression.* London: Methuen.

MACKINDER, H.J. 1919. *Democratic Ideals and Reality.* New York: Henry Holt.

MESSICK, D.M., and D.M. MACKIE. 1989. 'Intergroup Relations'. *Annual Review of Psychology* 40:45–81.

MONTAGUE, A. 1976. *The Nature of Human Aggression.* New York: Oxford University Press.

MORRIS, D. 1967. *The Naked Ape.* New York: McGraw-Hill.

O'LOUGHLIN, J. 1992. 'Ten Scenarios for a "New World Order"'. *Professional Geographer* 44:22–8.

OPENSHAW, S., P. STEADMAN, and O. GREENE. 1983. *Doomsday: Britain After Nuclear Attack.* Oxford: Blackwell.

OVERTON, J.D. 1981. 'A Theory of Exploration'. *Journal of Historical Geography* 7:53–70.

PADDISON, R. 1983. *The Fragmented State: The Political Geography of Power.* New York: St Martin's Press.

ROKKAN, S. 1980. 'Territories, Centres and Peripheries'. In *Centre and Periphery*, edited by J. Gottman, 163–204. London: Sage.

ROYLE, S. 1991. 'St Helena: A Geographical Summary'. *Geography* 76:266–8.

TAYLOR, P.J. 1989. *Political Geography: World Economy, Nation State and Locality*, 2nd ed. London: Longman.

WILKINSON, P. 1971. *Social Movement.* London: Macmillan.

WOLFE, R.I. 1962. 'Transportation and Politics: The Example of Canada'. *Annals, Association of American Geographers* 52:176–90.

WOOD, W.B. 2001. 'Geographic Aspects of Genocide: A Comparison of Bosnia and Rwanda'. *Transactions, Institute of British Geographers* NS26:57–75.

ZURICK, D. 1999. 'Lands of Conflict in South Asia'. *Focus* 45, 3:33–7.

CHAPTER 10

APPLETON, J.H. 1962. *The Geography of Communications in Great Britain.* Toronto: Oxford University Press.

BLAINEY, G. 1968. *The Tyranny of Distance.* London: Macmillan.

CARROTHERS, G.A.P. 1956. 'An Historical Review of the Gravity and Potential Concepts of Human Interaction'. *Journal of the American Institute of Planners* 22:94–102.

CLEARY, M., and R. BEDFORD. 1993. 'Globalisation and the New Regionalism: Some Implications for New Zealand Trade'. *New Zealand Journal of Geography* 20:19–22.

CLIFF, A.D., P. HAGGETT, and J.K. ORD. 1986. *Spatial Aspects of Influenza Epidemics.* London: Pion.

———, and M.R. Smallman-Raynor. 1992. 'The AIDS Pandemic: Global Geographical Patterns and Local Spatial Processes'. *Geographical Journal* 158:182–98.

CONKLING, E.C., and M.H. YEATES. 1976. *Man's Economic Environment.* Toronto: McGraw-Hill.

DICKEN, P. 1992. *Global Shift: The Internationalization of Economic Activity.* London: Paul Chapman.

————.1993. 'The Changing Organization of the Global Economy'. In *The Challenge for Geography: A Changing World, A Changing Discipline*, edited by R.J. Johnston, 31–53. Oxford: Blackwell.

FORER, P. 1978. 'A Place for Plastic Space?' *Progress in Human Geography* 3:230–67.

FREEMAN, D.B. 1985. 'The Importance of Being First: Preemption by Early Adopters of Farming Innovations in Kenya'. *Annals, Association of American Geographers* 75:17–28.

GARRISON, W.L. 1960. 'Connectivity of the Interstate Highway System'. *Regional Science Association, Papers and Proceedings* 6:121–37.

GOULD, P. 1993. *The Slow Plague: A Geography of the AIDS Pandemic*. Oxford: Blackwell.

HAGERSTRAND, T. 1951. 'Migration and the Growth of Culture Regions'. *Lund Studies in Geography* B, 3.

————. 1967. *Innovation Diffusion as a Spatial Process*. Translated by A. Pred. Chicago: University of Chicago Press.

HALL, E.T. 1966. *The Hidden Dimension*. Garden City: Doubleday.

HARVEY, D. 1969. *Explanation in Geography*. London: Arnold.

HILLING, D. 1977. 'The Evolution of a Port System: The Case of Ghana'. *Geography* 62:97–105.

HOAD, D. 2002. 'The World Trade Organisation, Corporate Interests, and Global Opposition: Seattle and After'. *Geography* 87:148–54.

HOYLE, B.S., and R.D. KNOWLES. 1992. *Modern Transport Geography*. London: Belhaven Press.

ISARD, W. 1956. *Location and Space Economy*. New York: Wiley.

————, et al. 1960. *Methods of Regional Analysis: An Introduction to Regional Science*. Cambridge, Mass.: MIT Press.

JANELLE, D.G. 1968. 'Central Place Development in a Time-Space Framework'. *Professional Geographer* 20:5–10.

————. 1969. 'Spatial Reorganization: A Model and Concept'. *Annals, Association of American Geographers* 59:348–64.

KANSKY, K. 1963. *The Structure of Transportation Networks*. Chicago: University of Chicago, Department of Geography, Research Paper No. 84.

KNIFFEN, F.B. 1951. 'The American Covered Bridge'. *Geographical Review* 41:114–23.

LOWE, J.C., and S. MORYADAS. 1975. *The Geography of Movement*. Boston: Houghton Mifflin.

MASSEY, D. 2002. 'Globalization: What Does it Mean for Geography?' *Geography* 87:293–96.

NORBERG, J. 2001. *In Defence of Global Capitalism*. Stockholm: Timbro.

OHMAE, K. 1993. 'The Rise of the Region State'. *Foreign Affairs* 76, no. 2:78–87.

PYLE, G. 1969. 'The Diffusion of Cholera in the United States in the Nineteenth Century'. *Geographical Analysis* 1:59–75.

RAITZ, K.B. 1973. 'Ethnicity and the Diffusion and Distribution of Cigar Tobacco Production in Wisconsin and Ohio'. *Tijdschrifte voor Economische en Sociale Geografie* 64:293–306.

REILLY, W.J. 1931. *The Law of Retail Gravitation*. New York: Free Press.

RIGG, J. 2001. 'Is Globalization Good?' *Geography Review* 14, 4:36–7.

ROGERS, E.M. 1962. *Diffusion of Innovations*. New York: Free Press.

ROSTOW, W.W. 1960. *The Stages of Economic Growth*. Cambridge: Cambridge University Press.

SMALLMAN-RAYNOR, M.R., A.D. CLIFF, and P. HAGGETT. 1992. *Atlas of AIDS*. Oxford: Blackwell.

TAAFE, E.J., R.L. MORRILL, and P. GOULD. 1963. 'Transport Expansion in Underdeveloped Countries: A Comparative Analysis'. *Geographical Review* 53:503–29.

ULLMAN, E.L. 1956. 'The Role of Transportation and the Bases for Interaction'. In *Man's Role in Changing the Face of the Earth*, edited by W.L. Thomas, Jr, 862–80. Chicago: University of Chicago Press.

WARNTZ, W. 1957. 'Transportation, Social Physics and the Law of Refraction'. *Professional Geographer* 9:2–7.

WATSON, J.W. 1955. 'Geography: A Discipline in Distance'. *Scottish Geographical Magazine* 71:1–13.

WILSON, A.G., and R.J. BENNETT, eds. 1986. *Mathematical Methods in Human Geography and Planning*. Chichester: Wiley.

ZIPF, G.K. 1949. *Human Behavior and the Principle of Least Effort*. Cambridge, Mass.: Addison Wesley.

CHAPTER 11

ARKELL, T. 1991. 'The Decline of Pastoral Nomadism in the Western Sahara'. *Geography* 76:162–6.

ATKINS, P., and Bowler, I. 2001. *Food in Society: Economy, Culture, Geography*. London: Arnold.

BARNES, T.J. 2001. 'Retheorizing Economic Geography: From the Quantitative Revolution to the " Turn"'. *Annals, Association of American Geographers* 91:546–65.

BARRETT, H., and A. BROWNE. 1991. 'Environmental and Economic Sustainability: Woman's Horticultural Production in the Gambia'. *Geography* 76:241–8.

BASSETT, T.J. 1988. 'The Political Ecology of Peasant-Herder Conflicts in the Northern Ivory Coast'. *Annals, Association of American Geographers* 78:453–72.

BINFORD, L. 1968. 'Post-Pleistocene Adaptations'. In *New Perspectives in Archaeology*, edited by L. Binford and S. Binford. Chicago: Aldine, 313–41.

BOSERUP, E. 1965. *The Conditions of Agricultural Change*. London: Allen and Unwin.

BOWLER, I.R. 1985. *Agriculture under the Common Agriculture Policy*. Manchester: Manchester University Press.

CARLYLE, W.J. 1994. 'Rural Population in the Canadian Prairies'. *Great Plains Research* 4:65–87.

CHISHOLM, M. 1962. *Rural Settlement and Land Use: An Essay in Location*. London: Hutchinson.

CONZEN, M.P. 1971. *Frontier Farming in an Urban Shadow*. Madison: State Historical Society of Wisconsin.

ERICKSON, R.A. 1989. 'The Influence of Economics in Geographic Inquiry'. *Progress in Human Geography* 13:223–49.

FERNANDEZ-ARMESTO. F. 2001. *Food: A History*. London: Macmillan.

FRIEDMANN, H. 1991. 'Changes in the International Division of Labour: Agri-Food Complexes and Export Agriculture'. In *Towards a New Political Economy of Agriculture*, edited by W. Friedland et al., 65–93. Boulder: Westview.

GRIFFIN, E. 1973. 'Testing the Von Thünen Theory in Uruguay'. *Geographical Review* 63:500–16.

GRIGG, D.B. 1974. *The Agricultural Systems of the World: An Evolutionary Approach*. New York: Cambridge University Press.

————. 1992. 'Agriculture in the World Economy: An Historical Geography of Decline'. *Geography* 77:210–22.

GROSSMAN, L.S. 1993. 'The Political Ecology of Banana Exports and Local Food Production in St Vincent, Eastern Caribbean'. *Annals, Association of American Geographers* 83:347–67.

HALL, P.G., ed. 1966. *Von Thünen's Isolated State*. Oxford: Pergamon.

HARRIS, D.R. 1996. 'Preface'. In *The Origins and Spread of Agriculture and Pastoralism in Eurasia*, edited by D.R. Harris, ix-xi. Washington DC: Smithsonian Institution Press.

HARVEY, D.W. 1973. *Social Justice and the City*. London: Arnold.

HORVATH, R.J. 1969. 'Von Thünen's Isolated State and the Area Around Addis Ababa, Ethiopia'. *Annals, Association of American Geographers* 59:308–23.

ISARD, W. 1956. *Location and Space-Economy*. New York: Wiley.

JOHNSON, H.B. 1962. 'A Note on Thünen's Circles'. *Annals, Association of American Geographers* 52:213–20.

JONASSON, O. 1925. 'Agricultural Regions of Europe'. *Economic Geography* 1:277–314.

KING, L.J. 1969. *Statistical Analysis in Geography*. Englewood Cliffs: Prentice-Hall.

LEAMAN, J.H., and E.C. CONKLING. 1975. 'Transport Change and Agricultural Specialization'. *Annals, Association of American Geographers* 65:425–37.

MCCALLUM, J. 1980. *Unequal Beginnings: Agriculture and Economic Development in Quebec and Ontario until 1870*. Toronto: University of Toronto Press.

MCDOWELL, L. 2000. 'Feminists Rethink the Economic: The Economics of Gender / The Gender of Economics'. In *The Oxford Handbook of Economic Geography*, edited by G.L. Clark, M.P. Feldman, and M.S. Gertler, 497–517. Toronto: Oxford University Press.

MORRISH, M. 1994. 'China Takes the Road to Market'. *Geographical Magazine* 67, no. 4:43–5.

MULLER, P.O. 1973. 'Trend Surfaces of American Agricultural Patterns: A Macro-Thünen Analysis'. *Economic Geography* 49:228–42.

NORTON, W., and E.C. CONKLING. 1974. 'Land Use Theory and the Pioneering Economy'. *Geografiska Annaler* 56B:44–56.

OVERTON, M. 1996. *Agricultural Revolution in England: The Transformation of the Agrarian Economy, 1500–1850*. Cambridge: Cambridge University Press.

PEET, J.R. 1969. 'The Spatial Expansion of Commercial Agriculture in the Nineteenth Century: A Von Thünen Interpretation'. *Economic Geography* 45:283–301.

RAITZ, K.B. 1973. 'Ethnicity and the Diffusion and Distribution of Cigar Tobacco Production in Wisconsin and Ohio'. *Tijdschrifte voor Economische en Sociale Geografie* 64:293–306.

REITSMA, H.J. 1971. 'Crop and Livestock Production in the Vicinity of the United States-Canada Border'. *Professional Geographer* 23:216–23.

SALAMON, S. 1985. 'Ethnic Communities and the Structure of Agriculture'. *Rural Sociology* 50:323–40.

SAUER, C.O. 1952. *Agricultural Origins and Dispersals*. New York: American Geographical Society.

SCHLEBECKER, J.T. 1960. 'The World Metropolis and the History of American Agriculture'. *Journal of Economic History* 20:187–208.

SEABORNE, A. 2001. 'Crop Diversification in Canada's Breadbasket: Land Use Changes in Saskatchewan's Agriculture'. *Geography* 86:151–8.

SINCLAIR, R. 1967. 'Von Thünen and Urban Sprawl'. *Annals, Association of American Geographers* 57:72–87.

SLATFORD, R., and Fishpool, I. 2002. "India and the Green Revolution'. *Geography Review* 15 (2):2–5.

SMIL, V. 2001. *Enriching the Earth: Fritz Haber, Carl Bosch and the Transformation of World Food Production*. Boston: MIT Press.

SMITH, J.R. 1925. *North America*. New York: Harcourt Brace and Co.

SMITH, W. 1984. 'The "Vortex" Model and the Changing Agricultural Landscape of Quebec'. *Canadian Geographer* 28:358–72.

SWYNGEDOUW, E. 2000. 'The Marxian Alternative: Historical–Geographical Materialism and the Political Economy of Capitalism'. In *A Companion to Economic Geography*, edited by E. Sheppard and T.J. Barnes, 41–59. Malden, Mass: Blackwell.

WOLPERT, J. 1964. 'The Decision Process in Spatial Context'. *Annals, Association of American Geographers* 54:537–58.

YOUNG, L.J. 1991. 'Agriculture Changes in Bhutan: Some Environmental Questions'. *Geographical Journal* 157:172–8.

ZIMMERER, K.S. 1991. 'Wetland Production and Smallholder Persistence: Agriculture Change in a Highland Peruvian Region'. *Annals, Association of American Geographers* 81:443–63.

CHAPTER 12

BASSETT, K., and J. SHORT. 1989. 'Development and Diversity in Urban Geography'. In *Horizons in Human Geography*, edited by D. Gregory and R. Walford, 175–93. London: Macmillan.

BERRY, B.J.L. 1967. *Geography of Market Centres and Retail Distribution*. Englewood Cliffs: Prentice Hall.

BORCHERT, J.R. 1967. 'American Metropolitan Evolution'. *Geographical Review* 57:301–32.

BRUSH, J.E. 1953. 'The Hierarchy of Central Places in South West Wisconsin'. *Geographical Review* 43:350–402.

BYLUND, E. 1960. 'Theoretical Considerations Regarding the Distribution of Settlement in Inner North Sweden'. *Geografiska Annaler* 42B:225–31.

CASTELLS, M., and P. HALL. 1994. *Technopoles of the World: The Making of the Twenty-first Century Industrial Complexes*. New York: Routledge.

CATER, J., and T. JONES. 1989. *Social Geography: An Introduction to Contemporary Issues*. New York: Arnold.

CHRISTALLER, W. 1966. *Central Places in Southern Germany*. Translated by C.W. Baskin. Englewood Cliffs: Prentice Hall.

CHURCH, R.L., and T.L. BELL. 1988. 'An Analysis of Ancient Egyptian Settlement Patterns Using Location-Allocation Covering Models'. *Annals, Association of American Geographers* 78:701–14.

COHEN, M. 1993. 'Megacities and the Environment'. *Finance and Development* 30, no. 2:44–7.

COOKE, P. 1990. 'Modern Urban Theory in Question'. *Transactions, Institute of British Geographers* 15:331–43.

COSGROVE, D. 1984. *Social Formation and Symbolic Landscape*. London: Croom Helm.

DAVIES, S., and M. YEATES. 1991. 'Exurbanization as a Component of Migration: A Case Study in Oxford County, Ontario'. *Canadian Geographer* 35:177–86.

DAVIS, M. 1992. *City of Quartz: Excavating the Future in Los Angeles*. London: Vintage Books.

DEAR, M.J., and S. FLUSTY. 1998. 'Postmodern Urbanism', *Annals of the Association of American Geographers* 88, 50–72.

DEAR, M.J., and J.R. WOLCH. 1987. *Landscapes of Despair*. Princeton: Princeton University Press.

DUNCAN, J.S. 1990. *The City as Text: The Politics of Landscape Interpretation in the Kandyman Kingdom*. New York: Cambridge University Press.

EARLE, C.V. 1977. 'The First English Towns of North America'. *Geographical Review* 67:34–50.

ENGLISH, P.W., and R.C. MAYFIELD, eds. 1972. *Man, Space and Environment*. New York: Oxford University Press, 601–10.

GODFREY, B.J. 1991. 'Modernizing the Brazilian City'. *Geographical Review* 81:18–34.

GREENBIE, B.B. 1981. *Spaces: Dimensions of the Human Landscape*. New Haven: Yale University Press.

GROSSMAN, D. 1971. 'Do We Have a Theory for Rural Settlement?' *Professional Geographer* 23:197–203.

HALL, P., et al. 1973. *The Containment of Urban England*. London: Allen and Unwin.

HARRIS, C.D., and E.L. ULLMAN. 1945. 'The Nature of Cities'. *Annals, American Academy of Political and Social Science* 37:7–17.

HARVEY, D.W. 1969. *Explanation in Geography*. London: Arnold.

———. 1973. *Social Justice and the City*. London: Arnold.

———. 1982. *The Limits to Capital*. Oxford: Blackwell.

———. 1989. *The Urban Experience*. Oxford: Blackwell.

———. 1996. *Justice, Nature and the Geography of Difference*. Oxford: Blackwell.

———. 2000. *Spaces of Hope*. Berkeley: University of California Press.

HOSKINS, W.G. 1955. *The Making of the English Landscape*. London: Hodder and Stoughton.

HOYT, H. 1939. *The Structure and Growth of Residential Neighbourhoods in American Cities*. Washington, DC: Federal Housing Administration.

HUDSON, J.C. 1969. 'A Location Theory for Rural Settlement'. *Annals, Association of American Geographers* 59:461–95.

KEDDIE, P.D., and A.E. JOSEPH. 1991. 'The Turnaround of the Turnaround? Rural Population Change in Canada, 1976 to 1986'. *Canadian Geographer* 35:367–79.

KNOX, P. 1992. 'Suburbia by Stealth'. *Geographical Magazine* 44, no. 8:26–9.

———. 1994. *Urbanization: An Introduction to Urban Geography*. Englewood Cliffs: Prentice-Hall.

LEY, D. 1992. 'Gentrification in Recession: Social Change in Six Canadian Inner Cities, 1981–1986'. *Urban Geography* 13:230–56.

———. 1993. *The New Middle Class and the Remaking of the Central City*. Oxford: Oxford University Press.

LINDEN, E. 1993. 'Megacities'. *Time*, 11 January: 30–40.

LÖSCH, A. 1954. *The Economics of Location*. Translated by W.H. Woglom. New Haven: Yale University Press.

LOWDER, S. 1994. 'Urban Development in Ecuador'. *Geography Review* 7, no. 4:2–6.

LYNCH, K. 1960. *The Image of the City*. Cambridge, Mass.: MIT Press.

MACMILLAN, B. 1996 'Chinese Cities'. *Geography Review* 9, no. 5:2–6.

MEYER, D.R. 1980. 'A Dynamic Model of the Integration of Frontier Urban Places into the United States System of Cities'. *Economic Geography* 56:120–40.

MIDDLETON, N. 1994. 'Environmental Problems in Mexico City'. *Geography Review* 7, no. 4:16–18.

MULLER, E.K. 1976. 'Selective Urban Growth in the Middle Ohio Valley, 1800–1860'. *Geographical Review* 66:178–99.

MURPHY, R.E., J.E. VANCE, and B.J. EPSTEIN. 1955. 'Internal Structure of the CBD'. *Economic Geography* 31:21–46.

NELSON, R.L. 1958. *The Selection of Retail Locations*. New York: Dodge.

OLDS, K. 2001. *Globalization and Urban Change: Capital, Culture, and Pacific Rim Mega-Projects*. New York: Oxford University Press.

PAHL, R.E. 1966. 'The Rural/Urban Continuum'. *Sociologia Ruralis* 6:299–327.

PARK, R.E., and E.W. BURGESS. 1921. *Introduction to the Science of Sociology*. Chicago: University of Chicago Press.

ROZMAN, G. 1978. 'Urban Networks and Historical Stages'. *Journal of Interdisciplinary History* 9:65–92.

SARGENT, C.G. 1975. 'Towns of the Salt River Valley, 1870–1930'. *Historical Geography Newsletter* 5, no. 2:1–9.

SASSEN, S. 1991. *The Global City: New York, London, Tokyo*. Princeton: Princeton University Press.

SJOBERG, G. 1960. *The Pre-Industrial City: Past and Present*. Glencoe: Free Press.

SMAILES, A.E. 1957. *The Geography of Towns*. London: Hutchinson.

SMITH, R.L. 1985. 'Activism and Social Status as Determinants of Neighbourhood Identity'. *Professional Geographer* 37:421–32.

VANCE, J.E. 1970. *The Merchant's World: The Geography of Wholesaling*. Englewood Cliffs: Prentice-Hall.

———. 1971. 'Land Assignment in Pre-Capitalist, Capitalist and Post-Capitalist Societies'. *Economic Geography* 47:101–20.

———. 1990. *The Continuing City: Urban Morphology in Western Civilization*. Baltimore: Johns Hopkins University Press.

WADE, R.C. 1959. *The Urban Frontier: 1790–1830*. Cambridge, Mass.: Harvard University Press.

WHEATLEY, P. 1971. *The Pivot of the Four Quarters: A Preliminary Enquiry into the Origins and Character of the Ancient Chinese City*. Chicago: Aldine.

WHITE, M. 1987. *American Neighborhoods and Residential Differentiation*. New York: Russell Sage Foundation.

WIRTH, C. 1938. 'Urbanism as a Way of Life'. *American Journal of Sociology* 44:46–63.

WOLCH, J., and DEVERTEUIL, G. 2001. 'Landscapes of the New Poverty Management.' In *Time Space*, edited by J. May and N.J. Thrift, 149–68. New York: Routledge.

CHAPTER 13

BAND, G.C. 1991. 'Fifty Years of UK Offshore Oil and Gas'. *Geographical Journal* 157:179–89.

BONE, R.M. 1992. *The Geography of the Canadian North*. Toronto: Oxford University Press.

BRITTON, J. 1996. *Canada and the Global Economy: The Geography of Structural and Technological Change*. Montreal: McGill-Queen's University Press.

CROMAR, P. 1979. 'Spatial Change and Economic Organization: The Tyneside Coal Industry (1751–1770)'. *Geoforum* 10:45–57.

DANIELS, P.W. 1985. *Service Industries: A Geographical Appraisal*. New York: Methuen.

DICKEN, P. 1992. *Global Shift: The Internationalization of Productive Activity*. London: Chapman.

EDWARDS, M. 1997. 'China's Gold Coast'. *National Geographic* 191, no. 3:2–31.

FENNELL, D.A. 1999. *Ecotourism*. London: Routledge.

FREEBERNE, M. 1993. 'The Northeast Asia Regional Development Area: Land of Metal, Wood, Water, Fire and Earth'. *Geography* 78:420–32.

FRIEDMANN, J. 1972. 'A General Theory of Polarized Development'. In *Growth Centers in Regional Economic Development*, edited by N.M. Hansen, 82–102. New York: Free Press.

FRIEDRICH, C., trans. 1929. *Alfred Weber's Theory of the Location of Industries*. Cambridge, Mass.: Harvard University Press.

GRAHAM, J., et al. 1988. 'Restructuring in U.S. Manufacturing: The Decline of Monopoly Capitalism'. *Annals, Association of American Geographers* 78:473–90.

GRUBEL, H.G., and K. WALKER. 1989. *Service Industry Growth: Causes and Effects*. Vancouver: The Fraser Institute.

HOOVER, E.M. 1948. *The Location of Economic Activity*. Toronto: McGraw-Hill.

HOPKINS, J.S.P. 1990. 'West Edmonton Mall: Landscape of Myths and Elsewhereness'. *Canadian Geographer* 34:2–17.

HOTELLING, H. 1929. 'Stability in Competition'. *Economic Journal* 39:40–57.

HUDSON, R., and D. SADLER. 1989. *The International Steel Industry: Restructuring, State Policies and Localities*. London: Routledge.

JACKSON, E.L., and D.B. JOHNSON, eds. 1991. 'The West Edmonton Mall and Mega Malls'. *Canadian Geographer* 35, no. 3.

JAKLE, J.A. 1985. *The Tourist: Travel in Twentieth-Century North America*. Lincoln: University of Nebraska Press.

KENNELLY, R.A. 1968. 'The Location of the Mexican Steel Industry'. In *Readings in Economic Geography: The Location of Economic Activity*, edited by R.H.T. Smith, E.J. Taafe, and L.J. King, 126–57. Chicago: Rand McNally.

LEY, D., and K. OLDS. 1988. 'Landscape as Spectacle: World's Fairs and the Culture of Heroic Consumption'. *Environment and Planning D: Society and Space* 6:191–212.

MASSEY, D. 1984. *Spatial Divisions of Labour: Social Structures and the Geography of Production*. New York: Methuen.

MELLINGER, A.D., J.D. SACHS, and J.L. GALLUP. 2000. 'Climate, Coastal Proximity, and Development'. In *The Oxford Handbook of Economic Geography*, edited by G.L. Clark, M.P. Feldman, and M.S. Gertler, 169–94. Toronto: Oxford University Press

MORRISH, M. 1994. 'China Takes the Road to Market'. *Geographical Magazine* 67, no. 4:43–5.

———. 1997. 'The Living Geography of China'. *Geography* 82:3–16.

MURGATROYD, T. 1993. 'The New Zealand Power Crisis'. *Geography* 78:433–7.

O'HARE, G., and H. BARRETT. 1993. 'The Fall and Rise of the Sri Lankan Tourist Industry'. *Geography* 78:438–42.

POLLARD, J.S. 1989. 'Gender and Manufacturing Employment: The Case of Hamilton'. *Area* 21:377–84.

PRED, A. 1966. *The Spatial Dynamics of U.S. Urban-Industrial Growth, 1800–1914*. Cambridge, Mass.: MIT Press.

ROBINSON, M. 1999. 'Cultural Conflicts in Tourism: Inevitability and Inequality'. In *Tourism and Cultural Conflicts*, edited by M. Robinson and P. Boniface, 1–32. New York: CABI Publishing.

ROSTOW, W.W. 1960. *The Stages of Economic Growth*. Cambridge: Cambridge University Press.

SHIELDS, R. 1989. 'Social Spatialisation and the Built Environment: The Example of West Edmonton Mall'. *Environment and Planning D: Society and Space* 7:147–64.

SQUIRE, S.J. 1994. 'Accounting for Cultural Meanings: The Interface Between Geography and Tourism Studies Re-examined'. *Progress in Human Geography* 18:1–16.

STAFFORD, H.A. 1985. 'Environmental Protection and Industrial Location'. *Annals, Association of American Geographers* 75:227–40.

WEBBER, M.J. 1984. *Industrial Location*. Beverly Hills: Sage.

WHEAT, S. 1994. 'Taming Tourism'. *Geographical Magazine* 67:16–19.

YEUNG, Y-M., and F-C. LO. 1996. 'Global Restructuring and Emerging Urban Corridors in Pacific Asia'. In *Emerging World Cities in Pacific Asia*, edited by F-C. Lo and Y-M. Yeung, 17–47. New York: United Nations University Press.

CONCLUSION

BUTTIMER, A. 1974. *Values in Geography*. Washington, DC: Association of American Geographers, Commission on College Geography, Resource Paper No. 24.

FOOTTIT, C. 1999. 'No Pride and Prejudice', *Geographical* 71, 11:20–24.

GREENLEE, J. 1988. 'Infinite Accretion'. *Geographical Magazine* 60, no. 11:24–6.

GREGORY, D. 1994. *Geographical Imaginations*. Oxford: Blackwell.

HANDBURY-TENISON, R., and C. SHANKEY. 1989. 'An Uphill Struggle for Survival'. *Geographical Magazine* 61, no. 9:29–33.

HARVEY, D.W. 1973. *Social Justice and the City*. London: Arnold.

JENKINS, S. 1992. 'Four Cheers for Geography'. *Geography* 77:193–7.

JOHNSTON, R.J. 1991. *A Question of Place: Exploring the Practice of Human Geography*. Oxford: Blackwell.

MARSH, G.P. [1864] 1965. *Man and Nature; or, Physical Geography as Modified by Human Action*. Cambridge, Mass: Belknap Press.

MORRILL, R.L. 1987. 'A Theoretical Imperative'. *Annals, Association of American Geographers* 77:535–41.

PEATTIE, R. 1940. *Geography as Human Destiny*. Port Washington, NY: Kennikat Press.

PRINCE, H.C. 1961. 'The Geographical Imagination'. *Landscape* 11, no. 2:22–5.

SCHNAIBERG, A., and K.A. GOULD. 1994. *Environment and Society: The Enduring Conflict*. New York: St Martin's Press.

GLOSSARY

NOTE: The number in parentheses indicates the page where each term is formally introduced.

accessibility
A variable quality of a location, expressing the ease with which it may be reached from other locations (60).

acculturation
The process by which an ethnic group is absorbed into a larger society while retaining aspects of its distinct identity (270); compare *assimilation*.

acid rain
The deposition on the earth's surface of sulphuric and nitric acids formed in the atmosphere as a result of fossil fuel and biomass burning; causes significant damage to vegetation and built environments (124).

adaptation
The process by which humans adjust to a particular set of circumstances; changes in behaviour that reduce conflict with the environment. See *cultural adaptation*.

agglomeration
The spatial grouping of humans or human activities to minimize the distances between them (60).

agricultural revolution
The slow transition in the human way of life, beginning about 12,000 years ago, from foraging to food production through plant and animal domestication (108).

alienation
The circumstance in which a person is indifferent to or estranged from nature or the means of production (258).

anarchism
A political philosophy that rejects the state and argues that social order is possible without a state (313).

animism
A general name for spiritual beliefs that attribute a spirit or soul to natural phenomena and inanimate objects (242).

apartheid
The South African policy by which four groups of people, as defined by the authorities, were spatially separated between 1948 and 1994 (272).

areal differentiation
A term favoured by Hartshorne; a synonym for 'regional geography' (34).

artifacts
Those elements of culture directly concerned with matters of livelihood (220).

assimilation
The process by which an ethnic group is absorbed into a larger society and loses its own identity (270); compare *acculturation*.

authority
The power or right, usually mutually recognized, to require and receive the obedience of others (271).

biomass
The mass of biological material present in an area, including both living and dead plant material (109).

capitalism
A social and economic system for the production of goods and services that is based on private enterprise; under capitalism the producer is separated from the means of production, and the power to exploit resources is limited to relatively few individuals (52). See also *competitive*, *organized*, and *disorganized capitalism*.

carrying capacity
The maximum population that can be supported by a given set of resources (157).

cartography
The conception, production, dissemination, and study of maps (66).

caste
A social rank, based solely on birth, to which an individual belongs for life and that limits interaction with members of other castes (336).

catastrophists
Those who argue that population increases and continuing environmental deterioration are leading to a nightmare future of food shortages, disease, and conflict (130).

ceiling rent
The maximum rent that a potential land user can be charged for use of a given piece of land (375).

census
The periodic collection and compilation of demographic and other data relating to all individuals in a given country at a particular time (180).

centrifugal forces
In political geography, forces that make it difficult to bind an area together as a effective state; in urban geography, forces that favour the decentralization of urban land uses (298).

centripetal forces
In political geography, forces that pull an area together as one unit to create a relatively stable state; in urban geography, forces that favour the concentration of urban land uses in a central area (298).

chain migration
A process of movement from one location to another through time that is sustained by social links of kinship or friendship; often results in the creation of distinct areas of ethnic settlement in rural or urban areas (269).

chorography
A Greek term revived by nineteenth-century German geographers who used it to refer to regional descriptions of large areas (118).

chorology
A Greek term revived by nineteenth-century German geographers as a synonym for 'regional geography' (31).

choropleth map
A thematic map using colour (or shading) to indicate density of a particular phenomenon in a given area (67).

civilization
A culture with agriculture and cities, food and labour surpluses, labour specialization, social stratification, and state organization (222).

class
A large group made up of individuals of similar social status, income, and culture (258).

Cold War
The period of non-military confrontation between western and communist regions that began shortly after the end of the

Second World War and lasted until the early 1990s; the central feature of international relations during this period, cold war politics had global implications (298).

collective consumption
The use of services that are produced and managed on a collective basis (481).

colonialism
The policy of a state or people seeking to establish and maintain authority over another state or people (201).

commercial agriculture
An agricultural system in which the production is primarily for sale (375).

commercial geography
An early term for economic geography, focusing on what was produced where (366).

competitive capitalism
The first of three phases of capitalism, beginning in the early eighteenth century (although some have identified signs of capitalism as early as the late sixteenth century); characterized by free market competition and laissez-faire economic development (258).

conservation
A general term referring to any form of environmental protection, including preservation (132).

constructionism
The school of thought according to which all our conceptual underpinnings (for example, ideas about identity) are socially constructed and therefore contingent and dynamic, not given or absolute (compare *essentialism*); because we are all born into an existing society, that social knowledge becomes a part of our worldview and ideology (256).

contextualism
Broadly, the idea that it is necessary to take into account the specific context, within which any research is conducted (264).

continental drift
The idea that the present continents were originally connected as one or two land masses that separated and drifted apart (83).

conurbation
A continuously built-up area formed by the coalescence of several expanding cities that were originally separate (436).

core/periphery
The concept that states are often unequally divided between powerful cores and dependent peripheries (308).

cornucopians
Those who argue that advances in science and technology will continue to create resources sufficient to support the growing world population (130).

cosmography
The science that maps and describes the entire universe, both heavens and earth; this Greek term is rarely used today (18).

counterfactual
An approach to historical analysis that compares what actually happened with what might have happened had some particular circumstance been different (296).

counterurbanization
A process of population decentralization that may be prompted by several factors, including the high cost of living in cities, improvements in personal spatial mobility, industrial deconcentration, and advances in information technologies (418).

creole
A pidgin language that assumes the status of a mother tongue for a group (238).

cultural adaptation
Changes in technology, organization, and ideology that permit sound relationships to develop between humans and their physical environment (229).

cultural regions
Areas in which there is a degree of homogeneity in cultural characteristics; areas with similar landscapes (224).

culture
The way of life of the members of a society (219).

cycle of poverty
The idea that poverty and deprivation are transmitted inter-generationally, reflecting home background and spatial variations in opportunities (439).

deconstruction
A method of critical interpretation applied to texts, including landscapes, that aims to show how the multiple positioning (in terms of class, gender, and so on) of an author or a reader affects the creation or reading of the text (263).

deglomeration
The spatial separation of humans or human activities so as to maximize the distances between them (60).

deindustrialization
Loss of manufacturing activity and related employment in a traditional manufacturing region in the more developed world (480).

democracy
A form of government involving free and fair elections, openness and accountability, civil and political rights, and the rule of law (312).

demography
The study of human populations (142).

density
A measure of the number of geographic facts (for example, people) per unit area (178).

dependence, dependency
In political contexts, a relationship in which one state or people is dependent on, and therefore dominated by, another state or people (201).

dependency theory
Centres on the relationship between dependence and under-development (293).

desertification
The process by which an area of land becomes a desert; typically involves the impoverishment of an ecosystem because of climatic change and/or human impact (119).

development
A term that should be handled with caution because it has often been used in an ethnocentric fashion; typically understood to refer to a process of becoming larger, more mature, and better organized; often measured by economic criteria (64).

developmentalism
Analysis of cultural and economic change that treats each country or region of the world separately in an evolutionary manner; assumes that all areas are autonomous and proceed through the same series of stages (185).

devolution
A process of transferring power from central to regional or local levels of government (302).

dictatorship
An oppressive, anti-democratic form of government in which the leader is often backed by the military (313).

diffusion
The spread of any phenomenon over space and its growth through time (62).

disorganized capitalism
The most recent form of capitalism, characterized by a process of disorganization and industrial restructuring (258).

distance
The spatial dimension of separation; a fundamental concept in spatial analysis (59).

distance decay
The declining intensity of any pattern or process with increasing distance from a given location (60).

distribution
The pattern of geographic facts (for example, people) within an area (59).

domestication
The process of making plants and/or animals more useful to humans through selective breeding (383).

ecology
The study of relationships between organisms and their environments (106).

economic base theory
A theory that tries to explain the growth or decline of particular regions or cities in terms of 'basic' and 'non-basic' economic activities: 'basic' goods and services are those produced for sale outside the city or region (423).

economic operator
A model of human behaviour in which each individual is assumed to be completely rational; economic operators maximize returns and minimize costs (368).

economic rent
The surplus income that accrues to a unit of land above the minimum income needed to bring a unit of new land into production at the margins of production (376).

ecosystem
An ecological system; comprises a set of interacting and interdependent organisms and their physical, chemical, and biological environment (106).

ecotourism
Tourism that is environmentally friendly and allows participants to experience a distinctive ecosystem (489).

empiricism
A philosophy of science based on the belief that all knowledge results from experience, and that therefore gives priority to factual observations over theoretical statements (48).

energy
The capacity of a physical system for doing work (107).

entropy
A measure of the disorder or disorganization in a system (339).

environmental determinism
The view that human activities are controlled by the physical environment (33).

equinox
Occurs twice each year, in spring and fall, when the sun is vertically overhead at the equator; on this day the periods of daylight and darkness are both twelve hours long all over the earth (82).

essentialism
Belief in the existence of fixed unchanging properties; attribution of 'essential' characteristics to groups (256); compare *constructionism*.

ethnic group
A group whose members perceive themselves as different from others because of a common ancestry and shared culture (268).

ethnocentrism
A form of prejudice or stereotyping that presumes that one's own culture is normal and natural and therefore that all other cultures are inferior (72).

ethnography
The study and description of social groups based on researcher involvement and first-hand observation in the field; a qualitative rather than quantitative approach (72).

existentialism
A humanistic philosophy according to which humans are responsible for making their own nature; stresses personal freedom, decision, and commitment in a world without absolute values outside of individuals (51).

exurbanization
The movement of households from urban areas to locations outside the urban area but within the commuting field (419).

fecundity
A biological term; the ability of a woman or man to produce a live child; refers to potential rather than actual number of live births (142).

federalism
A form of government in which power and authority are divided between central and regional governments (314).

feminism
The movement for or, more generally, advocacy of equal rights for women and men, and commitment to improve the position of women in society (215).

fertility
Generally, all aspects of human reproduction that lead to live births; also used specifically to refer to the actual number of live births produced by a woman (116).

feudalism
A social and economic system prevalent in Europe, prior to the industrial revolution, in which land was owned by the monarch, controlled by lords, and worked by peasants who were bound to the land and subject to the lords' authority (257).

fieldwork
A means of data collection; includes both qualitative (for example, observation) and quantitative (for example, questionnaire) methods (72).

first effective settlement
A concept based on the likely importance of the initial occupancy of an area in determining later landscapes (227).

flexible accumulation
Industrial technologies, labour practices, relations between firms, and consumption patterns that are increasingly flexible (480).

forces of production
A Marxist phrase that refers to the raw materials, tools, and workers that actually produce goods (52).

Fordism
A group of industrial and broader social practices introduced by Henry Ford, including use of the mass-production assembly line and development of a new relationship with workers characterized by higher wages and shorter working hours; dominant until recently in most industrial countries (258).

foreign direct investment
Direct investment by a government or multinational corporation in another country, often in the form of a manufacturing plant; raises employment and output in the host country, but is not always welcomed because it can be seen as an instance of economic colonialism (357).

formal region
A region that is identified as such because of the presence of some particular characteristic(s) (59).

friction of distance
A measure of the restraining effect of distance on human movement (332).

functional region
A region that is identified as such because it comprises a series of linked locations (59).

garden city
A planned settlement designed to combine the advantages of urban and rural living; an urban centre emphasizing spaciousness and quality of life (423).

Gemeinschaft
A term introduced by Tönnies; a form of human association based on loyalty, informality, and personal contact; assumed to be characteristic of traditional village communities (416).

gender
The social aspect of the relations between the sexes (260).

genocide
An organized, systematic effort to destroy a group defined in ethnic terms; usually the targeted group is seen as living in the 'wrong place' (303).

gentrification
A process of inner-city urban neighbourhood social change resulting from the in-movement of higher-income groups (439).

geographic information system (gis)
A computer-based tool that combines the storage, display, analysis, and mapping of spatially referenced data (70).

géographie Vidalienne
French school of geography initiated by Paul Vidal de la Blache at the end of the nineteenth century and still influential today, focusing on the study of human-made (cultural) landscapes (32).

geopolitics
The study of the importance of space in understanding international relations (296).

geopolitik
The study of states as organisms that need to expand in territory in order to fulfil their destinies as nation states (297).

gerrymandering
The creation of electoral boundaries designed to benefit a particular party (316).

Gesellschaft
A term introduced by Tönnies; a form of human association based on rationality and depersonalization; assumed to be characteristic of urban areas (416).

ghetto
A residential district in an urban area with a concentration of a particular ethnic group (269).

globalization
A complex combination of economic, political, and cultural changes that have long been evident but that have accelerated markedly since about 1980, bringing about a seemingly ever-increasing connectedness of both people and places; it is often claimed that this transformation, broadly understood, is the greatest challenge facing humans today (64).

green belt
A planned area of open, partially rural, land surrounding an urban area (423).

gross domestic product (GDP)
A monetary measure of the value at market prices of goods and services produced by a country over a given time period (usually one year); provides a better indication of domestic production than the GNP (198).

gross national product (GNP) or gross national income (GNI)
A monetary measure of the value at market prices of goods and services produced by a country, plus net income from abroad, over a given period (usually one year) (198).

heartland theory
A geopolitical theory of world power based on the assumption that the land-based state controlling the Eurasian heartland held the key to world domination (297).

historical materialism
An approach associated with Marxism that centres on the material basis of society and that attempts to understand social change by reference to historical changes in social relations (52).

Holocene
Literally, 'wholly recent': the postglacial period that began 10,000 years ago; preceded by the Pleistocene (98).

homeland
A cultural region that is especially closely associated with a particular cultural group; the term usually suggests a strong emotional attachment to place (268).

hominids
Bipedal primates; humans and their ancestors (97).

humanism
A philosophy that centres on such aspects of human life as value, quality, meaning, significance, and spirituality (49).

hypothesis
In positivist philosophy, a general statement deduced from theory but not yet verified (48).

iconography
The description and interpretation of visual images, including landscape, in order to uncover their symbolic meanings; the identity of a region as expressed through symbols (257).

idealism
A humanistic philosophy according to which human actions can be understood only by reference to the thought behind them (51).

ideal type
A term introduced by the sociologist Max Weber to refer to a hypothetical norm used to deepen understanding of the real world through comparison (258).

idiographic
Concerned with the unique and particular (54).

image
The perception of reality held by an individual or group (63).

imperialism
A relationship between states in which one is dominant over the other (293).

industrial revolution
The process that converted a fundamentally rural society into an industrial society beginning in England *c.*1750; primarily a technological revolution associated with the development of new energy sources (108).

informal sector
A part of a national economy involved in productive labour, but without any formal recognition, control, or remuneration (447).

infrastructure (base)
A Marxist concept that refers to the economic structure of a society, especially as it gives rise to political, legal, and social systems (53).

innovation
The introduction of a new invention or idea, especially one that leads to change in human behaviour or production processes (339).

interaction
The relationship or linkage between locations (60).

interaction theory

A body of theories that explain movements of goods and people between locations (367).

international division of labour

A term referring to the current tendency for high-wage and high-skill employment opportunities, often in the service sector, to be located in the more developed world, whereas low-wage and low-skill employment opportunities, often in the industrial sector, are located in the less developed world. (388)

irredentism

The view held by one country that a minority living in an adjacent country rightfully belongs to it (245).

isodapanes

In Weberian least-cost industrial location theory, lines of equal additional transport cost drawn around the point of minimum transport cost (337).

isopleth map

A map using lines to connect locations of equal data value (67).

isotims

In Weberian least-cost industrial location theory, lines of equal transport costs around material sources and markets (462).

landscape

A major concern of geographic study; the characteristics of a particular area especially as created through human activity (1).

landscape school

American school of geography initiated by Carl Sauer in the 1920s and still influential today; an alternative environmental determinism, focusing on human-made (cultural) landscapes (34).

Landschaftskunde

A German term, introduced in the late nineteenth century, best translated as 'landscape science'; refers to geography as the study of the landscapes of particular regions (34).

latitude

Angular distance on the surface of the earth, measured in degrees, minutes, and seconds, north and south of the equator (which is the line of 0° latitude); lines of constant latitude are called parallels (17).

law

In positivist philosophy, a hypothesis that has been proven correct and is taken to be universally true; once formulated, laws can be used to construct theories (48).

less developed world

All countries not classified as 'more developed' (109; see *more developed world*).

life cycle

The process of change experienced by individuals over their lifespans; often divided into stages (such as childhood, adolescence, adulthood, old age), each of which is associated with particular forms of behaviour (185).

limits to growth

The argument that, in the future, both world population and world economy may collapse because available world resources are inadequate (165).

lingua franca

An existing language used as a common means of communication between different language groups (238); compare *pidgin*.

lithosphere

The outer layer of rock on earth; includes crust and upper mantle (83).

locale

The setting or context for social interaction; a term introduced in structuration theory that has become popular in human geography as an alternative to 'place' (260).

location

A term that refers to a specific part of the earth's surface; an area where something is situated (56).

locational interdependence

The situation in which competing businesses base their location decisions on those of their rivals (463).

location theory

A body of theories explaining the distribution of economic activities (367).

longitude

Angular distance on the surface of the earth, measured in degrees, minutes, and seconds, east and west of a specific prime meridian (which is the line of 0° longitude); lines of constant longitude are called meridians (17).

malapportionment

A form of gerrymandering, involving the creation of electoral districts of varying population sizes so that one party will benefit (316).

malnutrition

A condition caused by a diet lacking some food necessary for health (202).

Maoism

The revolutionary thought and practice of Mao Zedong, based on protracted revolution to achieve power and socialist policies after power is achieved (313).

Marxism

The body of social and political theory developed by Karl Marx (1818–83), in which mode of production is seen as the key to understanding society. Marxism emphasizes the importance of labour; argues that throughout history the state has enabled a small dominant class to exploit the masses; and sees class struggle as the key to historical change (52).

material index

An index devised by Weber and used in industrial location theory to show the extent to which the least-cost location for a particular industrial firm will be either material- or market-oriented (461).

mental map

The individual psychological representation of space (63).

mentifacts

Those elements of culture that are attitudinal in character; the values held by members of a group (219).

mercantilism

A school of economic thought dominant in Europe in the seventeenth and early eighteenth centuries that argued for the involvement of the state in economic life so as to increase national wealth and power (421).

minority language

A language spoken by a minority group in a state in which the majority of the population speaks some other language; may or may not be an official language (236).

model

An idealized and structured representation of the real world (75).

mode of production

A Marxist term that refers to the organized social relations through which a human society organizes productive activity; human societies are seen as passing through a series of such modes (52).

modernism

A view that assumes the existence of a reality characterized by structure, order, pattern, and causality (262).

monarchy

The institution of rule over a state by the head of a hereditary family (313); *monarchists* are those who favour this system.

more developed world

(According to a United Nations classification) Europe, North America, Australia, Japan, New Zealand, and the former USSR (109).

multiculturalism

A policy that endorses the right of ethnic groups to remain distinct rather than be assimilated into a dominant society (270).

multilingual state

A state in which the population includes at least one linguistic minority (236).

nation

A group of people sharing a common culture and an attachment to some territory; a difficult term to define objectively (288).

nationalism

The political expression of nationhood or aspiring nationhood; reflects a consciousness of belonging to a nation (235).

nation state

A political unit that contains one principal national group that gives it its identity and defines its territory (288).

nativism

Intense opposition to an internal minority on the grounds that the minority is foreign (191).

neocolonialism

Economic relationships of dominance and subordination between countries without equivalent political relationships; such situations often develop after political colonialism ends and the former colony achieves independence, but may also occur without prior political colonialism (390).

nomothetic

Concerned with the universal and the general (55).

normative

Focusing on what ought to be rather than what actually is; in normative theory, the aim is to seek what is rational or optimal according to some given criteria (367).

nuptiality

The extent to which a population marries (145).

oligarchy

Rule by an élite group of people, typically the wealthy (313).

organized capitalism

The second phase of capitalism, beginning after the Second World War; characterized by increased growth of major (often transnational) corporations and increased involvement by the state (often in the form of public ownership) in the economy (258).

orientalism

A term used to refer to western views of the Orient, implying a view of the periphery from the centre; although the term is closely associated with postcolonial theory, especially the work of Edward Said, the idea of 'the Orient' as a foil for the west dates back to classical Greece (272).

other

Philosophically, an ambiguous term derived from Hegel and widely used in feminist and postcolonial theory. The 'other' (or 'Other') is usually defined in opposition to the same or to self. In cultural geography, 'other' often refers to subordinate groups as they are seen by and contrasted to dominant groups; implies both difference and inferiority (as, for example, when masculine identity is defined in terms of the exclusion of the feminine other) (272).

ozone layer

Layer in the atmosphere 16–40 km (10–25 miles) above the earth that absorbs dangerous ultraviolet solar radiation; ozone is a gas composed of molecules consisting of three atoms of oxygen (O_3) (82).

pandemic

A term used to designate diseases with very wide distribution (a whole country, or even the world); 'epidemic' diseases have more limited distribution (151).

participant observation

A qualitative method in which the researcher is directly involved with the subjects in question (72).

patriarchy

A social system in which men dominate, oppress, and exploit women (260).

perception

The process by which humans acquire information about physical and social environments (62).

phenomenology

A humanistic philosophy based on the ways in which humans experience everyday life and imbue activities with meaning (51).

phenotype

Any physical or chemical trait that can be observed or measured (99).

physiological density

Population per unit of cultivable land (179).

pidgin

A new language designed to serve the purposes of commerce between different language groups; typically has a limited vocabulary (238); compare lingua franca.

place

Location; in humanistic geography, 'place' has acquired a particular meaning as a context for human action that is rich in human significance and meaning (56).

placelessness

Homogeneous and standardized landscapes that lack local variety and character (58).

place utility

A measure of the satisfaction that an individual derives from a location relative to his or her goals (184).

Pleistocene

Geological time period, from about 1.5 million years ago to 10,000 years ago, characterized by a series of glacial advances and retreats (98); succeeded by the *Holocene*.

pollution

The release into the environment of substances that degrade one or more of land, air, or water (108).

population pyramid

A diagrammatic representation of the age and sex composition of a population (159).

positivism

A philosophy that contends that science is able to deal only with empirical questions (those with factual content), that scientific observations are repeatable, and that science progresses through the construction of theories and derivation of laws (48).

possibilism

The view that the environment does not determine either human history or present conditions; rather, humans pursue a course of action that they select from among a number of possibilities (34).

post-Fordism

A group of industrial and broader social practices evident in industrial countries since about 1970; involves more flexible production methods than those associated with Fordism (258).

postmodernism

A movement in philosophy, social science, and the arts based on the idea that reality cannot be studied objectively and that multiple interpretations are possible (262).

power
In general, the capacity to affect outcomes; more specifically, to dominate others by means of violence, force, manipulation, or authority (271).

pragmatism
A humanistic philosophy that focuses on the construction of meaning through the practical activities of humans (50).

primate city
The largest city in a country, usually the capital city, which dominates its political, economic, and social life (430).

principle of least effort
Considered to be a guiding principle in human activities; for human geographers, refers to minimizing distances and related movements (332).

projection
Any procedure employed to represent positions of all or a part of the earth's spherical (three-dimensional) surface onto a flat (two-dimensional) surface (67).

proxemics
The study of personal space, the invisible boundaries that protect and compartmentalize each individual (336).

public goods
Goods that are freely available to all or that are provided (equally or unequally) to citizens by the state (315).

qualitative methods
A set of tools used to collect and analyze data in order to subjectively understand the phenomena being studied; the methods include passive observation, participation, and active intervention (72).

quantitative methods
A set of tools used to collect and analyze data to achieve a statistical description and scientific explanation of the phenomena being studied; the methods include sampling, models, and statistical testing (75).

questionnaire
A structured and ordered set of questions designed to collect unambiguous and unbiased data (75).

race
A subspecies; a physically distinguishable population within a species (99).

raster
A method used in gis to represent spatial data; divides the area into numerous small cells and pixels, and describes the content of each cell (70).

rational choice theory
The theory that social life can be explained by models of rational individual action; an extension of the economic operator concept to other areas of human life (368).

recycling
The reuse of material and energy resources (132).

region
A part of the earth's surface that displays internal homogeneity and is relatively distinct from surrounding areas according to some criterion or criteria; regions are intellectual creations (2).

regionalization
A special kind of classification in which locations on the earth's surface are assigned to various regions, which must be contiguous spatial units (59).

reindustrialization
The development of new industrial activity in a region that has earlier experienced substantial loss of traditional industrial activity (480); see *deindustrialization*.

relations of production
A Marxist term referring to the ways in which the production process is organized, specifically the relationships of ownership and control (52).

remote sensing
A variety of techniques used for acquiring and recording data from points that are not in contact with the phenomena of interest (70).

renewable resources
Resources that regenerate naturally to provide a new supply within a human life span (109).

restructuring
In a capitalist economy, changes in or between the various component parts of an economic system resulting from economic change (401).

rimland theory
A geopolitical theory of world power based on the assumption that the state controlling the area surrounding the Eurasian heartland held the key to world domination (298).

sacred space
A landscape that is particularly esteemed by an individual or a group, usually (but not necessarily) for religious reasons (56).

sampling
The selection of a subset from a defined population of individuals to acquire data representative of that larger population (75).

satisficing behaviour
A model of human behaviour that rejects the rationality assumptions of the economic operator model, assuming instead that the objective is to reach a level of satisfaction that is acceptable (368).

scale
The resolution levels used in any human geographic research; most characteristically refers to the size of the area studied, but also to the time period covered and the number of people investigated (61).

scientific method
The various steps taken in a science to obtain knowledge; a phrase most commonly associated with a positivist philosophy (48).

sense of place
The deep attachments that humans have to specific locations such as home and also to particularly distinctive locations (56).

simulation
Representation of a real-world process in an abstract form for purposes of experimentation (340).

site
The location of a geographic fact with reference to the immediate local environment (56).

situation
The location of a geographic fact with reference to the broad spatial system of which it is a part (56).

slavery
Labour that is controlled through compulsion and is not remunerated; in Marxist terminology, one particular mode of production (187).

socialism
A social and economic system that involves common ownership of the means of production and distribution (259).

social physics
An approach to aggregate human movement and interaction based on the assumption that such phenomena are analogous to processes in the physical sciences (339).

society
Refers to the interrelationships that connect individuals as members of a culture (220).

sociofacts
Those elements of culture most directly concerned with

interpersonal relations; the norms that people are expected to observe (220).

solstice
Occurs twice each year when the sun is vertically overhead at the farthest distance from the equator, once 23.5°N and once 23.5°S of the equator; for example, when at 23.5°N, there is maximum daylight (longest day) in the Northern hemisphere and minimum daylight (shortest day) in the Southern hemisphere (82).

sovereignty
Supreme authority over the territory and population of a state, vested in its government; the most basic right of a state understood as a political community (289).

space
Areal extent; a term used in both absolute (objective) and relative (perceptual) forms (56).

spatial monopoly
The situation in which a single producer sells the entire output of a particular industrial good or service in a given area (463).

spatial preferences
Individual (sometimes group) evaluation of the relative attractiveness of different locations (185).

spatial separatism (or spatial fetishism)
A phrase critical of spatial analysis for treating space or distance as a cause without reference to humans (56).

species
A group of organisms that are able to produce fertile offspring among themselves but not with any other group (94).

spectacle
A term referring to places and events that are carefully constructed for the purposes of mass leisure and consumption (487).

squatter settlement
A shanty town, a concentration of temporary dwellings, neither owned nor rented, at the city's edge; related to large immigration into urban areas that are unable to cope with such population increases (449).

state
An area with defined and internationally acknowledged boundaries; a political unit (288).

state apparatus
The institutions and organizations through which the state exercises its power (315).

stock resources
Minerals and land that take a long time to form and hence, from a human perspective, are fixed in supply (109).

structuration
The conditions that govern the continuity and transformation of social structures (260).

subsistence agriculture
An agricultural system in which the production is not primarily for sale but is consumed by the farmer's household (375).

suburb
An outer commuting zone of an urban area; associated with social homogeneity and a lifestyle suited to family needs (271).

superorganic
An interpretation of culture that sees it as above both nature and individuals and therefore as the principal cause of the human world; a form of cultural determinism (219).

superstructure
A Marxist concept that refers to the political, legal, and social systems of a society (53).

sustainable development
A term popularized in the 1987 report of the World Commission on Environment and Development, referring to economic development that does not damage the environment (132).

symbolic interactionism
A group of social theories that see the social world as a social product with meanings resulting from interaction (256).

system
A set of interrelated components or objects that are linked together to form a unified whole (106).

tariff
A tax or customs duty charged by a country on imports from other countries (349).

technology
The ability to convert energy into forms useful to humans (107).

teleology
The doctrine that everything in the world has been designed by God; also refers to the study of purposiveness in the world (the idea that some phenomena are best explained in terms of ends, what they have become or what they achieve, rather than in terms of causes); sometimes used to refer to a recurring theme in history, such as progress or class conflict (19).

text
A term that originally referred to the written page, but that has broadened to include such products of culture as maps and landscape; postmodernists recognize that there may be any number of realities, depending on how a text is read (263).

theory
In positivist philosophy, an interconnected set of statements, often called assumptions or axioms, that deductively generates testable hypotheses (48).

time–space convergence
A decrease in the friction of distance between locations as a result of improvements in transport and communication technologies (334).

topography
A Greek term, revived by nineteenth-century German geographers to refer to regional descriptions of local areas (18).

topological map
A diagram that represents a network as a simplified series of straight lines (336).

toponym
Place name; such names usually had meaning for the first settlers, as they described a characteristic of the place or people; evidence provided by place names can be crucial in a historical study of movement and settlement if other sources of information are unavailable (238).

topophilia
The affective ties that humans have with particular places; literally, love of place (58).

topophobia
The feelings of dislike, anxiety, fear, or suffering associated with a particular landscape (58).

transnational
A large business organization that operates in two or more countries (65).

undernutrition
Diet inadequate to sustain normal activity (202).

urbanism
The urban way of life; associated with a declining sense of community and increasingly complex social and economic organization as a result of increasing size, density, and heterogeneity (420).

urbanization
The spread and growth of cities (420).

urban sprawl
The largely unplanned expansion of an urban area; typically discontinuous, leaving rural enclaves (418).

vector
A method used in GIS to represent spatial data; describes the data as a collection of points, lines, and areas, and describes the location of each of these (70).

vernacular region
A region that is identified as such on the basis of the perceptions held by people inside and outside the region (59).

verstehen
A research method, associated primarily with phenomenology, in which the researcher adopts the perspective of the individual or group under investigation; a German term, best translated as 'sympathetic' or 'empathetic understanding' (51).

welfare geography
An approach to human geography that maps and explains social and spatial variations (275).

well-being
The degree to which the needs and wants of a society are satisfied (274).

CREDITS

Every effort has been made to determine and contact copyright holders. In the event of errors or omissions, the publisher will be pleased to make suitable acknowledgement in future editions.

Academic Press Ltd: Figure 9.2 is adapted from 'A Theory of Exploration' by J.D. Overton from *Journal of Historical Geography* 7 (1981):57, by permission of Academic Press Ltd.

Figure 10.9 is reprinted from E. Pawson, *Transport and Economy*, Academic Press, 1977, p. 13, by permission of Academic Press Ltd.

Blackwell Publishing Ltd: Figure 4.2 is adapted from *The Human Impact: Man's Role in Environmental Change* by A. Goudie, Blackwell, 1981. By permission of Blackwell Publishers.

Figure 7.7 is adapted from 'The Mormon Cultural Region: Strategies nad Patterns in the Geography of the American West, 1847-1964', by D.W. Meinig from *The Annals of the Association of American Geographers* 55 (1965): 214. By permission of Blackwell Publishers.

Figure 8.2 is reprinted from W. Zelinsky, 'North American Vernacular Regions' from *The Annals of the Association of American Geographers* 70 (1980):14. Reprinted by permission of Blackwell Publishers.

Figure 9.13 from R. Paddison, ed., *The Fragmented State: The Political Geography of Power* (Oxford: Blackwell, 1983), p. 32. Reprinted by permission of Blackwell Publishing Ltd.

Figure 9.15 is adapted from 'The Identification and Evolution of Racial Gerrymandering' by J. O'Loughlin from *The Annals of the Association of American Geographers* 72 (1982):180. Reprinted by permission of Blackwell Publishers.

Figure 11.2 adapted from H.J. Reitsma, 'Crop and Livestock Production in the Vicinity of the United States-Canadian Border', *Professional Geographer*, Vol. 23, 1971, pp. 217, 219, 221. Reprinted by permission of Blackwell Publishing Ltd.

Table 11.4 is reprinted from T.J. Bassett, 'The Political Ecology of Peasant-Herder Conflicts in the Northern Ivory Coast' from *The Annals of the Association of American Geographers* 78 (1988):456. Reprinted by permission of Blackwell Publishers.

The British Petroleum Company plc: Table 13.1 from *BP Statistical Review of World Energy*, 2002.

Cambridge University Press: Table 4.2 is adapted from 'Forests' by M. Williams in B.L. Turner et al., eds, *The Earth as Transformed by Human Action*, (Cambridge: Cambridge University Press, 1990), p. 180. Reprinted by permission of Cambridge University Press.

Canadian Association of Geographers: Figure 2.9 is reprinted from R.I. Ruggles, 'Spatial Concepts of North America in 1763' from 'The West of Canada: Imagination and Reality' from *The Canadian Geographer*. Permission has been granted by the Canadian Association of Geographers.

Figure 10.3 is adapted from J.C. Muller, 'The Mapping of Travel Time in Alberta, Canada' from *The Canadian Geographer*. Permission has been granted by the Canadian Association of Geographers.

Table 12.2 is adapted from P.D. Keddie and A.E. Joseph, 'The Turnaround of the Turnaround? Rural Population Change in Canada, 1976 to 1986' from *The Canadian Geographer*. Permission has been granted by the Canadian Association of Geographers.

Table 12.3 is reprinted from S. Davies and M.H. Yeates, 'Exurbanization as a Component of Migration: A Case Study in Oxford County, Ontario', from *The Canadian Geographer*. Permission has been granted by the Canadian Association of Geographers.

Christopher Davies (Publishers) Limited: Figure 9.12 is reprinted from L. Kohr, *The Breakdown of Nations* (Swansea: Christopher Davies, 1957). Reprinted by permission.

Clark University: Figure 11.1 is adapted from O.E. Baker, 'The Potential Supply of Wheat' from *Economic Geography*, published by Clark University.

Current History, Inc.: Figure 4.7 is adapted with permission from *Current History* magazine (November 1996). Copyright © 1996, Current History, Inc.

Elsevier Science: Figure 12.17, from P.J. Taylor, 'World Cities and Territorial States under Conditions of Contemporary Globalization', *Political Geography*, Vol. 19 (2000), p. 25. Copyright ©2000. Reprinted with permission from Elsevier.

Encyclopædia Britannica, Inc.: Table 7.5 from *2002 Encyclopædia Britannica Book of the Year* (Chicago. Reprinted with permission from *Britannica Book of the Year* © 1996 by Encyclopædia Britannica, Inc.

Geographical Magazine: Table Conclu.1 is adapted from 'An Uphill Struggle for Survival' by R. Handbury-Tenison and C. Shankey, *Geographical Magazine*, 1989. Reprinted by permission.

Gordon Ewing and Pion Limited, London: Figure 10.4 is reprinted from G. Ewing and R. Wolfe, 'Toronto in Physical Space and Time Space' from 'Surface Feature Interpolation on Two-Dimensional Time-Space Maps' from *Environment and Planning* A9 (1977):429–37, figures 3 and 6. Reprinted by permission of Gordon Ewing and Pion Limited, London.

The Geographical Association: Figure 5.6 from J. Jowett, 'China's Population: 1,133,709, 738 and Still Counting', *Geography* 78 (1993):405.

Table 6.12 from M. Degg, 'Natural Disasters: Recent Trends and Future Prospects', *Geography* 77 (1992): 201.

Figure 6.13 from L. Young, 'World Hunger: A Framework for Analysis', *Geography* 81 (1996): 100.

A.L. Grove: Excerpt from F.P. Grove, *Settlers of the Marsh*. Reprinted by permission of A.L. Grove, Toronto.

Hodder Arnold: Table 6.5 from B. Moon, 'Paradigms in Migration Research: Exploring "Moorings" as a Schema', *Progress in Human Geography*, Vol. 19 (1995), p. 515.

Table 12.1 from H. Carter, *The Study of Urban Geography*, Fourth (London: Edward Arnold, 1995), pp. 10–12.

Houghton Mifflin Company: Figure 10.11 from John C. Lowe and S. Moryadas, *The Geography of Movement*. Copyright © 1975 by Houghton Mifflin Company. Used with permission.

International Energy Agency: Table 13.2 from <http://www.iea.org/statist/keyworld2002/key2002/keystats.htm>, International Energy Agency, 2003.

JIH Inc.: Table 12.4 is reprinted from G. Rozman, 'Urban Networks and Historical Stages' from the *Editions of the Journal of Interdisciplinary History* IX (1978) by permission.

Kluwer Academic Publishers: Table 13.7 from D.B. Weaver, 'Ecotourism in the Small Island Caribbean', *Geojournal*, Vol. 31 (1993), p. 458.

Mayfield Publishing Company: Table 7.1 is reprinted from *Human Antiquity* by K.L. Feder and M.A. Park with permission of Mayfield Publishing Company. Copyright © 1993, 1989 by Mayfield Publishing Company.

INDEX

Aboriginal peoples, 7, 9, 276, 292
abortion, 145, 146
accessibility, 60
acculturation, 270
acid rain, 124, 125
adaptation, 132; cultural, 229–32; preadaptation, 229–39
administration principle, 430, 431
Africa, 290, 350–1; fertility in, 144; HIV/AIDS in, 152–3; nation–state discordance, 299–302; refugees in, 197
African Union, 302
age, 143, 144, 159, 161–3, 276; age–sex structure, 160, 162
agency, human, 260, 262
agglomeration, 60, 462–3
aggression, 322
agricultural revolution, 108, 163–4, 222, 352; second, 385–6
agriculture, 87, 115–17, 122–4, 163–4, 204–5, 222–3, 352, 371–409; agribusiness, 399; and cities, 420; commercial, 375; core areas, 384; dairying, 394; evolution of, 382, 384–90; frontier, 385; large-scale grain, 395; location problem in, 371–9; Mediterranean, 392; mixed, 392–3; organic, 388; origins of, 382–4; pastoral nomadism, 392, 393; plantation, 394, 399; primitive subsistence, 391; ranching, 394–5; and region, 391–5; subsistence, 375; types of, 391–5; wet rice, 391–2; world, 382–99; see also farmers
AIDS (Acquired Immune Deficiency Syndrome), 150–4, 343
air, 106; transport by, 348
Albania, 504
Al-Idrisi, 21
alienation, 258
Alsace, 304
Amnesty International, 129
analysis: spatial, 2, 35, 37, 340–2; techniques of, 65–76
anarchism, 313
animals, 120–2, 247; see also domestication
animism, 242
anthropocentrism, 115
anthropogeography, 31, 46
anthropology, 219, 220
apartheid, 272–3
Apian, Peter, 24
Arabs, 295
Aral Sea, 124, 125
areal differentiation, 34, 36; see also regionalization

Aristotle, 288
artifacts, 220
Asia, 154; immigrants from, 191; industry in, 475–6
assimilation, 270
Association of South East Asian Nations (ASEAN), 350
asylum seekers, 193, 196
atmosphere, 106
Australia, 121, 314, 338, 386; urbanization in, 426
Australopithecus, 95, 97
authority, 271

Bahn, P., 108
Bangladesh, 128, 208, 210
Bangui, Central African Republic, 451
Banks, G., 276
Barnes, T.J., 370
Basque country, 304
Batwa people, 505–6
behaviourism, 184–5, 464–5
Belgium, 236, 237, 290
Belize, 489–90
Berry, Brian, 262
Bhopal, India, 112
Bhutan, 398
bias, electoral, 316–17
biochemical cycle, 107
biodiversity, 119, 122
biomass, 109
biosphere, 106
biotechnology, 390
Biswas, L., 246
Blainey, Geoffrey, 338
Blaut, J.M., 258–9
Bodin, J., 27
Bodoland, 307
Borchert, J.R., 425
boreal forests, 93
Boserup, Ester, 168
boundaries, 308–9; antecedent, 309; electoral, 316–17; subsequent, 309
Brandt Report, 197
Brasilia, 423
Brazil, 118, 160, 423, 450
Bretton Woods agreement, 354, 479
Britain, 404, 289, 494; energy in, 472; geography in, 31; industry in, 466–9; new towns in, 424, 425; rural settlement in, 417; transport in, 344–5
Brittany, 304
'brown agenda', 452
Brundtland Report, 112, 132, 134
Brush, J.E., 431
Buddhism, 240–2, 247

Buffon, G.-L., 27, 111
Bunge, W., 317, 320
Bylund, E., 415–16

Cairo, 451
Cajun, 266
Calcutta, 450
caliche, 387
Canada, 236, 237, 249, 270, 290, 296, 308; Aboriginal people in, 7; agriculture in, 374; development of, 493–4; geography in, 32–3; immigration to, 190–1; industry in, 474; labour in, 485; language in, 236, 237; population in, 149, 162; regions of, 228–9; rural settlement in, 414, 417–18
Canadian Shield, 86
canals, 344
capital, 461
capitalism, 52, 55, 112, 168, 201, 257–9, 312; competitive, 258, 422; disorganized, 258, 263, 421; ideal type, 258; and industry, 472; 'keno', 444; organized, 258, 422; and urban growth, 421–2, 438
carbon dioxide, 96, 119
Carey, H.C., 337–8
Carneiro, R., 223
Carrothers, G.A.P., 337–8
carrying capacity, 157
Carson, Rachel, 112
cartography, 29, 66–9; computer-assisted, 69–70; see also maps
caste, 305, 336
Castells, M., 436
Catalonia, 304
catastrophists, 130, 211
cattle, 247
ceiling rent, 375–6
Celtic languages, 234, 238
census, 180
central business district (CBD), 434–5
central place theory, 426–30, 431
centrifugal forces, 298–9
centripetal forces, 298–9
cereal crops, 397
Chamberlain, Houston Stewart, 100
Champlain, Samuel de, 296
Chandigarh, India, 423
Chernobyl, Ukraine, 110, 112
Childe, V.G., 222
China, 153, 158–9, 305, 313; agriculture in, 398–9; cities in, 448; early geography in, 20; industry in, 477–9; migrants from, 191; special economic zones, 478-9
chloro-fluorocarbons (CFCs), 113, 126–7

cholera, 340
chorography, 18
chorology, 31
choropleth maps, 67
Christaller, Walter, 368, 426–31
Christianity, 240–6, 249, 322
Churchill, Winston, 420
cities, 412; capital, 442; 'city beautiful'
    movement, 424; earliest, 420; 'edge',
    440; global, 443–6; 'garden', 423;
    growth of, 419–24; industrial, 421–3;
    internal structure of, 431–4; in less
    developed world, 446–51; in more
    developed world, 431–36; origins of,
    419–24; postindustrial, 438–9;
    preindustrial, 421; primate, 430; as
    service centres, 431; and urban
    problems, 442–3, 449–51; and
    symbolism, 440–2
civilization(s), 221–4, 420, 442; 'aborted',
    225; 'arrested', 225; clash of, 321–2;
    religion and, 244–5
Clark, A.H., 296
class, 223, 257–9, 280, 317; and cities, 437–8
Clayoquot Sound, 133
climate(s), 87, 89; and agriculture, 371; and
    development, 491; impacts on, 125–9;
    urban, 129
Club of Rome, 165, 211
coal, 465, 468, 469
coast, proximity to, 491
Cohen, S.B., 298, 323
Cold War, 298
colonialism, 201, 206–7, 291–3, 301, 394,
    399, 446, 475
commercialization, agricultural, 385
commodity flows, 349
communication, 256, 479
communism, 259, 298, 306
competition: and settlement patterns, 416;
    spatial, 428
compound unitary government, 314
computer-assisted cartography, 69–70
Comte, Auguste, 49
concentric zone model, 432
concepts, geographic, 55–65
conflict(s), 272–4, 289, 301, 315, 317–19,
    323; and agriculture, 402; categories of,
    318; and religion, 243–5
Congo, Democratic Republic, 451, 504
Conkling, E.C., 379
conservation, 109, 111, 132
constructionism, 256–7, 264
consumption: food, 403–4; mass, 488; oil,
    471
contextualism, 264
continental drift, 83–5
contraception, 145, 170
conurbations, 436
Cook, James, 23, 25
cooperation, 114; international, 125
Copernicus, Nicolaus, 23

cordilleran belts, 87
core regions, 201, 230, 302, 308, 492; see also
    peripheries
cornucopians, 130, 211
correlation analysis, 380
Corsica, 304
Cosgrove, D., 440
cosmography, 18
counterfactual approach, 296
counterurbanization, 418
creole (language), 238
crime, 275, 276
crops: cereals, 397; crop theory, 377–8; High
    Yielding Varieties (HYVs), 388–9
'Crow rate', 395
crude birth rate (CBR), 142
crude death rate (CDR), 148
'cultural turn', 264
culture(s), 217–85, 352, 505 ; and
    agriculture, 372–5; evolution of, 221–4;
    folk, 277–9; importance of, 264; major,
    321–2; material, 220, 222; national, 248;
    non-material, 219–20, 256; plurality of,
    256–7; popular, 277, 280–1; western,
    248
customs union, 350
cycle, biochemical, 107

dams, 110
Darwin, Charles, 100, 122
data: collection, 70–5; handling, 507
Davies, S. 419
Davis, M., 207, 442
Davis, William Morris, 32
Dear, M.J., 438, 443
deBlij, H., 298
debt, world, 208–9
decentralization, 481
decolonization, 292–3, 353
deconcentration, 418–19
deconstruction, 263
'deep ecology', 113, 115
deforestation, 108, 116–19, 123
deglomeration, 60, 462–3
deindustrialization, 480
Demeny, P., 147
democracy, 207, 249–50, 312–13, 324
demographic transition theory, 168–9
demography, 142; see also population
density: population, 178–81, 204, 434;
    physiological, 179
dependence, 201
dependency theory, 293
depopulation, 417–18
description, geographic, 24–5, 367
desertification, 119–21
deserts, 92
de Seversky, A.P., 298
determinism: cultural, 220; environmental,
    33, 46–8
Deutsch, K.W., 296
development: economic, 64, 291; 'global

partnership for', 212; human, 198–200;
    regional, 493–4; stages of, 491–2;
    'sustainable', 132–4; uneven, 491–5;
    zones, 201–2
developmentalism, 184, 199
devolution, 302–5
Dickenson, J., 197
dictatorship, 313
difference, human, 264, 270–1
diffusion, 62, 339–43
disappearing peoples, 505–6
disaster: ecological, 108; human, 209–10
discourse, 64
discovery, European, 22–4
discrimination, 353
disease, 150–4, 451; combatting, 211;
    diffusion of, 340, 343
dispersion, rural, 413, 414
distance, 59–61, 332–9, 461; and agriculture,
    368, 379–82; cognitive, 335–6; decay,
    60; economic, 335; and friction, 332–3;
    models of, 337–9; physical, 333; social,
    336; time, 334; transformations of, 337;
    'tyranny of', 339
distribution, 59–60; food, 204; language,
    233–5; population, 178–81; religion,
    241
districts (city), 440
diversity: cultural, 352; ethnic, 190; human,
    263, 281
Djibouti, 197
Doenges, C.E., 144
domain, 230
domestication, 383–4; animal, 120, 122;
    plant, 116–17
domestic space, 271
Doukhobors, 249
Doxiadis, C.A., 332
droughts, 112
Dutch language, 236, 237

earth: crust, 83–5; physical processes of,
    81–94; vital signs of, 130–2
Earth Day, 112
earthquakes, 83, 84
Easter Island, 108
Eberstadt, N., 153
ecology, 106, 351; 'deep' and 'shallow', 113,
    115; political, 401–3
economic base theory, 423
'economic operator' concept, 368, 432
economic rent, 431–4
economics: and agriculture, 375; and
    geography, 365–71; and migration, 182;
    schools of, 366
economic sectors, 460, 482; informal, 447
economy: and environment, 113; political,
    342–3; world, 396–9
ecosphere, 106
ecosystems, 106–7; human changes to,
    107–10
ecotourism, 489–90

'edge cities', 440
education, primary, 210
Ehrlich, P., 165, 211
ekistics, 332
elections, 315–17
élitist landscapes, 276-7
empire, 288, 291–3, 305, 306
empiricism, 48, 170, 257
employment, 481–2, 483, 484; and gender, 485; see also labour; unemployment
enclosure, 385–6, 413
energy, 106–10, 467, 472, 503; and civilization, 221–2; nuclear, 109, 110; renewable, 109–10; sources of, 469–72; see also power
Engels, Friedrich, 52, 54
engineering, group, 281
English language, 234–5, 236, 237
entitlements, 206–7
entropy, 339
environment(s): attractive, 180–1; environmental protection, 111, 132; global, 90–5; and industry, 485–6; and migration, 182–3; and states, 315; western concerns about, 110–13
environmental determinism, 33, 46–8
epidemics, 164
Epstein, B.J., 435
equality, 210–12, 275; gender, 210
equator, 82
equinox, 82–3
Eratosthenes, 17
erosion, 86–7, 123
Esperanto, 238
essentialism, 256
Estaville, L.E., Jr, 268
ethics, environmental, 110–13
Ethiopia, 197, 199
ethnicity, 268–70, 301; and agriculture, 372; ethnic areas, 269–70
ethnocentrism, 72
ethnography, 72
Europe, 22–4; expansion of, 352; industry in, 474, 475; and regional development, 494; regions of, 226; rural settlement in, 413; societies in, 257–9; states in, 288–91, 302–5
European Union, 310, 350
evaporation, 123
evolution: cultural, 281; human, 95–101; metropolitan, 425
existentialism, 51
exploration, 22–4, 29, 291–2
export-processing zones, 475
extinctions, 120–2
exurbanization, 419
Eyles, J., 265

factories, 467, 468–9
'fair trade', 388
famine, 207
farmers, yeoman vs. entrepreneur, 373

fascism, 313
favelas, 450
fear, 276, 277
fecundity, 142, 143; see also fertility
federalism, 314
feminism, 171–2, 257, 260; and economics, 369–70
Fernandez-Armesto, F., 403–4
fertility transition theory, 144, 169–70
fertility, 142–8, 157, 168–9, 171; measures of, 142–3; biological factors in, 143, economic factors in, 143–4; cultural factors in, 144–5, 147; declining, 147, 149
fertilizers, nitrogen, 386–7
fetishism, spatial, 56
feudalism, 257, 288–9
fieldwork, 72
fire, 115
First Nations, 9; see also Aboriginal peoples
Flanders, 304
Flemish, 237
Flenley, J., 108
flexible: accumulation, 480; production, 358, 422
flooding, 210
Flusty, S., 443
folk art, 245
food, 167–8, 202–8; aid, 203; consuming, 403–4; distribution, 204; 'ethnic', 404; genetically modified, 390; global trade in, 397; 'problem', 204–8; supply system, 400–1
Fordism, 258, 358, 479–80, 486
forests, 91, 93, 112, 117, 119–20, 133
Foucault, Michel, 64
Francaviglia, R.V., 231
France, 31, 289, 291, 304
Freeman, D.B., 342
French language, 234, 236, 237, 266
'friction of distance', 332–3
Friedmann, H., 400
frontier, 385, 386
fuels, 108, 109, 126

Gaeltacht, 238
'Gaia' hypothesis, 96
Gambia, 402–3
garden city, 423
Garrison, William, 345
gas, natural, 469
gasoline, 470
gazetteer, 18
Gee, M., 211
Gemeinschaft, 418
gender, 260, 369–70; and agriculture, 401–3; and employment, 485; and human development, 271–2; and religion, 240
General Agreement on Tariffs and Trade (GATT), 354–5, 356
general fertility rate (GFR), 142–3
genetically modified food, 390

genocide, 303
gentrification, 439
geo-economy, 356
geographical societies, 29
geographic information systems (GIS), 70, 72
géographie Vidalienne, 32, 34
geography: as academic discipline, 27, 30–3; agricultural, 365–409; applied, 38; analytic techniques, 65-75; basic concepts, 55–65; classical, 17–18; commercial, 366; contemporary, 35–8; cultural, 36, 339–40; and description, 24–5, 367; economic, 365–71; general, 26; history of, 16–35; human, 3–12, 219–21, 256–7, 259–64, 366, 501–10; landscape, 36; new cultural, 256–7; physical, 33, 36, 351; population, 141–75; preclassical, 16–17; qualitative methods, 72–5; quantitative methods, 75; regional, 31, 34–5, 36–7, 367; social, 36, 270–1; subdisciplines of, 11; transport, 343; universal, 27; welfare, 275
geometry, 333
geomorphology, Canadian, 86
geophagy, 279
geopolitics, 296–8
geopolitik, 297
German language, 234
Germany, 5, 30–1, 149, 289, 298
gerrymandering, 316
Gesellschaft, 418
Ghana, 346
ghetto, 269–70, 437
Giddens, Anthony, 260
glacial period, 125
glaciers, 84, 86
globalization, 8–9, 64–5, 129–30, 351–60, 443, 445, 479–86; cultural, 247–50; economic, 354–8; opposition to, 359; political, 324–5; support for, 359–60
global positioning system (GPS), 72
'global village', 360
global warming, 109, 119, 126–8
Gobineau, Joseph Arthur de, 100
'Gonneville' land, 25
goods, public, 315
Gorkhaland, 307
Gould, P., 185, 343, 345, 346
Gould, Steven Jay, 99
government: and agriculture, 398, 399; and development, 493–4; forms of, 312–15; and industry, 461, 462; and population control, 157-8; pronatalist, 156–7; state, 289; substate, 314–15; survey systems of, 413, 414; types, 314; see also state
gradation, 86
Grameen Bank, 208
graph theory, 345–6
grasslands, 93
gravity model, 337–8

grazing, commercial, 384–5
'Great American Desert', 239
Greece, ancient, geography in, 17–18
green belt, 423
Greenbie, B.B., 437
greenhouse effect, 96, 119; gases, 113, 126–7, 129; natural, 126
Green Plan, 112
green political parties, 112
'green revolution', 387–90, 391
Gregory, D., 508
Grigg, D.B., 384
Gritzner, Charles, 3
gross domestic product (GDP), 198, 357; and agriculture, 396–7
gross national income (GNI), 198
gross national product (GNP), 198
Grossman, D., 416
groups, 218–21; cultural, 289; dominant, 272–5; ethnic, 268–70, 303, 306; disadvantaged, 278
growth: economic, see development; limits to, 165
'growth poles', 493
gyres, 85

Hagerstrand, T., 62, 336, 340
Haiti, 203
Hale, R.F., 266
Hall, P., 436
Hamnett, C., 263–4
Hardin, G., 114
Harris, C.D., 432–3
Harris, M., 247
Hartshorne, Richard, 34, 35, 36, 50
Harvey, David, 248–9, 263, 432, 438, 509
Haushofer, K., 297
health, 211, 275–6
'hearth', 240–1
heartland theory, 297
Hettner, Alfred, 31
hierarchical effect, 62, 340, 341
High Yielding Varieties (HYVs), 388–9
Hinduism, 240–2, 245–6, 305
Hipparchus, 17–18
historical materialism, 52
HIV (Human Immunodeficiency Virus), 150–4; see also AIDS
Holocene period, 98
homelands, 267–8
homelessness, 442–3
hominids, 97
Homo erectus, 97–8
Homo habilis, 97
Homo sapiens sapiens, 98, 99
Horvath, R.J., 229
Hotelling, Harold, 465
housing, 436–9
Howard, Ebenezer, 423
Hoyt, H., 432
Hudson, J.C., 416
Hudson, R. 473

Human Development Index (HDI), 199–200; and gender, 271–2
humanism, 49–52, 171, 256, 259–60, 262
human(s): agency, 260, 262; and difference, 264, 270–1; environmental impacts, 105–39, 503; evolution, 95–101; origins, 351
Humboldt, Alexander von, 27–9, 36, 62, 111
hunger, 210
hunter gatherers, 114–15
Huntington, E., 47
Huntington, S.P., 303, 321
Hutterites, 249
Hutus, 301
hydraulic hypothesis, 222–3
hydroelectric power, 470, 472
hydrosphere, 106

ice, 84
iconography, 257, 264, 265
idealism, 51
identity, 30; dual, 248; ethnic, 268–70; human, 264; and language, 235–8; national, 265, 289, 292, 301, 302, 324; place, 265–6; and religion, 243–5; state, 298–9
idiographic approach, 54–5
images, mental, 63, 441
imagination, geographical, 508
immigration, 190, 269–70; policies, 189, 190, 191; see also migration
imperialism, 293
independence, political, 292–3
India, 153, 158–9, 242, 245, 247, 305; agriculture in, 389; industry in, 479
Indo-European language family, 233–5
industrialization, 421, 423, 469, 475–9; agricultural, 400–1
industrial revolution, 108, 352, 465–9
industry, 115, 124; decaying, 495; and environment, 485–6; and globalization, 479–86; location of, 435–6, 460–5; origins of, 465–9; service, 481–4; world patterns of, 469–79
inequality, 270–1, 281; and economic development, 491–5
infant mortality rate (IMR), 148, 151, 451
informal sector, 448
information, exchanging, 481
infrastructure, 53, 442
Innis, Harold, 338
innovation: adoption, 341–2; diffusion of, 339–43; resistance to, 341
integration, economic, 298, 350; European, 310–12
intensity theory, 378
interaction, 60; interaction theory, 367–8
interdependence, locational, 463, 465
internally displaced persons (IDPs), 193
International Monetary Fund (IMF), 354
investment, foreign direct, 357, 358
Iraq, 471
Ireland, 238

iron, 467
'irrational choices', 247
irredentism, 299
Isard, Walter, 339, 349
Islam, 240–5, 305, 307, 322
Islamic world, early geography in, 21–2
isodapanes, 462, 463
'isolated state', 376–9, 426–7
isopleth maps, 67
isotims, 462, 463
Israel, 294, 295
Italy, 289
Ivory Coast, 402

Jackson, P., 256, 280
James, Preston, 218
Japan, 445, 475, 502–3
Jews, 295
Jharkand, 307
Johnston, R.J., 317, 505
Jones, S.B., 294
Jordan, T.G., 226, 230, 239
Joseph, A.E., 417–18
Jowett, J., 158
Judaism, 242, 245

Kansky, K., 346
Kant, Immanuel, 27
Kaplan, R.D., 130, 211
Karan, P., 245
Kashmir, 305, 307
Kaups, M., 230, 239
Keddie, P.D., 417–18
Kennelly, R.A., 464
'keno capitalism', 444
Kenya: Green Belt Movement, 119
Keynes, John Maynard, 479
Khalistan, 307
Kinshasa, Democratic Republic of Congo, 451
Kjellen, R., 297
Kneiffen, F.B., 225
Knox, P., 439, 440
Knox, Thomas, 314
Kohr, L., 311
Kosovo, 303, 504
Krebheil, E., 315–16
Kroeber, Alfred, 219, 220
Kropotkin, Peter, 313, 423
Kurds, 299
Kuwait, 471
Kyoto protocol, 113, 127

labour, 461, 462, 483, 485, 486; and agriculture, 397; indentured, 188; international division of, 357–8; and service industry, 484; see also employment
Lagos, Nigeria, 451
Lake Erie, 124, 125
land, 83; competition for, 375–6; impacts on, 122–3; use, 379–82, 431–4; value of, 379–82

landforms, 86–90, 122
landmarks, 440
Landsat, 71–2
landscape(s), 1–2; agricultural, 371–2;
    cultural, 46–8, 219, 229–32; élitist,
    276–7; human, 502–6; iconography of,
    265; industrial, 468–9; and language,
    238–40; pariah, 277; as place, 264–6;
    religious, 245–7, 249; school, 34, 219,
    220, 230, 256, 257; as socially
    constructed, 257; of stigma, 277;
    urban, 440–3
Landschaftskunde, 34
language(s), 221, 232–40, 248; artificial, 238;
    families, 233–5; and identity, 235–8;
    minority, 236; new words, 239–40l;
    number of, 232–3; official, 236, 237,
    238; and territory, 237; universal, 238
Las Vegas, 487
latitude, 17–18, 82, 87, 88–9
Laurentian Shield, 86
law, 48
Leaman, J.H., 379
least-cost theory, 460–3, 464
least effort, principle of, 332
Le Corbusier, 423
Le Play, Pierre, 31
Léros, Greece, 278
Lesotho, 300
less developed world, 109,147, 197–212,
    348; industry in, 477–9, 488, 493;
    relations with more developed, 201–4;
    spatial patterns in, 380–1; transport in,
    345; urbanization in, 420, 446–51
Libya, 471
life, 94–5; human, 95–101
life cycle, 185–6
life expectancy (LE), 148–9, 151
limits to growth, 165
lingua franca, 238
lithosphere, 83, 106
locales, 260
localism, 248–9
location, 2, 56; agricultural, 371–82;
    industrial, 460-5, 481; movement
    between locations, 337–9; urban,
    424–30; *see also* settlement
locational figures, 462
locational interdependence, 463, 465
location theory, 367–8
logging, 117, 133
Lomborg, Bjorn, 131, 211
London, 382, 445
longitude, 17–18, 23, 82
Lorenz, Konrad, 322
Los Angeles, 441, 442, 443, 444
Lösch, August, 427, 463–4
Louisiana, 266
Love Canal, 125
Lovelock, James, 96
Lowe, J.C., 345–6
Lynch, K., 440

McDowell, L., 369–70
Mackinder, Halford J., 31, 297
Maghreb Union, 312, 393
malapportionment, 316
Malaysia, 157
malls, shopping, 280
malnutrition, 202, 205
Malte-Brun, Conrad, 27
Malthus, T.R., 27, 111, 165, 167–8
manufacturing, 460; *see also* industry
Maoism, 313
maps: biases of, 69; early, 16–25; 'of
    meaning', 256; mental, 63–4, 185, 441;
    population, 180–1; Portolano charts,
    19–20; T–O, 19; topological, 336; *see
    also* cartography
maquiladoras, 476
market: central, 377–9; common, 350;
    regional, 379; world, 381
market-area analysis, 463–4
marketing principle, 429
marriage, 144–5
Marsh, George Perkins, 111, 507
Martinique, 346, 347
Marx, Karl, 52, 168, 460
Marxism, 52–4, 55, 168, 170, 257–9, 259–60,
    263, 438, 484; and agriculture, 396–401;
    and civilization, 223; and economics, 369
Massey, D., 356, 484, 485
material index, 461–2
materialism, historical, 52
Mauritius, 188
Mediterranean areas, 92
Meinig, D.W., 231
Meitzen, August, 415
mental illness, 278
mentifacts, 219
mercantilism, 421; mercantile model, 424–5
Mercator, Gerardus, 23
metropolitan evolution, 425
Mexico, 464, 471
Mexico City, 450
migration, 98, 99, 142, 181–96; chain,
    269–70; forced, 187–8; free, 187, 188–9;
    illegal, 187, 189–92; laws of, 183;
    reasons for, 181–86; primitive, 187;
    selectivity of, 185–6; types of, 186–92;
    see also immigration
minority, 268–9
Mithila, India, 245
mobility transition, 184
models, spatial, 75
modernism, 262–3
modernization, 6–8
monarchy, 313
Mongolia, 393
Monmonier, M., 58
monopoly, spatial, 463
monsoon areas, 91
Montague, Ashley, 322
Monte Carlo simulation, 340
Montesquieu, C., 27

Moon, B., 186
'moorings', 186
more developed world, 109, 149, 348;
    industry in, 472–6; urbanization in,
    419–24, 431–46
Mormons, 230–1, 247, 249
Morrill, R.L., 345, 346, 508
Morris, Desmond, 322
mortality, 148–54, 168–9; child, 211, 451
Moryadas, S., 346
mountains, 83–4, 86, 87
Muller, E.K., 425
multiculturalism, 270
multilingual states, 235–6
multinationals, 394, 399; *see also*
    transnationals
Münster, Sebastian, 24
Murphy, R.E., 435

names, place, 238–9
nation state, 288–90; predecessors of, 288–9
nation, 288, 290; *see also* state
nationalism, 235, 289, 308
National Topographic System, 66–7
Native peoples, 7, 276, 292
nativism, 191
nature, 257; *see also* environment
Neanderthals, 98
nearest neighbour analysis, 428
neighbourhoods, 437–8; neighbourhood
    effect, 62, 340–1
neocolonialism, 390
Netherlands, 128
networks, transport, 345
New France, 413
New Haven, Connecticut, 442–3
newly industrializing countries (NICs),
    475–6
Newman, J.L., 144
New York, 445
New Zealand, 472
Niagara peninsula, 418–19
Nigeria, 180
nitrogen fertilizers, 386–7
nodes, 345–6, 440
nomadism, 392, 393
nomothetic approach, 55
normative theory, 368
North America: agriculture in, 385; industry
    in, 473–4; regions of, 226–9; rural
    settlement in, 413; urban location in,
    424–6, 432–3
North Atlantic Treaty Organization (NATO),
    298
Northern Ireland, 304
North Korea, 313
North Sea, 472
'north/south', 197, 198
Nostrand, R.L., 268
nuclear energy, 470
nucleation, rural, 413, 414
nuptiality, 144

objectivity, 263
observation, participant, 72
oceans, 85, 125, 127
Ohmae, K., 325
oil, 129, 469–71, 480; conflicts over, 323; spills, 125
Olduvai Gorge, 96
oligarchy, 313
Ontario, 379, 418–19
optimists, 211
organization(s): cultural, 181; human, 222–3; regional, 354–5, 357
Organization of African Unity, 302
Organization of Petroleum Exporting Countries (OPEC), 470, 479–80
orientalism, 272
Ortelius, Abraham, 25
Osborne, B.S., 265
'others', 272
overpopulation, 168, 204
Owen, Robert, 423
ozone layer, 82, 94, 112, 126, 129

Paddison, R., 314
Pahl, R.E., 416
Pakistan, 305
Palestine, 295
pandemic, 151, 340
Park, C., 232
participant observation, 72
Pashtunistan, 307
paths (city), 440
patriarchy, 260, 270
patterns: point, 60; settlement, 412–16; spatial, 59–60
Pawson, E., 276
Peace, W., 265
peace, 317–20; 'perpetual', 320–1
Pearl, R., 165, 167
Pearl River, 478
Peattie, R., 508
perception, 62–4
peripheries, 201–2, 302, 308, 492
permafrost, 93
pessimists, 211
pesticides, 112
Petersen, W., 186–7
phenomenology, 51
phenotypes, 99
philosophy, 46–55; and economic geography, 366–7; and scale, 221
pidgin (language), 238
pipelines, 129
place(s), 56, 58; artificial, 487–8; and meaning, 264–5; names, 238–9; sacred, 245–7; sense of, 267–8; as social creations, 264–6; and voting, 317
placelessness, 58, 265–6, 280
place utility, 184–5
planning: industrial, 495; urban, 423–4
plates, tectonic, 83–5
Plato, 288

Pleistocene, 98
point patterns, 60
Poland, 298
polar areas, 94, 127–8
political ecology, 401–3
politics, 207, 287–329, 503; and agriculture, 372–5; and environment, 112–13; international, 112–13; and migration, 182
Pollard, J.S., 485
pollution, 108–9, 124–5, 450
population, 141–75, 177–215, 502; aging, 159–63; declines in, 166; doubling time, 156; government policies on, 156–7; growth, 353; natural increase of, 154–6; projections of, 165–6; pyramid, 158, 159, 160, 162, 169; rural, 417–18; surplus, 168
Portolano charts, 19–20
ports, 346
positivism, 48–9, 170–1, 259, 262, 367–8
possibilism, 34, 46–8, 219
post-Fordism, 258, 358, 422, 479–80, 486
postmodernism, 257, 262–4, 443, 370
'postsuburbia', 438–9, 440
potential model, 338–9
poverty, 210; cycle of, 439; urban, 449–51
power: lack of, 207; political, 287; relations, 261, 270–7; solar, 110; state, 315; wind, 110, 111; see also energy
pragmatism, 50–1
preadaptation, 229–39
precipitation, 123
'predisposition, culturally habituated', 229
preferences, spatial, 185
primate city, 430, 448
prime meridian, 82
Prince, H.C., 508
principle of least effort, 332
privatization, 478
processes: physical, 81–94; structuration, 261
production: 'flexible', 358, 422; forces of, 52; mode of, 52, 352; relations of, 52
'profit maximizers', 372, 463
projection, 67–8; conformal, 68; equal area, 68; Mercator, 23, 68; Mollweide, 68; Robinson, 68
protectionism, 354, 359
protest, 314
proxemics, 336
psychogeography, 267, 268
Ptolemy, 18
public goods, 315
push–pull logic, 181–3, 186

qualitative methods, 72–4, 75
quantitative methods, 75
questionnaire, 75

rabbits, 120, 121
race, 99–100, 268–9
racism, 100

railways, 343, 344, 347–8, 467
rain forest: temperate, 117, 133; tropical, 91, 112, 119–20
Raitz, R.B., 268
range (of good or service), 428
rank size rule, 430
rape, 276
raster, 70
rate of natural increase (RNI), 154–6
rational choice theory, 368
Ratzel, Friedrich, 30, 31, 46, 294, 296–7
Ravelstein, E.G., 183, 338
Réclus, Elisée, 31, 313, 423
recreation, 486–91
recycling, 132
Redman, Charles, 222
refugees, 192–6, 197
region(s), 2, 58–9; cultural, 224–9, 267; ethnic, 269–70, 300, 311; European, 226; formal, 59; functional, 59; global environmental, 90–5; industrial, 473; language, 233–5; North American, 226–9, 267; regional development policies, 493–4; and religions, 240–3; vernacular, 59, 266–8; world, 225–6
regionalism, 302–5, 308
regionalization, 2, 59; cultural, 225–6
regional science, 367
region state, 325
regression analysis, 380
regularity, spatial, 332
reindustrialization, 480
relief: and agriculture, 371–2; land, 181
religion, 145, 147, 230–2, 240–7, 288, 289; and agriculture, 372; classification of, 240–3; and conflict, 243–5; ethnic, 242; and gender, 240; and land use, 246; universalizing, 242–3
renewable resources, 109
rent: ceiling, 375–6, economic, 375, 377, 431–4; land, 432
repatriation, 196
repopulation, 417–18, 419
resettlement, 196
residential areas, 436–9
resistance, landscapes of, 272–4
resources: natural, 109–10; renewable, 109; stock, 109
restructuring, 443, 445; agricultural, 401–3; industrial, 479, 480–1
retailing, 434–5
'returnees', 193
revolution, 82–3; agricultural, 108, 163–4, 222, 352, 385–6; 'green', 387–90; industrial, 108, 164, 352;
Rhine River, 124, 125
Ricardo, David, 376
Richthofen, Ferdinand von, 31
Rigg, J., 356
rights, equal, 260
rimland theory, 298
Rio de Janeiro, 450

Ritter, Carl, 27–9, 36
rivalries, international, 318
roads, 348; turnpike, 344
Rogers, E.M., 341
Rokkan, S., 302
Romania, 157
Rooney, J.F., Jr., 280
Rostow, W.W., 347, 491–2
rotation (earth), 82
Rotestein, Frank, 165
routes, 345–6
Rozman, G., 426
Ruhr Valley, 474
'rural idyll', 417
rural–urban fringe, 418–l9
Russell, R.J., 225
Russia, 154, 153, 305, 471; *see also* USSR
Rwanda, 301, 504

sacred space, 56, 58
Sadler, D., 473
Sahara, 393
Sahel, 119, 120
St Helena, 291
sampling, 75
São Paulo, Brazil, 450
satellites, 71–2, 127
satisficing behaviour, 368
Saudi Arabia, 471
Sauer, Carl, 34, 35, 219, 220, 256, 383–4
savannas, 92
scale, 61–2, 66–7; continental, 381; group, 221; human, 221; local, 382; social, 62; spatial, 61, 178; temporal, 62; world, 381
Scandinavia, 189
Schaefer, F.W., 35, 50
Schlüter, Otto, 34
science: regional, 339; social, 259
scientific method, 48
Scotland, 304
seas, epeiric, 86; rising levels, 128
seasons, 82–3
sector model, 432
segregation, spatial, 436
semiperiphery, 201
Semple, Ellen Churchill, 32
Sen, Amartya, 207
sensing, remote, 70–2, 114, 117, 118, 124
separatism, spatial, 56, 60
Serbia, 303, 504
service industry, 481–4
settlement(s) , 332, 411–57; 'first effective', 227, 267 local, 196; patterns of, 412–16; rural, 412–19; squatter, 449; theories of, 415–18; urban, 412, 424–51
sexism, 485
sex ratio, 158, 159
sexuality, 260, 274, 280
Simon, J., 131, 211
simulation, Monte Carlo, 340
Singapore, 156–7

site, 56
situation, 56
Sjoberg, G., 421
'skeptical environmentalist', 131, 211
slavery, 187–8
slums, 439
Smallman-Raynor, M.R., 343
Smil, V., 109, 122–3, 130–1, 208, 221–2
Smith, Adam, 460
Smith, D.M., 274–5
Smith, George, 231
Smith, S.J., 256
Smith, W., 400
smog, 129
socialism, 259, 312–13, 495
social physics, 339
social structure, 260
society, 218, 219, 220, 502; and industry, 484–6; types of, 257–9
sociofacts, 220
sociolinguistics, 240
sociology, 219, 220, 256
soil, 87, 88, 106, 181; and agriculture, 371–2; impacts on, 122–3
solstice, 82–3
Somalia, 197
South Africa, 191, 272–3, 385, 426
South America, 350
South Asia, 305, 307–8
South Korea, 475
sovereignty, 289
space(s), 56; concepts of, 333–9; control of, 280; domestic, 271; proxemic/distemic, 437; sacred, 56, 58
spatial analysis, 2, 35, 37, 340–2; and economic geography, 367
spatial competition, 428
spatial fetishism, 56
spatial monopoly, 463
spatial preferences, 185
spatial segregation, 436
spatial separatism, 56, 60
spatial social indicators, 274–5
special economic zones, 478
species, 94, 99, 100; introduced, 120–2
spectacles, 487
Spencer, J.E., 229
sphere, 230
sport, 280
Spykman, N., 298
squatter settlements, 449
Sri Lanka, 201, 305, 307, 489
S–shaped curve, 62, 166–7, 341
Stafford, H.A., 486
staple theory, 424–5, 492
state(s), 223, 260, 352, and agriculture, 372–5; apparatus, 315; and culture, 270; creation of, 288–96; decline of, 325; divided, 310; growth of, 294, 296–8; groupings of, 310–12; internal division of, 299–310; multinational, 290; nation, 248, 288–90; region, 325; role of,

312–15; stability of, 298–312; *see also* government; nation
steel, 464, 467, 473
stigma, landscapes of, 277
stock resources, 109
Strabo, 17
structuration theory, 260–2, 369
subsistence, 391, 392
subspecies, 99, 100
suburbanization, 419
suburbs, 271, 440
Sudan, 129–30
sun, 106–7
superorganic, 219, 220, 256
superpower, 323–4
superstructure, 53–4
supply and demand, 375
survey, township, 413, 414
Süssmilch, Johann Peter, 211
sustainability, 113, 132–4, 212
Switzerland, 235–6, 290
symbolic interactionism, 256
symbols, religious, 246–7
systems, 106–7, 134

Taafe, E.J., 345, 346
Talisman Oil, 129–30
Tamil-Eelam, 307
Tansley, A.G., 106
tariff, 349
Taylor, Griffith, 33
technology, 107–10, 131–2, 358, 480; information, 353; and urbanization, 436
tectonic plates, 83–5
teleology, 19
temperature, 126–8, 181
territorialization, 237
terrorism, 244, 245, 321, 322–3
textile industry, 467–8
theory, 48, 367–8; social, 259–64
'Third World', 197–8; *see also* less developed world
Thomas, W.L., 112
Three Gorges dam, 110
threshold (in central place theory), 428
Thünen, Johann Heinrich von, 368, 376–82, 426–31
time–space convergence, 334
toad, cane, 121
Tobler, W., 60
Tokyo, 445, 502–3
tool-making, 221
topography, 18
topological maps, 336
toponyms, 238–9
topophilia, 58, 265
topophobia, 58, 265
total fertility rate (TFR), 142–3, 144
tourism, 486–91; alternative, 489; ecotourism, 489–90; industry, 490–1; in less developed world, 488; mass, 488
'tourist gaze', 490–1

Toynbee, A.J., 225, 244–5

trade, 348–51; 'fair', 388; free, 350, 355; protectionism, 354, 359; and regional integration, 349–51; theories, 349; and urban location, 424–5

tragedy of the commons, 114

transnationals, 65, 129, 354, 357–8, 445, 472, 484; *see also* multinationals

transportation, 343–8, 349, 425; costs of, 461–2, 464; modes of, 346–8; networks, 345–6; principle, 429, 430

Trépanier, C., 266

tribalism, 325; in Africa, 301

Trudeau, P.E., 270

Tuan, Yi–Fu, 51, 58

tundra, 93–4

Tutsis, 301

Tylor, E., 219

Tyneside coalfield, 468

Tyrol, 304

Ullman, Edward, 349, 432–3

uncertainty, 461

undernutrition, 202, 205

underpopulation, 166

unemployment, 442

United Arab Emirates, 471

United Nations (UN), 8; and conflict management, 312, 318, 319, 324; High Commission for Refugees (UNHCR), 193–4; population projections, 165

United States, 191, 248, 270, 290, 293, 310, 323–4, 354, 471; agriculture in, 374; geography in, 32–4; regions of, 226–8, 230–1

urban: ecology, 432; problems, 442–3, 449–51; sprawl, 418, 436; *see also* cities

urbanism, 420; postmodern, 443

urban location theories, 431-2: central place theory, 426–30; mercantile model, 424; metropolitan evolution, 425; selective urban growth, 425

urbanization, 223, 353, 420, 421, 502–3; 'premature', 449; premodern, 426; urban growth, 417–18, 421, 422–31, 447–8

Uruguay, 381

USSR, 305, 306, 310, 474–5; *see also* Russia

Uttarkhand, 307

vacation, annual, 487

values, human, 109–10

Vance, James, 420, 421, 424, 435

vapour transport, 123

Varenius, Bernhardus, 25–6

vector, 70

vegetation: human impacts on, 115–20; natural, 87, 89

Venezuela, 471

Venice, 424–5

vernacular region, 59, 266–8

verstehen, 51

Vidal de la Blache, Paul, 31, 32, 34, 219, 220

violence, 260, 277, 301

volcanoes, 85

voting, 315–17

Wade, R.C., 425

Wagner, P.L., 256

Wales, 238, 304

Wallerstein, I., 201

Wallonia, 304

war, 317–20, 393; civil, 293, 318–19; cold, 298, 320–1; costs of, 320; nuclear, 320

water, 83, 84, 106, 181; and civilization, 222–3; conflicts over, 323; global cycle, 123–4; impacts on, 123–5; transport by, 346–7

Watson, Wreford, 332

weather, 90

weathering, 86–7

Webber, M.J., 462

Weber, Alfred, 368, 426, 427, 460–3, 464

Weber, Max, 258

well-being, 274–5

White, M., 434

White, R., 185

Whittlesey, D., 391

Wilson, E.O., 119, 122

Wirth, C., 416

Wittfogel, K., 222

Wolch, J.R., 438

women, 171–2, 208, 271–2, 276; and agriculture, 401–3; and religion, 240

Woodstock, Ontario, 419

World Bank, 110, 119, 199, 210, 212, 354

World Development Report, 199

World Health Organization (WHO), 278

world systems theory, 201–4, 206, 271, 293

World Trade Organization (WTO), 354–5, 356

Wright, Frank Lloyd, 423

Yeates, M., 419

Young, L., 206

Yugoslavia, 303, 503–4

Zelinsky, W., 184, 227, 266–7

zones, development, 201–2

Greenland

3700

BROOKS RANGE

Baffin
Bay

Yukon

Mt McKinley • 6194

Davis strait

• 5951
• 5489

Mackenzie

Great
Bear Lake

British
Isles

Great
Slave Lake

Lake Athabasca

Hudson
Bay

CANADIAN SHIELD

Newfoundland

NORTH
ATLANTIC
OCEAN

PYRENEES

ROCKY MOUNTAINS

Iberia

Fraser

L.Winnipeg

Mt Rainier • 4392

Columbia

L. Superior

COAST RANGES

• 3427

Missouri

The
Great
Lakes

St. Lawrence

ATLAS

• 3187

• 4418

Ohio

4165

Colorado

APPALACHIAN MOUNTAINS

PACIFIC
OCEAN

Sahara

SIERRA MADRE

Rio Grande

Mississippi

C. Falso

Gulf of
Mexico

West Indies

5452 • • 5099

Niger

CARIBBEAN SEA

GUINEA
PLATEAU

Orinoco

GUIANA
HIGHLANDS

2579 •

Cotopaxi

Negro

5896

Japurá

Amazon

AMAZON BASIN

SOUTH
ATLANTIC
OCEAN

ANDES

Purus

Madera

Tapajós

Xingu

Tocantins

ALTIPLANO

BRAZILIAN HIGHLANDS

6155 •

Paraguay

Paraná

• 2787

Aconcagua

6960

Paraná

Paraná

ENTRE RIOS

ANDES

PAMPAS

PATAGONIA

698

Cape Horn

	Over 3,000 m
	2,000–3,000 m
	1,000–2,000 m
	500–1,000 m
	200–500 m
	0–200 m